MOHLING, FRANZ
STATISTICAL MECHANICS: METHODS
000416757

QC174.

WITHDRAWN FROM STOCK
The University of Liverpool

TABLE OF CONSTANTS

Speed of light	$c = 2.998 \times 10^{8}$ m/s
Electronic charge	$e = 1.602 \times 10^{-19}$ C
Planck's constant	$h = 6.626 \times 10^{-34}$ J.s.
	$= 4.136 \times 10^{-15}$ eV.s
	$\hbar = 1.055 \times 10^{-34}$ J.s
Boltzmann's constant	$\kappa = 1.381 \times 10^{-23}$ J/K
	$= 8.617 \times 10^{-5}$ eV/K
Gas constant	$R = N_{A}\kappa = 8.314$ J/K.mole
Avogadro's constant	$N_{A} = 6.023 \times 10^{23}$ mole^{-1}
Electron rest mass	$m_e = 9.1091 \times 10^{-31}$ kg
Proton rest mass	$m_p = 1.6725 \times 10^{-27}$ kg
Bohr magneton	$\mu_B = \dfrac{eh}{2m_e} = 9.273 \times 10^{-24}$ J/T
Standard Pressure and Temperature (STP)	$P_o = 1$ atmosphere
	$= 1.013 \times 10^{5}$ N/m^2 at
	$T_o = 273$ K
Normal volume perfect gas (at STP)	$\Omega_o = 2.24 \times 10^{-2}$ m^3/mole
Freezing point of water	$0°C = 273.15°K$
Gravitational acceleration, standard value	$g = 9.81$ m/s^2

Franz Mohling

Published by
Publishers Creative Services Inc.

Distributed by
HALSTED PRESS Division of
John Wiley and Sons, Inc.
New York Toronto London Sydney

UNIVERSITY LIBRARY
LIVERPOOL

STATISTICAL MECHANICS

Methods and Applications

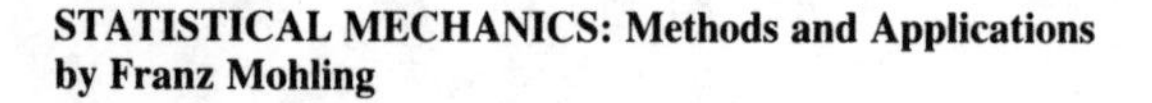

STATISTICAL MECHANICS: Methods and Applications
by Franz Mohling

Copyright © 1982 by Publishers Creative Services Inc.
All rights reserved.

Printed in the United States of America

No part of this publication may be reproduced, stored in a retrieval system, or transmitted, in any form or by any means, electronic, mechanical, photo-copying, recording or otherwise, without the prior written permission of the publisher:

PUBLISHERS CREATIVE SERVICES INC.
89-31 161 Street; Suite 611
Jamaica, Queens, N.Y. 11432

Distributed by the HALSTED PRESS Division
of JOHN WILEY & SONS, INC., New York

ISBN 0-470-27340-2

Book Design by Sidney Solomon with Raymond Solomon

REMEMBERING FRANZ MOHLING

It is a difficult experience to lose a friend—and Franz Mohling was a friend, even though I had never met him in person. My first contact with him came in the form of a manuscript that reached me in the mail, namely his book *STATISTICAL MECHANICS: Methods and Applications*. Professor Mohling dealt with this manuscript, the culmination of ten years' work, in a modest manner, with a helpful, cooperative approach, and an eagerness to blend into the fascinating world of book publishing. It appears that we hit it off early on.

A man of tremendous intellectual achievements, he obviously never limited his interests and his desires to his special academic accomplishments. Tragically, it was his work as a mountaineer which resulted in his death. He was killed by an avalanche as he led a climbing party to the heights of Mt. Logan in Canada in June 1982.

In our many phone conversations we had often spoken of meeting in person, to cement a relationship of mutual trust and confidence. I shall regret that this meeting never took place.

Before he went on this fatal expedition on Mt. Logan, he wrote me that he would take care of all the necessary remaining work on the preparation of his book. With the most meticulous attention to detail, he sent in all that was asked of him, including a beautifully prepared index. We had a long telephone conversation a few days before he departed, in which he expressed his great joy at the prospect of seeing his life-work come to fruition in the publication of his *STATISTICAL MECHANICS: Methods and Applications*. His memory will be well served by this book.

SIDNEY SOLOMON
Publisher

CONTENTS

003613

ACKNOWLEDGMENTS

The maturing of this text from rough first draft to published form was a process to which many contributed, and my gratitude to friends and colleagues who assisted me is enormous. The list begins with the name of Boris Jacobsohn, who was a source of inspiration to me in Statistical Mechanics when I was a graduate student at the University of Washington. Next are past graduate students at the University of Colorado, who studied with me and who at some time made a significant contribution. These former students, now colleagues, are W. Tom Grandy, Jim Rainwater, I. Ramarao, C. Ray Smith and Elizabeth Tuttle. A major contribution was made by my colleague Joe Dreitlein, and several important suggestions came from two former students, Bruce Ackerson and John Lemmon, who used early versions of the text. I am indebted to Stephen G. Brush for his assistance with the Historical Notes.

Preparation of the manuscript proceeded with the able typing of Edith Krause and Karen Danville, under the helpful guidance of Jeannette Royer. Careful drafting of the figures was done by Gloria Kroeger and Dennis Ehmsen. I am also grateful to reviewers of early drafts for their suggestions, and in particular, to the unknown reviewer from Oxford University Press whose thorough and constructive critique was most helpful.

Finally, the entire publication effort was supervised by a most able editor and wonderful gentleman, Sidney Solomon. To him and his fine staff at Publishers Creative Services Inc. I am especially appreciative.

FRANZ MOHLING
Boulder, Colorado

PREFACE

The scope of statistical mechanics is interdisciplinary, ranging from its foundations in pure mathematics to its applications in the applied sciences. Relative to the learning of statistical mechanics, we may consider an undergraduate student who, upon being introduced to the notions of thermodynamics, inquires as to the why's of its laws. This student may be satisfied only after an introduction to the subject's microscopic, or statistical, basis. At the other extreme in the educational process are the theoretical chemists. Their needs in statistical mechanics are not for its mathematical foundations, but rather require comprehensive exposition of statistical-mechanical methods as they apply to the chemical systems ordinarily encountered. The physicist's interests normally lie between these extremes, and it is primarily physicists for whom this book was written.

The level of this text has been set for graduate physics students in their second or third years, and thus it presumes a prior introduction to statistical thermodynamics. Chapter I presents probability concepts and counting problems in statistical physics. The concept of statistical entropy from the information theory viewpoint is introduced in Section 4. Then in Chapter II the discipline of thermodynamics is established, starting from the maximum entropy principle, along lines used in the wonderful old text by Landau and Lifshitz. This treatment, indeed a sophisticated approach, seems to be quite suitable for a graduate-level text on statistical mechanics. In these two introductory chapters frequent references to applications are made. Also, a number of instructive exercises and problems have been included, a pattern that is followed for the most part throughout the text.

From a broader viewpoint, Chapters I–V are primarily source material on the methods of statistical mechanics, and these methods are then applied in the succeeding chapters. Here it should be stressed that statistical mechanics is a very subjective discipline pedagogically. After completing a course introduction based on the first two chapters, some professors will feel that kinetic theory (Chapter III) provides the next logical introductory step. Others will prefer to proceed with a treatment of the quantum-mechanical many-body problem (Chapter IV), whereas in a third camp will be those who can only initiate their course properly by discussing the foundations of ensemble theory (Chapter V). Probably the most popular doorway to statistical mechanics is this third approach combined with common applications of ensemble theory (Chapters VI and VII). In recognition of this diversity in the discipline, Chapters III, IV, and V can be read independently. Although some cross-referencing does occur, one can proceed through them in various orders and with desired omissions.

Material of an applied nature is distributed throughout the text, but those topics that a physicist would label as 'standard application' are included mostly in Chapters VI, VII, and XI. Therefore, for a one-semester course in statistical mechanics it would be a mistake to proceed sequentially and comprehensively through this text. Most students would *not* want initially to cover *all* of the topics in the first five chapters. I suggest that professors guide students into the methods of quantum statistical mechanics according to their interests, and allow ample time to pursue the more applied material of subsequent chapters, beginning with Chapter VI. Two useful features of this text are that sections of advanced or more difficult material have been inserted throughout to stimulate the more devoted student, and the extensive collection of exercises and problems includes some that are quite challenging.

I have considered that a good text is one that carries its interests beyond the classroom, and therefore this text has been written to serve also as a source book for practicing physicists. What may appear to be overly pedantic to a student, or as an unnecessary footnote to someone else, may be the missing link in a postdoctoral's chain of reasoning. Moreover, there are many topics of considerable interest that have not been included previously in most standard textbooks. In addition to the "Historical Notes," the following Sections are more or less unique to this text:

12—A general derivation of the macroscopic equations of hydrodynamics from the Boltzmann transport equation.
19—A careful development of Landau's Fermi-fluid theory.
37—The cluster-integral expansion of the grand potential in quantum statistics.
44—Nuclear spin relaxation.
45—Methods by which the approach to equilibrium is studied (Kubo's method, Chapman–Kolmogorov equation, master equation, Ehrenfests' urn model).
49—Explicit treatment of long-range order in crystals and of fluctuations near the critical point.

50—Graphical expansion in perturbation theory of the one-particle thermodynamic Green's function.
Appendix E—Wigner distribution function.

The material in Chapters VIII, IX, and X is mostly of an advanced nature. These chapters are intended to introduce topics of contemporary interest and to carry the development up to the point where the specialist begins. The extensive footnotes included here may direct the interested reader toward further thinking and to more recent developments in the field. Treatments of irreversible processes (Chapter IX) and few-particle distribution functions (Chapter X) are rarely found in standard texts. Featured in this latter chapter is an emphasis on the important linear response function for scattering processes. Of course, it is difficult for one person to maintain an acute awareness of all the latest developments in a diverse collection of topics, and I welcome suggestions for up-to-date modifications, corrections, or additions to my treatment of contemporary topics.

Historical Notes

That matter can be resolved ultimately into atoms and molecules is common knowledge today. Yet every initiate to physical science must be convinced anew, by relevant experiments, because the atomistic nature of matter is not apparent from everyday experience. Historically, atomic concepts first became important to quantitative physical science, not among the Greeks, but from experiments conducted with air during the 17th century.

The physical nature of air was established by Torricelli, Pascal, and Boyle, whose experiments and theoretical reasoning served to establish that the earth is surrounded by a sea of air that exerts pressure in much the same way that water does. Robert Boyle (1627–1691) attempted to show that air has elasticity by demonstrating that it can support a column of water or mercury. He introduced pressure as a physical quantity, and his experiments led to the discovery that the pressure of a gas is inversely proportional to the volume of space in which it is confined. After these experiments it was no longer necessary to explain many phenomena as being due to "nature's abhorrence of a vacuum." The generalization that pressure is proportional to absolute temperature at constant volume, although realized soon after, was not established firmly until the experiments of Gay-Lussac around the beginning of the 19th century.

Both Boyle and Newton attempted to explain the elasticity of air, by imagining that in some manner atoms interact elastically with each other through repulsive forces. Most scientists accepted an atomic model of this kind, in which the atoms of a gas were suspended in the ether, up until about the middle of the 19th century. In 1738 Daniel Bernoulli developed the first kinetic theory of gases, according to which the gas laws were derived from a billiard-ball model—a method still used in elementary textbooks today. The assumption that atoms could move freely through space until they collide as billiard balls do, and that heat is nothing more than atomic motion, was a full century ahead of its time, and Bernoulli's kinetic theory was unacceptable to the scientists of his day.

One of the earliest subsequent kinetic theories was advanced by Herapath in 1820–1821. This work later inspired J. P. Joule, in 1848, to revive kinetic theory as an illustration of the equivalence of heat and work, or motion.

In addition to investigation of the gas laws, attention was also directed during the 18th century to the correlation of chemical and thermal phenomena. The laws of thermodynamics were developed and based on the assumption that heat is a substance, called "caloric." It was realized that this caloric must, in some sense, be related to atomic motion. Thus, in the quantitative theory of Laplace and Poisson the repulsion between atoms was attributed to the action of "atmospheres" of caloric surrounding the atoms.

By the middle of the 19th century thermodynamics was a well-developed branch of physics and the gas laws were fully established. The background for a kinetic theory of gases, in which macroscopic laws could be related to the microscopic world of atoms, was available. Several notable physicists took up the challenge to develop such a theory.

Kinetic Theory of Dilute Gases

The objective of classical kinetic theory was to reduce the explanation of macroscopic phenomena to atomic models in which atoms move deterministically according to the laws of Newtonian mechanics. One of the earliest kinetic theories in the middle 1800s was put forth by Kronig in 1856. During this period the general principle of conservation of energy was also established by the researches and arguments of Joule, Mayer, and others. Then substantial progress toward the modern kinetic theory was made by Rudolf Clausius (1822–1888), who in 1858 introduced the idea of the "mean free path" of a molecule between successive collisions.[1] Subsequently, James Clerk Maxwell (1831–1879) and others used this concept to develop the theory of diffusion, viscosity, and heat conduction. Acceptance of this first true kinetic theory of transport phenomena was promoted by the experimental confirmation of Maxwell's surprising prediction (1859) that the viscosity of a gas should increase with temperature while remaining independent of the density.[2]

The kinetic theory of gas viscosity was used by J. Loschmidt in 1865 to make the first reliable estimate of atomic size. By this means scientists for the first time were able to establish absolute values for the scale of "atomic weights," previously known only relative to hydrogen, and to calculate "Avogadro's number." This synthesis of theory with observation was one of the most remarkable achievements of 19th century kinetic theory. We must also mention here that in the 1859 paper Maxwell introduced his celebrated velocity distribution function,[3] a conceptual development stemming from earlier statistical theories of Gauss, Quetelet, and others.

In a memoir published in 1866, Maxwell set forth the foundations of the modern theory of transport phenomena by writing down a general equation for the rate of change of any physical quantity as it develops in time due to molecular motions and collisions. Comparison of his "microscopic" equations with corresponding "macroscopic" equations of hydrodynamics, such as the Navier–Stokes equation, then made it possible to compute transport coefficients (e.g., the coefficients of viscosity and heat conductivity) on a microscopic basis.[4] This was the framework within which Ludwig Boltzmann (1844–1906) began his extensive researches. The Boltzmann transport equation,[5] derived in 1872, is a special case of Maxwell's general differential equation, in which the physical quantity of interest is the velocity distribution function for real gases. Once this function was determined, then all transport coefficients could be computed. Much of modern research in the field still focuses on attempts to solve either the Boltzmann transport equation or similar equations for other distribution functions.

It was not until 1917 that a definitive solution of the Boltzmann equation was achieved by D. Enskog.[6] A similar method of solution, applied to Maxwell's original transport equations, was deduced earlier in 1916 by S. Chapman. These solutions provide a quantitative understanding of those nonequilibrium phenomena in dilute gases that are referred to as steady-state processes.

In an application of his own equation, Boltzmann deduced that a quantity, which he called H, must always decrease or remain constant. Proof of the time development of Boltzmann's H-function, which was identified as the negative of the entropy, made possible a molecular interpretation of the law of increasing entropy and, therefore, of the irreversibility of macroscopic phenomena. Such a conclusion, subject to much of Boltzmann's attentions, brought both widespread acknowledgment and attack. The possibility of explaining irreversibility by the time-reversible Newtonian mechanics seemed inherently contradictory.[7] Central to Boltzmann's proof was the two-part assumption that, for a gas at rest, in a given volume element containing a sufficiently large number of molecules, all directions of velocity are represented with equal frequency and that this condition applies also to two colliding groups of molecules. The second part of this assumption, called *Stosszahlansatz* in the literature,[8] was itself the subject of considerable debate (as reviewed in the Ehrenfests' article). But independent of this complex question one firm and important conclusion remained; namely, that under the influence of collisions every velocity distribution function in a gas goes eventually, at equilibrium, to the Maxwell–Boltzmann distribution.[3]

It was a significant conceptual development to suggest that an irreversible process, such as the flow of heat in a gas from high to low temperature, could be interpreted on the microscopic level as a molecular mixing process. The associated connection between thermodynamic entropy and 'molecular disorder' that was established by Boltzmann had already been anticipated qualitatively by Maxwell. An ever-intriguing aspect of this connection has been the impossibility of perpetual-motion machines that violate the second law of thermodynamics. In his 1871 book *Theory of Heat* Maxwell suggested a microscopic mechanism for violating the second law by imagining a molecule-sorting fictitious demon: "Maxwell's demon."

It should be emphasized that the successes of the kinetic theory did not imply a widespread acceptance of reality for the atomic model. Boltzmann himself asserted that "the theory agrees in so many respects with the facts, that we can hardly doubt that in gases certain entities, the number and size of which can roughly be determined, fly about pell-mell." But then he asks rhetorically: "Can it be seriously expected that they will behave exactly as aggregates of Newtonian centres of force, or as the rigid bodies of our Mechanics?" There were many physicists, whose most famous leader was Ernst Mach (1838–1916), who argued that the use of hypothetical concepts, such as atoms or ether, to describe natural phenomena was a violation of science's basic function, which is to achieve an economy of thought through quantitative laws about nature. It was not until a few years after Boltzmann's death that Perrin's experiments on and Einstein's interpretation of Brownian motion, and other accumulated evidence, led to a general belief in the validity of the atomic world.

Classical Statistical Mechanics

Boltzmann's investigations of irreversibility and studies of his famous *H*-function opened up a field of research in what came to be called the "foundations of statistical mechanics." The term "statistical mechanics" was actually coined by J. Willard Gibbs (1839–1903) in 1901, and it was he who first pursued the general use of ensemble theory. Boltzmann, however, first introduced the microcanonical ensemble, with systems of constant energy, for

THE FOUNDING FATHERS OF STATISTICAL MECHANICS

James Clerk Maxwell
(1831–1879)

J. Willard Gibbs
(1839–1903)

Ludwig Boltzmann
(1844–1906)

Albert Einstein
(1879–1955)

(PHOTO OF EINSTEIN FROM BURNDY LIBRARY.
PHOTOGRAPHS COURTESY AMERICAN INSTITUTE OF PHYSICS NIELS BOHR LIBRARY)

which he used the term "ergode." Later, the Ehrenfests, in their famous review article, adopted the term ergodic to refer to a single isolated system whose trajectory passes through (or arbitrarily close to) every point on the energy surface.[7] For Boltzmann's ergode, an aggregate or ensemble of systems whose positions and velocities are distributed over all possible sets of values on the energy surface in $6N$-dimensional space, the Ehrenfests used the term "ergodic distribution" . . . fortunately superseded by the ensemble terminology of Gibbs.[9]

Thus it developed that the distinction between a microcanonical ensemble and an ergodic system was recognized and a program to probe the equivalence of their macroscopic properties, the ergodic theorem, was stated.[7] Included in a developing awareness about the deeper problems embedded in the foundation of his theory was more and more of a reliance by Boltzmann on statistical or probability considerations. Thus, to answer objections directed toward the *Stosszhlansatz,* Boltzmann asserted that one can only expect this assumption to hold most of the time and that deviations from this condition are overwhelmingly improbable. To clarify this situation definitively the Ehrenfests outlined a detailed program for proving Boltzmann's H-theorem, which was based on a statistical approach involving the use of the microcanonical ensemble.[7]

Most of the early work in kinetic theory centered around derivations of the ideal-gas law and the transport properties of a dilute gas in which the intermolecular potential could be neglected, except insofar as it makes collisions possible. In 1868 Boltzmann generalized the Maxwell velocity distribution to include potential energy by means of the "Boltzmann factor" $e^{-U/\kappa T}$.[10] Although U referred originally only to external fields such as gravity, it was subsequently applied, by Boltzmann and others, to the treatment of intermolecular forces for real gases.

Specific consideration of the nonideal gas can be traced to the original work of Clausius (1870) in which the "virial" theorem was derived.[11] By use of this theorem the equation of state for an imperfect gas was related to averages over intermolecular forces. Subsequent developments included the introduction by J. D. Van der Waals (1873) of his famous equation of state[12] with its many successful empirical applications, such as the law of corresponding states. One of the most significant contributions made by Van der Waals was the idea that the same molecular model can describe different macroscopic states of matter as well as transitions between them. Previously it had been thought that the gaseous, liquid, and solid states of matter might each be ascribed to different properties of the constituent molecules.

L. S. Ornstein was the first physicist to attempt (1908) a derivation of the equation of state using ensemble theory. Then, in 1927, H. D. Ursell developed the principles of a cluster expansion of the partition function—a method that was pursued successfully by Mayer, Uhlenbeck, and Kahn, and others.[13] The term "cluster" was first introduced by J. E. Mayer (1937) in his derivation of the cluster-integral expansion of the equation of state. The use of such procedures in quantum statistical mechanics, with an associated reduction of the dynamics of N interacting particles to sums of integrals involving two-body cluster functions, was first studied systematically by Lee and Yang (1959).[14] An underlying motivation in much of this recent work has been a desire to demonstrate the existence of phase transitions within the framework of a single molecular model, or partition function.

Quantum Statistical Mechanics

The failure of the classical kinetic theory of dilute gases to give a satisfactory explanation of the specific heat capacities of polyatomic gases, thereby implying failure of the fundamental equipartition theorem, stimulated much theoretical research toward the end of the 19th century. Clarification by Boltzmann of statistical considerations in the kinetic theory, and his associated statistical interpretation of entropy,[15] was one such effort. In turn, the problem of understanding the puzzling laws of black-body radiation[16] was attacked by both Lord Rayleigh and Max Planck (1858–1947), among others.

Prior to 1900 Planck had been quite skeptical of the atomistic-statistical approach toward a theory of black-body radiation, and he attempted with great effort to arrive at a theory using only macroscopic thermodynamics and electromagnetic theory. It was only after Planck had discovered the correct radiation law empirically (1900) that he departed from these strictly classical considerations and turned to the use of Boltzmann's statistical theory of entropy. In this connection there is even a question as to whether Planck's equation $E = h\nu$ was originally more than a necessary division of phase space for combinatorial calculations in his derivation of the radiation law. Certainly, recognition of the physical importance of this hypothesis as an "energy quantization of the radiation field" was first made several years later by Einstein and Ehrenfest.

One other significant failure of classical statistical mechanics was embodied in Gibbs' paradox, whereby the entropy of an ideal gas fails to be an extensive quantity when calculated microscopically.[17]

An understanding of the limitations of classical mechanics from the viewpoint of the developing quantum theory was as striking in statistical physics as it was in atomic physics. Not surprisingly, the atomic domain of the new quantum concepts had important consequences in the microscopic interpretation of many-particle phenomena. Thus the low-temperature departure of the heat capacities of simple crystals from the classical law of Dulong and Petit was explained by P. Debye (1912) in one of the first truly successful applications of the quantum theory.[18] Debye's analysis was preconditioned by Einstein (1879–1955), who showed how energy-level quantization could explain *qualitatively* the drop in the heat capacities that occurs in solids at low temperatures.[18] The quantization of phase space, first invoked by Planck, coupled with a correct understanding of particle indistinguishability, led to a resolution of Gibbs' paradox.[17]

The realization that phase space is quantized evolved to the concept of quantum-mechanical states for free particles. The interpretation of Planck's radiation law in terms of a correct counting of the possibilities for occupation of quantum-mechanical states by photons was first achieved by S. N. Bose in 1924, and later in the same year Einstein applied this procedure to an ideal gas.[19] Hence the terminology "Bose–Einstein statistics" developed, and the theory of black-body radiation became truly understood statistically. One of the remarkable consequences of this new statistics, was its prediction of a "Bose-Einstein condensation" for Bose (integral-spin) particles into their lowest single-particle energy state.[20] This hypothetical phenomenon, a condensation in momentum space rather than in the familiar position space of most phase transitions, was later associated by F. London (1938) with the so-called λ-transition of liquid helium four from its normal fluid state into its superfluid state below 2K.[20] Bose–Einstein condensation (with sophisticated contemporary modifications) is indeed a reality in nature.

Following the introduction of the exclusion principle by W. Pauli (1900–1958) in 1925, E. Fermi (1901–1954) applied a second kind of statistics, or counting procedure, to statistical discussions.[19] The connection between particle statistics and wave mechanics was then investigated extensively by P. A. M. Dirac (1902–) in 1926. The second kind of particle statistics was referred to subsequently as "Fermi–Dirac statistics." Whereas in its original form the Pauli exclusion principle accounted for the filling of energy levels by atomic electrons, as the atomic number increases throughout the periodic table, application of the generalized Fermi–Dirac statistics has led to an understanding of such diverse phenomena as atomic nuclei, the conduction electrons in metals,[21] liquid helium three near absolute zero, white dwarf stars,[22] and neutron stars. In the limit that Planck's constant $h \to 0$, both of these new statistics reduce to the Boltzmann-statistics case,[23] thereby satisfying Bohr's correspondence principle as is required of all quantum-mechanical formulas.

In the treatment of quantum statistical mechanics, one must be careful to make a clear distinction between the probability aspects inherent to quantum mechanics and the solely statistical considerations introduced by ensemble theory.[24] Thus, the concept of the "pure" quantum-mechanical state developed, in which ordinary (classical) statistical aspects are omitted. The study of pure states involves only the Schrödinger or Heisenberg formulations of wave mechanics. That classical statistical considerations must also be introduced into the general analysis of quantum-mechanical systems, to give the so-called "mixed" states, was realized and treated by both L. D. Landau and J. von Neumann in 1927. The description of mixed quantum-mechanical states by the statistical, or density, matrix[24] was introduced by them and also by Dirac. In the classical limit ($h \rightarrow 0$), the statistical matrix becomes the ensemble probability-density function of classical statistical mechanics.[25]

Irreversible Processes

Fundamental to the understanding of irreversible processes are fluctuation phenomena, of which the prototype is Brownian motion, discovered in 1827 by the nauturalist Robert Brown. Carbonelle (1874) was the first to consider a theory of this phenomenon based on random molecular impulses impinging on the macroscopically small particles which undergo Brownian motion in a fluid. It was not until 1905, however, that the fundamental papers of Einstein and Smoluchowski (1906) presented a modern kinetic-theoretical explanation of the fluctuation process.[26]

Boltzmann himself had been surprisingly unaware of the significance of Brownian motion, and he had assumed that fluctuations could never be observed. Einstein's major breakthrough in this area was his demonstration that it is not the average velocity of a Brownian particle that must be measured, but rather, its mean square displacement as a function of time. The subsequent experiments by Jean-Baptiste Perrin (1908) on Brownian motion were guided by Einstein's theory; 3/4 of a century of earlier measurements were of almost no value because observers had not focussed attention on an appropriate physical quantity. Perrin also measured the variation of concentration with height for Brownian particles in suspension, thereby verifying directly the Boltzmann factor.[10]

Other fluctuation phenomena subsequently investigated were the Schottky effect (1918), in which random thermionic emission in a vacuum tube produces circuit noise upon amplification, and Johnson noise (1928), analyzed theoretically by Nyquist, in which circuit noise is generated by the thermal fluctuations in any conductor.

A general basis for the study of irreversible processes and fluctuation phenomena, or noise, is provided by the fluctuation-dissipation theorem,[27] derived originally by Callen and Welton in 1951. In this theorem, which relies upon the statistical definition of entropy, fluctuations in a system are related to their dissipation and hence to the tendency of any system to go to equilibrium. The theorem is a generalization of earlier work by L. Onsager (1931) in which the Onsager reciprocal relations[28] between dissipation coefficients were deduced.

The approach to thermodynamic equilibrium, i.e., irreversible processes, and departures from the equilibrium condition can be understood theoretically in terms of the fluctuation-dissipation theorem. Boltzmann's H-theorem also deals with irreversibility for the particular case of the velocity distribution function. However, the complete scope of contemporary research into the subject of irreversibility ranges much further. Thus a general equation which governs the time development of system probability functions, whether classical or quantum mechanical, is the master equation, first derived heuristically by W. Pauli in 1928.[29] Theoretical research relating to its derivation continues to the present, because the master equation governs directly the general approach to equilibrium of a system and, also, a form of Boltzmann's H-theorem can be deduced from it. Similar remarks apply to the fluctuation-dissipation theorem which is, in application, closely related to the master equation.[30]

Many other aspects of statistical mechanics, such as the general theory of distribution functions or the theory of phase transitions, have interesting historical facets and significant contemporary developments. The temptation to be comprehensive is resisted here, because the purpose of providing a brief historical perspective of statistical mechanics would then not be achieved.

Bibliography

1. Ludwig Boltzmann, "Lectures on Gas Theory"; translated by Stephen G. Brush (Univ. of California, 1964).
   A comprehensive exposition of the kinetic theory of gases by one of its foremost creators, and a text whose study continues to be valuable even today. The translator's introduction contains an historical sketch of the theory. The bibliography of historically significant papers is extensive.
2. Stephen G. Brush, "The Kind of Motion We Call Heat: A History of the Kinetic Theory of Gases in the Nineteenth Century", edited by E. W. Montroll and J. L. Lebowitz (North Holland, 1976)
   This sixth volume in a series "Studies in Statistical Mechanics" provides an unusually comprehensive account of the development of kinetic theory and statistical mechanics up to the beginning of the 20th century.
3. Stephen G. Brush, "Kinetic Theory Volume 3, The Chapman-Enskog Solution of the Transport Equation for Moderately Dense Gases" (Pergamon Press, 1972).
   The introduction to this text provides a historical survey of the major applications of Chapman-Enskog theory in the 20th century to dense-gas phenomena. It includes reprints of basic works of historical importance in the kinetic theory of dense gases.
4. R. Bowling Barnes and S. Silverman, "Brownian Motion as a Natural Limit to All Measuring Process" in Rev. Mod. Phys. *6* (1934), p. 162.
   In which the history and general description of Brownian Motion, and its relation to various other fluctuation phenomena, is discussed extensively.
5. Sydney Chapman, "The Kinetic Theory of Gases Fifty Years Ago" in "Lectures

in Theoretical Physics, Kinetic Theory," Vol. IXC, edited by W. E. Brittin (Gordon and Breach, 1967).
Reminiscences by Professor Chapman of the famous and elegant solutions of the (microscopic) transport equations for dilute gases, deduced independently by himself and Enskog.
6. Paul and Tatiana Ehrenfest, "The Conceptual Foundations of the Statistical Approach in Mechanics", translated by Michael J. Moravcsik (Cornell Univ., 1959).
A celebrated review of the foundations of statistical mechanics, published originally in 1912 in the German "Encyclopaedia of Mathematical Sciences."
7. D. ter Haar, "Elements of Statistical Mechanics" (Rinehart, 1954).
This text and its second edition, "Elements of Thermostatistics," has historical notes and references included throughout.

Notes to Pertinent Sections of the Text

1. This concept is introduced in the present text in Sec. 12.
2. Equation (13.33) contains this result.
3. Kinetic-theory derivations of the Maxwell-Boltzmann distribution function are given in Sec. 11.
4. This basic procedure is used in Sec. 13 on "Viscous Hydrodynamics."
5. See Eq. (11.27) and Sec. 12.
6. Section 14.
7. See Sec. 25 on "Macroscopic Irreversibility."
8. A mathematical statement of the Stosszahlansatz is Eq. (11.13).
9. See Sec. 26, "Ensemble Theory, a General Discussion."
10. See Eq. (11.29).
11. See Sec. 28 on "Classical Systems."
12. Section 39.
13. See Sec. 37, "Cluster-Integral Expansion of Grand Potential."
14. One part of the methods developed by Lee and Yang is included as App. F.
15. The statistical definition of entropy, in contemporary terms, is given in Sec. 4.
16. Section 35.
17. See the discussion at the end of Sec. 4.
18. Section 30, "Heat Capacity of Simple Crystals."
19. See Sec. 5.
20. Section 34.
21. Section 33.
22. Section 54.
23. See the discussion associated with Eq. (31.25). The distinction between classical statistics and Boltzmann statistics is clarified at the end of Sec. 28.
24. See Sec. 21, "Formal Theory of Statistical Matrix."
25. See Sec. 23.
26. A simplified version of this theory is given in Sec. 15.
27. Section 41.

28. See Sec. 42 on "Irreversible Thermodynamics."
29. See Sec. 45, "Approach to Equilibrium."
30. That macroscopic irreversibility need not be inconsistent with the time reversibility of microscopic processes is discussed in Sec. 25.

Text Notation

Vectors: boldface type
Unit vectors: boldface type with caret: $\hat{\mathbf{i}}$
Quantum-Mechanical Operators: Roman symbols and numbers are sans serif type: H, $\mathbf{\mathsf{A}}$;
Greek symbols are underscored: $\underline{\psi}$, $\underline{\boldsymbol{\sigma}}$

Dyads: Indicated by a double arrow above: $\overleftrightarrow{\mathbf{B}}$
Important Equations: boldface numbering

CHAPTER I

STATISTICAL PHYSICS

Statistical methods are employed in physics whenever problems involving large numbers occur. Particularly relevant are applications to the microscopic world of atoms and molecules when the goal is to understand macroscopic phenomena. In this chapter methods of probability theory are reviewed within the context of defining and deriving the most commonly encountered distribution functions. Typical applications in statistical physics are described, and the statistical definition of entropy is given.

1 Binomial Distribution, Random Walk, and Other Applications

The notion of probability requires the specification of conditions for *events,* or trials, under which a given experiment can be repeated many times. Often we imagine that there is an assembly, or *ensemble,* of M identically prepared experiments for which the frequency of various possible outcomes can be measured. In the simple coin-tossing experiment, for which the outcome is either heads or tails, a collection of M experiments, each with N tosses, would be called the ensemble.

Experimentally calculated probabilities for the occurrence of the various outcomes can be determined from the frequencies measured for any given ensemble. In the limit of large numbers ($M \to \infty$) it is expected that these probabilities will "agree" with those predicted theoretically for the given experimental conditions. Comparison can then yield information about either theory or experiment, or both. Of course, it makes no sense to refer to the probability for the occurrence of an isolated event; probabilities must always refer to a suitably defined ensemble.

In this first section we introduce the most basic theoretical probability distribution, the binomial distribution of probabilities for experiments in which only two outcomes are possible. Several applications in physics will also be given. Then in following sections we shall show how other probability distributions can be related to, or derived from, the binomial distribution.

We begin by considering the simple coin-tossing experiment. Now, to be sure, a good coin will have *equal a priori probabilities* for tossing heads and tails. But we will consider a loaded coin for which the probability p_1 of tossing a head differs from the probability p_2 for tossing a tail. Clearly, $p_1 + p_2 = 1$; for a perfect coin $p_1 = p_2 = 1/2$.

In successive trials of tossing the coin the probabilities p_1 and p_2 will remain constant; in this case the probabilities are *independent* throughout the trials. If n_1 heads are tossed and n_2 tails are tossed, then the (theoretical) probability for any one sequence of heads and tails will be $p_1^{n_1}p_2^{n_2}$.

Although the probability for the occurrence of a given sequence of n_1 heads and n_2 tails can be of interest, usually we wish only to know the probability for all possible sequences of n_1 heads and n_2 tails, that is, the unordered probability. To compute this probability we need only multiply the above single-sequence probability by the total number of distinct sequences,[1]

$$\binom{N}{n_1} \equiv \frac{N!}{n_1!\,n_2!} \qquad (N = n_1 + n_2) \tag{1.1}$$

which number is called the binomial coefficient. Then the (unordered) probability $W_N(n_1)$ for tossing n_1 heads and n_2 tails will be

$$W_N(n_1) = \binom{N}{n_1} p_1^{n_1} p_2^{n_2} \qquad (N = n_1 + n_2) \tag{1.2}$$

That these probabilities are properly normalized, that is, with sum equal to unity, is verified by the binomial expansion

$$1 = (p_1 + p_2)^N = \sum_{n_1=0}^{N} \binom{N}{n_1} p_1^{n_1} p_2^{n_2} \qquad (N = n_1 + n_2) \tag{1.3}$$

The theoretical probability distribution $W_N(n_1)$ is the binomial distribution. When $p_1 = p_2 = 1/2$ it reduces to

$$W_N(n_1)\Big|_{p_1=1/2} = 2^{-N}\binom{N}{n_1} \tag{1.2a}$$

Suppose now that one tosses a coin N times and records the results. In order to make a meaningful comparison with Eq. (1.2) it would be necessary to repeat this N-toss experiment many times, say M, where M is a very large number. Experimentally determined probabilities $W_N^{(M)}(n_1)$ could then be related to the frequencies $f_N^{(M)}(n_1)$ with which n_1 out of N heads are tossed by the definition

$$W_N^{(M)}(n_1) \equiv \frac{f_N^{(M)}(n_1)}{M} \tag{1.4}$$

In the limit $M \to \infty$, we would expect $W_N^{(M)}(n_1)$ to approach the theoretical probability distribution (1.2) or (1.2a), that is,

$$\lim_{M\to\infty} W_N^{(M)}(n_1) = W_N(n_1) \tag{1.4a}$$

In Figure I.1 we compare the results of $N = 10$ coin tosses repeated $M = 250$ times with the theoretical expression (1.2a) for $W_{10}(n_1)$.

A much less accurate, but also much simpler, alternative to making complete comparisons of the type shown in Figure I.1 is to define quantities that measure the "goodness of fit" of Equation (1.2) or (1.2a) to experiment. Two parameters that are normally used as goodness-of-fit criteria are the *average values* $\langle n_1 \rangle$ and $\langle n_1^2 \rangle$. For a binomial distribution the average value $\langle n_1 \rangle$ can be calculated as follows:

$$\begin{aligned}\langle n_1 \rangle = \sum_{n_1=0}^{N} n_1 W_N(n_1) &= p_1 \frac{\partial}{\partial p_1} \sum_{n_1=0}^{N} W_N(n_1) \\ &= p_1 \frac{\partial}{\partial p_1}(p_1 + p_2)^N = Np_1\end{aligned} \tag{1.5}$$

where for $p_1 = 1/2$, we have $\langle n_1 \rangle = N/2$. The particular method by which $\langle n_1 \rangle$ has been evaluated, by employing the generating function $(p_1 + p_2)^N$, is

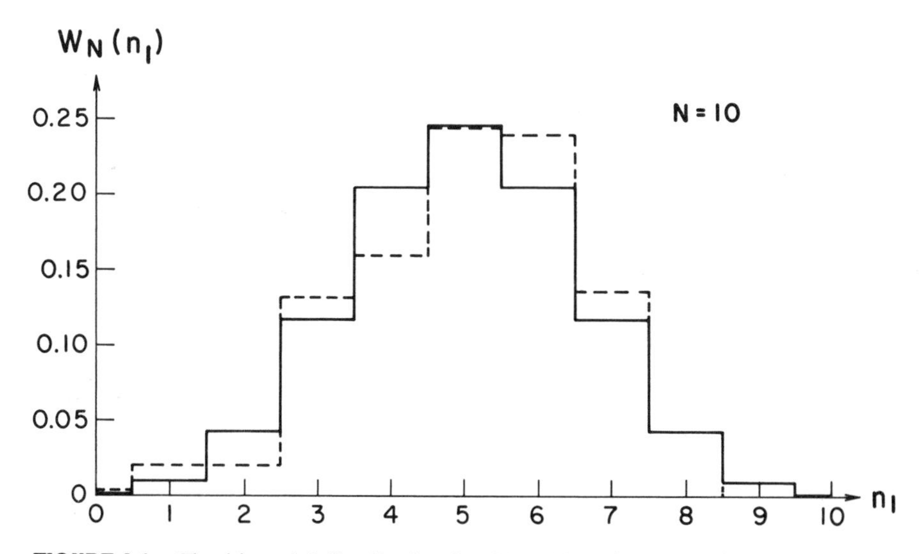

FIGURE I.1 The binomial distribution for $N = 10$ and an experimental histogram (dashed line) for 250 sets of 10 coin tosses.

very commonly used in probability theory (see the end of Section 5). That the average value $\langle n_1 \rangle$ is given by the product Np_1 should be intuitively clear.

Equation (1.5) can also be used to determine the probabilities p_1 and $p_2 = 1 - p_1$ for any given experiment. Thus for an experiment whose outcome can be described theoretically by a binomial distribution we may define p_1 to be the ratio $\langle n_1 \rangle / N$ in the limit $M \to \infty$. Experimentally, $\langle n_1 \rangle$ is defined by using $W_N^{(M)}(n_1)$ in the first equality of (1.5).

The average value of n_1^2 can be calculated in a manner similar to that for $\langle n_1 \rangle$, and one finds for the binomial distribution

$$\begin{aligned} \langle n_1^2 \rangle &= \sum_{n_1=0}^{N} n_1^2 W_N(n_1) = \left(p_1 \frac{\partial}{\partial p_1} \right)^2 \sum_{n_1=0}^{N} W_N(n_1) \\ &= (Np_1)^2 + Np_1 p_2 \end{aligned} \tag{1.6}$$

Now, the quantity that is even more interesting than $\langle n_1^2 \rangle$ is the *mean-square deviation* (or *variance* or *dispersion*) of n_1 from $\langle n_1 \rangle$,

$$\langle (\Delta n_1)^2 \rangle \equiv \langle (n_1 - \langle n_1 \rangle)^2 \rangle = \langle n_1^2 \rangle - \langle n_1 \rangle^2 \tag{1.7}$$

where $\langle n_1 \langle n_1 \rangle \rangle = \langle n_1 \rangle^2$.[2] In passing, we note from this equation, since $\langle (\Delta n_1)^2 \rangle > 0$, that $\langle n_1^2 \rangle$ is always greater than $\langle n_1 \rangle^2$, a result that is true for any averaged quantity. According to Equations (1.5) and (1.6) we have for $\langle (\Delta n_1)^2 \rangle$ the result

$$\langle (\Delta n_1)^2 \rangle = Np_1 p_2 \xrightarrow[p_1 = 1/2]{} (1/4)N \tag{1.8}$$

Therefore the *root-mean-square* (rms) deviation of n_1 from its average value is given by $\sqrt{N}/2$ when $p_1 = 1/2$. This is a general result of statistical physics, namely, that rms deviations, or *fluctuations,* in statistical quantities are usually proportional to $\sqrt{N}$. For the experimental histogram of Figure I.1 $\langle n_1 \rangle = 5.060$ and $\langle n_1^2 \rangle = 2.464$, whereas for the binomial distribution, when $N = 10$ and $p_1 = 1/2$, $\langle n_1 \rangle = 5$ and $\langle n_1^2 \rangle = 2.5$.

We consider next another quantity of interest, which is the difference d between the number (n_1) of heads tossed and the number (n_2) of tails tossed.

$$d \equiv n_1 - n_2 = 2n_1 - N \tag{1.9}$$

Exercise 1.1 Verify the following average values for the binomial distribution:

$$\begin{aligned} \langle d \rangle &= N(p_1 - p_2) \xrightarrow[p_1 = 1/2]{} 0 \\ \langle d^2 \rangle &= \langle d \rangle^2 + 4Np_1 p_2 \xrightarrow[p_1 = 1/2]{} N \end{aligned} \tag{1.10}$$

The quantity d occurs in several of the physics problems, which we now discuss.

Random Walk and Brownian Motion

In the one-dimensional random-walk problem, steps of length (a) in both the positive and negative directions are taken randomly. Assuming the probabilities p_+ and p_- to be equally likely ($p_+ = p_- = ½$), then the distances (da) traveled after N steps, for a large number of experiments all starting from the origin, will be expected to follow the binomial distribution (1.2a) with $n_1 = ½(N + d)$. According to the first of Equations (1.10), the average value $\langle d \rangle a$ for many experiments with N steps will be zero. But this result should not be confused with the fact that after N steps the rms distance from the origin will, on the average, have the value,

$$(\Delta d)_{\text{rms}} \cdot a = \langle d^2 \rangle^{1/2} a = \sqrt{N} a \tag{1.11}$$

Figure I.2 shows an example of random walk in two dimensions. In colloidal suspensions there is a random zigzag motion of the colloidal particles, which, except for the small influence of gravity on their concentration, is essentially three-dimensional random walk. The causes of the random walk are to be found, ultimately, in the fluctuating momentum transfers due to molecular collisions between the colloidal particles and the fluid molecules. The end result of the motion is that in any given small time interval τ a colloidal particle will travel the random vector distance $\boldsymbol{\ell}$. Equation (1.11) then leads us to guess that after a time $t = N\tau$ the colloidal particle will be expected to have traveled a distance

$$\langle \mathbf{r}_N^2 \rangle^{1/2} = \sqrt{N}\, \ell = \sqrt{3}\, \alpha\, t^{1/2} \tag{1.12}$$

Simple kinetic-theory considerations are required to derive the coefficient α.[3] To show that the first equality in Equation (1.12) is indeed correct, we use the fact that in the time interval τ, after time $(N - 1)\tau$, the colloidal particle will travel a random vector distance

$$\boldsymbol{\ell} = \mathbf{r}_N - \mathbf{r}_{N-1}$$

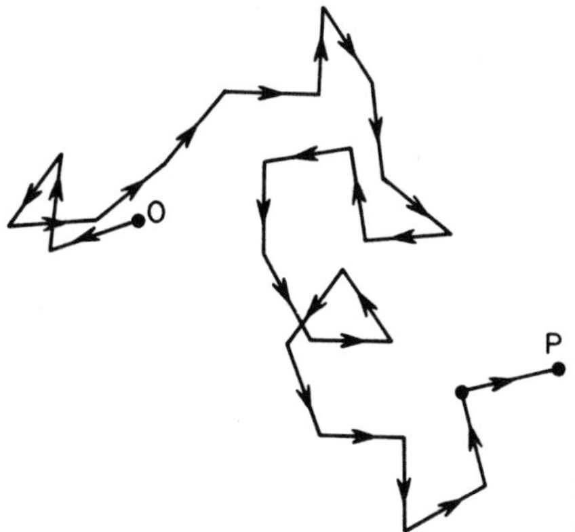

FIGURE I.2 An example of random walk in two dimensions, from 0 to P in 25 equal-length steps.

The average $\langle \mathbf{r}_N^2 \rangle$ will then be given by

$$\langle \mathbf{r}_N^2 \rangle = \langle \mathbf{r}_{N-1}^2 \rangle + \langle \boldsymbol{\ell} \cdot \mathbf{r}_{N-1} \rangle + \langle \ell^2 \rangle = \langle \mathbf{r}_{N-1}^2 \rangle + \ell^2$$

which leads to Equation (1.12) by an induction proof.

Polymers and Rubber Bands

In the simplest model of deformable polymer chains or rubber bands we consider a one-dimensional chain of N units each of length a. The units can fold on each other in all possible ways so as to give an effective length to the polymer, which we denote by $(d \cdot a)$. See Figure I.3. To investigate the question as to what elastic force the polymer exerts when it is deformed, that is, when d is varied, we first need to know how many possible configurations can occur for a given value of d. The answer, of course, is just the binomial coefficient of Equation (1.1), with $n_1 = \frac{1}{2}(N + d)$. This idealized problem can now be solved with the aid of simple thermodynamic considerations.[4]

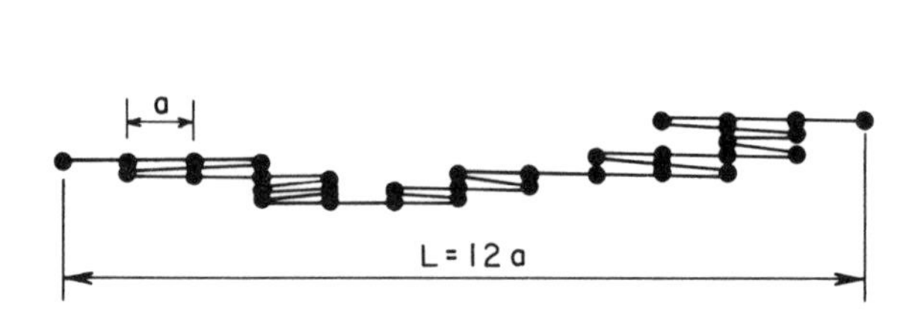

FIGURE I.3 The one-dimensional folding model of a chain molecule, shown with a vertical spread for clarity. Starting from the left end, there are $n_+ = 23$ units directed to the right and $n_- = 11$ units directed to the left, so that the net length of the chain is $L = (n_+ - n_-)a = da = 12a$.

Ferromagnetism

In ferromagnetic metals each atom has a net spin and magnetic moment, and it is the spontaneous alignment of the atomic magnets that results in bulk ferromagnetic properties. For the simplest model of a ferromagnet we consider a regular arrangement of N spin-½ atoms in a magnetic field, whose direction (z) defines the axis of quantization for the magnetic quantum numbers $m = \pm\frac{1}{2}$.

We represent the spin state $m = +\frac{1}{2}$ by $\uparrow$ and $m = -\frac{1}{2}$ by $\downarrow$ and attach a subscript α to denote the state of the αth atom. Since there are two states for each atom, there will be 2^N quantum states in all, each of which can be represented symbolically by a state vector Ψ_i of the form

$$\Psi_i = \uparrow_1 \uparrow_2 \downarrow_3 \uparrow_4 \downarrow_5 \cdots \downarrow_N$$

Clearly, these Ψ_i are eigenfunctions of the total spin-angular-momentum operator S_z, where

$$\mathsf{S}_z = \sum_{\alpha=1}^{N} \mathsf{S}_{\alpha,z}$$

If we denote by $n_\pm$ the number of atoms with spin state $m = \pm\frac{1}{2}$, then the eigenfunctions of S_z will be given by $\frac{1}{2}\,d\hbar$,

$$\mathsf{S}_z\,\Psi_i = \left(\frac{d}{2}\right)\hbar\,\Psi_i \tag{1.13}$$

where $d = n_+ - n_- = 2n_+ - N$.

In the original version of Heisenberg's theory of ferromagnetism,[5] one needed to enumerate the various possible spin states for N atoms. For the simple model of spin-½ atoms, the number of ways in which a given eigenvalue of S_z can occur is, of course, $\binom{N}{n_+}$ of Equation (1.1), where $n_+ = \frac{1}{2}(N + d)$. Of further interest is the enumeration of eigenfunctions of $\mathsf{S}^2 = \mathbf{S}\cdot\mathbf{S}$. These eigenfunctions, denoted by χ_n, are linear combinations of the Ψ_i, and they can be simultaneously eigenfunctions of S_z.

Consider the special case $N = 4$. In Table I.1 we have listed all $2^4 = 16$ eigenfunctions χ_n in rows according to their eigenvalues of S_z, Equation (1.13). Since each eigenvalue $s(s + 1)\hbar^2$ of S^2 must be associated with $(2s + 1)$ S_z eigenvalues, we see that the columns of Table I.1 can be arranged to list all the eigenfunctions of S_z for each s value. In this example, then, there are five eigenfunctions with $s = 2$, nine with $s = 1$, and two with $s = 0$. Equivalently, we can say that the degeneracy of s values (for the lowest eigenvalue of $|\mathsf{S}_z|$) is one for $s = 2$, three for $s = 1$, and two for $s = 0$.

Table I.1 Eigenfunctions of S^2 for Four Spin-½ Atoms, Arranged in Rows According to Their S_z Eigenvalues.

		$(\mathsf{S}_z\chi_i = d/2\hbar\chi_i;\ \mathsf{S}^2\chi_i = s(s+1)\hbar^2\chi_i)$					
$d/2$	$\binom{N}{n_+}$	$s = 2$	$s = 1$	$s = 1$	$s = 1$	$s = 0$	$s = 0$
2	1	χ_1					
1	4	χ_2	χ_6	χ_9	χ_{12}		
0	6	χ_3	χ_7	χ_{10}	χ_{13}	χ_{15}	χ_{16}
−1	4	χ_4	χ_8	χ_{11}	χ_{14}		
−2	1	χ_5					

$(n_\pm = 2 \pm d/2)$

We now generalize the results of the preceding paragraph and Table I.1. Let

$$\mathcal{N}(\mathsf{S}_z) = \binom{N}{n_+} = \binom{N}{\frac{1}{2}(N+d)}$$

be the number of eigenfunctions χ_n (or Ψ_i) with eigenvalue $(d/2)\hbar$, and let $\mathcal{N}_D(s)$ be the degeneracy or number of eigenfunctions χ_n for $s = |d/2|$. By referring to Table I.1 for the special case $N = 4$, one can deduce that the general formula for $\mathcal{N}_D(s)$ is given by

$$\mathcal{N}_D\left(\frac{N}{2}\right) = \binom{N}{N} = 1$$

$$\mathcal{N}_D(s) = \binom{N}{n_+} - \binom{N}{n_+ + 1}, \quad \left(\frac{N}{2} \leqq n_+ < N\right) \tag{1.14}$$

in which n_+ and d are related by $n_+ = \frac{1}{2}(N + d)$ and $s = d/2$. By this procedure we have enumerated all possible s values and their degeneracies for any given N.

Exercise 1.2 List all of the eigenvalues of S_z and S^2 for eight spin-½ atoms. What is the degeneracy $\mathcal{N}_D(s)$ for each s value?

Now when N is large, the binomial coefficient $\binom{N}{n_+}$ will always be much larger than $\binom{N}{n_+ + 1}$, except for n_+ values around $n_+ \cong N/2$. For example, when $N = 24$, so that there are $2^{24} = 16{,}777{,}216$ possible spin states, we can write down the list of n_+ values and corresponding binomial coefficients given as in Table I.2. For large n_+ values it is clear that in Equation (1.14)

Table I.2 Binomial Coefficients for $N = 24$.

n_+	$d/2$	$\binom{N}{n_+}$
24	12	1
23	11	24
22	10	276
21	9	2,024
20	8	10,626
19	7	42,504
18	6	134,596
17	5	346,104
16	4	735,471
15	3	1,307,504
14	2	1,961,256
13	1	2,496,144
12	0	2,704,156

Source: M. Abramowitz and I. A. Stegun, *Handbook of Mathematical Functions* (Dover, 1965), p. 828.

we may set

$$\mathcal{N}_D(s) \cong \binom{N}{n_+} \qquad (n_+ \sim N >> 1) \tag{1.15}$$

The magnetization M associated with the spins is determined by the eigenvalues of S_z, Equation (1.13). If g is the intrinsic g factor for the atoms (e.g., $g \cong 2$ for electrons), then M is given by

$$M = g\left(\frac{d}{2}\right)\left(\frac{e\hbar}{2mc}\right) \tag{1.16}$$

When combined with a simple dynamical theory of the valence electrons in ferromagnetic substances, this result, together with the preceding statistical considerations, becomes Heisenberg's theory of ferromagnetism.[5] In particular, one can use the approximate expression (1.15) for the number of spin states with spin $s\hbar$ in the calculation of the expected magnetization $\langle M \rangle$. (An alternative formulation of this theory is given in Section 52.)

Exercise 1.3 The number of eigenvalues of S_z associated with a given spin value $s = |d/2|$ is $2s + 1 = |d| + 1$. Show, using Equation (1.14), that

$$\sum_{n_+ \geqq N/2}^{N} (d + 1)\mathcal{N}_D\left(\frac{d}{2}\right) = 2^N \tag{1.17}$$

Is this result physically reasonable?

2 Limit $N \to \infty$, Poisson and Normal Distributions

Statistical physics is characterized by problems involving large numbers. In this section we show that in the limit $N \to \infty$ the binomial distribution can be made to approach either a Poisson distribution or a normal distribution, according to how the limit is taken. For large, but finite, N values these latter distributions can be good approximate representations of the binomial distribution. For appropriate experimental conditions the Poisson and the normal (or Gaussian) distributions may be valid independently for describing the outcome of probability measurements.

To proceed we need an asymptotic expression, that is, Stirling's formula, for $N!$ when $N >> 1$. Thus we are motivated to study the gamma function $\Gamma(N)$,[6] which is an integral representation of $(N - 1)!$, defined for $\text{Re}(z) > 0$ by

$$\Gamma(z) = \int_0^\infty dt\, t^{z-1}e^{-t} \xrightarrow[z=N>0]{} (N - 1)! \tag{2.1}$$

It is easy to show, after an integration by parts, that $\Gamma(z)$ satisfies the recursion relation

$$\Gamma(z + 1) = z\Gamma(z) \tag{2.2}$$

and, therefore, that the second line of (2.1) is correct for z = (positive integer). We see from Equation (2.1) that, since $\Gamma(1) = 1$, the value of 0! is unity.

When z is 1/2, we can calculate $\Gamma(\frac{1}{2})$ by the following manipulations:

$$\begin{aligned}[\Gamma(1/2)]^2 &= \left(\int_0^\infty dt(t^{-1/2})e^{-t}\right)^2 = \left(2\int_0^\infty dx\, e^{-x^2}\right)^2 \\ &= \int_{-\infty}^{+\infty} dx\, e^{-x^2} \cdot \int_{-\infty}^{+\infty} dy\, e^{-y^2} \\ &= \int_0^{2\pi} d\phi \int_0^\infty r\, dr\, e^{-r^2} = \pi\end{aligned}$$

where $t = x^2$ and we have switched from the cartesian coordinates (x, y) to the polar coordinates (r, ϕ) in going to the last line of this equation. Therefore, since $\Gamma(1/2)$ is a positive number, we may conclude that

$$\Gamma\left(\frac{1}{2}\right) = \sqrt{\pi} \tag{2.3}$$

Exercise 2.1 Verify, for all positive integers n, the two definite integrals:

$$\begin{aligned}\int_0^\infty x^{2n+1}\, e^{-ax^2}\, dx &= (1/2)n!/a^{n+1} \\ \int_0^\infty x^{2n}\, e^{-ax^2}\, dx &= \sqrt{\frac{\pi}{a}}\,\frac{(2n-1)!!}{2^{n+1}a^n}\end{aligned} \tag{2.4}$$

where $(2n - 1)!! = (2n)!/2^n n!$. [Thus $(-1)!! = (1)!! = 1$, $(3)!! = 3$, $(5)!! = 15$, etc.] (*Suggestion:* First verify these results for the case $n = 0$, and then differentiate the $n = 0$ answers with respect to a.)

Stirling's Formula

We now derive Stirling's formula, which is an approximate expression for $N!$ valid in the limit of large N. The technique we shall use is based on the method of steepest descents, which is described in some detail in Appendix B. We can understand this method intuitively by observing that the integrand of Equation (2.1) for $\Gamma(N + 1)$ has a sharp maximum, when N is large, for some positive t value. See Figure I.4. In fact, the maximum is so sharp when $N >> 1$ that it is far more accurate to write the integrand in exponential form and then to expand the argument of the exponent about the maximum value than it is to expand the integrand itself directly about its maximum value. Thus we write

$$\begin{aligned} t^N e^{-t} &= \exp(-t + N \ln t) \equiv e^{g(t)} \\ g(t) &= N \ln t - t \end{aligned} \tag{2.5}$$

The maximum of both the integrand and $g(t)$ occurs at the value $t_o = N$, so that we can write

$$\begin{aligned} g(t) &= g(t_o) - (1/2N)(t - t_o)^2 + O(t - t_o)^3 \\ &\cong N \ln N - N - (2N)^{-1}(t - t_o)^2 \end{aligned}$$

We see that the integrand (2.5) does indeed have an extremely sharp maximum for large N; because of the exponential form it becomes negligible

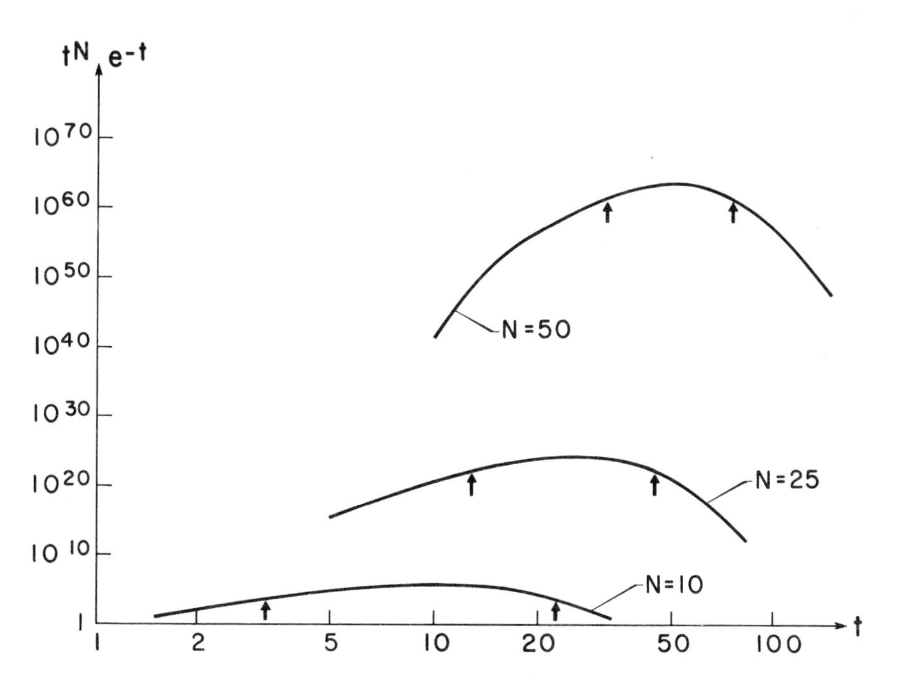

FIGURE I.4 Integrand of the gamma function for three N values, showing how the sharp maximum develops with increasing N. The arrows indicate where the integrand has fallen to 1% of its maximum value.

when $|t - t_o| >> \sqrt{N}$.[7] Therefore, for large N we may write Equation (2.1) to good accuracy as

$$\Gamma(N+1) = N! \underset{N\to\infty}{\longrightarrow} e^{N \ln N - N} \int_0^{\infty} dt\, e^{-(t-t_o)^2/2N}$$

$$\cong \sqrt{2N}\, e^{N \ln N - N} \int_{-\infty}^{+\infty} dx\, e^{-x^2}$$

or

$$N! = \sqrt{2\pi N}\left(\frac{N}{e}\right)^N \qquad (N \to \infty) \tag{2.6}$$

The last equality of Equation (2.6) is Stirling's formula for $N!$. It can be shown that an improvement in the accuracy of this formula is obtained when the factor $[1 + 1/12N]$ is included on the right side.[8]

Poisson Distribution

The limit $N \to \infty$ can be taken for the binomial distribution with $p_1 \to 0$ such that $Np_1 \equiv \lambda$ remains finite. The result is called the *Poisson distribution* $P_\lambda(n_1)$ of n_1 values. To derive the expression for $P_\lambda(n_1)$ we first use Equation (2.6) to show, for $n << N$, that

$$\ln \frac{N!}{(N-n)!} \underset{N\to\infty}{\longrightarrow} n \ln N + \frac{n}{2N} + O\left(\frac{n^2}{N}\right)$$

Thus $N!/(N-n)! \cong N^n$ for very large N values. We may show similarly that $p_2^{n_2} = (1-p_1)^{N-n_1} = \exp[(N-n_1)\ln(1-p_1)] \cong e^{-Np_1} = e^{-\lambda}$ in the limit $N \to \infty$ and $p_1 \to 0$, with $\lambda \equiv Np_1 =$ finite. Then, when we use these two limiting expressions the binomial distribution (1.2) becomes[9]

$$P_\lambda(n_1) \equiv \lim_{\substack{N\to\infty \\ p_1\to 0 \\ (Np_1=\lambda)}} [W_N(n_1)] = e^{-\lambda}\left(\frac{\lambda^{n_1}}{n_1!}\right) \tag{2.7}$$

The Poisson distribution is normalized "properly" in the sense that $\Sigma_{n_1=0}^{n_1=\infty} P_\lambda(n_1) = 1$, with large n_1 values ($n_1 \gtrsim N$) contributing negligibly to the sum. [For large n_1 values, $P_\lambda(n_1) \sim e^{-\lambda}(\lambda e/n_1)^{n_1}$.]

Exercise 2.2 Show that the average value $\langle n_1 \rangle$ equals λ for the Poisson distribution, in agreement with Equation (1.5) for the binomial distribution.

Although the Poisson distribution is used frequently as a good approximation to the binomial distribution (to simplify calculations), it is also a probability distribution valid in its own right. For example, radioactive disintegrations obey a Poisson distribution. The criteria for valid application of the Poisson distribution are that: (1) experimental conditions remain constant, in time usually, and (2) information concerning the number of randomly occurring events in one interval does not affect the probability for observing events in any other interval. Also, (3) there must be negligible probability for observing more than one event in the limit that the interval size goes to zero. To clarify these points, we now discuss the example of radioactive decay.

Consider the α particles emitted by a long-lived radioactive source during some short time interval t. One can imagine this time interval to be subdivided into many very small intervals, each of length Δt. Since the α particles are emitted at random times, the probability of a radioactive disintegration occurring during any one time interval Δt is completely independent of whatever disintegrations occur at other times. Furthermore, Δt can be imagined to be chosen small enough so that the probability for more than one disintegration occurring in the time interval Δt is negligibly small. This means that there is a probability p of one disintegration occurring during a time Δt (with $p \ll 1$, when Δt is chosen small enough) and a probability $(1 - p)$ of no disintegration occurring during the same time. Each time interval Δt can be regarded as an independent trial, there being a total of $N = t/\Delta t$ such trials during a time t.

From the preceding argument we may conclude that the probability for observing n_1 α-decays in N time intervals Δt will follow a binomial distribution for sufficiently small, but finite, Δt values. By writing $p = \lambda_\alpha (\Delta t)$ and letting $\Delta t \to 0$ we satisfy the conditions established for the derivation of the Poisson distribution (2.7). In the limit $\Delta t \to 0$ we obtain Equation (2.7) exactly, with λ replaced by $\lambda_\alpha t$.

Note that the particular case $n_1 = 0$, in which no decays occur, corresponds to the well-known radioactive decay law for parent nuclei. Although they are one and the same, one can distinguish conceptually between the decay constant λ_α for parent nuclei decays, considered alone, and the time constant $\langle n_1 \rangle t^{-1} = \lambda_\alpha$ for the complete Poisson distribution (2.7), as considered above, when multiple decays in a time t are included.

Normal Distribution

When N is large, the binomial distribution $W_N(n_1)$ is appreciable only for (large) n_1 values close to the average value Np_1. Table I.2 of Section 1 shows this fact, when $p_1 = 1/2$, even for the comparatively small value $N = 24$. The fact that $W_N(n_1)$ is appreciable only near its maximum value, when $N \gg 1$, can be exploited to derive an approximate form in the large-N limit. This limiting form of the binomial distribution, called the *normal or Gaus-*

sian distribution, describes continuous distributions of variables. The normal distribution occurs frequently in physics.

If N is large and we consider regions near the maximum of W_N, where n_1 is also large, then the fractional change in W_N when n_1 changes by unity will be quite small,[10] that is,

$$|W_N(n_1 + 1) - W_N(n_1)| \ll W_N(n_1), \qquad \text{(near max of } W_N) \tag{2.8}$$

Thus, to good approximation, W_N can be considered as a continuous function of the continuous variable n_1, even though only integral values of n_1 are of physical relevance. Moreover, since $W_N(n_1)$ exhibits a sharp maximum when $N \ggg 1$, we can determine an exponential form for $W_N(n_1)$. We are thereby led to investigate the expansion of

$$\begin{aligned} \ln W_N(n_1) &= \ln(N!) - \ln(N - n_1)! - \ln(n_1!) \\ &\quad + n_1 \ln p_1 + (N - n_1)\ln p_2 \end{aligned} \tag{2.9}$$

about its maximum value at $n_1 = \bar{n}_1$.

Using Stirling's formula (2.6), we rewrite Equation (2.9) in the approximate form, for $n_1 \sim N \gg 1$, as

$$\begin{aligned} \ln W_N(n_1) &= N \ln N - (N - n_1)\ln(N - n_1) - n_1 \ln n_1 - \frac{1}{2}\ln 2\pi \\ &\quad + \frac{1}{2}[\ln N - \ln(N - n_1) - \ln n_1] + n_1 \ln p_1 \\ &\quad + (N - n_1)\ln p_2 \end{aligned} \tag{2.10}$$

The most probable value $\bar{n}_1$ is then determined by the condition

$$\begin{aligned} \left.\frac{d \ln W_N(n_1)}{dn_1}\right|_{n_1 = \bar{n}_1} &= \ln(N - \bar{n}_1) - \ln \bar{n}_1 + \frac{1}{2}\left[\frac{1}{N - \bar{n}_1} - \frac{1}{\bar{n}_1}\right] \\ &\quad + \ln p_1 - \ln p_2 = 0 \end{aligned}$$

Since $p_1 + p_2 = 1$, one finds for large $(N, \bar{n}_1)$ the value

$$\bar{n}_1 = Np_1 = \langle n_1 \rangle \tag{2.11}$$

Thus in the continuous-distribution limit the mean value $\langle n_1 \rangle$ of n_1 is identical with the most probable value $\bar{n}_1$.

If we expand $\ln W_N(n_1)$ about $\bar{n}_1$, then we find the Taylor-series expression

$$\begin{aligned} \ln W_N(n_1) &= \ln W_N(\bar{n}_1) - \frac{(n_1 - \bar{n}_1)^2}{2Np_1p_2} + \frac{1}{6}\frac{(p_2^2 - p_1^2)}{(Np_1p_2)^2}(n_1 - \bar{n}_1)^3 + \cdots \\ &\cong \ln W_N(\bar{n}_1) - \frac{1}{2}\frac{(n_1 - \bar{n}_1)^2}{\Delta n_{\text{rms}}^2} \end{aligned} \tag{2.12}$$

where we have used Equation (1.8) for the rms value of Δn_1 to obtain the second line of this equation. We may drop the third term in the first line of Equation (2.12) in the region of n_1 values

$$|n_1 - \bar{n}_1| \ll Np_1p_2 \tag{2.13}$$

which condition is equivalent to (2.8), since the latter inequality can be rewritten as

$$\left|\frac{d \ln W_N(n_1)}{dn_1}\right|_{n_1 \sim \bar{n}_1} \ll 1$$

To verify the equivalence, differentiate Equation (2.12).

On using Equations (2.10)–(2.12), the explicit form for $W_N(n_1)$ in the limit $N, n_1 \to \infty$ becomes the normal distribution $P(n_1)$

$$P(n_1) \equiv \lim_{N,n_1 \to \infty} [W_N(n_1)] = \frac{1}{\sqrt{2\pi \, \Delta n_{\text{rms}}^2}} \exp\left\{-\frac{1}{2}\left[\frac{(n_1 - \bar{n}_1)^2}{\Delta n_{\text{rms}}^2}\right]\right\} \tag{2.14}$$

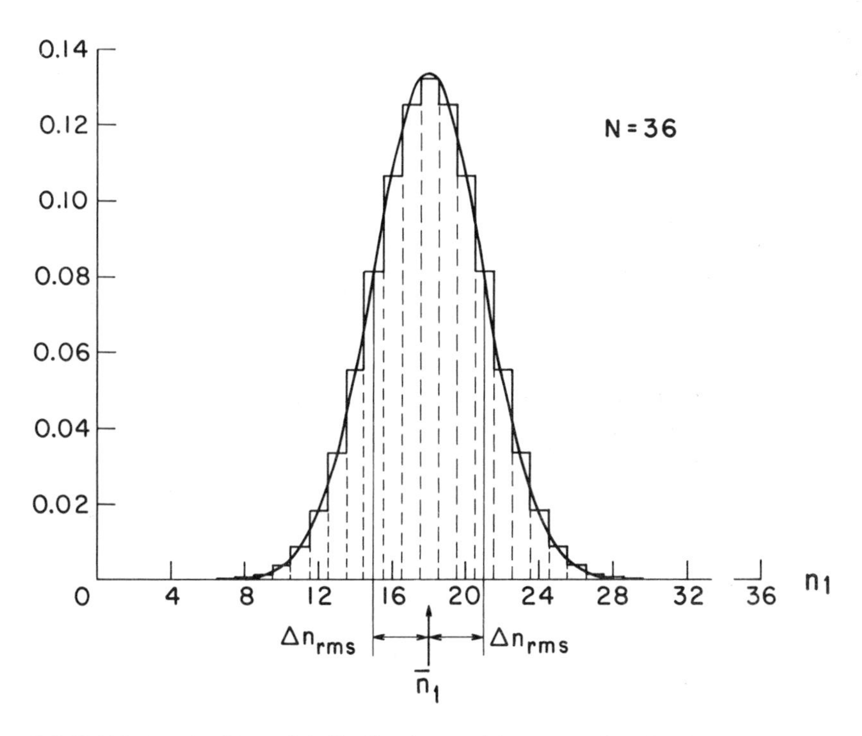

FIGURE 1.5 The binomial distribution and its approximation by a normal distribution for $N = 36$ and $p_1 = p_2 = \frac{1}{2}$.

This distribution falls rapidly to zero for n_1 values such that

$$|n_1 - \bar{n}_1| >> \Delta n_{rms} = \sqrt{Np_1p_2} \tag{2.15}$$

which condition must not conflict with (2.13) if Equation (2.14) is to be accurate for all n_1 values where it is appreciable. But conditions (2.13) and (2.15) are consistent if $Np_1p_2 >> 1$, which implies, for $p_1 \sim p_2 \sim 1$, that $N >> 1$. Since the condition of large N is inherent in the derivation of Equation (2.14) from Equation (2.9), we may conclude that the entire calculation has been self-consistent.

In Figure I.5 we show the binomial distribution and its normal-distribution approximation for the case $N = 36$ and $p_1 = p_2 = 1/2$, for which $\bar{n}_1 = 18$ and $\Delta n_{rms} = 3$.

Exercise 2.3 Verify directly from Equation (2.14) that:

$$\int_0^\infty dn_1\, P(n_1) = 1$$

$$\langle n_1 \rangle \equiv \int_0^\infty dn_1 \cdot n_1 \cdot P(n_1) = \bar{n}_1$$

$$\langle (n_1 - \langle n_1 \rangle)^2 \rangle = \Delta n_{rms}^2$$

Here we must assume that $\bar{n}_1 \to \infty$, so that the use of continuous n_1 values can be justified.

3 Continuous Distributions, Central Limit Theorem

The normal distribution introduced in the preceding section is an example of a *continuous probability distribution*. In this section we first clarify the meaning of continuous distributions for one or more variables, and then we introduce the notion of stochastic independence. Following this, we discuss the problem of finding an unknown continuous probability distribution $\mathcal{P}(x)$ when x is a randomly changing variable. Under certain general conditions the solution to this problem is provided by the central limit theorem, which we investigate only for the simple example of one-dimensional random walk. More generally, the central limit theorem shows that $\mathcal{P}(x)$ will usually be a normal probability distribution.

A continuous distribution $P'(n)dn$ can always be related to some corresponding discrete distribution $W(n)$ by the relation

$$W(n) \cong P'(n)\, dn \tag{3.1}$$

The quantity $P'(n)$ is called a *probability density*, and dn is an infinitesimal interval that is characterized by a single value of the variable n. Here it should be noticed that when the number of intervals, or trials, becomes very large, the corresponding probability $W(n)$ for any one interval value n becomes extremely small, as is suggested by Equation (3.1).

We can understand the continuum limit in the following sense. A macroscopic, but very small, interval δn can be characterized by one negligibly varying probability value $W(n)$. Multiplying by the number $(\delta n/dn)$ of discrete values of n in this interval gives, according to Equation (3.1), the result

$$W(n)\left(\frac{\delta n}{dn}\right) \cong P'(n)\,\delta n$$

The left side of this equation is the probability for finding any n value in the interval δn, so that $P'(n)$ is, indeed, a probability density.

Assuming that the probability distribution $W(n)$ is normalized to unity, we have in the continuum limit the result

$$1 = \sum_n W(n) \cong \int P'(n)\,dn \tag{3.2}$$

The integral in Equation (3.2) is a definite integral with limits of integration determined by the range of values of n. Similarly, the average value of some function $f(n)$ of n can be calculated in the continuum limit by the relation

$$\langle f(n)\rangle = \sum_n f(n)W(n) \cong \int f(n)P'(n)\,dn \tag{3.3}$$

In the limit in which the number of intervals becomes infinite the approximate equalities in Equations (3.1) to (3.3) become exact, and the integration limits span an infinite interval. In these integrals we may also transform to some other variable $s = s(n)$, in which case we must use different probability densities $P(s)$, defined by

$$P(s) = P'(n)\left|\frac{dn}{ds}\right| \tag{3.4}$$

The quantity $|dn/ds|$ is the density of states for the interval ds.

Distributions Involving Several Variables

The probability density P may be a function of two or more independent variables, corresponding in the discrete case to two or more sets of integers that characterize the probability distribution W. For example, the random-walk process in three dimensions involves a distribution function of three independent variables. We shall not write down the discrete case; in fact, we discuss only the sufficiently general case of a continuous probability distribution $P(s, t)\, ds\, dt$ of two variables s and t. The normalization in this case is determined by the definite integral

$$\int P(s, t)\, ds\, dt = 1 \tag{3.5}$$

where the double integration limits range over the complete span of possible s and t values. We can also write down the generalization of Equation (3.3) that gives the average value of a function $f(s, t)$. By integrating over all allowed values of one variable, say, t, the probability density $P(s, t)$ can be reduced to the probability density of a single variable, discussed above:

$$P(s) = \int P(s, t)\, dt \tag{3.6}$$

Thus, for example, the average value of a function $f(s)$ does not depend on the distribution of t values given by $P(s, t)$, so that we obtain

$$\langle f(s) \rangle = \int f(s)P(s, t)\, ds\, dt = \int f(s)P(s)\, ds$$

Stochastic Independence

Two variables s and t are said to be *stochastically independent*, or statistically independent, if the corresponding probability function $P(s, t)$ [or $W(n_1, n_2)$ in the discrete case] satisfies the equality

$$P(s, t) = P(s)P(t) \tag{3.7}$$

where $P(s)$ is defined by Equation (3.6), and similarly for $P(t)$. Thus, when there is stochastic independence, an event characterized by s has a probability distribution $P(s)$ that is completely independent of the variable t.

The stochastic independence of three variables s_1, s_2, and s_3 requires not only that they be pairwise stochastically independent, as with Equation (3.7), but also that

$$P(s_1, s_2, s_3) = P(s_1)P(s_2)P(s_3) \tag{3.8}$$

Ordinarily, the three pairwise conditions of the type (3.7) imply condition (3.8), and vice versa, but there are examples, mostly nonphysical, of exceptions to this rule.[11] The definition of stochastic independence for distributions involving more than three variables should be clear. The random-walk problem in three dimensions involves a distribution function of three independent variables that are stochastically independent, unless there is a constraint requiring, for example, that the random walk be restricted to the surface of a sphere.

The concepts of multivariable probability distributions and stochastic independence can be applied to give a more general description of single-variable processes, such as the one-dimensional random-walk problem. We imagine that each trial in the process (displacement from the preceding location in one-dimensional random walk) involves a distribution function. Thus after i steps the probability density distribution will be $P(s_1, s_2, \ldots, s_{i-1}, s_i)$, where s_i is the variable characterizing the ith trial. If there are N trials in all, then the complete distribution function will be $P(s_1, s_2, \ldots, s_i, \ldots, s_N)$. Moreover, if the N trials are all stochastically independent, then we will have

$$P(s_1, s_2, \ldots, s_i, \ldots, s_N) = P(s_1)P(s_2) \cdots P(s_i) \cdots P(s_N) \tag{3.9}$$

with the function $P(s)$ being the same for every trial. The right side of (3.9) can be interpreted as the probability density distribution for N successive, independent, and identical experiments. In probability theory the associated N-dimensional space is called a *product space*.

It might seem that the above description of a single-variable process is unnecessarily cumbersome, but as we show below, this viewpoint provides an extremely powerful tool for the general analysis of random events. To appreciate this fact, note that $P(s)$ in Equation (3.9) permits step sizes other than the uniform length a of the random-walk problem discussed in Section 1. We can, of course, describe the fixed-step-length random-walk problem in one dimension with the above mathematical apparatus by setting

$$P(s) = p_+\delta(s - a) + p_-\delta(s + a) \tag{3.10}$$

Exercise 3.1 In the N-step one-dimensional random-walk process the net displacement x is defined by

$$x = \sum_{i=1}^{N} s_i \tag{3.11}$$

where s_i is the displacement during the ith step.

(a) Show that the average value $\langle x \rangle$ is given by $\langle x \rangle = N\langle s \rangle$ and that

$$(\Delta x)^2_{\text{rms}} \equiv \langle (x - \langle x \rangle)^2 \rangle = \left\langle \sum_{i,j=1}^{N} \Delta s_i \Delta s_j \right\rangle = N(\Delta s)^2_{\text{rms}}$$

where $\Delta s_i = s_i - \langle s \rangle$.

(b) Using Equation (3.10) for the special case of the fixed-step-length problem, show that $\langle x \rangle = (p_+ - p_-)Na$ and $(\Delta x)^2_{\text{rms}} = 4Np_+p_-a^2$, in agreement with the results of Section 1, where the particular choice $p_+ = p_- = 1/2$ was made for this problem.

In the general single-variable random process, a question that can be asked is: What final probability density function $\mathcal{P}_N$ can be expected after N identical experiments are repeated independently? Here, the distribution function $\mathcal{P}_N$ represents the probability density of the cumulative effect of individual processes, each governed by some unknown distribution function P. For the special case of one-dimensional random walk we would want to know the "most likely" probability distribution $\mathcal{P}_N(x)\,dx$ to be expected for the variable x defined by Equation (3.11).

Now the Dirac δ function $\delta(x - x_o)$ is the probability density for finding x at the known value x_o. Therefore, to find $\mathcal{P}_N(x)$ for one-dimensional random walk, one must average $\delta[x - \Sigma_{i=1}^{N} s_i]$ over all possible values of the N variables s_i using the probability density function (3.9). One obtains

$$\mathcal{P}_N(x) = \int \delta\left[x - \sum_{i=1}^{N} s_i\right] P(s_1)P(s_2) \cdots P(s_N)\, ds_1\, ds_2 \cdots ds_N \tag{3.12}$$

in which an N-dimensional integral must be evaluated over the entire ranges of the variables s_i.

Central Limit Theorem

Despite the very complicated appearance of Equation (3.12), we can use it to determine $\mathcal{P}_N(x)$ for large N by making a few very plausible assumptions concerning the behavior of $P(s)$. The result is a special case of the *central limit theorem* for the probability distribution of a randomly occurring variable. One version of this theorem can be stated as follows.

Suppose that we have a sequence ($i = 1, 2, \ldots, N$) of mutually independent random variables s_i that are governed by probability density functions $P(s_i)$, and let the Fourier transform $Q(k)$ be defined by

$$Q(k) \equiv \int ds\, e^{-iks} P(s) \tag{3.13}$$

Then, subject to conditions on the moments of $P(s)$, such as those to be discussed below,[12] we obtain in the limit of large N a normal form for the probability density function $\mathcal{P}_N(x)$ of the variable $x = \Sigma_{i=1}^{N} s_i$.

$$\mathcal{P}_N(x) = \frac{1}{\sqrt{2\pi\sigma^2}} e^{-(x-x_o)^2/2\sigma^2}\left[1 + \mathrm{O}\left(\frac{1}{\sqrt{N}}\right)\right] \tag{3.14}$$

where

$$x_o = N\langle s\rangle$$
$$\sigma^2 = N(\Delta s)^2_{\text{rms}} \tag{3.15}$$

The central limit theorem determines, in essence, the most probable probability distribution for the variable x.

We now return to the special case of one-dimensional random walk and consider Equation (3.12). First we introduce the Fourier transform of the δ function.

$$\delta\left(x - \sum_{i=1}^{N} s_i\right) = \frac{1}{2\pi}\int_{-\infty}^{+\infty} dk \exp\left[ik\left(-\sum_{i=1}^{N} s_i + x\right)\right] \tag{3.16}$$

Substitution of this expression into Equation (3.12) then yields for $\mathcal{P}_N(x)$ the alternative form

$$\mathcal{P}_N(x) = \frac{1}{2\pi}\int_{-\infty}^{+\infty} dk\, e^{ikx}[Q(k)]^N \tag{3.17}$$

in which definition (3.13) has been used.

Exercise 3.2 Substitution of Equation (3.10) for the constant-step-length random-walk problem into Equation (3.13) yields

$$Q(k) = p_+ e^{-ika} + p_- e^{ika}$$

Insert this result into Equation (3.17) for $\mathcal{P}_N(x)$, and show that

$$\mathcal{P}_N(x) = \sum_{n_+=0}^{N} \binom{N}{n_+} p_+^{n_+} p_-^{(N-n_+)}\, \delta[x - (2n_+ - N)a]$$

where $(2n_+ - N)$ is the difference d of Equation (1.9). Therefore, the discrete probability distribution

$$W_N(n_+) \equiv \int_{(2n_+-N)a-\varepsilon}^{(2n_+-N)a+\varepsilon} \mathcal{P}_N(x)\, dx$$

agrees with our earlier result (1.2). Here ε is a small positive length.

Consider next any interval Δs in the integration region of Equation (3.13) such that

$$\left|\frac{dP(s)}{ds}\right|_{s=s_o} \Delta s \ll P(s_o) \tag{3.18}$$

where $P(s)$ and its first derivative are evaluated at some arbitrary point s_o in the interval Δs. We also consider sufficiently large k values so that

$$k(\Delta s) \gg 1 \tag{3.19}$$

Under these conditions, for which $P(s)$ is slowly varying and the exponential factor e^{-iks} is rapidly varying, we have

$$\int_{\Delta s} ds\, e^{-iks} P(s) \cong P(s_o) \int_{\Delta s} ds\, e^{-iks} \cong 0$$

But for very large k values this result will be true for all intervals Δs, with each Δs governed by inequality (3.18). Therefore, we may surmise, by combining inequalities (3.18) and (3.19), that $Q(k)$ is negligible (i.e., falls off rapidly with increasing k) at sufficiently large k values such that for all s_o,

$$\frac{1}{k}\left|\frac{dP(s)}{ds}\right|_{s=s_o} \ll \left|\frac{dP(s)}{ds}\right|_{s=s_o} \cdot \Delta s \ll P(s_o) \tag{3.20}$$

Note that this smoothly varying condition on $P(s)$ will not be satisfied by probability densities, such as (3.10), which are not truly random. But even for such distributions the final result (3.14) of the central limit theorem may be true in the limit of large N, as Exercise 3.2 and the derivation of the normal distribution at the end of Section 2 demonstrate for the special case (3.10).

The above discussion suggests that we may often be able to derive a useful approximate expression for $Q(k)$ by making a Taylor-series expansion about $k = 0$.[13] More important, since $Q(k)$ falls rapidly to zero for large k values, as determined by (3.20), $[Q(k)]^N$ will, when $N >>> 1$, fall to zero extremely rapidly for large k values. So we may with good confidence in the general applicability of the approximation proceed to write

$$\begin{aligned} Q(k) &= \int ds\, P(s)\left[1 - iks - \frac{1}{2}k^2s^2 + \cdots\right] \\ &= 1 - ik\langle s\rangle - \frac{1}{2}k^2\langle s^2\rangle + \cdots \end{aligned} \tag{3.21}$$

For large N values we then use this assumed result in an expansion of the logarithm of $[Q(k)]^N$, which exhibits a sharp maximum around $k = 0$, to obtain

$$\begin{aligned} \ln[Q(k)]^N &= N \ln\left[1 - ik\langle s\rangle - \frac{1}{2}k^2\langle s^2\rangle + \cdots\right] \\ &= N\left[-ik\langle s\rangle - \frac{1}{2}k^2(\Delta s)^2_{\text{rms}} + \cdots\right] \end{aligned}$$

Thus, to good approximation,

$$[Q(k)]^N \cong \exp\left(-iNk\langle s\rangle - \frac{1}{2}Nk^2(\Delta s)^2_{\text{rms}}\right) \tag{3.22}$$

Upon substituting Equation (3.22) into Equation (3.17), the final result for $\mathcal{P}_N(x)$ becomes

$$\mathcal{P}_N(x) \cong \frac{1}{2\pi}\int_{-\infty}^{+\infty} dk \exp[ik(x - N\langle s\rangle)] \exp\left(-\frac{1}{2}Nk^2(\Delta s)^2_{\text{rms}}\right)$$

$$= \frac{1}{2\pi}\int_{-\infty}^{+\infty} dk \exp\left(-\frac{N}{2}(\Delta s)^2_{\text{rms}}\right)\left[k + i\left(\frac{N\langle s\rangle - x}{N(\Delta s)^2_{\text{rms}}}\right)\right]^2 \exp\left(-\frac{(x - N\langle s\rangle)^2}{2N(\Delta s)^2_{\text{rms}}}\right)$$

$$= \frac{1}{\sqrt{2\pi\sigma^2}} \exp(-(x - x_o)^2/2\sigma^2) \qquad \textbf{(3.23)}$$

which is Equation (3.14) for the special case of one-dimensional random walk. Here we have used the second integral of Exercise 2.1 to derive the final equality.

By explicit calculation (see Exercise 2.3) one can show easily that x_o is the average value $\langle x\rangle$ and σ is the rms value $(\Delta x)_{\text{rms}}$. Therefore, we have derived in a quite general manner that after a large number N of random steps in one dimension the distribution of cumulative displacements x from the origin will follow the normal distribution (2.14). This is the central limit theorem for the special case of a one-dimensional random walk. The theorem applies to a wide range of randomly occurring physical phenomena, such as errors in measurement. It is also used in mathematical studies of the foundations of statistical mechanics. Investigations of this type are not pursued in this text, however.

Exercise 3.3 Consider the next term in the expansions of $Q(k)$ and $\ln[Q(k)]^N$ in order to find the leading correction to Equation (3.23) for $\mathcal{P}_N(x)$. Define $k_o \equiv i(\Delta x/\sigma^2)$, where $\Delta x = x - N\langle s\rangle = x - \langle x\rangle$, and argue that the spread $\Delta k = (k - k_o)$ of significantly contributing k values to the integral for $\mathcal{P}_N(x)$ is proportional to $1/\sqrt{N}$. Then derive the expression

$$\mathcal{P}_N(x) = \frac{1}{\sqrt{2\pi\sigma^2}} \exp\left[-\frac{1}{2}\left(\frac{\Delta x}{\sigma}\right)^2\right]\left\{1 - \frac{1}{6\sqrt{N}}\left(\frac{\Delta x}{\sigma}\right)\left[3 - \left(\frac{\Delta x}{\sigma}\right)^2\right]\right.$$
$$\left.\times\left[\frac{\langle s^3\rangle - 3\langle s^2\rangle\langle s\rangle + 2\langle s\rangle^3}{(\Delta s)^3_{\text{rms}}}\right] + O\left(\frac{1}{N}\right)\right\}$$

4 Maxwell–Boltzmann Distribution for Free Particles, Statistical Entropy

The Maxwell–Boltzmann distribution for the velocities or momenta of molecules in an ideal gas is a normal distribution. This fact suggests that the central limit theorem of the preceding section might be utilized to give a derivation of the Maxwell–Boltzmann distribution. We first pursue this idea to arrive at the distribution function for the total energy of an ideal gas, and then we indicate the nature of Khinchin's actual derivation along these lines of the Maxwell–Boltzmann distribution.

The derivation of the Maxwell–Boltzmann distribution for an ideal gas which we choose to outline in this section is a popular one, in which the probability distribution for occupation of single-particle states is maximized subject to energy- and particle-number constraints. We discuss the Maxwell–Boltzmann distribution in some detail, and then, after introducing the statistical definition of entropy, we show that this derivation is equivalent to maximizing the entropy. The section is concluded with a short discussion of the Gibbs paradox.

We begin with a brief characterization of the *ideal gas* of classical mechanics by considering a box of volume Ω in which there are N noninteracting atoms or molecules. Now, in any real gas the particles are continually undergoing collisions. To have collisions, in which kinetic energy is transferred from one molecule to another, intermolecular forces (if only hard-sphere repulsions) must be operative in the system. The magnitude of these forces, or the laws governing them, need not be known for the present discussion. However, it is essential that the forces be negligibly small, except at distances of approach between the molecules that are very small compared with the average distance between them. Thus if we denote by r_o the average range of the intermolecular forces, we must have

$$r_o \ll \ell \qquad \text{(for almost-ideal gases)} \tag{4.1}$$

where $\ell^3 = \Omega/N$. Only under this condition is the potential energy in the gas negligible for all probable positions of the molecules in the system. Stated differently, condition (4.1) implies that at any instant only an infinitesimal fraction of the molecules will be in the process of undergoing a collision. The gas is then one with very weakly (or rarely) interacting molecules. In the limit $r_o \rightarrow 0$, the gas is an ideal gas.

We also neglect gravitational effects and consider sufficiently high temperatures so that quantum effects can be ignored. This latter condition is satisfied if $\lambda_T \ll \ell$, where the quantum-mechanical thermal wavelength, which is a measure of wave-function overlap in the gas, is defined by

$$\lambda_T = (2\pi\hbar^2/m\kappa T)^{1/2} \tag{4.2}$$

Here T is the temperature of the gas and m is the mass of a single molecule whose average momentum is $\sim(m\kappa T)^{1/2}$. (The factor of 2π is inserted arbitrarily for later convenience.)

For the classical ideal gas, the probability density function $P(\mathbf{k}_1, \mathbf{k}_2, \ldots, \mathbf{k}_N)$, where we use the symbol $\mathbf{k}$ to denote a single-particle momentum, will be expected to show stochastic independence. We expect it to be of the form (3.9).

$$P(\mathbf{k}_1, \mathbf{k}_2, \ldots, \mathbf{k}_N) = P(\mathbf{k}_1)P(\mathbf{k}_2)P(\mathbf{k}_3) \cdots P(\mathbf{k}_N) \tag{4.3}$$

But our primary interest here is not in the probability density for all the momenta, $\mathbf{k}_1 \cdots \mathbf{k}_N$. Rather, we wish first to determine the "most likely" probability density $\mathcal{P}(\mathbf{K})$ for the total momentum $\mathbf{K}$ of the system, where

$$\mathbf{K} = \sum_{i=1}^{N} \mathbf{k}_i \tag{4.4}$$

We use a three-dimensional version of the development at the end of the preceding section and write the density $\mathcal{P}(\mathbf{K})$, in analogy with Equation (3.12), as

$$\mathcal{P}(\mathbf{K}) = \int d^3\mathbf{k}_1\, d^3\mathbf{k}_2 \cdots d^3\mathbf{k}_N\, \delta\left[\mathbf{K} - \sum_{i=1}^{N} \mathbf{k}_i\right] P(\mathbf{k}_1)P(\mathbf{k}_2) \cdots P(\mathbf{k}_N) \tag{4.5}$$

To determine $\mathcal{P}(\mathbf{K})$, we follow the development that begins with Equation (3.16) and write

$$\mathcal{P}(\mathbf{K}) = \frac{1}{(2\pi)^3} \int_{-\infty}^{+\infty} d^3\mathbf{R}\, e^{-i\mathbf{K}\cdot\mathbf{R}}[Q(\mathbf{R})]^N \tag{4.6}$$

where the Fourier transform $Q(\mathbf{R})$ of the probability density $P(\mathbf{k})$ is defined by

$$Q(\mathbf{R}) \equiv \int d^3\mathbf{k}\, e^{i\mathbf{k}\cdot\mathbf{R}} P(\mathbf{k}) \tag{4.7}$$

The subsequent derivation of Equation (3.23) applies here, with changes required only to incorporate the vector notation for a three-dimensional problem. Thus in place of Equation (3.21) we write

$$Q(\mathbf{R}) = 1 + i\langle \mathbf{k} \rangle \cdot \mathbf{R} - \frac{1}{2} \sum_{i,j=1}^{3} \langle k_i k_j \rangle R_i R_j + \cdots \tag{4.8}$$

Now, for a system in which the particles are moving isotropically we must have $\langle k_i k_j \rangle = \langle k_i^2 \rangle\, \delta_{ij}$. Moreover, if the system is at rest we will have $\langle \mathbf{k} \rangle = 0$ and $\langle k_i^2 \rangle = \frac{1}{3}\langle \mathbf{k}^2 \rangle$. Under these conditions we obtain:

$$Q(\mathbf{R}) = 1 - \frac{1}{6}\langle k^2 \rangle R^2 + \cdots$$

For $[Q(\mathbf{R})]^N$ we then get

$$[Q(\mathbf{R})]^N = e^{-(N/6)\langle k^2 \rangle R^2} \tag{4.9}$$

Substitution of Equation (4.9) into Equation (4.6) yields, finally, the result

$$\begin{aligned}
\mathcal{P}(\mathbf{K}) &= \frac{1}{(2\pi)^3}\int_{-\infty}^{+\infty} d^3\mathbf{R}\,\exp(-i\mathbf{K}\cdot\mathbf{R})\exp(-N\langle k^2\rangle R^2/6) \\
&= \frac{1}{4\pi^2 iK}\int_0^{\infty} R\,dR\,\exp(-N\langle k^2\rangle R^2/6)[e^{iKR} - e^{-iKR}] \\
&= \frac{1}{4\pi^2 iK}\int_{-\infty}^{\infty} R\,dR\,\exp\left(-\frac{N}{6}\langle k^2\rangle\left[R - \frac{3iK}{N\langle k^2\rangle}\right]^2\right) e^{-3K^2/2N\langle k^2\rangle} \\
&= \frac{3}{4\pi^2 N\langle k^2\rangle} e^{-3K^2/2N\langle k^2\rangle}\int_{-\infty}^{+\infty} dx\,\exp(-N\langle k^2\rangle x^2/6) \\
&= \left(\frac{3}{2\pi N\langle k^2\rangle}\right)^{3/2} e^{-3K^2/2N\langle k^2\rangle} \qquad (4.10)
\end{aligned}$$

The result (4.10) for $\mathcal{P}(K)$ can be simplified somewhat by defining

$$E_{\text{c.m.}} = \frac{K^2}{2Nm} \qquad (4.11)$$

and

$$\beta = \frac{3}{2}\left(\frac{2m}{\langle k^2\rangle}\right) \qquad (4.12)$$

Clearly, $E_{\text{c.m.}}$ is the energy of the center-of-mass (c.m.) motion of the particles in the gas, and $\frac{3}{2}\beta^{-1}$ is their average kinetic energy ($\beta^{-1} = \kappa T$, as we know but discuss later). With these definitions we obtain for $\mathcal{P}(K)$ the expression

$$\mathcal{P}(K) = \left(\frac{\beta}{2\pi Nm}\right)^{3/2} e^{-\beta E_{\text{c.m.}}} \qquad (4.13)$$

We might inquire at this point as to why we did not start in Equation (4.3) with probability densities for the individual particle energies $\omega_i = k_i^2/2m$. Indeed we could have, and then, with luck, we would have followed the lengthy path of Khinchin's derivation of the Maxwell–Boltzmann distribution for the molecules of an ideal gas.[14] We prefer to give an alternative derivation, one that is both popular and illustrative of another important method in statistical physics.

Maxwell–Boltzmann Distribution

Equation (4.13) shows, when the average momentum $\langle\mathbf{K}\rangle$ is zero, that the c.m. energy of a system is distributed according to the famous Maxwell–Boltzmann law:

$$P \propto e^{-\beta E} \qquad (4.14)$$

For a macroscopic system this distribution of energies varies only over an extremely small range on any macroscopic energy scale. Equation (4.13) can be interpreted in terms of fluctuation motions for the system about its rest position.

Of greater interest, usually, is the distribution function $P(\mathbf{k})$ for a single particle. Intuitively, we might expect Equation (4.14) to apply to a free particle as well as to the entire ideal gas. We defer to Equation (4.18) and Exercise 4.1 for a derivation of this anticipated result and first explore some of its implications. Thus we write for the probability density $P(\mathbf{k}_i)$ the expression

$$P(\mathbf{k}_i) = \mathcal{N} e^{-\beta\omega_i} \tag{4.15}$$

where $\omega_i = \mathbf{k}_i^2/2m$. This expression is, of course, the Maxwell–Boltzmann distribution for individual particle momenta, or velocities, in a classical ideal gas. Its normalization constant $\mathcal{N}$ is the same as that for $\mathcal{P}(K)$ in Equation (4.13), except that $N = 1$ for $P(\mathbf{k}_i)$. The fraction of particles in the ith momentum interval $d^3\mathbf{k}_i$, with momentum $\mathbf{k}_i$, is then

$$\frac{\bar{n}_i}{N} = P(\mathbf{k}_i)\, d^3\mathbf{k}_i = \left(\frac{\beta}{2\pi m}\right)^{3/2} e^{-\beta\omega_i}\, d^3\mathbf{k}_i \tag{4.16}$$

Consider for this paragraph the language of quantum mechanics, in which it is meaningful to talk about *momentum states* and to refer to the density of momentum states $d^3\mathbf{n}/d^3\mathbf{k} = \Omega/(2\pi\hbar)^3$. Then Equation (4.16) can also be written as the number of particles in the ith momentum state $d^3\mathbf{n}_i$ with momentum $\mathbf{k}_i$, that is

$$\bar{n}_i = (n\lambda_T^3)\, e^{-\beta\omega_i}\, d^3\mathbf{n}_i \tag{4.17}$$

Here n is the particle density N/Ω, and we have introduced the thermal wavelength λ_T of Equation (4.2), assuming $\beta = (\kappa T)^{-1}$. For the particular choice $d^3\mathbf{n}_i$ = one quantum state, we obtain in the high-temperature (or classical) limit where $n\lambda_T^3 \ll 1$, the result $\bar{n}_i \ll 1$, as one would expect. Classical mechanics is characterized by a low probability for the occupation of single quantum states.

Let us consider next macroscopic cells $\delta^3\mathbf{n}_i$ in momentum space, normalized as above and of size such that $\delta^3\mathbf{n}_i \gg 1$, but still sufficiently small that the probability distribution (4.16) varies negligibly over all momentum values in the cell. (Here we are using the concept of a continuous distribution as discussed at the beginning of Section 3.) Under these conditions the advertised derivation of Equation (4.17) can be given.[15] We let Z_i be the size, or degeneracy, of the ith momentum cell $\delta^3\mathbf{n}_i$, that is, we set $Z_i = \delta^3\mathbf{n}_i$. Then the probability $W_N(n_1, n_2, \ldots)$ of finding the N free particles, with n_i of them in the ith cell ($i = 1, 2, \ldots$), is given by

$$W_N(n_1 \cdots n_N) = CN! \prod_i \frac{Z_i^{n_i}}{n_i!} \tag{4.18}$$

The product Π_i ranges over the entire (denumerably infinite) set of (distinguishable) momentum cells Z_i, and C is a normalization constant to be determined below.[16]

To derive Equation (4.18) one must use a generalization of the combinatorial number $\binom{N}{n_1}$ of Equation (1.1) and the assumption that the probability for finding a particle in the ith cell is proportional to the size Z_i of that cell. This latter condition is a mathematical statement of the assumption that there are *equal a priori probabilities* for finding particles in various quantum states. The assumption is a fundamental one in statistical physics, and its validity is tested only indirectly when the predictions of statistical mechanics are verified experimentally. The derivation of Equation (4.17) from Equation (4.18) is the subject of the following important exercise.

Exercise 4.1 Assume (for $N >>> 1$) that the significantly contributing n_i values in Equation (4.18) are all large and that $W_N(n_i)$ exhibits a sharp maximum as a function of each of these n_i. Then derive Equation (4.17), for the most probable number $\overline{n}_i$ of particles in the ith cell, by maximizing $\ln W_N(n_i)$ subject to the two constraints that the total number of particles,

$$N = \sum_i n_i \tag{4.19}$$

and their total energy,

$$E = \sum_i n_i \omega_i \tag{4.20}$$

are both constant. [*Suggestion:* Use Lagrange's method of undetermined multipliers (see Problem II.1), with unknown multipliers α and $-\beta$.] [With $Z_i = \delta^3\mathbf{n}_i$, one can use the constraint equation (4.19) to show that the normalization constant e^α is equal to $n\lambda_T^3$. Or, without the language of quantum mechanics, one can instead obtain Equation (4.16) by assuming that $Z_i \propto \delta^3\mathbf{k}_i$—an assumption due originally to Gibbs.]

Information Theory, Statistical Entropy, and the Gibbs Paradox

One notion, mentioned only briefly at the start of Section 1, although included implicitly throughout our discussions of probability distributions, is the concept of the ensemble. Whenever we describe a situation from a statistical point of view, that is, in terms of probabilities, we must imagine that there is an assembly, or *ensemble*, that consists of a very large number M of identically prepared systems. The probability of occurrence of a particular event (for example, a particular set of occupation numbers n_i in a gas) can then be defined operationally with respect to the ensemble and is given, for $M \to \infty$, by the fraction of systems in the ensemble that are characterized by the occurrence of the specified event. In this connection it is important to define accurately the ensemble being considered, such as by restricting the ensemble to identical fermion systems or to systems containing all possible ionized states of a fixed number of N_2 molecules, and so on. It makes no sense to speak simply of the probability that one isolated system or particle will have some particular energy.

For the ideal gas of classical mechanics we can identify an ensemble with one system itself and refer to each of the N noninteracting molecules as a member of the ensemble. We can then inquire as to the probability $P_i Z_i$ for finding n_i particles in the ith momentum cell Z_i. Here P_i is defined as the corresponding probability per unit cell size (or quantum-mechanical state). In terms of the notation used above, the unknown probabilities P_i are defined operationally by

$$P_i = \frac{n_i}{NZ_i} \tag{4.21}$$

Note the distinction between cell sizes Z_i and probabilities P_i per unit cell size for their occupation (see Figure I.6). We note in passing that in the absence of any constraints on the gas, the P_i for large N would be expected to be independent of i; that is, they would all be equal (and extremely small).

In the limit $N \rightarrow \infty$ we can expect that Equation (4.21) for operationally defined probabilities P_i will agree with *a priori*, or intrinsic, probabilities $\overline{P}_i$, characteristic of the ideal-gas system. Alternatively, an average set of probability values $\langle P_i \rangle$, computed by applying Equation (4.21) to a large number of similar ensembles, would also be expected to characterize the ideal-gas system. We would expect that $\overline{P}_i = \langle P_i \rangle$.[17] Henceforth, we shall choose either one of these two procedures (usually the former) for arriving at a set of *a priori*, or intrinsic, probabilities. To justify the former choice we now show that it is equivalent to a general procedure of information theory, in which the statistical entropy is maximized. Thus, for example, the procedure of Exercise 4.1 for determining the $\overline{n}_i$ (for the "most likely" probabilities $\overline{P}_i$) can be characterized as a procedure of information theory in which the *uncertainty of our information*, that is, the entropy, is maximized.

Suppose that for some situation n possibilities can occur. Information theory is concerned not with the information that is known, such as N and E in Exercise 4.1, but with the unknown or missing information denoted by $I = I(n)$. If we assume that each of the n possibilities is equally probable, then it is possible to specify $I(n)$ to within a constant after making a few simple postulates:

$$I(n) > I(m), \qquad \text{if } n > m \tag{4.22a}$$

$$I(1) = 0 \tag{4.22b}$$

$$I(nm) = I(n) + I(m) \tag{4.22c}$$

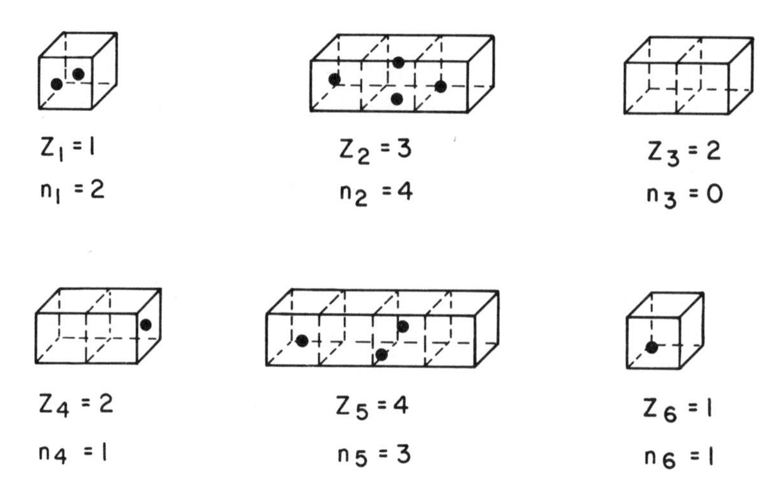

FIGURE I.6 Arbitrary (but integral-sized) cells Z_i, in which Z_1 and Z_6 are unit cells. There are n_i particles in the ith cell and n_i/Z_i particles per unit cell size in the ith cell. The operationally defined probability P_i, per unit cell size, for occupation of the ith cell is given by $P_i = n_i/NZ_i$, where $N = \Sigma_i n_i$. Note that in the derivation of Exercise 4.1 one must choose all $Z_i >> 1$.

The first of these postulates says that as n increases, the amount of missing information also increases. The second postulate says that there is no missing information when $n = 1$. Finally, with the third postulate it is required that missing information be additive. That is, when (missing) information is accumulated or grouped into m mutually exclusive sets, each set itself characterized by n possibilities, then the total amount of information must be equal to the sum of the information conveyed by choosing first from among the m sets and then from among the n possibilities in the selected set.

It can be shown (see Problem I.7) that the requirements (4.22) lead to a unique expression for $I(n)$ that is valid as a continuous function of n; namely, that the missing information is given by

$$I(n) = \kappa(\ln n) \tag{4.23}$$

where κ is a positive constant to be determined by the particular situation under consideration. In statistical physics we call the missing information (or uncertainty in our knowledge about systems) S, instead of I, and the arbitrary constant κ is identified eventually as Boltzmann's constant. From the information-theory point of view, Equation (4.23) is taken to be the definition of *statistical entropy* S.[18]

For the ideal gas we can evaluate the entropy, or missing information, explicitly. We must, of course, consider N to be very large so that the "ensemble'' will be properly representative of the molecules in the gas. Then, on including the assumption (or "given information'') that the probability for occupation of the ith cell is proportional to the cell size (or degeneracy) Z_i, we conclude that the number of possibilities n in Equation (4.23) is precisely $W_N(n_i)$ of Equation (4.18) with $C = 1$.

$$S = \kappa \ln W_N(n_1, n_2, \cdots)\big|_{C=1} \tag{4.24}$$

In the justification of Equation (4.24), one must be careful to distinguish between the degeneracies Z_i (which may vary from cell to cell) and the equally likely possibilities for different arrangements of particles within the cells (for a given set of n_i values).

The expression (4.24) can be rewritten in terms of the P_i of Equation (4.21) by using Stirling's formula (2.6) for the significantly contributing n_i, assuming all $Z_i >> 1$. [The $\sqrt{2\pi n_i}$ factors must be dropped so that (2.6) will also apply to all the vanishingly small n_i.] One finds for the statistical entropy per particle, when $N >>> 1$, the result

$$S_{\text{system}} \equiv \frac{1}{N} S = -\kappa \sum_i Z_i [P_i \ln P_i] \tag{4.25}$$

This expression itself is quite general, and it is often used in entropy discussions, usually with $Z_i = 1$ so that i ranges over all possible states of a system.[19]

Exercise 4.2

(a) Argue that Equation (4.24) follows from Equation (4.23), with $I \rightarrow S$, by considering a two-step process similar to that described below Equations (4.22). First consider all possible ways of arranging the N particles into a given set of n_i values. Then in the second step assume for the ith cell that there are Z_i equal compartments into which each of the n_i particles can be placed.

(b) Derive Equation (4.25) from Equation (4.24).

We can now see, from the information-theory point of view, that Exercise 4.1 determines the most probable n_i values by maximizing the statistical entropy (4.24) subject to the constraints (4.19) and (4.20). But maximum entropy means maximum missing information or maximum uncertainty. Any other entropy value would be equivalent to less uncertainty, that is, to more knowledge, about the system than occurs for the maximum entropy condition. It is natural to assume, then, that in nature undisturbed (or closed) many-body systems will evolve until our ignorance about them is maximal, with experimentally available information about the system inserted as constraints on the maximum condition. Thus we use the *maximum entropy prin-*

ciple in statistical physics to determine equilibrium situations; nonequilibrium situations always correspond to smaller values of the entropy. At equilibrium, then, S takes the value

$$\bar{S} = \kappa \ln [W_N(n_i)]_{\max} \tag{4.26}$$
$$= \kappa \ln W_N(\bar{n}_i)$$

For the classical ideal gas this expression is correct, provided that we do not set $C = 1$ in Equation (4.18), as we now discuss.

The fact that the normalization constant C should not be set equal to unity in Equations (4.24) and (4.26) is simply a consequence of the indistinguishability of identical particles. In effect, the indistinguishability of the particles is known information that must be included in the entropy expression. We may argue (as did Gibbs) that, since the N identical particles are indistinguishable, each of the possibilities $W_N(n_i)$ must be divided by $N!$, the number of possible N-particle permutations. Accordingly, we set $C = (N!)^{-1}$ for identical-particle systems, even in the limit in which classical mechanics applies.[20] Stated differently, with $C = 1$ in Equation (4.26) we must add an entropy constant $-\kappa \ln N! \cong \kappa N(1 - \ln N)$ to the right side. This added term, also required in order that $\bar{S}/N$ be an intensive quantity (see Exercise 4.3), was not understood at first from the point of view of classical physics, and hence the need for it was originally called the *Gibbs paradox*.

Exercise 4.3 Set $Z_i = \delta^3\mathbf{n}_i$ in Equation (4.25), and show for the continuum limit that the entropy per particle is

$$\frac{1}{N}\bar{S} \xrightarrow[C=1]{} \kappa\left[\frac{3}{2} - \ln(n\lambda_T^3) + \ln N\right] \qquad (\neq \text{intensive quantity})$$
$$\xrightarrow[C=(N!)^{-1}]{} \kappa\left[\frac{5}{2} - \ln(n\lambda_T^3)\right]$$

Note that $\bar{P}_i = (\lambda_T^3/\Omega)\, e^{-\beta\omega_i}$, according to Equations (4.17) and (4.21). This second result for S is known as the Sackur–Tetrode equation for the entropy of a classical ideal gas.

5 Counting Problems, Statistical Entropy of Quantum Ideal Gases

At the end of the preceding section we explained, for a system of identical particles described classically, that we must include a factor of $(N!)^{-1}$ in the probability function $W_N(n_1, n_2, \ldots, n_i, \ldots)$. The proper origin of this factor is a treatment of the indistinguishability of identical-particle systems from a quantum-mechanical viewpoint. From the viewpoint of classical mechanics, the $(N!)^{-1}$ must be included so that the thermodynamic entropy per particle is an intensive quantity. When the factor $(N!)^{-1}$ is included in classical mechanics, then we say that the particles are described by *Boltzmann statistics*.

In this section we begin by elaborating on the entropy expression for the classical ideal gas using Boltzmann statistics, and then we define the counting procedures that correspond to a correct quantum-mechanical description of identical-particle systems. These procedures lead to expressions for the statistical entropy of ideal gases of identical particles, described either by *Fermi–Dirac statistics* or by *Bose–Einstein statistics*, which expressions can be maximized to find the most probable distributions of n_i values.

At the end of the section we treat an entirely different aspect of counting problems, namely, the identification of generating functions to represent the analytic continuation of sums encountered in probability theory. The usefulness of generating functions to evaluate sums for which there are constraints is illustrated with an example.

For the classical ideal gas that satisfies Boltzmann statistics we must use $C = (N!)^{-1}$ in Equation (4.24). If in addition we introduce a quantity ν_i, defined as the number of molecules n_i per unit cell (or per state), by

$$\nu_i \equiv \frac{n_i}{Z_i} = NP_i \tag{5.1}$$

then the entropy expression (4.25) can be conveniently rewritten for the Boltzmann-statistics case. Still considering $Z_i = \delta^3\mathbf{n}_i >> 1$, we obtain instead of Equation (4.25) the result

$$S = -\kappa \sum_i Z_i[\nu_i \ln \nu_i - \nu_i] \qquad \text{(for Boltzmann statistics)} \tag{5.2}$$

As we have already shown in connection with Equation (4.17), the most probable occupation numbers $\overline{\nu}_i$ are all very small compared with unity in the high-temperature, or classical, limit. Their values are

$$\overline{\nu}_i = (n\lambda_T^3)\, e^{-\beta\omega_i} \tag{5.3}$$

Consider next the quantum-mechanical ideal gas. In this case we must discuss separately the two possibilities of Fermi–Dirac statistics and Bose–Einstein statistics. We continue to use the concept of cells in momentum space, with the ith cell containing Z_i quantum states. Then, without considering the symmetry characteristics of the many-body wave function for either Fermi or Bose systems (see Section 16), we focus here on the distinctive feature of state occupation by identical Fermi-particle systems, which results from the Pauli exclusion principle. Thus, according to the

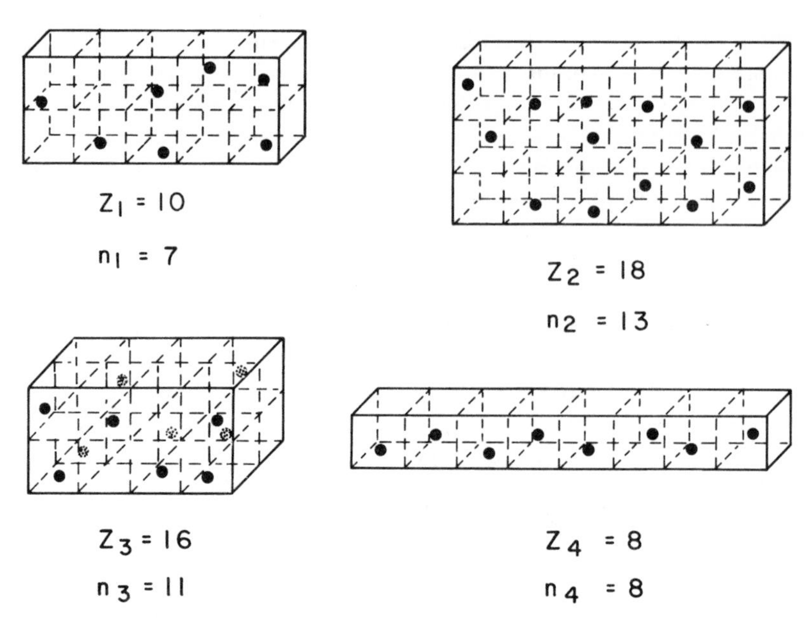

FIGURE I.7 Integral-sized cells Z_i in momentum space to illustrate the Fermi-system case in which $n_i \leqq Z_i$, due to the Pauli principle. Note that in each quantum-mechanical state, represented by a unit-cell subivision, there can only be 0 or 1 fermion (cf. Figure I.6).

Pauli exclusion principle, the *i*th cell in momentum space can contain no more than Z_i fermions, that is, $n_i \leqq Z_i$ (see Figure I.7). By contrast, a system of identical Bose particles may have $n_i > Z_i$ (as well as $n_i \leqq Z_i$) for the *i*th cell, there being no constraining Pauli principle for bosons. In either case we assume that $n_i \sim Z_i >> 1$.

For a Fermi gas we consider the probability function $W_{FD}(n_i)$, for the *i*th cell, which is the number of ways of occupying n_i out of Z_i states:

$$W_{FD}(n_i) = \frac{Z_i!}{n_i!(Z_i - n_i)!} \tag{5.4}$$

With $n_i \sim Z_i >> 1$, the entropy of a Fermi gas then becomes, according to Equations (4.22c) and (4.23),

$$S_{FD} = \kappa \ln\left[\prod_i W_{FD}(n_i)\right]$$

$$S_{FD} = -\kappa \prod_i Z_i\{\nu_i \ln \nu_i + (1 - \nu_i)\ln(1 - \nu_i)\} \tag{5.5}$$

where ν_i is defined by Equation (5.1). To derive the second line of Equation

(5.5) we have used Stirling's formula (2.6), as in the derivation of Equation (4.25) from Equation (4.24).

An expression for the probability function $W_{BE}(n_i)$, for the ideal Bose gas, is only slightly more difficult to derive than Equation (5.4) for $W_{FD}(n_i)$. In the case of Bose statistics there can be any number of particles in each quantum state, and we must allow for this fact when we compute the number of ways of distributing n_i bosons among Z_i states. The answer to the counting problem is

$$W_{BE}(n_i) = \frac{(Z_i + n_i - 1)!}{n_i!(Z_i - 1)!} \tag{5.6}$$

To arrive at this expression, imagine arranging the n_i particles along a straight line with $(Z_i - 1)$ partitions between them (see Figure I.8). The problem is equivalent to the number of ways of distributing n_i identical balls into Z_i identical boxes. Altogether there will be $(n_i + Z_i - 1)$ objects that must be arranged in all possible (distinguishable) ways, that is, into $W_{BE}(n_i)$ different ways as given by Equation (5.6). Then, in analogy with the derivation of Equation (5.5), one can determine the entropy of an ideal Bose gas to be

$$S_{BE} = \kappa \ln\left[\prod_i W_{BE}(n_i)\right]$$

$$S_{BE} = -\kappa \sum_i Z_i\{\nu_i \ln \nu_i - (1 + \nu_i)\ln(1 + \nu_i)\} \tag{5.7}$$

Although the entropy expressions (5.5) and (5.7) have been derived assuming $Z_i >> 1$ for all i, we shall momentarily consider the case $Z_i = 1$ for the purpose of discussion. Thus if $Z_i = 1$, then in Equation (5.5) the occu-

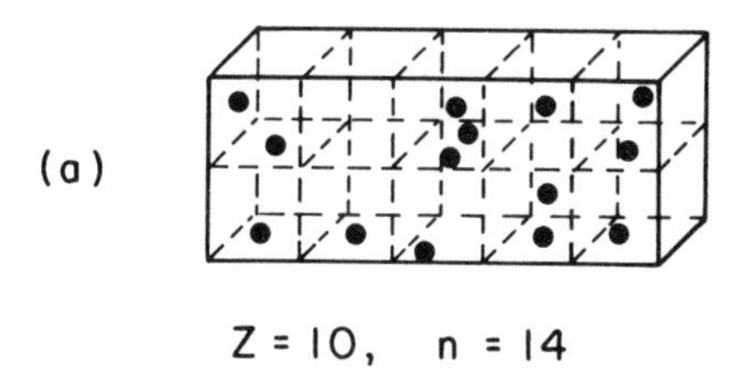

(b) |• •| • | | • |•••| • | • |• •|••| • ¦

FIGURE I.8 (a) A momentum-space cell Z occupied by n bosons. (b) The corresponding counting problem with the n particles separated by $Z - 1$ partitions.

pation number ν_i is restricted to the values 0 and 1, according to the Pauli principle, whereas in Equation (5.7) ν_i may take on all possible integral values. It is readily verified, for the general Z_i case again, that both expressions reduce to Equation (5.2) in the classical limit, in which $\nu_i \ll 1$ (on the average) for all i.

The ideal-gas statistical-entropy expressions we have deduced here are valid for the general nonequilibrium case. When most probable ν_i values are inserted into these expressions, they describe ideal gases at thermodynamic equilibrium. Now, according to the maximum entropy principle enunciated below Equation (4.25), the most probable values $\bar{\nu}_i$ can be determined by maximizing the entropy, subject to constraints of known information. For a system at rest the constraints usually are only those of fixed particle number N and total energy E, as given by Equations (4.19) and (4.20), with $n_i = Z_i\nu_i$.

Exercise 5.1 Show that a functional variation with respect to the occupation numbers ν_i of the entropy expressions (5.5) and (5.7) yields

$$\delta S = -\kappa \sum_i Z_i \ln\left(\frac{\nu_i}{1 + \varepsilon\nu_i}\right) \delta\nu_i \tag{5.8}$$

where $\varepsilon = +1$ for Bose statistics and $\varepsilon = -1$ for Fermi statistics. Similarly, write for the functional variation of the constraints (4.19) and (4.20):

$$\delta N = \sum_i Z_i \, \delta\nu_i = 0$$

$$\delta E = \sum_i Z_i\omega_i \, \delta\nu_i = 0 \tag{5.9}$$

Exercise 5.2 The maximum entropy of an ideal gas, with respect to variation of the occupation numbers ν_i, can be determined from Equation (5.8) by setting $\kappa^{-1}\,\delta S = 0$. Use Lagrange's method of undetermined multipliers (see Problem II.1) to determine the most probable values $\bar{\nu}_i$ from this condition, subject to the constraints (5.9). Show that

$$\bar{\nu}_i = [e^{\beta(\omega_i - g) - \varepsilon}]^{-1} \tag{5.10}$$

where the unknown multipliers are $(-\beta)$ for the total-energy constraint and (βg) for the total-particle-number constraint.

Generating Functions

Complicated sums occur frequently in probability theory, and it is usually quite difficult by direct inspection to identify answers in closed form.

Although the method of induction can be useful on occasion, a powerful technique for evaluating sums involves the use of generating functions. In a characteristic case we may readily know the result of a related simpler sum whose manipulation can then lead to the desired answer. Consider, for example, the generating function

$$1 = (p_1 + p_2)^N = \sum_{n_1=0}^{N} W_N(n_1)$$

that was used in Equation (1.5). Derivatives of this function with respect to p_1, say, are closely related to the average values $\langle n_1^s \rangle$, which are, at best, tedious to evaluate solely by examination of the defining expressions.

In general, a *generating function* is an analytic function that, when written out as a finite or infinite series, is closely related to a particular series being examined. The best example in statistical mechanics is the partition function Z, introduced and used in Chapter VI. Its use is quite similar to that of $(p_1 + p_2)^N$ in the above example. Another important application of generating functions in statistical mechanics is to facilitate the evaluation of sums that are subject to constraints, such as those of constant total particle number and energy [see, e.g., Equations (4.19) and (4.20)]. Often the constraints make the summation difficult to evaluate. The problem, then, becomes tractable by introduction of a suitable generating function, whose expanded form does not include all the constraints but in which it is possible to identify the desired sum as a contributing part. This technique is illustrated in practice by the Darwin–Fowler method of Appendix A. There, by way of illustration, the sum

$$P_N \equiv \sum_{\{n_i\}} W_N(n_i)$$

where $W_N(n_i)$ is defined by Equation (4.18), is to be evaluated, subject to the constraints (4.19) and (4.20). In this case an associated generating function is $[u(\mathfrak{z})]^N$, where $u(\mathfrak{z})$ is defined by Equation (A.7) [see also Equation (A.8)]. The usefulness of generating functions for problems with constraints is illustrated below by the solution to a tricky counting problem for a system of mixed composition.

Consider a system in which there are N oxygen atoms, which may occur in either of the molecular states O_2 and O_3 (ozone) in addition to the single-atom state. Let n_1, n_2, and n_3 be the numbers of these three possible molecular components in the gas, so that

$$N = n_1 + 2n_2 + 3n_3 \tag{5.11}$$

The counting question we ask is: How many different mixtures, that is, sets $\{n_1, n_2, n_3\}$, can be made from the N atoms?

Consider next the generating function $G(x)$ and its power series expansion, defined by

$$G(x) = \left(\frac{1}{1-x}\right)\left(\frac{1}{1-x^2}\right)\left(\frac{1}{1-x^3}\right) = \sum_{N=0}^{\infty} C_N x^N \tag{5.12}$$

In the power series expansion of $G(x)$, the number C_N is the coefficient of x^N. For example, one contribution to C_N comes from the Nth term in the expansion $\Sigma_{n=1}^{\infty} x^n$ of $(1-x)^{-1}$. In general, C_N will be equal to the sum of all such contributions when various powers of x, x^2, and x^3 are all combined. In other words, C_N is precisely the number required in the specified counting problem of all possible oxygen-atom mixtures. The utility of the generating function, for this particular example, is that the coefficient C_N can be calculated alternatively from the closed form of $G(x)$ by use of the formula

$$C_N = \left[\frac{1}{N!}\left(\frac{d}{dx}\right)^N G(x)\right]_{x=0} \tag{5.13}$$

The counting problem has thereby been reduced to the explicit evaluation of the right side of Equation (5.13).

To evaluate the derivatives of $G(x)$, as required in Equation (5.13), it is easiest first to write $G(x)$ as a sum of partial fractions. For this purpose it is necessary to know the three roots of $x^3 = 1$. In addition to the value unity, these are w and w^*, where

$$w \equiv -e^{i\pi/3} \tag{5.14}$$

and w^* is the complex conjugate of w. The partial fraction decomposition of $G(x)$ is then found to be

$$G(x) = \frac{1}{72}\left\{\frac{17}{(1-x)} + \frac{18}{(1-x)^2} + \frac{12}{(1-x)^3} + \frac{9}{(1+x)} + \frac{8w}{(w-x)} + \frac{8w^*}{(w^*-x)}\right\} \tag{5.15}$$

One now finds easily, using Equation (5.13), that C_N is given by

$$\begin{aligned} C_N &= \frac{1}{72}\{17 + 18(N+1) + 6(N+2)(N+1) \\ &\quad + [9(-1)^N + 8(w)^N + 8(w^*)^N]\} \\ &= \frac{1}{12}(N+1)(N+5) + \frac{1}{8}[1 + (-1)^N] \\ &\quad + \frac{1}{9}[(w)^N + (w)^{2N} + (w)^{3N}] \end{aligned} \tag{5.16}$$

where $w^2 = w^*$ has been used to obtain the second line of this result. Clearly, the use of generating functions can be extremely expedient. For $N = 25$, one finds from Equation (5.16) that $C_{25} = 65$.

Exercise 5.3

(a) Show from Equation (5.16) that

$$C_1 = 1, C_2 = 2, C_3 = 3, C_4 = 4, C_5 = 5, C_6 = 7$$

$$C_7 = 8, C_8 = 10, C_9 = 12, \text{ and } C_{10} = 14$$

Do these values agree with simple counting procedures? [Note that the term that involves the w's in the second line of Equation (5.16) is one-third if N is divisible by 3, and zero otherwise.]

(b) Show that C_N is also given by

$$C_N = \sum_{n=0}^{N/3} \sum_{m=0}^{\frac{1}{2}(N-3n)} 1 = \sum_{n=0}^{N/3} \left[\frac{1}{2}(N - 3n) + 1 - \frac{1}{4}(1 - (-1)^{N-3n}) \right]$$

In the first equality each sum terminates with the integer that is $\leqq$ to the indicated upper limit. Therefore, in the second equality we must include the subtracted term $-\frac{1}{4}[1 - (-1)^{N-3n}]$, which equals $-\frac{1}{2}$ when $N - 3n$ is an odd number.

Bibliography

Texts

R. Baierlein, *Atoms and Information Theory* (Freeman, 1971). The meaning of the term "probability" is discussed extensively in Chapter 2, and a careful application of probability concepts in information theory is given in Chapter 3.

W. Feller, *An Introduction to Probability Theory and Its Applications*, Vol. I (Wiley, 1968), Chaps. I, V, and VI. Covers basic notions of probability theory with a mathematician's care, including abundant examples.

B. V. Gnedenko, *The Theory of Probability* (Chelsea, 1963), translated from the Russian edition by B. D. Seckler. An excellent and quite readable text, with careful explanations and clear notation.

A. Katz, *Principles of Statistical Mechanics* (Freeman, 1967). Chapter 2 outlines the principles of information theory and their application to statistical physics, which application was first developed in detail by E. T. Jaynes, *op. cit.*

A. I. Khinchin, *Mathematical Foundations of Statistical Mechanics* (Dover, 1949), translated from the Russian edition by G. Gamow. The appendix of this monograph contains a rather rigorous derivation of the central limit theorem using arguments similar to, although more elaborate than, those invoked at the end of Section 3. See also Section 1 of Chapter 10 in the text by Feller.

L. D. Landau and E. M. Lifshitz, *Statistical Physics* (Addison-Wesley, 1958), translated by E. Peierls and R. F. Peierls. The first part of our Section 5 is adapted from the treatment in Sections 40 and 54 of this superb reference.

F. Reif, *Fundamentals of Statistical and Thermal Physics* (McGraw-Hill, 1965), Chap. 1. A sophisticated yet detailed introduction to statistical methods in physics, some parts of which have been adapted for the presentations given here.

Original Paper

E. T. Jaynes, "Information Theory and Statistical Mechanics," *Phys. Rev.* **106** (1957), p. 620.

Notes

1. This is the number of distinct ways in which N objects of two different kinds can be arranged in a straight line, when there are n_1 (indistinguishable) objects of the first kind and $n_2 = N - n_1$ objects of the second kind.

2. We consider $\langle(\Delta n)^2\rangle$, because $\langle(\Delta n)\rangle = \langle(n_1 - \langle n_1\rangle)\rangle \equiv 0$ and $\langle|\Delta n|\rangle$ is mathematically more intractable.

3. See F. W. Sears, *Thermodynamics* (Addison-Wesley, 1953), Sec. 17.2. In the second equality we have used the fact that $t \propto N$; the factor $\sqrt{3}$ is introduced by convention, because it appears in the three-dimensional problem. The one-dimensional version of this result is derived in our Section 15.

4. This problem and its three-dimensional generalization is treated in Section 55. See also R. Kubo, *Statistical Mechanics* (Interscience, 1965), pp. 58 and 77. A brief discussion of rubber elasticity is given in Section 15.1 of D. ter Haar, *Elements of Statistical Mechanics* (Rinehart, 1954).

5. J. Van Vleck, *The Theory of Electric and Magnetic Susceptibilities* (Oxford, Clarenden, 1932), Secs. 76 and 77.

6. M. Abramowitz and I. A. Stegun, *Handbook of Mathematical Functions* (Dover, 1965), Chap. 6.

7. Since $t_o = N$, note how small would be the radius of convergence of the integrand (2.5) were it instead of its logarithm expanded about $t = t_o$. This expansion would be

$$e^{g(t)} = e^{g(t_o)}[1 - (1/2N)(t - t_o)^2 + \cdots]$$

For another example, consider the expansion of $(1 + y)^{-N} = \exp[-N \ln(1 + y)]$ about the point $y = 0$.

8. F. Reif, *op. cit.*, App. A.6.

9. An independent derivation of the Poisson distribution is given as Problem I.2.

10. This inequality is verified explicitly below Equation (2.12).

11. W. Feller, *op. cit.*, p. 127.

12. A. I. Khinchin, *op. cit.* The appendix gives a full statement of conditions for the central limit theorem in the form in which we have given it.

13. Even when condition (3.20) holds, it is possible that $\langle s^2\rangle$ may not be finite. For

such cases, expansion (3.21) is not valid, and the central limit theorem fails. See, for example, Problem 1.26 in F. Reif, *loc. cit.*

14. A. I. Khinchin, *loc. cit.*, Chap. 5. Khinchin's derivation includes a procedure of ensemble theory, which we discuss in connection with Figure VI.1 in Section 26.

15. This derivation is taken from D. ter Haar, *loc. cit.*, p. 23.

16. A general method for calculating the normalization of $W_N(n_i)$ in ensemble theory when there are constraints on the n_i and which uses the Darwin–Fowler method of mean values is given in Appendix A. See, in particular, Exercise A.2.

17. This equality is derived explicitly for an ensemble of ideal-gas systems in Appendix A. In the simple coin-tossing experiment of Section 1, we have $Z_1 = Z_2 = 1$ and the *a priori* probabilities p_1 and p_2 were defined below Equation (1.5) to be the average values $\langle n_1 \rangle / N$ and $\langle n_2 \rangle / N$ in the limit $M \to \infty$.

18. This definition shows that entropy, whether we refer to statistical entropy or to the *maximum* statistical entropy identified as thermodynamic entropy (Chapter II), depends on our specification of the macroscopic state of a system. Inasmuch as such a specification in terms of known information can be operational in part, entropy is (in the words of E. P. Wigner) an anthropomorphic concept. A physical system does not carry entropy intrinsically.

19. When applied to a nonideal classical gas the single-particle energies ω_i' to be used in the Maxwell–Boltzmann distribution function for the most likely probabilities $\bar{P}_i \propto \bar{n}_i$ must include (both internal and external) potential-energy contributions. See J. C. Rainwater and R. F. Snider, *Phys. Rev.* **A13** (1976), p. 1190.

20. The most satisfactory justification of this fact occurs when the classical limit is taken in quantum statistics (Section 28). The other factors in Equation (4.18) for W_N are retained, even for indistinguishable particles, because they refer to the different possible arrangements of the particles into the Z_i "compartments," or states, of the ith cell. See also Section 5.

21. Rigorous details, supplementary to this proof, are given in A. Katz, *op. cit.*

Problems

I.1. (a) What is the probability of tossing three dice to get a 1 and two 3s?
(b) What is the probability of tossing three dice sequentially to get the numbers (1, 3, 3)?
(c) What is the probability of drawing any full house from a full deck of cards?
(*Suggestion:* Consider 13 piles with four cards in each pile.)

I.2. *Poisson distribution for dots along a line*

Let λ be the mean density of dots along a line, so that the chance that a randomly chosen infinitesimal element dx of the line will contain a dot is $\lambda\, dx$. Define the Poisson distribution $P_{\lambda x}(n)$, for $n = 0, 1, 2, \ldots$, to be the probability for finding n and only n dots in the interval of length x.

(a) Show for the case $n = 0$, in which no dots occur, that

$$P_{\lambda(x+dx)}(0) = P_{\lambda x}(0)(1 - \lambda\, dx)$$

and therefore that

$$P_{\lambda x}(0) = e^{-\lambda x}$$

(b) Argue that the probability for finding the first dot along the line of length x, between y and $(y + dy)$, is equal to

$$P_{\lambda y}(0) \cdot \lambda\, dy = e^{-\lambda y} \lambda\, dy \qquad (y < x)$$

Show therefore that the probability for finding n, and only n, dots in the interval x can be written as the recursion relation

$$P_{\lambda x}(n) = \lambda \int_0^x e^{-\lambda y}\, dy\, P_{\lambda(x-y)}(n - 1)$$

(c) Prove, in agreement with Equation (2.7), that

$$P_{\lambda x}(n) = \frac{(\lambda x)^n}{n!} e^{-\lambda x}$$

***I.3.** A metal is evaporated in vacuum from a hot filament. The resultant metal atoms are incident upon a quartz plate some distance away and form a thin metallic film there. This quartz plate is maintained at a low temperature, so that any metal atom incident on it sticks at its place of impact without further migration. The metal atoms can be assumed to impinge with equal likelihood on any area of the plate.

If one considers an element of substrate area b^2 (where b is the metal atom diameter), show that the number of metal atoms piled up on this area should be distributed approximately according to a Poisson distribution. Suppose that one evaporates enough metal to form a film of mean thickness corresponding to six atomic layers. What fraction of the substrate area then is not covered by metal at all? What fraction is covered, respectively, by metal layers three atoms thick and six atoms thick?

***I.4.** A set of telephone lines is to be installed to connect town A to town B. The town A has 2000 telephones. If each of the telephone users of A were to be guaranteed instant access to make calls to B, 2000 telephones lines would be needed. This would be rather extravagant. Suppose that during the busiest hour of the day each subscriber in A requires, on the average, a telephone connection to B for two minutes, and that these telephone calls are made at random. Find the minimum

---

*Problems taken from Chapter 1 of *Fundamentals of Statistical and Thermal Physics* by F. Reif. Copyright © 1965 by McGraw-Hill, Inc. Used with the permission of McGraw-Hill Book Company.

number M of telephone lines that must be installed to B so that on the average only 1% of the callers in town A will fail to have immediate access to a telephone line to B. *Answer:* $M = 85$ or 86 lines.

***I.5.** Radar signals have been reflected from the planet Venus. Suppose that in such an experiment a pulse of electromagnetic radiation of duration τ is sent from the earth toward Venus. At a time t later (which corresponds to the time necessary for light to go from the earth to Venus and back again) the receiving antenna on the earth is turned on for a time τ. The returning echo ought then to register on the recording meter, placed at the output of the electronic equipment following the receiving antenna, as a very faint signal of definite amplitude a_s. But a fluctuating random signal (due to the inevitable fluctuations in the radiation field in outer space and to current fluctuations always existing in the sensitive receiving equipment itself) also registers as a signal of amplitude a_n on the recording meter. This meter thus registers a total amplitude

$$a = a_s + a_n$$

Although $\langle a_n \rangle = 0$ on the average, since a_n is as likely to be positive as negative, there is considerable probability that a_n attains values considerably in excess of a_s; that is, the rms amplitude $\langle a_n^2 \rangle^{1/2}$ can be considerably greater than the signal a_s of interest. Suppose that $\langle a_n^2 \rangle^{1/2} = 1000\ a_s$. Then the fluctuating signal a_n constitutes a background of "noise" that makes observation of the desired echo signal essentially impossible.

On the other hand, suppose that N such radar pulses are sent out in succession and that the total amplitudes a, picked up at the recording equipment after each pulse, are added together before being displayed on the recording meter. The resulting amplitude must then have the form $A = A_s + A_n$, where A_n represents the resultant noise amplitude (with $\langle A_n \rangle = 0$), and $\langle A \rangle = A_s$ represents the resultant echo-signal amplitude. How many pulses must be sent out before $\langle A_n^2 \rangle^{1/2} = A_s$, so that the echo signal becomes detectable? Explain.

***I.6.** Consider the random walk of a particle in three dimensions, and let $P(\mathbf{s})\, d^3\mathbf{s}$ denote the single-step probability that its displacement $\mathbf{s}$ lies in the range between $\mathbf{s}$ and $(\mathbf{s} + d\mathbf{s})$. Let $\mathcal{P}_N(\mathbf{r})\, d^3\mathbf{r}$ denote the probability that the total displacement $\mathbf{r}$ of the particle after N steps lies in the range between $\mathbf{r}$ and $(\mathbf{r} + d\mathbf{r})$.

(a) Generalize the derivation of Equation (3.17) to three dimensions, and show that

$$\mathcal{P}_N(\mathbf{r}) = \frac{1}{(2\pi)^3} \int d^3\mathbf{k}\, \exp(i\mathbf{k} \cdot \mathbf{r})[Q(\mathbf{k})]^N$$

where

$$Q(\mathbf{k}) = \int d^3\mathbf{s} \exp(-i\mathbf{k} \cdot \mathbf{s})P(\mathbf{s})$$

(b) Using an appropriate Dirac δ function, find the probability density $P(\mathbf{s})$ for displacements of uniform length (a), but in any random direction of three-dimensional space. (*Suggestion:* Use spherical coordinates.) Then calculate $Q(\mathbf{k})$.

(c) Using the results of parts (a) and (b), show for $N > 1$ that

$$\mathcal{P}_N(\mathbf{r}) = \mathcal{P}_N(r) = \frac{N(N-1)}{2\pi r(2a)^N} \sum_{n=0}^{N/2} \frac{(-1)^n}{n!(N-n)!}$$
$$\theta[(N-2n)a - r][(N-2n)a - r]^{N-2}$$

where $\theta(x) = 1$ if $x > 0$, $\theta(x) = 0$ if $x \leqq 0$.
(To obtain this particular expression, the radial k integral has been evaluated as a closed contour in the complex-k plane; other, equivalent expressions can be derived.)

Write out explicit expressions for the cases $N = 2$ and $N = 3$, and sketch these two results graphically.

(d) Verify from the general expression of part (c) that $\mathcal{P}_N(\mathbf{r})$ is correctly normalized.
(*Suggestion:* The generating function

$$(\sinh y)^N = 2^{-N} \sum_{n=0}^{N} \binom{N}{n} (-1)^n \exp[(N-2n)y]$$

can be applied to evaluate the sum that is encountered.)

(e) Use the result of part (c) to show that $\langle \mathbf{r}^2 \rangle = Na^2$.

I.7. (a) *Generalization of the information-theory postulate (4.22c)*
Begin by introducing the missing-information function $M(P_1, P_2, \ldots, P_N)$, in which there are N missing-information probabilities P_i with $\Sigma_{i=1}^{N} P_i = 1$. Then assume that this function is equal to a sum of $(m + 1)$ missing-information functions, m of which are associated with probabilities P'_α, such that $\Sigma_{\alpha=1}^{m} P'_\alpha = 1$ as determined by the two-step process described below (4.22c). Finally, argue that your resulting equation must also apply to a special case (4.22c) of equal probabilities, $P_i = 1/N$ and $N = mn$, in which case $M(P_1, P_2, \ldots, P_N)$ can be denoted by $I(N) = I(mn)$.

(b) Give a derivation of Equation (4.23) from the postulates (4.22).[21]
(*Suggestion:* Consider the more general case of nonintegral values $n \geqq 1$.)

I.8. *Photon random-walk problem*

Consider a model of radiation transfer inside the sun in which photons are scattered primarily by classical Thomson scattering from free electrons. Let $\lambda = (n_e \sigma_T)^{-1}$ be the estimated mean free path of photons scattered in this way, where n_e is the electron number density and σ_T is the total Thomson scattering cross section.

(a) Obtain a number ($\sim$1 cm) for λ from the mean density of the sun, $\rho \cong 1.42\ \text{gm/cm}^3$.

(b) Estimate the average number N of random-walk steps, of uniform length λ, a photon must take to get from the center of the sun to the perimeter, a distance R away. Use $R = 2.32$ light-sec.

(c) Calculate the time T for light to travel the distance $N\lambda$, using the value of N estimated in part (c). *Answer:* $T \sim 3000$ years.

I.9. In a game, five dice are tossed, and they each land randomly on one of the five sections as shown. Two sections are red and three are green. Explain all answers.

(a) What are the number of possibilities (n) and corresponding (missing) information $I = \kappa \ln(n)$ for tossing 3 sixes and 2 fours? (The figure shows only one such toss.) *Answer:* $n = 100$.

(b) To clarify your answer to part (a), view the toss as a two-step process, and explain the addition of information when probabilities multiply.

(c) What is the probability of the particular toss shown, in which both fours land on red sections? *Answer:* 1 in 6^5.

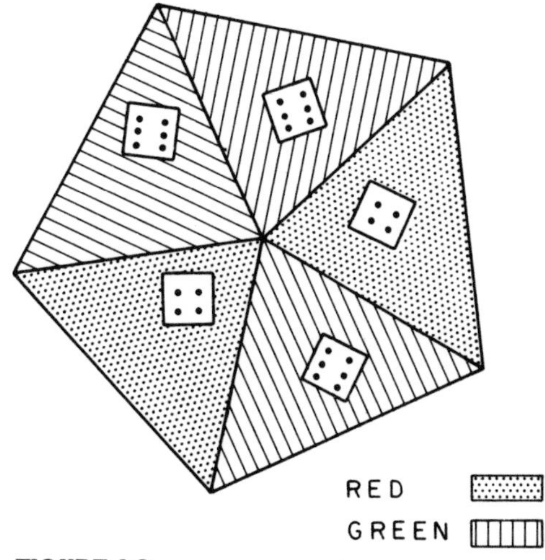

FIGURE I.9

CHAPTER II THERMODYNAMICS

We are interested eventually in the relationship between microscopic and macroscopic descriptions of nature. Equilibrium thermodynamics, the subject of this chapter, provides a framework for correlating physical laws in the macroscopic domain. Because its roots are intertwined in the microscopic world, thermodynamics is one of the most profound branches of physics. In Section 6 we give recognition to the microscopic foundations of equilibrium thermodynamics by initiating our development from the maximum entropy principle. Then in Section 7 we establish a precise relationship for any system between its thermodynamic description and the result of performing statistical averages over its microscopic description, which may be either classical or quantum mechanical. The remainder of the chapter pursues the development of thermodynamics, not comprehensively, but with emphasis on those features of this science that are most useful to an understanding of statistical mechanics.

6 Maximum Entropy Principle

Any microscopic description of our macroscopic world, whether classical or quantum mechanical, must involve the use of countless variables. From the outset we must appreciate that these variables will almost all be continually and randomly varying. The randomness, inevitably to be averaged on some macroscopic time or distance scale, is the key to passage from a microscopic description to the macroscopic description of nature. On any macroscopic level statistical regularities occur, and departures or fluctuations from these regularities are normally small.

The physical quantities associated with statistical regularities are called *thermodynamic variables*. These variables can be used to describe or define *macroscopic states* of a system, each macroscopic state being somehow related, in general, to many, many microscopic states or configurations. Some thermodynamic variables, such as total energy and volume, have a purely mechanical significance as well as a thermodynamic one. There are, in addition, quantities that are due solely to the existence of statistical regularities; temperature, for example, is one such variable. The science of thermodynamics deals with the laws that relate thermodynamic variables and govern macroscopic bodies.

One way to characterize the intrinsic randomness of the microscopic behavior of a large system is to refer to our lack of information, that is, to our *incomplete information,* about that system. To proceed in this direction, we assume that every allowable piece of microscopic information we might have can occur with equal likelihood. This is the assumption of equal *a priori* probabilities for microscopic states. Given this assumption, one can follow the development in Section 4 and introduce statistical entropy S as a quantitative measure of our incomplete information or "ignorance" about a system. We then assume, quite plausibly, that thermodynamic equilibrium corresponds to the condition of maximum ignorance (subject to constraints of known information) or, in other words, to a maximum entropy $\overline{S}(=\langle S\rangle)$.[1] Thus we are led to apply the *maximum entropy principle* to characterize systems in (or approaching) thermodynamic equilibrium, and in this section we use this principle to introduce the discipline of thermodynamics.

We begin by giving careful attention to the definition of a system and, in particular, to its external conditions. The simplest case, and also the most idealized one, is the *closed system* for which external conditions are held absolutely constant in time. The rest of the universe has no varying influence on a closed system. Its total energy is constant, but it may or may not be considered totally *isolated.*

The possible categories of systems that are isolated from external influences include one very important case for statistical considerations, namely, the *thermally isolated* (or insulated) *system.* Such a system may be subject to a variety of external influences, such as physical fields and varying pressure, but there is no microscopic energy flow, or *heat,* into or out of it. Its total energy can, therefore, be defined entirely in terms of external parameters and the time (see Section 7).

In contrast with the preceding two examples of restricted systems is the *open system,* which can be influenced by any and all external conditions, including heat flow. Many possibilities are allowed by the above terminology. For example, a system can be mechanically isolated, but open to thermal influence (heat flow). See Figure II.1.

Heat flow from one system into another occurs only when there is a temperature difference between the systems. We define the thermodynamic variable temperature precisely subsequently; for the present it suffices to use the intuitive notion that the relative terms hotter and colder are correlated with higher and lower temperatures, respectively, of systems. When

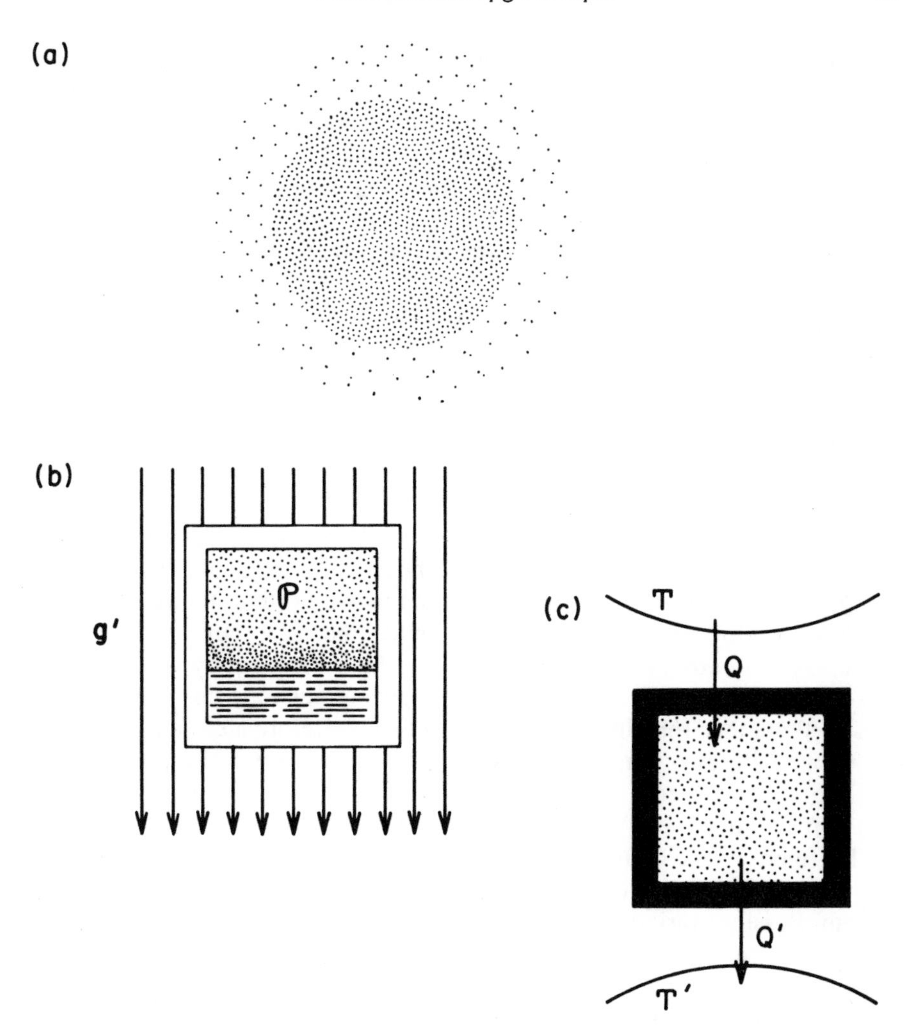

FIGURE II.1 (a) A closed system, completely isolated (ideally) from the rest of the universe; (b) a thermally isolated liquid–vapor system subject to two external conditions, an applied pressure $\mathcal{P}$ and a uniform gravitational field $\mathbf{g}'$; (c) a mechanically isolated system, with heat flow Q in from a reservoir at temperature T and heat flow Q' out into a reservoir at temperature T'.

one system is neither hotter nor colder than another system, there is no heat flow between the two systems and they are said to be in *thermal equilibrium*. This concept of thermal equilibrium is formalized operationally by a postulate called the *zeroth law of thermodynamics:* If systems A and B are each in thermal equilibrium with a third body C (say, a thermometer), then A and B are in thermal equilibrium with each other.

A system may be balanced, or in equilibrium, not only thermally, but also with respect to either mechanical effects or chemical reactions. A system that is in equilibrium with respect to all possible (thermal, mechanical, and chemical) small disturbances, or fluctuations, is said to be in *thermodynamic equilibrium*. We mention, in passing, that all possible processes that systems undergo must be consistent with overall conservation of energy, a constraint referred to in thermodynamics as the *first law of thermodynamics* (see Section 7).

With the above preliminaries in mind, we focus attention on the maximum entropy principle. In macroscopic terms, the maximum entropy principle states that the entropy of a closed system will never decrease. At thermodynamic equilibrium, the entropy is, therefore, a maximum for given external conditions. The entropy of a closed system can increase, of course, and this leads us to consider various possible processes by which the macroscopic states of systems can change.

It is usual to divide all processes that occur in macroscopic bodies into *irreversible* and *reversible* ones. By the first we understand processes accompanied by an increase in entropy of the whole system; the reverse processes cannot occur without external influence, because they would involve a decrease in entropy. Slow processes in which the entropy of the whole closed system remains constant (the entropies of separate parts of the system need not also remain constant), and which therefore also can take place in the opposite direction, are called *reversible processes*.[2] A strictly reversible process represents an ideal limiting case; processes actually occurring in nature can only be reversible to some limiting degree of accuracy.

Thermodynamic equilibrium corresponds to a condition in which the entropy is maximal with respect to variations in the macroscopic state of a system. Possible variations, whether real or virtual, must be imagined, however, to be carried out subject to fixed external conditions; the external parameters must all be held constant. On the other hand, for a real process (as opposed to a variation) the external conditions may vary. Thus for a reversible process the entropy remains constant and the system is at all times in thermodynamic equilibrium; at any particular moment, however, the entropy is a maximum with respect to virtual variations that are subject to the momentary external conditions. Contrasted with this situation is the irreversible, or nonequilibrium, process for which the entropy of a whole system always increases. The tendency for the entropy of whole closed systems to increase, which has its basis in the statistical or probabilistic definition of entropy, can be called the *law of increasing entropy*. As such it becomes one form of the *second law of thermodynamics*.

If we were to apply the principle of maximum entropy to the universe as a whole, then we might conclude that the universe ought to be in a state of total statistical equilibrium, assuming that the relaxation processes for all mechanical "disordering" phenomena are finite. But the universe is not in equilibrium; rather, it is a quite ordered structure. The reason for this contradiction is due in part to the importance of the gravitational field, which maintains "order" in the role of an external condition,[3] thereby facilitating the burning of nuclear fuel in stars. Without gravitational ordering, the burning of nuclear fuel would be a process with an essentially infinite relaxation time. In the sense that the burning of nuclear fuel is actually quite "rapid," it may be that the entropy of the universe is, in fact, still increasing.

It is one thing to assert that the most probable (equilibrium) condition for a closed system is the one of maximum entropy. It is quite another to understand the time evolution of an arbitrarily prepared system to the state of thermodynamic equilibrium. As a matter of fact, irreversibility, that is, the law of increasing entropy, has never been successfully formulated on a microscopic basis. The underlying reason for the difficulty is that in the microscopic world all physical processes seem to be completely time reversible. Moreover, it may be that the explanation of macroscopic irreversibility can only be found ultimately in the laws of quantum mechanics.[3] We shall return to more detailed discussions of irreversibility in Sections 25 and 45.

Temperature

We now use the maximum entropy principle to introduce the concept of temperature. It is assumed here that the reader is familiar with a standard introductory discussion of temperature based on the operational definition of an ideal gas and on the Carnot cycle.[4] Thus we consider two bodies in thermal equilibrium with each other, the bodies together forming a closed system. See Figure II.2. Then the entropy S of this system will have its maximum possible value for a given value E of the total energy of the system. We use the fact that $E = E_1 + E_2 =$ constant, where the subscripts 1 and 2 refer to the two bodies, choose E_1 to be an independent variable, and examine the variation of S with respect to E_1. Now, since the probability for any microscopic state of the whole system is the product of the probabilities for each part, we can, according to Equation (4.22c), write the total entropy

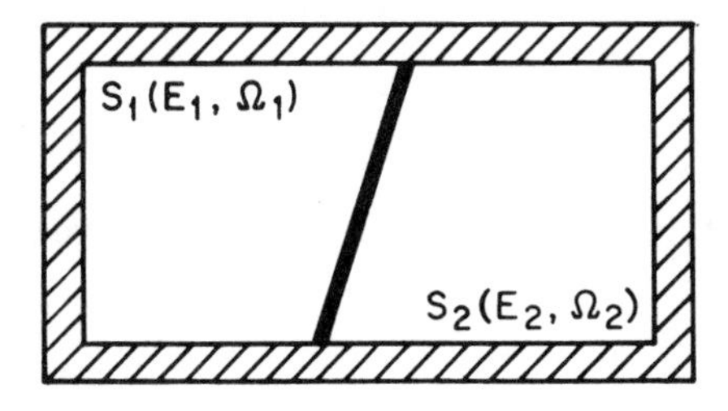

FIGURE II.2 A closed system consisting of two bodies or subsystems, each with a fixed volume and between which there can be energy transfer. At thermal equilibrium there is a common temperature T defined by Equation (6.3).

S additively as[5]

$$S = S_1(E_1) + S_2(E_2) \tag{6.1}$$

The maximum entropy principle then says that at thermal equilibrium (when microscopic energy flow has ceased),

$$\frac{dS}{dE_1} = 0 = \frac{dS_1}{dE_1} - \frac{dS_2}{dE_2}$$

where $E_2 = E - E_1$. We may conclude that for two bodies in thermal equilibrium,

$$\frac{dS_1}{dE_1} = \frac{dS_2}{dE_2}$$

a result that can be generalized easily to the case of an arbitrary number of bodies in thermal equilibrium with each other. [One may consider the entire system to be divided arbitrarily into two pieces and then argue that dS_i/dE_i is characteristic of the system rather than of the arbitrary piece i. See also Problem II.1(c).]

We must be careful because the above derivatives might be taken subject to any specified conditions of the external parameters of the system. We really meant to specify that each body is not directly in contact with any other bodies, or surrounded by an external medium. If, however, we allow the bodies to be in contact with each other, or if one body is immersed in an external medium, then we can eliminate associated mechanical effects, which might occur when the energy is varied, by specifying that the above derivatives are all to be taken at constant volume Ω. Then we obtain the more precise result, of interest here,

$$\left(\frac{\partial S_1}{\partial E_1}\right)_{\Omega_1} = \left(\frac{\partial S_2}{\partial E_2}\right)_{\Omega_2} \tag{6.2}$$

In general, we must denote which thermodynamic variables are being held constant in the partial derivatives of thermodynamics, since the choice of independent variables always involves some arbitrariness.

When several bodies are in thermal equilibrium, then the quantity

$$\left(\frac{\partial S_i}{\partial E_i}\right)_{\Omega_i}$$

is the same for each of them, and therefore it is a parameter for the entire system. This conclusion leads us to define the *temperature* T of a body as

$$T \equiv \left(\frac{\partial E}{\partial S}\right)_{\Omega} \tag{6.3}$$

This definition is consistent with our intuitive notion, used previously in the discussion of the zeroth law of thermodynamics, about energy flow when two bodies not in thermal equilibrium are brought together. Thus for processes at constant volume the entropy variation with respect to time is given by

$$\frac{\partial S}{\partial t} = \left(\frac{\partial S_1}{\partial E_1} - \frac{\partial S_2}{\partial E_2}\right)\frac{\partial E_1}{\partial t} = \left(\frac{1}{T_1} - \frac{1}{T_2}\right) \cdot \left(\frac{\partial E_1}{\partial t}\right) > 0$$

Therefore, if T_2 is greater than T_1 (body 2 is "hotter" than body 1), then the energy of body 1 will increase. Correspondingly, $(\partial E_2/\partial t) < 0$, and energy flows from a body with higher temperature to one with lower temperature.

We discuss with Equation (7.12) the fact that definition (6.3) of absolute temperature is equivalent to the more familiar thermodynamic definition.

Macroscopic Motion

Motions in which separate macroscopic parts of a body move as independent units are called macroscopic, as distinguished from microscopic motion. Most microscopic motion is a disordered or random motion; when macroscopic motion occurs, then, clearly, the microscopic motion is ordered in part. We now show that a closed system in equilibrium can perform only uniform translation and rotation as a whole, as internal macroscopic motion is impossible in a state of thermodynamic equilibrium.

Imagine a body to be divided into a large number of small (but macroscopic) parts, and let m_i, E_i, and $\mathbf{P}_i$ denote the mass, energy, and momentum of the ith part. Here we note that to be macroscopic, a small part must be sufficiently large, so that only the average over microscopic motions is important for the part as a whole; the macroscopic dimensions must be large compared with the mean free path[6] of molecular motions. Next, we emphasize that the internal energy of the ith part is merely the difference

$$\left(E_i - \frac{P_i^2}{2m_i}\right)$$

between its total energy and its macroscopic translational energy, because in a finely divided body rotational motion of one small part relative to adjoining parts of the body cannot occur. Now the entropy S_i of the ith part is a function of its internal energy only, since the number of its quantum states (and hence the number of possibilities for identifying missing information) must be the same for all inertial frames of reference, including the rest frame. Therefore, the total entropy can be written as

$$S = \sum_i S_i \tag{6.4}$$

where

$$S_i = S_i\left(E_i - \frac{P_i^2}{2m_i}\right)$$

Assume that the body is a closed system. Then, in addition to its total internal energy, other conserved quantities will be its total momentum **P** and angular momentum **L**, where

$$\mathbf{P} = \sum_i \mathbf{P}_i$$
$$\mathbf{L} = \sum_i \mathbf{r}_i \times \mathbf{P}_i \tag{6.5}$$

Here, the radius vectors $\mathbf{r}_i$ are all assumed to be drawn from some inertial origin. In the equilibrium state the total entropy of the body as a function of the momenta $\mathbf{P}_i$ will be a maximum subject to these three constraints. Upon using Lagrange's method of undetermined multipliers (see Problem II.1), we obtain

$$\frac{\partial}{\partial \mathbf{P}_i}\sum_j \{S_j + (\mathbf{u}/T)\cdot\mathbf{P}_j + (\mathbf{\Omega}/T)\cdot(\mathbf{r}_j \times \mathbf{P}_j)\}_\Omega = 0$$

where $\mathbf{u}/T$ and $\mathbf{\Omega}/T$ are constant unknown vectors. The factors of T^{-1} are introduced for convenience, with $\mathbf{u}$ and $\mathbf{\Omega}$, so that factors of T do not appear in Equation (6.6). Note that whereas the $\mathbf{P}_i$ occur functionally in Equation (6.4), the total internal energy itself does not depend on the $\mathbf{P}_i$.

Denoting by $\mathbf{v}_i$ the velocity of the ith part in the body ($\mathbf{v}_i = \mathbf{P}_i/m_i$), we have for the derivative of the first term in the last equation:

$$\frac{\partial S_j}{\partial \mathbf{P}_i} = \frac{1}{T}\left(-\frac{\mathbf{P}_i}{m_i}\right)\delta_{ij} = -\left(\frac{\mathbf{v}_i}{T}\right)\delta_{ij}$$

Therefore the maximum entropy condition gives for $\mathbf{v}_i$ the result

$$\mathbf{v}_i = \mathbf{u} + (\mathbf{\Omega} \times \mathbf{r}_i) \tag{6.6}$$

which says that the velocities of all parts of the body are given by the same expression. A closed body moves as a whole in thermodynamic equilibrium, with uniform velocity **u** and constant angular velocity **Ω**, internal macroscopic motion being impossible. Internal macroscopic motion can occur, of course, and we treat such motion in Chapter III in connection with the condition of local equilibrium. Our conclusion here applies only to closed systems in complete thermodynamic equilibrium. Henceforth we consider systems to be at rest unless we specify otherwise.

7 Relation of Thermodynamics to Statistical Physics

The primary objective of statistical physics is to explain macroscopic phenomena in microscopic terms. The most straightforward procedure for achieving this goal is to average various physical quantities over the microscopic domain. Such averages must be statistical averages, because they involve the enormous numbers of molecules in any macroscopic system. Thus most of this text after the present chapter is concerned with averaging techniques and their application. In particular, we begin in Chapter V with the specific problem of the statistical averaging of thermodynamic quantities.

It is important to show, as we do in the present section, that the statistical calculation of the equilibrium properties for a system gives results that, with proper identification, are always in agreement with those derived by thermodynamic methods. Our development leads us into a discussion of the thermodynamic identity and the first law of thermodynamics. The discussion is extended to include a derivation of the Clausius inequality.

To relate the statistical averages of a variable to its calculation by thermodynamic methods, we consider a general *adiabatic process,* in which the external conditions to which a system is subjected must vary slowly. By slowly varied external conditions we mean that the conditions are changed sufficiently slowly so that at every instant the system can be considered to be in the equilibrium state corresponding to the momentary external conditions. Thus the adiabatic process refers to a thermally isolated system for which the total energy is a function only of time-dependent external parameters. As we shall clarify below, the adiabatic process is also a reversible one.[7] Because there is no heat exchange with the external world, the temperature is not a variable of primary interest for adiabatic processes.

For the system to remain in thermodynamic equilibrium during an adiabatic process, the times for changes must be slow compared with relevant relaxation times in the system. Of course, these slow times may be, in fact, rather fast, as in the expansion of a gas, for which "slow" would require only that boundary velocities be small compared with the velocity of sound in the gas. On the other hand, the adiabatic process must be sufficiently fast so that heat cannot be exchanged with the surrounding medium.

The reason for the latter speed requirement is that only by considering bodies in thermal isolation can we specify uniquely their equilibrium state in terms of external parameters. Thus a thermally isolated body can be described by a Hamiltonian function $H(\lambda_1, \lambda_2, \cdots)$ in which the external conditions are taken into account entirely by the use of position- and time-dependent parameters $\lambda_i(\mathbf{r}, t)$ [λ_1 might be the volume of the system, λ_2 might be an external magnetic field, etc.]. If the system interacted directly with other systems, as with each system in Figure II.2, then it would not, itself, have a Hamiltonian function $H(\lambda_i)$.

It must be stressed that the maximum entropy principle holds for thermally isolated bodies as well as for closed ones, inasmuch as the external conditions in both cases may be regarded as purely mechanical functions of position (and time) rather than as statistical objects. From the discussions of Section 6, this fact suggests that the adiabatic process must be a reversible one; we now demonstrate this fact unambiguously.

We first observe that the time variation of the entropy must be due entirely to time variations of the parameters λ_i. Since for an adiabatic process $d\lambda_i/dt$ is expected to be a "small" quantity, with the system always maintained at thermodynamic equilibrium, we assume that we can expand dS/dt as a Taylor series in $d\lambda_i/dt$, keeping only the first few terms. By considering henceforth only one parameter λ, for simplicity, we then have

$$\frac{dS}{dt} = \left(\frac{dS}{dt}\right)_0 + \left(\frac{dS}{dt}\right)_0^{(1)} \frac{d\lambda}{dt} + \frac{1}{2}\left(\frac{dS}{dt}\right)_0^{(2)} \left(\frac{d\lambda}{dt}\right)^2 + \cdots \tag{7.1}$$

where the superscript (n) indicates the nth derivative of dS/dt with respect to $d\lambda/dt$, and the subscript 0 means that this derivative is evaluated at $d\lambda/dt = 0$. But $(dS/dt)_0 = 0$, because when $d\lambda/dt = 0$, the body becomes a closed system in thermodynamic equilibrium with constant entropy. Moreover, the second term in Equation (7.1) changes its sign with $d\lambda/dt$ whereas dS/dt must be a positive quantity according to the maximum entropy principle. Therefore, $(dS/dt)_0^{(1)} = 0$, and we are left with the approximate expression

$$\frac{dS}{dt} = A\left(\frac{d\lambda}{dt}\right)^2 + \cdots$$

where A is a positive quantity. This equation can be rewritten as

$$\frac{dS}{d\lambda} = A\left(\frac{d\lambda}{dt}\right) + \cdots \tag{7.2}$$

which shows that $dS/d\lambda$ vanishes in the limit $d\lambda/dt = 0$. Hence an adiabatic process is a reversible one (in the limit $d\lambda/dt \to 0$), its entropy remaining unchanged as the external conditions are varied.

Consider next the Hamiltonian function $H(\lambda)$ of the body. Since the total time derivative of the Hamiltonian is equal to the partial time derivative, we may write for an adiabatic process

$$\frac{dH(\lambda)}{dt} = \frac{\partial H(\lambda)}{\partial t} = \frac{\partial H}{\partial \lambda} \cdot \frac{d\lambda}{dt}$$

Here the partial time derivative involves all time dependence that is due explicitly to variations of the external parameters, and not to any time variations associated explicitly with dynamical variables internal to the system of interest. Now the calculation by statistical methods of the equilibrium value of a physical quantity O is indicated as $\langle O \rangle$ or $\overline{O}$, by definition.[1] In addition, the average value of dH/dt for an adiabatic process is denoted by dE/dt, where E is the thermodynamic energy function.[8] The statistical average of the last equation then gives

$$\frac{dE}{dt} = \left\langle \frac{\partial H}{\partial \lambda} \right\rangle \cdot \frac{d\lambda}{dt} \tag{7.3}$$

where the statistical average on the right side must be calculated for a statistical distribution of systems whose most probable state corresponds to the equilibrium macroscopic state determined by the specified external conditions (value of λ).

As we discuss further below, the independent thermodynamic variables that characterize a macroscopic system at equilibrium are not unique. However, straightforward application of Equation (7.3) is restricted to use of the independent variables S and λ. Thus the thermodynamic prescription for calculating dE/dt for an adiabatic process is to vary the external parameters at constant entropy, that is,

$$\frac{dE}{dt} = \left(\frac{\partial E}{\partial \lambda} \right)_S \frac{d\lambda}{dt} \qquad \text{(for adiabatic processes)} \tag{7.4}$$

Therefore, by comparing Equations (7.3) and (7.4) we obtain the important formula

$$\left\langle \frac{\partial H(\lambda)}{\partial \lambda} \right\rangle = \left(\frac{\partial E}{\partial \lambda} \right)_S \tag{7.5}$$

which relates statistically averaged derivatives of the energy to the corresponding quantities calculated by thermodynamic methods. In quantum mechanical calculations of the left side of Equation (7.5), the Hamiltonian $H(\lambda)$ and its derivative are operators.

As it stands, the statistical average on the left side of Equation (7.5) is to be calculated for an ensemble of (identically prepared) thermally isolated systems, the *microcanonical ensemble* of statistical mechanics (see Section 26). It is also permissible, however, to calculate this average value for equilibrium ensembles corresponding to other given external situations, despite our particular specification that Equation (7.3) is restricted to thermally isolated systems. For example, the average value in Equation (7.5) can be calculated for systems at a definite temperature T (i.e., in a temperature bath), but in this case the right side of Equation (7.5) becomes more meaningful if it is expressed in terms of the Helmholtz free energy [see Exercise 8.2(b)]. The uniqueness of the adiabatic process is that the time variation of the energy function can be expressed by way of Equation (7.3), but this uniqueness does not extend to Equation (7.5), which also applies when other external conditions are considered for systems in thermodynamic equilibrium.

Pressure

In the simplest example of a thermally isolated body the energy depends on only one external parameter, namely, the volume Ω of the body. Shape dependence of the energy occurs for solid bodies, of course, but we do not consider this complication here. Similarly, surface effects, important for small systems, are ignored (see, however, the end of Section 10). Then, in the absence of other external parameters, the macroscopic state of a stationary body is defined completely by only two independent quantities, for example, volume and entropy. One aspect of the dependence of the energy on the volume and entropy has already been explored in Section 6, and the principal result of this study was the definition of temperature by Equation (6.3). In the present discussion we deduce a second relationship [Equation (7.8) below] between energy, volume, and entropy by setting $\lambda = \Omega$ in Equation (7.5).

Consider the average force $\langle \mathbf{F} \rangle$ with which a body acts on a particular surface element $d\mathbf{a}$ located at point $\mathbf{r}$ of its bounding surface,

$$\langle \mathbf{F}(\mathbf{r}) \rangle = -\langle \nabla H(\mathbf{r}) \rangle \tag{7.6}$$

The sign on the right side of this expression is such that a body acts to minimize its own energy. By using identity (7.5), with $\lambda = \mathbf{r}$, we find for this average force $\langle \mathbf{F} \rangle$ the expression

$$\langle \mathbf{F}(\mathbf{r}) \rangle = -[\nabla E(\mathbf{r})]_S = -\left(\frac{\partial E}{\partial \Omega}\right)_S \nabla \Omega$$

where $\nabla \Omega = d\mathbf{a}$, since at the point $\mathbf{r}$ we can write $d\Omega = d\mathbf{a} \cdot d\mathbf{r}$. Thus

$$\langle \mathbf{F}(\mathbf{r}) \rangle = -\left(\frac{\partial E}{\partial \Omega}\right)_S d\mathbf{a} \tag{7.7}$$

a result telling us, as expected, that the average force acting on a surface element is directed along the normal to it and is proportional to its area.

The force per unit area, in absolute magnitude, is defined to be the *pressure* $\mathcal{P}$.

$$\mathcal{P} = -\left(\frac{\partial E}{\partial \Omega}\right)_S = -\left\langle \frac{\partial H(\Omega)}{\partial \Omega} \right\rangle \tag{7.8}$$

the second equality following again from Equation (7.5). Of course, the pressure is only completely meaningful for a fluid. For solids we would have to discuss, in addition, various stress moduli. The minus sign in Equation (7.8) assures us that the pressure is a positive quantity, a fact we discuss further in Section 9. As with Equation (7.5), the pressure can be calculated for conditions other than that of thermal isolation [see Exercise 8.2(b)].

Exercise 7.1 Bodies totally in equilibrium with each other have equal pressures, neglecting surface effects, since thermodynamic equilibrium includes mechanical equilibrium in which the forces between bodies balance along their common boundaries. The equality of pressures for bodies in equilibrium can also be deduced from the maximum entropy principle. Demonstrate that this latter statement is correct by considering a closed system composed of two parts in thermal equilibrium and with constant total volume. Neglect specific surface effects. (*Note:* The energy of each part of the system must be held constant for this proof. How then does the proof fail for a system in a position-dependent external potential?)

Thermodynamic Identity and First Law of Thermodynamics

The complete description of a system in thermodynamic equilibrium requires the use of a minimum number of independent thermodynamic variables. For thermally isolated systems, the "natural" choice of independent variables is the entropy S and the external parameters λ_i. Moreover, for the simple case of a fluid in thermal isolation, with no external fields present, the only external parameter is the volume Ω. The thermodynamic energy depends in this case on only two independent variables; it can be differentiated to give either the temperature by Equation (6.3) or the pressure by Equation (7.8). With the choice of volume and entropy as the independent variables, these two relations can be combined to give the *thermodynamic identity* for variations dE at thermodynamic equilibrium.

$$\begin{aligned} d\langle H\rangle \equiv dE &= \left(\frac{\partial E}{\partial S}\right)_\Omega dS + \left(\frac{\partial E}{\partial \Omega}\right)_S d\Omega \\ &= T\,dS - \mathcal{P}\,d\Omega \end{aligned} \tag{7.9}$$

Consider next the work $đW$ done by a body on its surroundings, when the volume is the only external parameter, for an infinitesimal process during which mechanical equilibrium is maintained. According to Equations (7.7) and (7.8) this work is given by

$$đW = \oint \langle \mathbf{F}(\mathbf{r})\rangle \cdot d\mathbf{r} = \mathcal{P}\,d\Omega \tag{7.10}$$

where the integration is a surface integral over the entire outer surface of the body. The quantity $đW$ is an inexact differential (as indicated by the bar on d), which means that it depends on the external conditions that are maintained during the infinitesimal process. Thus $\mathcal{P}\, d\Omega$ for an adiabatic process (constant S) differs from $\mathcal{P}\, d\Omega$ for an isothermal process (constant T).

The total energy of a system can change in an infinitesimal process, not only through the mechanism of mechanical work, but also by the addition of *heat* $đQ$. As we discussed at the beginning of Section 6, heat is energy transfer by nonmechanical means in which there is microscopic energy flow into or out of a body. We note, in particular, that heat does not include energy transfer due directly to macroscopic motion, or current.[9] Of course, $đQ$ is also an inexact differential.

The relation between total-system energy changes dE and infinitesimal thermal and mechanical energy transfers, $đQ$ and $đW$, is expressed by the conservation-of-energy principle. The macroscopic form of this principle for a body,

$$dE = đQ - đW \tag{7.11}$$

is called the *first law of thermodynamics*. Here there is a minus sign associated with $đW$, because $đW$ is defined to be positive when work is done by the body. The distinction between Equations (7.9) and (7.11) should be noted carefully. The thermodynamic identity refers only to reversible processes for a system, those in which the system is maintained in thermodynamic equilibrium, whereas the first law of thermodynamics applies also to those irreversible processes in which the initial and final states are both states of thermodynamic equilibrium [in which case it is likely that $đW$ cannot be given by an equation of the type (7.10)].

The notation of Equation (7.11) indicates that both $đW$ and $đQ$ are inexact differentials. For the particular case of reversible processes, comparison of Equations (7.9)–(7.11) yields the relation for infinitesimal heat transfer,[10]

$$đQ = T\, dS \qquad \text{(for reversible processes)} \tag{7.12}$$

This result shows that for reversible processes T^{-1} is the integrating factor for the inexact differential $đQ$. In integrated form for a cyclic process [see Equation (7.15)], this fact is used in conventional treatments of thermodynamics to introduce entropy as a thermodynamic variable.

Equation (7.12) can be applied to a thermally isolated system that consists of two bodies at different temperatures (temperature reservoirs), which are each kept at a constant volume. One can show (see Appendix C) that the maximum work that can be extracted from this system in the process of bringing the two bodies to thermal equilibrium is obtained ideally with an engine operating on a *Carnot cycle*. It can then be concluded, since the Carnot cycle is the basis of the thermodynamic temperature scale, that definition (6.3) of absolute temperature T is equivalent to the more common thermodynamic definition.[11]

The Clausius Inequality

We now consider infinitesimal irreversible processes between two equilibrium states of a body. Two examples of such processes are an infinitesimal Joule–Thomson process with a small pressure difference (see Exercise 8.1) and a small transfer of heat to a body from a system at a slightly higher temperature. We show that Equation (7.12) corresponds to the *maximum heat* that can be added to a body to take it from one equilibrium state to another state infinitesimally removed. Corresponding to this maximum heat, and for fixed energy change dE, is a *maximum work* done by the body according to Equation (7.11). We note that the irreversible processes being considered may include ones for which the temperature and pressure are uniform throughout the body at every instant (thermal and mechanical equilibrium is maintained), such as chemical reactions in a homogeneous mixture of interacting substances. Examples of this latter type are best treated by applying the "minimal conditions for free energies" of Section 9, which are derived from the present result (7.14).

Let us imagine that the heat transfer to the body is the result of interaction between the body and a surrounding medium. Further, we let dS be the entropy change in the body and dS_o be the entropy change in the medium. Then for an irreversible process, in which the body plus medium together form a closed system, the total entropy will increase:

$$dS + dS_o > 0$$

Now assume that the medium is very large, so that its temperature T_o remains constant during the process. Then since the medium (now a heat reservoir) remains in thermodynamic equilibrium during the heat exchange, the heat transferred to the body will be given by

$$đQ = -\, T_o\, dS_o < T_o\, dS \tag{7.13}$$

If, finally, we let the temperature T of the body (which is changed negligibly by an infinitesimal process) be equal to T_o, then we obtain the important inequality

$$đQ \leqq T\, dS \tag{7.14}$$

Thus Equation (7.12) provides an upper limit for the heat that can be added to a body in an infinitesimal process between two equilibrium states.

In the final result (7.14) the artificial role of the medium has disappeared, as might have been expected. Of course, without a medium the system under consideration must be sufficiently large that its temperature T is well defined; otherwise inequality (7.13) applies. Inequality (7.14) is, in fact, quite general.

The preceding derivation enables us to derive a further result for a body that is passed through a closed cycle, returning eventually to its initial equilibrium state. If in each infinitesimal part of the cycle for which the temperature of the body is not well defined we use inequality (7.13), where T_o is the temperature of some associated heat reservoir, then we can derive the *Clausius inequality:*

$$\oint \frac{đQ}{T_o} \leqq \oint dS = 0 \tag{7.15}$$

Here T_o can be replaced by T in parts of the cycle for which the temperature of the body is well defined, and equality holds when the entire cyclic process is reversible. When derived without reference to entropy considerations, the equality can be used to introduce the concept of entropy from the macroscopic point of view. (Also required in such presentations is the fact that a closed-line integral vanishes generally only when its integrand is an exact differential.)

Exercise 7.2

(a) Explain how to calculate the entropy change $[S(B) - S(A)]$ for a body that is subjected to a finite irreversible process between two equilibrium states A and B. Show for all processes between equilibrium states A and B that

$$\int_A^B \frac{đQ}{T_o} \leqq [S(B) - S(A)] \tag{7.16}$$

(b) Show, using only (7.16), that the entropy of a thermally isolated system never decreases. Consider only processes between equilibrium states.

8 Energy Functions of Thermodynamics

The macroscopic state of a system in thermodynamic equilibrium is described completely by only a small number of independent thermodynamic variables. For a uniform fluid in the absence of external fields this number is two. We have already observed for this case, in which the volume is the only external parameter, that the system is characterized completely once the entropy is also specified. Indeed, for reversible processes at constant S or Ω

the change in the internal (thermodynamic) energy function E is readily determined by the thermodynamic identity (7.9). But it frequently occurs that a reversible process takes place at constant $\mathcal{P}$ or at constant T, in which case the internal-energy change, although calculable, is not so readily determined. There is, in fact, no unique set of independent thermodynamic variables to be used in the description of reversible processes. Various processes may be most easily described thermodynamically in terms of various other energy functions, different from, but all simply related to, the basic energy function E. It is the purpose of this section to construct these other functions. Our development includes the introduction of the particle number N as an additional independent thermodynamic variable and, at the end, a discussion of heat capacities.

Heat Function W

If the volume of a body remains constant during an infinitesimal process, then $đQ = dE$; that is, the quantity of heat gained by the body is equal to the change in its internal energy. If, however, the process takes place at constant pressure, then the quantity of heat can be written as the differential of the *heat function* W, also called the *enthalpy*, which is defined by[12]

$$W \equiv E + \mathcal{P}\Omega \tag{8.1}$$

The differential of the heat function is given by

$$\begin{aligned} dW &= dE + \mathcal{P}d\Omega + \Omega d\mathcal{P} \\ &= T\,dS + \Omega d\mathcal{P} \end{aligned} \tag{8.2}$$

whence it follows that

$$\begin{aligned} T &= \left(\frac{\partial W}{\partial S}\right)_{\mathcal{P}} \\ \Omega &= \left(\frac{\partial W}{\partial \mathcal{P}}\right)_S \end{aligned} \tag{8.3}$$

The transformation (8.1) to (8.3) from the energy function to the enthalpy W is known mathematically as a *Legendre transformation*. Geometrically, the function $E(S, \Omega)$ has a slope at constant S given by $\mathcal{P} = -(\partial E/\partial\Omega)_S$. The tangent line of E as a function of Ω, therefore, has the intercept W of Equation (8.1) along the E-axis in a plane at constant S. In principle we can invert this expression for $\mathcal{P}$ and solve for the functional dependence $\Omega(S, \mathcal{P})$ on the surface E. The result of expressing W, rather than E, in terms of S and slope $-\mathcal{P}$ is Equations (8.2) and (8.3).

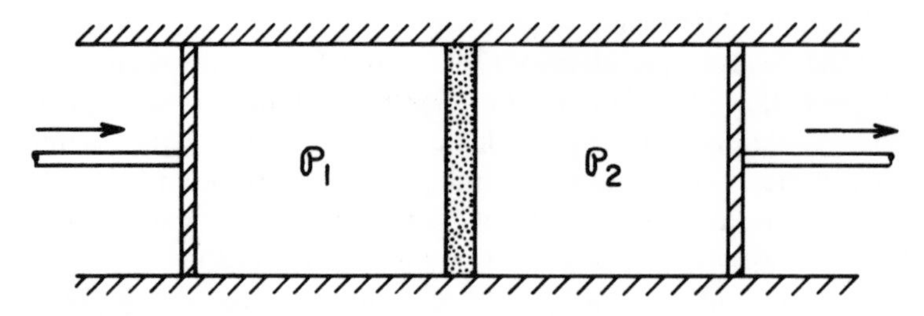

FIGURE II.3 Joule–Thomson process idealized.

Exercise 8.1 In the Joule–Thomson process a fluid at pressure $\mathcal{P}_1$ is transferred steadily to a vessel where the pressure is $\mathcal{P}_2 < \mathcal{P}_1$, with $\mathcal{P}_1$ and $\mathcal{P}_2$ held constant throughout the process. The process can be idealized, as shown in Figure II.3, by considering a gas that is transferred steadily through a porous partition in a tube that is itself thermally isolated from any external medium. The pressures are maintained by suitably moving pistons, together with frictional forces in the partition. These frictional forces also destroy any macroscopic gas velocities that tend to occur because of the pressure difference across the partition. The process is clearly an irreversible one between two equilibrium states of the gas. Show that the heat function per particle remains constant in the Joule–Thomson process. (*Suggestion:* Consider the work done to transfer a mass of gas through the partition.)

Helmholtz Free Energy Ψ

The work done on a body in an infinitesimal, isothermal reversible change in its state can be written as the differential of the *Helmholtz free energy* Ψ, defined by the Legendre transformation

$$\Psi \equiv E - TS \tag{8.4}$$

Thus the differential of the Helmholtz free energy is readily found to be

$$\begin{aligned} d\Psi &= dE - T\,dS - S\,dT \\ &= -S\,dT - \mathcal{P}\,d\Omega \end{aligned} \tag{8.5}$$

and therefore

$$\begin{aligned} S &= -\left(\frac{\partial \Psi}{\partial T}\right)_{\Omega} \\ \mathcal{P} &= -\left(\frac{\partial \Psi}{\partial \Omega}\right)_{T} \end{aligned} \tag{8.6}$$

Exercise 8.2

(a) Let $\beta = (\kappa T)^{-1}$, where κ is Boltzmann's constant, and show that the energy E can be derived from the Helmholtz free energy by the expression

$$E = \left.\frac{\partial(\beta\Psi)}{\partial\beta}\right|_{\Omega,N} \tag{8.7}$$

In Chapter VI we encounter this equation as a prescription for calculating the average energy in a macrocanonical ensemble [see Equations (27.7) and (27.17)].

(b) Show, for an arbitrary external parameter λ, that

$$\left(\frac{\partial E}{\partial\lambda}\right)_S = \left.\frac{\partial}{\partial\lambda}(\Psi + TS)\right|_S = \left(\frac{\partial\Psi}{\partial\lambda}\right)_T \tag{8.8}$$

Thus, as was discussed below Equation (7.5), the right side of Equation (7.5) can be evaluated for a system at constant temperature if the energy E is replaced by Ψ. When $\lambda = \Omega$, then substitution of Equation (8.8) into Equation (7.8) yields Equation (8.6) for the pressure $\mathcal{P}$.

Thermodynamic Potential G

The energy function for which the natural thermodynamic variables are temperature and pressure is the *thermodynamic potential* G, also known as the *Gibbs free energy*. It is defined by

$$G = E - TS + \mathcal{P}\Omega = \Psi + \mathcal{P}\Omega \tag{8.9}$$

and its differential form is given by

$$dG = -S\,dT + \Omega\,d\mathcal{P} \tag{8.10}$$

with

$$S = -\left(\frac{\partial G}{\partial T}\right)_{\mathcal{P}}$$

$$\Omega = \left(\frac{\partial G}{\partial \mathcal{P}}\right)_T \tag{8.11}$$

Because the conditions of constant temperature and pressure are the easiest ones to maintain in the laboratory, the thermodynamic potential is of central importance to the chemist. Equilibrium conditions in the theory of chemical reactions, and also in the theory of coexisting phases of a substance, are characterized by $\delta G = 0$ [see also inequality (9.15)].

The first derivatives of the thermodynamic potential G are given by Equations (8.11). Inasmuch as

$$\frac{\partial^2 G}{\partial T\,\partial \mathcal{P}} = \frac{\partial^2 G}{\partial \mathcal{P}\,\partial T}$$

we may conclude from Equations (8.11) that

$$\left(\frac{\partial \Omega}{\partial T}\right)_{\mathcal{P}} = -\left(\frac{\partial S}{\partial \mathcal{P}}\right)_T \tag{8.12}$$

Simple partial-derivative identities such as this one can be derived from the differential forms of each of the energy functions. These identities are called Maxwell relations, and each involves cross-derivatives between the pairs of variables (S, T) and $(\Omega, \mathcal{P})$. Further analysis of partial-derivative relations is given in the following section.

Exercise 8.3. Establish the three Maxwell relations:

$$\begin{aligned}
\left(\frac{\partial T}{\partial \Omega}\right)_S &= -\left(\frac{\partial \mathcal{P}}{\partial S}\right)_\Omega \\
\left(\frac{\partial T}{\partial \mathcal{P}}\right)_S &= \left(\frac{\partial \Omega}{\partial S}\right)_{\mathcal{P}} \\
\left(\frac{\partial S}{\partial \Omega}\right)_T &= \left(\frac{\partial \mathcal{P}}{\partial T}\right)_\Omega
\end{aligned} \tag{8.13}$$

When external parameters λ_i other than the volume are needed to define the state of a system, then additional work terms

$$-\sum_i \Lambda_i \, d\lambda_i$$

must be included in the thermodynamic identity (7.9) for dE. Appropriate generalizations of the other energy functions can then be made according to whether Λ_i or λ_i is constant in the process being studied.

Chemical Potential g

We have indicated that for the simplest systems in thermodynamic equilibrium there are only two independent thermodynamic variables. If, however, the number N of particles in the system is allowed to vary, then a third macroscopic variable is needed for the full description of even a uniform fluid. By these considerations one is led to introduce the *chemical potential* g, defined as the average energy change associated with the addition of a single particle to a system in thermodynamic equilibrium.

Let us consider first the *extensive* property of the energy functions. The energy, entropy, and volume are all extensive quantities, which means that at fixed density they are all proportional to the volume Ω, or to N, of the system.[13] Correspondingly, the temperature and the pressure are *intensive* quantities, which means that for fixed density they are independent of the size of the system. Referring to the preceding definitions of the energy functions W, Ψ, and G, we may conclude that the following functions are all intensive quantities.

$$\begin{aligned}
\frac{E}{N} &\equiv e = e\left(\frac{S}{N}, \frac{\Omega}{N}\right) \\
\frac{W}{N} &\equiv w = w\left(\frac{S}{N}, \mathcal{P}\right) \\
\frac{\Psi}{N} &\equiv \psi = \psi\left(T, \frac{\Omega}{N}\right) \\
\frac{G}{N} &\equiv g' = g'(T, \mathcal{P})
\end{aligned} \tag{8.14}$$

Here we have used the fact, in each of the second equalities, that an intensive quantity can only be a function of (other) intensive variables.

We now consider variation of the energy functions with particle number. It is necessary, for this purpose, to introduce into each of Equations (7.9), (8.2), (8.5), and (8.10) a term proportional to dN. Since dN takes on only integral values, such a term is meaningful, of course, only when N is extremely large, as it always is in thermodynamic considerations. As explained above, the proportionality coefficient of dN is defined to be the chemical potential g, so that one obtains the differential forms

$$\begin{aligned}
dE &= T\,dS - \mathcal{P}\,d\Omega + g\,dN \\
dW &= T\,dS + \Omega\,d\mathcal{P} + g\,dN \\
d\Psi &= -S\,dT - \mathcal{P}\,d\Omega + g\,dN \\
dG &= -S\,dT + \Omega\,d\mathcal{P} + g\,dN
\end{aligned} \tag{8.15}$$

The chemical potential is given by the derivative of each of the energy functions with respect to particle number when the appropriate other thermodynamic variables are held constant. Thus

$$g = \left(\frac{\partial E}{\partial N}\right)_{S,\Omega} = \left(\frac{\partial W}{\partial N}\right)_{S,\mathcal{P}} = \left(\frac{\partial \Psi}{\partial N}\right)_{T,\Omega} = \left(\frac{\partial G}{\partial N}\right)_{T,\mathcal{P}} \tag{8.16}$$

Each of these partial-derivative equalities can be verified independently by the method with which Equation (8.8) was proved.

The first of the equalities (8.16) shows, in keeping with its definition, that g is the "removal energy" of a particle at constant S and Ω. The final equality, together with the fourth of Equations (8.14), shows that $g' = g$. Therefore, the chemical potential of a system of identical particles is also its thermodynamic potential per particle.[14]

$$G = Ng \tag{8.17}$$

When expressed in terms of the independent variables T and $\mathscr{P}$, the differential form of g is found easily, from the fourth of Equations (8.15) and Equation (8.17), to be

$$dg = -\left(\frac{S}{N}\right) dT + \left(\frac{\Omega}{N}\right) d\mathscr{P} \tag{8.18}$$

Suppose that we consider a closed system in thermodynamic equilibrium and divided into two parts between which there is particle exchange. For single-component systems, in which there is only one kind of particle, application of the maximum entropy principle yields the condition that the chemical potential must be uniform throughout [Problem II.5(a)]. A special case of this result is the two-phase equilibrium condition (10.6) derived in Section 10.

Grand Potential *f*

Yet another energy function occurs in statistical analyses when, as in Chapter VIII, grand canonical ensembles are treated. The motivation in this case is to consider the chemical potential to be an independent variable instead of the particle number. We choose T and Ω to be the other two independent variables, so that the energy function F we introduce now is related to the Helmholtz free energy Ψ [see the third of Equations (8.15)] by the definition:

$$F \equiv -(\Psi - G) = \mathscr{P}\Omega \tag{8.19}$$

where we have used Equation (8.9) to obtain the second equality. The differential dF is found from Equation (8.17) and the third of Equations (8.15) to be

$$dF = S\,dT + \mathscr{P}\,d\Omega + N\,dg \tag{8.20}$$

so that the particle number N can be derived from F by the relation

$$N = \left(\frac{\partial F}{\partial g}\right)_{T,\Omega} = \Omega\left(\frac{\partial \mathscr{P}}{\partial g}\right)_T \tag{8.21}$$

Note, from Equation (8.18), that $\mathcal{P}$ depends on only two independent variables, say, g and T.

We also define the intensive quantity f in terms of F by the expression, with $\beta = (\kappa T)^{-1}$:

$$f \equiv \beta(\Omega^{-1}F) = \beta\mathcal{P} \tag{8.22}$$

The quantity f is called the *grand potential,* and in Section 36 Ωf will appear in statistical mechanics as the logarithm of the grand partition function. According to Equation (8.21) the density n of a system of identical particles is derived from the grand potential by the relation

$$n \equiv \frac{N}{\Omega} = \left(\frac{\partial \mathcal{P}}{\partial g}\right)_T = \beta^{-1}\left(\frac{\partial f}{\partial g}\right)_T \tag{8.23}$$

Since the grand potential differs from the pressure $\mathcal{P}$ only by the factor β, it is a function of two independent variables for simple single-component systems.

Heat Capacity C

The quantity of heat that, when added to a body in thermodynamic equilibrium, raises its temperature by one degree is called its heat capacity. When the heat capacity is calculated for a unit quantity (mass, number of particles, etc.) of the body, then the heat added is called *specific heat capacity* or merely specific heat. It is obvious that the heat capacity of a body depends on the process by which it is warmed up. We distinguish between the heat capacity C_V at constant volume and the heat capacity C_P at constant pressure, both energy-transfer processes being conducted reversibly (and at constant N).

$$\begin{aligned} C_V &\equiv \frac{dQ_V}{dT} = T\left(\frac{\partial S}{\partial T}\right)_\Omega \\ C_P &\equiv \frac{dQ_P}{dT} = T\left(\frac{\partial S}{\partial T}\right)_\mathcal{P} \end{aligned} \tag{8.24}$$

Here we used Equation (7.12) to write the second equality in each case. Since $đQ$ is an inexact differential, there is an infinity of possible warming processes for a given body in a specified thermodynamic state. A few processes other than those at constant volume and pressure are of interest (see, e.g., Problem II.2).

Upon using the thermodynamic identity (7.9), we see that C_V can also be written as

$$C_V = \left(\frac{\partial E}{\partial T}\right)_\Omega \tag{8.25}$$

Similarly, Equation (8.2) for dW, taken at constant pressure, shows that C_P can also be written as

$$C_P = \left(\frac{\partial W}{\partial T}\right)_{\mathcal{P}} \tag{8.26}$$

In both of these last two expressions it is assumed that the particle number N is held constant [see Equations (8.15)]. We usually suppress the subscript indicating this constraint.

Exercise 8.4 Use the result of Exercise 4.3 to show for the monatomic classical ideal gas at constant volume that the specific heat per particle is $(3/2)\kappa$. We may conclude, therefore, that the proportionality constant κ introduced into the entropy expressions following Equation (4.23) is indeed Boltzmann's constant. For in the ideal-gas limit, which can be identified experimentally using, for example, a constant-volume gas thermometer, Boltzmann's constant is defined operationally by using the equivalent ideal-gas result:

$$\frac{1}{N}\langle E\rangle = \frac{3}{2}\kappa T$$

9 Partial Derivatives in Thermodynamics, Thermodynamic Inequalities, Nernst's Theorem

The partial derivatives that arise in thermodynamic calculations are not always directly measurable quantities, as are the ones of Equations (8.3), say. In the first part of this section we give some standard procedures for expressing partial derivatives in terms of recognizable measurable quantities. We then pursue this analysis by applying it to an investigation of fundamental thermodynamic inequalities, all derivable from the maximum entropy principle. The concluding topic of this section is Nernst's theorem, which deals with the entropy constant of a body at absolute zero temperature.

We begin by listing the most directly measurable thermodynamic quantities of interest in this section. In addition to T, S, Ω, $\mathcal{P}$, and the heat capacities C_V and C_P, we shall consider some commonly occurring relations between T, S, Ω, and $\mathcal{P}$. Of course, the equation of state for a substance summarizes all possible relations, but we focus attention on the three particular quantities:

$$\alpha \equiv \Omega^{-1}\left(\frac{\partial \Omega}{\partial T}\right)_{\mathscr{P}} \qquad \text{(coefficient of thermal expansion)} \tag{9.1}$$

$$\kappa_T \equiv -\Omega^{-1}\left(\frac{\partial \Omega}{\partial \mathscr{P}}\right)_T \qquad \text{(isothermal compressibility} \tag{9.2}$$

$$\kappa_S \equiv -\Omega^{-1}\left(\frac{\partial \Omega}{\partial \mathscr{P}}\right)_S \qquad \text{(adiabatic compressibility)} \tag{9.3}$$

Relations between partial derivatives are derived in the elementary theory of partial differentiation. Let x, y, z be quantities satisfying a functional relation $f(x, y, z) = 0$, and let w be some function of any two of the variables x, y, z. Then three simple relations are

$$\left(\frac{\partial y}{\partial x}\right)_z^{-1} = \left(\frac{\partial x}{\partial y}\right)_z \tag{9.4a}$$

$$\left(\frac{\partial z}{\partial x}\right)_w = \left(\frac{\partial z}{\partial y}\right)_w \left(\frac{\partial y}{\partial x}\right)_w \tag{9.4b}$$

$$\left(\frac{\partial x}{\partial y}\right)_z \cdot \left(\frac{\partial y}{\partial z}\right)_x \cdot \left(\frac{\partial z}{\partial x}\right)_y = -1 \tag{9.4c}$$

Exercise 9.1 Starting from Equation (8.7), show that the energy per particle can be derived from the grand potential by the relation

$$\frac{E}{N} = g - \frac{1}{n}\left(\frac{\partial f}{\partial \beta}\right)_g \tag{9.5}$$

The transition from Equation (8.7) to Equation (9.5) reflects a shift from the independent variables (β, n) to (β, g). In Section 36 Equation (9.5) reappears as Equation (36.17).

Sometimes it is possible to deduce a recognizable form for a partial derivative by using one or more of the Maxwell relations. In the simplest example, Equations (8.12) and (9.1) give directly,

$$\left(\frac{\partial S}{\partial \mathscr{P}}\right)_T = -\left(\frac{\partial \Omega}{\partial T}\right)_{\mathscr{P}} = -\Omega\alpha \tag{9.6}$$

Another, more complicated example is to evaluate $(\partial E/\partial \Omega)_{T,N}$ by considering the first of Equations (8.15) and the third of Equations (8.13). One finds (for constant N)

$$\left(\frac{\partial E}{\partial \Omega}\right)_T = T\left(\frac{\partial S}{\partial \Omega}\right)_T - \mathcal{P}$$

$$= T\left(\frac{\partial \mathcal{P}}{\partial T}\right)_\Omega - \mathcal{P} \tag{9.7}$$

$$= \frac{T\alpha}{\kappa_T} - \mathcal{P}$$

where Equations (9.1), (9.2), (9.4a), and (9.4c) have all been used to derive the third equality.

In the example of Equation (9.7) the two independent variables Ω and T are involved. Often we want to switch independent variables, say, from E and T to Ω and T. In such cases there is a convenient generalization of Equation (9.4c) that involves the Jacobian $\partial(u, v)/\partial(x, y)$, defined as a determinant by

$$\frac{\partial(u, v)}{\partial(x, y)} = \begin{vmatrix} \dfrac{\partial u}{\partial x} & \dfrac{\partial u}{\partial y} \\ \dfrac{\partial v}{\partial x} & \dfrac{\partial v}{\partial y} \end{vmatrix} = -\frac{\partial(v, u)}{\partial(x, y)} \xrightarrow[v\,=\,y]{} \left(\frac{\partial u}{\partial x}\right)_y \tag{9.8}$$

In this definition u and v are considered functions of the independent variables x and y. The generalization of Equation (9.4c) is, then,

$$\frac{\partial(u, v)}{\partial(x, y)} = \frac{\partial(u, v)}{\partial(s, t)} \frac{\partial(s, t)}{\partial(x, y)} \tag{9.9}$$

as one can verify by setting $v = y$, $s = x$, and $t = u$. Equation (9.9) can be proved by expanding each partial derivative on the left side in terms of the variables s and t; for example,

$$\frac{\partial u}{\partial x} = \frac{\partial u}{\partial s}\frac{\partial s}{\partial x} + \frac{\partial u}{\partial t}\frac{\partial t}{\partial x}$$

To illustrate the application of Equation (9.9), we shall derive an expression for $(C_P - C_V)$ by writing C_P in terms of the independent variables Ω and T instead of $\mathcal{P}$ and T. We write

$$C_P = T\left(\frac{\partial S}{\partial T}\right)_{\mathcal{P}} = T\frac{\partial(S, \mathcal{P})}{\partial(T, \mathcal{P})}$$

$$= T\frac{\partial(S, \mathcal{P})}{\partial(\Omega, T)} \Big/ \frac{\partial(T, \mathcal{P})}{\partial(\Omega, T)}$$

(Continued on p. 73)

$$= -T\left\{\left(\frac{\partial S}{\partial \Omega}\right)_T\left(\frac{\partial \mathcal{P}}{\partial T}\right)_\Omega - \left(\frac{\partial S}{\partial T}\right)_\Omega\left(\frac{\partial \mathcal{P}}{\partial \Omega}\right)_T\right\} \Big/ \left(\frac{\partial \mathcal{P}}{\partial \Omega}\right)_T$$

$$= C_V - T\left(\frac{\partial \mathcal{P}}{\partial T}\right)_\Omega^2\left(\frac{\partial \Omega}{\partial \mathcal{P}}\right)_T$$

where we have again used the third of the Maxwell relations (8.13) to obtain the final equality. This is a well-known result, and it can be rewritten, using Equations (9.1), (9.2), and (9.4c), as

$$C_P - C_V = \frac{\Omega T \alpha^2}{\kappa_T} \tag{9.10}$$

Exercise 9.2

(a) Show that C_P and C_V can each be expressed in terms of α, κ_T, and κ_S by deriving the relation

$$\frac{C_P}{C_V} = \frac{\kappa_T}{\kappa_S} \tag{9.11}$$

[*Suggestion:* Follow the method used to derive Equation (9.10).]

(b) In the free expansion of a thermally isolated gas (Figure II.4) the thermodynamic energy E is conserved. For a small volume change $\Delta\Omega$ between two equilibrium states the temperature change ΔT is given by

$$\Delta T = (\partial T/\partial \Omega)_E \, \Delta\Omega$$

Show that

$$\left(\frac{\partial T}{\partial \Omega}\right)_E = -\frac{1}{C_V}\left(\frac{\partial E}{\partial \Omega}\right)_T \tag{9.12}$$

where $(\partial E/\partial \Omega)_T$ is given by Equation (9.7). What is the value of this partial derivative for the classical ideal gas?

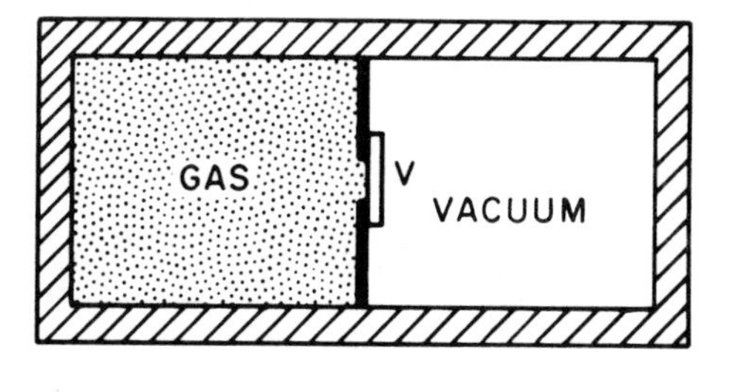

FIGURE II.4 The free expansion of a thermally isolated gas occurs when valve V is opened.

Minimal Conditions for Free Energies

We now state a theorem concerning the minimal conditions for the free energies Ψ and G: In a state of thermodynamic equilibrium, the Helmholtz free energy Ψ and the thermodynamic potential G of a body are minimal, the first with respect to small irreversible changes of state at constant T and Ω and the second with respect to small irreversible changes at constant T and $\mathcal{P}$.

The proof of this theorem starts from the basic inequality (7.14), which applies to irreversible and infinitesimal processes for which the temperature is essentially well defined. We may substitute the first law of thermodynamics, Equation (7.11), into this inequality to obtain

$$đW \leqq -(dE - T\,dS) = -d\Psi_T \tag{9.13}$$

where definition (8.4) of the Helmholtz free energy has been used. Thus we have a preliminary result that the maximum possible work a system can do in an isothermal change of state is given by the negative of the Helmholtz-free-energy change. For processes at constant temperature and volume $đW = 0$, and inequality, (9.13) becomes

$$d\Psi_{T,\Omega} \leqq 0 \tag{9.14}$$

Thus an irreversible process at constant temperature and volume results in a decrease of the Helmholtz free energy.[15] Corresponding to the maximum entropy condition at thermodynamic equilibrium, an isolated system kept at constant temperature and volume will be, at equilibrium, in a state in which the Helmholtz free energy is a minimum.

For processes, such as chemical reactions, that take place at constant temperature and pressure, substitution of Equation (7.10) for $đW$ into inequality (9.13) yields

$$d(\Psi + \mathcal{P}\Omega)_{T,\mathcal{P}} = dG_{T,\mathcal{P}} \leqq 0 \tag{9.15}$$

which says that an irreversible process at constant temperature and pressure results in a decrease in its thermodynamic potential. We may conclude that an isolated system kept at constant temperature and pressure will have a minimum Gibbs free energy at thermodynamic equilibrium. We discuss in Section 10 (especially with reference to Figure II.7) how this condition gives an understanding of the phase changes of materials.

Thermodynamic Inequalities

Thermodynamic equilibrium is dictated by the maximum entropy principle. The second law of thermodynamics, which asserts an increasing tendency for the total entropy of closed systems, encompasses the rest of this important principle. We now show how thermodynamic inequalities result from examining both necessity and sufficiency in the identification of the maximum entropy condition with thermodynamic equilibrium.

We begin by examining the implications (necessity) of the maximum entropy condition and show that in the equilibrium state of a system both the temperature and the pressure must be positive quantities. We can prove this fact after solving the thermodynamic identity (7.9) for the entropy differential

$$dS = \frac{1}{T}\,dE + \left(\frac{\mathcal{P}}{T}\right)d\Omega \qquad (9.16)$$

From this expression we obtain

$$\left(\frac{\partial S}{\partial E}\right)_{\Omega} = \frac{1}{T}$$

which is our original definition (6.3) of absolute temperature. If T were negative, then the entropy would increase with decreasing energy. The lowest internal energy state of a system would then not correspond to the lowest entropy, in violation of our probabilistic notions about entropy. Furthermore, in order to increase its entropy, every part of a body would tend to fly apart, thereby minimizing its internal energy.[16] In other words, macroscopic bodies could not exist at negative temperatures.[17]

We also have from Equation (9.16) that

$$\left(\frac{\partial S}{\partial \Omega}\right)_{E} = \frac{\mathcal{P}}{T}$$

for bodies in thermodynamic equilibrium. But $(\partial S/\partial \Omega)_E > 0$ always, because otherwise a body at constant energy would tend to contract spontaneously in order to increase its entropy. By contrast, the spontaneous expansion of fluids is physically expectable and is prevented only by the existence of external forces, such as are provided by gravity or a container. Thus, since T is a positive quantity, so is the pressure $\mathcal{P}$.[18]

We now derive some other, less obvious thermodynamic inequalities. We have already seen that a necessary condition that the entropy of a system increase for infinitesimal irreversible processes is given by inequality (7.14), which is equivalent to the condition that at thermodynamic equilibrium the thermodynamic potential G must be a minimum for processes at constant T and $\mathcal{P}$. We now examine the sufficiency condition for the maximum entropy principle by requiring that the second-order variation of G about its equilibrium value be positive for virtual processes at constant T and $\mathcal{P}$.[19] Thus we write

$$\delta G = \delta E - T\,\delta S + \mathscr{P}\,\delta\Omega$$
$$= \left(\frac{\partial E}{\partial S}\right)_{\Omega} \delta S + \left(\frac{\partial E}{\partial \Omega}\right)_{S} \delta\Omega - T\,\delta S + \mathscr{P}\,\delta\Omega \tag{9.17}$$
$$+ \frac{1}{2}\left[\frac{\partial^2 E}{\partial S^2}(\delta S)^2 + 2\frac{\partial^2 E}{\partial S\,\partial\Omega}\delta S\,\delta\Omega + \frac{\partial^2 E}{\partial\Omega^2}(\delta\Omega)^2\right] + \cdots$$

and require $\delta G > 0$ when the first line of the second equality vanishes. This latter requirement means that for variations with $\delta\Omega = 0$ we must have

$$\left(\frac{\partial^2 E}{\partial S^2}\right)_{\Omega} = \left(\frac{\partial T}{\partial S}\right)_{\Omega} = \frac{T}{C_V} > 0 \tag{9.18}$$

which implies that the constant-volume heat capacity C_V must be a positive quantity. Similarly, for variations with $\delta S = 0$ we must have

$$\left(\frac{\partial^2 E}{\partial \Omega^2}\right)_{S} = -\left(\frac{\partial \mathscr{P}}{\partial \Omega}\right)_{S} = (\Omega\kappa_S)^{-1} > 0 \tag{9.19}$$

which implies that the adiabatic compressibility κ_S is a positive quantity.

For a general variation of G, the entire bracket term in Equation (9.17) must be a positive quantity. If it could vanish, then the relation between the variations δS and $\delta\Omega$ would be given by

$$\frac{\partial S}{\partial \Omega} = \left(\frac{\partial^2 E}{\partial S^2}\right)^{-1}\left[-\left(\frac{\partial^2 E}{\partial S\,\partial\Omega}\right) \pm \sqrt{\left(\frac{\partial^2 E}{\partial S\,\partial\Omega}\right)^2 - \frac{\partial^2 E}{\partial S^2}\frac{\partial^2 E}{\partial\Omega^2}}\right]$$

For there to be no real zeros of this bracket term, the radical in this last equation must be an imaginary quantity. Therefore, we must have

$$\left(\frac{\partial^2 E}{\partial S^2}\right)_{\Omega} \cdot \left(\frac{\partial^2 E}{\partial \Omega^2}\right)_{S} - \left(\frac{\partial^2 E}{\partial S\,\partial\Omega}\right)^2 = \frac{\partial\left[\left(\frac{\partial E}{\partial S}\right)_{\Omega}, \left(\frac{\partial E}{\partial \Omega}\right)_{S}\right]}{\partial(S, \Omega)} > 0$$

where we have used definition (9.8) of the Jacobian. But this inequality is equivalent to

$$\frac{\partial(T, \mathscr{P})}{\partial(S, \Omega)} < 0$$

which with the aid of Equation (9.9) can be rewritten as

$$\frac{\partial(T, \mathscr{P})}{\partial(T, \Omega)} = \frac{\dfrac{\partial(T, \mathscr{P})}{\partial(T, \Omega)}}{\dfrac{\partial(S, \Omega)}{\partial(T, \Omega)}} = \frac{\left(\dfrac{\partial \mathscr{P}}{\partial \Omega}\right)_T}{\left(\dfrac{\partial S}{\partial T}\right)_{\Omega}} = -\frac{T}{\Omega\kappa_T C_V} < 0 \tag{9.20}$$

Inequalities (9.18) and (9.20), taken together, imply that the isothermal compressibility κ_T is a positive quantity. But then, according to Equation (9.11) and inequality (9.19), the constant-pressure heat capacity C_P must also be a positive quantity. In fact, Equation (9.10), together with inequalities (9.18) and (9.20), implies that

$$C_P > C_V > 0 \tag{9.21}$$

for all substances.

It is truly impressive that the thermodynamic inequalities can all be derived from one simple principle, namely, that the entropy of closed systems tends to increase, becoming a maximum at thermodynamic equilibrium.[20] Although these inequalities are mostly obvious intuitively, it is of value to have shown that the laws of thermodynamics are consistent with this intuition.

Nernst's Theorem

We initiate this next discussion by arguing heuristically. The fact that the heat capacity C_V is a positive quantity means, according to Equation (8.25), that at constant volume the energy is a monotonic increasing function of temperature. Conversely, as the temperature falls the energy decreases monotonically, and hence at the lowest possible temperature (absolute zero) a body must be in the state with the lowest possible energy. But the lowest possible energy state is the ground state of the system, which is characteristically assumed to be nondegenerate. Moreover, a nondegenerate state has a statistical weight of unity in the statistical-entropy definition (4.23), which implies that $S = 0$ for such a system in its ground state. If we generalize this conclusion to allow for degenerate ground states, then we arrive plausibly at a statement of *Nernst's theorem:* The entropy of every body must approach a constant Ns_0 at absolute zero.

$$\lim_{T\to 0} S(T) = Ns_0 \tag{9.22}$$

where s_0 is independent of all external parameters in the system. Nernst's theorem is also known as the *third law of thermodynamics*.

Let us now approach a justification of Nernst's theorem with greater care. We consider entropy from the probabilistic viewpoint and assume that Equation (4.25) applies (as indeed it does) when the states i are states of a thermally isolated system instead of a particle. Taking $(1/N)S$ in (4.25) to be the entropy S of the system, with all $Z_i = 1$, we obtain

$$S = -\kappa \sum_i P_i \ln P_i \underset{T\to 0}{\longrightarrow} -\kappa \sum_0 P_0 \ln P_0 \tag{9.23}$$

The sum in the second line is over all degenerate states 0, corresponding to the ground-state energy of the system, each with probability P_0. Here we have presumed, as above, that at $T = 0$ a system would be in one of its ground states.

Lacking information to the contrary, we may assume that each of the g_0 degenerate states occurs with equal probability so that $P_0 = g_0^{-1}$. Equation (9.23) then becomes

$$\lim_{T\to 0} S = \kappa \sum_{0} g_0^{-1} \ln g_0 = \kappa \ln g_0 \tag{9.24}$$

Thus if g_0 is a small number ($\ll N$), then the result (9.24) implies that $s_0 = O(N^{-1})$ in Nernst's theorem; this result corresponds to the common statement that the entropy (per particle) of every body must vanish at $T = 0$. Consider a counterexample, however. If $g_0 = 2^N$, as would be the case if each atom or molecule of a system had a doubly degenerate ground state and interparticle forces were absent, then we would have

$$\lim_{T\to 0} S = N\kappa \ln 2 \neq 0$$

We may conclude that the common form of Nernst's theorem is a statement about the quantum mechanical and dynamical properties of systems, because it asserts that g_0 must be a small number. The above counterexample is normally invalidated by the universal presence in nature of intermolecular forces, whether strong or ever so weak. However, there are systems for which g_0 is apparently not a small number, such that $N^{-1} \ln g_0 = \kappa^{-1} s_0 = O(1)$.[21]

An alternative formulation of the third law of thermodynamics sometimes is given by the statement that absolute zero is unattainable in nature.[22] This aspect of Nernst's theorem is operational in part, as it deals with the practical matter of attaining absolute zero in the limiting region where every heat transfer process is, according to $dQ \leqq T\,dS$, negligible. Thus Nernst's theorem implies that the heat capacity vanishes at $T = 0$, since from Equations (8.24)

$$C = T\frac{\partial S}{\partial T} = \frac{\partial S}{\partial \ln T}$$

and $\ln T \to -\infty$, in conjunction with the constancy of S, as $T \to 0$.

10 Phases in Equilibrium

So far we have restricted our discussions to single-phase, single-component systems. Of course, systems in nature can be multicomponent (i.e., composed of more than one kind of particle), and they may also include several different interacting phases. In this section we focus attention on the latter case.

The phases of materials are those macroscopic conditions we call solid, liquid, gas, and plasma, and we first investigate the equilibrium conditions for coexisting phases. Then as an application of this condition we derive the Clausius–Clapeyron equation. Finally, we deduce an expression for the pressure difference that can occur at the surface separating two phases.

When the system under consideration consists of several phases and/or components α, all in thermodynamic equilibrium, then each of the energy functions of Section 8 must be expressed more generally. The simplest case occurs when, ideally, there are no interactions between the various phases and particle components. For this case the total energy functions can each be written as a sum over components and phases α. But even when there are interactions between particle components, the energy differentials (8.15) may remain quite simple. Thus, as we argued in Section 6, the temperature at thermodynamic equilibrium will be uniform throughout a system. Moreover, in the absence of surface effects and position-dependent external potentials, the pressure will also be uniform throughout (Exercise 7.1). For this still simple but realistic case the only change necessary in Equations (8.15), for the differentials of the energy functions, is that the term $g\,dN$ must be generalized to $\Sigma_\alpha\, g_\alpha\, dN_\alpha$, where g_α is the chemical potential of the component or phase α.[23]

Exercise 10.1 Show, for systems in which the temperature and pressure are uniform throughout, that the generalization of Equation (8.17) for the thermodynamic potential is

$$G = \sum_\alpha N_\alpha g_\alpha \tag{10.1}$$

[*Suggestion:* Let

$$G = \sum_\alpha N_\alpha g'_\alpha = N \sum_\alpha x_\alpha g'_\alpha$$

with $g'_\alpha = g'_\alpha(T, \mathcal{P}, x_\alpha, x_\beta, x_\gamma, \cdots)$ [cf., the fourth of Equations (8.14)] and $\Sigma_\alpha\, x_\alpha = 1$. Then show that

$$\sum_\alpha N_\alpha g_\alpha = \sum_\alpha N_\alpha g'_\alpha$$

where

$$g_\alpha \equiv \left(\frac{\partial G}{\partial N_\alpha}\right)_{T,\mathcal{P}} \qquad (10.2)]$$

We now consider the example of systems with two phases in equilibrium. For a two-phase system the total entropy S will be given by the sum of the entropies of each phase and the intervening surface separately:

$$S = S_1 + S_2 + S_S \tag{10.3}$$

The reason is that the probability distributions for the two phases will multiply, even for identical-particle systems. For this latter case the particles in different phases will be, on the average, in energetically different single-particle states, so that exchange effects (due specifically to particle identity) will be negligible.[24] Similarly, for phases separated in space, the total thermodynamic energy will be given by

$$E = E_1 + E_2 + E_S \tag{10.4}$$

where the only interaction energy that must be considered between the two phases is a (mechanical) surface energy E_S. Here we have included surface energy to show its relevance in the following development and, more particularly, because it is necessary for the final topic of this section. The surface entropy term S_S included in Equation (10.3) is treated in Exercise 10.3.

We can apply the maximum entropy principle to the two-phase system at temperature T. If we hold all bounding surfaces fixed and assume that the entropy of each phase α is a function only of its energy E_α, volume Ω_α, and particle number N_α, then the maximum of the total entropy with respect to particle number N_1 will giv

$$\frac{\partial S}{\partial N_1} = \left(\frac{\partial S_1}{\partial N_1}\right)_{E_1,\Omega_1} - \left(\frac{\partial S_2}{\partial N_2}\right)_{E_2,\Omega_2} = 0$$

in which the total particle number $N = N_1 + N_2$ is assumed constant. Inasmuch as the first of Equations (8.15) holds for each phase separately, we then obtain quite generally

$$\left(\frac{\partial S_\alpha}{\partial N_\alpha}\right)_{E_\alpha,\Omega_\alpha} = -\frac{g_\alpha}{T} \tag{10.5}$$

Thus the maximum entropy condition gives, for a two-phase system,

$$g_1(T, \mathcal{P}_1) = g_2(T, \mathcal{P}_2) \tag{10.6}$$

In thermodynamic equilibrium the chemical potentials of the two phases will be equal. When surface effects can be neglected and there are no position-dependent external potentials, then their pressures will also be equal ($\mathcal{P}_1 = \mathcal{P}_2$), as we have already discussed.

When surface effects cannot be neglected, then $\mathcal{P}_1 \neq \mathcal{P}_2$, and we must modify the work terms in Equations (8.15). Instead of Equation (7.10) we write

$$\text{đ}W = \mathscr{P}_2\, d\Omega_1 + \mathscr{P}_2\, d\Omega_2 - \sigma\, da \tag{10.7}$$

where da is an element of the surface between phases 1 and 2 and σ is the *coefficient of surface tension* for the substance. The surface energy $dE_S = \sigma\, da$, that is, the work done in reversibly changing the surface area by da, is best understood empirically and macroscopically; it is not easily related to microscopic considerations by an equation such as (7.8) for the pressure. Nevertheless, some elaboration on our physical understanding of surface energy will be given before we write Equation (10.13). The surface-energy term remains the same for each of the energy-function differentials (8.15).

Clausius–Clapeyron Equation

The equation of state for a substance is a relation between $\mathscr{P}$, Ω, and T corresponding to all possible macroscopic states for which it exists in thermodynamic equilibrium. Figure II.5 depicts the equation of state of a substance that contracts on freezing, showing that the equation of state is a surface when plotted against the three rectangular coordinates $\mathscr{P}$, Ω, and T. The $\mathscr{P}$–T and $\mathscr{P}$–Ω projections of Figure II.5 are shown in Figure II.6. The term *vapor* is usually applied to a gas in equilibrium with its liquid or solid (i.e., a saturated vapor) or to a gas at a temperature below the critical temperature T_c, but the properties of a vapor differ in no essential respect from those of a gas.

We now focus attention on the parts of the $\mathscr{P}$–Ω–T surface where phases coexist in equilibrium. These parts are represented as lines in the $\mathscr{P}$–T diagram of Figure II.6. As we have already discussed, along these lines the phases in equilibrium are characterized by equal temperatures, equal pressures (when surface effects are neglected), and equal chemical potentials.

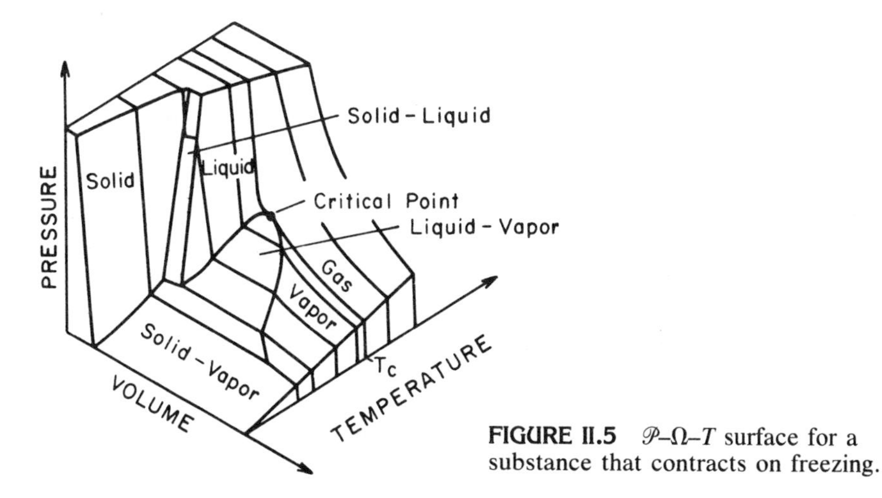

FIGURE II.5 $\mathscr{P}$–Ω–T surface for a substance that contracts on freezing.

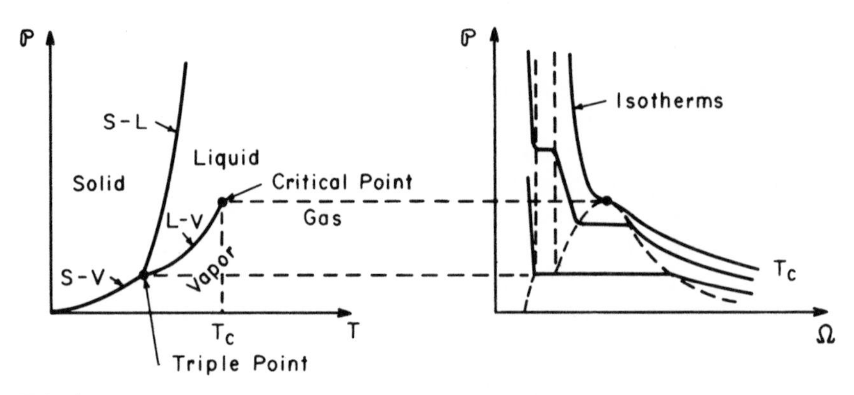

FIGURE II.6 $\mathcal{P}$–T and $\mathcal{P}$–Ω projections of Figure II.5.

Therefore, comparison of definition (8.9) for G with Equations (10.1), (10.3), (10.4), and $\Omega = \Omega_1 + \Omega_2$ gives for each phase the result

$$g_\alpha = \left(\frac{E_\alpha}{N_\alpha}\right) - T\left(\frac{S_\alpha}{N_\alpha}\right) + \mathcal{P}\left(\frac{\Omega_\alpha}{N_\alpha}\right) \qquad (10.8)$$
$$= w_\alpha - Ts_\alpha$$

neglecting surface effects, where w_α is the enthalpy per particle and s_α is the entropy per particle of the phase α. On equating the chemical potentials for two phases in equilibrium, and using the fact that the latent heat ℓ per molecule, absorbed by a body in changing phases, is defined by[25]

$$\ell \equiv w_2 - w_1 \qquad (10.9)$$

we obtain the result

$$\ell = T(s_2 - s_1) \qquad (10.10)$$

Equation (10.10) can also be derived by direct integration of $đQ = T\,dS$.

We must emphasize that the phase-equilibrium condition (10.6) applies only for certain ($\mathcal{P}_o$, T_o) relations corresponding to the curves in the $\mathcal{P}$–T projection of Figure II.6. Moreover, as we know from Equation (8.18) for dg and have already indicated by the notation of Equation (10.6), the chemical potentials g_α are functions of only two independent variables, say, T and $\mathcal{P}$. Suppose 1 denotes the liquid phase and 2 the gaseous phase. Then we can expect that Equation (10.8) is valid for two functions $g_1(T, \mathcal{P})$ and $g_2(T, \mathcal{P})$, which when plotted against T for a constant pressure $\mathcal{P}_o$ would appear as shown in Figure II.7. By referring to this figure we can understand how the minimal condition (9.15) for the thermodynamic potential causes a substance to change phases at some pressure $\mathcal{P}_o$ and temperature T_o. Equation (10.6) then determines precisely where this change occurs.

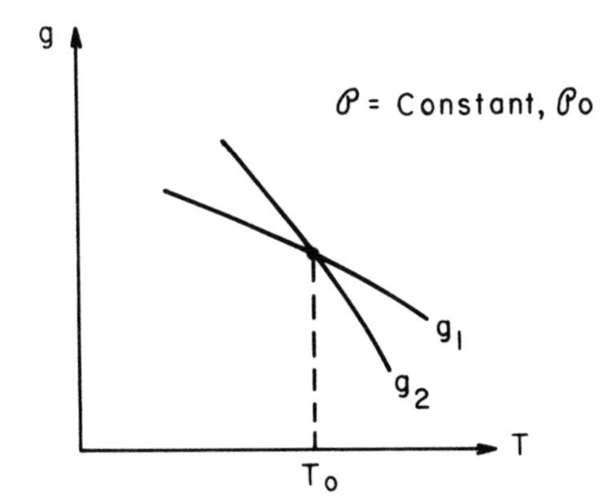

FIGURE II.7 Chemical potentials versus T for liquid (1) and gaseous (2) phases of a substance (schematic).

We now derive an explicit relation between $\mathcal{P}_o$ and T_o along the liquid–vapor curve in the $\mathcal{P}$–T projection of Figure II.6. Using the fact that along this curve $dg_1 = dg_2$, and dropping the subscript o, we have from Equation (8.18),

$$-s_1\,dT + v_1\,d\mathcal{P} = -s_2\,dT + v_2\,d\mathcal{P} \tag{10.11}$$

where $v_\alpha \equiv \Omega_\alpha/N_\alpha$ is the volume per particle in the phase α. This equation, together with Equation (10.10), is readily solved to give for the slope of the vapor pressure line,

$$\frac{d\mathcal{P}}{dT} = \frac{\ell}{T(v_2 - v_1)} \tag{10.12}$$

which is the *Clausius–Clapeyron equation*. In application, we can often neglect $v_1(<<v_2)$ and use the ideal-gas law $v_2 = \kappa T/\mathcal{P}$ for the vapor phase.

Exercise 10.2 Verify for the summit of Mt. Evans in Colorado, where the atmospheric pressure is 6/10 that at sea level, that water boils at approximately 86°C.

From Figure II.7 we see that a liquid–vapor phase change can be characterized by a continuous value of the chemical potential, but that at the point $(\mathcal{P}_o, T_o)$ there will be discontinuous first derivatives

$$\left(\frac{\partial g_1}{\partial T}\right)_{\mathcal{P}} \neq \left(\frac{\partial g_2}{\partial T}\right)_{\mathcal{P}} \quad \text{and} \quad \left(\frac{\partial g_1}{\partial \mathcal{P}}\right)_T \neq \left(\frac{\partial g_2}{\partial \mathcal{P}}\right)_T$$

The first inequality here is equivalent to $s_1 \neq s_2$ and the second inequality to $v_1 \neq v_2$. Phase changes with discontinuous first derivatives of the chemical potential are called *first-order phase transitions*.[26]

Surface Pressure

The existence of surface tension or surface energy causes a pressure difference to exist at the boundary between surfaces in equilibrium. We have already included a surface energy term $\sigma\, da$ in Equation (10.7). One can show for any contour bounding some part of the surface between two phases that σ is a force per unit length directed tangentially to the surface along the inward-pointing normal to the contour. For positive σ the surface tends to contract, thereby decreasing the thermodynamic energy of the system. In fact, for phases to exist at all, σ must be a positive number, since otherwise the surfaces would tend to stretch out in order to achieve a lower energy state. That the surface energy term in Equation (10.4) is positive can also be understood microscopically. The term E_S corrects for the fact that particles near the surface do not get to attract other particles in every direction, as do particles far away from the surface; that is, surface particles cannot "saturate" their attraction capabilities.

We can deduce the pressure difference at a point on a surface by minimizing the Helmholtz free energy there at constant temperature and total volume element $\Omega = \Omega_1 + \Omega_2$ [see Equation (9.14)]. Thus we write, using the third of Equations (8.15) and Equation (10.7),

$$\begin{aligned} d\Psi &= -\mathcal{P}_1\, d\Omega_1 - \mathcal{P}_2\, d\Omega_2 + \sigma\, da + g_1\, dN_1 + g_2\, dN_2 \\ &= -(\mathcal{P}_1 - \mathcal{P}_2)\, d\Omega_1 + \sigma\, da = 0 \end{aligned} \tag{10.13}$$

To derive the second equality, we have assumed that the total particle number $N = N_1 + N_2$ is also fixed and have used Equation (10.6). The pressure difference at the element δa, therefore, is given by

$$\mathcal{P}_1 - \mathcal{P}_2 = \sigma\left(\frac{da}{d\Omega_1}\right) = \sigma\left(\frac{1}{\rho} + \frac{1}{\rho'}\right) \tag{10.14}$$

where when the volume element Ω_1 changes by the amount $d\Omega_1 = \delta a \cdot dr$, then the area δa changes by the amount $d(\delta a) = (\rho^{-1} + \rho'^{-1})\, \delta a\, dr$, ρ and ρ' being the principal radii of curvature. See Figure II.8. The radii of curvature are taken as positive when they are directed into phase 1 and negative otherwise. Equation (10.14) is known as Lord Kelvin's formula.

If the first phase is a sphere immersed in the second phase (a drop of liquid in a vapor or a bubble of vapor in a liquid), then the surface of separation is spherical, and its radii of curvature are everywhere equal to the radius r of the sphere. In this case we obtain for the pressure $\mathcal{P}_1$ inside the sphere, relative to the outside pressure $\mathcal{P}_2$, the expression

$$\mathcal{P}_1(r) = \mathcal{P}_2 + \frac{2\sigma}{r} \tag{10.15}$$

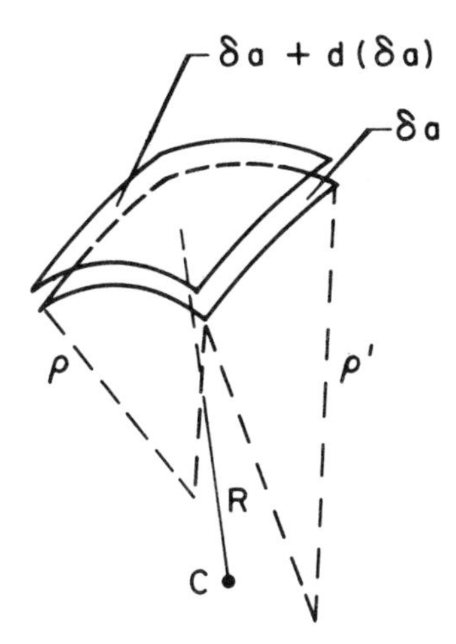

FIGURE II.8 The mean curvature R for a surface element δa, where ρ and ρ' are its minimum and maximum radii of curvature, respectively, is given by $R^{-1} = \frac{1}{2}(\rho^{-1} + \rho'^{-1})$.[27] In an expansion of the surface outward by a distance dr the area element changes by an amount $d(\delta a) = (2/R)\,\delta a \cdot dr$.

which is the same for both a liquid drop within vapor and for a bubble in liquid. In the limit of a plane surface ($r \to \infty$) $\mathcal{P}_1 = \mathcal{P}_2$. If for a given temperature $\mathcal{P}_o$ is the common pressure value of the two phases when the surface is plane, then one can show that the difference between a bubble and a drop is that $\mathcal{P}_1$ and $\mathcal{P}_2$ are both greater than $\mathcal{P}_o$ for a drop and both less than $\mathcal{P}_o$ for a bubble.[28,29]

Exercise 10.3 Argue that the surface contribution to the Helmholtz free energy is given by

$$\Psi_S = \sigma A$$

where A is the total surface area of the boundary between two phases. Then assume $\sigma = \sigma(T)$, with no dependence on g, and determine the surface contribution to the entropy to be

$$S_S = -A\frac{d\sigma}{dT}$$

Hence find an expression for the surface energy

$$E_S = \Psi_S + TS_S$$

Bibliography

K. Huang, *Statistical Mechanics* (Wiley, 1967), Chaps. 1 and 2. Includes an outline of the laws of thermodynamics, as they are conventionally presented, with emphasis on those results most relevant for the microscopic considerations of statistical mechanics. Several interesting applications are treated.

L. D. Landau and E. M. Lifshitz, *Statistical Physics* (Addison-Wesley, 1958), translated by E. Peierls and R. F. Peierls. An excellent source for relationships between statistical physics and thermodynamics. Our treatment of thermodynamics follows the approach presented in Chapter II of this text.

M. W. Zemansky, *Heat and Thermodynamics* (McGraw-Hill, 1968). A good standard introduction to classical thermodynamics presented entirely from a macroscopic viewpoint.

Notes

1. Our convention in this text is that $\langle O \rangle$ denotes the average value of O, whereas $\overline{O}$ is its maximum, or most probable, value. For the extensive quantities of large systems these two entities are essentially identical, as is shown in Appendix A.

2. One example of a reversible process is the adiabatic process treated in detail in Section 7. Clarification of the designation *slow process* is given at the beginning of Section 7.

3. A provocative discussion of the law of increasing entropy is given in the text by Landau and Lifshitz, *op. cit.*, Section 8.

4. See, for example, the elementary presentation of these ideas in the text by R. Resnick and D. Halliday, *Physics, Part I* (Wiley, 1966), Sections 21-5 and 25-6. See also Appendix C.

5. In Chapter VI (see Exercise 27.2), we show that the statistical definition (4.24) of entropy is consistent with the thermodynamic definition (7.12).

6. The molecular mean free path is defined in Section 12.

7. We use the term *adiabatic* here as it is defined in classical mechanics and in quantum mechanics and not as in conventional thermodynamic discussions, where the term also applies to thermally isolated irreversible processes.

8. See Equation (23.5). In other parts of this text we also use the letter E for microscopically calculated system energies. Confusion should not occur, since it is usually clear in context whether the considerations are microscopic or thermodynamic ones.

9. This statement is expressed mathematically by Equation (43.14) in Chapter IX. We shall include energy transfer due to currents when the subject of irreversible processes is treated (Chapter IX). Such energy transfers are omitted from Equation (7.11), although not from Equations (8.15).

10. Of course, the total entropy change for the system of interest plus a connecting heat reservoir must be zero for a reversible process.

11. This equivalence is demonstrated in Appendix C.

12. Confusion between the use of W for work and W for the heat function should not occur, because the context will always clarify which quantity is being used. At any rate, we do not make frequent reference to the heat function.

13. Strictly, the extensive property of system variables can be defined only for a sufficiently large system in which surface effects can be neglected. In subsequent chapters we refer to the limiting case of an infinite system, for which surface effects truly vanish, as the *thermodynamic limit*. By this limit we mean $(N, \Omega) \rightarrow \infty$ in such a way that the particle density, $n = N/\Omega$, remains constant (and finite).

14. We show explicitly in Section 31 that the quantity g in Equation (5.10) is also the chemical potential. Indeed, this fact becomes obvious if one compares the first of Equations (8.15) with the derivation of Exercise 5.2, in which Equation (5.10) is determined by the vanishing of $\kappa^{-1}\,\delta S - \beta\,\delta E + \beta g\,\delta N$. (Set Ω = constant and $\beta^{-1} = \kappa T$.)

15. Huang, *op. cit.*, p. 24, discusses an example of such a process that is virtual. An application of the equilibrium condition (9.14) is given at the end of Section 10 in connection with Equation (10.13).

16. Recall from Equation (6.4) that the argument of S_i, for the ith part of a body, is its internal energy $(E_i - P_i^2/2m_i)$. Of course, both the total momentum and angular momentum of Equations (6.5) would, in general, remain constant in such a disintegration.

17. The fact that spin systems in crystals can exist in metastable states at negative temperatures does not violate these considerations. See L. D. Landau and E. M. Lifshitz, *loc. cit.*, Section 70, for a discussion of this special case.

18. A special case can occur in that superheated liquids near the liquid–vapor transition region may, under certain circumstances, exist in metastable states with negative pressures. *Ibid.*, Section 79.

19. In other words, sufficiency requires that when a system is at equilibrium, then G must be at a minimum for processes at constant T and $\mathcal{P}$. Note the distinction between the virtual variations from the minimum value of G at thermodynamic equilibrium, being considered here, and the real variations in G for an irreversible process, as governed by inequality (9.15).

20. A generalization, called Le Chatelier's principle, of the results derived from Equation (9.17) is discussed in Section 22 of the text by Landau and Lifshitz.

21. Some substances, such as ice, seem to have a residual entropy constant s_0 describable in terms such as we have elaborated with this counterexample. See the introduction of an article by E. H. Lieb, *Phys. Rev.* **162** (1967), p. 167.

22. Considerable historical and thermodynamic discussion of Nernst's postulate, as it is also called, is given in P. S. Epstein, *Textbook of Thermodynamics* (Wiley, 1947), Sections 88-92 and 97. See also D. ter Haar, *Elements of Statistical Mechanics* (Rinehart, 1954), App. III, and E. Schrödinger, *Statistical Thermodynamics* (Cambridge University, 1947), Chapter III. The unattainability of absolute zero is examined in some detail in Section 1.7 of the text by Huang.

23. When the pressure is not uniform throughout, because of the presence of a position-dependent external potential and/or surface effects, then the work terms in Equations (8.15) must also be modified, say, by using Equation (10.7).

24. A many-body wave function, or state vector, can always be resolved as a linear combination of (properly symmetrized) single-particle product states.

25. Recall the discussion preceding Equation (8.1).

26. Similarly, a *second-order phase change* is one at which both s and v are continuous, and instead, second-order derivatives of g are discontinuous. The nomenclature is attributable to P. Ehrenfest.

27. J. J. Stoker, *Differential Geometry* (Wiley, 1969), pp. 93 and 107.

28. L. D. Landau and E. M. Lifshitz, *loc. cit.*, p. 465.

29. Further treatment of surface effects, supplementary to that of Section 10, can be found in L. D. Landau and E. M. Lifshitz, *loc. cit.*, Chapter XV, and P. S. Epstein, *loc. cit.*, Chapter XII.

Problems

II.1. Lagrange's *method of undetermined multipliers* states that "the restricted maximum of a function S of N variables $(x_1, \cdots, x_n)$ subject to M constraint equations $C_\alpha(x_i)$, with $\alpha = 1, 2, \cdots, M (M < N)$, can be found by using the unrestricted maximum condition

$$dS + \sum_{\alpha=1}^{M} \lambda_\alpha \, dC_\alpha = 0$$

where the numbers λ_α are unknown multipliers.

(a) Write:

$$dS = \sum_{i=1}^{N} S_i \, dx_i = 0, \qquad S_i = S_i(x_1 \cdots x_N)$$

$$dC_\alpha = \sum_{i=1}^{N} C_{\alpha,i} \, dx_i = 0, \qquad C_{\alpha,i} = C_{\alpha,i}(x_1 \cdots x_N)$$

and prove Lagrange's method for $M = 1$. Be careful to show that λ_1 is independent of the label i.

(b) Construct a general proof of Lagrange's method for arbitrary $M < N$, for example, by using an induction hypothesis.

(c) Consider N systems in thermodynamic equilibrium with each other, each with constant volume and particle number, and with constant total energy:

$$E = \sum_{i=1}^{N} E_i$$

Prove that in thermodynamic equilibrium

$$\left(\frac{\partial S_i}{\partial E_i}\right)_{\Omega_i, N_i}$$

is independent of i, where S_i is the entropy of the ith system.

II.2. (a) Show that $T\,dS = C_P\,dT - \Omega \alpha T\,d\mathcal{P}$ where α is the coefficient of thermal expansion.

(b) Prove that the heat capacity C_{Sat} of a vapor, for the process in which liquid (phase 1) remains at all times in equilibrium with its saturated vapor (phase 2), is given by

$$C_{Sat} = C_P - \frac{\Omega_2 \alpha_2 \ell}{v_2 - v_1}$$

where ℓ is the latent heat in the phase transition and $(v_2 - v_1)$ is the volume per particle change. See the $\mathcal{P}$–T projection of Figure II.6.

Simplify the second term by neglecting v_1 in comparison with v_2 and assuming that the vapor is an ideal gas.

(c) According to Equation (4.20), $\overline{E} = \Sigma_i \, \overline{n}_i \omega_i$ for the ideal gas, where $\overline{n}_i$ is given classically by Equation (4.16). Substitute this result into Equation (8.25), and show by direct calculation that $C_V = \frac{3}{2} N\kappa$.

II.3. (a) A gas does work when it expands adiabatically. What is the energy source for this work? Consider an adiabatic curve and an isothermal curve plotted for a gas on a $\mathcal{P}$–Ω diagram. Give a general argument, either physical or mathematical, to prove that the adiabatic curve must always be steeper than the isothermal one where they cross.

(b) Forty-four grams of CO_2, taken to be an ideal gas, are used to operate a Carnot heat engine between the isotherms $T_1 = 273$K and $T_2 = 373$K, and the adiabatics $S_1 = 197.52$ joules/K and $S_2 = 222.46$ joules/K. See Appendix C. In one cycle, how much work does the engine do, and what is its efficiency? By what factor does the volume change during either isothermal segment of the cycle?

II.4. For macroscopic systems in which the energy fluctuations are small, use

$$S = -\kappa \sum_i P_i \ln P_i$$

where P_i is the probability for finding the system in its ith microscopic energy state, to show that the maximum entropy is given approximately by

$$\overline{S} \cong \kappa \ln[n(\overline{E}) \, \delta E]$$

Here $n(\overline{E})$ is the density of states for the system, evaluated at $E = \overline{E}$, and δE is the rms spread of energy states about the most probable value $\overline{E}$. See Figure II.9. Argue then that the number of energy states in a unit interval of energy for a many-body system increases exponentially with the entropy.

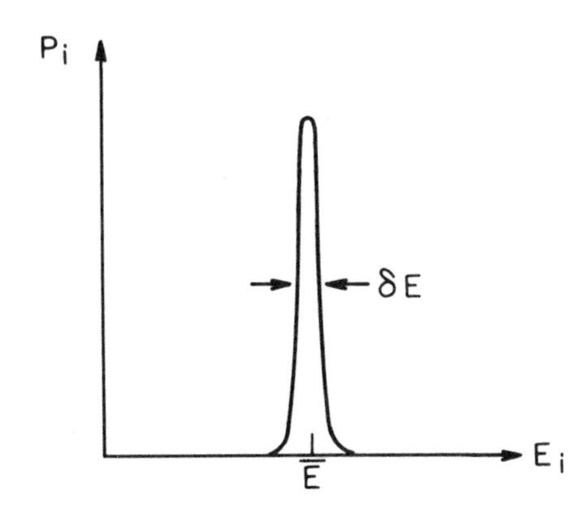

FIGURE II.9 The probability P_i for finding a macroscopic system in its ith energy state is a sharp maximum peaked at the most probable energy $\overline{E}$. The curve is normalized according to $\Sigma_i \, P_i = 1$.

II.5. (a) Consider a thermally isolated system in a time-independent external field, which is represented by a potential energy per molecule $U(\mathbf{r})$. Use the maximum entropy principle to show that at thermodynamic equilibrium the temperature and chemical potential, but not necessarily the pressure, are uniform throughout the system. If $g_o(T, \mathcal{P})$, with $\mathcal{P} = \mathcal{P}(\mathbf{r})$, is the chemical potential in the absence of an external field, then argue that

$$g = g_o(T, \mathcal{P}) + U(\mathbf{r}) = \text{uniform throughout system}$$

(b) In a uniform gravitational field, $U(\mathbf{r}) = mg'z$, where m is the (average) mass of a molecule, g' is the gravitational acceleration, and z is the vertical coordinate, we may assume that the pressure varies slowly and uniformly with z. Hence show that

$$\frac{d\mathcal{P}}{n} = -mg'\,dz$$

where n is the particle density and $\mathcal{P}$ is the pressure.

(c) Assume that the result in part (b) applies to the atmosphere, considered to be an ideal gas with constant temperature, and derive the *barometer formula* (valid for moderate elevation changes),

$$n(z) = n_0 \exp(-mg'z/\kappa T)$$

where n_0 is the density at $z = 0$.

II.6. *Adiabatic processes*

(a) Show for an adiabatic process that

$$d\mathcal{P} = \left(\frac{\gamma}{\gamma - 1}\right)\left(\frac{\alpha}{\kappa_T}\right) dT \xrightarrow[\text{ideal gas}]{} \left(\frac{\gamma}{\gamma - 1}\right)\left(\frac{\mathcal{P}}{T}\right) dT$$

where $\gamma \equiv C_P/C_V$. Use this result together with that of Problem II.5(b) to calculate dT/dz in degrees per kilometer for the atmosphere. Set $\gamma = 1.4$ and $N_o m = 29$ gm/mole, where N_o is Avogadro's number.

(b) A method of measuring γ for a gas is as follows. A gas, assumed ideal, is confined within a vertical cylindrical container, and it supports a freely moving piston of mass M. The piston and container both have the same cross-sectional area A. Let $\mathcal{P}_o$ denote the external atmospheric pressure, g' the acceleration of gravity, and Ω_o the equilibrium volume of the gas. When the piston is displaced from its equilibrium position it is found to oscillate with a frequency ν; the corresponding gas oscillations occur *adiabatically*. Express γ in terms of M, g', A, $\mathcal{P}_o$, Ω_o, and ν.

(c) Determine the variation of pressure with height for an adiabatic atmosphere and compare your answer with the isothermal result of Problem II.5(c).

II.7. *Fluctuations*

(a) Apply definition (4.24) of the entropy to an ensemble of N systems and argue that the probability for finding a fluctuation ΔS_t in the total entropy of any one system is given by the proportionality

$$P(\Delta S_t) \propto e^{\Delta S_t/\kappa}$$

where $\Delta S_t \equiv (1/N)(S - \overline{S})$ with $\overline{S}$ given by Equation (4.26).

(b) Consider a small part of a large system, for example, the system selected in part (a), which small part is characterized by fluctuations ΔS, ΔE, and $\Delta\Omega$. Show, if the total system energy and volume are constant, that the probability for finding fluctuations in the subsystem is given by

$$P \propto \exp\left\{-\frac{1}{\kappa T}[\Delta E + \mathcal{P}\,\Delta\Omega - T\,\Delta S]\right\}$$

(*Suggestion:* Treat the large part of the system as a surrounding medium for the small subsystem, and write $\Delta S_t = \Delta S + \Delta S_{\text{med}}$.)

(c) Expand ΔE as a series in small variations and show that

$$[\Delta E + \mathcal{P}\,\Delta\Omega - T\,\Delta S] = \frac{1}{2}[\Delta S \cdot \Delta T - \Delta\mathcal{P} \cdot \Delta\Omega]$$

(d) Choosing T and Ω to be the independent variables, show with the aid of part (c) that the result of part (b) can be rewritten as

$$P \propto \exp\left\{-\frac{1}{2\kappa T}\left[\frac{C_V}{T}(\Delta T)^2 + (\Omega\kappa_T)^{-1}(\Delta\Omega)^2\right]\right\}$$

and hence that, for example,

$$(\Delta\Omega)_{\text{rms}} = (\kappa T\Omega\kappa_T)^{1/2}$$

when N = constant in the subsystem.

II.8. A gas has a Helmholtz free energy per particle $\psi = \Psi/N$ given by *either* $\psi_1 = -aT^2v^2$ or $\psi_2 = -aT^2v^{1/2}$, where a is a dimensioned constant and $v = \Omega/N$.

(a) For each case calculate:

Entropy per particle $s = S/N$
Pressure, $\mathcal{P}$
Internal energy per particle, $e = E/N$
Specific heat capacity at constant volume, $c = C_V/N$

(b) Does this substance obey the third law of thermodynamics (consider each case)? What must be the sign of a?

(c) Prove that one of the two given cases cannot describe a real gas.

CHAPTER III

KINETIC THEORY OF GASES

An introduction to the microscopic theory of real gases is provided in this chapter by the kinetic theory of classical physics. All microscopic theories must necessarily involve the dynamics of interacting particles. Thus the kinetic theory of gases predicts macroscopic properties from a detailed knowledge of molecular collisions, which, in turn, depend on the dynamics. After a careful derivation of the Maxwell–Boltzmann distribution of particle energies in a dilute gas, we focus attention on the very important Boltzmann transport equation and its solution. The basis this equation provides for the macroscopic equations of hydrodynamics is pursued in detail. Finally, an elementary outline of the theory of noise is given.

11 Collisions and the Maxwell–Boltzmann Distribution

In Section 4 we introduced the one-particle distribution function $P(\mathbf{k})$ for a gas of weakly interacting molecules. Classical kinetic theory employs a position- and time-dependent generalization of this function, which we write as $P(\mathbf{r}, \mathbf{k}, t)$.[1] Whereas $P(\mathbf{r}, \mathbf{k}, t)$ is the probability density function for finding a single molecule with momentum $\mathbf{k}$ at the position $\mathbf{r}$ and time t, the function $P_2(\mathbf{r}_1, \mathbf{r}_2; \mathbf{k}_1, \mathbf{k}_2; t)$ is the corresponding probability density for two molecules. A central concept of kinetic theory is that two-particle correlations in a system, as governed by P_2, influence the time evolution of P itself. Thus we begin in this section by studying the time rate of change of P due to collisions. This quantity, $(\partial P/\partial t)_{\text{coll}}$, is related to P_2 by an integration, and hence it is popularly referred to as the collision integral.

Thermodynamic equilibrium is characterized in the kinetic-theory view by the vanishing of the collision integral. By this identification one is led to a derivation of the Maxwell–Boltzmann distribution function $P_0(\mathbf{r}, \mathbf{k})$, which is the particular form $P(\mathbf{r}, \mathbf{k}, t)$ takes at equilibrium. The derivation requires the assumption of stochastic independence for P_2 (at equilibrium), an assumption referred to historically as the *stosszahlansatz*.

After a discussion of the Maxwell–Boltzmann distribution we examine an alternate derivation of it from the Boltzmann transport equation. This equation, derived heuristically here and more rigorously in Section 23, is simply an explicit expression for the total time rate of change of P. In the alternative derivation of the Maxwell–Boltzmann distribution for a gas we are able to identify position dependence in P_0 on any single-particle potential energies.

Classical kinetic theory applies in the domain where classical mechanics is valid. For a gas an important criterion of this domain is that the thermal wavelength λ_T of Equation (4.2) be $\ll \ell$, the average interparticle spacing. When this condition is satisfied, then the particles of the gas can be regarded as being distinguishable, and we may treat their interactions classically.[2]

In this chapter we always consider a gas of large volume Ω with N identical, but distinguishable, point particles. The single-particle distribution function P, which describes the typical behavior of any one of these particles, is assumed to vary negligibly over a "small" region $d\omega \equiv d^3\mathbf{r}\, d^3\mathbf{k}$ of the total available position and momentum space, collectively called *phase space* (or μ space). The normalization of $P(\mathbf{r}, \mathbf{k}, t)$ is chosen arbitrarily to be the number N instead of unity; that is,

$$\int d\omega P(\mathbf{r}, \mathbf{k}, t) = N \qquad \textbf{(11.1)}$$

For the normalization integral (11.1) to be meaningful we must be able to choose the tiny phase-space volume elements $d\omega$ (or $\delta\omega$) to be sufficiently "large" that, on the average, they contain many molecules. Otherwise, the concept of a continuous probability density (Section 3) would not be meaningful. Consider what this means for the size of $\delta^3\mathbf{r}$. At standard conditions (273K and atmospheric pressure) 1 cm^3 of gas contains approximately 3×10^{19} molecules. Thus if we make the reasonable choice $\delta^3\mathbf{r} \sim 10^{-9}$ cm^3, then there will be $\sim 10^{10}$ molecules in $\delta^3\mathbf{r}$.

More generally, we must choose the $\bar{n}_i$ of Equation (4.17) to be much greater than unity in the region of interest where $\beta\omega_i \lesssim 1$. This condition implies (since $\delta\omega = h^3\, \delta^3 n$) that we must choose $\delta\omega$ so that, on the average,

$$\delta\omega \gg \left(\frac{\ell h}{\lambda_T}\right)^3 \qquad (11.2)$$

Since the average interparticle spacing ℓ is related to the particle density $n = N/\Omega$ by $n = \ell^{-3}$, we find for standard conditions that $\ell \sim 30$ Å. The following exercise then provides an estimate of the minimum velocity interval δv, defined by $\delta^3\mathbf{k} = (m\,\delta v)^3$.

Exercise 11.1

(a) Show that the thermal wavelength (4.2) is given, in angstroms, approximately by

$$\lambda_T \sim \left(\frac{300}{AT}\right)^{1/2}$$

where A is the molecular weight of a molecule in units of the hydrogen atom's mass.

(b) Show that condition (11.2) gives for hydrogen gas ($A = 2$) at standard conditions, with the choice $d^3\mathbf{r} \sim 10^{-9}\ \mathrm{cm}^3$, the condition on velocity intervals in any one dimension,

$$\delta v \gg 1\ \mathrm{m/sec}$$

Collision Integral $(\partial P/\partial t)_{\mathrm{coll}}$

The single-particle distribution function P changes with time as a result of numerous causes. Its position dependence changes because of particle velocities, which in turn are influenced by both external and internal forces. There will be explicit time dependence whenever nonsteady-state conditions occur.

Finally, interparticle collisions cause erratic changes in particle trajectories, which, when averaged, produce time variations in P. We use the notation $(\partial P/\partial t)_{\mathrm{coll}}$ and the terminology *collision integral* to indicate the time rate of change of P due to collisions. The development of this chapter begins with an elementary derivation of an expression for this important quantity.

Consider the collision $\{\mathbf{k}_1, \mathbf{k}_2\} \rightarrow \{\mathbf{k}_1', \mathbf{k}_2'\}$, in which two molecules initially with momenta $\mathbf{k}_1$ and $\mathbf{k}_2$ are scattered finally into states with momenta $\mathbf{k}_1'$ and $\mathbf{k}_2'$. Because of symmetry considerations, the dynamical description of this collision is simplest in the center-of-momentum (CM) coordinate system; the collision integral can be expressed in terms of its coordinates. Thus in the CM system the total molecular momentum

$$\mathbf{K} = \mathbf{k}_1 + \mathbf{k}_2 = \mathbf{k}_1' + \mathbf{k}_2'$$

is zero, and we need only consider the relative momenta $\mathbf{k}_{12}$ (before) and $\mathbf{k}_{12}'$ (after), defined by

$$\begin{aligned} \mathbf{k}_{12} &\equiv \frac{1}{2}(\mathbf{k}_1 - \mathbf{k}_2) \\ \mathbf{k}_{12}' &\equiv \frac{1}{2}(\mathbf{k}_1' - \mathbf{k}_2') \end{aligned} \tag{11.3}$$

For elastic collisions[3] we have $|\mathbf{k}'_{12}| = |\mathbf{k}_{12}|$. Since $\mathbf{k}_1 = \frac{1}{2}\mathbf{K} + \mathbf{k}_{12}$, $\mathbf{k}_2 = \frac{1}{2}\mathbf{K} - \mathbf{k}_{12}$, and so on, the collision may be viewed in the CM system as shown in Figure III.1. Because of the symmetry of the scattering process in this coordinate system, we need only focus attention on the scattering of one particle (particle 2, say) relative to the origin O for various impact parameters b.

It is clear from Figure III.1 that to determine completely the final momenta in the scattering process we must specify the scattering angles (θ_c, ϕ_c) in the CM system. These two angles are the most convenient ones to specify; moreover, they are essentially the same for the inverse scattering process $\{\mathbf{k}'_1, \mathbf{k}'_2\} \rightarrow \{\mathbf{k}_1, \mathbf{k}_2\}$. They define the direction of the momentum-transfer vector

$$\mathbf{q} \equiv (\mathbf{k}_{12} - \mathbf{k}'_{12}) = \mathbf{k}'_2 - \mathbf{k}_2 \tag{11.4}$$

The corresponding solid angle is denoted by $d\omega_q$. There is a one-to-one correspondence, in classical mechanics, between the scattering angle θ_c and the impact parameter b, which correspondence is determined for any given collision by the value of $|\mathbf{k}_{12}|$ and the dynamics, or interparticle potential V_{12}.

We next introduce the differential scattering cross section $\sigma(k_{12}, \hat{\mathbf{q}})$ for a two-particle collision in the CM system. Let $I(\mathbf{r}, \mathbf{k}_{12})$, the incident flux at $\mathbf{r} = \mathbf{r}_2 - \mathbf{r}_1$, be the number of particles 2 with momentum $-\mathbf{k}_{12}$ incident per unit time and per unit area normal to a beam. Then $\sigma(k_{12}, \hat{\mathbf{q}})$ is defined operationally by

$$I(\mathbf{r}, \mathbf{k}_{12})\sigma(k_{12}, \hat{\mathbf{q}})\, \delta\omega_q \equiv \left[\begin{array}{l}\text{Number of molecules with incident} \\ \text{momentum } -\mathbf{k}_{12} \text{ deflected, per} \\ \text{second, into a direction lying within} \\ \text{the solid-angle-element } \delta\omega_q\end{array}\right] \tag{11.5}$$

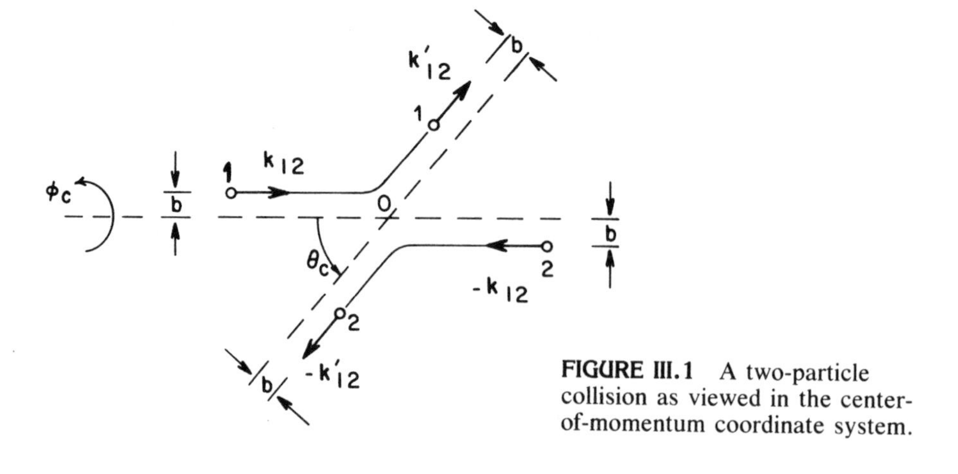

FIGURE III.1 A two-particle collision as viewed in the center-of-momentum coordinate system.

We also now specify the collision integral in the following manner:

$$\left(\frac{\partial P}{\partial t}\right)_{\text{coll}} = B - A \tag{11.6}$$

where

$$A(\mathbf{r}_1, \mathbf{k}_1, t)\,\delta\omega_1 \equiv \begin{bmatrix}\text{Number of collisions occurring per}\\ \text{unit time in which one of the}\\ \text{molecules is initially in the}\\ \text{phase-space volume element } \delta\omega_1\end{bmatrix}$$

$$B(\mathbf{r}_1, \mathbf{k}_1, t)\,\delta\omega_1 \equiv \begin{bmatrix}\text{Number of collisions occurring per}\\ \text{unit time in which one of the}\\ \text{molecules is finally in the}\\ \text{phase-space volume element } \delta\omega_1\end{bmatrix} \tag{11.7}$$

Clearly, $B - A$ gives the net gain of molecules per second in $\delta\omega_1$ due to collisions.

To derive an explicit expression for the collision integral we need to relate the above expressions (11.6) and (11.7) to definition (11.5). The following assumptions are involved in the derivation:

(a) Only two-particle (binary) collisions are taken into account, an approximation that is valid for a dilute gas.
(b) The walls of the container reflect incident molecules elastically and, therefore, can be ignored.
(c) The effect of external forces on $\sigma(k_{12}, \hat{\mathbf{q}})$ can be ignored.[4]
(d) Inequality (11.2) for $\delta\omega$ is satisfied.

With these assumptions in mind we proceed by observing that since $I(\mathbf{r}, \mathbf{k}_{12})$ is the flux at $\mathbf{r}$ of particles 2 in a momentum volume element $\delta^3\mathbf{k}_2$ about $\mathbf{k}_2$, incident on any one particle 1 with momentum $\mathbf{k}_1$, we may write

$$P(\mathbf{r}_1, \mathbf{k}_1, t)I(\mathbf{r}, \mathbf{k}_{12}) = P_2(\mathbf{r}_1, \mathbf{r}_2; \mathbf{k}_1, k_2; t)\,\delta^3\mathbf{k}_2\left(\frac{2k_{12}}{m}\right)$$

In words, at time t the phase-space density of particles 1 times the flux of particles 2 in $\delta^3\mathbf{k}_2$ equals the two-particle distribution function P_2 times $\delta^3\mathbf{k}_2$ times the relative velocity of particles 2. Here we have used assumption (d). To derive an expression for the quantity A of Equations (11.7) we must multiply this expression by $\sigma(k_{12}, \hat{\mathbf{q}})\,\delta\omega_q$, according to Equation (11.5), and integrate the result over all $\delta^3\mathbf{k}_2\,\delta\omega_q$.

$$A(\mathbf{r}_1, \mathbf{k}_1, t) = \int d^3\mathbf{k}_2\, d\omega_q P_2(\mathbf{r}_1, \mathbf{r}_2; \mathbf{k}_1, \mathbf{k}_2; t)\left(\frac{2k_{12}}{m}\right)\sigma(k_{12}, \hat{\mathbf{q}}) \tag{11.8}$$

The vector $\mathbf{r}_2 \cong \mathbf{r}_1$ for particles 2 in the volume element $\delta^3\mathbf{r}_1$ at time t. Assumptions (a) and (c) have also been used here. Similarly, we may argue that the quantity B of Equations (11.7) can be deduced from the inverse-collision expression

$$B(\mathbf{r}_1, \mathbf{k}_1, t)\, d^3\mathbf{k}_1 = \int_{\mathbf{k}_1 = \text{fixed}} d^3\mathbf{k}_1'\, d^3\mathbf{k}_2'\, d\omega_{q'} P_2(\mathbf{r}_1, \mathbf{r}_2'; \mathbf{k}_1', \mathbf{k}_2', t) \left(\frac{2k_{12}'}{m}\right) \sigma(k_{12}', \hat{\mathbf{q}}')$$

But $|\mathbf{k}_{12}'| = |\mathbf{k}_{12}|$, $\mathbf{q}' = -\mathbf{q}$ and $d^3\mathbf{k}_1'\, d^3\mathbf{k}_2'\, d\omega_{q'} = d^3\mathbf{k}_1\, d^3\mathbf{k}_2\, d\omega_q$[5]. Therefore, with $\mathbf{r}_2' \cong \mathbf{r}_1$, for particles 2 in the volume element $\delta^3\mathbf{r}_1$ at time t, we may conclude that B is given by

$$B(\mathbf{r}_1, \mathbf{k}_1, t) = \int d^3\mathbf{k}_2\, d\omega_q P_2(r_1, \mathbf{r}_2'; \mathbf{k}_1', \mathbf{k}_2'; t) \left(\frac{2k_{12}}{m}\right) \sigma(k_{12}, -\hat{\mathbf{q}}) \tag{11.9}$$

To complete the derivation of an expression for the collision integral, it is only necessary to substitute Equations (11.8) and (11.9) into Equation (11.6). We obtain

$$\left[\frac{\partial}{\partial t} P(\mathbf{r}_1, \mathbf{k}_1, t)\right]_{\text{coll}} = \int d^3\mathbf{k}_2\, d\omega_q [P_2(\mathbf{r}_1, \mathbf{r}_2'; \mathbf{k}_1', \mathbf{k}_2'; t) - P_2(\mathbf{r}_1, \mathbf{r}_2; \mathbf{k}_1, \mathbf{k}_2; t)] \times R(\mathbf{k}_1, \mathbf{k}_2, \hat{\mathbf{q}}) \tag{11.10}$$

where we have also introduced the transition probability R, defined by

$$R(\mathbf{k}_1, \mathbf{k}_2, \hat{\mathbf{q}}) \equiv \sigma(k_{12}, \hat{\mathbf{q}}) \left(\frac{2k_{12}}{m}\right) = \sigma(k_{12}, -\hat{\mathbf{q}}) \left(\frac{2k_{12}}{m}\right) \tag{11.11}$$

The second equality in this last equation, which applies to the inverse collision $\{\mathbf{k}_1', \mathbf{k}_2'\} \to \{\mathbf{k}_1, \mathbf{k}_2\}$, is proved in Appendix D [see Equation (D.6)]. The units of R are (volume/time).

Stosszahlansatz and the Maxwell – Boltzmann Distribution

The two-particle distribution function $P_2(\mathbf{r}_1, \mathbf{r}_2; \mathbf{k}_1, \mathbf{k}_2; t)$ is the probability density at time t for finding one particle in the phase-space volume element $d\omega_1$ about the point $(\mathbf{r}_1, \mathbf{k}_1)$ and a second particle in the phase-space volume element $d\omega_2$ about the point $(\mathbf{r}_2, \mathbf{k}_2)$. Its normalization is defined by the equation

$$\int d\omega_1\, d\omega_2 P_2(\mathbf{r}_1, \mathbf{r}_2; \mathbf{k}_1, \mathbf{k}_2; t) = N(N - 1) \cong N^2 \tag{11.12}$$

which is consistent with Equation (11.1) for the normalization of the one-particle distribution function P.

We now discuss the famous *stosszahlansatz*,[6] which reduces the collision integral (11.10) to a more calculable form. In modern terminology the *stosszahlansatz* is equivalent to assuming stochastic independence [recall Equation (3.7)] for the coordinates 1 and 2 in P_2', that is, we now assume

$$P_2(\mathbf{r}_1, \mathbf{r}_2; \mathbf{k}_1, \mathbf{k}_2; t) \cong P(\mathbf{r}_1, \mathbf{k}_1, t)P(\mathbf{r}_2, \mathbf{k}_2, t) \tag{11.13}$$

Approximation (11.13) deserves some discussion, for although it is ordinarily a plausible assumption if particles 1 and 2 are widely separated, its relevant application to the collision integral (11.10) is not so evident. Thus we have derived Equations (11.8) and (11.9) by using the picture of a two-particle collision, during which particles 1 and 2 experience a close encounter. But when two particles in a gas are within the range r_o of their mutual interaction potential, then their positions and momenta will be correlated; stochastic independence will not then be a valid approximation for P_2. The approximation is validated, in part, only by assumption (d) below Equations (11.7), which enables us to choose phase-space volume elements for which the *Stosszahlansatz* can be valid on the average. Beyond this argument one can only resort to saying that Equation (11.13) provides a good first approximation for the kinetic-theory calculations of real gases. On the other hand, use of the correct collision integral, instead of either Equation (11.10) or Equation (11.14), improves the derivation from microscopic considerations of the equations of hydrodynamics, as we discuss in Section 12 and Appendix D.

Having dealt with the apologies, we now substitute Equation (11.13) into Equation (11.10) to obtain for the collision integral the approximate form

$$\left(\frac{\partial P}{\partial t}\right)_{\text{coll}} \cong J(P \mid P) \tag{11.14}$$

where the more general function $J(P \mid Q)$ is defined by

$$J(P \mid Q) \equiv \int d^3\mathbf{k}_2 \, d\omega_q R(\mathbf{k}_1, \mathbf{k}_2, \hat{\mathbf{q}}) \cdot [P(\mathbf{r}_1, \mathbf{k}_1', t)Q(r_1, \mathbf{k}_2', t) - P(\mathbf{r}_1, \mathbf{k}_1, t)Q(\mathbf{r}_1, \mathbf{k}_2, t)] \tag{11.15}$$

Here we have set $\mathbf{r}_2 = \mathbf{r}_2' = \mathbf{r}_1$, inasmuch as all three positions lie within the same position-space volume element $\delta^3\mathbf{r}$ and after application of the *Stosszahlansatz* (11.13) distinguishing between them is no longer important. The functional generalization $J(P \mid Q)$ is needed in Section 14.

We now discuss the case of thermodynamic equilibrium. This condition is characterized by a time-independent single-particle distribution function and, in particular, a vanishing collision integral. For every position-space element $\delta^3\mathbf{r}$ at equilibrium, the time rate at which molecules enter a given momentum state because of collisions is equal to the time rate at which molecules leave this same state because of inverse collisions. If we represent the one-particle distribution function for a dilute gas at equilibrium by

$P_0(\mathbf{r}, \mathbf{k})$, then the collision integral (11.14) will vanish at equilibrium when we set

$$P_0(\mathbf{r}, \mathbf{k}_1)P_0(\mathbf{r}, \mathbf{k}_2) = P_0(\mathbf{r}, \mathbf{k}_1')P_0(\mathbf{r}, \mathbf{k}_2') \tag{11.16}$$

Equation (11.16) is a sufficiency equilibrium condition; it can also be shown to be a necessary condition. The inverse proof, in which Equation (11.16) is derived from $(\partial P/\partial t)_{\text{coll}} = 0$, makes use of the Boltzmann transport equation.[7]

We have assumed that only elastic two-particle collisions need be considered in a dilute real gas, and this condition (together with other assumptions) has led us to Equation (11.16) for the equilibrium distribution P_0. Thus the momenta in Equation (11.16) must satisfy the conservation laws for both total momentum and total kinetic energy. These two quantities, together with the total angular momentum of the colliding molecules, are the only known conserved additive quantities involving the momenta. Therefore, upon writing Equation (11.16) in the form

$$\ln P_0(\mathbf{r}, \mathbf{k}_1) + \ln P_0(\mathbf{r}, \mathbf{k}_2) = \ln P_0(\mathbf{r}, \mathbf{k}_1') + \ln P_0(\mathbf{r}, \mathbf{k}_2')$$

one can see that the most general function $P_0(\mathbf{r}, \mathbf{k})$ that satisfies Equation (11.16) and the above three conservation laws is

$$\ln P_0(\mathbf{r}, \mathbf{k}) = \alpha - \beta\omega(\mathbf{k}) + \beta\mathbf{u} \cdot \mathbf{k} + \beta\mathbf{\Omega} \cdot \mathrm{L} \tag{11.17}$$

Here α, β, $\beta\mathbf{u}$, and $\beta\mathbf{\Omega}$ are unknown quantities that are independent of $\mathbf{k}$ but not necessarily of $\mathbf{r}$, and $\mathbf{L} = \mathbf{r} \times \mathbf{k}$ is the angular momentum of a (spinless) particle with respect to some inertial origin.[8] We show below, in fact, that the only position dependence in P_0, besides that in $\mathbf{L}$, is in $\alpha = \alpha(\mathbf{r})$. The parameter β is introduced as a factor of $\mathbf{u}$ and $\mathbf{\Omega}$ so that these latter quantities can be interpreted directly as a uniform velocity and angular velocity, respectively, for the system as a whole. [See Equation (6.6) and associated discussion.]

Upon defining $\mathbf{k}_o = m\mathbf{u}$, we derive from Equation (11.17) the expression

$$P_0(\mathbf{r}, \mathbf{k}) = \mathcal{N}(\mathbf{r}) \exp\left\{-\beta\left[\frac{1}{2m}(\mathbf{k} - \mathbf{k}_o)^2 - \mathbf{\Omega} \cdot \mathbf{L}\right]\right\}$$

where $\ln \mathcal{N} = \alpha(\mathbf{r}) + \beta\omega(k_o)$. The normalization constant $\mathcal{N}(\mathbf{r})$ is determined by Equation (11.1); this equation also motivates us to define the position-space distribution function $P_0(\mathbf{r})$ by

$$P_0(\mathbf{r}) \equiv \int d^3\mathbf{k}\, P_0(\mathbf{r}, \mathbf{k}) \tag{11.18}$$

We may then refer to this quantity as the local particle density, inasmuch as when $\mathbf{\Omega} = 0$ and in the absence of external forces it reduces to $n = N/\Omega$. If it is recalled that in Equation (4.16) $P(\mathbf{k})$ is normalized to unity, then the identification $\mathcal{N}(r) = P_0(\mathbf{r})(\beta/2\pi m)^{3/2}$ can readily be made for the case

$\mathbf{\Omega} = 0$, which result holds for any momentum $\mathbf{k}_o$. This conclusion applies also when $\mathbf{\Omega} \neq 0$, if we define $\mathbf{k}_o = m(\mathbf{u} + \mathbf{\Omega} \times \mathbf{r})$. We obtain, therefore,

$$P_0(\mathbf{r}, \mathbf{k}) = P_0(\mathbf{r})\left(\frac{\beta}{2\pi m}\right)^{3/2} \exp\left[-\frac{\beta}{2m}(\mathbf{k} - \mathbf{k}_o)^2\right] \tag{11.19}$$

which is the *Maxwell–Boltzmann distribution* for a dilute real gas. Henceforth we consider only the case $\mathbf{\Omega} = 0$.

The position-dependent quantity $P_0(\mathbf{r})$ is determined below in terms of the external potential $U(\mathbf{r})$. For the present we set $U(\mathbf{r}) = 0$, in which case $P_0(\mathbf{r}) \rightarrow n$, and study some of the implications of the simpler distribution function. Thus the average value for the momentum of a single molecule in the gas is given by the equation

$$\langle \mathbf{k} \rangle = n^{-1} \int d^3\mathbf{k}\, \mathbf{k}\, P_0(\mathbf{r}, \mathbf{k}) = \mathbf{k}_o = m\mathbf{u} \tag{11.20}$$

which shows, as already suggested, that $\mathbf{u}$ must be identified as the velocity with which the whole system is moving. For systems at rest, with $\mathbf{u} = 0$, we may compute the average energy of a single molecule by using one of the integrals (2.4). We find

$$\langle \omega(k) \rangle = n^{-1} \int d^3\mathbf{k}\left(\frac{k^2}{2m}\right) P_0(\mathbf{r}, \mathbf{k}) = \frac{3}{2}\beta^{-1} = \frac{3}{2}\kappa T \tag{11.21}$$

where we have assumed that $\beta = (\kappa T)^{-1}$.[9] According to this last result, the rms velocity of a molecule in a dilute gas at rest is given by

$$v_{\text{rms}} = \langle v^2 \rangle^{1/2} = \left(\frac{3\kappa T}{m}\right)^{1/2} \tag{11.22}$$

Similarly, the average speed of a molecule is computed to be

$$\langle v \rangle = (nm)^{-1} \int d^3\mathbf{k}\, k P_0(\mathbf{r}, \mathbf{k}) = \left(\frac{8\kappa T}{\pi m}\right)^{1/2} < v_{\text{rms}} \tag{11.23}$$

For hydrogen gas at standard conditions $\langle v \rangle \sim 1500$ m/sec, which is a result showing that the estimate for δv in Exercise 11.1(b) is tolerable.

The probability distribution density of speeds, normalized according to $\int F(v)\, dv = 1$, is given by

$$\begin{aligned} F(v) &\equiv 4\pi k^2\left(\frac{m}{n}\right) P_0(\mathbf{r}, \mathbf{k}) \\ &= 4\pi v^2\left(\frac{m}{2\pi\kappa T}\right)^{3/2} \exp\left(-\frac{m}{2\kappa T}v^2\right) \end{aligned} \tag{11.24}$$

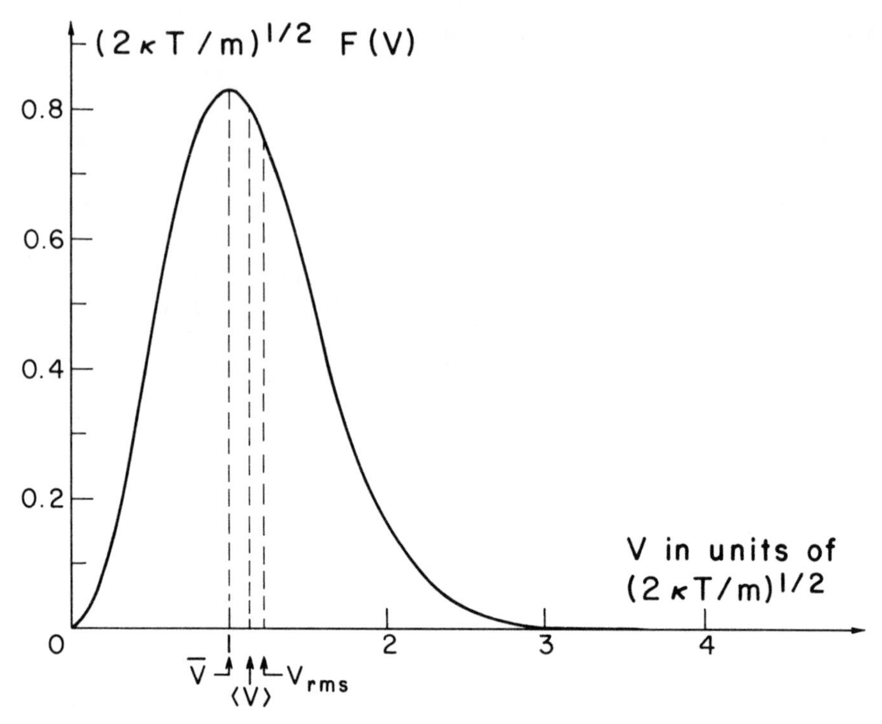

FIGURE III.2 The Maxwell–Boltzmann distribution of molecular speeds in a dilute gas.

It has its maximum value, the most probable speed $\bar{v}$, at the value where $\frac{1}{2}mv^2 = \kappa T$, that is,

$$\bar{v} = \left(\frac{2\kappa T}{m}\right)^{1/2} < \langle v \rangle \tag{11.25}$$

The velocities $v_{\rm rms}$, $\langle v \rangle$, and $\bar{v}$ are all indicated in Figure III.2, which shows $F(v)$ plotted versus v.

Exercise 11.2 Show that the number of molecules that escape per unit time and per unit area, with speed v in the internal dv, from a hole in an oven containing a gas at temperature T and density n is given by

$$\frac{1}{4} nvF(v)\, dv$$

This result can be used to show that the rms speed of escaping molecules is $(4\kappa T/m)^{1/2}$, instead of $(3\kappa T/m)^{1/2}$, and hence why evaporation is a cooling process in this example.

Boltzmann Transport Equation

The Maxwell–Boltzmann distribution can be derived in full generality, with an expression for the particle density $P_0(\mathbf{r})$ included, from Boltzmann's transport equation. We give here an elementary derivation of the Boltzmann equation, deferring a more rigorous treatment to Section 23. We consider only systems at rest and external forces $\mathbf{F}_{ext}$ that are velocity-independent. Our interest is in the total time rate of change of the one-particle distribution function $P(\mathbf{r}, \mathbf{k}, t)$. For this purpose we focus attention on molecules with coordinates $(\mathbf{r}, \mathbf{k})$ in the phase-space volume element $\delta\omega$ at time t. At a slightly later time $(t + \delta t)$ these molecules will have the coordinates $(\mathbf{r} + \mathbf{k}\,\delta t/m, \mathbf{k} + \mathbf{F}_{ext}\,\delta t)$, apart from the effect of collisions. Although it will be distorted, the phase-space volume element will remain constant in magnitude.[10] Thus, on including the effect of molecular collisions [Equations (11.6) and (11.7)] we obtain

$$P\left(\mathbf{r} + \frac{\mathbf{k}\,\delta t}{m}, \mathbf{k} + \mathbf{F}_{ext}\,\delta t, t + \delta t\right) = P(\mathbf{r}, \mathbf{k}, t) + \left(\frac{\partial P}{\partial t}\right)_{coll} \delta t \tag{11.26}$$

Note that the collision term is included on the right side of this equation, because it is a contribution to the resulting distribution function P at the time $t + \delta t$.

We now expand the left side of Equation (11.26) to first order in δt. This gives an equation of motion, the *Boltzmann transport equation,* for $P(\mathbf{r}, \mathbf{k}, t)$.

$$\left[\frac{\partial}{\partial t} + \left(\frac{\mathbf{k}}{m}\right) \cdot \nabla_r + \mathbf{F}_{ext} \cdot \nabla_k\right] P(\mathbf{r}, \mathbf{k}, t) = \left(\frac{\partial P}{\partial t}\right)_{coll} \tag{11.27}$$

where ∇_r and ∇_k are gradient operators that operate on position and momentum coordinates, respectively, and the collision integral is given (approximately) by Equation (11.10) or (11.14).[11]

We now investigate the consequence of Equation (11.27) for systems at thermodynamic equilibrium. We have already argued that the collision integral must vanish at equilibrium, which condition implies here that the Boltzmann equation for the time-independent equilibrium distribution function $P_0(\mathbf{r}, \mathbf{k})$ takes the form, for systems at rest,

$$[(\mathbf{k}/m) \cdot \nabla_r + \mathbf{F}_{ext} \cdot \nabla_k] P_0(\mathbf{r}, \mathbf{k}) = 0 \tag{11.28}$$

We have also demonstrated that $P_0(\mathbf{r}, \mathbf{k})$ has the exponential behavior (11.19). It is then easy to verify, for $\mathbf{k}_o = 0$, that this analytical form satisfies the differential equation (11.28) with

$$P_0(\mathbf{r}) = n'\, e^{-\beta U(\mathbf{r})} \tag{11.29}$$

where $U(\mathbf{r})$ is the general external potential energy corresponding to $\mathbf{F}_{ext}$.

The quantity n' is related to the average particle density n by the integral $n = \Omega^{-1} \int d^3\mathbf{r} P_0(\mathbf{r})$, so that $n' = n$ when $U = 0$. Expression (11.29) agrees with the special case of the barometer formula derived in Problem II.5(c); it gives the famous Boltzmann factor $e^{-\beta U}$ of statistical physics.

12 Boltzmann Transport Equation and Hydrodynamics

The Boltzmann transport equation (11.27) is the fundamental equation of kinetic theory. It makes possible the quantitative calculation of macroscopic phenomena from a microscopic basis, including application to systems with velocity-independent external forces. It gives, from first principles essentially, a basis for the macroscopic equations of transport phenomena, such as the equations of hydrodynamics. Finally, it applies to nonequilibrium phenomena as well as to systems in equilibrium; in particular, it applies to steady-state transport.

In this section we show how the Boltzmann transport equation provides a microscopic basis for the equations of hydrodynamics. After a preliminary discussion, which deals quantitatively with the concepts of mean free path, collision time, and collision frequency, we attack our central objective by deriving three basic conservation laws. With the aid of results from Appendix D, the derivation becomes complete and applies to fluids quite generally.

On transformation to the local rest system at point $\mathbf{r}$, the conservation laws begin to resemble the equations of hydrodynamics. This identification is completed in the following section when the conditions of viscous motion are inserted, but in the present section the hydrodynamic equations for nonviscous flow already can be extracted. We complete this section by using the condition of local equilibrium to arrive at the nonviscous-flow equations and by making several illustrative applications.

In any presentation of hydrodynamics from the microscopic view we must have a conceptual understanding of the approach to equilibrium, both locally and throughout the system of interest. Initially, the transition to equilibrium is determined at the local level by local conditions; the transition from local equilibrium to complete thermodynamic equilibrium then relies on macroscopic transport phenomena. Of course, interplay between these two processes must occur, as, for example, when changes in local conditions are caused by macroscopic transport.

To describe the approach to local equilibrium when it is governed by molecular collisions, as is the case for fluids, we introduce a length λ, called the *mean free path,* and a time τ, called the *collision time*. Also, as will become apparent in the discussions of this chapter (and in Chap. IX), the transition region in which description in the microscopic world changes to description in the macroscopic world is best characterized on a numerical scale by these two quantities. They can be estimated to good accuracy for a dilute gas by referring to the quantity $A(\mathbf{r}, \mathbf{k}, t)$ of Equations (11.7) and (11.8).

At equilibrium, the integral of $A(\mathbf{r}, \mathbf{k}, t)$ over all momenta defines a number Zn,

$$\begin{aligned} Zn &\equiv \int d^3\mathbf{k}_1 A(\mathbf{r}, \mathbf{k}_1, t) \\ &\cong \int d^3\mathbf{k}_1\, d^3\mathbf{k}_2 \sigma(k_{12})\left(\frac{2k_{12}}{m}\right)P_0(\mathbf{r}, \mathbf{k}_1)P_0(\mathbf{r}, \mathbf{k}_2) \end{aligned} \tag{12.1}$$

where $\sigma(k_{12}) = \int d\omega_q \sigma(k_{12}, \hat{\mathbf{q}})$ is the total scattering cross section for collisions between molecules with momenta $\mathbf{k}_1$ and $\mathbf{k}_2$. Here we have also used the *stosszahlansatz* (11.13), with $\mathbf{r}_1 = \mathbf{r}_2 = \mathbf{r}$ as is discussed below Equation (11.15). Clearly, Zn is the number of collisions per unit volume and per unit time (at $\mathbf{r}$) in the dilute gas. The quantity Z itself is then the number of collisions per molecule per second, that is, the *collision frequency,* and its reciprocal is the collision time τ.

For a dilute gas at equilibrium, $P_0(\mathbf{r}, \mathbf{k})$ in (12.1) is the Maxwell–Boltzmann distribution (11.19), with $\mathbf{k}_0 = 0$ and $P_0(\mathbf{r}) = n$ when there are no external forces. We can make an order-of-magnitude estimate of Z by assuming that $\sigma(k_{12})$ is not a sensitive function of k_{12} and may be replaced by a constant of order πa^2, where a is the molecular diameter. From Equations (12.1) and (11.19) we then obtain for the collision frequency Z,

$$\begin{aligned} Z &\cong 2\sigma(nm)^{-1} \int d^3\mathbf{k}_1\, d^3\mathbf{k}_2 k_{12} P_0(\mathbf{r}, \mathbf{k}_1)P_0(\mathbf{r}, \mathbf{k}_2) \\ &= \frac{2\sigma n}{m}\left(\frac{\beta}{2\pi m}\right)^3 \int d^3\mathbf{k}_1\, d^3\mathbf{k}_2 k_{12} \exp\left[-\frac{\beta}{2m}(k_1^2 + k_2^2)\right] \\ &= \frac{2\sigma n}{m}\left(\frac{\beta}{2\pi m}\right)^3 \int d^3\mathbf{K}\, d^3\mathbf{k}_{12} k_{12} \exp\left[-\frac{\beta}{2m}\left(\frac{1}{2}K^2 + 2k_{12}^2\right)\right] \\ &= \sqrt{2} n\sigma\langle v\rangle \end{aligned} \tag{12.2}$$

where we have transformed to CM coordinates [see Equations (11.3)] to get the third line of this expression. The last line is derived by using Equations (2.4) and (11.23). An estimate of the collision time in a dilute gas is, therefore,

$$\tau = Z^{-1} = (\sqrt{2} n\sigma\langle v\rangle)^{-1} \tag{12.3}$$

The mean free path λ, defined as the average distance between collisions, is given by the product of the average speed $\langle v \rangle$ and the collision time[12]

$$\lambda \equiv \langle v \rangle \tau = (\sqrt{2} n\sigma)^{-1} \tag{12.4}$$

The numbers (12.3) and (12.4) are good estimates for a dilute gas not far from equilibrium, which is the case of primary interest in this chapter. For H_2 gas in interstellar space, where the density is ~ 1 molecule/cm^3, one readily calculates, for example, that the mean free path is $\sim 10^{15}$ cm $\cong 10^{-3}$ light-years.

Exercise 12.1 Consider a gas of H_2 molecules at standard conditions and take the molecular diameter a to be ~ 1 Å. Show for this case that

$$\lambda \sim 10^{-4} \text{ cm}$$

$$\tau \sim 10^{-9} \text{ sec}$$

The Conservation Laws

In the investigation of nonequilibrium phenomena using classical kinetic theory we must solve the Boltzmann transport equation, with given initial conditions, to obtain the distribution function $P(\mathbf{r}, \mathbf{k}, t)$. We may then calculate average values of physical interest using the derived distribution function. Some rigorous properties of any solution to the Boltzmann equation may be deduced from the fact that in a molecular collision there are dynamical quantities that are rigorously conserved. We now derive the associated *conservation laws,* of mass, momentum, and energy, which laws are equivalent to the basic equations of hydrodynamics.

Let $\chi(\mathbf{r}, \mathbf{k}_1, t)$ be some physical quantity associated with the ith molecule, with momentum $\mathbf{k}_1$ and located in $\delta^3\mathbf{r}$, such that in any collision $\{\mathbf{k}_1, \mathbf{k}_2\} \rightarrow \{\mathbf{k}_1', \mathbf{k}_2'\}$ taking place at $\mathbf{r}$ we have

$$\chi(\mathbf{r}, \mathbf{k}_1, t) + \chi(\mathbf{r}, \mathbf{k}_2, t) = \chi(\mathbf{r}, \mathbf{k}_1', t) + \chi(\mathbf{r}, \mathbf{k}_2', t) \tag{12.5}$$

We call any $\chi(\mathbf{r}, \mathbf{k}, t)$ that satisfies Equation (12.5) a conserved property, and we are interested in three particular ($\mathbf{r}$- and t-independent) conserved properties for elastic collisions; namely,

$\chi = m$ (mass)

$\chi = \mathbf{k}$ (momentum)

$\chi = k^2/2m$ (kinetic energy)

We now state in detail the three associated conservation laws in the form in which we derive them.

Let $A(\mathbf{r}, \mathbf{k}, t)$ be any molecular physical quantity such as the functions χ defined above. We define momentum-space average values $\langle A \rangle$ by

$$\langle A(\mathbf{r}, t) \rangle \equiv \frac{1}{n(\mathbf{r}, t)} \int d^3\mathbf{k} A(\mathbf{r}, \mathbf{k}, t) P(\mathbf{r}, \mathbf{k}, t) \tag{12.6}$$

where $P(\mathbf{r}, \mathbf{k}, t)$ is the one-particle distribution function that when integrated over all momenta gives the position-space density function $n(\mathbf{r}, t)$ [set $A = 1$ in (12.6)]. For equilibrium distributions, $n(\mathbf{r}, t)$ reduces to the function $P_0(\mathbf{r})$ of Equation (11.18). We define, in addition, the density ρ_χ and the current density $\mathbf{j}_\chi$ by

$$\begin{aligned} \rho_\chi &\equiv n\langle \chi \rangle \\ \mathbf{j}_\chi &\equiv n\langle \chi \mathbf{k}/m \rangle \end{aligned} \tag{12.7}$$

The simplest form of the conservation laws might then appear to be the continuity equations,

$$\frac{\partial \rho_\chi}{\partial t} + \nabla \cdot \mathbf{j}_\chi = 0 \qquad \text{(incorrect, in general)}$$

Although this guess is correct for mass, the continuity equations for momentum and kinetic energy need to be modified because of molecular interactions and external forces. The correct conservation laws are as follows.

Mass: When $\chi = m$ we do obtain the continuity equation, which can also be written in the familiar form

$$\frac{\partial \rho}{\partial t} + \nabla \cdot (\rho \mathbf{u}) = 0 \tag{12.8}$$

where the local mass density $\rho = \rho_m$ is defined by

$$\rho(\mathbf{r}, t) = mn(\mathbf{r}, t) \tag{12.9}$$

The local macroscopic velocity $\mathbf{u}$ is defined by $\mathbf{j}_m = \rho\mathbf{u}$, or, more directly, as an average over molecular velocities $\mathbf{v} = \mathbf{k}/m$ by

$$\mathbf{u}(\mathbf{r}, t) = \langle \mathbf{v} \rangle \tag{12.10}$$

Momentum: With $\chi = \mathbf{k}$, the first of Equations (12.7) gives $\rho_\mathbf{k} = \rho\mathbf{u} = \mathbf{j}_m$. The current density $\mathbf{j}_\mathbf{k}$ becomes the velocity tensor $(\mathbf{j}_\mathbf{k})_{ij} = \rho\langle v_i v_j \rangle$. However, the tensor that appears in the correct form of the momentum conservation law is the modified velocity tensor

$$\Pi_{ij}(\mathbf{r}, t) \equiv \rho\langle v_i v_j \rangle + B_{ij}(\mathbf{r}, t) \tag{12.11}$$

where, with $\mathbf{r}_{21} \equiv \mathbf{r}_2 - \mathbf{r}_1$,

$$B_{ij}(\mathbf{r}, t) \equiv -\frac{1}{2}\int d^3\mathbf{r}_1\, d^3\mathbf{r}_2 P(\mathbf{r}_1, \mathbf{r}_2, t)\, \delta^{(3)}(\mathbf{r}_1 - \mathbf{r}) V'(r_{21}) r_{21}^{-1} (r_{21})_i (r_{21})_j \qquad \textbf{(12.12)}$$

Here $V'(r)$ denotes the derivative of the two-particle (central) interaction potential $V(r)$, and $P(\mathbf{r}_1, \mathbf{r}_2, t)$ is the *pair distribution function* in position space, defined in terms of the two-particle probability density function P_2 by the integral

$$P(\mathbf{r}_1, \mathbf{r}_2, t) \equiv \int d^3\mathbf{k}_1\, d^3\mathbf{k}_2 P_2(\mathbf{r}_1, \mathbf{r}_2; \mathbf{k}_1, \mathbf{k}_2; t) \qquad \textbf{(12.13)}$$

The tensor B_{ij} gives a contribution to the velocity tensor due to molecular interactions. In terms of the symmetric tensor Π_{ik} the momentum conservation law can be stated as

$$\frac{\partial \mathbf{j}}{\partial t} + \nabla \cdot \overleftrightarrow{\Pi} = \rho(\mathbf{F}_{\text{ext}}/m) \qquad \textbf{(12.14)}$$

where $\mathbf{j} = \mathbf{j}_m = \rho\mathbf{u}$ and $\overleftrightarrow{\Pi}$ is the dyadic notation for Π_{ij}.

Energy: It is not sufficient to consider only the kinetic energy when molecular interactions are present. Thus, instead of $\chi = k^2/2m$ in Equations (12.7), we need to examine the quantity

$$\chi = \frac{k^2}{2m} + n^{-1}\phi(\mathbf{r}, t)$$

where ϕ is an average potential energy density that is defined in terms of the pair distribution function (12.13) by

$$\phi(\mathbf{r}, t) \equiv \frac{1}{2}\int d^3\mathbf{r}_1\, d^3\mathbf{r}_2 P(\mathbf{r}_1, \mathbf{r}_2, t)\, \delta^{(3)}(\mathbf{r}_1 - \mathbf{r}) V(r_{21}) \qquad \textbf{(12.15)}$$

Then the energy density ρ_E and energy flux $\mathbf{j}_E$ are given by

$$\begin{aligned} \rho_E(\mathbf{r}, t) &= \frac{1}{2}\rho\langle v^2\rangle + \phi \\ \mathbf{j}_E(\mathbf{r}, t) &= \frac{1}{2}\rho\langle v^2\mathbf{v}\rangle + \phi\mathbf{u} \end{aligned} \qquad \textbf{(12.16)}$$

where $\langle\phi\rangle = \phi$. Again, the correct form of the energy conservation law involves yet an additional contribution to $\mathbf{j}_E$. We define the more general energy flux $\mathbf{Q}$ by the equation

$$\mathbf{Q}(\mathbf{r}, t) \equiv \mathbf{j}_E + \overleftrightarrow{\mathbf{B}} \cdot \mathbf{u} + \mathbf{q}_V \qquad \textbf{(12.17)}$$

where B_{ij} is the tensor (12.12) and the potential-energy flux vector $\mathbf{q}_V$ is defined by the two-particle phase-space average

$$\mathbf{q}_V(\mathbf{r}, t) \equiv \frac{1}{2} \int d\omega_1\, d\omega_2 P_2(\mathbf{r}_1, \mathbf{r}_2; \mathbf{k}_1, \mathbf{k}_2; t)\, \delta^{(3)}(\mathbf{r}_1 - \mathbf{r}) \times (\mathbf{k}_1/m - \mathbf{u}) \cdot [V(r_{21})\overleftrightarrow{1} - V'(r_{21}) r_{21}^{-1} \mathbf{r}_{21}\mathbf{r}_{21}] \tag{12.18}$$

Here the square-brackets term includes the unit dyad and the dyad $\mathbf{r}_{21}\mathbf{r}_{21}$. Then, in terms of the flux vector $\mathbf{Q}$ the energy conservation law can be stated as

$$\frac{\partial \rho_E}{\partial t} + \nabla \cdot \mathbf{Q} = \mathbf{j} \cdot (\mathbf{F}_{\text{ext}}/m) \tag{12.19}$$

Derivation of the Conservation Laws

The conservation laws (12.8), (12.14), and (12.19) have indeed been complicated to state, despite the simplicity of their final forms, due essentially to Irving and Kirkwood.[13] We now deduce them in part, reserving for Appendix D the derivation of the flux correction terms $\overleftrightarrow{B}$ and $\mathbf{q}_V$. In the derivation we start with the Boltzmann transport equation (11.27).

We must be careful when we apply the Boltzmann equation to nonequilibrium systems with macroscopic motion. Thus if a system is not translationally invariant, that is, if it is not moving uniformly with velocity $\mathbf{u}$, then it is not proper to include in Equation (11.27) the replacement $\mathbf{k} \to \mathbf{k} - m\mathbf{u}$ of footnote 11.

The derivation proceeds generally if we multiply the Boltzmann equation by any conserved quantity χ and integrate over all momentum space to get the result (for velocity-independent external forces $\mathbf{F}_{\text{ext}}$),

$$\int d^3\mathbf{k}\chi(\mathbf{r}, \mathbf{k}, t)\left[\frac{\partial}{\partial t} + \mathbf{v} \cdot \nabla_r + \mathbf{F}_{\text{ext}} \cdot \nabla_k\right] P(\mathbf{r}, \mathbf{k}, t) = \int d^3\mathbf{k}\chi(\mathbf{r}, \mathbf{k}, t)\left(\frac{\partial P}{\partial t}\right)_{\text{coll}} \tag{12.20}$$

After an integration by parts this result can be rewritten, using definition (12.6), in the abstract form

$$\frac{\partial}{\partial t}(n\langle\chi\rangle) - n\left\langle\frac{\partial \chi}{\partial t}\right\rangle + \nabla \cdot n\langle \mathbf{v}\chi\rangle - n\langle(\mathbf{v} \cdot \nabla_r)\chi\rangle = n\mathbf{F}_{\text{ext}} \cdot \langle \nabla_k \chi\rangle + \int d^3\mathbf{k}\chi(\mathbf{r}, \mathbf{k}, t)\left(\frac{\partial P}{\partial t}\right)_{\text{coll}} \tag{12.21}$$

The desired conservation laws are deduced as special cases of this basic conservation theorem. In these applications we do not require the second and fourth terms on the left side.

Exercise 12.2

(a) Establish the theorem for conserved properties χ,

$$\int d^3\mathbf{k}_1 \chi(\mathbf{r}, \mathbf{k}_1, t)[J(P \mid Q) + J(Q \mid P)] = 0 \tag{12.22}$$

where $J(P \mid Q)$ is defined by Equation (11.15).
[*Suggestion:* Use Equation (11.11) and the kinematical identities preceding (11.9) to cast the left side of (12.22) into a form in which Equation (12.5) can be used.]

This theorem shows that when the approximate form (11.14) is used for the collision integral, then it gives no contribution to the conservation theorem (12.21). In this approximation, the quantities $\overleftrightarrow{B}$ and $\mathbf{q}_V$ in the conservation laws vanish [see Equations (12.11) and (12.17)]. We must use an exact expression for the collision integral to derive Equations (12.12) and (12.18) for $\overleftrightarrow{B}$ and $\mathbf{q}_V$ respectively.

(b) Argue that the potential-energy flux vector $\mathbf{q}_V$ of Equation (12.18) vanishes when there are no position-momentum correlations in P_2. This result shows that the nonvanishing of $\mathbf{q}_V$ depends on intricate correlations in a fluid that may well be negligible normally in real gases. By contrast, the quantity $\overleftrightarrow{B}$ of Equation (12.12) gives an important contribution for real fluids to the velocity tensor $\overleftrightarrow{\Pi}$ and the energy flux vector $\mathbf{Q}$.

The continuity equation for mass, Equation (12.8), follows easily from the basic theorem (12.21) after setting $\chi = m$ and using definitions (12.9) and (12.10). Even the exact collision integral gives no contribution in this case, since

$$\int d^3\mathbf{k}\left(\frac{\partial P}{\partial t}\right)_{\text{coll}} = 0 \tag{12.23}$$

quite generally, a result that is proved in Appendix D [see Equation (D.8)].

We next prove the momentum conservation law (12.14) by substituting $\chi = \mathbf{k}$ into Equation (12.21). Upon using the definitions $\mathbf{j} = \rho\mathbf{u}$ and (12.11) for $\overleftrightarrow{\Pi}$, Equation (12.14) is obtained provided that we set

$$-\int d^3\mathbf{k}\,\mathbf{k}\left(\frac{\partial P}{\partial t}\right)_{\text{coll}} = \nabla \cdot \overleftrightarrow{B} \tag{12.24}$$

This identification is verified in Appendix D as Equation (D.13).

The energy conservation law (12.19) is more subtle to demonstrate. We begin by substituting $\chi = k^2/2m$ into the general conservation theorem

(12.21). Apart from the collision-integral contribution, this step yields the kinetic-energy terms in Equation (12.19) [cf. the first terms on the right sides of Equations (12.16)]. Thus the correct energy conservation law requires proof of the following rather complex result:

$$-\int d^3\mathbf{k}(k^2/2m)\left(\frac{\partial P}{\partial t}\right)_{\text{coll}} = \frac{\partial \phi}{\partial t} + \nabla \cdot (\phi \mathbf{u}) + \nabla \cdot (\overleftrightarrow{B} \cdot \mathbf{u} + \mathbf{q}_V) \tag{12.25}$$

This proof is outlined in Appendix D [see Equation (D.15)].[14]

In passing, we note that all of the quantities that appear in the three basic conservation laws can be defined macroscopically. For example, $(\mathbf{F}_{\text{ext}}/m)$, the external force per unit mass, is a field quantity. On the one hand, we have given microscopic expressions for each of the densities and current densities; on the other hand, the conservation laws can be transformed into forms that are familiar as the basic (macroscopic) equations of hydrodynamics.

Transformation to Local Rest System

We now wish to separate explicitly all contributions due to microscopic and macroscopic motion in the momentum and energy conservation laws. In so doing we shall be able to identify pressure, internal energy, and heat as averages over microscopically defined functions. The resulting differential equations then serve to relate these three quantities on a macroscopic scale.

The continuity equation for mass is unaffected by the present analysis. However, we do rewrite it in terms of the total time derivative

$$\frac{d}{dt} \equiv \frac{\partial}{\partial t} + \mathbf{u} \cdot \nabla \tag{12.26}$$

which is the time derivative for an observer moving with the local average velocity $\mathbf{u}(\mathbf{r}, t)$.[15] With this notation, Equation (12.8) becomes

$$\frac{d\rho}{dt} = -\rho(\nabla \cdot \mathbf{u}) \qquad \text{(continuity equation for mass)} \tag{12.27}$$

To proceed, we now introduce the microscopic velocity $\mathbf{U}(\mathbf{r}, t)$ in the local rest system by

$$\mathbf{U}(\mathbf{r}, t) = \mathbf{v} - \mathbf{u}(\mathbf{r}, t) \tag{12.28}$$

where $\mathbf{v} = \mathbf{k}/m$. We can then make the following definitions in terms of this velocity:

$$\mathcal{P}_{ij}(\mathbf{r}, t) \equiv \rho\langle U_i U_j\rangle + B_{ij} \qquad \text{(pressure tensor)} \tag{12.29}$$

$$e(\mathbf{r}, t) \equiv \frac{1}{2}\langle U^2\rangle + \phi/\rho \qquad \text{(internal energy per unit mass)} \tag{12.30}$$

$$\mathbf{q}_K(\mathbf{r}, t) \equiv \frac{1}{2}\rho\langle U^2\mathbf{U}\rangle \qquad \text{(kinetic-energy flux vector)} \tag{12.31}$$

$$\Lambda_{ij}(\mathbf{r}, t) \equiv \frac{1}{2}\left(\frac{\partial u_i}{\partial x_j} + \frac{\partial u_j}{\partial x_i}\right) \tag{12.32}$$

Both $\mathscr{P}_{ij}$ and Λ_{ij} are symmetric tensors; the pressure tensor is so named because one-third of its trace is the fluid pressure.[16] The trace of Λ_{ij} is given by $\Sigma_i\Lambda_{ii} = \nabla \cdot \mathbf{u}$.

The momentum conservation law (12.14) and the energy conservation law (12.19) can be rewritten in terms of the preceding four quantities as:

$$\rho\frac{du}{dt} = \rho(\mathbf{F}_{\text{ext}}/m) - \nabla \cdot \overleftrightarrow{\mathscr{P}} \qquad \text{(momentum conservation law)} \tag{12.33}$$

$$\rho\frac{de}{dt} = -\nabla \cdot (\mathbf{q}_K + \mathbf{q}_V) - \overleftrightarrow{\mathscr{P}} \cdot \overleftrightarrow{\Lambda} \qquad \text{(energy conservation law)} \tag{12.34}$$

In this form the laws have been transformed to the local rest system in which microscopic velocities are given by Equation (12.28). In Equation (12.34) we note that the local energy flux, or *heat,* has both a kinetic-energy part, $\mathbf{q}_K$, and a potential-energy contribution, $\mathbf{q}_V$, of Equation (12.18). Also, whereas external forces influence the flow of total energy in a fluid [see Equation (12.19)], they have no direct bearing on the flow of internal energy.

Exercise 12.3 Derive Equations (12.33) and (12.34) from Equations (12.14) and (12.19) respectively. In both cases the continuity equation (12.8) must be used; to derive Equation (12.34) the first result (12.33) is also required.

Nonviscous Flow

The conservation laws (12.27), (12.33), and (12.34) have been derived here for fluids in general. They can also be derived by using general heuristic arguments that involve only macroscopic concepts.[17] We now return to the case of real gases, and, in particular, we investigate the simple limiting case of a dilute gas.

For real gases we henceforth neglect the potential-energy flux vector $\mathbf{q}_V$ in Equation (12.34), an omission made plausible by the argument of Exercise 12.2(b). For dilute real gases, we also neglect the potential-energy contribution $\overleftrightarrow{B}$, in $\overleftrightarrow{\mathscr{P}}$, which means that molecular-interaction effects will not be included in either the pressure tensor or the equation of state.[18] There is also a potential-energy contribution to de/dt, which we treat by writing, for systems near equilibrium,

$$
\begin{aligned}
\rho\frac{de}{dt} &= \frac{1}{\Omega}\left(\frac{\partial E}{\partial T}\right)_{\Omega}\frac{dT}{dt} + \rho\left(\frac{\partial E}{\partial \Omega}\right)_{T}\frac{d(1/\rho)}{dt} \\
&= c_V\frac{dT}{dt} + \rho^{-1}[\mathcal{P} - T\alpha/\kappa_T]\frac{d\rho}{dt} \\
&= c_V\frac{dT}{dt} - (\mathcal{P} - T\alpha/\kappa_T)\nabla\cdot\mathbf{u} \\
&\cong c_V\frac{dT}{dt}
\end{aligned}
\tag{12.35}
$$

where $T(\mathbf{r}, t)$ is the local absolute temperature of a small (constant) gas mass with volume Ω. The second equality is derived by using Equations (8.25) and (9.7), and the third by using Equation (12.27). The quantity $c_V \equiv C_V/\Omega$ is the constant-volume heat capacity per unit volume of the gas. With the first three lines of this equation we have made a general macroscopic identification of local internal-energy variations in terms of local temperature and density variations. Now in a dilute gas without internal molecular structure $e = \frac{3}{2}\kappa T/m$ and $c_V/\rho = \frac{3}{2}\kappa/m$, where κ is Boltzmann's constant. Neglect of the contribution ϕ to e is, therefore, equivalent to writing the final equality in Equation (12.35).

The concept of a local temperature $T(\mathbf{r}, t)$ can only be meaningful for systems that are locally at equilibrium, a condition that is achieved ordinarily after several collision times τ. Furthermore, the approximations of the preceding paragraph imply that we can treat the dilute gas that is locally at equilibrium by assuming that the ideal gas law

$$
\mathcal{P}(\mathbf{r}, t) = n(\mathbf{r}, t)\kappa T(\mathbf{r}, t) \tag{12.36}
$$

holds locally. The corresponding one-particle distribution function $P \cong f^{(0)}$ is, according to Equation (11.19),

$$
f^{(0)}(\mathbf{r}, \mathbf{k}, t) = n(2\pi m\kappa T)^{-3/2}\exp\left(-\frac{mU^2}{2\kappa T}\right) \tag{12.37}
$$

where $\mathbf{U}$ is defined by Equation (12.28). Here, each of the quantities n, T, and $\mathbf{u}$ is assumed to be a slowly varying function of $\mathbf{r}$ and t.[19] When no corrections to this approximation for $P(\mathbf{r}, \mathbf{k}, t)$ are included, then the three conservation laws become the hydrodynamic equations for *nonviscous flow* in which flow patterns, as determined by boundary conditions, persist indefinitely; there is no dissipation of macroscopic order.

For the nonviscous flow of dilute gases the pressure tensor is diagonal, $\mathcal{P}_{ij}^{(0)} = \mathcal{P}\,\delta_{ij}$, as we show below, and the ideal-gas law (12.36) holds locally.

The conservation laws (12.27), (12.33), and (12.34) then become

$$\frac{d\rho}{dt} + \rho\nabla \cdot \mathbf{u} = 0 \tag{12.38a}$$

$$\frac{d\mathbf{u}}{dt} + \frac{1}{\rho}\nabla\mathcal{P} = \mathbf{F}_{\text{ext}}/m \tag{12.38b}$$

$$\frac{dT}{dt} + \frac{2}{3}T\nabla \cdot \mathbf{u} = 0 \tag{12.38c}$$

where we have used $me = \frac{3}{2}\kappa T$ to derive the final equation. Also, as we verify velow, there is no heat transfer ($\mathbf{q}_K = 0$) in the nonviscous flow.

We now employ the local-equilibrium function $f^{(0)}$ in definition (12.6) to calculate average values for nonviscous flow. First we note that the quantities n, T, and $\mathbf{u}$, as they appear in Equation (12.37) for $f^{(0)}$, are precisely the average quantities defined: (1) by using Equation (12.6) with $A = 1$ to calculate $n(\mathbf{r}, t)$; (2) by using Equation (12.30) with $\phi = 0$ to calculate $e(\mathbf{r}, t) = \frac{3}{2}\kappa T/m$; and (3) by using Equation (11.20) or (12.10) to calculate $\mathbf{u}(\mathbf{r}, t)$.

Next we calculate the pressure tensor $\mathcal{P}_{ij}^{(0)}$ for nonviscous flow. On substituting Equation (12.37) into Equation (12.29) and setting $B_{ij} = 0$, we obtain

$$\begin{aligned}\mathcal{P}_{ij}^{(0)}(\mathbf{r}, t) &= mn(2\pi m\kappa T)^{-3/2}\int d^3\mathbf{k}U_iU_j\exp\left(-\frac{mU^2}{2\kappa T}\right)\\ &= \delta_{ij}\mathcal{P}(\mathbf{r}, t)\end{aligned} \tag{12.39}$$

It is easy to verify that the expression for $\mathcal{P}(\mathbf{r}, t)$ is precisely the ideal-gas law (12.36). We have omitted superscripts (0) from the quantities n, T, $\mathbf{u}$, and $\mathcal{P}$, because, as we shall find in the next two sections, these quantities are unaffected for dilute real gases by corrections (in P) to $f^{(0)}$. The underlying reason for this fact is that these four quantities characterize local equilibrium self-consistently. Other physical quantities we calculate depend, via the Boltzmann transport equation, on macroscopic variations with $\mathbf{r}$ and t of the above four local-equilibrium parameters. They therefore vary with the approximation to $P(\mathbf{r}, \mathbf{k}, t)$. Thus one readily verifies from Equation (12.31) for $\mathbf{q}_K$ that in nonviscous flow there is no heat transfer, that is,

$$\mathbf{q}_K^{(0)}(\mathbf{r}, t) = \frac{1}{2}mn(2\pi m\kappa T)^{-3/2}\int d^3\mathbf{k}\mathbf{U}U^2\exp\left(-\frac{mU^2}{2\kappa T}\right) = 0 \tag{12.40}$$

The vanishing of $\mathbf{q}_K^{(0)}(\mathbf{r}, t)$ is, of course, not a general result.

In general, the average values that give physical quantities such as $\mathcal{P}_{ij}$ and $\mathbf{q}_K$ can be calculated only after the Boltzmann transport equation (11.27) has been solved in any particular application. The solution to the hydrodynamic equations, on the other hand, is a boundary-value problem. These microscopic and macroscopic solutions must be combined, as suggested above Equation (12.1), in any complete treatment of a dilute real gas. In the next two sections we describe methods for solving the Boltzmann transport equation. Some applications of the hydrodynamic equations for nonviscous flow complete the present section.

Examples of Nonviscous Flow

Equations (12.38) are the hydrodynamic equations for nonviscous flow in dilute gases. In any solution of the hydrodynamic equations for a fluid, we must know the (equilibrium) equation of state for the fluid. For the present case of dilute gases we have assumed consistently that the equation of state is the ideal-gas law (12.36); we have, in fact, used this equation to arrive at the particular form (12.38c) of the third nonviscous-flow equation.

We have seen with Equation (12.40) that there is no heat flow associated with nonviscous motion. See Figure III.3. We now show that Equations (12.38a) and (12.38c) can be combined to express this fact as an adiabatic flow condition on $\mathcal{P}$, ρ, and T. This condition is:

$$\mathcal{P}\rho^{-\gamma} = \text{constant (along a streamline)} \qquad (12.41)$$

where $\gamma = 5/3$ for simple dilute gases. To prove Equation (12.41), multiply Equation (12.38c) by $-(3\rho/2T)$ and then add it to Equation (12.38a) to obtain $(d/dt)(\rho T^{-3/2}) = 0$. With the aid of the ideal-gas law (12.36), this last expression is readily shown to be equivalent to Equation (12.41).

Steady-State Flow: We consider next Equation (12.38b), which is known as *Euler's equation of motion* in hydrodynamics, and examine the case of *steady-state flow* ($\partial \mathbf{u}/\partial t = 0$) under the influence of a conservative external potential $U(\mathbf{r})$. With the aid of the ideal-gas law, the expression $\mathbf{F}_{\text{ext}} = -\nabla U(r)$ and the vector identity

$$(\mathbf{u} \cdot \nabla)\mathbf{u} = \frac{1}{2}\nabla(u^2) - \mathbf{u} \times (\nabla \times \mathbf{u})$$

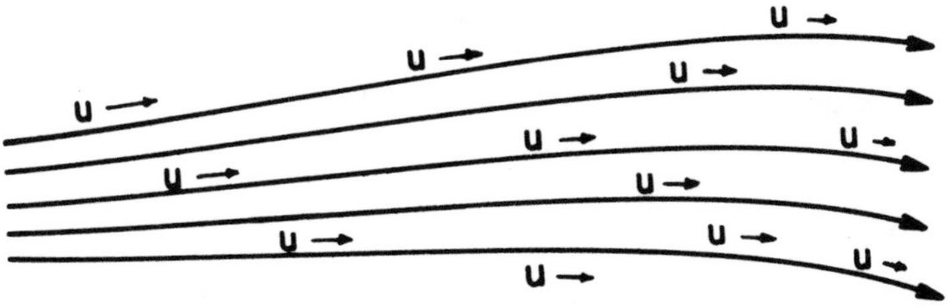

FIGURE III.3 Two-dimensional view of streamlines in a fluid. Flow is along streamlines, and the velocity is greatest where the streamlines are closest. In nonviscous flow there is no heat transfer.

we can rewrite Equation (12.38b) for steady-state flow as

$$\nabla\left(\frac{1}{2}u^2 + \frac{\mathcal{P}}{\rho} + \frac{U}{m}\right) = \mathbf{u} \times (\nabla \times \mathbf{u}) - \left(\frac{\kappa T}{m\rho}\right)\nabla\rho \tag{12.42}$$

In the case of uniform temperature $\nabla T = 0$ and irrotational flow $\nabla \times \mathbf{u} = 0$, this equation reduces to the simpler form

$$\nabla\left(\frac{1}{2}u^2 + \frac{1}{m}U\right) = -\frac{\kappa T}{m}\nabla(\ln \rho)$$

Integration then gives a macroscopic version of the Maxwell–Boltzmann distribution (11.19) and (11.29); namely,

$$\rho = \rho_o \exp\left[-\frac{1}{\kappa T}\left(\frac{1}{2}mu^2 + U\right)\right] \tag{12.43}$$

An important difference between this last result and the Maxwell–Boltzmann distribution is that the velocity $\mathbf{u}(\mathbf{r}, t)$ is a macroscopic quantity that is subject to the theorems of vector calculus. For example, if T is not only uniform but time independent, then Equation (12.38c) shows that $\mathbf{u}$ is solenoidal, $\nabla \cdot \mathbf{u} = 0$, in addition to being irrotational. The velocity $\mathbf{u}$ is, in this case, the gradient of a scalar function; for very large systems it is a constant vector.[20]

Exercise 12.4 The second of the nonviscous equations, (12.38b), follows from the general conservation law (12.33) whenever the pressure tensor is diagonal. Therefore, Equation (12.42) can be applied generally along the streamlines of a uniform (or incompressible) fluid, for which $\nabla\rho = 0$. Derive Bernoulli's equation for steady-state flow from Equation (12.42) for this case.

Sound Propagation: We complete this section by deriving the linear wave equation for sound in a dilute gas. We can understand the propagation of sound by considering small deviations from equilibrium in a system at rest with no external forces present. Under these circumstances, Equations (12.38) can be linearized by assuming that $\mathbf{u}$ and the derivatives of ρ, T, $\mathcal{P}$, and $\mathbf{u}$ are all small quantities whose products with each other can be neglected. In this approximation we obtain the set of equations

$$\frac{\partial\rho}{\partial t} + \rho\,\nabla \cdot \mathbf{u} = 0$$

$$\rho\frac{\partial\mathbf{u}}{\partial t} + \nabla\mathcal{P} = 0 \tag{12.44}$$

$$\frac{3}{2}\rho\frac{\partial T}{\partial t} = -\rho T(\nabla \cdot u) = T\frac{\partial\rho}{\partial t}$$

where the last equation is equivalent to the adiabatic flow condition (12.41). The first two equations can be combined, neglecting products of small quantities again, to give the second-order partial differential equation

$$\frac{\partial^2 \rho}{\partial t^2} - \nabla^2 \mathcal{P} = 0 \tag{12.45}$$

If we now use the adiabatic flow condition in the form (entropy S = constant), then the gradient of $\mathcal{P}$ can be written as

$$\nabla \mathcal{P} = \left(\frac{\partial \mathcal{P}}{\partial \rho}\right)_S \nabla \rho = (\rho \kappa_S)^{-1} \nabla \rho$$

The adiabatic compressibility κ_S is given by Equation (9.3), which, for an ideal monatomic gas, becomes

$$\kappa_S = \rho^{-1}\left(\frac{\partial \rho}{\partial \mathcal{P}}\right)_S = (\gamma \mathcal{P})^{-1} = \frac{3}{5}\left(\frac{m}{\rho \kappa T}\right) \tag{12.46}$$

Thus Equation (12.45) can be written in the linear approximation as

$$\rho \kappa_S \frac{\partial^2 \rho}{\partial t^2} - \nabla^2 \rho = 0 \tag{12.47}$$

which is a wave equation for ρ. Since sound is propagated by density fluctuations, we may conclude that Equation (12.47) describes the transmission of sound with a velocity of propagation v_S given, for a dilute monatomic gas, by

$$v_S = (\rho \kappa_S)^{-1/2} = \left(\frac{5 \kappa T}{3 m}\right)^{1/2} = \frac{\sqrt{5}}{3} v_{\text{rms}} \tag{12.48}$$

It is not surprising that the velocity of sound should be less than the rms molecular velocity (11.22).

13 Viscous Hydrodynamics

The development of the preceding section is concluded here by the phenomenological introduction of viscous coefficients to characterize heat flow and viscosity in fluids. The standard equations of viscous hydrodynamics are then derived from the conservation laws of Section 12 after making the simplifying assumption that spatial variations of $\mathbf{u}(\mathbf{r}, t)$ and the viscous coefficients are small.

In the second half of this section we show how the Boltzmann transport equation provides a microscopic basis for the viscous coefficients. First we investigate the approach to local equilibrium, and then we discuss qualitatively how corrections to the local equilibrium function $f^{(0)}$ must vary on a macroscopic scale. Finally, we make a rough calculation of the leading correction term $f^{(1)}$ and of the viscous coefficients.

The macroscopic equations of viscous hydrodynamics are derived from the conservation laws (12.27), (12.33), and (12.34) by substitution of particular forms for the total heat-flux vector $\mathbf{q}$ and the pressure tensor $\overleftrightarrow{\mathcal{P}}$. We assume, for an isotropic system, that $\mathbf{q} = \mathbf{q}_K + \mathbf{q}_V$ is proportional to the temperature gradient, that is,

$$\mathbf{q} \equiv \mathbf{q}_K + \mathbf{q}_V = -K\,\nabla T \tag{13.1}$$

in which K is the *coefficient of thermal conductivity*. We also assume that $\mathcal{P}_{ij}$ is given by the expression

$$\mathcal{P}_{ij} = \mathcal{P}\,\delta_{ij} - 2\mu\Lambda_{ij} + \left(\frac{2}{3}\,\mu - \phi\right)\delta_{ij}(\nabla\cdot\mathbf{u}) \tag{13.2}$$

where μ is called the *coefficient of shear viscosity* and ϕ the *coefficient of bulk viscosity*. The trace of this pressure tensor is given by

$$\sum_i \mathcal{P}_{ii} = 3(\mathcal{P} - \phi\,\nabla\cdot u) \tag{13.3}$$

where we have used Equation (12.32) for Λ_{ij}. Finally, for an incompressible fluid (one with ρ = constant), the continuity equation (12.27) gives

$$\frac{d\rho}{dt} = -\rho(\nabla\cdot\mathbf{u}) = 0 \qquad \text{(incompressible fluid)} \tag{13.4}$$

in which case the coefficient ϕ is unnecessary. The quantities K, μ, and ϕ are known as *viscous,* or *kinetic, coefficients,* so-called because they provide for currents or fluxes that dissipate macroscopic order.[21]

Using Equations (13.1) and (13.2), we now show that Equations (12.33) and (12.34) can be recast into the forms:

$$\rho\frac{d\mathbf{u}}{dt} = \rho(\mathbf{F}_{\text{ext}}/m) - \nabla\mathcal{P} + \mu\,\nabla^2\mathbf{u} + \left(\frac{\mu}{3} + \phi\right)\nabla(\nabla\cdot\mathbf{u}) + \mathbf{R} \tag{13.5}$$

$$\begin{aligned} c_V\frac{dT}{dt} &= K\,\nabla^2 T + \nabla K\cdot\nabla T - \frac{T\alpha}{\kappa_T}(\nabla\cdot\mathbf{u}) - \left(\frac{2}{3}\,\mu - \phi\right)(\nabla\cdot\mathbf{u})^2 \\ &\quad + \mu[\nabla^2(u^2) - 2\mathbf{u}\cdot\nabla^2\mathbf{u} - (\nabla\times\mathbf{u})^2] \end{aligned} \tag{13.6}$$

where in Equation (13.5) we have introduced the vector $\mathbf{R}$, defined by

$$R_i \equiv 2\sum_{j=1}^{3}\left(\frac{\partial\mu}{\partial x_j}\right)\Lambda_{ij} - \left(\frac{2}{3}\frac{\partial\mu}{\partial x_i} - \frac{\partial\phi}{\partial x_i}\right)\nabla\cdot\mathbf{u} \tag{13.7}$$

In Equation (13.6) α and κ_T are the thermodynamic coefficients defined by Equations (9.1) and (9.2) respectively.

The derivation of Equation (13.5) from (12.33) is straightforward. To arrive at Equation (13.6) we first inserted the third equality of Equation (12.35) into (12.34). The derivation then requires a reduction of the last term in Equation (12.34), which, with the aid of Equations (13.2) and (12.32), can be written in the form

$$-\sum_{i,j=1}^{3} \mathcal{P}_{ij}\Lambda_{ij} = -\mathcal{P}(\nabla \cdot \mathbf{u}) - \left(\frac{2}{3}\mu - \phi\right)(\nabla \cdot \mathbf{u})^2 + 2\mu \sum_{i,j=1}^{3} \Lambda_{ij}\Lambda_{ij}$$

where

$$\begin{aligned} 2\sum_{i,j=1}^{3} \Lambda_{ij}\Lambda_{ij} &= \sum_{i,j=1}^{3} \frac{\partial u_i}{\partial x_j}\left(\frac{\partial u_i}{\partial x_j} + \frac{\partial u_j}{\partial x_i}\right) \\ &= 2\sum_{i,j=1}^{3} \left(\frac{\partial u_i}{\partial x_j}\right)^2 + \sum_{i,j=1}^{3} \left[\frac{\partial u_i}{\partial x_j}\frac{\partial u_i}{\partial x_i} - \left(\frac{\partial u_i}{\partial x_j}\right)^2\right] \end{aligned}$$

In the second line of this last expression the two terms on the right side can be written further as follows:

$$\begin{aligned} 2\sum_{i,j=1}^{3} \left(\frac{\partial u_i}{\partial x_j}\right)^2 &= 2\sum_{i,j=1}^{3} \frac{\partial}{\partial x_j} u_i\left(\frac{\partial u_i}{\partial x_j}\right) - 2\sum_{i,j=1}^{3} u_i\left(\frac{\partial^2 u_i}{\partial x_j^2}\right) \\ &= \nabla^2(u^2) - 2\mathbf{u} \cdot \nabla^2 \mathbf{u} \end{aligned}$$

and

$$\begin{aligned} \sum_{i,j=1}^{3} \left[\frac{\partial u_i}{\partial x_j}\frac{\partial u_j}{\partial x_i} - \left(\frac{\partial u_i}{\partial x_j}\right)^2\right] &= \sum_{i,j=1}^{3} \left(\frac{\partial u_i}{\partial x_j} - \frac{\partial u_j}{\partial x_i}\right) \cdot \left(\frac{\partial u_j}{\partial x_i} - \frac{\partial u_i}{\partial x_j}\right) \\ &\quad + \sum_{i,j=1}^{3} \left[\left(\frac{\partial u_j}{\partial x_i}\right)^2 - \frac{\partial u_j}{\partial x_i}\frac{\partial u_i}{\partial x_j}\right] \\ &= -(\nabla \times \mathbf{u})^2 \end{aligned}$$

Substitution of these last four equations into Equation (12.34) yields directly the final set of terms in Equation (13.6).

Useful approximate forms for Equations (13.5) and (13.6) can be deduced from these two equations by: (1) neglecting the spatial variations of K, μ, and ϕ; and (2) considering velocities sufficiently slowly varying that quadratic products of derivatives of $\mathbf{u}$ can be omitted. In this case Equations (13.5) and (13.6), together with Equation (12.27), simplify to the equations of macroscopic hydrodynamics in a standard form:

$$\frac{d\rho}{dt} = -\rho(\nabla \cdot \mathbf{u}) \qquad \textbf{(13.8)}$$

$$\rho \frac{d\mathbf{u}}{dt} = \rho(\mathbf{F}/m) - \nabla \mathcal{P} + \mu \nabla^2 \mathbf{u} + \left(\frac{\mu}{3} + \phi\right) \nabla(\nabla \cdot \mathbf{u}) \tag{13.9}$$

$$c_V \frac{dT}{dt} = K \nabla^2 T - \frac{T\alpha}{\kappa_T}(\nabla \cdot \mathbf{u}) \tag{13.10}$$

where the total time derivative d/dt is defined by (12.26). These three equations are correct for fluids in general. For dilute real gases, one can use the ideal-gas law (12.36) to replace $T\alpha/\kappa_T$ by $\mathcal{P}$ in Equation (13.10).

Equation (13.9) is called the *Navier–Stokes equation,*[22] and Equation (13.10) is called the *heat-conduction equation.* If $\mathbf{u} = 0$, then this latter equation reduces to the familiar diffusion equation for heat conduction

$$c_V \frac{\partial T}{\partial t} - K \nabla^2 T = 0 \tag{13.11}$$

There are four independent variables, ρ, $\mathcal{P}$, T, and $\mathbf{u}$ in Equations (13.8) to (13.10). The general solution of these equations therefore requires that one have an additional relation among the variables, such as the equation of state for the material of the system.

Deviations from Local Equilibrium

Our next objective is to deduce explicit expressions for the coefficients of heat conduction and viscosity for a dilute gas by deriving Equations (13.1) and (13.2). We assume in the derivation that the one-particle distribution function $P(\mathbf{r}, \mathbf{k}, t)$ differs only slightly from the local-equilibrium distribution (12.37), and we show under what conditions this assumption should be valid. But first let us examine the local equilibrium condition itself.

Here and in the following section we are interested not in the most general solution to the Boltzmann transport equation, but rather in a particular type of solution in which time is contained only implicitly through variations of the local density, velocity and temperature (n, $\mathbf{u}$, and T). Thus it is assumed that the mechanism of collisions rapidly produces a local equilibrium condition that is characterized by the single-particle distribution function $f^{(0)}(\mathbf{r}, \mathbf{k}, t)$ of Equation (12.37).

We now show that any explicit time dependence of $P(\mathbf{r}, \mathbf{k}, t)$ can be expected to damp out in a characteristic time τ_R, which is of order the collision time (12.3). We write

$$P(\mathbf{r}, \mathbf{k}, t) = f(\mathbf{r}, \mathbf{k}, t) + g(\mathbf{r}, \mathbf{k}, t) \tag{13.12}$$

where $f(\mathbf{r}, \mathbf{k}, t)$ is a solution to the Boltzmann equation (11.27), which has no explicit time dependence, and $g(\mathbf{r}, \mathbf{k}, t)$ is any part of $P(\mathbf{r}, \mathbf{k}, t)$, which has explicit time dependence. Then

$$\frac{\partial P}{\partial t} = \frac{\partial f}{\partial t} + \left(\frac{\partial g}{\partial t}\right)_{\text{explicit}} + \left(\frac{\partial g}{\partial t}\right)_{\text{implicit}} \tag{13.13}$$

where the derivative $\partial f/\partial t$ involves only implicit time dependence.

We next define the differential operator $\mathscr{D}$ by the equation

$$\mathscr{D}(f) \equiv (\mathbf{v} \cdot \nabla_r + \mathbf{F}_{\text{ext}} \cdot \nabla_k)f \tag{13.14}$$

Then the Boltzmann equation (11.27) can be written, using this definition and Equation (13.13), as

$$\frac{\partial f}{\partial t} + \left(\frac{\partial g}{\partial t}\right)_{\text{explicit}} + \left(\frac{\partial g}{\partial t}\right)_{\text{implicit}} + \mathscr{D}(f) + \mathscr{D}(g) = \left(\frac{\partial P}{\partial t}\right)_{\text{coll}} \tag{13.15}$$

We also introduce an approximation, to be used in this section, but not in Section 14, that the collision integral (11.10) or (11.14) can be estimated by writing

$$\left(\frac{\partial P}{\partial t}\right)_{\text{coll}} \sim -\frac{\delta P}{\tau_R} \cong -\frac{1}{\tau_R}(f^{(1)} + g) \tag{13.16}$$

where τ_R is called the *relaxation time* and $f^{(1)}$ is the leading correction term in f, to $f^{(0)}$, that is,

$$f(\mathbf{r}, \mathbf{k}, t) - f^{(0)}(\mathbf{r}, \mathbf{k}, t) \cong f^{(1)}(\mathbf{r}, \mathbf{k}, t) \tag{13.17}$$

We shall presume that the relaxation time of Equation (13.16) and the collision time (12.3) are essentially the same. We note that the local equilibrium function $f^{(0)}(\mathbf{r}, \mathbf{k}, t)$ satisfies Equation (11.16), and therefore, it makes no contribution to the collision integral (11.14). Therefore, $(f^{(1)} + g)$ is the relevant quantity to use on the right side of approximation (13.16). The minus sign is introduced there because we expect on the average that variations $(\partial P/\partial t)_{\text{coll}}$, as time increases, will always be opposite in sign to the deviations $(f^{(1)} + g)$ themselves. In other words, we assume that systems tend to be restored by collisions to the equilibrium condition.

By definition, $f(\mathbf{r}, \mathbf{k}, t)$ is a solution to the Boltzmann equation

$$\frac{\partial f}{\partial t} + \mathscr{D}(f) = \left(\frac{\partial f}{\partial t}\right)_{\text{coll}} \cong -\frac{f^{(1)}}{\tau_R} \tag{13.18}$$

and Equation (13.15) may therefore be written as

$$\left(\frac{\partial g}{\partial t}\right)_{\text{explicit}} + \left(\frac{\partial g}{\partial t}\right)_{\text{implicit}} + \mathscr{D}(g) \cong -\frac{g}{\tau_R} \tag{13.19}$$

But if we assume that the first term on the left side of this equation dominates the other two, then we obtain the simple differential equation

$$\frac{\partial g}{\partial t} \cong -\frac{g}{\tau_R} \tag{13.19a}$$

which has the solution

$$g \cong g_o\, e^{-t/\tau_R} \xrightarrow[t \gg \tau_R]{} 0 \tag{13.20}$$

It is because of this last, heuristically derived result that it is meaningful to focus attention only on solutions $f(\mathbf{r}, \mathbf{k}, t)$ to the Boltzmann equation. The explicit time dependence of any solution, as contained in the part $g(\mathbf{r}, \mathbf{k}, t)$, damps out rapidly and is unimportant on a macroscopic time scale.

We consider next the function $f^{(1)}(\mathbf{r}, \mathbf{k}, t)$, defined by Equation (13.17). We can obtain an expression for $f^{(1)}$ by substituting Equation (13.14) into Equation (13.18) and then making the approximation $f \cong f^{(0)}$. This procedure gives

$$f^{(1)}(\mathbf{r}, \mathbf{k}, t) \cong -\tau_R\left[\frac{\partial}{\partial t} + \mathbf{v} \cdot \nabla_r + \mathbf{F}_{\text{ext}} \cdot \nabla_k\right] f^{(0)}(\mathbf{r}, \mathbf{k}, t) \tag{13.21}$$

We now use this estimate of $f^{(1)}$ to discuss deviations from local equilibrium on a macroscopic scale.

We may assume that $f^{(0)}$ varies in space by a significant amount (i.e., of the order of itself) only when $|\mathbf{r}|$ varies by some distance L. Then Equation (13.21) furnishes the estimate

$$\frac{f^{(1)}}{f^{(0)}} \sim \tau_R \frac{\langle v \rangle}{L} \sim \frac{\lambda_R}{L} \cdot \frac{u}{\langle U \rangle} < \frac{\lambda_R}{L}$$

where $\lambda_R \equiv \tau_R \langle U \rangle$ is a length of order the mean free path (12.4). We may conclude from this estimate that $f^{(0)}(\mathbf{r}, \mathbf{k}, t)$ should be a good approximation to $f(\mathbf{r}, \mathbf{k}, t)$ if the local density, temperature, and macroscopic velocity all have characteristic wavelengths L that are much larger than λ_R. The fractional corrections to $f^{(0)}$ will then be $< \lambda_R/L \ll 1$.

Viscous Flow in a Dilute Gas

We now turn to the calculation of the viscous coefficients for a dilute gas. Our procedure is to calculate the heat-flux vector $\mathbf{q}$, neglecting $\mathbf{q}_V$, and the pressure tensor $\mathcal{P}_{ij}$ and to cast these quantities into the form of Equations (13.1) and (13.2). This method enables us to identify expressions for the coefficients K and μ, the coefficient of bulk viscosity ϕ being zero in the approximate calculations of this section and the next.

A systematic procedure for calculating the deviations of $f(\mathbf{r}, \mathbf{k}, t)$ from the local-equilibrium function $f^{(0)}(\mathbf{r}, \mathbf{k}, t)$ is given by the Chapman–Enskog method, which is outlined in the next section. Here we make only a rough calculation of these deviations by using the estimate (13.21) for $f^{(1)}(\mathbf{r}, \mathbf{k}, t)$, which is based on approximation (13.16).

We must emphasize that the local density, temperature, and velocity are determined self-consistently by the condition of local equilibrium. This is a requirement for the calculation of $f(\mathbf{r}, \mathbf{k}, t)$ that is motivated physically by our interpretation of "local equilibrium" as a condition characterized by local values of n, T, and $\mathbf{u}$. So it is, for a dilute gas, that we set

$$n(\mathbf{r}, t) \equiv \int d^3\mathbf{k} f(\mathbf{r}, \mathbf{k}, t) = \int d^3\mathbf{k} f^{(0)}(\mathbf{r}, \mathbf{k}, t) \tag{13.22}$$

$$T(\mathbf{r}, t) \equiv \frac{1}{3}\left(\frac{m}{n\kappa}\right) \int d^3\mathbf{k} U^2 f(\mathbf{r}, \mathbf{k}, t) = \frac{1}{3}\left(\frac{m}{n\kappa}\right) \int d^3\mathbf{k} U^2 f^{(0)}(\mathbf{r}, \mathbf{k}, t) \tag{13.23}$$

$$\mathbf{u}(\mathbf{r}, t) \equiv \frac{1}{mn} \int d^3\mathbf{k}\,\mathbf{k} f(\mathbf{r}, \mathbf{k}, t) = \frac{1}{mn} \int d^3\mathbf{k}\,\mathbf{k} f^{(0)}(\mathbf{r}, \mathbf{k}, t) \tag{13.24}$$

where the local microscopic velocity $\mathbf{U}$ is defined by (12.28) and $n = \rho/m$. These three average values are always to be calculated by the second equalities, as explained above Equation (12.39). For each case the first equality becomes a condition on the general function $f(\mathbf{r}, \mathbf{k}, t)$.

To calculate $f^{(1)}(\mathbf{r}, k, t)$ with Equation (13.21), we first define the differential operator D by

$$\begin{aligned} D(A) &\equiv \left(\frac{\partial}{\partial t} + \mathbf{v} \cdot \nabla_r\right) A \\ &= \frac{dA}{dt} + (\mathbf{U} \cdot \nabla_r) A \end{aligned} \tag{13.25}$$

where we have used (12.26) for the total time derivative d/dt and Equation (12.28) for $\mathbf{U}(\mathbf{r}, t)$. Upon substituting Equation (12.37) for $f^{(0)}$ into Equation (13.21) and writing $\mathbf{F}_{\text{ext}} = \mathbf{F}$, we obtain

$$\begin{aligned} f^{(1)}(\mathbf{r}, \mathbf{k}, t) &= -\tau_R\left[D - \frac{1}{\kappa T}\mathbf{F} \cdot \mathbf{U}\right] f^{(0)}(\mathbf{r}, \mathbf{k}, t) \\ &= -\tau_R f^{(0)}\left\{\rho^{-1} D(\rho) + T^{-1}\left[\frac{m}{2\kappa T}U^2 - \frac{3}{2}\right] D(T)\right. \\ &\qquad \left. + \left(\frac{m}{\kappa T}\right) \sum_{j=1}^{3} U_j D(u_j) - (\kappa T)^{-1}\mathbf{F} \cdot \mathbf{U}\right\} \end{aligned} \tag{13.26}$$

To evaluate the derivatives in Equation (13.26) we use the nonviscous flow equations (12.38), which may be rewritten, using the second equality of Equation (13.25), as

$$D(\rho) = -\rho(\nabla \cdot \mathbf{u}) + \mathbf{U} \cdot \nabla\rho$$

$$D(T) = -\frac{2}{3}T(\nabla \cdot \mathbf{u}) + \mathbf{U} \cdot \nabla T$$

$$D(u_j) = -\rho^{-1}\frac{\partial \mathcal{P}}{\partial x_j} + m^{-1}F_j + (\mathbf{U} \cdot \nabla)u_j$$

Substitution of these relations into Equation (13.26) then gives for $f^{(1)}(\mathbf{r}, \mathbf{k}, t)$ the expression

$$\begin{aligned} f^{(1)}(\mathbf{r}, \mathbf{k}, t) = & -\tau_R f^{(0)}\Bigg\{ -\nabla \cdot \mathbf{u} + \rho^{-1}\mathbf{U} \cdot \nabla\rho + \Theta^{-1}\left(\frac{m}{2\Theta}U^2 - \frac{3}{2}\right) \\ & \times \left(-\frac{2}{3}\Theta\,\nabla \cdot \mathbf{u} + \mathbf{U} \cdot \nabla\Theta\right) \\ & + m\Theta^{-1}\left[-\rho^{-1}(\mathbf{U} \cdot \nabla)\mathcal{P} + m^{-1}\mathbf{U} \cdot \mathbf{F} + \sum_{i,j=1}^{3} U_iU_j\frac{\partial u_j}{\partial x_i}\right] \\ & - \Theta^{-1}\mathbf{F} \cdot \mathbf{U}\Bigg\} \end{aligned}$$

where $\Theta \equiv \kappa T$ and $\mathcal{P} = \rho\Theta/m$. After some manipulations, this expression simplifies to

$$\begin{aligned} f^{(1)}(\mathbf{r}, \mathbf{k}, t) = & -\tau_R\Theta^{-1}f^{(0)}(\mathbf{r}, \mathbf{k}, t)\Bigg\{\left(\frac{m}{2\Theta}U^2 - \frac{5}{2}\right)(\mathbf{U} \cdot \nabla\Theta) \\ & + m\sum_{i,j=1}^{3}\Lambda_{ij}\left(U_iU_j - \frac{1}{3}\delta_{ij}U^2\right)\Bigg\} \end{aligned} \quad (13.27)$$

where Λ_{ij} is defined by Equation (12.32).

Exercise 13.1 Use Equation (12.37) for $f^{(0)}$ and the second of the integrals (2.4) to show that $f^{(1)}$, as given by Equation (13.27), makes no contribution to the average values n, T, and $\mathbf{u}$ defined by Equations (13.22) to (13.24). Note that $d^3\mathbf{k} = m^3d^3\mathbf{U}$.

We now wish to calculate the macroscopic variables $\mathbf{q}_K$ and $\mathcal{P}_{ij}$, defined by Equations (12.31) and (12.29), respectively, with $f(\mathbf{r}, \mathbf{k}, t)$ substituted for $P(\mathbf{r}, \mathbf{k}, t)$ in definition (12.6). As explained above Equation (12.35), for the dilute real gas we neglect $\mathbf{q}_V$ and the contribution B_{ij} to the pressure tensor. With the aid of Equations (13.17), (12.40), and (13.27), it is then easy to verify that $\mathbf{q}_K$ and $\mathcal{P}_{ij}$ can be written as

$$\begin{aligned} \mathbf{q}_K(\mathbf{r}, t) & \cong \frac{1}{2}m\int d^3\mathbf{k}U^2\mathbf{U}f^{(1)}(\mathbf{r}, \mathbf{k}, t) \\ & = -\frac{1}{2}\tau_R m\Theta^{-1}\sum_{j=1}^{3}\frac{\partial\Theta}{\partial x_j}\int d^3\mathbf{k}U^2\mathbf{U}U_j\left(\frac{m}{2\Theta}U^2 - \frac{5}{2}\right)f^{(0)}(\mathbf{r}, \mathbf{k}, t) \end{aligned} \quad (13.28)$$

$$\mathcal{P}_{ij}(\mathbf{r}, t) \cong m \int d^3\mathbf{k} U_i U_j (f^{(0)} + f^{(1)}) \\ = \mathcal{P}\,\delta_{ij} + \mathcal{P}'_{ij} \tag{13.29}$$

where the scalar pressure $\mathcal{P} = \rho\Theta/m$ is given by Equation (12.39) and

$$\mathcal{P}'_{ij}(\mathbf{r}, t) \cong -\left(\frac{\tau_R m^2}{\Theta}\right) \sum_{k,\ell=1}^{3} \Lambda_{k\ell} \int d^3\mathbf{k} U_i U_j \left(U_k U_\ell - \frac{1}{3}\delta_{k\ell} U^2\right) f^{(0)}(\mathbf{r}, \mathbf{k}, t) \tag{13.30}$$

We can see from Equation (13.28) that the heat-flux vector is proportional to the temperature gradient, as was assumed more generally by Equation (13.1), with the coefficient of thermal conductivity K given, for a dilute gas, by

$$K(\mathbf{r}, t) = \left(\frac{m\tau_R}{6T}\right) \int d^3\mathbf{k} U^4 \left(\frac{m}{2\Theta}U^2 - \frac{5}{2}\right) f^{(0)}(\mathbf{r}, \mathbf{k}, t) \\ = \frac{5}{2} n\tau_R \kappa^2 T/m \tag{13.31}$$

To derive the second equality of this expression we have used Equations (12.37) and (2.4).

To evaluate $\mathcal{P}'_{ij}$, we note first from Equation (13.30) that this quantity is a symmetric tensor with zero trace. Since it depends linearly on Λ_{ij}, whose trace is $\nabla \cdot \mathbf{u}$, we may, therefore, write $\mathcal{P}'_{ij}$ in the form

$$\mathcal{P}'_{ij} = -2\mu\left(\Lambda_{ij} - \frac{1}{3}\delta_{ij}\nabla \cdot \mathbf{u}\right) \tag{13.32}$$

which agrees with Equation (13.2) when $\phi = 0$. The coefficient of viscosity μ for a dilute gas can then be calculated by setting $i = 1$ and $j = 2$ in Equation (13.30), to get

$$\mu(\mathbf{r}, t) = \left(\frac{\tau_R m^2}{\Theta}\right) \int d^3\mathbf{k} U_1^2 U_2^2 f^{(0)}(\mathbf{r}, \mathbf{k}, t) = n\tau_R \kappa T \tag{13.33}$$

where again we have used Equations (12.37) and (2.4) to arrive at the second equality. If we set $\tau_R \cong \tau$, then according to expression (12.3) for the collision time τ, both K and μ will be density-independent for a dilute gas, a fact that was first noticed by Maxwell and is confirmed by observation. This fact is indeed surprising, for Equation (13.33) suggests at first sight and in agreement with intuition that μ must increase with density; however, the combination $n\tau_R$ has no density dependence. Using Equations (13.32) and (13.33) we can also verify that $|\mathcal{P}'_{ij}|/\mathcal{P} \sim (\lambda_R/L)(u/\langle U \rangle)$ in agreement with our general considerations below Equation (13.21).

For a dilute gas without internal molecular structure, $c_V/\rho = 3/2(\kappa/m)$, where c_V is the constant-volume heat capacity per unit volume. From Equations (13.31) and (13.33) we therefore have

$$\frac{K}{\mu(c_V/\rho)} = \frac{5}{3} \tag{13.34}$$

for a dilute real gas. Since the relaxation time drops out of this ratio, we might expect the result to be of quantitative significance, and, in fact, the ratio is ~2.5 for the noble gases at 0°C. See Problem III.5(c).

Equation (13.33) for the viscosity coefficient μ can be understood generally for a fluid by starting from the elementary definition of μ in terms of momentum transfer perpendicular to the flow direction. In such a transport process the steady-state flux of particles perpendicular to the flow direction is the same both ways, to and fro, being $\sim\frac{1}{3}n\langle U\rangle$. Since most molecules that cross a given plane parallel to the flow direction (x-axis) originate within a mean free path λ, the net transfer of momentum per particle (in the y direction) will be proportional to the velocity gradient $\partial u_x/\partial y$ with a constant of proportionality $\sim m\lambda$. Now according to Equation (13.2) the viscosity coefficient μ is defined to be the ratio of the force per unit area F_{xy} to the velocity gradient $\partial u_x/\partial y$. Therefore, we can deduce from

$$F_{xy} = (\text{particle flux}) \times (\text{momentum transfer per particle})$$

that $u \sim \frac{1}{3}nm\lambda\langle U\rangle \sim \frac{1}{3}nm\tau\langle U\rangle^2 \sim n\tau\kappa T$, in agreement with Equation (13.33).

In passing, we note that the approximation (13.16) we have used in this section as

$$\left(\frac{\partial f}{\partial t}\right)_{\text{coll}} \cong -\frac{f^{(1)}}{\tau_R}$$

satisfies the dilute-gas conservation theorem (12.22) in the form

$$\int d^3\mathbf{k}\chi(\mathbf{k})\left(\frac{\partial f}{\partial t}\right)_{\text{coll}} \cong \int d^3k\chi(\mathbf{k})[-f^{(1)}/\tau_R] = 0$$

where $\chi(\mathbf{k})$ is any of the three conserved properties m, $\mathbf{k}$, or $k^2/2m$ listed below Equation (12.5). The proof of this claim follows from Exercise 13.1. Approximation (13.16) is therefore remarkably consistent for giving a first orientation to the calculation of f and the viscous coefficients K and μ for a dilute gas. We turn now to a more accurate approximation procedure, the Chapman–Enskog method, for calculating $f(\mathbf{r}, \mathbf{k}, t)$.

14 Chapman–Enskog Method

The purpose of the Chapman–Enskog method is to solve the Boltzmann transport equation (11.27) for a dilute real gas by a scheme of successive approximations without relying on the heuristic approximation (13.16). The method is directed toward calculating the deviations of $f(\mathbf{r}, \mathbf{k}, t)$ from the local equilibrium function $f^{(0)}(\mathbf{r}, \mathbf{k}, t)$ of Equation (12.37); it pays no attention to the short-time, or transient, function $g(\mathbf{r}, \mathbf{k}, t)$, discussed below Equation (13.12). In this section we only outline the Chapman–Enskog method and then develop the explicit equations whose solution gives the leading correction term $f^{(1)}(\mathbf{r}, \mathbf{k}, t)$.

We wish to solve the Boltzmann transport equation (11.27), which, with definition (13.14), can be written for the distribution function $f(\mathbf{r}, \mathbf{k}, t)$ as

$$\frac{\partial f}{\partial t} + \mathscr{D}(f) = J(f \mid f) \tag{14.1}$$

where we have used approximation (11.14) for the dilute-gas collision integral. The general integral $J(f \mid h)$ is defined by Equation (11.15) and satisfies theorem (12.22). As we discussed above Equation (13.12), time dependence of $f(\mathbf{r}, \mathbf{k}, t)$ is included only implicitly through the variations with time of n, T, and $\mathbf{u}$. It is assumed further now that these three quantities vary negligibly over a spatial distance comparable to the mean free path and also over a time interval comparable to the collision time.[23]

We now outline the Chapman–Enskog method for solving the approximate Boltzmann equation (14.1). The zeroth approximation in this method of solution is the local equilibrium function $f^{(0)}(\mathbf{r}, \mathbf{k}, t)$ of Equation (12.37). Moreover, we shall continue to assume that Equations (13.22) to (13.24) give the quantities n, T, and $\mathbf{u}$ quite generally, providing thereby conditions on the distribution function $f(\mathbf{r}, \mathbf{k}, t)$ as it is determined by the Chapman–Enskog method. We shall also continue to set the molecular-interaction quantities $\mathbf{q}_V$ and $\overleftrightarrow{B}$ of Equations (12.18) and (12.12), respectively, equal to zero, as we explained above Equation (12.35).

Chapman–Enskog Expansion

We introduce formally a parameter ζ, eventually to be set equal to unity, and write $f(\mathbf{r}, \mathbf{k}, t)$ as an expansion in powers of ζ,

$$f(\mathbf{r}, \mathbf{k}, t) = \sum_{n=0}^{\infty} \zeta^{n-1} f^{(n)}(\mathbf{r}, \mathbf{k}, t) \tag{14.2}$$

where the intention of the expansion is that $f^{(n)}$ gets smaller as n increases. The parameter ζ has no physical significance and is used merely to keep track of the order of terms in the series. We associate $f^{(0)}$ with a factor ζ^{-1}, so that the lowest order approximation to the Boltzmann equation (14.1) becomes simply

$$J(f^{(0)} \mid f^{(0)}) = 0 \tag{14.3}$$

As is discussed in Section 11, this last equation is equivalent to Equation (11.16) for determining the zeroth-order distribution function $f^{(0)}$ of Equation (12.37), or (11.19).

We have already indicated that the purpose of the Chapman–Enskog expansion (14.2) is to provide a method for calculating $f^{(n)}$ that is both consistent and practical. Now, according to Equations (13.22) to (13.24) we can and shall require that the $f^{(n)}$ satisfy the general condition

$$\int d^3\mathbf{k}\chi(\mathbf{r}, \mathbf{k}, t) f^{(n)}(\mathbf{r}, \mathbf{k}, t) = 0 \qquad (n \neq 0) \tag{14.4}$$

where $\chi(\mathbf{r}, \mathbf{k}, t)$ is one of the three conserved properties: $\chi = 1$, $\chi = \mathbf{U}$, and $\chi = U^2$. As demonstrated with Exercise 13.1, our approximate solution for $f^{(1)}$ in Section 13 satisfies condition (14.4), and we may anticipate no great difficulty in maintaining this condition more generally.

If the expansion (14.2) is substituted into Equations (12.31) and (12.29), for $\mathbf{q}_K$ and $\overleftrightarrow{\mathcal{P}}$, respectively, (with $B_{ij} = 0$), then we obtain

$$\begin{aligned} \mathbf{q}_K(\mathbf{r}, t) &= \sum_{n=0}^{\infty} \zeta^{n-1} q^{(n)}(\mathbf{r}, t) \\ \mathcal{P}_{ij}(\mathbf{r}, t) &= \sum_{n=0}^{\infty} \zeta^{n-1} \mathcal{P}_{ij}^{(n)}(\mathbf{r}, t) \end{aligned} \tag{14.5}$$

where according to Equation (12.6) with $P \rightarrow f$,

$$\mathbf{q}^{(n)} \equiv \frac{1}{2} m \int d^3\mathbf{k} f^{(n)} \mathbf{U} U^2 \tag{14.6}$$

$$\mathcal{P}_{ij}^{(n)} \equiv m \int d^3\mathbf{k} f^{(n)} U_i U_j \tag{14.7}$$

If Equations (13.22) to (13.24) are utilized also, then the conservation laws (12.27), (12.33), and (12.34) can be written in the following manner:

$$\frac{\partial \rho}{\partial t} + \nabla \cdot (\rho \mathbf{u}) = 0 \tag{14.8}$$

$$\rho\left(\frac{\partial}{\partial t} + \mathbf{u} \cdot \nabla\right)\mathbf{u} = \rho \mathbf{F}_{\text{ext}}/m - \sum_{n=0}^{\infty} \zeta^n \nabla \cdot \overleftrightarrow{\mathcal{P}}^{(n)} \tag{14.9}$$

$$c_V\left(\frac{\partial}{\partial t} + \mathbf{u} \cdot \nabla\right)T = -\sum_{n=0}^{\infty} \zeta^n [\nabla \cdot \mathbf{q}^{(n)} + \overleftrightarrow{\Lambda} \cdot \overleftrightarrow{\mathcal{P}}^{(n)}] \tag{14.10}$$

The fourth equality of Equation (12.35) has been used to obtain Equation (14.10). As explained in Section 12, this substitution is equivalent to the ideal-gas approximation for the energy per unit mass ($e = {}^3\!/\!_2\kappa T/m$) and for the heat capacity per unit volume ($c_V = {}^3\!/\!_2\kappa\rho/m$), where the absolute temperature T is defined microscopically by Equation (13.23). The powers of ζ in Equations (14.8) to (14.10) can be reconciled if one agrees to extract a factor of ζ^{-1} from Equation (13.22) for the particle density n.

In zeroth order we can utilize the development associated with Equation (12.37) to see that the conservation laws (14.8) to (14.10) reduce again to the nonviscous flow equations (12.38). We shall use the latter set of equations or, equivalently, the set below Equation (13.26), in the first-order solution to the Boltzmann equation discussed below.

We must be careful, in the general case, not to lose the nth order terms on the right sides of Equations (14.9) and (14.10). Thus although the three space- and time-dependent quantities ρ, T, and $\mathbf{u}$ must self-consistently satisfy Equations (13.22) to (13.24) or, equivalently, Equation (14.4), these quantities are determined ultimately through the solution to the Boltzmann equation by Equations (14.8) to (14.10). To this end we prescribe formally that whenever the partial-derivative operator $\partial/\partial t$ is applied to ρ, Θ, or u_i it shall be considered to be the following infinite series of operators,

$$\frac{\partial}{\partial t} \equiv \sum_{n=0}^{\infty} \zeta^n \frac{\partial_n}{\partial t} \tag{14.11}$$

Substitution of this expression into Equations (14.8) to (14.10) then yields the definition of ∂_n/∂_t when operating on each of the three functions ρ, Θ, and u_i. For example, $(\partial_n/\partial t)\rho \equiv 0$ when $n \neq 0$. With no effect on the end result we could just as well have chosen formally, with Equation (14.11), to decompose the operator $(\partial/\partial t + \mathbf{u} \cdot \nabla)$ instead of $\partial/\partial t$ alone.

We return finally to the Boltzmann transport equation itself. With expansions (14.2) and (14.11), Equation (14.1) becomes

$$\left(\sum_{m=0}^{\infty} \zeta^m \frac{\partial_m}{\partial t} + \mathscr{D}\right)\left(\sum_{n=0}^{\infty} \zeta^{n-1} f^{(n)}\right) = \sum_{m,n=0}^{\infty} \zeta^{m+n-2} J(f^{(m)} \,|\, f^{(n)}) \tag{14.12}$$

We derive an infinite set of integral equations for determining the $f^{(n)}$ from this one equation by matching the coefficients of equal powers of ζ on both sides. Thus the lowest order equation, which is the coefficient of ζ^{-2}, is Equation (14.3) for determining $f^{(0)}$.

First-Order Calculation

We now specialize our considerations to the determination of $f^{(1)}(\mathbf{r}, \mathbf{k}, t)$ and refer to the interested reader to the bibliography at the end of the chapter for references to more general calculations of $f^{(1)}$ and $f^{(2)}$. From Equation (14.12) the coefficient of ζ^{-1} yields the integral equation

$$\frac{\partial_0 f^{(0)}}{\partial t} + \mathscr{D}(f^{(0)}) = J(f^{(0)} \,|\, f^{(1)}) + J(f^{(1)} \,|\, f^{(0)}) \tag{14.13}$$

The condition for the existence of a solution to this inhomogeneous linear equation,[24] with $J(f \mid h)$ defined by Equation (11.15), is that both Equations (12.22) and (14.4) hold and that the conservation laws (14.8) to (14.10) are used in the consistent manner discussed in connection with definition (14.11). We do not repeat the existence proof here; rather, we assume that the solution exists and proceed directly to outline the Chapman–Enskog method for calculating $f^{(1)}$ from Equation (14.13).

We begin by introducing a function $\phi(\mathbf{r}, \mathbf{k}, t)$, which is related to $f^{(1)}(\mathbf{r}, \mathbf{k}, t)$ by the definition

$$f^{(1)}(\mathbf{r}, \mathbf{k}, t) \equiv f^{(0)}(\mathbf{r}, \mathbf{k}, t)\phi(\mathbf{r}, \mathbf{k}, t) \tag{14.14}$$

We also introduce the integral operator $\mathscr{I}$ by

$$\mathscr{I}(\phi) \equiv \frac{1}{f^{(0)}}[J(f^{(0)} \mid f^{(1)}) + J(f^{(1)} \mid f^{(0)})]$$

$$\mathscr{I}(\phi) = \int d^3\mathbf{k}_2\, d\omega_q R(\mathbf{k}_1, \mathbf{k}_2, \hat{\mathbf{q}}) f^{(0)}(\mathbf{k}_2)[\phi(\mathbf{k}_1') + \phi(\mathbf{k}_2') - \phi(\mathbf{k}_1) - \phi(\mathbf{k}_2)] \tag{14.15}$$

in which dependence on $\mathbf{r}$ and t has been suppressed in the notation and Equations (11.15) and (11.16) have both been used, the latter with $P_0 \rightarrow f^{(0)}$. On substituting Equation (14.15) into Equation (14.13) and comparing the left side of (14.13) with Equations (13.14) and (13.21), we obtain

$$\Theta^{-1}\left\{\left(\frac{m}{2\Theta}U^2 - \frac{5}{2}\right)(\mathbf{U} \cdot \nabla\Theta) + m \sum_{i,j=1}^{3} \Lambda_{ij}\left(U_iU_j - \frac{1}{3}\delta_{ij}U^2\right)\right\} = \mathscr{I}(\phi) \tag{14.16}$$

where Equation (13.27) has also been used. We recall that to derive the result (13.27) the zeroth-order form of the conservation laws (14.8) to (14.10) was used, as discussed below these equations.

Our previous solution (13.27) for $f^{(1)}$, which was based on approximation (13.16), can be recovered from Equation (14.16) by setting

$$\mathscr{I}(\phi) \cong -\tau_R^{-1}\phi(\mathbf{k}_1)$$

Examination of Equation (14.16) reveals that, since $\mathscr{I}(\phi)$ is linear in ϕ, a particular solution for the scalar quantity $\phi(\mathbf{k}_1)$ must be composed of a term linear in $\partial\Theta/\partial x_i$ and a term linear in Λ_{ij}

$$\phi_{\text{part}}(\mathbf{k}_1) = \Theta^{-1}\left[\sum_{i=1}^{3} R_i(\mathbf{k}_1)\frac{\partial\Theta}{\partial x_i} + m \sum_{i,j=1}^{3} S_{ij}(\mathbf{k}_1)\Lambda_{ij}\right] \tag{14.17}$$

where $R_i(\mathbf{k}_1)$ and $S_{ij}(\mathbf{k}_1)$ are, respectively, vector and tensor functions of Θ, ρ, and U_i. On substituting Equation (14.17) into Equation (14.16), one finds that R_i and S_{ij} satisfy separately the equations

$$\mathscr{I}(R_i) = \left(\frac{m}{2\Theta}U^2 - \frac{5}{2}\right)U_i$$

$$\mathscr{I}(S_{ij}) = \left(U_iU_j - \frac{1}{3}\,\delta_{ij}U^2\right)$$

Since the only momentum-dependent vector available is **U** (or **k**), and since S_{ij} must be a symmetric traceless tensor (see below), we may conclude that particular solutions R_i and S_{ij} must be of the form

$$R_i = -U_iF(\rho, \Theta, U^2) \tag{14.18}$$
$$S_{ij} = -\left(U_iU_j - \frac{1}{3}\,\delta_{ij}U^2\right)G(\rho, \Theta, U^2)$$

The unknown scalar functions F and G, therefore, satisfy the integral equations

$$\mathscr{I}(U_iF) = -\left(\frac{m}{2\Theta}U^2 - \frac{5}{2}\right)U_i \tag{14.19}$$

$$\mathscr{I}\left[\left(U_iU_j - \frac{1}{3}\,\delta_{ij}U^2\right)G\right] = -\left(U_iU_j - \frac{1}{3}\,\delta_{ij}U^2\right) \tag{14.20}$$

Inasmuch as the trace of the right side of this last equation vanishes, any function $\delta_{ij}G'(\rho, \Theta, U^2)$, added to S_{ij} in Equation (14.18), must satisfy the homogeneous integral equation $\mathscr{I}(G') = 0$. Such a function must therefore be included with the homogeneous solutions to Equation (14.16). But the second equality of Equation (14.15) for $\mathscr{I}(\phi)$ shows that the homogeneous solutions must, in turn, be conserved quantities $\chi(\mathbf{r}, \mathbf{k}, t)$. We now show that such solutions can be omitted.

Consider the general homogeneous solution to $\mathscr{I}(\phi) = 0$:

$$\phi_{\text{homo}}(\mathbf{k}) = \alpha_0 + \boldsymbol{\alpha}_1 \cdot \mathbf{U} + \alpha_2U^2 \tag{14.21}$$

where 1, **U**, and U^2 are the conserved properties that satisfy Equation (12.5). The complete solution for ϕ is then

$$\phi(\mathbf{k}) = \phi_{\text{part}}(\mathbf{k}) + \phi_{\text{homo}}(\mathbf{k}) \tag{14.22}$$

This solution must satisfy the constraint equation (14.4) for $n = 1$, where $f^{(1)}$ is given by Equation (14.14) and the $\chi(\mathbf{r}, \mathbf{k}, t)$ are, again, 1, **U**, and U^2 (or any linear combination of these quantities). It is straightforward to show that the three constraint equations, with $\phi(\mathbf{k})$ given by Equation (14.22), are

$$\int d^3\mathbf{U}f^{(0)}[\alpha_0 + \alpha_2U^2] = 0$$
$$\int d^3\mathbf{U}f^{(0)}\left[-\Theta^{-1}\frac{\partial\Theta}{\partial x_i}F + \alpha_{1,i}\right]U^2 = 0 \tag{14.23}$$
$$\int d^3\mathbf{U}f^{(0)}[\alpha_0 + \alpha_2U^2]U^2 = 0$$

where the tensor S_{ij} of Equations (14.18) satisfies (14.4) directly.

Exercise 14.1

(a) Use the first and third of Equations (14.23) to show that

$$\int d^3\mathbf{U} f^{(0)}(\alpha_0 + \alpha_2 U^2)^2 = 0$$

and argue, therefore, that $\alpha_0 = \alpha_2 = 0$.

(b) Use the second of Equations (14.23) to argue that the term $\boldsymbol{\alpha}_1 \cdot \mathbf{U}$ in $\phi_{\text{homo}}(\mathbf{k})$ can be incorporated into $\phi_{\text{part}}(\mathbf{k})$ if we replace condition (14.4) for $n = 1$ by the requirement

$$\int d^3\mathbf{U} f^{(0)} F U^2 = 0 \tag{14.24}$$

The complete solution for $f^{(1)}$ can now be written as

$$f^{(1)}(\mathbf{k}) = -\Theta^{-1} f^{(0)}(\mathbf{k}) \Bigg[(\mathbf{U} \cdot \nabla\Theta) F(\rho, \Theta, U^2) + m \sum_{i,j=1}^{3} \Lambda_{ij} \left(U_i U_j - \frac{1}{3}\delta_{ij} U^2\right) G(\rho, \Theta, U^2) \Bigg] \tag{14.25}$$

Comparison with our earlier solution (13.27) for $f^{(1)}$ then reveals that the relaxation-time approximation (13.16) is equivalent in Equation (14.25) to the approximations

$$F \cong \left(\frac{m}{2\Theta} U^2 - \frac{5}{2}\right) \tau_R \sim \tau_R$$
$$G \cong \tau_R \tag{14.26}$$

If Equation (14.25) for $f^{(1)}$ is substituted into the first lines of both Equations (13.28) and (13.29), then we can derive expressions for the coefficients of thermal conductivity K and viscosity μ. On referring to Equations (13.1) and (13.2) and to the argument associated with Equation (13.32), we find

$$K = \left(\frac{m}{6T}\right) \int d^3\mathbf{k} f^{(0)} U^4 F(\rho, \Theta, U^2) \tag{14.27}$$

$$\mu = \left(\frac{m^2}{\Theta}\right) \int d^3\mathbf{k} f^{(0)} U_1^2 U_2^2 G(\rho, \Theta, U^2)$$
$$= \left(\frac{m^2}{15\Theta}\right) \int d^3\mathbf{k} f^{(0)} U^4 G(\rho, \Theta, U^2) \tag{14.28}$$

These two results reduce to (13.31) and (13.33), respectively, when the approximations (14.26) are substituted.

Techniques for solving Equations (14.19) and (14.20) subject to the constraint (14.24) are given in the text by Chapman and Cowling. The solutions for F and G as functions of the velocity U are expressed in terms of Sonine polynomials. Equation (14.15) shows, as would be expected, that these solutions depend, via the transition probability of Equation (11.11), on the nature of the intermolecular potential. Calculations employing realistic potentials are in excellent agreement with experiments on dilute gases.[25] Figure III.4 shows experimental curves of K and μ for dilute samples of the noble gases.

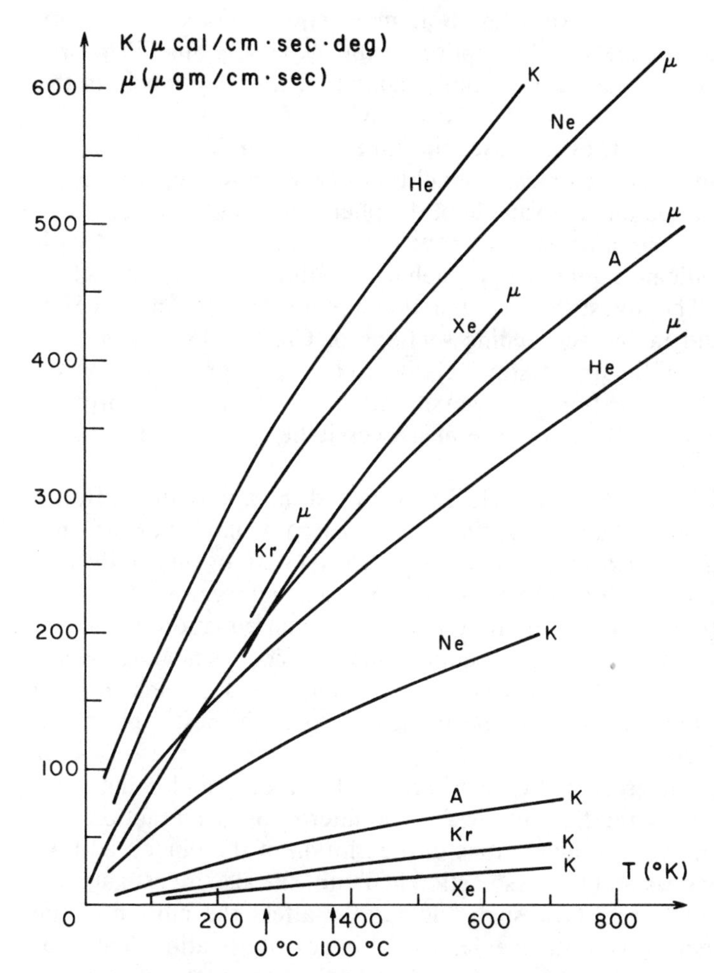

FIGURE III.4 Experimental curves for pure dilute gases showing the temperature dependence of the coefficients of thermal conductivity K and viscosity μ. Recall [see below Equation (13.33)] that both K and μ are density-independent for dilute gases. Data taken from J. O. Hirschfelder, C. F. Curtiss, and R. B. Bird, *Molecular Theory of Gases and Liquids* (Wiley, 1954), Tables 8.4-2 and 8.4-9, and from *Handbook of Chemistry and Physics* (Chemical Rubber Co., 1970), pp. F-43–46.

15 Noise and Brownian Motion

It is impressive that the Boltzmann transport equation, which we have derived only approximately from microscopic theory in Section 11, leads to quantitative agreement of calculations with a wide variety of measurements of transport coefficients. By contrast, the approximate Boltzmann equation cannot be used unambiguously in statistical mechanics to study its foundations, given that physics is time-reversible in the truly microscopic world. In other words, whereas the approximate Boltzmann equation of kinetic theory seems to provide an excellent description of the "coarse-grained" macroscopic world for dilute gases, it certainly cannot be taken seriously in the "fine-grained" microscopic world.[26] To emphasize for measurements the distinction between short, or microscopic, times (comparable with the collision time τ_{coll}) and macroscopically small times (with, however $t >> \tau_{\text{coll}}$), we consider in this section an example of the phenomenon of "noise."

The example is one-dimensional Brownian motion, but we intend with this example to indicate a general approach to the study of noise, or fluctuation phenomena. The investigation is pursued in considerable detail in Section 41. There and in the succeeding sections of Chapter IX we shall be forced to dodge the ultimate question relative to transport processes, which is the microscopic explanation of irreversibility. We shall, however, provide detailed discussions about the nature of irreversibility in both Sections 25 and 45.

In Section 13, below Equation (13.12), we used the approximate Boltzmann equation to investigate short-time, or transient, behavior of the one-particle distribution function. Although the result (13.20) of this study was satisfying, and undoubtedly correct, microscopic time intervals ($t \sim \tau_{\text{coll}}$) are precisely those for which the approximate Boltzmann equation cannot be justified. We are left with a need to understand transient behavior, even if only nonrigorously, in order, at least, to "explain" the approach to local equilibrium. The study of fluctuation phenomena gives us the best explanation.

Imagine a tiny macroscopic object immersed in a fluid and being bombarded continually by the fluid particles. The microscopic interactions are associated intimately with fluctuations in the motion of the object, and we now attempt to understand this association in detail. We also investigate, for the one-dimensional case, how such fluctuations affect the motion of the macroscopic object through the mechanism of energy dissipation. Thus our problem is to understand in detail the random walk of the macroscopic particle, that is, its Brownian motion, in one dimension.

So we consider a particle of mass m whose center-of-mass coordinate at time t is designated by $x(t)$ and whose corresponding velocity is $u \equiv dx/dt$. This particle is immersed in a fluid that is assumed to be at the absolute temperature T. Although we do not intend to consider the detailed microscopic behavior of the fluid molecules, considering them instead to act as a temperature reservoir, we allow for their dynamic effect on the macroscopic particle by introducing a net force $F(t)$ on the particle at position x. This force $F(t)$, which accounts for the microscopic interactions between fluid particles and the macroscopic particle, is some rapidly fluctuating function of time that varies in a highly irregular fashion. We may call it the "noise force," and it cannot, in general, be disregarded. Despite the fact that one cannot specify the precise functional dependence of F on t, one can calculate its average effects using statistical methods, and, of course, it is only the average effects that one measures.

In addition to the noise force $F(t)$ the macroscopic particle may also be subject to some external forces $F_{\text{ext}}(t)$, such as gravitational or electromagnetic forces. We treat the macroscopic particle using Newtonian mechanics, reserving until Section 44 the more elaborate machinery of quantum mechanics for problems involving fluctuation phenomena. We therefore apply Newton's second law to the particle, in the form

$$m\frac{du}{dt} = F_{\text{ext}}(t) + F(t) \tag{15.1}$$

Since $F(t)$ is a random force, it is appropriate to average this last expression over an ensemble of identical systems, all at time t and each having a macroscopic particle immersed at position $x(t)$ with velocity $u(t)$. We then assume that both $u(t)$ and the external force vary slowly over a macroscopically small time interval τ, where $\tau >> \tau_{\text{coll}}$, and integrate the ensemble-averaged Equation (15.1) over the time interval τ. The result is

$$m(\langle u(t + \tau)\rangle - \langle u(t)\rangle) = F_{\text{ext}}(t) \cdot \tau + \int_t^{t+\tau} dt' \langle F(t')\rangle \tag{15.2}$$

where the brackets $\langle\ \ \rangle$ indicate ensemble averages.

At first sight one might expect that $\langle F(t)\rangle = 0$ for the rapidly varying noise force. Indeed, when $\langle u(t)\rangle = 0$ this is the case. But when $\langle u(t)\rangle \neq 0$, then there is a slowly varying part of $F(t)$ that depends linearly (for small velocities) on the slowly varying part of $u(t)$. One way of seeing that $\langle F(t)\rangle \neq 0$ when $\langle u(t)\rangle \neq 0$ is to observe that in this case the ensemble-averaging process itself is time-dependent, because as the particle slows down energy is transferred to the fluid, thereby changing its thermodynamic state. One can show[27] that this effect has the primary consequence of changing the time-dependent ensemble average of $F(t')$ to a time-independent average over $F(t')F)t'')$, which does not vanish. For the present we do not examine further the reasons why $\langle F(t)\rangle \neq 0$ and assume merely that the time integral over $\langle F(t')\rangle$ has the form, for small velocities,

$$\int_t^{t+\tau} dt' \langle F(t')\rangle = -\alpha\langle u(t)\rangle\tau \qquad (\tau >> \tau_{\text{coll}}) \tag{15.3}$$

where α is a damping constant. That $\langle F \rangle$ and $\langle u \rangle$ have opposite signs is physically reasonable, as the natural consequence of innumerable microscopic interactions between the macroscopic particle and the fluid must be damping of the macroscopic particle's motion. Damping must also be related macroscopically to the viscosity μ of the fluid, and, in fact, we cite below a particular relationship between α and μ.

Substitution of Equation (15.3) into Equation (15.2) gives the results

$$m[\langle u(t + \tau) \rangle - \langle u(t) \rangle] = [F_{\text{ext}}(t) - \alpha \langle u(t) \rangle]\tau$$

which can be rewritten for small τ as

$$m\frac{d\langle u(t) \rangle}{dt} = F_{\text{ext}}(t) - \alpha \langle u(t) \rangle \tag{15.4}$$

Here, as in Equation (15.2), we may use the fact that for a slowly varying variable, such as $u(t)$ in this case, the operations d/dt and $\langle \quad \rangle$ commute:

$$\frac{d}{dt}\langle A \rangle = \left\langle \frac{dA}{dt} \right\rangle \qquad \text{[for slowly varying } A(t)] \tag{15.5}$$

We now apply the concepts we introduced in the preceding two paragraphs to our first equation (15.1) so that it can be used as a starting point for subsequent developments. Thus upon separating the noise force $F(t)$ into two parts,

$$F(t) = -\alpha u(t) + F'(t) \tag{15.6}$$

where $F'(t)$ is a rapidly fluctuating part whose ensemble average vanishes irrespective of the position or velocity of the small particle, we obtain from Equation (15.1) the so-called *Langevin equation* for one-dimensional Brownian motion,

$$m\frac{du}{dt} = F_{\text{ext}}(t) - \alpha u(t) + F'(t) \tag{15.7}$$

This equation is applied below to the calculation of quantities of physical interest.

If the macroscopically small particle under consideration is spherical, with radius a, then for an incompressible fluid the damping coefficient α is related to the coefficient of viscosity μ by *Stoke's law*

$$\alpha = 6\pi\mu a \tag{15.8}$$

assuming that $\rho\langle u \rangle a << \mu$. The derivation of Stoke's law is given as Problem III.6.

Exercise 15.1

(a) A simple electrical example of Equation (15.7) occurs for a particle with electric charge q within a medium that is placed in a uniform electric field $\mathscr{E}$. Perform an ensemble average of Equation (15.7) for this example and consider the steady-state situation. Show that the electrical mobility, defined by

$$\mu_e \equiv \frac{\langle u \rangle}{\mathscr{E}}$$

is given by

$$\mu_e = \frac{q}{\alpha}$$

(b) A free electron in a wire moves under the influence of a constant electric field $\mathscr{E}$. Although the electron is not a macroscopic particle, it will have an ordered (macroscopic) motion in such a situation. Then according to part (a) the average velocity of the electron under steady-state conditions is equal to $-e\mathscr{E}/\alpha$. Show that the conductivity σ of the wire is given by

$$\sigma = \frac{ne^2}{\alpha}$$

To estimate α we can equate $\alpha\langle u \rangle$ with the average rate of momentum transfer, $\sim m\langle u \rangle \tau_i^{-1}$, to an electron in ion collisions, where τ_i is the average time between ion collisions. Thus $\alpha \sim m/\tau_i$.[28]

The electrical example of Exercise 15.1 can be pursued further, with analogies to both Equations (15.1) and (15.4). If an electrical conductor has a self-inductance L and carries a current I, then the analogy to Equation (15.1) is

$$L\frac{dI}{dt} = \mathscr{V}(t) + V(t)$$

where $\mathscr{V}(t)$ denotes any applied electromotive force (emf) and $V(t)$ is the effective fluctuating emf. The fluctuating potential represents the interaction energy per unit charge between a conduction electron and all random microscopic influences on it in the conductor. The ensemble average of this last equation is, in analogy with Equation (15.4),

$$L\frac{d\langle I \rangle}{dt} = \mathscr{V} - R\langle I \rangle$$

where R is the electrical resistance.

Calculation of $\langle x^2 \rangle$

We now consider a special case of Equation (15.7) in which the total system (fluid plus particle) is at thermodynamic equilibrium with no external forces present. If we choose the origin of the x-axis to coincide with the location of the macroscopically small particle at $t = 0$, then by symmetry we must have $\langle x(t) \rangle = 0$. To calculate the mean-square displacement $\langle x^2 \rangle$ we first multiply Equation (15.7), with $F_{\text{ext}} = 0$, by $x(t)$ to get

$$mx\frac{d^2x}{dt^2} = m\left[\frac{d}{dt}\left(x\frac{dx}{dt}\right) - \left(\frac{dx}{dt}\right)^2\right] = -\alpha x\frac{dx}{dt} + xF'(t) \tag{15.9}$$

We then perform an ensemble average of this equation, for which the last term vanishes as was discussed below Equation (15.6).[29] Upon using Equation (15.5) for particle variables and the equipartition theorem for one-dimensional motion,[30]

$$\frac{1}{2}m\left\langle\left(\frac{dx}{dt}\right)^2\right\rangle = \frac{1}{2}\kappa T$$

we can write the ensemble average of Equation (15.9) as

$$m\frac{d}{dt}\left\langle x\frac{dx}{dt}\right\rangle = m\left\langle\frac{d}{dt}\left(x\frac{dx}{dt}\right)\right\rangle = \kappa T - \alpha\left\langle x\frac{dx}{dt}\right\rangle \tag{15.10}$$

Equation (15.5) can be used freely for particle variables, x and its time derivatives, because these quantities all vary slowly on a microscopic, or macroscopically small, time scale.

The solution to the simple differential equation (15.10) is

$$\left\langle x\frac{dx}{dt}\right\rangle = Ce^{-\gamma t} + \frac{\kappa T}{\alpha}$$

where $\gamma \equiv (\alpha/m)$. Thus γ^{-1} is a characteristic time constant for the system. The constant of integration C can be determined by assuming, as above, that $x = 0$ at $t = 0$. Then

$$\frac{1}{2}\frac{d}{dt}\langle x^2\rangle = \left\langle x\frac{dx}{dt}\right\rangle = \frac{\kappa T}{\alpha}\left(1 - e^{-\gamma t}\right) \tag{15.11}$$

which can be integrated once again to produce the final result

$$\begin{aligned}\langle x^2\rangle &= \frac{2\kappa T}{\alpha}[t - \gamma^{-1}(1 - e^{-\gamma t})] \\ &\to \left(\frac{\kappa T}{m}\right)t^2 \qquad \text{for } t \ll \gamma^{-1} \\ &\to \left(\frac{2\kappa T}{\alpha}\right)t \qquad \text{for } t \gg \gamma^{-1}\end{aligned} \tag{15.12}$$

From the second line of the result (15.12) we see that for short time intervals the small macroscopic particle moves as though it were a free particle with the constant thermal velocity $(\kappa T/m)^{1/2}$. On the other hand, after long times such that $t >> \gamma^{-1}$, the particle is displaced by an rms distance $x_{rms} \propto t^{1/2}$, which is the one-dimensional version of the random-walk result (1.12). Using Stoke's law, (15.8), the constant α in Equation (1.12) is readily found from the third line of Equation (15.12) to be $(\kappa T/3\pi\mu a)^{1/2}$.

In this section we have seen, with Equations (15.2) and (15.3), an example of how random microscopic events (collisions) can be averaged to give measurable consequences at the "coarse-grained level." Continually fluctuating microscopic influences on a macroscopic motion result in its dissipation, promoting thereby a more complete equilibrium condition. We might even visualize, more generally, how such a randomizing process is associated with the rapid approach of a fluid to local equilibrium. Alternatively, random fluctuations away from the equilibrium condition of a system constitute a background of "noise," which imposes limitations on the possible accuracy of delicate physical measurements. In Chapter IX we give a more general treatment of the fluctuation-dissipation process and its physical consequences.

Bibliography

S. Chapman and T. G. Cowling, *The Mathematical Theory of Non-Uniform Gases.* (Cambridge University Press, 1939), Chap. 7 ff. In this text the Chapman–Enskog method is developed from its original perspective, with explicit solutions included.

K. Huang, *Statistical Mechanics* (Wiley, 1967), Chaps. 3, 5, and 6. An excellent introduction to the problems and methods of kinetic theory as it applies to the very dilute gas. Much of our treatment of the subject in this chapter is adapted from Huang's coverage.

F. Reif, *Fundamentals of Statistical and Thermal Physics* (McGraw-Hill, 1965), Chaps. 12–14. An introduction to transport phenomena with considerable analysis based on the Boltzmann equation. Our treatment of noise and Brownian motion in Section 15 is taken from Sections 15.5–15.7 of this text.

G. E. Uhlenbeck and G. W. Ford, *Lectures in Statistical Mechanics* (American Mathematical Society, 1963), Chap. VI. An outline of the Chapman–Enskog method, with emphasis on points of mathematical interest.

G. H. Wannier, *Statistical Physics* (Wiley, 1966), Part III. A different presentation of kinetic theory, with emphasis on the relaxation-time concept and comparisons with experiment.

Notes

1. In Chapter V, where some of the foundations of kinetic theory are outlined, we use the notation $P_1(\mathbf{r}, \mathbf{k}, t)$ instead of $P(\mathbf{r}, \mathbf{k}, t)$ for the one-particle distribution function. The notation $f(\mathbf{r}, \mathbf{v}, t)$ is also frequently used, but in this text we reserve the symbol f for the slowly-varying part of P (see Sections 13 and 14). We continue to use $\mathbf{k} = m\mathbf{v}$ for a single-particle momentum value.

2. Of course, the theoretical explanation for particle interactions is quantum mechanical in part. The consequence of particle indistinguishability in the classical domain is discussed at the end of Section 4; there is no direct effect on classical kinetic theory.

3. We ignore inelastic processes in this chapter. In general, these only occur at temperatures much higher than room temperature.

4. See, in this connection, K. R. Symon, *Mechanics* (Addison-Wesley, 1964), p. 180, where it is shown that only constant external forces that are directly proportional to the particle mass cancel out in the CM equation of motion.

5. An explicit proof of the equality of these volume elements, that is, that the Jacobian of the transformation is unity, is given in D. ter Haar, *Elements of Thermostatics* (Holt, Rinehart and Winston, 1966), Section 1.4. The equality follows, more generally, because it is an "integral invariant" of Poincaré. See H. Goldstein, *Classical Mechanics* (Addison-Wesley, 1965), Sections 8-3 and 9-1.

6. The literal translation of *Stosszahlansatz* is collision-number hypothesis. This hypothesis received considerable attention in discussions at the turn of the century, by Boltzmann and others, of the foundations of kinetic theory. See P. and T. Ehrenfest, *The Conceptual Foundations of the Statistical Approach in Mechanics,* translated by M. J. Moravesik (Cornell University, 1959).

7. This "necessity" proof is given in Section 4.1 of K. Huang, *op. cit.*

8. When $\mathbf{u} \neq 0$, we can remove an uninteresting contribution to the angular momentum and write, with equal validity in Equation (11.17), $\mathbf{L} = \mathbf{r} \times (\mathbf{k} - m\mathbf{u})$. See also Problem V.2.

9. Actually, our earlier treatment in Section 4, ending with Exercises 4.3 and 8.4, has already demonstrated that $\beta = (\kappa T)^{-1}$. More directly, one can make a kinetic calculation of the pressure, to show for an ideal gas that $\mathcal{P} = \frac{2}{3}n\langle\omega\rangle$, and then make the final identification in Equation (11.21) by using the ideal-gas law $\mathcal{P} = n\kappa T$. See F. W. Sears, *Thermodynamics* (Addison-Wesley, 1959), Section 11.4.

10. See the second reference in footnote 5.

11. When $\mathbf{k}_o = m\mathbf{u} \neq 0$, then in Equation (11.27) we must make the replacement $\mathbf{k} \rightarrow \mathbf{k} - m\mathbf{u}$, because our derivation considered a system at rest. See also Problem III.1.

12. See R. Resnick and D. Halliday, *Physics, Part I* (Wiley, 1966), Section 24.1, for an elementary derivation of this result. The $\sqrt{2}$ factor occurs when one takes into account, as we have done, the relative motion of colliding molecules.

13. J. H. Irving and J. G. Kirkwood, *J. Chem. Phys.* **18** (1950), p. 817.

14. Equations (12.12) for $\overleftrightarrow{B}(\mathbf{r}, t)$ and (12.18) for $\mathbf{q}_V(\mathbf{r}, t)$ are exact only for classical fluids that are near equilibrium. The (generally excellent) approximation involved for nonequilibrium systems is discussed below Equation (D.12) in Appendix D.

15. Transformation to a local (primed) coordinate system involves the macroscopic transformation equation $\mathbf{r}' = \mathbf{r} - \mathbf{u}t$. In this coordinate system the second term of (12.26) eliminates time-derivative contributions due to the $-\mathbf{u}t$ terms. The operator (12.26) is also referred to as a material derivative.

16. Problem X.1(c) gives this identification of the pressure, that is, that $\mathcal{P}(\mathbf{r}, t) = \frac{1}{3}[\rho\langle U^2\rangle + Tr(B_{ii})]$, by an independent derivation.

17. See D. F. Fitts, *Nonequilibrium Thermodynamics* (McGraw-Hill, 1962), Chapter 2, for multicomponent generalizations of these equations derived using macroscopic concepts.

18. The equation of state for imperfect gases is treated in Sections 38 and 39.

19. We do not require the explicit expression (11.29) for $P_0(\mathbf{r})$. Instead we write the quantity $P_0(\mathbf{r})$ as the local density $n(\mathbf{r}, t)$ [see Equation (11.18)].

20. H. B. Phillips, *Vector Analysis* (Wiley, 1949), Sections 80 and 83.

21. A microscopic basis for kinetic coefficients, which emphasizes their connection to macroscopic irreversibility in nature, is presented in Section 42.

22. A phenomenological derivation of the Navier–Stokes equation, assuming that the bulk-viscosity coefficient $\phi = 0$, is given in K. Huang, *loc. cit.*, Section 5.7.

23. This assumption was discussed below Equation (13.21). It can be used to justify the replacement of Equation (13.19) by Equation (13.19a).

24. The general existence proof for the determination of $f^{(n)}$ from the integral equation that is the coefficient of ζ^{n-2} in Equation (14.12) is given in Section 6.3 of K. Huang, *loc. cit.*

25. Extensive calculations of transport coefficients and comparisons with experiment for dilute gases are summarized in the comprehensive text by J. O. Hirschfelder, C. F. Curtiss, and R. B. Bird, *Molecular Theory of Gases and Liquids* (Wiley, 1954).

26. The exact Boltzmann transport equation is derived in Section 23 and used in Appendix D to derive the rigorous conservation laws of Section 12. In Sections 14 and the second half of 13, only consequences of the approximate Boltzmann equation (14.1) have been explored.

27. See Equation (41.3) and the proof preceding it in Section 41.

28. We can use Equation (12.3) without the $\sqrt{2}$ factor, since the ions are essentially fixed (see footnote 12), to make the estimate $\tau_i^{-1} = n\sigma_i\langle u\rangle$, where σ_i is the average electron–ion collision cross section. If electron–electron collisions predominate, then the entire calculation must be done as at the end of Section 13.4 in F. Reif, *op. cit.*

29. Although the ensemble average of $xF'(t)$ vanishes, it is not true that all averages of the rapidly fluctuating noise term $F'(t)$ vanish. In particular, one can integrate the Langevin equation (15.7), square the result, and then average over an equilibrium ensemble to show that

$$\langle F'(t)F'(t')\rangle_0 = 2\alpha\kappa T\,\delta(t - t')$$

The proof uses Equations (41.17) and (41.18) for a particle with $\langle u\rangle = 0$.

30. The equipartition theorem of classical statistical mechanics is derived in Section 28. The special case of interest here is Equation (28.16) with $Ns = 1$.

31. L. D. Landau and E. M. Lifshits, *Statistical Physics* (Addison-Wesley, 1958), p. 73.

32. The total scattering cross section of hard spheres at low energies is four times its classical value. See L. I. Schiff, *Quantum Mechanics*, 3rd ed. (McGraw-Hill, 1968), p. 125.

33. W. Pospišil, *Ann. Phys.* **83** (1927), p. 735.

Problems

III.1. Consider a system rotating with angular velocity $\mathbf{\Omega}$, but at rest translationally ($\mathbf{u} = 0$). Its internal motion can be described by using, in addition to the molecular interaction terms, a single-particle Hamiltonian of the form[31]

$$H_o = K + U(r) - \mathbf{\Omega} \cdot (\mathbf{r} \times \mathbf{k})$$

where $\mathbf{r}$ and $\mathbf{k} = m\mathbf{v}$ are laboratory coordinates, $K = k^2/2m$, and $U(\mathbf{r})$ is the external potential energy.

(a) Use Hamilton's equations to deduce the rotating-system replacements for the Boltzmann transport equation (11.27):

$$\frac{\mathbf{k}}{m} = \mathbf{v} \rightarrow \mathbf{v} + (\mathbf{r} \times \mathbf{\Omega})$$

$$\mathbf{F}_{\text{ext}} \rightarrow \mathbf{F}_{\text{ext}} + (\mathbf{k} \times \mathbf{\Omega})$$

(b) Use the result of part (a) to deduce the Maxwell–Boltzmann distribution for rotating systems with $\mathbf{u} = 0$. Assume an exponential form.

III.2. *Radiometer*

A square vane of area 1 cm^2, painted white on one side and black on the other, is attached to a vertical axis and can rotate freely about it as shown. Assume that the vane can support a temperature difference between its black and white sides. Suppose the arrangement is placed in He gas at room temperature and sunlight is allowed to shine on the vane. Explain qualitatively why:

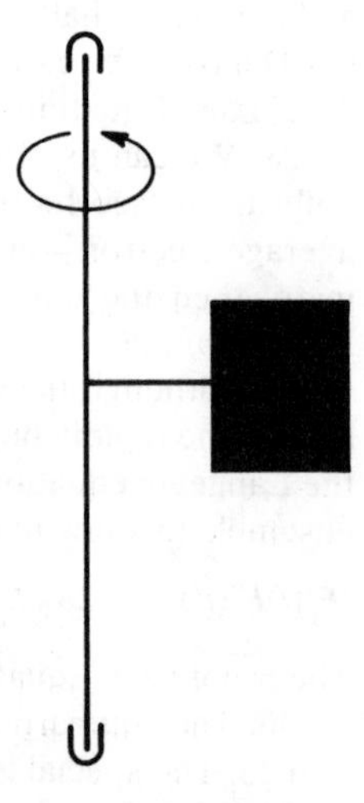

(a) At extremely small densities the vane rotates.

(b) At some intermediate (very low) density the vane rotates in a sense opposite to that in (a); estimate this intermediate density and the corresponding pressure.

(c) At a higher (but still low) density the vane stops.

III.3. *Electrical conduction*

The results of Exercise 15.1 can be derived from the approximate Boltzmann transport equation (13.18) as follows:

(a) Assume a steady-state distribution function f, which is position-independent, and an external force $\mathbf{F}_{\text{ext}} = q\mathscr{E}\hat{\imath}_z$. Show that

$$f \cong f^{(0)} - q\mathscr{E}\tau_R \frac{\partial f^{(0)}}{\partial k_z}$$

where the relaxation time τ_R can here be identified as m/α (see argument at the end of Exercise 15.1).

(b) Use the definition of electrical mobility μ_e in Exercise 15.1, along with Equation (12.6), to prove that

$$\mu_e = \frac{q}{m}\left\langle \frac{\partial}{\partial k_z}(\tau_R k_z) \right\rangle \xrightarrow[\tau_R = \text{const.}]{} \frac{q\tau_R}{m}$$

a result that does not depend on the form, that is, on the statistics, of the equilibrium distribution $f^{(0)}$.

III.4. Equation (12.3) provides an estimate of the relaxation time τ_R for a dilute gas. Deduce two different estimates for τ_R by substituting the relaxation-time approximations (14.26), that is,

$$F(\rho, \theta, U^2) \cong \left(\frac{m}{2\theta}U^2 - \frac{5}{2}\right)\tau_F(\rho, \theta)$$

$$G(\rho, \theta, U^2) \cong \tau_G(\rho, \theta)$$

into Equations (14.19) and (14.20) of the Chapman–Enskog method. To deduce explicit expressions for τ_F and τ_G, multiply Equations (14.19) and (14.20) through by suitable functions of $\mathbf{U}(\mathbf{k}_1)$ and integrate over all $d^3\mathbf{k}_1$ so that Equations (13.31) and (13.33) can be utilized for the right-side integrations. The left-side integrations are tedious but can be performed in part by transforming to CM and relative coordinates [see Equations (11.3)] and then arbitrarily selecting particular laboratory components i and j. It is also convenient to choose the (ξ, η, ζ) relative-coordinates system of Figure III.5 and to express

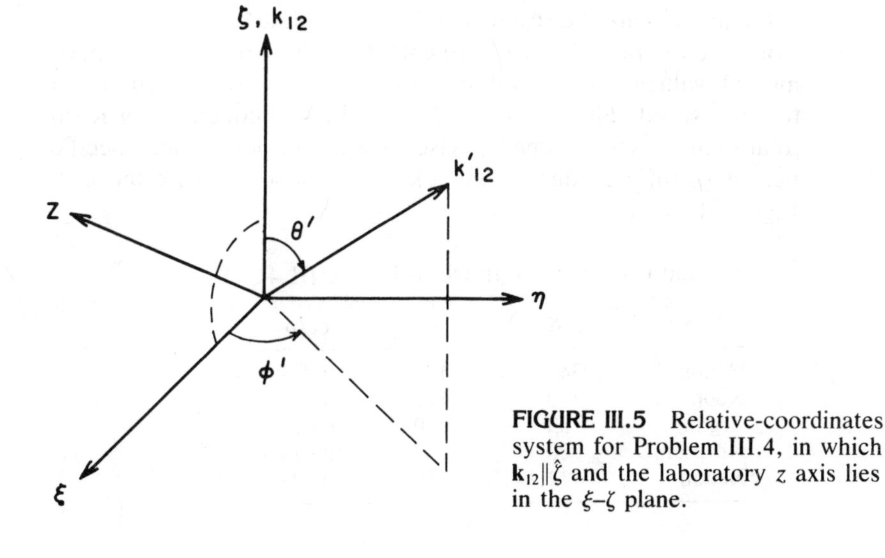

FIGURE III.5 Relative-coordinates system for Problem III.4, in which $\mathbf{k}_{12} \| \hat{\zeta}$ and the laboratory z axis lies in the ξ–ζ plane.

laboratory components in this system. Note that the angle Θ_q of $\hat{\mathbf{q}}$ is related to the angle Θ' of $\mathbf{k}'_{12}$ by $\Theta' = \pi - 2\Theta_q$.

The final answers are

$$\tau_F = \frac{15}{8}(\sqrt{2}n\bar{\sigma}\langle v\rangle)^{-1} \quad \text{and} \quad \tau_G = \frac{5}{4}(\sqrt{2}n\bar{\sigma}\langle v\rangle)^{-1}$$

where $\bar{\sigma}$ is an average total scattering cross section defined in terms of the differential cross section $\sigma(k_{12}, \hat{\mathbf{q}})$ of Equations (11.5) and (11.11) by

$$\bar{\sigma} \equiv 8\int_0^\infty x^7\, dx\, e^{-2x^2} \int d\omega_q \sin^2\Theta' \sigma(\sqrt{2m\kappa T}x, \hat{\mathbf{q}})$$

$$= 8\int_0^\infty x^7\, dx\, e^{-2x^2} \int d\omega' \sin^2\Theta' \sigma(\sqrt{2m\kappa T}x, \Theta')$$

III.5. The result of Problem III.4 gives an improvement in the approximate expressions (13.31) and (13.33), for the coefficients of thermal conductivity K and viscosity μ respectively. We can replace τ_R by τ_F in the former expression and τ_R by τ_G in the latter one.

(a) For a dilute gas of hard-sphere molecules of diameter a the average total scattering cross section of Problem III.4 is $\bar{\sigma} = \sigma_o \equiv 4\pi a^2$.[32] Show that the viscosity for this model gas is given by

$$\mu \cong \frac{5}{16}(\pi m\kappa T)^{1/2}\sigma_o^{-1}$$

(b) Use the answers to Problem III.4 to calculate the ratio $K/[\mu(c_V/\rho)]$, and compare the result with Equation (13.34).

(c) Compare the ratio $K/[\mu(c_V/\rho)]$ calculated in part (b) with experimental values for several monatomic gases. Appended are a few measured values at $T = 273$K of thermal conductivity K (in μcal · cm^{-1} · sec^{-1} · deg^{-1}), viscosity μ (in μpoise), and specific heat c_V/ρ (in cal · deg^{-1} · gm^{-1}). The data sources are those of Figure III.4.

The data sources are those of Figure III.4.

Gas	K	μ	c_V/ρ
Helium	343	186	0.753
Neon	110	302	0.150
Argon	39	210	0.076
Krypton	20	233	0.036
Xenon	12.4	210	0.023

III.6. *Stoke's law*

The damping force $\mathbf{F}$ with which an incompressible fluid of viscosity μ retards a sphere of radius a moving with velocity $\mathbf{u}_o$ is given by *Stoke's law*

$$\mathbf{F} = -6\pi\mu a u_o\left[1 + \frac{3}{8}\eta + O(\eta^2)\right]$$

assuming that $\eta \equiv \rho u_o a/\mu << 1$.

(a) Show for an incompressible fluid that the Navier–Stokes equation can be written, under steady-state conditions, as

$$\nabla^2 u - \frac{1}{\mu}\nabla\mathcal{P} = \frac{\rho}{\mu}(\mathbf{u}\cdot\nabla)\mathbf{u}$$

with

$$\nabla^2\mathcal{P} = -\rho\,\nabla\cdot(\mathbf{u}\cdot\nabla)\mathbf{u}$$

Argue that the right sides of these two equations can be neglected, in first approximation, when $\eta << 1$.

(b) Solve the coupled differential equations of part (a) to leading order when $\eta << 1$. Choose the coordinate system in which the sphere is at rest with a boundary condition that $\mathbf{u}(\mathbf{r}) = 0$ at the surface of the sphere. Show that

$$\mathcal{P}(r, w) = \mathcal{P}_o - \frac{3}{2}(\mu u a)\left(\frac{w}{r^2}\right) \xrightarrow[r>>a]{} \mathcal{P}_o$$

$$\mathbf{u}(r, w) = u_o\left[\hat{\imath}_z\left(1 - \frac{a}{r}\right) + \frac{1}{4}(r^2 - a^2)\nabla_r\left(\frac{w}{r^2}\right)\right] \xrightarrow[r>>a]{} u_o\hat{\imath}_z$$

where $w = \cos\Theta$ and (r, Θ) are spherical coordinates. The gradient operator is $\nabla_r = \hat{\imath}_r a\partial/\partial r$.

(c) Prove that the fluid force $\mathbf{f}(\mathbf{r})$ per unit area acting on the sphere is equal to

$$\mathbf{f}(\mathbf{r}) = -\hat{\imath}_r\cdot\overleftrightarrow{\mathcal{P}} = -\hat{\imath}_r\mathcal{P} + \frac{\mu}{r}[\nabla(\mathbf{r}\cdot\mathbf{u}) - \mathbf{u} + (\mathbf{r}\cdot\nabla)\mathbf{u}]$$

where the pressure tensor $\overleftrightarrow{\mathcal{P}}$ is given by Equation (13.2). Then use $\mathbf{F} = \oint d\mathbf{a}\cdot\mathbf{f}$ and the results of part (b) to derive the leading term of Stoke's law.

III.7. Pospišil[33] observed the Brownian motion of soot particles of radius 0.4×10^{-4} cm. The particles were immersed in a water–glycerin solution having a viscosity of 0.0278 gm $\cdot$ cm$^{-1}\cdot$ sec^{-1} at the temperature 18.8°C of the experiment. The observed mean-square (horizontal) x component of displacement in a 10-second time interval was $\langle x^2\rangle = 3.3 \times 10^{-8}$ cm^2.

(a) Use these data to calculate Boltzmann's constant.

(b) Compute the sedimentation velocity. Would you expect to be able to observe the barometer formula [see Problem II.5(c)] for these suspended particles?
Answer: 1.2×10^{-5} cm/sec.

III.8. *Diffusion*

(a) Consider the Boltzmann transport equation for a dilute gas in the absence of external forces, and make the relaxation-time approximation for the collision integral. Assume that there are density gradients at $t = 0$ and write in zeroth approximation

$$P^{(0)}(\mathbf{r}, \mathbf{k}) \cong n(\mathbf{r})P_0(\mathbf{k}) \qquad (t = 0)$$

where $P_0(\mathbf{k})$ is the Maxwell–Boltzmann distribution normalized to unity in momentum space and

$$\int d^3\mathbf{r} n(\mathbf{r}) = N$$

Derive an approximate expression for $P^{(1)} \equiv P - P^{(0)}$.

(b) Show that the drift velocity $\mathbf{V}$ by which molecules reduce density gradients in the gas is given by

$$n\mathbf{V} \cong -\tau_R\left(\frac{\kappa T}{m}\right)\nabla n$$

where τ_R is the relaxation time.

(c) Now let the density n be a function of t and use the continuity equation and the result of (b) to derive the diffusion equation

$$\frac{\partial n}{\partial t} = D\,\nabla^2 n$$

where $D = \tau_R(\kappa T/m)$ is the diffusion constant. *Note:* Assuming that n is time-dependent does not change the answer to part (b).

CHAPTER IV

MANY-BODY PROBLEM IN QUANTUM MECHANICS

A quantum mechanical description of the microscopic world of matter is necessitated by the existence of observables that cannot be understood from the classical viewpoint. When the particles of a system are numerous, the calculation of system wave functions and energy levels, starting from an appropriate Schrödinger equation, is referred to as the *many-body problem*. Quite apart from purely statistical considerations, we are interested in solutions to the many-body problem, if only to be able to enumerate energy levels and to characterize symmetries of the associated wave functions. The many methods that have been developed to solve these problems all have their origins in ordinary time-independent perturbation theory, the subject of Section 18. To simplify application of this and other formalisms, we first develop the notation of the occupation-number representation in Fock space. The notation and methods are illustrated by application to a nonideal Fermi gas. Finally, in Section 20 we introduce single-particle Green's functions for describing many-body systems quantum mechanically.

16 Indistinguishability of Identical Particles, State Descriptions

We begin with a discussion of identical particles and how they can be characterized. The quantum-mechanical concept that identical particles must always be treated as being indistinguishable in many-body systems is emphasized. One important consequence of indistinguishability for systems of identical particles is that many-body wave functions must be totally symmetric or antisymmetric in all of the particle coordinates. This feature of the many-body problem is introduced from the outset, and it leads naturally to use of the occupation-number representation in the many-body problem. A discussion of the correct method for counting many-body states when single-particle state descriptions are used is given at the end of the section.

Identical particles are particles that cannot be distinguished by any intrinsic property, such as mass, charge, or spin. Sharply defined trajectories, although sufficient to identify or distinguish classical particles, do not exist in the quantum domain. In quantum mechanics finite size and the spreading of wave packets result in overlap of particle paths, especially if the particles interact appreciably. Moreover, the indistinguishability of identical particles has important consequences in many-body systems when, as for quantum fluids, the so-called "particle statistics" becomes important. An important measure of wave-function overlap is provided in this case by the thermal wavelength λ_T of Equation (4.2). Because the thermal wavelength increases as the temperature is lowered, the use of quantum-mechanical descriptions and the occurrence of specific quantum phenomena are ordinarily restricted to many-body systems at very low temperatures.

The definition of a *particle* itself is subject to arbitrariness. Thus, apart from associated quantum numbers due to internal structure, the composite nature of particles such as atoms or molecules may be unimportant for the description of many-body systems. An extreme example is the many-body description of an atomic nucleus in which the effective role of mesons can be incorporated into the two-nucleon interaction to good approximation. It is unnecessary to consider three-, or more, body interaction forces in this example, because nucleons in nuclei are, on the average, sufficiently separated that their single-particle character is maintained. In closer proximity the overlapping meson clouds of, say, three nucleons would result in interaction forces that could not be represented merely as the sum of three pairs of two-body forces.

To give an explicit representation of the Hilbert space for a many-body system we should have to solve the Schrödinger equation for N interacting particles. But definitive solutions, even for the case $N = 3$, have never been written down. Since it is impossible to characterize the N-particle Hilbert space explicitly in this fashion, we must proceed from a different vantage point. We assume, in fact, that single-particle product functions can always be used to span the complete N-particle space. The choice of the appropriate single-particle functions, with quantum numbers μ, will, of course, vary from system to system.

The quantum number μ is used here to denote a composite of all translational and internal quantum numbers for a single particle. Often we are interested in infinite fluids (gases and liquids), for which there are no preferred spatial sites for the constituent particles. Momentum space is then the natural single-particle space for such systems. Thus for fluids in the *thermodynamic limit* (density $n = N/\Omega$ = finite as N, $\Omega \to \infty$, where Ω = volume of system), μ represents three momentum values (k_x, k_y, k_z) in addition to spin and any other intrinsic quantum numbers.

The notation must now be expanded. We first attach a subscript i to the quantum number μ, to denote the ith single-particle state of the μ representation. In addition, we shall have to use a superscript (α) to specify that the state $\mu_i^{(\alpha)}$ is associated with the αth particle, where $\alpha \varepsilon \{1, 2, \cdots, N\}$. Of course, as soon as a complete notation for describing identical-particle systems has been outlined, then it will no longer be necessary, or even appropriate, to retain the superscript (α).

We consider now a system of N identical particles, reserving until Chapter VIII the further complication of multicomponent systems in which there are different kinds of particles. Using the Dirac notation for state vectors, we could write an N-particle product state as

$$|\mu_1^{(1)}\rangle|\mu_2^{(2)}\rangle \cdots |\mu_N^{(N)}\rangle$$

in which the αth particle is in the ith single-particle state and both labels α and i have been numbered sequentially from the left. However, in view of the identity of the N particles, we could, with equal justification, write the N-particle product state as

$$U_P|\mu_1^{(1)}\rangle \cdots |\mu_N^{(N)}\rangle$$

where U_P is the *unitary permutation operator* corresponding to the permutation $\mathcal{P}$ of the N particle labels α.[1] More particularly, the fact that U_P commutes with the N-body Hamiltonian $H^{(N)}$, that is, $[H^{(N)}, U_P] = 0$, implies that all permutations of a given unsymmetrized solution to the Schrödinger equation give equally valid solutions for possible motions of the system. As we discuss below, however, only two linear combinations of these permuted solutions need be considered for the description of systems of identical particles.

A general permutation $\mathcal{P}$ can be constructed from a sequence of binary interchanges among the particles.[2] Although this sequence will not be unique, the number of binary interchanges will always be either even or odd for a given $\mathcal{P}$, so that $\mathcal{P}$ itself can always be characterized as even or odd. We let the number P be 0 (zero) for even permutations and 1 (unity) for odd permutations of N objects, and we consider the double-valued quantity $\varepsilon = \pm 1$. Then the two linear combinations of interest are

$$\sum_{\mathcal{P}} \varepsilon^P U_P|\mu_1^{(1)}\rangle|\mu_2^{(2)}\rangle \cdots |\mu_N^{(N)}\rangle$$

where the sum is over all $N!$ permutations $\mathscr{P}$ of the N (superscript) particle labels. Thus

$$|\mu_1\mu_2 \cdots \mu_N\rangle^{(S)} \equiv \sum_{\mathscr{P}} U_P|\mu_1^{(1)}\rangle|\mu_2^{(2)}\rangle \cdots |\mu_N^{(N)}\rangle \tag{16.1a}$$

is a totally symmetric state vector under interchange of the particles, whereas

$$|\mu_1\mu_2 \cdots \mu_N\rangle^{(A)} \equiv \sum_{\mathscr{P}} (-1)^P U_P|\mu_1^{(1)}\rangle|\mu_2^{(2)}\rangle \cdots |\mu_N^{(N)}\rangle \tag{16.1b}$$

is a totally antisymmetric state vector under interchange of the particles. Inasmuch as the sums in Equations (16.1) are over all permutations of the particles, the superscript particle labels α can, and should, be suppressed on the left sides.

Each fundamental particle in nature appears to be associated with a unique value of ε that determines its statistics. Thus particles for which $\varepsilon = +1$ are said to be Bose–Einstein particles (or bosons) and to satisfy Bose–Einstein (or merely Bose) statistics, whereas particles for which $\varepsilon = -1$ are said to be Fermi–Dirac particles (or fermions) and to satisfy Fermi–Dirac (or merely Fermi) statistics.[3]

We now make the assumption that the only appropriate state vectors to use for describing identical-particle systems are those in the restricted Hilbert space of totally symmetric state vectors (for systems of bosons) or of totally antisymmetric state vectors (for systems of fermions).[4] Thus we consider only the N-particle product state vectors

$$|\mu_1\mu_2 \cdots \mu_N\rangle^{(S)} \text{ and } |\mu_1\mu_2 \cdots \mu_N\rangle^{(A)}$$

One can show, in fact, that these state vectors are the only ones that are invariant (apart from a factor ε) under an arbitrary permutation.[5] Although symmetry characters other than these two would also be preserved in time by a symmetric Hamiltonian, such possibilities do not seem to correspond to any systems found in nature.

We note for antisymmetric state vectors $|\mu_1\mu_2 \cdots \mu_N\rangle^{(A)}$ that no two μ_i can be equal; otherwise $|\mu_1\mu_2 \cdots \mu_N\rangle^{(A)}$ reduces to the null vector. This is the exclusion principle—first postulated by Pauli as an explanation for the periodic system of the chemical elements—which specifies that identical fermions must all be in different single-particle states. Further experimental evidence for the Fermi statistics of electrons is given by the spectrum of helium, which would be altogether different if the two atomic electrons were distinguishable. Also, the scattering of any two identical particles gives rise to observable interference effects, which are associated in the transition matrix with the occurrence of exchange integrals. Exchange integrals are due to the cross terms in the product of two symmetrized (or antisymmetrized) state vectors.

There is a connection between the spins of fundamental particles and their statistics. Particles with integral spin satisfy Bose statistics, and particles with half-integral spin satisfy Fermi statistics. There is good reason theoretically, based on relativistic quantum mechanics, to believe that this connection between spin and statistics holds universally.[6] On the experimental side the alternating intensities in the band (or rotational) spectra of homonuclear diatomic molecules establishes this relation for many atomic nuclei.

Occupation-Number Representation

We now introduce the occupation-number representation for systems of identical particles. By virtue of the overall symmetry of the state vectors (16.1), in which we are unable to specify which particle is in which single-particle quantum state, we may characterize the state vectors by numbers n_i only. Thus we say that the quantum state μ_i is occupied by n_i identical particles. Equations (16.1a) and (16.1b) together then become

$$|n_1 n_2 \cdots n_i \cdots \rangle = \mathcal{N} \sum_{\mathcal{P}} \varepsilon^P U_P |\mu_1^{(1)}\rangle |\mu_2^{(2)}\rangle \cdots |\mu_N^{(N)}\rangle \qquad \textbf{(16.2)}$$

where $\varepsilon = +1$ for Bose particles, $\varepsilon = -1$ for Fermi particles, and $\mathcal{N}$ is a normalization constant to be specified below. Of course, as remarked above, for Fermi systems the exclusion principle restricts the numbers n_i to values 0 or 1.

Equation (16.2) defines state vectors in the occupation-number representation, in which N-particle states are characterized only by sets of the numbers n_i. This equation needs some clarification, because on the left side we must assume some sequential ordering for the entire denumerable set of state labels i. Therefore, for consistency we must now adopt this same ordering on the right side; this ordering then defines, essentially, the identity permutation for the superscript particle labels.

Given a set of occupation numbers n_i, we can inquire as to how many different ways N particles can be distributed into the corresponding quantum states. For distinguishable particles the answer is (N!) divided by the number of ways, for each i, of arranging n_i particles in the state μ_i. Thus thinking classically we would obtain

$$C(n_1, n_2, \cdots, n_i, \cdots) = \frac{N!}{n_1!\, n_2! \cdots n_i! \cdots} = N! \prod_i \frac{1}{(n_i!)} \qquad (16.3)$$

ways of distributing N (identical but distinguishable) particles into the quantum states μ_i for a given set of n_i values. It is to be emphasized, however, that all these $C(n_i)$ distributions, for a given set of occupation numbers, correspond to the same many-particle quantum state. Properly symmetrized many-body product states are characterized by their occupation numbers n_i only. The combinatorial number $C(n_i)$ was encountered previously in Equation (4.18); it is quite familiar in statistical physics.

The normalization constant $\mathcal{N}$ in Equation (16.2) bears a similarity to the number $C(n_i)$. To calculate $\mathcal{N}$, we can write

$$\mathcal{N}^{-2} = \sum_{\mathcal{P}',\mathcal{P}''} \varepsilon^{P'+P''} \langle \mu_N^{(N)}| \cdots \langle \mu_2^{(2)}|\langle \mu_1^{(1)}|U_{P'}^{\dagger} U_{P''}|\mu_1^{(1)}\rangle|\mu_2^{(2)}\rangle \cdots |\mu_N^{(N)}\rangle \tag{16.4}$$

One then finds from this equation that

$$\mathcal{N} = \left[N! \prod_i (n_i!) \right]^{-1/2} \tag{16.5}$$

apart from an overall complex phase factor.

Exercise 16.1

(a) Apply Equation (16.4) to the case of two identical bosons in states μ_1 and μ_2 and show explicitly, in agreement with Equation (16.5) for this case, that $\mathcal{N} = 1/\sqrt{2}$ when $\mu_1 \neq \mu_2$ and $\mathcal{N} = ½$ when $\mu_1 = \mu_2$.

(b) Derive the general result (16.5) from Equation (16.4). (*Suggestion:* Use the group property of the permutation operators to replace $\mathcal{P}''$ by $\mathcal{P}$ and $U_{P''}$ by U_P through the definition $\mathcal{P}'' \equiv \mathcal{P}\mathcal{P}'$, which implies $U_{P''} = U_P U_{P'}$ and $P'' = P + P'$ with $\varepsilon^{2P'} = 1$.)

In classical statistical mechanics we use the so-called Boltzmann statistics in which identical particles are regarded as being distinguishable (see Section 5). Physical systems for which it is proper to use Boltzmann statistics can be characterized by referring to the preceding discussion. We observe that $C(n_i)$ reduces to the simple distinguishable-particle result $N!$ when all the n_i are zero or one. Equation (4.17) shows that the most probable values $\bar{n}_i$ will, in fact, all be negligibly small when $n\lambda_T^3 \ll 1$, that is, when

$$\lambda_T \ll \ell \equiv n^{-1/3} \qquad \text{(for Boltzmann statistics)} \tag{16.6}$$

This is the condition that must be satisfied for the correct application of Boltzmann statistics to fluids. The thermal wavelength λ_T is defined by Equation (4.2). But even when Boltzmann statistics are applied in the proper domain, we must still omit the combinatorial factor $N!$, as in the quantum case discussed below (16.3). This point was explained previously for entropy calculations in connection with the discussion at the end of Section 4 of the Gibbs paradox in classical statistical mechanics.

17 Fock Space

The development of Section 16 is extended to systems with an indefinite number of particles, whose corresponding Hilbert space is called Fock space. In this space it is quite natural to use the occupation-number representation of Equation (16.2). Although we consider few systems in this text in which particle creation and destruction actually occur, it is nevertheless convenient in this section to introduce corresponding construction operators that create and destroy particles in the occupation-number representation. With this notation the use of Fock space and the number representation offers the big advantage for many-body theory, say, in quantum statistical mechanics, in that it readily permits consideration of systems with variable particle number N. It is especially convenient to use this representation when calculations are made with the grand-canonical-ensemble formulation of statistical mechanics.

After writing the many-body Hamiltonian in construction-operator language the section is concluded with some observations about the consequence of translational invariance for two-particle interactions.

Variable particle number does not imply that there must be particle annihilation and creation, because particle transfer may occur at the boundaries of a system. Even when the total particle number N of a system is strictly constant, it can still be convenient to use a construction-operator notation. Thus we begin by introducing operators a_i, called *annihilation operators,* and their Hermitian conjugates $\mathsf{a}_i^\dagger$, called *creation operators*. These construction operators are defined below in terms of their effects on the state vectors (16.2) in the occupation-number representation. In the description of real processes for which N is constant the number of annihilation operators in any physical operator must be equal to the number of creation operators.

We next recall the discussion below Equation (16.2), in which an ordering convention was assumed for the particle states in the occupation-number representation. Not only is such an ordering convention convenient, but now for the case of Fermi systems it becomes essential in order to reproduce correctly the so-called anticommutation relations for Fermi-particle operators. With each single-particle state in a Fermi system we can, to this end, associate a number

$$\Theta_i \equiv \sum_{j=1}^{i-1} n_j \tag{17.1}$$

where $n_j = 0$ or 1, and a corresponding sign $(-1)^{\Theta_i}$.

The particle annihilation operator a_i for a system of identical particles is now defined, for both statistics, as follows:

$$\mathsf{a}_i|n_1n_2 \cdots n_i \cdots\rangle \equiv \varepsilon^{\Theta_i}\sqrt{n_i}|n_1n_2 \cdots (n_i - 1) \cdots\rangle \tag{17.2}$$

where $\varepsilon = +1$ for Bose systems and $\varepsilon = -1$ for Fermi systems. The factor $\sqrt{n_i}$ assures a vanishing result when $n_i = 0$. Moreover, for Bose systems this factor is required to preserve the state-vector normalization. Similarly, the creation operator $\mathsf{a}_i^\dagger$ is defined by the relation

$$\mathsf{a}_i^\dagger|n_1n_2 \cdots n_i \cdots\rangle \equiv \varepsilon^{\Theta_i}\sqrt{1 + \varepsilon n_i}|n_1n_2 \cdots (1 + n_i) \cdots\rangle \tag{17.3}$$

a definition that gives, as it should, zero on the right side for a Fermi system when $n_i = 1$. Again, for a Bose system the factor $\sqrt{1 + n_i}$ is necessary because of normalization considerations. That $\mathsf{a}_i^\dagger$ is indeed the Hermitian adjoint of a_i can be checked by the equation

$$\langle n_1n_2 \cdots n_i \cdots|\mathsf{a}_i|n_1n_2 \cdots (1 + n_i) \cdots\rangle = [\mathsf{a}_i^\dagger|n_1n_2 \cdots n_i \cdots\rangle]^\dagger|n_1n_2 \cdots (1 + n_i) \cdots\rangle \tag{17.4}$$

We next define the *number operator* for particles in the ith state by

$$\mathsf{n}_i \equiv \mathsf{a}_i^\dagger\mathsf{a}_i \tag{17.5}$$

Then, as can readily be verified, the result of n_i operating on the state vector (16.2) gives n_i times the same state vector. The total number operator is clearly

$$\mathsf{N} \equiv \sum_i \mathsf{n}_i = \sum_i \mathsf{a}_i^\dagger\mathsf{a}_i \tag{17.6}$$

From their definitions it follows for Bose particles that the operators a_i and $\mathsf{a}_i^\dagger$ satisfy the well-known *commutation relations*

$$\begin{aligned} [\mathsf{a}_i, \mathsf{a}_j^\dagger] &\equiv (\mathsf{a}_i\mathsf{a}_j^\dagger - \mathsf{a}_j^\dagger\mathsf{a}_i) = \delta_{ij}1 \\ [\mathsf{a}_i, \mathsf{a}_j] &= [\mathsf{a}_i^\dagger, \mathsf{a}_j^\dagger] = 0 \end{aligned} \tag{17.7}$$

Similarly, for Fermi particles a careful application of Equations (17.2) and (17.3) yields the *anticommutation relations*

$$\begin{aligned} [\mathsf{a}_i, \mathsf{a}_j^\dagger]_+ &\equiv (\mathsf{a}_i\mathsf{a}_j^\dagger + \mathsf{a}_j^\dagger\mathsf{a}_i) = \delta_{ij}1 \\ [\mathsf{a}_i, \mathsf{a}_j]_+ &= [\mathsf{a}_i^\dagger, \mathsf{a}_j^\dagger]_+ = 0 \end{aligned} \tag{17.8}$$

Here the sign factors $(-1)^{\Theta_i}$ in Equations (17.2) and (17.3), for the case $\varepsilon = -1$, are essential to produce the anticommutators. The Pauli exclusion principle follows from the second of Equations (17.8), in the form $(\mathsf{a}_i^\dagger)^2 = 0$; successive creation of two particles in the same state cannot occur.

Exercise 17.1

(a) Verify for all allowed sets of occupation numbers that the anticommutation relations (17.8) are a consequence of definitions (17.2) and (17.3).

(b) Show for both statistics that

$$[\mathrm{n}_i, \mathrm{a}_j] = -\mathrm{a}_i\,\delta_{ij} \qquad [\mathrm{n}_i, \mathrm{a}_j^\dagger] = \mathrm{a}_i^\dagger\,\delta_{ij} \tag{17.9}$$

Hamiltonian in Terms of Construction Operators

We now wish to write the N-particle Hamiltonian $H^{(N)}$ in terms of construction operators, in which it takes an N-independent form. With the Hamiltonian expressed in terms of annihilation and creation operators we also say that it has been written in second-quantized form. In a rough sense one may say that with "first" quantization, physical variables associated with the wave function for a particle are required to satisfy certain quantum conditions. With second quantization the particulate nature of matter is itself exhibited as a quantum condition on more fundamental wave fields.

The Hamiltonian $H^{(N)}$ in general consists of a kinetic-energy part $K^{(N)}$, a single-particle potential $U^{(N)}$, and a two-particle interaction $V_2^{(N)}$. We write the first two of these as

$$K^{(N)} + U^{(N)} = \sum_{\alpha=1}^{N} H_\alpha^{(1)} \tag{17.10a}$$

and $V_2^{(N)}$ as

$$V_2^{(N)} = \frac{1}{2}\sum_{\substack{\alpha,\beta=1\\(\alpha\neq\beta)}}^{N} V_{\alpha\beta}^{(2)} \tag{17.10b}$$

where the sums are over the N identical particles.

Since the Hamiltonian is most commonly written as an operator in position space (the $\mathbf{r}$ representation), we derive our standard result by using this representation. To this end we require the unitary transformation elements that transform the state vectors (16.2) to position space.

$$\langle \mathbf{r}_1 \cdots \mathbf{r}_N \mid n_1 n_2 \cdots n_i \cdots \rangle$$
$$= (N!)^{-1/2} \prod_i (n_i!)^{-1/2} \sum_{\mathcal{P}} \varepsilon^P U_P[\phi_{\mu_1}(\mathbf{r}_1)\phi_{\mu_2}(\mathbf{r}_2) \cdots \phi_{\mu_N}(\mathbf{r}_N)] \tag{17.11}$$

Here we have used Equation (16.5), and the permutation operator U_P now operates on the N position coordinates $\mathbf{r}_\alpha$. (We suppress notation for any internal coordinates such as spin degrees of freedom.) The $\phi_\mu(\mathbf{r})$ belong to the complete set of orthonormalized wave functions for the states μ (in the Hilbert space of $H^{(1)}$) or, alternatively, they are the unitary transformation elements $\langle \mathbf{r} \mid \mu\rangle$ for a single particle. Also, we continue to adhere to the particle-state ordering convention introduced below Equation (16.2), which determines the identity permutation for the N coordinates $\mathbf{r}_\alpha$. For systems of identical bosons the ordering of the coordinates will be arbitrary whenever two or more bosons are in the same single-particle state.

Our first objective is to derive an explicit expression for $(K^{(N)} + U^{(N)})$ in terms of construction operators, so we consider the effect of this operator [or any single-particle operator of the form (17.10a)] on N-particle state vectors in the number representation; that is, we consider $(K^{(N)} + U^{(N)}) \times |n_1 n_2 \cdots n_i \cdots\rangle$. We show in detail below that $(K^{(N)} + U^{(N)})$ can be written in the N-independent form

$$\mathsf{K} + \mathsf{U} = \sum_{i,j} \langle j|H^{(1)}|i\rangle \mathsf{a}_j^\dagger \mathsf{a}_i \tag{17.12}$$

where the label i refers to the ith state in the set μ so that $|i\rangle$ is a simplified notation for $|\mu_i\rangle$.

Derivation of Equation (17.12): We first utilize the transformation matrix elements (17.11) in the following manipulations, in which $(K^{(N)} + U^{(N)}) \times |n_1 \cdots n_i \cdots\rangle$ is transformed to the position representation.

$$\begin{aligned}
&\langle \mathbf{r}_1\mathbf{r}_2 \cdots \mathbf{r}_N|(K^{(N)} + U^{(N)})|n_1 n_2 \cdots n_i \cdots\rangle \\
&= \int d^3\mathbf{r}_1' \cdots d^3\mathbf{r}_N' \langle \mathbf{r}_1\mathbf{r}_2 \cdots \mathbf{r}_N|(K^{(N)} + U^{(N)})|\mathbf{r}_1' \cdots \mathbf{r}_N'\rangle \\
&\quad \times \langle \mathbf{r}_1' \cdots \mathbf{r}_N' \mid n_1 n_2 \cdots n_i \cdots\rangle \\
&= \sum_{\alpha=1}^{N} \int d^3\mathbf{r}_1' \cdots d^3\mathbf{r}_N' \left[\prod_{\substack{\beta=1 \\ (\beta\neq\alpha)}}^{N} \delta^{(3)}(\mathbf{r}_\beta - \mathbf{r}_\beta') \right] \langle \mathbf{r}_\alpha|H^{(1)}|\mathbf{r}_\alpha'\rangle \qquad (17.13) \\
&\quad \times \left[N! \prod_i (n_i!) \right]^{1/2} \sum_{\mathcal{P}} \varepsilon^P U_P[\phi_{\mu_1}(\mathbf{r}_1') \cdots \phi_{\mu_N}(\mathbf{r}_N')] \\
&= \left[N! \prod_i (n_i!) \right]^{-1/2} \sum_{\mathcal{P}} \varepsilon^P \sum_{\alpha=1}^{N} \int d^3\mathbf{r}_\alpha' \langle \mathbf{r}_\alpha|H^{(1)}|\mathbf{r}_\alpha'\rangle \\
&\quad \times U_P[\phi_\mu(\mathbf{r}_1) \cdots \phi_{\mu_\alpha}(\mathbf{r}_\alpha') \cdots \phi_{\mu_N}(\mathbf{r}_N)]
\end{aligned}$$

At this stage of the derivation, position-space matrix elements of the single-particle operator $H^{(1)}$ occur. It is well to be familiar with these matrix elements in a few simple cases. For example, for a point potential U we have

$$\langle \mathbf{r}|U^{(1)}|\mathbf{r}'\rangle = U(r)\, \delta^{(3)}(\mathbf{r} - \mathbf{r}') \tag{17.14a}$$

whereas for the kinetic-energy operator K we obtain

$$\langle \mathbf{r}|K^{(1)}|\mathbf{r}'\rangle = -(\hbar^2/2m)\,\nabla'^2\,\delta^{(3)}(\mathbf{r} - \mathbf{r}') \tag{17.14b}$$

We now observe that the U_P and the symmetric function $\Sigma_{\alpha=1}^{N} \int d^3\mathbf{r}'_\alpha\langle \mathbf{r}_\alpha| \times H^{(1)}|\mathbf{r}'_\alpha\rangle$ commute. With the operator U_P to the left of this function, we can then perform the following manipulations:

$$\begin{aligned}
\int d^3\mathbf{r}'_\alpha\langle \mathbf{r}_\alpha|H^{(1)}|\mathbf{r}'_\alpha\rangle\phi_{\mu_\alpha}(\mathbf{r}'_\alpha) &= \langle \mathbf{r}_\alpha|H^{(1)}|\mu_\alpha\rangle \\
&= \sum_{\mu'_\alpha} \langle \mathbf{r}_\alpha|\mu'_\alpha\rangle\langle\mu'_\alpha|H^{(1)}|\mu_\alpha\rangle \\
&= \sum_{\mu'_\alpha} \langle\mu'_\alpha|H^{(1)}|\mu_\alpha\rangle\phi_{\mu'_\alpha}(\mathbf{r}_\alpha)
\end{aligned}$$

Upon substituting this result into the final equality of Equation (17.13) we obtain, after noticing that $\Sigma_{\alpha=1}^{N}$ is equivalent to $\Sigma_{\mu_\alpha} n_\alpha$ on the right side,

$$\begin{aligned}
&\langle \mathbf{r}_1\mathbf{r}_2 \cdots \mathbf{r}_N|(K^{(N)} + U^{(N)})|n_1 n_2 \cdots n_i \cdots\rangle \\
&= \left[N!\prod_i (n_i!)\right]^{-1/2} \sum_{\mathscr{P}} \varepsilon^P \sum_{\mu_\alpha\mu'_\alpha} n_\alpha\langle\mu'_\alpha|H^{(1)}|\mu_\alpha\rangle \\
&\quad\times U_P[\phi_{\mu_1}(\mathbf{r}_1)\cdots\phi_{\mu'_\alpha}(\mathbf{r}_\alpha)\cdots\phi_{\mu_N}(\mathbf{r}_N)] \qquad (17.15)\\
&= \sum_{\mu_\alpha,\mu'_\alpha} \langle\mu'_\alpha|H^{(1)}|\mu_\alpha\rangle n_\alpha \varepsilon^{\Theta_\alpha^{(0)}}\varepsilon^{\Theta_{\alpha'}^{(f)}}\left(\frac{1 + \varepsilon n'_\alpha}{n_\alpha}\right)^{1/2} \\
&\quad\times \langle \mathbf{r}_1\mathbf{r}_2\cdots\mathbf{r}_N \mid n_1 n_2 \cdots (n_\alpha - 1)\cdots(n'_\alpha + 1)\cdots\rangle
\end{aligned}$$

We can explain the second equality of this expression by noting first that the inserted sign factors make it possible, for Fermi systems, to relocate the function $\phi_{\mu'_\alpha}(\mathbf{r}_\alpha)$ according to our ordering convention for the states μ. The superscript (0) refers to Θ_α for the state μ_α in the initial wave function (17.11), and the superscript (f) refers to $\Theta_{\alpha'}$ for the final wave function in (17.15). The factor

$$\left(\frac{1 + \varepsilon n_{\alpha'}}{n_\alpha}\right)^{1/2}$$

allows for the normalization change in these two wave functions; when $\mu_{\alpha'} = \mu_\alpha$, this factor must be replaced by 1. Finally, for Fermi systems the resulting factor $[n_\alpha(1 - n_{\alpha'})]^{1/2}$ deals properly with the requirements of the Pauli exclusion principle.

To complete our derivation it is now necessary only to utilize definitions (17.2) and (17.3) to obtain from Equation (17.15) the final expression

$$\langle \mathbf{r}_1\mathbf{r}_2 \cdots \mathbf{r}_N|(K^{(N)} + U^{(N)})|n_1n_2 \cdots n_i \cdots \rangle$$
$$= \langle \mathbf{r}_1\mathbf{r}_2 \cdots \mathbf{r}_N| \sum_{\mu_i,\mu_j} \langle \mu_j|H^{(1)}|\mu_i\rangle \mathsf{a}_j^\dagger \mathsf{a}_i|n_1n_2 \cdots n_i \cdots \rangle \quad (17.16)$$

We conclude that the N-independent operator expression (17.12) is correct. It is also a quite general result, as part (b) of the following exercise demonstrates.

Exercise 17.2

(a) Show, using the representation in which the kinetic-energy operator is diagonal, that

$$\mathsf{K} = \sum_i \mathsf{n}_i\omega_i \quad (17.17)$$

where n_i is defined by Equation (17.5) and

$$\langle j|K^{(1)}|i\rangle = \omega_i\,\delta_{ij}$$

(b) Argue that the position-space density operator

$$n(\mathbf{r}) \equiv \sum_{\alpha=1}^{N} \delta^{(3)}(\mathbf{r} - \mathbf{r}_\alpha) \quad (17.18a)$$

can be written in terms of construction operators as

$$\mathsf{n}(\mathbf{r}) = \sum_{i,j} \phi_j^*(\mathbf{r})\phi_i(\mathbf{r})\mathsf{a}_j^\dagger \mathsf{a}_i \quad \mathbf{(17.18b)}$$

In a manner quite similar to the above derivation for single-particle Hamiltonians, one can prove that two-particle interaction Hamiltonians $V_2^{(N)}$, of the form (17.10b), can be written as the N-independent operators

$$\mathsf{V}_2 = \frac{1}{2}\sum_{i,j,k,l} \langle ij|V^{(2)}|kl\rangle \mathsf{a}_j^\dagger \mathsf{a}_i^\dagger \mathsf{a}_k \mathsf{a}_l \quad (17.19a)$$

The different ordering of the labels i and j in the two parts of the summand is important for Fermi systems.

Exercise 17.3 Verify Equation (17.19a) using the method of Equations (17.13)–(17.16).

Equation (17.19a) can be rewritten to exhibit the matrix elements of the two-particle interaction $V^{(2)}$ in a useful symmetrized form. Thus with the aid of a commutation relation from either Equation (17.7) or (17.8) one can readily verify that

$$\mathsf{V}_2 = \frac{1}{4}\sum_{i,j,k,l} \langle ij|V^{(S)}|kl\rangle \mathsf{a}_j^\dagger \mathsf{a}_i^\dagger \mathsf{a}_k \mathsf{a}_l \quad \mathbf{(17.19b)}$$

where the superscript (S) denotes either a symmetric or an antisymmetric combination defined as follows:

$$\begin{aligned}\langle ij|V^{(S)}|kl\rangle &\equiv \langle ij|V^{(2)}|kl\rangle + \varepsilon\langle ij|V^{(2)}|lk\rangle \\ &= \varepsilon\langle ij|V^{(S)}|lk\rangle = \varepsilon\langle ji|V^{(S)}|kl\rangle\end{aligned} \tag{17.20}$$

For Bose systems $\varepsilon = +1$, and for Fermi systems $\varepsilon = -1$. The decision as to which of Equations (17.19) is to be used in an application often depends merely on whether or not it is desired to exhibit a result in terms of the symmetrized two-particle interaction matrix elements.

Having expressed the Hamiltonian in terms of construction operators, it is now also appropriate to have an explicit expression for the state vector (16.2) in this notation. If we represent the no-particle, or vacuum, state by $|0\rangle$, then one can show that state vectors in the number representation can be written in the N-independent form

$$|n_1 n_2 \cdots n_i \cdots\rangle = \left[\prod_j (n_j!)\right]^{-1/2} (\mathsf{a}_1^\dagger)^{n_1}(\mathsf{a}_2^\dagger)^{n_2} \cdots (\mathsf{a}_i^\dagger)^{n_i} \cdots |0\rangle \tag{17.21}$$

where for Fermi systems the creation operators must be ordered according to the convention established below Equation (16.2). This expression can be verified by proving that the number operator (17.5) is diagonal in the states $|n_i\rangle$ and that these state vectors are correctly normalized. The proof is easy for Fermi systems.

Exercise 17.4 Prove Equation (17.21) for Bose systems by using the commutation relations (17.7) and (17.9).

The complete set of single-particle states μ (or μ_i) that has been used in the above development is restricted only by the condition that it span the Hilbert space of the single-particle Hamiltonian $H^{(1)}$. A special case of frequent interest is the position representation, normally considered to be continuous. For this case we use the notation $\underline{\psi}(\mathbf{r})$ and $\underline{\psi}^\dagger(\mathbf{r})$ for the construction operators, instead of $\mathsf{a}_\mathbf{r}$ and $\mathsf{a}_\mathbf{r}^\dagger$. The operators $\underline{\psi}(\mathbf{r})$ and $\underline{\psi}^\dagger(\mathbf{r})$ are often referred to as field operators, because they can be identified with a matter field. We define them here by the equations

$$\begin{aligned}\underline{\psi}(\mathbf{r}) &= \sum_i \phi_i(\mathbf{r})\mathsf{a}_i \\ \underline{\psi}^\dagger(\mathbf{r}) &= \sum_i \phi_i^*(\mathbf{r})\mathsf{a}_i^\dagger\end{aligned} \tag{17.22}$$

The interpretation is given that $\underline{\psi}(\mathbf{r})$ is an annihilation operator for a particle at point $\mathbf{r}$, whereas $\underline{\psi}^\dagger(\mathbf{r})$ is a creation operator for a particle at point $\mathbf{r}$.

On using Equation (17.7) or (17.8) and the completeness relation for the set of single-particle functions $\phi_i(\mathbf{r})$, one can readily establish commutation

(−) or anticommutation (+) relations for the operators $\underline{\psi}(\mathbf{r})$ and $\underline{\psi}^\dagger(\mathbf{r})$ as

$$[\underline{\psi}(\mathbf{r}), \underline{\psi}^\dagger(\mathbf{r}')]_\pm = \delta^{(3)}(\mathbf{r} - \mathbf{r}')\mathsf{1}$$
$$[\underline{\psi}(\mathbf{r}), \underline{\psi}(\mathbf{r}')]_\pm = [\underline{\psi}^\dagger(\mathbf{r}), \underline{\psi}^\dagger(\mathbf{r}')]_\pm = 0 \tag{17.23}$$

Similarly, by using the orthogonality relation for the functions $\phi_i(\mathbf{r})$ one can show that the total number operator (17.6) can be written as

$$\mathsf{N} = \int d^3r\,\underline{\psi}^\dagger(\mathbf{r})\underline{\psi}(\mathbf{r}) \tag{17.24}$$

The single-particle and two-particle interaction Hamiltonians of Equations (17.12) and (17.19a), when written in terms of the field operators $\underline{\psi}(\mathbf{r})$ and $\underline{\psi}^\dagger(\mathbf{r})$, become

$$\mathsf{K} + \mathsf{U} = \int d^3\mathbf{r}\, d^3\mathbf{r}'\,\underline{\psi}^\dagger(\mathbf{r})\langle\mathbf{r}|H^{(1)}|\mathbf{r}'\rangle\underline{\psi}(\mathbf{r}') \tag{17.25}$$

$$\mathsf{V}_2 = \frac{1}{2}\int d^3\mathbf{r}_1\, d^3\mathbf{r}_2\, d^3\mathbf{r}_1'\, d^3\mathbf{r}_2'\,\underline{\psi}^\dagger(\mathbf{r}_2')\underline{\psi}^\dagger(\mathbf{r}_1') \times \langle\mathbf{r}_1'\mathbf{r}_2'|V^{(2)}|\mathbf{r}_1\mathbf{r}_2\rangle\underline{\psi}(\mathbf{r}_1)\underline{\psi}(\mathbf{r}_2) \tag{17.26}$$

Translational Invariance and the Two-Particle Interaction

We are frequently interested in very large, or infinite, systems, characterized by the thermodynamic limit $(N, \Omega) \to \infty$ with the particle density $n = N/\Omega$ remaining finite. Such a system ordinarily is assumed to be *translationally invariant,* which means, apart from a qualification that must be made for crystalline solids, that there is no position point in the system, such as a center of force, that has a special significance. In the absence of position-space localization, the natural single-particle representation is the complementary one implied by the Heisenberg uncertainty principle, namely, the momentum representation with quantum numbers $\mathbf{k}_j$.

For a fluid with no associated periodic potential, as for the electron gas in a metal, the momentum eigenstates will be plane-wave states. The single-particle wave function $\phi_j(\mathbf{r})$ becomes

$$\langle\mathbf{r}\,|\,j\rangle = \phi_j(\mathbf{r}) = \Omega^{-1/2}\, e^{i\mathbf{k}_j\cdot\mathbf{r}}\chi_{m_j}(\xi) \tag{17.27}$$

where we can use box normalization to enumerate the states $\mathbf{k}_j$ (or merely the states j) before going to the thermodynamic limit. The $\chi_m(\xi)$ are spin-wave functions for particles with spin, with m being the magnetic spin quantum number and ξ the spin coordinate (in a column matrix, say).

In the thermodynamic limit the discrete set of states (j) becomes a continuum of states and, correspondingly, sums over momentum states are replaced by integrals:

$$\sum_{\mathbf{k}_j} \rightarrow \int d^3\mathbf{n} \qquad (\Omega \rightarrow \infty)$$

The integral is over all three-dimensional elements $d^3\mathbf{n}$ in number space. Of course, in practice we evaluate such integrals in wave-number space [here, $\mathbf{k}$ = (momentum)/$\hbar$], so that we need to know the density of states

$$\frac{d^3\mathbf{n}}{d^3\mathbf{k}} = \frac{dn_x}{dk_x} \cdot \frac{dn_y}{dk_y} \cdot \frac{dn_z}{dk_z} = \left(\frac{L_x}{2\pi}\right) \cdot \left(\frac{L_y}{2\pi}\right) \cdot \left(\frac{L_z}{2\pi}\right) = \frac{\Omega}{(2\pi)^3} \tag{17.28}$$

where the "rectangular-box" dimensions of the system are L_x, L_y, and L_z. After we also consider spin degrees of freedom for the single particles we write

$$\sum_j \rightarrow \frac{\Omega}{(2\pi)^3} \sum_m \int d^3\mathbf{k} \tag{17.29}$$

For spin-independent integrands the sum over spin states can be replaced by $(2S + 1)$, where S is the particle spin.

For translationally invariant two-particle interactions $V^{(2)}$, the position-space matrix elements must satisfy the identity (with spin coordinates ξ_i suppressed)

$$\langle \mathbf{r}_1\mathbf{r}_2|V^{(2)}|\mathbf{r}_3\mathbf{r}_4\rangle = \langle \mathbf{r}_1 + \mathbf{a}, \mathbf{r}_2 + \mathbf{a}|V^{(2)}|\mathbf{r}_3 + \mathbf{a}, \mathbf{r}_4 + \mathbf{a}\rangle$$

where $\mathbf{a}$ is an arbitrary displacement vector. Transformation of both sides of this equality to the momentum representation using (17.27) shows that momentum is conserved for translationally invariant interactions; that is, we can write

$$\begin{aligned} \langle \mathbf{k}_1\mathbf{k}_2|V^{(2)}|\mathbf{k}_3\mathbf{k}_4\rangle &= \langle \mathbf{k}_1\mathbf{k}_2|V^{(2)}|\mathbf{k}_3\mathbf{k}_4\rangle\, \delta_{KR}(\mathbf{k}_1 + \mathbf{k}_2, \mathbf{k}_3 + \mathbf{k}_4) \\ &= \Omega^{-1}\langle \mathbf{k}_{12}|V^{(2)}|\mathbf{k}_{34}\rangle_{\mathbf{K}_{12}}\, \delta_{KR}(\mathbf{K}_{12}, \mathbf{K}_{34}) \end{aligned} \tag{17.30}$$

where $\delta_{KR}(\ )$ is a Kronecker-δ function. In the second equality we have transformed to relative- and total-momentum coordinates, using Equations (11.3) and extracted a factor Ω^{-1} due to (17.27). One can also exhibit conservation of angular momentum for two-particle interactions that are rotationally invariant by using the above procedure. The method applies as well when the particles are not identical.

We have attached a subscript $\mathbf{K}_{12}$ to the matrix element of $V^{(2)}$ in the second equality of Equation (17.30), and we now show that this subscript is unnecessary; that is, these matrix elements do not depend on the total momentum of the two interacting particles. This assertion is demonstrated by inverting Equation (17.30), again using the functions (17.27), and then transforming to the center-of-momentum coordinate system. One finds the result (suppressing all spin coordinates):

$$\langle \mathbf{r}_1\mathbf{r}_2|V^{(2)}|\mathbf{r}_3\mathbf{r}_4\rangle = \sum_{\mathbf{k}_1\mathbf{k}_2\mathbf{k}_3\mathbf{k}_4} \langle \mathbf{r}_1 \mid \mathbf{k}_1\rangle\langle \mathbf{r}_2 \mid \mathbf{k}_2\rangle\langle \mathbf{k}_1\mathbf{k}_2|V^{(2)}|\mathbf{k}_3\mathbf{k}_4\rangle\langle \mathbf{k}_3 \mid \mathbf{r}_3\rangle\langle \mathbf{k}_4 \mid \mathbf{r}_4\rangle$$

$$= \Omega^{-3} \sum_{\mathbf{k}_{12},\mathbf{k}_{34},\mathbf{K}} \langle \mathbf{k}_{12}|V^{(2)}|\mathbf{k}_{34}\rangle_{\mathbf{K}} \exp[i\mathbf{K}\cdot(\mathbf{R}_{12} - \mathbf{R}_{34})]$$

$$\times \exp[i(\mathbf{k}_{12}\cdot\mathbf{r}_{12} - \mathbf{k}_{34}\cdot\mathbf{r}_{34})]$$

where $\mathbf{R}_{12} = \frac{1}{2}(\mathbf{r}_1 + \mathbf{r}_2)$, $\mathbf{r}_{12} = (\mathbf{r}_1 - \mathbf{r}_2)$, and so on. We now use the argument that the position-space matrix elements of $V^{(2)}$ cannot result in a shift in the center of mass $\mathbf{R}_{12}$ of the two interacting particles; that is, we must have $\mathbf{R}_{12} = \mathbf{R}_{34}$. But this can only be true if the momentum-space matrix elements are independent of the total momentum $\mathbf{K}$. We obtain, therefore, the two results (with spin coordinates now included):

$$\langle k_1k_2|V^{(2)}|k_3k_4\rangle = \Omega^{-1}\langle \mathbf{k}_{12}; m_1m_2|V^{(2)}|\mathbf{k}_{34}; m_3m_4\rangle \times \delta_{KR}(\mathbf{k}_1 + \mathbf{k}_2, \mathbf{k}_3 + \mathbf{k}_4) \quad \textbf{(17.31)}$$

where by k_i we mean $(\mathbf{k}_i, m_i)$, and

$$\langle r_1r_2|V^{(2)}|r_3r_4\rangle = \langle \mathbf{r}_{12}; \xi_1\xi_2|V^{(2)}|\mathbf{r}_{34}; \xi_3\xi_4\rangle\, \delta^{(3)}(\mathbf{R}_{12} - \mathbf{R}_{34}) \quad (17.32)$$

where by r_i we mean $(\mathbf{r}_i, \xi_i)$. In Equation (17.32) we have also used the fact, due again to translational invariance, that the position-space matrix elements on the right side cannot depend on the center-of-mass coordinate $\mathbf{R}_{12}$.

The result (17.32) applies to a general, nonlocal two-particle interaction $V^{(2)}$ whose relative-coordinates matrix elements depend on both $\mathbf{r}_{12}$ and $\mathbf{r}_{34}$. For a local, or point, interaction Equation (17.32) becomes, neglecting spin considerations,

$$\langle \mathbf{r}_1\mathbf{r}_2|V^{(2)}|\mathbf{r}_3\mathbf{r}_4\rangle = V(\mathbf{r}_{12})\, \delta^{(3)}(\mathbf{R}_{12} - \mathbf{R}_{34})\, \delta^{(3)}(\mathbf{r}_{12} - \mathbf{r}_{34}) \quad (17.33)$$

The relative-coordinates momentum-space matrix elements of Equation (17.31) are then related to $V(\mathbf{r})$ by the equation

$$\langle \mathbf{k}_{12}|V^{(2)}|\mathbf{k}_{34}\rangle = \int d^3\mathbf{r}\, e^{-i\mathbf{q}\cdot\mathbf{r}}V(\mathbf{r}) = V_{\mathbf{q}} \quad (17.34)$$

where $\mathbf{q}$ is the momentum-transfer vector of Equation (11.4), that is, $\mathbf{q} = \mathbf{k}_{12} - \mathbf{k}_{34} = \mathbf{k}_1 - \mathbf{k}_3$.

In the infinite-volume limit, the Kronecker-δ functions in Equations (17.30) and (17.31) must be replaced by Dirac-δ functions, according to the relation

$$\delta_{KR}(\mathbf{k}_i, \mathbf{k}_j) \rightarrow \frac{(2\pi)^3}{\Omega}\, \delta^{(3)}(\mathbf{k}_i - \mathbf{k}_j) \qquad (\Omega \rightarrow \infty) \quad \textbf{(17.35)}$$

The three-dimensional δ function $\delta^{(3)}(\mathbf{k})$ represents a product of three δ functions $\delta(k_x)\,\delta(k_y)\,\delta(k_z)$.

Exercise 17.5 Prove that the replacement (17.35) is correct by using (17.29).

18 Rayleigh–Schrödinger Perturbation Theory

Ordinary time-independent perturbation theory in quantum mechanics occupies a fundamental position in the theoretical understanding of many-body systems. It is a tool that when applied to systems of many identical particles can provide some of our simplest insights into their microscopic behavior. Moreover, an understanding of the mathematical structure of time-independent perturbation theory in simple applications serves as a useful stepping stone to the further assimilation of more elaborate many-body theories such as time-dependent perturbation theory, quantum statistical mechanics, and those in which Green's-function techniques are applied.

In this chapter we are interested only in thermally isolated closed systems, those that can be associated uniquely with a Hamiltonian operator as described in Section 7. This is the so-called "pure-states" case for which it is possible to ignore the statistical considerations of Chapter I. But even when our incomplete knowledge of a macroscopic body is properly accounted for, the solutions to the pure-state problem are still necessary. Thus in Section 21 of the next chapter, where we introduce the proper "mixed-state" description using the statistical matrix, we shall see that both the present pure-state considerations and statistical methods from Section 4 must be combined to give a correct microscopic treatment of real macroscopic systems.

In this section we first review the so-called *Rayleigh–Schrödinger perturbation theory,* which is the version of time-independent perturbation theory most easily applied to many-body systems. Then we illustrate the use of this method by applying it straightforwardly to a nonideal Fermi gas.

Rayleigh–Schrödinger perturbation theory applies to the eigenstates of a physical system whose Hamiltonian operator H can be separated into two Hermitian parts:

$$H = H_o + gV \tag{18.1}$$

Of these, H_o will be regarded as the unperturbed part and gV as the perturbation. By considering the parameter g to be a small quantity we can justify an expansion procedure of the many-body state vector $|\Psi_n\rangle$ and its associated energy E_n in powers of g. Eventually, though, we set g equal to unity.

Here n refers to the nth energy level, assumed discrete, of the many-body system, and $|\Psi_n\rangle$ usually reduces in the unperturbed case ($g = 0$) to a state vector of the form (17.21).

We assume that as g increases from zero to unity $|\Psi_n\rangle$ and E_n develop smoothly from unperturbed values $E_n^{(0)}$ and $|\Psi_n^{(0)}\rangle$; systems for which this "adiabatic hypothesis" is valid are called *normal* systems. Moreover, for systems such as nonideal gases, for which the eigenstates of H and H_o do not differ greatly, we may expect that even when g is unity E_n decreases rapidly as n increases. Roughly speaking, this is the convergence condition for the perturbation theory in its approximated, or unsummed, form.[7] We have then for normal systems

$$H|\Psi_n\rangle = E_n|\Psi_n\rangle \tag{18.2}$$

$$H_o|\Psi_n^{(0)}\rangle = E_n^{(0)}|\Psi_n^{(0)}\rangle \tag{18.3}$$

We now assume that the energy levels $E_n^{(0)}$ are *nondegenerate,* an assumption that can pose difficulty in some applications to many-body systems. With this qualification one can expect expansions of E_n and $|\Psi_n\rangle$ in powers of g to be valid for normal systems. Therefore, we write

$$\begin{aligned} E_n &= E_n^{(0)} + gE_n^{(1)} + g^2E_n^{(2)} + \cdots \\ &= \sum_{k=0}^{\infty} g^k E_n^{(k)} \end{aligned} \tag{18.4}$$

$$\begin{aligned} |\Psi_n\rangle &= |\Psi_n^{(0)}\rangle + g|\Psi_n^{(1)}\rangle + g^2|\Psi_n^{(2)}\rangle + \cdots \\ &= \sum_{k=0}^{\infty} g^k |\Psi_n^{(k)}\rangle \end{aligned} \tag{18.5}$$

It should be emphasized that unless the normalization of $|\Psi_n\rangle$ is specified precisely, the expansion (18.5) is nonunique. The normalization we choose in the present section is

$$\langle\Psi_n^{(0)} \mid \Psi_n\rangle = 1 \tag{18.6}$$

which implies, in general, that $|\Psi_n\rangle$ is not normalized to unity. The normalization constant corresponding to the choice $\langle\Psi_n \mid \Psi_n\rangle = 1$ may then have to be calculated for some applications, such as that in Problem IV.4.

When Equations (18.4) and (18.5) for E_n and $|\Psi_n\rangle$, respectively, are substituted into the wave equation (18.2), and the coefficients of like powers in g are collected together and matched on both sides, then one obtains in addition to Equation (18.3), for the zeroth-power case, the following set of equations:

$$H_o|\Psi_n^{(1)}\rangle + V|\Psi_n^{(0)}\rangle = E_n^{(0)}|\Psi_n^{(1)}\rangle + E_n^{(1)}|\Psi_n^{(0)}\rangle \qquad (k = 1)$$

as the result for coefficients of g; and

$$H_o|\Psi_n^{(k)}\rangle + V|\Psi_n^{(k-1)}\rangle = \sum_{l=0}^{k} E_n^{(l)}|\Psi_n^{(k-l)}\rangle \qquad (k > 1)$$

as the result for coefficients of g^k when $k > 1$. These equations can be rearranged in the forms

$$[H_o - E_n^{(0)}]|\Psi_n^{(1)}\rangle = [E_n^{(1)} - V]|\Psi_n^{(0)}\rangle \qquad (k = 1)$$

$$[H_o - E_n^{(0)}]|\Psi_n^{(k)}\rangle = [E_n^{(1)} - V]|\Psi_n^{(k-1)}\rangle + E_n^{(k)}|\Psi_n^{(0)}\rangle + \sum_{l=2}^{k-1} E_n^{(l)}|\Psi_n^{(k-l)}\rangle \qquad (k \geq 2) \tag{18.7}$$

where the last term in the second equation vanishes when $k = 2$. We may conclude that $|\Psi_n^{(k)}\rangle$ satisfies an inhomogeneous linear equation.

For nondegenerate states the solution to Equations (18.7) can be written in an operator form as[8]

$$|\Psi_n^{(k)}\rangle = C_n^{(k)}|\Psi_n^{(0)}\rangle - \left(\frac{1 - P_n^{(0)}}{E_n^{(0)} - H_o}\right) \times \left[(E_n^{(1)} - V)|\Psi_n^{(k-1)}\rangle + \Theta(k - 2^+) \sum_{l=2}^{k-1} E_n^{(l)}|\Psi_n^{(k-l)}\rangle\right] \tag{18.8}$$

where $k \geqslant 1$. The second term inside the square brackets vanishes for $k = 1$ and $k = 2$, as is indicated by the step function $\Theta(k - 2^+)$. The quantity $P_n^{(0)}$ is a projection operator that projects any state vector onto the unperturbed state $|\Psi_n^{(0)}\rangle$, that is,

$$P_n^{(0)} = |\Psi_n^{(0)}\rangle\langle\Psi_n^{(0)}| \tag{18.9}$$

Therefore $[1 - P_n^{(0)}]$ is a projection operator away from the state $|\Psi_n^{(0)}\rangle$. When use is made of the closure relation for the unperturbed states i, that is, $\Sigma_i|\Psi_i^{(0)}\rangle\langle\Psi_i^{(0)}| = 1$, in conjunction with the operator $(1 - P_n^{(0)})$, then the effect is to eliminate the term $i = n$. Since this particular term is precisely the one that leads to a vanishing energy denominator $(E_n^{(0)} - H_o)$, it can be seen how the operator $(1 - P_n^{(0)})$ in Equation (18.8) eliminates possible singularities when the unperturbed states n are nondegenerate.

If in Equation (18.8) we choose the arbitrary constant $C_n^{(k)}$ to be zero, then we obtain immediately for $k \neq 0$ that

$$\langle\Psi_n^{(0)} \mid \Psi_n^{(k)}\rangle = 0$$

since by definition

$$\langle\Psi_n^{(0)}|\left[\frac{1 - P_n^{(0)}}{E_n^{(0)} - H_o}\right] = 0 \tag{18.10}$$

The choice of $C_n^{(k)} = 0$ is therefore equivalent to the required normalization condition (18.6), and Equation (18.8) becomes, for $k \geqslant 1$,

$$|\Psi_n^{(k)}\rangle = \left(\frac{1 - P_n^{(0)}}{E_n^{(0)} - H_o}\right)\left[V|\Psi_n^{(k-1)}\rangle - \Theta(k - 1^+)\sum_{l=1}^{k-1} E_n^{(l)}|\Psi_n^{(k-l)}\rangle\right] \tag{18.11}$$

We return now to Equations (18.7) and take the inner product of each equation with $\langle\Psi_n^{(0)}|$. One obtains readily, for all $k \geqslant 1$, the result

$$E_n^{(k)} = \langle\Psi_n^{(0)}|V|\Psi_n^{(k-1)}\rangle \tag{18.12}$$

which together with Equation (18.11) can be used for explicit calculations of the perturbation corrections to $E_n^{(0)}$ due to the perturbation V. With the aid of Equations (18.4) and (18.5) one can finally derive, for $g = 1$, a well-known energy-shift expression

$$\Delta E_n \equiv E_n - E_n^{(0)} = \langle\Psi_n^{(0)}|V|\Psi_n\rangle \tag{18.13}$$

The simplicity of this final result is due, in part, to the normalization choice (18.6). To third order in V, Equations (18.11) to (18.13) yield the explicit result

$$\begin{aligned}\Delta E_n = {} & \langle\Psi_n^{(0)}|V|\Psi_n^{(0)}\rangle + \langle\Psi_n^{(0)}|V\left(\frac{1 - P_n^{(0)}}{E_n^{(0)} - H_o}\right)V|\Psi_n^{(0)}\rangle \\ & + \langle\Psi_n^{(0)}|V\left(\frac{1 - P_n^{(0)}}{E_n^{(0)} - H_o}\right)[V - \langle\Psi_n^{(0)}|V|\Psi_n^{(0)}\rangle]\left(\frac{1 - P_n^{(0)}}{E_n^{(0)} - H_o}\right)V|\Psi_n^{(0)}\rangle \\ & + O(V^4)\end{aligned} \tag{18.14}$$

The preceding development outlines a form of Rayleigh–Schrödinger perturbation theory that can be readily applied to weakly interacting many-body systems. It is straightforward to show with the above expression that for normal systems each of the three terms in Equation (18.14) is proportional to N (or Ω) in the thermodynamic limit.[9] A general argument makes use of the Ω dependences in Equations (17.29) and (17.31) when sums over intermediate states are inserted between the interaction operators in Equation (18.14) and all momentum-conserving δ functions are used to eliminate dependent momentum variables. This linear N dependence is to be expected so long as the interparticle potentials $V_{\alpha\beta}$ are short-ranged, that is, fall off faster than $1/r_{\alpha\beta}$.

When the energy E_n is proportional to N, that is, when E_n is an extensive quantity, then we say that the interparticle forces *saturate,* which means that any one particle exerts an appreciable force only on a cluster of particles in its immediate neighborhood. For long-range forces, that is, for $1/r$ potentials, any one particle exerts an appreciable force on distant neighbors, and the energy E_n has an $N(N - 1) \cong N^2$ dependence. Thus Coulomb forces, for example, require special analysis of the perturbation series, and it is necessary to examine the entire infinite series of terms in ΔE_n to extract meaningful results in the thermodynamic limit.[10] For systems with overall charge neutrality, however, one finds, again, that the total energy per particle is independent of N in the large-N limit. The reason is that for neutral systems electrical screening occurs around each charged particle, thereby reducing effectively the range of its influence on neighboring charges.

There is an alternative formulation of time-independent perturbation theory called the *Brillouin–Wigner form*. This form is not readily applicable to many-body systems, because one cannot extract proper N dependence in the thermodynamic limit from individual terms in the series expression for ΔE_n.[11]

We have observed that the above development of Rayleigh–Schrödinger perturbation theory applies only, strictly speaking, to nondegenerate states. One can see why this is so by referring to Equation (18.11), which shows that vanishing energy denominators occur for degenerate states n. Here it should be mentioned that in quantum mechanics it is assumed that the *ground state* (the state $n = 0$) of any system is usually nondegenerate, as we discussed previously at the end of Section 9 in connection with Nernst's theorem. For this particular state, then, there is no degeneracy problem. The ground state is of particular interest theoretically, because it corresponds in thermodynamics to the zero-temperature limit.

Symmetry considerations can eliminate part or all of the difficulty associated with energy degeneracy. Thus unperturbed states that with the state $|\Psi_n^{(0)}\rangle$ have vanishing matrix elements of the perturbation V will not contribute to $|\Psi_n^{(k)}\rangle$, nor to $E_n^{(k)}$, when sums over intermediate states are inserted between interaction operators, that is, when the closure relation is used, in Equation (18.11) or (18.14). The occurrence of vanishing matrix elements is usually associated with symmetry properties of the single-particle Hilbert space and, therefore, often can be considered from a general viewpoint. When there still are unresolved degeneracies, then for the many-body problem it is customary to specify in the thermodynamic limit that *principal values,* or some other prescription, of the energy denominators are to be computed when they vanish.

Prescriptions for the evaluation of energy denominators can be justified by resorting to more elaborate theories such as those that use Green's functions. The question of the proper treatment of energy denominators is a difficult one, but once a prescription has been given, the degeneracy problem usually creates no further difficulties for infinite systems. Formulations of degenerate perturbation theory, appropriate for application to finite systems, have also been developed.[12]

Nonideal Fermi Gas

We now consider the example of a gas of weakly interacting identical fermions, that is, a nonideal Fermi gas, for which the system energy can be calculated from Equation (18.14). At very low temperatures it is indeed appropriate to make a quantum-mechanical calculation, because the probability for occupation of low-lying single-particle states is quite high. The Fermi statistics of the particles as manifested, say, by the Pauli exclusion principle is, therefore, an important consideration, and we shall allow the occupation numbers n_i of single-particle states to be 0 and 1 only. The label n, introduced at the beginning of this section, refers in the unperturbed state to the entire set of occupation numbers $(n_1 n_2 \cdots n_i \cdots)$.

We first write down the unperturbed energy $E_n^{(0)}$, which in the absence of any single-particle potential U will be the total microscopic kinetic energy in the system. Thus $E_n^{(0)}$ will be the nth eigenvalue of $\mathsf{H}_o = \mathsf{K}$; according to Equation (17.17) it is

$$E_n^{(0)} = \sum_i n_i \omega_i \tag{18.15}$$

where $\omega_i = \hbar^2 k_i^2/2m$.

We proceed next to the calculation of perturbation-theory corrections to $E_n^{(0)}$. In first order we have from either Equation (18.12) or (18.14), with the perturbation V given by Equation (17.19b),[13]

$$\begin{aligned} E_n^{(1)} &= \langle n_1 n_2 \cdots n_t \cdots | \mathsf{V}_2 | n_1 n_2 \cdots n_t \cdots \rangle \\ &= \frac{1}{4} \sum_{ijkl} \langle ij | V^{(A)} | kl \rangle \langle n_1 n_2 \cdots n_t \cdots | \mathsf{a}_j^\dagger \mathsf{a}_i^\dagger \mathsf{a}_k \mathsf{a}_l | n_1 n_2 \cdots n_t \cdots \rangle \\ &= \frac{1}{2} \sum_{ij} \langle ij | V^{(A)} | ij \rangle \langle n_1 n_2 \cdots n_t \cdots | \mathsf{a}_j^\dagger \mathsf{a}_i^\dagger \mathsf{a}_i \mathsf{a}_j | n_1 n_2 \cdots n_t \cdots \rangle \end{aligned} \tag{18.16}$$

The third equality is derived by observing that, since the state vectors (17.21) form a complete orthonormal set, we must have either $(k, l) = (i, j)$ or $(k, l) = (j, i)$. In the second case Equation (17.20) and one of the anticommutation relations (17.8) must also be used. To evaluate the matrix element of annihilation and creation operators in the last line of Equation (18.16), we can use Equation (17.5) and either of Equations (17.9) to obtain

$$E_n^{(1)} = \frac{1}{2} \sum_{i,j} n_i n_j \langle ij | V^{(A)} | ij \rangle - \frac{1}{2} \sum_i n_i \langle ii | V^{(A)} | ii \rangle$$

$$E_n^{(1)} = \frac{1}{2} \sum_{\substack{i,j \\ (i \neq j)}} n_i n_j \langle ij | V^{(A)} | ij \rangle \tag{18.17}$$

$$= \frac{1}{2} \sum_{\substack{i,j \\ (i \neq j)}} n_i n_j (\langle ij | V^{(2)} | ij \rangle - \langle ij | V^{(2)} | ji \rangle)$$

where $\langle ii | V^{(A)} | ii \rangle \equiv 0$ according to Equation (17.20).

It is instructive to represent the first-order energy $E_n^{(1)}$ by a picture. For this purpose we introduce the concept of a *Fermi sea,* which is defined as the collection of occupied single-particle states (those with $n_i = 1$) in the unperturbed many-fermion state. The unoccupied states (those with $n_i = 0$) are then said to be above the Fermi sea, although only for the ground state ($n = 0$) will the concept "above" correspond to a state-ordering of increasing energy. Since we have considered an arbitrary pure state n, the Fermi sea, which we represent in Figure IV.1 by fluid in a box, may include any N single-particle states.

If we represent the occupied states i and j by points in the Fermi sea and the interaction $V^{(2)}$ by two directed lines from the initial $|\ \ \rangle$ to the final $\langle\ \ |$ states, then the direct and exchange terms in the final equality of Equation (18.17) can be pictured as shown in Figure IV.1. Without presenting any formal rules for their construction, we can use such pictures as a guide to the calculation of higher order perturbation-theory contributions to ΔE_n. These pictures are intended only as visual aids to give direction to tedious calculations, and we now illustrate their value by calculating the second-order energy $E_n^{(2)}$.

The first-order energy $E_n^{(1)}$ is merely the sum over all pairs of occupied single-particle states, in the unperturbed many-body state n, of the diagonal, but antisymmetrized, matrix elements of $V^{(2)}$. To calculate $E_n^{(2)}$ we must, according to Equation (18.14), consider off-diagonal matrix elements of $V^{(2)}$; when we insert a sum over intermediate states between the two interaction

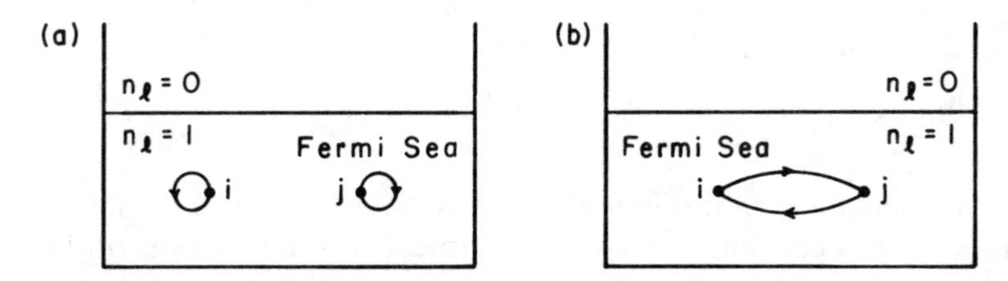

FIGURE IV.1 A pictorial representation of the direct (a) and exchange (b) contributions to the first-order energy $E_n^{(1)}$ of Equation (18.17) for a system of identical fermions. The Fermi sea here refers to the collection of occupied single-particle states in the unperturbed many-body state.

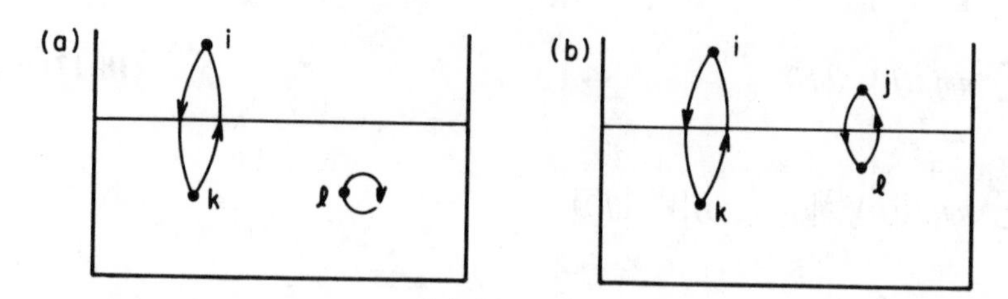

FIGURE IV.2 A pictorial representation of the two possible contributions from V_2 to the second-order energy $E_n^{(2)}$. Only the direct terms are shown, with one-particle excitations depicted in part (a) and two-particle excitations in part (b).

operators of the second-order term, then the projection operator $(1 - P_n^{(0)})$ causes the state n to be excluded. The intermediate, or excited, single-particle states must, therefore, include excitations above the Fermi sea, in agreement with the Pauli exclusion principle. Moreover, since the perturbation V_2 of Equation (17.19b) involves, at most, only two-particle excitations, we need consider only those terms represented pictorially in Figure IV.2, where for reasons of simplicity we have represented only the direct terms of the antisymmetric matrix elements of $V^{(2)}$.

Figure IV.2 shows immediately that the second-order perturbation energy in Equation (18.14) can be written, using the fact that V_2 is Hermitian, in the form

$$
\begin{aligned}
E_n^{(2)} &= \langle n|V_2\left(\frac{1 - P_n^{(0)}}{E_n^{(0)}1 - H_o}\right)V_2|n\rangle \\
&= \sum_{i,k} |\langle n_1 n_2 \cdots (n_k - 1) \cdots (n_i + 1) \cdots |V_2|n_1 n_2 \cdots \rangle|^2 \\
&\quad \times (\omega_k - \omega_i)^{-1}\, \delta_{n_k,1}\, \delta_{n_i,0} \\
&\quad + \frac{1}{4} \sum_{\substack{ijkl \\ (i\neq j),(k\neq l)}} |\langle n_1 n_2 \cdots (n_k - 1) \cdots (n_l - 1) \cdots (n_i + 1) \cdots (n_j + 1) \\
&\quad \cdots |V_2|n_1 n_2 \cdots \rangle|^2 (\omega_k + \omega_l - \omega_i - \omega_j)^{-1}\, \delta_{n_k,1}\, \delta_{n_l,1}\, \delta_{n_i,0}\, \delta_{n_j,0}
\end{aligned}
\tag{18.18}
$$

The energy denominators in this expression have been computed by noting that the differences between $E_n^{(0)}$ and the intermediate-state energies must involve only those single-particle states that are excited out of the Fermi sea. The restrictions $i \neq j$ and $k \neq l$ in the second term reflect the Pauli-principle restriction that no two fermions can occupy the same single-particle state. The factor of ¼ then occurs, due to these two pairs of states, because the sum over intermediate states must be a sum over different many-body states.

We next evaluate the matrix elements in Equation (18.18). For this purpose it is now necessary to substitute the explicit expression (17.19b). If one recalls the properties (17.20) of the antisymmetrized matrix elements of $V^{(2)}$, then one obtains, with the aid of Equations (17.2), (17.3), and (17.8), the results

$$\langle n_1 n_2 \cdots (n_k - 1) \cdots (n_i + 1) \cdots |V_2| n_1 n_2 \cdots \rangle \, \delta_{n_k,1} \, \delta_{n_i,0}$$
$$= \sqrt{n_i(1 - n_i)} \sum_{l \neq k} n_l \langle il|V^{(A)}|kl\rangle \, \varepsilon^{\Theta_i^{(f)}} \, \varepsilon^{\Theta_k^{(0)}} \tag{18.19a}$$

$$\langle n_1 n_2 \cdots (n_k - 1) \cdots (n_l - 1) \cdots (n_i + 1) \cdots (n_j + 1) \cdots |V_2| n_1 n_2 \cdots \rangle$$
$$\times \, \delta_{n_k,1} \, \delta_{n_l,1} \, \delta_{n_i,0} \, \delta_{n_j,0}$$
$$= \sqrt{n_l n_k (1 - n_i)(1 - n_j)} \langle ij|V^{(A)}|kl\rangle \, \varepsilon^{\Theta_j^{(f)}} \, \varepsilon^{\Theta_i^{(f)}} \, \varepsilon^{\Theta_k^{(0)}} \, \varepsilon^{\Theta_l^{(0)}} \tag{18.19b}$$

Here $\Theta^{(f)}$ refers to the intermediate, or excited, state and $\Theta^{(0)}$ refers to the (initial) state n. In Equation (18.19b), $\Theta_k^{(0)}$ is computed by using Equation (17.1) with the single-particle state l unoccupied, and similarly, $\Theta_i^{(f)}$ is computed with the excited state j unoccupied. For all of the sign factors we must set $\varepsilon = -1$, although as soon as the matrix elements are squared for substitution into Equation (18.18), the signs become irrelevant. It is not until one calculates the third-order energy $E_n^{(3)}$ (see Problem IV.3) that such sign factors become important in final energy expressions. The particular occupation-number coefficients on the right sides of Equations (18.19) assure that $n_k = 1$, $n_i = 0$, and so on, when only values 0 and 1 are allowed in the unperturbed state n.

Exercise 18.1 Derive Equations (18.19a) and (18.19b).
[*Suggestion:* To avoid confusion with the state labels in these equations, it is convenient to relabel the single-particle states (i, j, k, l) in Equation (17.19b) as, say, (q, r, s, t). For each matrix element there will then be four nonvanishing possibilities for identifying the set (q, r, s, t) with the excited-state labels of Equations (18.19).]

Substitution of Equations (18.19) into Equation (18.18) yields for $E_n^{(2)}$ the expression

$$E_n^{(2)} = \frac{1}{4} \sum_{\substack{ijkl \\ (i \neq j),(k \neq l)}} n_l n_k (1 - n_i)(1 - n_j) |\langle ij|V^{(A)}|kl\rangle|^2 (\omega_k + \omega_l - \omega_i - \omega_j)^{-1}$$
$$+ \sum_{i,k} n_k (1 - n_i) \sum_{l \neq k} n_l |\langle il|V^{(A)}|kl\rangle|^2 (\omega_k - \omega_i)^{-1} \tag{18.20}$$
$$\xrightarrow[\Omega \to \infty]{} \frac{1}{4} \sum_{ijkl} n_k n_l (1 - n_i)(1 - n_j) |\langle ij|V^{(A)}|kl\rangle|^2 (\omega_k + \omega_l - \omega_i - \omega_j)^{-1}$$

The expression for $E_n^{(2)}$ in the infinite-volume, or thermodynamic, limit does not include the one-particle excitation term, because momentum conservation [Equation (17.31)] occurs in this limit, and this is impossible when i and k are required to be different states. For a finite system the single-particle eigenstates of H_o will not be plane waves, so that this momentum-conservation argument cannot then be applied. The infinite-volume limit for $E_n^{(2)}$ exhibits the Pauli principle in a characteristic fashion. The states k and l represent, via the factor $n_k n_l$, all possible pairs of occupied single-particle unperturbed states that can be excited. The states i and j to which excitation occurs must be, according to the exclusion principle, unoccupied states, and this condition is guaranteed by the factor $(1 - n_i)(1 - n_j)$.

The $i \neq j$ and $k \neq l$ restrictions on the state summations in Equations (18.17) and (18.20) are actually unnecessary, because these restrictions are automatically realized when one takes account of the antisymmetrized matrix elements of $V^{(2)}$. Thus when $\varepsilon = -1$ Equation (17.20) gives zero identically for these matrix elements if $i = j$ or $k = l$. For the ground state, in which all excited-state energies (ω_i and ω_j) are higher than all unexcited-state energies (ω_k and ω_l), we note that the second-order energy $E_0^{(2)}$ is negative definite. Second-order perturbation theory always lowers the unperturbed ground-state energy.

With the aid of Equations (17.29) and (17.31) we can immediately deduce the N dependence in the thermodynamic limit of the explicit perturbation energies $E_n^{(1)}$ and $E_n^{(2)}$. After recalling that the density $n = N/\Omega$ is finite in the thermodynamic limit, one concludes that both $E_n^{(1)}$ and $E_n^{(2)}$ are proportional to N (for short-ranged forces) in this limit, as has already been indicated in the discussion below Equation (18.14). This property is also true for the unperturbed, or kinetic, energy $E_n^{(0)}$ of Equation (18.15).

19 The Fermi Fluid

The *Fermi fluid* is a quantum fluid whose identical particles are fermions. A *quantum fluid* is a gas or liquid whose identical particles have overlapping wavelengths, that is, a fluid for which $\lambda_T \gtrsim \ell$, where λ_T is the thermal de-Broglie wavelength (4.2) and $\ell = n^{-1/3}$ is the interparticle spacing. Quantum-mechanical effects, such as particle statistics and properly symmetrized wave functions, are important in the temperature–density regions that characterize quantum fluids.

Inasmuch as particle wave packets are necessarily spread out in position space for quantum fluids, we can expect, because of the Heisenberg uncertainty relation, that wave packets can be well defined in momentum space. In fact, there can be *momentum-space ordering* when $\lambda_T >> \ell$. Momentum-space ordering contrasts with the more familiar position-space ordering, or localization, that characterizes models of many-particle systems in classical physics. The Fermi sea, introduced in the preceding section, is a primary example of momentum-space ordering for Fermi fluids.

In this section we explore some of the concepts associated with the Fermi sea. We first study the ground state of the ideal Fermi gas, whose Fermi sea is particularly simple to characterize. Then in the rest of the section we extend this notion by developing Landau's theory of a normal Fermi fluid, a theory whose success enables us to view the Fermi sea as a common, low-temperature phenomenon characteristic of such diverse substances as liquid helium three, the conduction electrons in metals, large atomic nuclei, and the electrons in white dwarf stars.

In the preceding section we calculated the energy of a nonideal Fermi gas to second order in perturbation theory. In this section we explore implications of this model calculation in considerable detail, and we begin by focusing attention on the unperturbed, or ideal, Fermi gas to its ground state ($n = 0$), which corresponds to zero absolute temperature ($T = 0$).

We have already observed below Equation (18.14) that the ground states of many-body systems are of particular theoretical interest, in part because their unperturbed states can usually be characterized quite easily. Thus the unperturbed ground state of a weakly interacting normal Fermi gas is characterized by a Fermi sea in which all of the fermions are in the lowest possible single-particle energy states. Of course, the Pauli exclusion principle dictates that the fermions must all be in different single-particle states, so that the energetically highest occupied state at $T = 0$ will be at the top of the

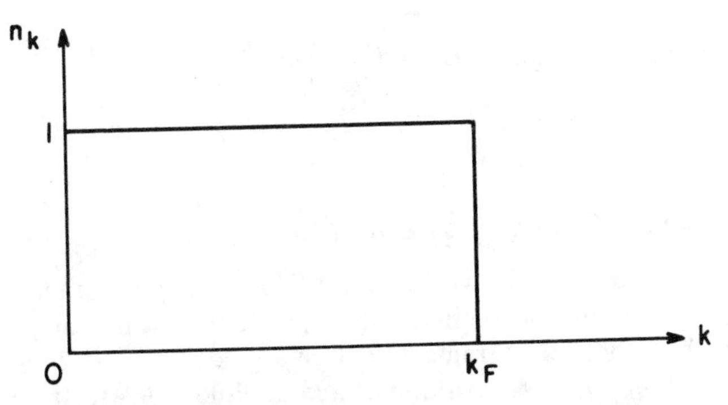

FIGURE IV.3 The occupation numbers n_k for the ground state ($T = 0$) of an ideal Fermi gas. The occupied states $k \leq k_F$ comprise the Fermi sea in this case. One could just as well use an energy scale for the abscissa, in which case k_F would be replaced by the Fermi energy E_F. This latter alternative lends itself readily to the Fermi-fluid generalization (19.15).

Fermi sea. For an infinite system the highest occupied state is a momentum state whose eigenvalue is called the *Fermi momentum* $\hbar k_F$. The occupation numbers for an infinite system, therefore, are (see Figure IV.3):

$$\begin{aligned} n_i = n(k_i) &= 1 \quad \text{if} \quad k_i \leq k_F \\ &= 0 \quad \text{if} \quad k_i > k_F \end{aligned} \tag{19.1}$$

We can determine the Fermi momentum in the thermodynamic limit by using Equation (17.6) together with (17.29).

$$\begin{aligned} n = \Omega^{-1}\langle \mathrm{N} \rangle &= (2\pi)^{-3}(2S + 1) \int d^3\mathbf{k}\, n(k) \\ &\xrightarrow[T\to 0]{} (2\pi)^{-3}(2S + 1) \int_{k<k_F} d^3\mathbf{k} = \frac{(2S + 1)k_F^3}{6\pi^2} \end{aligned} \tag{19.2}$$

With $n = \ell^{-3}$, where ℓ is the average interparticle spacing, we have for the common case of spin $S = 1/2$ fermions,

$$k_F^3 = 3\pi^2/\ell^3 \tag{19.3}$$

Thus the Fermi momentum is essentially equal to the reciprocal of the interparticle spacing ($k_F \sim 3/\ell$), as indeed it must be from a dimensional argument. In applications one frequently uses Equation (19.3) to define the Fermi momentum for nonideal Fermi fluids at $T = 0$, and also for all Fermi fluids when $T \neq 0$.

With the aid of Equation (19.2) one can show that the unperturbed ground-state energy (18.15) in a Fermi gas is given by

$$\frac{1}{\langle \mathrm{N} \rangle} E_0^{(0)} \xrightarrow[\Omega\to\infty]{} \frac{3}{5} E_F \tag{19.4}$$

where the *Fermi energy* E_F is defined by $E_F = \hbar^2 k_F^2/2m$.

Exercise 19.1 Prove Equation (19.4).

Landau's Theory of a Fermi Fluid

We have indicated in Section 17, above Equation (17.27), that a momentum description of large systems can be more appropriate than a position-space description, simply by virtue of translational invariance. At very low temperatures the momentum-space description of a large fluid gains further significance when momentum-space ordering occurs. Thus the Fermi sea described above characterizes the ordering of momentum values in an ideal Fermi gas. Indeed, the calculation at the end of Section 18 illustrates the importance of a Fermi-sea concept for the description of a nonideal Fermi gas. The essence of Landau's theory, which we now develop, is that Fermi fluids are characterized quite generally by an ordering of single-particle energy values into a Fermi sea. For translationally invariant fluids these energy values correspond to momentum, or plane-wave, eigenstates.

The Landau theory gives a semiphenomenological description of a normal Fermi liquid or gas. Its microscopic basis is contained in perturbation theory, yet the theory stands on its own in providing a basis for the understanding of low-temperature observations in Fermi fluids. With regard to the perturbation-theoretic basis of Landau's theory our study of the nonideal Fermi gas in Section 18 is instructive, for even in the second-order energy expression (18.20) one can see important general features, as we discuss below. Central to this theory is the fact that physical or thermodynamic quantities can be regarded as functions (or functionals in the limit $\Omega \to \infty$) of the occupation numbers n_i.

We use the qualifying adjective normal, as in Section 18, to designate those Fermi fluids whose many-body states are in one-to-one correspondence with the corresponding unperturbed states of an ideal Fermi gas. Moreover, it is assumed for a normal Fermi liquid that there exist single-particle-like quantities, called *quasi-particles,* that are in one-to-one correspondence with the particles of the corresponding noninteracting Fermi gas. That is to say, if we imagine the particle interactions to be turned on gradually from the noninteracting case, then in this process the classification of single-particle states with associated quantum numbers remains essentially unchanged. Of course, the labels i used to describe the noninteracting system cannot have precisely their same meaning when used to label effective single-particle states in the real Fermi fluid. In an infinite homogeneous system, however, translational invariance would prevail as the interactions were turned on, and the k_i labels would continue to indicate plane-wave states. In the case of infinite systems, then, the conceptual problem is only to relate the meaning of momentum-dependent quasi-particle quantities of the real interacting Fermi fluid to their noninteracting counterparts.

Associated with each quasi-particle state i in the Fermi fluid is a quasi-particle energy ω_i',[14] which corresponds in the unperturbed state to the energy ω_i (usually the kinetic energy, $\hbar^2 k_i^2/2m$). The difference between ω_i' and ω_i is a single-particle interaction energy U_i which takes into account, for the dynamics of a quasi-particle in state i, the inertial effect or "drag" of all the other quasi-particles in the system. We specify this energy U_i quantitatively below, but for the moment we note only that U_i depends generally in some very complicated manner on the particle interactions $V^{(2)}$ in the system.

In addition to their energies, ω_i', quasi-particles are also characterized by their occupation numbers, n_i'. Here we rely on the procedure of perturbation theory, wherein for any pure quantum state the same set of occupation numbers n_i is used to characterize, or label, both the perturbed state and the unperturbed state, even though the perturbation does indeed alter the expectation values of these occupation numbers. For pure states, then, we set $n_i' = n_i$. Moreover, no matter what values the occupation numbers $n_i' = n_i$ take, the system energy E_n retains its same functional dependence on the n_i' for all pure states, as can be seen in the first two orders of perturbation theory from Equations (18.17) and (18.20). Thus the numbers n_i' become a set of undetermined variables in the present view, and the question to be asked below is: What actual values do the n_i' take for a statistically averaged Fermi fluid (one in thermodynamic equilibrium)?

We now study the functional dependence of the quasi-particle energies ω_i' on their occupation numbers n_i'. Here we observe that the change in the total energy E (suppressing the subscript n) due to the removal of a particle from the state i, in a virtual process that does not allow any other quasi-particles to change states, must be the single-particle energy ω_i'. Therefore we must have for functional variations $\delta n_i'$ the expression

$$\delta E = \sum_i \omega_i' \, \delta n_i' \tag{19.5}$$

This relation shows that the energy ω_i' is given by a partial derivative of the system energy E

$$\omega_i' = \omega_i + U_i = \frac{\partial E}{\partial n_i'} \tag{19.6}$$

In the limit $N \to \infty$, the partial derivative becomes a functional derivative.[15]

The result (19.6) is consistent with the above discussion of the interaction energy U_i, since according to Equation (18.15) ω_i arises in Equation (19.6) from $E^{(0)}$ whereas U_i arises from perturbation corrections to $E^{(0)}$. The single-particle potential U_i is, in fact, defined by the partial derivative (19.6). In contemporary literature it is called the *Hartree–Fock potential*.

It must be stressed that the quasi-particle energies ω_i' have not been defined to be effective single-particle eigenstates of the many-body Hamiltonian such that E is the sum over all i of $n_i'\omega_i'$. In fact, for interacting systems with the ω_i' defined by Equation (19.6),

$$E \neq \sum_i n_i' \omega_i'$$

The physical reason why this inequality is so can be appreciated by referring to the above interpretation of U_i as an energy due to the effect of all other quasi-particles on the one in the state i. We can exhibit this effect to leading order in perturbation theory by deriving from Equations (19.6) and (18.17) the term $U_i^{(1)}$

$$U_i^{(1)} = \sum_j n_j' \langle ij | V^{(A)} | ij \rangle \tag{19.7}$$

It is then easy to verify that the sum over all i of $n_i' U_i$ gives twice the total interaction energy $E^{(1)}$, a result that occurs because all interacting pairs are counted twice. [One finds, more generally, that ΔE of Equation (18.13) is not precisely equal to $\frac{1}{2} \Sigma_i n_i' U_i$, because of many-particle correlations. This fact can be checked in second order by calculating $U_i^{(2)}$ from Equation (18.20).]

What, then, is the advantage of definition (19.6)? The answer is that its use as a constraint equation, similar to the second of Equations (5.9), leads to a meaningful expression for the n_i' of statistically averaged Fermi fluids. In analogy with the ideal Fermi gas, in which Equations (5.8) and (5.9) together resulted in the average value (5.10), we derive below an expression for the $\bar{n}_i'$. For the present we shall pursue other implications of Equation (19.6).

To use the prescription (19.6) in a microscopic calculation we must first have an expression for the system energy E. Even without a specific expression, however, we can extract useful conclusions from the following general consideration. Thus perturbation theory via Equations (18.17) and (18.20) (see also Problem IV.3.) suggests that the energy of a Fermi fluid when calculated microscopically must have the general form

$$E = \sum_i n_i' \omega_i + \frac{1}{2} \sum_{i,j} n_i' n_j' E(i, j) \tag{19.8}$$

in which a symmetric "interaction energy" $E(i, j) = E(j, i)$, itself a functional of n', is defined. The origin of the first term is quite clear; it is the kinetic energy (18.15). The second term, which can be associated with quasi-particle interactions but occurs because of real-particle interactions, must contain at least two occupation-number factors to represent interacting quasi-particles. In the example of the nonideal Fermi gas, an expression for $E(i, j)$ valid to second order in perturbation theory can be derived from Equations (18.17) and (18.20). One readily obtains (for $\Omega \rightarrow \infty$)

$$\begin{aligned} E(i, j) &= \langle ij|V^{(A)}|ij\rangle \\ &\quad + \frac{1}{2} \sum_{k,l} (1 - n_k')(1 - n_l')|\langle kl|V^{(A)}|ij\rangle|^2(\omega_i + \omega_j - \omega_k - \omega_l)^{-1} \\ &\quad + O(V^3) \end{aligned} \tag{19.9}$$

Equation (19.9) shows explicitly that the interaction energy $E(i, j)$ depends functionally on the occupation numbers $n_s' = n_s$ (see also Problem IV.5.). Therefore, when the general expression (19.8) is substituted into Equation (19.6) two distinct terms for U_i result.

$$U_i = \sum_j n_j' E(i, j) + \frac{1}{2} \sum_{k,l} n_k' n_l' \frac{\partial}{\partial n_i'} E(k, l) \tag{19.10}$$

The first term, easily interpretable in terms of pairs of interacting quasi-particles, has as its leading contribution in perturbation theory the quantity $U_i^{(1)}$ of Equation (19.7). The second term, known as the *rearrangement energy,* arises: (1) because of many (i.e., more than two) particle excitations of the fluid's unperturbed state vector, and (2) because of the Pauli exclusion principle.

In analogy with Equation (19.5), one can also consider functional variations of ω_i' itself with respect to the occupation numbers:

$$\delta\omega_i' = \sum_j U_2(i, j)\, \delta n_j' \tag{19.11}$$

Here the weighting function $U_2(i, j)$ for a virtual change $\delta n_j'$ must be the interaction energy between two quasi-particles in states i and j. To verify this assertion interpret the effect on ω_i' of the simple virtual change $\delta n_j' = \delta_{jl}$. We must emphasize, however, that the interaction energy $U_2(i, j)$ differs significantly from the energy $E(i, j)$ in Equation (19.8), because the latter quantity does not include "rearrangement-energy" contributions.

Equation (17.31) shows that both of the symmetric interaction energies $E(i, j)$ and $U_2(i, j)$ must have a volume dependence Ω^{-1} for large Fermi fluids. In the Landau theory the volume-independent quantity $\Omega U_2(i, j)$ is labeled as $f(i, j)$. It is the quantity that most characterizes the distinction between real Fermi fluids and the corresponding free Fermi gas of quasi-particles. In principle, the calculation of $f(i, j)$ from microscopic theory requires use of the general prescription

$$U_2(i, j) = \Omega^{-1} f(i, j) = \frac{\delta^2 E}{\delta n_j'\, \delta n_i'} \tag{19.12a}$$

which follows from Equations (19.11) and (19.6). One can also write, equivalently,

$$U_2(i, j) = \frac{\partial U_i}{\partial n_j'} = \frac{\partial U_j}{\partial n_i'} \tag{19.12b}$$

Statistically Averaged Fermi Fluid

The preceding considerations apply quite generally to the pure states of a normal Fermi fluid for any set of the occupation numbers $n_i' = n_i$. Now we wish to determine the most probable set of occupation numbers $\bar{n}_i'$ for a real Fermi fluid, that is, one that corresponds to the statistical average of many Fermi fluids in pure states. For this purpose we must refer to a statistical-entropy expression for the Fermi fluid.

The analysis of the entropy of the ideal Fermi gas given in Section 5 applies to normal Fermi fluids in general. As surprising as this might seem at first, reflection reveals that only the enumeration of single-particle states enters into the derivation of Equation (5.5). And as we have carefully specified above Equation (19.5), the quasi-particle occupation numbers n_i' are identical for pure states to the unperturbed (or ideal-gas) occupation numbers n_i.[16] Thus on defining $v_i' \equiv n_i'/Z_i$ [see Equation (5.1)], we obtain immediately from Equation (5.5) the entropy expression

$$S = -\kappa \sum_i Z_i[\nu_i' \ln \nu_i' + (1 - \nu_i')\ln(1 - \nu_i')] \tag{19.13}$$

where $Z_i = \delta^3\mathbf{n}_i' >> 1$ is a cell in the space of single-particle quantum numbers i. The derivation of an expression for the most probable occupation numbers $\overline{n}_i'$ is the subject of the following exercise.

Exercise 19.2 Verify by analogy with the development in Exercises 5.1 and 5.2 that the most probable occupation numbers $\overline{n}_i'$ for a Fermi fluid are given by

$$\overline{n}_i' = \delta^3 n_i'[\exp\{\beta(\omega_i' - g)\} + 1]^{-1} \tag{19.14}$$

$$\xrightarrow[\delta^3\mathbf{n}'_i = 1]{} [\exp\{\beta(\omega_i' - g)\} + 1]^{-1}$$

where β is related to the absolute temperature T by $\beta = (\kappa T)^{-1}$ and g is the chemical potential of the Fermi fluid.[17]

The simple result (19.14) is remarkable, as it shows that the quasi-particles of a normal Fermi fluid are distributed in energy in precisely the same way as are the particles in an ideal gas. Yet according to Equation (19.11) quasi-particles also interact. In the zero-temperature limit, $\beta \to \infty$, $g \to g_0$ and

$$\begin{aligned} \overline{n}_i' &= 1 \quad \text{if} \quad \omega_i' \leq g_0 \\ &= 0 \quad \text{if} \quad \omega_i' > g_0 \end{aligned} \tag{19.15}$$

The concept of a Fermi sea, Equation (19.1), applies not only to the ideal Fermi gas but also to the quasi-particles of a normal Fermi fluid. See Figure IV.3.

Inasmuch as the microscopic energy expression (19.8) applies to an arbitrary pure state of a Fermi fluid, with an arbitrary set of occupation numbers n_i', it also applies when the most probable set of occupation numbers $\overline{n}_i'$ (with $\delta^3\mathbf{n}_i' = 1$) is used. Thus the pure-state analysis also applies to the real-fluid case.

Exercise 19.3 Show that the constant-volume heat capacity of a Fermi fluid (for constant total particle number N) is given by

$$C_V = \sum_i \omega_i' \frac{d\overline{n}_i'}{dT} = \sum_i (\omega_i' - g)\frac{d\overline{n}_i'}{dT} \tag{19.16}$$

[*Suggestion:* Use Equation (19.5).]
The merit of the second equality is that the quasi-particle energies can be expressed relative to the chemical potential g, which as $T \to 0$ becomes the energy at the top of the Fermi sea.

In the introduction to this section we cited several examples of Fermi fluids in nature. These are not always normal. In particular, the electron gas in metals can become superconducting for some substances at very low temperatures. In the superconducting state the single-particle energy levels of the fermions no longer satisfy the adiabatic condition of perturbation theory, and a gap can appear near the Fermi surface. Special treatment of this nonnormal case is required.

Bose fluids also occur in nature, and these also exhibit specific quantum phenomena. Three known examples are blackbody radiation, phonons in a crystal, and liquid helium four. Each has its own distinctive characteristics, in addition to some common features; discussions of them are given in Sections 30, 34, and 35.

20 Green's Functions for Many-Particle Systems

The use of Green's functions in classical mechanics and in electromagnetic theory is well known as a powerful mathematical tool for the solution of physical problems. These functions also find extensive application in quantum theory; in particular, they are the "propagators" of quantum field theory. In this section we show how Green's functions can be applied to the nonrelativistic many-body problem in quantum mechanics.

We consider only one-particle Green's functions in this section and define a first one, $\tilde{G}(x_2, x_1)$, in terms of both position and time variables, where $x = (\mathbf{r}, \tau)$. Then we introduce a second, thermodynamic Green's function, $G(\mathbf{r}_2 t_2; \mathbf{r}_1 t_1)$, in terms of position and temperature variables, where $0 < t < \beta = (\kappa T)^{-1}$. The thermodynamic Green's function is more useful for applications to macroscopic systems, and we show how it can be used to calculate equilibrium averages; it is also the easier function to calculate in a perturbation series using Feynman-like diagrams (see Section 50). Notwithstanding, there are important physical interpretations of physical systems that involve their single-particle excitation energies and corresponding lifetimes,[18] and these are best expressed using the space–time Green's function. It is, therefore, fortunate that an intimate connection between these two Green's functions can be established using the so-called *Lehmann representation,* which we develop in the second half of this section.

We begin by introducing the single-particle space–time Green's function $\tilde{G}(x_2, x_1)$, for nonrelativistic problems, in terms of the Fock-space field operators $\underline{\psi}(\mathbf{r})$ and $\underline{\psi}^\dagger(\mathbf{r})$ of Equations (17.22). We first express these operators in the Heisenberg picture[19] as $\underline{\psi}_H(x)$ and $\underline{\psi}_H^\dagger(x)$, using the definitions

$$\begin{aligned} \underline{\psi}_H(x) &\equiv e^{i\mathsf{H}\tau/\hbar}\underline{\psi}(\mathbf{r})\, e^{-i\mathsf{H}\tau/\hbar} \\ \underline{\psi}_H^\dagger(x) &\equiv e^{i\mathsf{H}\tau/\hbar}\underline{\psi}^\dagger(\mathbf{r})\, e^{-i\mathsf{H}\tau/\hbar} \end{aligned} \tag{20.1}$$

where H is the many-body Hamiltonian of, say, Equations (17.25) and (17.26):

$$\mathsf{H} = \mathsf{K} + \mathsf{U} + \mathsf{V}_2 \tag{20.2}$$

The one-particle Green's function $\tilde{G}(x_2, x_1)$ is then defined by

$$\begin{aligned} i\tilde{G}(x_2, x_1) &\equiv \langle P\{\underline{\psi}_H(x_2)\underline{\psi}_H^\dagger(x_1)\}\rangle \\ &= \Theta(\tau_2 - \tau_1)\langle\underline{\psi}_H(x_2)\underline{\psi}_H^\dagger(x_1)\rangle \\ &\quad + \varepsilon\Theta(\tau_1 - \tau_2)\langle\underline{\psi}_H^\dagger(x_1)\underline{\psi}_H(x_2)\rangle \end{aligned} \tag{20.3}$$

In this definition P is a chronological ordering operator, as defined by the second line, such that when the operators $\underline{\psi}_H$ and $\underline{\psi}_H^\dagger$ are interchanged the exchange factor ε is included, with $\varepsilon = +1$ for Bose particles and $\varepsilon = -1$ for Fermi particles. The variable τ is a time coordinate here, and $\Theta(\tau)$ is a step function defined by $\Theta(\tau) = +1$ for $\tau > 0$ and $\Theta(\tau) = 0$ for $\tau \leqslant 0$. The average-value brackets $\langle\ \ \rangle$ are assumed to be those of a statistically averaged quantum-mechanical state.[20]

We now show the sense in which $\tilde{G}(x_2, x_1)$ is a Green's function. Consider the equation of motion for operators in the Heisenberg picture,[19] that is,

$$i\hbar\frac{\partial}{\partial\tau}\underline{\psi}_H(x) = [\underline{\psi}_H(x), \mathsf{H}] \tag{20.4}$$

with a similar equation for $\underline{\psi}_H^\dagger(x)$. The commutator bracket in this equation can be expressed alternatively by using Equations (17.23), (17.25), and (17.26). Thus it is straightforward to show for either statistics that

$$\begin{aligned} [\underline{\psi}(\mathbf{r}), \mathsf{H}_o] &= \int d^3\mathbf{r}'\langle\mathbf{r}|H^{(1)}|\mathbf{r}'\rangle\underline{\psi}(\mathbf{r}') \\ &\equiv \mathsf{H}_o\underline{\psi}(\mathbf{r}) \\ [\underline{\psi}(\mathbf{r}), \mathsf{V}_2] &= \int d^3\mathbf{r}_1\, d^3\mathbf{r}_2\, d^3\mathbf{r}_2'\underline{\psi}^\dagger(\mathbf{r}_2')\langle\mathbf{r}\mathbf{r}_2'|V^{(2)}|\mathbf{r}_1\mathbf{r}_2\rangle\underline{\psi}(\mathbf{r}_2)\underline{\psi}(\mathbf{r}_1) \\ &\equiv \mathsf{V}_2\underline{\psi}(\mathbf{r}) \end{aligned} \tag{20.5}$$

where $\mathsf{H}_o \equiv \mathsf{K} + \mathsf{U}$ and the second lines of these two equations are defined in each case by the first equalities.

Exercise 20.1 Verify the first equalities of Equations (20.5).

On substituting Equations (20.5) into the equation of motion (20.4), we obtain the formal result

$$i\hbar\frac{\partial}{\partial\tau}\underline{\psi}_H(x) = e^{i\mathsf{H}\tau/\hbar}[\underline{\psi}(\mathbf{r}), \mathsf{H}]\, e^{-i\mathsf{H}\tau/\hbar} = \mathsf{H}\underline{\psi}_H(x) \tag{20.6}$$

We must stress that this last equality is a formal result that can be written out explicitly only by referring to Equations (20.5). We next derive from Equation (20.3) the differential equation,

$$\begin{aligned} i\hbar\frac{\partial}{\partial\tau_2}\tilde{G}(x_2, x_1) &= \hbar\,\delta(\tau_1 - \tau_2)\langle[\underline{\psi}_H(x_2)\underline{\psi}_H^\dagger(x_1) - \varepsilon\underline{\psi}_H^\dagger(x_1)\underline{\psi}_H(x_2)]\rangle \\ &\quad + \hbar\left\langle P\left\{\frac{\partial}{\partial\tau_2}\underline{\psi}_H(x_2)\underline{\psi}_H^\dagger(x_1)\right\}\right\rangle \\ &= \hbar\,\delta(\tau_1 - \tau_2)\,\delta^{(3)}(\mathbf{r}_1 - \mathbf{r}_2) + \mathsf{H}(2)\tilde{G}(x_2, x_1) \end{aligned}$$

where the second equality follows after using Equation (20.6) and the first of Equations (17.23). The operator $\mathsf{H}(2) = [\mathsf{H}_o(2) + \mathsf{V}_2(2)]$ is defined to operate on $\underline{\psi}(\mathbf{r}_2)$ according to Equations (20.5). Rearrangement of this last equation now yields formally the familiar differential equation for a Green's function in four dimensions,[21]

$$\left\{i\hbar 1\frac{\partial}{\partial\tau_2} - \mathsf{H}(2)\right\}\tilde{G}(x_2, x_1) = \hbar\,\delta(\tau_1 - \tau_2)\,\delta^{(3)}(\mathbf{r}_1 - \mathbf{r}_2) \tag{20.7}$$

Thermodynamic Green's Function

We introduce the temperature variable t, where $0 \leqslant t \leqslant \beta = (\kappa T)^{-1}$, and define the *thermodynamic Green's function* for temperature variables by

$$G(\mathbf{r}_2 t_2; \mathbf{r}_1 t_1) \equiv \langle P\{\underline{\psi}_H(\mathbf{r}_2, t_2)\underline{\psi}_H^\dagger(\mathbf{r}_1, t_1)\}\rangle \tag{20.8}$$

where, similar to Equations (20.1), we also write

$$\begin{aligned} \underline{\psi}_H(\mathbf{r}, t) &\equiv e^{t\mathsf{H}}\underline{\psi}(\mathbf{r})\, e^{-t\mathsf{H}} \\ \underline{\psi}_H^\dagger(\mathbf{r}, t) &\equiv e^{t\mathsf{H}}\underline{\psi}^\dagger(\mathbf{r})\, e^{-t\mathsf{H}} \end{aligned} \tag{20.9}$$

Then, in complete analogy with the derivation of Equation (20.7), one can show that the thermodynamic Green's function satisfies formally the diffusionlike equation,

$$\left\{ 1\frac{\partial}{\partial t_2} + \mathsf{H}(2) \right\} G(\mathbf{r}_2 t_2; \mathbf{r}_1 t_1) = \delta(t_1 - t_2)\, \delta^{(3)}(\mathbf{r}_1 - \mathbf{r}_2) \tag{20.10}$$

Instead of the Green's function (20.8), it is useful to consider its transform, which for large fluid systems is defined with the aid of Equations (17.22) and (17.27) as

$$G(\mathbf{r}_2 t_2; \mathbf{r}_1 t_1) = \frac{1}{\Omega} \sum_{\mathbf{k}_1 m_1, \mathbf{k}_2 m_2} \exp[i(\mathbf{k}_2 \cdot \mathbf{r}_2 - \mathbf{k}_1 \cdot \mathbf{r}_1)] \\ \chi_{m_2}(\xi_2) \chi^{\dagger}_{m_1}(\xi_1) G(k_2 t_2; k_1 t_1) \tag{20.11}$$

For generality we have included the spin-transform function $\chi_m(\xi)$, and

$$G(k_2 t_2; k_1 t_1) \equiv \langle P\{\mathsf{a}_H(k_2, t_2)\mathsf{a}^{\dagger}_H(k_1, t_1)\}\rangle \tag{20.12}$$

with $k_i = (\mathbf{k}_i, m_i)$. Of course, for large translationally invariant systems we must obtain from this average value that $\mathbf{k}_2 = \mathbf{k}_1$.[22]

We now anticipate a result from Section 21, in the next chapter, that the statistical average of a physical operator A is calculated by computing the trace $\mathrm{Tr}(\underline{\rho}\mathsf{A})$, where $\underline{\rho}$ is called the statistical matrix. See Equation (21.9).

$$\langle \mathsf{A} \rangle = \mathrm{Tr}(\underline{\rho}\mathsf{A}) \tag{20.13}$$

For systems at thermodynamic equilibrium we also know [Equation (23.3)] that the statistical matrix commutes with the Hamiltonian, that is, $[\mathsf{H}, \underline{\rho}] = 0$. Moreover, for a system characterized by stationary-state occupation probabilities, that is, by probabilities for occupying the eigenstates of H, we have $\langle n|\underline{\rho}|n'\rangle = \rho_n\, \delta_{nn'}$; when the system occupies one definite, or pure, eigenstate n_o, then $\rho_n = \delta_{nn_o}$. See Equations (21.15) and (21.7b).

We may apply the above facts to the Green's function (20.12), when the average value is evaluated for an equilibrium ensemble, to show for large fluid systems that[23]

$$G(kt_2; kt_1) = S(k, t_2 - t_1) \qquad \text{(equilibrium ensembles)} \tag{20.14}$$

To establish this property that the thermodynamic Green's function depends only on the difference $(t_2 - t_1)$ of the temperature variables, we need to use the cyclic property of the trace represented by the average value brackets in Equation (20.12).

Exercise 20.2

(a) Use Equations (20.9) and $[\mathsf{H}, \underline{\rho}] = 0$ to verify property (20.14) of the one-particle thermodynamic Green's function, and show that $S(k, t)$

can be written as

$$S(k, t) = \theta(t)S^{(+)}(k, t) + \theta(-t)S^{(-)}(k, t) \tag{20.15}$$

where (on suppressing the subscripts H)

$$\begin{aligned} S^{(+)}(k, t) &= \mathrm{Tr}[\underline{\rho}\mathsf{a}(k, t)\mathsf{a}^{\dagger}(k)] \\ S^{(-)}(k, t) &= \varepsilon\mathrm{Tr}[\underline{\rho}\mathsf{a}^{\dagger}(k)\mathsf{a}(k, t)] \end{aligned} \tag{20.16}$$

(b) The transform (20.11) can be defined in an identical manner for the space–time Green's function (20.3). Show that property (20.14) also holds for the space–time transformed Green's function $\tilde{G}(k\tau_2; k\tau_1)$, so that one can also write Equation (20.15), with $t \to \tau$, for the equilibrium function $\tilde{S}(k, \tau)$, where

$$\begin{aligned} i\tilde{S}^{(+)}(k, \tau) &= \mathrm{Tr}[\underline{\rho}\mathsf{a}(k, \tau)\mathsf{a}^{\dagger}(k)] \\ i\tilde{S}^{(-)}(k, \tau) &= \varepsilon\mathrm{Tr}[\underline{\rho}\mathsf{a}^{\dagger}(k)\mathsf{a}(k, \tau)] \end{aligned} \tag{20.17}$$

Equilibrium Ensemble Averages

The utility of many-particle Green's functions lies in the fact that they can be used to calculate essentially all observable properties of macroscopic systems. For example, we may apply definition (17.5) of the number operator, with $i \to k$ for large fluid systems, to establish that

$$\begin{aligned} \langle \mathsf{n}(k) \rangle &= \varepsilon G(k0^-; k0) \\ &\to \varepsilon S^{(-)}(k, 0) \qquad \text{(for equilibrium ensembles)} \end{aligned} \tag{20.18}$$

Here we have used the thermodynamic Green's functions (20.12) and (20.16). The quantity $\langle \mathsf{n}(k) \rangle$ is called the momentum distribution (see Sections 24 and 46). The result (20.18) suggests that one-particle Green's functions in statistical mechanics can be referred to as generalized distribution functions.

We now show how Equation (20.10) can be used to derive an expression for the average energy $\langle E \rangle$ of a large fluid system at equilibrium. We have already indicated that the formal equations (20.7) and (20.10) can be somewhat misleading. This point is emphasized when one notes in the first equality of Equation (20.5) for $[\underline{\psi}(\mathbf{r}), \mathsf{V}_2]$ that the factor of 1/2 that occurs in Equation (17.26) is missing. If we make use of this fact, then it is straightforward to show from Equations (20.5), (20.8), and (20.10), with $t_1 \neq t_2$, that

$$-\varepsilon \sum_{\xi} \int d^3\mathbf{r} \lim_{t_1 \to t_2^+} \frac{\partial}{\partial t_2} G(\mathbf{r}t_2; \mathbf{r}t_1) = \langle \mathsf{H}_o \rangle + 2\langle \mathsf{V}_2 \rangle \tag{20.19}$$

where H_o and V_2 are given by Equations (17.25) and (17.26) and we have here included spin coordinates by using the notation $r = (\mathbf{r}, \xi)$.

Substitution of Equation (20.11) into this last result leads to the conclusion that we may also write, for large fluid systems:

$$
\begin{aligned}
-\varepsilon \sum_k \lim_{t_1 \to t_2^+} \frac{\partial}{\partial t_2} G(kt_2; kt_1) &= \langle \mathsf{H}_o \rangle + 2\langle \mathsf{V}_2 \rangle \\
&= 2\langle E \rangle - \varepsilon \sum_k \omega_k G(k0^-; k0)
\end{aligned} \tag{20.20}
$$

Here the second equality is derived with the aid of (17.12) and (20.18), assuming that H_o has single-particle eigenvalues ω_k. We may then finally use Equation (20.14) to obtain for the average energy of a large fluid system at equilibrium:

$$
\langle E \rangle = \frac{\varepsilon}{2} \sum_k \lim_{t_1 \to t_2^+} \left\{ \omega_k - \frac{\partial}{\partial t_2} \right\} S(k, t_2 - t_1) \tag{20.21}
$$

Exercise 20.3 According to Equations (8.19) and (8.20), the thermodynamic (or maximum) entropy is given by the partial derivative

$$
\langle S \rangle = \left. \frac{\partial(\mathscr{P}\Omega)}{\partial T} \right|_{\Omega, g}
$$

Use this result and $\mathscr{P}\Omega = \langle N \rangle g - \langle E \rangle + T\langle S \rangle$ to show that

$$
\left. \frac{\partial \langle S \rangle}{\partial T} \right|_{\Omega, g} = \frac{1}{T} \frac{\partial}{\partial T} (\langle E \rangle - \langle N \rangle g) \bigg|_{\Omega, g} \tag{20.22}
$$

The entropy can be calculated by integration of this expression at constant Ω and g. Moreover, since the chemical potential g is determined implicitly by the density equation $n = \Omega^{-1} \Sigma_k \langle n(k) \rangle$,[24] we can also calculate the pressure, and therefore all thermodynamic quantities, from the Green's function (20.14) for an equilibrium ensemble.

Lehmann Representation

The Fourier transform in time of the space–time Green's function $\tilde{S}(k, \tau)$ is defined by the equations

$$
\tilde{S}(k, \tau) = \frac{1}{2\pi} \int_{-\infty}^{+\infty} d\omega \, e^{-i\omega\tau} \tilde{S}(k, \hbar\omega) \tag{20.23}
$$

$$
\tilde{S}(k, \hbar\omega) = \int_{-\infty}^{+\infty} d\tau e^{i\omega\tau} \tilde{S}(k, \tau) \tag{20.24}
$$

By contrast, for the thermodynamic Green's function we introduce a spectrum of energies ω_n,

$$\omega_n = \left[2n + \frac{1}{2}(1 - \varepsilon)\right]\pi/\beta \qquad \textbf{(20.25)}$$

where $n = 0, \pm 1, \pm 2, \ldots$, and expand $S(k, t)$ as the Fourier series

$$S(k, t) = \frac{1}{\beta}\sum_{n=-\infty}^{+\infty} S(k, \omega_n)\, e^{-i\omega_n t} \qquad \textbf{(20.26)}$$

We show below that the frequencies ω_n arise as a consequence of certain periodicity properties of the thermodynamic Green's function. The Fourier coefficients $S(k, \omega_n)$ can then be defined by a temperature integral over the interval $(-\beta, \beta)$ by the inverse equation

$$S(k, \omega_n) = \frac{1}{2}\int_{-\beta}^{+\beta} dt\, S(k, t)\, e^{i\omega_n t} \qquad \textbf{(20.27)}$$

The *Lehmann representations* of the Fourier transforms $\tilde{S}(k, \hbar\omega)$ and $S(k, \omega_n)$ are explicit expressions for these two quantities in terms of the states of the fluid system under consideration. The remarkable fact, which we shall demonstrate using these representations, is that each of these quantities can be written as a simple state-independent integral over the same function $\rho(k, \omega)$. This new quantity $\rho(k, \omega)$, called the *spectral density function,* is defined in terms of sums over system states μ and ν by

$$\rho(k, \omega) \equiv 2\pi \sum_{\mu,\nu} \rho_\mu |\langle\mu|\mathrm{a}(k|\nu\rangle|^2\, \delta(\omega + E_\mu - E_\nu)(1 - \varepsilon e^{-\beta(\omega - g)}) \qquad \textbf{(20.28)}$$

where g is the chemical potential.

The derivation of the Lehmann representation of the space–time Green's function $\tilde{S}(k, \hbar\omega)$ is straightforward. We appeal to Equations (20.15) and (20.17) for $\tilde{S}(k, \tau)$. We also require the integral representation of the step function $\theta(\tau)$.

$$\theta(\tau) = \frac{i}{2\pi}\int_{-\infty}^{+\infty} d\omega \frac{e^{-i\omega\tau}}{\omega + i\hbar^{-1}\eta} \qquad (20.29)$$

where $\hbar^{-1}\eta = 0^+$ is a positive infinitesimal. Substitution of these three equations into definition (20.24) then yields the expression

$$\begin{aligned}\tilde{S}(k, \hbar\omega) = \hbar \sum_{\mu,\nu} |\langle\mu|\mathrm{a}(k)|\nu\rangle|^2 \{&\rho_\mu(\hbar\omega + E_\mu + g - E_\nu + i\eta)^{-1} \\ &- \varepsilon\rho_\nu(\hbar\omega + E_\mu + g - E_\nu - i\eta)^{-1}\}\end{aligned} \qquad (20.30)$$

where Equations (20.1) and (20.13) have also been used.[25,26] Equation (20.30) is the Lehmann representation of $\tilde{S}(k, \hbar\omega)$.

Exercise 20.4 Derive Equation (20.30), starting from definition (20.24).

We now use the fact [see Equation (27.18) or (36.15)] that the energy dependence of the statistical matrix element ρ_μ is proportional to the Boltzmann factor $e^{-\beta E_\mu}$. Then[25,26]

$$\rho_\nu = \rho_\mu \, e^{\rho(E_\mu + g - E_\nu)} \tag{20.31}$$

and we can rewrite Equation (20.30) in terms of the (real) spectral density function (20.28) as

$$\tilde{S}(k, \omega) = \frac{\hbar}{2\pi} \int_{-\infty}^{+\infty} d\omega' \, \rho(k, \omega')\{[1 + \varepsilon\nu(\omega')](\omega + g - \omega' + i\eta)^{-1} - \varepsilon\nu(\omega')(\omega + g - \omega' - i\eta)^{-1}\} \tag{20.32}$$

where the distribution function $\nu(\omega)$ is defined by [see also Equation (5.10)]

$$\nu(\omega) \equiv [e^{\beta(\omega - g)} - \varepsilon]^{-1} \tag{20.33}$$

Equations (20.28) and (20.32) show that in the Lehmann representation of $\tilde{S}(k, \omega)$ all microscopically derived properties of the space–time single-particle Green's function are determined by the spectral density function. We shall show that a similar statement applies to the thermodynamic single-particle Green's function.

Inasmuch as the variable t in the thermodynamic Green's function $S(k, t)$ represents the difference between two temperature variables, $t_2 - t_1$, its full range of variation is $-\beta \leqslant t \leqslant \beta$, and not just 0 to β. More particularly, $S^{(+)}(k, t)$ of Equations (20.16) is defined over the positive temperature interval $(0, \beta)$, and $S^{(-)}$ is defined over the negative interval $(-\beta, 0)$. By using the fact that the equilibrium statistical matrix $\underline{\rho}$ is proportional to $e^{-\beta \mathsf{H}}$ [see Equation (27.23)],[25] we can relate the two Green's functions $S^{(+)}$ and $S^{(-)}$ as follows. On referring to Equation (20.13) and using the cyclic property of the trace, we obtain from Equations (20.16)

$$\begin{aligned} S^{(+)}(k, t) &= \mathrm{Tr}[\underline{\rho} e^{t\mathsf{H}} \mathsf{a}(k) \, e^{-t\mathsf{H}} \mathsf{a}^\dagger(k)] \\ &= \mathrm{Tr}[\underline{\rho} \mathsf{a}^\dagger(k) \, e^{(t-\beta)\mathsf{H}} \mathsf{a}(k) \, e^{-(t-\beta)\mathsf{H}}] \\ &= \varepsilon S^{(-)}(k, t - \beta) \end{aligned} \tag{20.34a}$$

a result that can also be written as

$$S^{(-)}(k, t) = \varepsilon S^{(+)}(k, t + \beta) \tag{20.34b}$$

Equations (20.34a) and (20.34b) can be inserted into Equation (20.15) to give the single relation, when both t and $(t \pm \beta)$ are within the interval $(-\beta, \beta)$,

$$S(k, t) = \varepsilon S(k, t \pm \beta) \tag{20.35}$$

When t is outside of the interval $(-\beta, \beta)$, then $S(k, t)$ is undefined, and we can assign it any value. If, however, we arbitrarily take $S(k, t)$ to be periodic in t with period 2β, which is consistent with Equation (20.35),[27] then $S(k, t)$ will be defined for all values of t. Moreover, Equations (20.26) and (20.27) will then comprise an acceptable definition of the Fourier coefficients $S(k, \omega_n)$ for this periodicity if $\omega_n = n(\pi/\beta)$, where n is any integer.

In fact, we must choose ω_n according to Equation (20.25) because of the more specific condition (20.35). The identity

$$\exp(\pm i\omega_n\beta) = \varepsilon \tag{20.36}$$

shows that Equations (20.25) and (20.26) satisfy condition (20.35) quite generally. We can easily understand expression (20.25) for ω_n by considering separately the two cases $\varepsilon = \pm 1$. Thus for Bose systems ($\varepsilon = +1$), condition (20.35) shows that $S(k, t)$ is periodic with a period β, instead of 2β, and this smaller periodicity interval can be accommodated by choosing the ω_n to be even-integral multiples of π/β. On the other hand, for systems of fermions ($\varepsilon = -1$) the oddness characteristic of condition (20.35) shows that even-integral multiples of π/β must be excluded from the list of ω_n values; that is, for this case the ω_n must be odd-integral multiples of π/β. Finally, we can use Equations (20.15), (20.34b), and (20.36) to simplify expression (20.27) for the Fourier coefficients as follows:

$$\begin{aligned} S(k, \omega_n) &= \frac{1}{2}\int_0^\beta dt\, S^{(+)}(k, t)\, e^{i\omega_n t} + \frac{1}{2}\int_{-\beta}^0 dt\, S^{(-)}(k, t)\, e^{i\omega_n t} \\ &= \frac{1}{2}\int_0^\beta dt\, S^{(+)}(k, t)\, e^{i\omega_n t} + \frac{1}{2}\int_{-\beta}^0 dt\, e^{i\omega_n\beta} S^{(+)}(k, t + \beta)\, e^{i\omega_n t} \\ &= \int_0^\beta dt\, S^{(+)}(k, t)\, e^{i\omega_n t} \end{aligned} \tag{20.37}$$

We now substitute the first of Equations (20.16) into Equation (20.37) for the Fourier coefficient $S(k, \omega_n)$. On using Equations (20.13), (20.9), and (20.36) we derive the Lehmann representation of these coefficients:

$$\begin{aligned} S(k, \omega_n) &= \sum_{\mu,\nu} \rho_\mu |\langle\mu|\mathrm{a}(k)|\nu\rangle|^2[1 - \varepsilon \exp\{-\beta(E_\nu - E_\mu - g)\}] \\ &\quad \times (E_\nu - E_\mu - g - i\omega_n)^{-1} \end{aligned}$$

$$S(k, \omega_n) = \frac{1}{2\pi}\int_{-\infty}^{+\infty} d\omega'\, \rho(k, \omega')(\omega' - g - i\omega_n)^{-1} \tag{20.38}$$

where μ and ν are labels for the eigenstates of the fluid Hamiltonian. The spectral density function (20.28) has again appeared in the Lehmann representation. The fact that $\rho(k, \omega)$ has appeared in both of the expressions

(20.32) and (20.38) shows that there is an intimate connection between the space–time and thermodynamic one-particle Green's functions. Thus if one of these functions were to be calculated, say, $S(k, \omega_n)$, then the spectral density function could be identified by using Equation (20.38). Equation (20.32) would then provide a ready prescription for calculating the other function $\tilde{S}(k, \omega)$. Of course, such a calculation for a real fluid system can only be approximate at best.[28]

Bibliography

Texts

A. L. Fetter and J. D. Walecka, *Quantum Theory of Many-Particle Systems* (McGraw-Hill, 1971), Chapters 1–3. A comprehensive formal introduction to the many-body problem in quantum mechanics.

L. D. Landau and E. M. Lifshitz, *Statistical Physics* (Addison-Wesley, 1958), translated by E. Peierls and R. F. Peierls, Section 68. An elementary treatment of Landau's Fermi liquid theory.

N. H. March, W. H. Young, and S. Sampanthar, *The Many-Body Problem in Quantum Mechanics* (Cambridge University Press, 1967), Sections 1.8, 3.1–3.7, and 4.1–4.3, and Chapter 10. Includes development and applications of many-body perturbation theory. The Brillouin–Wigner form of perturbation theory is discussed. A rather complete introduction to the role of Green's functions in the many-body problem, with notation that can be readily assimilated, is given in Chapter 10.

E. Merzbacher, *Quantum Mechanics* (Wiley, 1970), Chapters 17 and 20. Included are excellent outlines of the principles of Raleigh–Schrödinger perturbation theory and the method of second quantization for identical-particle systems.

P. Roman, *Advanced Quantum Theory* (Addison-Wesley, 1965), Chapter 1. The discussion of second quantization is comprehensive.

L. I. Schiff, *Quantum Mechanics,* 3rd ed. (McGraw-Hill, 1968), Sections 40 and 55. Salient features in the theory of identical particles and second quantization are covered.

Original Papers

V. Fock, "Konfigurationsraum und Zweite Quantelung," *Zeitz. Phys.* **75** (1932), p. 622.

L. D. Landau, "The Theory of a Fermi Liquid," *J. Exp. Theoret. Phys.* **30** (1956), p. 1058 [English translation: *Soviet Phys.—JEPT* **3** (1957), p. 920].

Notes

1. We could, instead, have considered U_p to operate on the N state labels i; in the symmetrized forms (16.1) these two possibilities become equivalent. We have chosen to permute particle labels with U_p to preserve the particle-state ordering for manipulation in Equation (16.2).

2. E. P. Wigner, *Group Theory and Its Applications, . . .* (Academic Press, 1959), p. 126.

3. The proper counting of all distinct ways for distributing particles into single-particle states according to each of these statistics is explained in Section 5.

4. See, in this connection, M. D. Girardeau, *J. Math. Phys.* **10** (1969), p. 1302.

5. E. P. Wigner, *loc. cit.*, p. 127. Problem 1 in Chapter 10 of the text by Schiff, *op. cit.*, can be solved to show explicitly that no other one-dimensional, or unit-rank, representations of the permutation (or symmetry) group of three objects exist.

6. See, for example, P. Roman, *Theory of Elementary Particles* (North Holland, 1961), pp. 206–214, who follows the original explanation due to W. Pauli.

7. Formal convergence conditions of the general perturbation theory are derived with a mathematician's care in T. Kato, *Perturbation Theory for Linear Operators* (Springer-Verlag, 1966). In practice, the convergence of many-body perturbation theory is not well understood. See, for example, G. A. Baker, *Rev. Mod. Phys.* **43** (1971), p. 479.

8. E. Merzbacher, *op. cit.*, Chapter 17, Sections 2 and 3. The first term in Equation (18.8) is, of course, the solution to the homogeneous set of Equations (18.7). A necessary condition for the existence of such solutions, for $k \geqslant 1$ and when we choose $C_n^{(k)} = 0$, is Equation (18.12).

9. We demonstrate this property of the perturbation series explicitly with the calculation to second order, which is given at the end of this section. See also N. H. March, W. H. Young, and S. Sampanthar, *op. cit.*, Section 4.3.

10. A rigorous multistaged proof of "the stability of matter" and many excellent references on this subject are provided by E. H. Lieb, *Rev. Mod. Phy* **48** (1976), p. 553.

11. Leading terms in both the Rayleigh–Schrödinger and Brillouin–Wigner forms of perturbation theory are displayed in the text by N. H. March, W. H. Young, and S. Sampanthar, *loc. cit.*, Sections 4.2 and 4.3.

12. For a theory that applies to finite nuclei, see B. H. Brandow, "Reaction-Matrix Theory of Nuclei and Other Hard-Core Fermi Liquids," in *Lectures in Theoretical Physics*, Vol. XIB (Univ. of Colorado, 1968).

13. For the present calculation we shall replace the superscript (S) in Equation (17.19b) by (A), to emphasize that we are using antisymmetrized matrix elements (17.20) with $\varepsilon = -1$.

14. In the literature the notation ε_i is frequently used for quasi-particle energies.

15. The definition of a functional derivative entails the elimination of an integration over the variable that becomes k_i. A density-of-states factor must also be removed in this operation. See L. I. Schiff, *loc. cit.*, p. 494. The classic reference on functionals is by Vito Votterra, *Theory of Functionals and of Integral and Integro-Differential Equations* (Dover, 1959).

16. In the present discussion we must temporarily use the language of Sections 4 and 5, in which the label i refers not to a single quantum state but rather to a collection of single-particle quantum states $Z_i = \delta^3\mathbf{n}_i'$. See, for example, above Equation (4.18). Then for Equation (19.5) to apply we must assume, quite reasonably, that ω_i' does not vary appreciably within any given cell Z_i.

17. See note 14 in Chapter II.

18. We shall not explore these interesting properties of the space–time Green's

function in this text. See, however, Problem X.6 and N. H. March, W. H. Young, and S. Sampanthar, *loc. cit.*, Section 10.3.

19. L. I. Schiff, *loc. cit.*, pp. 170–171.

20. See Section 21.

21. One can use this formal result to derive an equation of motion for $\tilde{G}(x_2, x_1)$ in which particle interactions are governed by a two-particle Green's function. See, for example, A. L. Fetter and J. D. Walecka, *op. cit.*, Problems 3.3 and 3.4. This equation of motion has as its analogue in classical kinetic theory the Boltzmann transport equation (23.27).

22. Recall the argument above Equation (17.30).

23. The function $S(k, t_2 - t_1)$ is another notation for the Green's function (20.12) when large fluid systems at thermodynamic equilibrium are considered. For convenience we neglect any spin-dependent interactions, in which case the one-particle Green's function will be independent of the spin projections m_i.

24. The density can be calculated by using Equation (20.18). Moreover, the statistical matrix $\underline{\rho}$ to be used in the calculation of $S(k, t)$ will usually be chosen for the variable-N case of the grand canonical ensemble, with the system parameters of the grand potential f, Equation (8.22); namely, T, Ω, and g.

25. We must remark here that use of the more appropriate grand canonical ensemble (Section 36) is equivalent to the replacement of H in $\underline{\rho}$ by the effective Hamiltonian $\mathsf{H}' = \mathsf{H} - g\mathsf{N}$, where g is the chemical potential and N is the total number operator (17.6). Because of the identities

$$\exp(-i\tau_2 g\mathsf{N}/\hbar)\mathsf{a}(k)\exp(i\tau_2 g\mathsf{N}/\hbar) = \mathsf{a}(k)\exp(i\tau_2 g/\hbar)$$

$$\exp(-i\tau_1 g\mathsf{N}/\hbar)\mathsf{a}^\dagger(k)\exp(i\tau_1 g\mathsf{N}/\hbar) = \mathsf{a}^\dagger(k)\exp(-i\tau_1 g/\hbar)$$

which can be proved by using Equations (17.9), the effective Hamiltonian H′ can readily be used in definitions (20.1) and (20.9). Note that $[\mathsf{N}, \mathsf{H}] = 0$ (see Problem IV.1). The definition of the space–time Green's function $\tilde{S}(k, \tau)$ then changes only by an overall factor $\exp(i\tau g/\hbar)$. A similar statement applies to the thermodynamic Green's function $S(k, t)$. Thus either Hamiltonian H or H′ can be used in the definitions of the single-particle Green's functions; our development beginning with the Lehmann representation is completely correct only when H′ is used. The eigenvalues of H′ are of the form $(E_\mu - Ng)$.

26. The chemical potential g occurs in this expression, because the number of particles in state ν is one more than in state μ.

27. For this definition to be consistent the step function in Equation (20.15) must be defined so that, with t in the interval $(-\beta, \beta)$, $\theta(t + 2n\beta) = \theta(t)$ for any integer n.

28. An example is provided by Problem X.6, and the ideal-gas limit is examined in Problem IV.7.

Problems

IV.1. Show that the total number operator N commutes with $(\mathsf{K} + \mathsf{U})$ as well as with the total Hamiltonian $(\mathsf{K} + \mathsf{U} + \mathsf{V}_2)$. Consider both

Fermi- and Bose-statistics systems. [*Suggestion:* Use Equations (17.6), (17.9), (17.12), and (17.19a).]

IV.2. Consider an infinite, weakly interacting normal Fermi gas as in Section 18. The object of this problem is to derive an expression for the state vector $|\Psi_n\rangle$ to second order in perturbation theory, that is, to find $|\Psi_n^{(1)}\rangle$ and $|\Psi_n^{(2)}\rangle$.

(a) Follow the example of Figure IV.2(b) and give a pictorial representation of $|\Psi_n^{(1)}\rangle$, which is a two-particle excited state. Then write down an explicit expression for $|\Psi_n^{(1)}\rangle$.

(b) The second-order state-vector $|\Psi_n^{(2)}\rangle$ consists of two-particle, three-particle, and four-particle excited states. Represent all possible contributions to $|\Psi_n^{(2)}\rangle$ pictorially, noting that single-line loops as in Figure IV.2(a) can occur in this case even for an infinite system.

(c) Derive explicit expressions for the pictures of part (b) from the general expression (18.11) for $|\Psi_n^{(2)}\rangle$. You will discover that the second, $E_n^{(1)}$, term cancels many of the two-particle excitations of the first term.

IV.3. Consider the nonideal Fermi gas of Problem IV.2 and show that the third-order perturbation-theory energy $E_n^{(3)}$ is given by

$$
\begin{aligned}
E_n^{(3)} &= \frac{1}{8}\sum_{1\cdots 6} n_1 n_2 (1-n_3)(1-n_4) n_5 n_6 \\
&\quad \times \frac{\langle 12|V^{(A)}|34\rangle\langle 34|V^{(A)}|56\rangle\langle 56|V^{(A)}|12\rangle}{(\omega_1+\omega_2-\omega_3-\omega_4)(\omega_5+\omega_6-\omega_3-\omega_4)} \\
&\quad + \frac{1}{8}\sum_{1\cdots 6} n_1 n_2 (1-n_3)(1-n_4)(1-n_5)(1-n_6) \\
&\quad \times \frac{\langle 12|V^{(A)}|34\rangle\langle 34|V^{(A)}|56\rangle\langle 56|V^{(A)}|12\rangle}{(\omega_1+\omega_2+\omega_3-\omega_4)(\omega_1+\omega_2-\omega_5-\omega_6)} \\
&\quad - \sum_{1\cdots 6} n_1 n_2 n_4 (1-n_3)(1-n_4)(1-n_5)(1-n_6) \\
&\quad \times \frac{\langle 12|V^{(A)}|36\rangle\langle 34|V^{(A)}|52\rangle\langle 56|V^{(A)}|14\rangle}{(\omega_1+\omega_2-\omega_3-\omega_6)(\omega_1+\omega_4-\omega_5-\omega_6)} \\
&\quad + \frac{1}{2}\sum_{1\cdots 5} n_1 n_2 (1-n_3)(1-n_4) n_5 \frac{\langle 12|V^{(A)}|34\rangle\langle 34|V^{(A)}|12\rangle\langle 35|V^{(A)}|35\rangle}{(\omega_1+\omega_2-\omega_3-\omega_4)^2} \\
&\quad - \frac{1}{2}\sum_{1\cdots 5} (1-n_1)(1-n_2) n_3 n_4 n_5 \frac{\langle 12|V^{(A)}|34\rangle\langle 34|V^{(A)}|12\rangle\langle 35|V^{(A)}|35\rangle}{(\omega_1+\omega_2-\omega_3-\omega_4)^2}
\end{aligned}
$$

The numbers 1, 2, 3, ... represent states $i, j, k, \ldots$. Some rearrangements and relabeling may be required to derive this expression. [*Sug-*

gestion: One method for obtaining this answer is to use Equation (18.12) and the results of Problem IV.2(c), in which case note that only the two-particle excitations in $|\Psi_n^{(2)}\rangle$ are needed.]

IV.4. Refer to the Fermi gas, Problem IV.2.

(a) Use the result of part (a) to calculate the normalization constant $\langle\Psi_n \mid \Psi_n\rangle^{-1/2}$ to second order in perturbation theory.

(b) Derive an expression to second order in perturbation theory for the momentum distribution

$$\langle \mathsf{n}_k \rangle \equiv \langle \mathsf{a}_k^\dagger \mathsf{a}_k \rangle$$

Does $\langle \mathsf{N} \rangle = \Sigma_k \langle \mathsf{n}_k \rangle$ (as it should) by explicit calculation using your second-order result? Recall Equation (17.6).

(c) In the context of the Landau theory of Section 19 can you explain why $\langle \mathsf{n}_k \rangle \neq n_k' = n_k$?

IV.5. Consider again the infinite Fermi gas of Problem IV.2. Show that the last two terms (called self-energy terms) in the expression for $E_n^{(3)}$ of Problem IV.3 can be combined with the second-order energy term $E_n^{(2)}$ of Equation (18.20) by redefining the single-particle energies that appear in the energy denominator. Use the notation

$$\omega_i \rightarrow \omega_i' \equiv \omega_i + U_i$$

Does the expression for U_i so obtained agree with the first-order Landau energy (19.7)?

IV.6. *Kinetic equation for a normal Fermi fluid*

In analogy with the Boltzmann transport equation for a dilute classical real gas, Equations (11.27) and (11.14), one can write down a general transport, or kinetic, equation for a normal Fermi fluid. The equation used in Landau's theory, with $k = (\mathbf{k}, m)$, $\mathbf{k}$ = momentum and m = spin projection, is

$$\frac{\partial n'}{\partial t} + \nabla_{\mathbf{r}} n' \cdot \nabla_{\mathbf{k}} \omega' - \nabla_{\mathbf{k}} n' \cdot \nabla_{\mathbf{r}} \omega' = \left(\frac{\partial n}{\partial t}\right)_{\text{coll}}$$

where

$$\left(\frac{\partial n_1'}{\partial t}\right)_{\text{coll}} = (2h^3)^{-1} \sum_{m_1, m_2, m_3} \int d^3\mathbf{k}_1'\, d^3\mathbf{k}_2'\, d^3\mathbf{k}_2\, w(k_1, k_2; k_1', k_2')$$
$$\times\, \delta^{(3)}(\mathbf{k}_1 + \mathbf{k}_2 - \mathbf{k}_1' - \mathbf{k}_2')\, \delta(\omega_1' + \omega_2' - \omega_{1'}' - \omega_{2'}')$$
$$\times\, [n_{1'}' n_{2'}'(1 - n_1')(1 - n_2') - n_1' n_2'(1 - n_{1'}')(1 - n_{2'}')]$$

(a) Arrive at this kinetic equation by giving careful heuristic arguments. Your considerations should include: discussion of the analogies $P \rightarrow n'$, $H_o \rightarrow \omega'$; discussion of $\mathbf{r}$, t, and spin dependences in $\omega'(\mathbf{r}, k, t)$ and $n'(\mathbf{r}, k, t)$; the analogy to Equation (11.1); conservation of energy and momentum for quasi-particle interactions; influence of Equation (19.11) on ω' for nonequilibrium situations; role of the Pauli exclusion principle in determining $(\partial n'/\partial t)_{\text{coll}}$; relation of the transition probability w to R of Equation (11.11).

(b) Derive a basic conservation theorem similar to Equation (12.21) from the above kinetic equation, and show that the collision-integral term vanishes for conserved properties χ.

IV.7. *Green's functions for an ideal gas*

In this problem we use the effective free-particle Hamiltonian $\mathsf{H}_o' = \mathsf{H}_o - g\mathsf{N}$ (see note 25).

(a) Verify the two identities, with $\mathsf{H} \rightarrow \mathsf{H}_o'$ in Equations (20.9),

$$\mathsf{a}(k, t) = \mathsf{a}(k) \exp[-t(\omega_k - g)]$$

$$\mathsf{a}^\dagger(k, t) = \mathsf{a}^\dagger(k) \exp[t(\omega_k - g)]$$

(b) Show that the momentum distribution (20.18) can also be written as

$$\langle \mathsf{n}(k) \rangle = S(k, \beta)$$

(c) Use Equations (20.15) and (20.16) to show that

$$S_o(k, t) = \exp[-t(\omega_k - g)][\theta(t) + \varepsilon \langle \mathsf{n}(k) \rangle_o]$$

and hence, using the result (b), that

$$\langle \mathsf{n}(k) \rangle_o = \nu(\omega_k)$$

where $\nu(\omega)$ is defined by Equation (20.33).

(d) Show that

$$S_o(k, \omega_n) = (\omega_k - g - i\omega_n)^{-1}$$

and

$$\rho_o(k, \omega') = 2\pi\, \delta(\omega' - \omega_k)$$

(e) Derive explicit expressions for the space–time free-particle Green's functions $\tilde{S}_o(k, t)$ and $\tilde{S}_o(k, \omega)$.

CHAPTER V

STATISTICAL MATRIX AND CLASSICAL DISTRIBUTION FUNCTION

The introduction of probabilistic notions into the microscopic description of large systems is accomplished with the use of the statistical matrix $\underline{\rho}$ in quantum mechanics and with its counterpart the distribution function ρ in classical mechanics. Only when our uncertainty in knowledge about the occupation of microscopic states by real systems is incorporated into their description can we expect to arrive at a fundamental understanding of the macroscopic behavior of real systems. The time evolution of this behavior is governed by an equation of motion for $d\underline{\rho}/dt$ quantum mechanically and by Liouville's theorem for $\partial\rho/\partial t$ classically; both cases are derived in Section 23. Also, the Boltzmann transport equation of classical kinetic theory is derived rigorously from Liouville's equation. Reduced distribution functions are related systematically to ρ (or $\underline{\rho}$) in Section 24. Finally, Section 25 is devoted to a discussion of various attempts to understand irreversibility in nature.

21 Formal Theory of Statistical Matrix

We adopt the quantum-mechanical view of statistics, and this forces us to generalize the purely statistical probability distribution functions introduced in Sections 3 and 4. We begin by explaining certain peculiarities which in quantum mechanics distinguish macroscopic bodies from those containing a comparatively small number of particles. We then define the statistical matrix $\underline{\rho}$, which is the most useful quantum-mechanical generalization of a classical distribution function.

In Sections 16 to 18 we discussed methods for investigating stationary-state solutions to the Schrödinger-wave equation. These are the *pure states* of quantum mechanics. When our incomplete knowledge about the occupation of pure states by any real macroscopic system is combined with its pure-state description, we arrived at a *mixed-state* description of the system. The statistical matrix $\underline{\rho}$ provides a mixed-state description of real systems. After giving a general formulation of $\underline{\rho}$, the calculation of quantum-mechanical average values using $\underline{\rho}$ is demonstrated.

The main quantum-mechanical feature that distinguishes a macroscopic body from a few-particle system is the extraordinarily high density of levels in the energy spectrum of the macroscopic body. The reason for this high density is easily understood if one notices that, owing to the immense number of particles in the body, any energy can, roughly speaking, be "distributed" among different particles in an innumerable number of distinct ways. The connection between this fact and the high density of the energy levels becomes particularly clear if one considers, for example, a macroscopic body that is a gas of N completely noninteracting particles confined in a certain volume. The energy levels of such a system represent simply the sums of the energies of the separate particles, each ranging over an infinite series of discrete values (when box normalization is used). Inasmuch as the values of the N terms in the total-energy sum can be chosen in many, many possible ways, it is clear that in any appreciable energy interval there will be an enormous number of possible values for the energy of the system, which values consequently must lie very close to one another.[1] In fact, it can be shown generally (see Problem II.4) that, except near the ground state, the number of levels in any given finite interval of the energy spectrum of a macroscopic body increases exponentially with the number N, the level spacing becoming, correspondingly, exponentially smaller with N.

Suppose that the energy of a macroscopic system could have some definite value. It is evident that this value would be "spread out" by an amount equal to the interaction energy of the system with its surroundings. But because of the extraordinary density of levels, this interaction energy would also be extremely large in comparison with the separation of the levels, not only for a closed subsystem but also for systems that could be considered to be rigorously isolated from every other point of view. There exist in nature no rigorously isolated systems whose interaction with every other body is zero. Thus even the smallest residual interactions, for example, those so small that they have negligible influence on all other properties of a system, will be large in comparison with the vanishingly small spacing between energy levels.

We may conclude that a macroscopic body can never be brought to a strictly stationary state. This fact stems not only from the existence of system interactions, it also has a fundamental origin in quantum mechanics. We distinguish here between the energy value E that the system would have in the absence of interactions and the value E' that results in the presence of interaction with a measuring device. Then the indeterminacy $\Delta(E' - E)$ of the perturbation energy $(E' - E)$ is connected with the duration Δt of the interaction process by the uncertainty relation[2]

$$|\Delta(E' - E)| \sim \frac{\hbar}{\Delta t} \tag{21.1}$$

Moreover, for the state with energy E' to be considered stationary, the energy uncertainty $|\Delta(E' - E)|$ would have to be small in comparison with the separation of adjacent levels. Owing to the extreme smallness of the level spacing, we see then that an incredibly long time Δt would be required to bring the macroscopic state, even with negligibly small measurement-interaction energy, into a definite quantum state. In other words, it is impossible for a real macroscopic body to exist exactly in a stationary state.

Despite the foregoing negative evaluation of stationary-state descriptions as they apply to real macroscopic systems in nature, there is absolutely no reason why every macroscopic system cannot be characterized by a suitably qualified Hamiltonian function with a corresponding spectrum of stationary states. What is essential is to utilize such stationary-, or pure-, state descriptions in conjunction with the complete and realistic mixed-state description, which we consider next.

The Mixed-State Description

It is essential for any microscopic description of a macroscopic system to be valid that we incorporate our vast incomplete knowledge about the system into this description. By incomplete knowledge we mean missing information, as discussed in connection with Equation (4.23), and missing information here means the absence of kinematical detail relating to microscopic motions within the system or, equivalently, the absence of well-defined system energies in the sense described above. We use the terminology *incomplete-information probabilities* to describe this missing information. Despite the fact that its origins are quantum mechanical in part, incomplete-information probabilities are basically classical, in that they can be described by conventional probability theory. But in quantum mechanics even a pure state, with no associated incomplete-information probabilities, is inherently probabilistic. We call the probabilities of a pure state, due entirely to quantum mechanics, *potential probabilities*. These probabilities are the potentialities for occurring that various values of some physical quantity have if a suitable measurement is made.

We can illustrate the concept of potential probabilities by considering the simple example of the spin-½ particle.[3] Suppose we consider measurements, using the Stern–Gerlach experiment, of the z component of the spin of many such particles and form an ensemble by choosing only particles for which the value found is $+\frac{1}{2}\hbar$. We will then have an ensemble for which the (potential) probability of getting the value $S_z = \frac{1}{2}\hbar$ is unity. If next we measure the value of S_x, the potential probability is ½ for each of the two possible values $+\frac{1}{2}\hbar$ and $-\frac{1}{2}\hbar$. Whether we now consider either the entire original ensemble (after measuring S_x) or one of the two subensembles sorted out according to the values of S_x, we now find on measuring S_z that it does not have the value $+\frac{1}{2}\hbar$ for all systems, but only for half of them. The measurement of S_x has disturbed the systems so that they no longer have a definite value of S_z. The potential probabilities of pure states depend, of course, on how these states are prepared, but they also can be changed by the process of measurement. Potential probabilities have no classical analogue.

In the general case we have to consider both kinds of probabilities in our description of macroscopic bodies. We must combine the inherently quantum-mechanical potential probabilities of pure states with the incomplete-information probabilities. The resulting quantum-mechanical description is in terms of *mixed states,* and the most useful tool for characterizing the mixed-state ensemble is the *statistical matrix* or *density matrix.*

Let us characterize the (pure) states of a many-body system by state vectors $|i\rangle$, where i denotes all of the quantum numbers needed to specify one state completely. We then imagine that we can form an ensemble of M identical systems such that, with the i values ordered ($i = 1, 2, \ldots$), there will be

$w_1 M$ systems in the state $|1\rangle$

$w_2 M$ systems in the state $|2\rangle$

and so forth, with

$$\sum_i w_i = 1 \tag{21.2}$$

The w_i are the incomplete-information probabilities of the various pure states in the mixed-state ensemble.

We now introduce the statistical matrix $\underline{\rho}$ by the definition

$$\underline{\rho} \equiv \sum_i w_i \mathsf{P}_i \tag{21.3}$$

where P_i is the projection operator onto the state i, defined by

$$\mathsf{P}_i \equiv |i\rangle\langle i| \tag{21.4}$$

For orthonormal states the projection operator satisfies

$$\mathsf{P}_i\mathsf{P}_j = \delta_{ij}\mathsf{P}_i \tag{21.5}$$

and for a complete set of states we have

$$\sum_i \mathsf{P}_i = 1 \qquad \text{(completeness relation)} \tag{21.6}$$

It must be emphasized that the w_i in Equation (21.3) are (classical) probabilities and not quantum-mechanical amplitudes.

Whereas the w_i in the definition of $\underline{\rho}$ give the incomplete-information probabilities in the ensemble, the projection operators P_i relate to the potential probabilities, that is, to purely quantum-mechanical aspects, of the system in the ensemble. Thus the statistical matrix $\underline{\rho}$ provides a complete description of a system in a mixed state, any one real system being characterized by the ensemble. The sum in definition (21.3) shows that for mixed states the two kinds of probabilities are inseparably intertwined; their combined effect cannot be represented as the result of applying consecutively a purely quantum-mechanical averaging process and a purely statistical averaging process. Note, however, that for a system in a pure state j we have $w_i = \delta_{ij}$, and the statistical matrix reduces to

$$\underline{\rho} = \mathsf{P}_j = |j\rangle\langle j| \qquad \text{(pure state)} \tag{21.7a}$$

Thus in the diagonal representation, that is, the i representation in which the probabilities w_i are given, the matrix $\underline{\rho}$ for a pure-state description has only one nonvanishing matrix element:

$$\underline{\rho}_{jj} = w_j = 1 \qquad \text{(pure state } j\text{)} \tag{21.7b}$$

In actual calculations it is usually more convenient to use some other representation $|n\rangle$ of many-body states, such as the free-particle state vectors (17.21), instead of the system eigenstates $|i\rangle$. The matrix elements of $\underline{\rho}$ in the n representation will then be

$$\begin{aligned}\langle n|\underline{\rho}|n'\rangle &= \sum_i \langle n \mid i\rangle w_i\langle i \mid n'\rangle \\ &= \sum \Phi_i(n) w_i \Phi_i^\dagger(n')\end{aligned} \tag{21.8}$$

where $\Phi_i(n) = \langle n \mid i\rangle$ is the many-body wave function for the state i in the n-representation (which will often be either the momentum or the position representation).

Calculation of Average Values

We now justify Equation (21.3) by demonstrating that the matrix elements (21.8) are those one encounters naturally in the calculation of an expression for the average value $\langle \mathsf{A} \rangle$ of some physical quantity, or observable, A. Intuitively one would expect to have

$$\langle \mathsf{A} \rangle = \sum_i w_i \langle i|\mathsf{A}|i\rangle \tag{21.9a}$$

which is the correct mixed-state expression for $\langle \mathsf{A} \rangle$. But the matrix elements of A are often more usefully evaluated in some other representation n, in which case this expression becomes

$$\begin{aligned}\langle \mathsf{A} \rangle &= \sum_i \sum_{n,n'} w_i \langle i \mid n'\rangle\langle n'|\mathsf{A}|n\rangle\langle n \mid i\rangle \\ &= \sum_{n,n'} \langle n|\underline{\rho}|n'\rangle\langle n'|\mathsf{A}|n\rangle\end{aligned}$$

$$\langle \mathsf{A} \rangle = \mathrm{Tr}(\underline{\rho}\mathsf{A}) \tag{21.9b}$$

The second line of this equation shows generally how to calculate $\langle \mathsf{A} \rangle$ for mixed states, and the derivation shows that the statistical matrix (21.3) or (21.8) does indeed provide a proper characterization of mixed states in quantum mechanics.

It is easy to verify using Equation (21.9) with $\mathsf{A} = 1$ that the trace of the statistical matrix is unity

$$\mathrm{Tr}(\underline{\rho}) = 1 \tag{21.10}$$

a property corresponding to Equation (21.6) for projection operators. On the other hand, a property of the type (21.5) is not, in general, satisfied by the statistical matrix. Indeed,

$$\begin{aligned}\mathrm{Tr}(\underline{\rho}^2) &= \sum_{n,n'} \langle n|\underline{\rho}|n'\rangle\langle n'|\underline{\rho}|n\rangle \\ &= \sum_i w_i^2 = 1\end{aligned} \tag{21.11}$$

where the equality holds only for pure states, as can be seen by using Equation (21.7b).

Let us now consider the expansion of the stationary-state wave function $\Phi_i(\mathbf{r})$ in terms of the complete set of basis wave functions $\phi_n(\mathbf{r})$, with expansion coefficients $c_n^{(i)}$,

$$\Phi_i(\mathbf{r}) = \sum_n c_n^{(i)} \phi_n(\mathbf{r}) \tag{21.12a}$$

Written in Dirac notation this equation is

$$\langle \mathbf{r} \mid i\rangle = \sum_n \langle \mathbf{r} \mid n\rangle\langle n \mid i\rangle \tag{21.12b}$$

and therefore $\langle n \mid i \rangle = c_n^{(i)}$. It is then easy to see that the statistical-matrix elements (21.8) can also be written as

$$\begin{aligned} \langle n|\underline{\rho}|n'\rangle &= \sum_i w_i c_n^{(i)} c_{n'}^{(i)*} \\ &= \langle n'|\underline{\rho}|n\rangle^* \end{aligned} \tag{21.13}$$

Inasmuch as the probabilities w_i are real (and positive), the statistical matrix is Hermitian, a fact we could easily have observed sooner.

We can gain further understanding of the statistical matrix by applying Equations (21.9) and (21.13) to the familiar pure-state case characterized by Equation (21.7b). Then, with the superscript label (j) suppressed,

$$\langle n|\underline{\rho}|n'\rangle = c_n c_{n'}^* \qquad \text{(pure-state case)} \tag{21.14}$$

For pure states, a diagonal matrix element gives the potential probability $|c_n|^2$ for finding a system in the state $|n\rangle$. The expectation value of the physical quantity A for the pure state Φ is given by

$$\begin{aligned} \langle \mathsf{A} \rangle = \mathrm{Tr}(\underline{\rho}\mathsf{A}) &= \sum_{n,n'} \langle n|\underline{\rho}|n'\rangle\langle n'|\mathsf{A}|n\rangle \\ &= \sum_{n,n'} c_n c_{n'}^* \int \mathbf{d}^3\mathbf{r}\phi_{n'}(\mathbf{r})\mathsf{A}\phi_n(\mathbf{r}) \\ &= \int d^3\mathbf{r}\Phi^*(\mathbf{r})\mathsf{A}\Phi(\mathbf{r}) \end{aligned}$$

The second and third lines of this equation are, of course, familiar quantum-mechanical expressions. This evaluation of the expectation value $\langle \mathsf{A} \rangle$ shows that a knowledge of the statistical matrix is equivalent for pure states to a knowledge of the wave function $\Phi(\mathbf{r})$ to within an overall phase factor.

Exercise 21.1 Show for pure states that

$$\langle n|\underline{\rho}^2|n'\rangle = \langle n|\underline{\rho}|n'\rangle \qquad \text{(pure state)}$$

in agreement with Equations (21.11) and (21.10).

In the original i representation, as characterized by Equations (21.2) and (21.3), the statistical matrix has no off-diagonal elements. This fact follows, by definition, because of the classical nature of incomplete-information probabilities.[4]

$$\langle i|\underline{\rho}|j\rangle = w_i\,\delta_{ij} \tag{21.15}$$

Therefore, the usual program of statistical mechanics, in which probabilities w_i are determined by standard statistical procedures, can be combined with definition (21.3) to determine the statistical matrix of quantum statistical mechanics. It is precisely this procedure that is used in Section 27.

Alternative Treatment

There is an alternative, commonly used formulation of the statistical matrix $\underline{\rho}$ that is completely equivalent to Equation (21.3). It applies Equation (21.14) to an ensemble of M systems, each described by a pure-state statistical matrix $\underline{\rho}_\mu$.

We define $\underline{\rho}$ to be the ensemble average

$$\underline{\rho} \equiv \frac{1}{M} \sum_{\mu=1}^{M} \underline{\rho}_\mu \tag{21.16}$$

In the n representation of Equation (21.14) the matrix elements of $\underline{\rho}$ are then given by

$$\begin{aligned} \langle n|\underline{\rho}|n'\rangle &= \frac{1}{M} \sum_{\mu=1}^{M} \langle n|\underline{\rho}_\mu|n'\rangle \\ &= \frac{1}{M} \sum_{\mu=1}^{M} c_{\mu,n} c^*_{\mu,n'} \end{aligned} \tag{21.17}$$

Now in the diagonal (i) representation for the $\underline{\rho}_\mu$, we may describe the mixed-state ensemble by saying that there are

$w_1 M$ systems in state 1

$w_2 M$ systems in state 2

and so forth, with $\Sigma_i w_i = 1$. Therefore, Equation (21.17) can be rewritten in the form (21.13), which is equivalent to definition (21.3).

It is instructive to apply Equation (21.17) to a calculation of $\langle n|\underline{\rho}^2|n'\rangle$. One finds

$$\begin{aligned} \langle n|\underline{\rho}^2|n'\rangle &= \sum_m \langle n|\underline{\rho}|m\rangle\langle m|\underline{\rho}|n'\rangle \\ &= \frac{1}{M^2} \sum_{\mu,\nu=1}^{M} \sum_m c_{\mu,n} c^*_{\mu,m} c_{\nu,m} c^*_{\nu,n'} \\ &\neq \langle n|\underline{\rho}|n'\rangle \qquad \text{(mixed state)} \end{aligned}$$

It is because, generally, the μth and νth systems are in different states, not coherently related, that we get this last result. Only when all systems are in the same quantum-mechanical state with the same phases do we get the result of Exercise 21.1 for pure states. The absence, in general, of coherent phase relations between the systems in an ensemble, and therefore in the statistical matrix (21.17), is one reason why the mixed-state statistical matrix successfully describes real systems in nature.

22 Example of Spin-½ Particles

In ordinary quantum-mechanical applications the statistical matrix is called the density matrix. To gain insight into the usefulness and meaning of the density-matrix formalism we apply it in this section to the description of polarization measurements, using spin-½ particles. We show first how to express the polarization vector in terms of the density matrix, and vice versa, and then we give an application of the formalism to scattering phenomena.

Although our interest in this text is primarily in many-particle systems, in this section we are considering a one-particle system that is part of a beam of many such particles. In this example the beam is an ensemble of identical particles, which should not be confused with a system of identical particles where, for example, wave-function symmetrization is important. It is important to realize, however, that a typical particle in the beam will generally be in a mixed state.

The density matrix that characterizes a single particle will have its rows and columns labeled by all the possible single-particle states. Here we are not interested in the translational states or in other internal degrees of freedom, but only in the spin degrees of freedom (magnetic quantum numbers S_z) of the single particle with spin S. Therefore we consider only the $(2S + 1)$ by $(2S + 1)$ submatrix of the full density matrix and suppress other quantum numbers in the notation. In particular we investigate only the simple case of spin-½ particles with S_z values $\pm\frac{1}{2}\hbar$. The corresponding density matrix is labeled as

$$\underline{\rho} = \begin{pmatrix} \rho_{++} & \rho_{+-} \\ \rho_{-+} & \rho_{--} \end{pmatrix} \tag{22.1}$$

where

$$\rho_{ij} = \langle i|\underline{\rho}|j\rangle$$

Consider now the spin part of the wave function for a single particle with spin-½.

$$\Phi = c_+\begin{pmatrix} 1 \\ 0 \end{pmatrix} + c_-\begin{pmatrix} 0 \\ 1 \end{pmatrix}$$

where the $c_{\pm}$ can be functions of the time, momentum, and other quantum numbers of the particle. Then for a pure state,

$$\langle +|\underline{\rho}|+\rangle = c_+c_+^*, \qquad \langle -|\underline{\rho}|+\rangle = c_-c_+^*, \qquad \text{etc.}$$

We next write down the Pauli spin matrices $\underline{\sigma}_+$, $\underline{\sigma}_-$, $\underline{\sigma}_z$, and $\mathsf{1}$:

$$\begin{aligned}
\underline{\sigma}_+ &= \frac{1}{2}(\underline{\sigma}_x + i\underline{\sigma}_y) = \begin{pmatrix} 0 & 1 \\ 0 & 0 \end{pmatrix} \\
\underline{\sigma}_- &= \frac{1}{2}(\underline{\sigma}_x - i\underline{\sigma}_y) = \begin{pmatrix} 0 & 0 \\ 1 & 0 \end{pmatrix} \\
\underline{\sigma}_z &= \begin{pmatrix} 1 & 0 \\ 0 & -1 \end{pmatrix}, \qquad \mathsf{1} = \begin{pmatrix} 1 & 0 \\ 0 & 1 \end{pmatrix}
\end{aligned} \tag{22.2}$$

These matrices are four linearly independent 2×2 matrices, and there cannot be more than four such matrices. Their expectation values are readily computed with the aid of Equation (21.9b). On using the 2×2 density matrix (22.1), one finds for the average of a general spin operator O

$$\begin{aligned}
\langle \mathsf{O} \rangle &= \mathrm{Tr}(\underline{\rho}\mathsf{O}) \\
&= \langle -|\underline{\rho}|-\rangle\langle -|\mathsf{O}|-\rangle + \langle -|\underline{\rho}|+\rangle\langle +|\mathsf{O}|-\rangle \\
&\quad + \langle +|\underline{\rho}|-\rangle\langle -|\mathsf{O}|+\rangle + \langle +|\underline{\rho}|+\rangle\langle +|\mathsf{O}|+\rangle
\end{aligned} \tag{22.3}$$

Therefore

$$\begin{aligned}
\langle \underline{\sigma}_+ \rangle &= \langle -|\underline{\rho}|+\rangle, \\
\langle \underline{\sigma}_- \rangle &= \langle +|\underline{\rho}|-\rangle = \langle \underline{\sigma}_+ \rangle^* \\
\langle \underline{\sigma}_z \rangle &= \langle +|\underline{\rho}|+\rangle - \langle -|\underline{\rho}|-\rangle \\
\langle \mathsf{1} \rangle &= \langle +|\underline{\rho}|+\rangle + \langle -|\underline{\rho}|-\rangle
\end{aligned} \tag{22.4}$$

which for the pure-state case yields

$$\begin{aligned}
\langle \underline{\sigma}_+ \rangle &= c_-c_+^* = \langle \underline{\sigma}_- \rangle^* \\
\langle \underline{\sigma}_z \rangle &= |c_+|^2 - |c_-|^2 \\
\mathsf{1} &= |c_+|^2 + |c_-|^2
\end{aligned} \tag{22.5}$$

We next introduce the polarization vector $\boldsymbol{P}$, which is defined for a spin-½ particle by

$$\boldsymbol{P} \equiv \langle \underline{\boldsymbol{\sigma}} \rangle \tag{22.6}$$

It is easily shown, using Equations (22.5) for pure states, that

$$P^2 = \langle \underline{\sigma}_x \rangle^2 + \langle \underline{\sigma}_y \rangle^2 + \langle \underline{\sigma}_z \rangle^2$$
$$= \langle \underline{\sigma}_z \rangle^2 + 4\langle \underline{\sigma}_+ \rangle \langle \underline{\sigma}_- \rangle = 1 \tag{22.7}$$

A pure state for spin-½ particles has the polarization value unity and can never be considered unpolarized. If we wish to describe an unpolarized or partially polarized beam of particles, then we must use a mixed-state density matrix, in which case Equations (22.5) must be replaced by expressions of the form (21.17). The generalization of definition (22.6) for the polarization vector of a beam of N spin-½ particles becomes

$$\mathbf{P} = \langle \underline{\boldsymbol{\sigma}} \rangle = \frac{1}{N} \sum_{\alpha=1}^{N} \langle \underline{\boldsymbol{\sigma}} \rangle \alpha$$

Inasmuch as Equations (22.4) hold in general, we can invert these equations to express the density-matrix elements in terms of the polarization-vector components. Thus the problem is abruptly turned around, and instead of attempting to calculate mixed-state expressions for the density-matrix elements using Equation (21.17) we now adopt a common operational method in which these matrix elements are expressed in terms of observable quantities. One obtains

$$\langle + | \underline{\rho} | + \rangle = \frac{1}{2}(1 + P_z)$$
$$\langle - | \underline{\rho} | - \rangle = \frac{1}{2}(1 - P_z)$$
$$\langle - | \underline{\rho} | + \rangle = P_+ = \frac{1}{2}(P_x + iP_y) \tag{22.8}$$
$$\langle + | \underline{\rho} | - \rangle = P_- = \frac{1}{2}(P_x - iP_y)$$

The polarization components in Equations (22.8) may all be functions of time, momentum, and other quantum numbers of the particles. Since the terms beam and ensemble are equivalent terms in this example, the density matrix will be time-independent if the beam is constant in time, and so on. Equations (22.8) show that the general mixed-state density matrix for a beam of spin-½ particles can be written as the matrix operator

$$\underline{\rho} = \begin{pmatrix} \tfrac{1}{2}(1 + P_z) & P_- \\ P_+ & \tfrac{1}{2}(1 - P_z) \end{pmatrix}$$

$$\underline{\rho} = \frac{1}{2}[1 + \mathbf{P} \cdot \underline{\boldsymbol{\sigma}}] \tag{22.9}$$

This result can also be derived easily as the most general scalar that can be constructed using the spin matrices (22.2).[5] Note that for the pure-state case $P = P_z = 1$, Equation (22.9) reduces to

$$\underline{\rho} \xrightarrow[P=P_z=1]{} \begin{pmatrix} 1 & 0 \\ 0 & 0 \end{pmatrix}$$

in agreement with (21.7b).

Exercise 22.1 Let $c_+ = \cos\beta e^{i\delta}e^{-i\gamma}$ and $c_- = \sin\beta e^{i\delta}e^{i\gamma}$ for a spin-½ particle in a pure state, and define the unit vector

$$\hat{\mathbf{n}} = \hat{\mathbf{i}}_x \sin\theta\cos\phi + \hat{\mathbf{i}}_y \sin\theta\sin\phi + \hat{\mathbf{i}}_z\cos\theta$$

(a) With the z-axis chosen as the axis of quantization for S_z, find the direction in which the particle's spin "points," that is, find θ and ϕ in terms of (β, δ, γ), given that

$$\mathbf{P}\cdot\hat{\mathbf{n}} = 1$$

(b) What are θ and ϕ for the spinor function

$$\Phi = \begin{pmatrix} \sqrt{1/5} \\ i\sqrt{4/5} \end{pmatrix}?$$

For a completely unpolarized beam, $\mathbf{P} = 0$, and the density matrix becomes

$$\underline{\rho} = \frac{1}{2}\begin{pmatrix} 1 & 0 \\ 0 & 1 \end{pmatrix} = \frac{1}{2}\cdot 1 \qquad \text{(unpolarized beam)} \tag{22.10}$$

Let us write the wave-function amplitudes c_+ and c_- for the αth particle in terms of an amplitude and a phase factor as

$$c_{\alpha,+} = |c_{\alpha,+}|\exp(i\theta_{\alpha,+}), \qquad c_{\alpha,-} = |c_{\alpha,-}|\exp(i\theta_{\alpha,-})$$

Then for an unpolarized beam of N particles we have from Equations (21.17) and (22.10)

$$\begin{aligned} 0 = \langle -|\underline{\rho}|+\rangle &= \langle +|\underline{\rho}|-\rangle^* \\ &= \frac{1}{N}\sum_{\alpha=1}^{N} c_{\alpha,-}c^*_{\alpha,+} \\ &= \frac{1}{N}\sum_{\alpha=1}^{N} |c_{\alpha,-}|\cdot|c_{\alpha,+}|\exp[i(\theta_{\alpha,-} - \theta_{\alpha,+})] \end{aligned}$$

We see very explicitly from this last expression that there can be no general phase relationships between the (+) and (−) states for different particles in an unpolarized beam. An unpolarized beam is characterized by random phase relationships. This conclusion is in accord with Equation (21.15), which assumes that only probabilities w_i need be specified for the general mixed-state ensemble, the phase relationships between systems of the ensemble being random.

Elastic Scattering of Particles with Spin

We wish to describe the elastic scattering of an unpolarized beam of particles by some target. One way to do this using pure states is to "sum over final states (after scattering) and average over initial states (before scattering)." This procedure, although equivalent to a mixed-state description, is often quite cumbersome; the most expedient way to treat such problems, and also to deal with arbitrary polarization values, is to introduce the mixed-state density matrix for a large number of particles (or, more precisely, for one representative particle from the ensemble). We now elucidate the procedure.

Consider a single elastic-scattering process in the center-of-momentum coordinate system for which the (unnormalized) asymptotic form of the scattering wave function is given, for a particle with spin quantum numbers (S, m), by

$$\phi_{\mathbf{k},m}(\mathbf{r}) \underset{r\to\infty}{\longrightarrow} e^{i\mathbf{k}\cdot\mathbf{r}}\chi_m + \frac{e^{ikr}}{r}\sum_{m'}\langle m'|\mathsf{f}|m\rangle\chi_{m'} \tag{22.11}$$

The scattering matrix f is a function of both the initial (m) and final (m') magnetic-spin quantum numbers as well as the scattering angles (θ, ϕ).[6] The notation only exhibits the $(2S + 1)$ by $(2S + 1)$ submatrix $\langle m'|\mathsf{f}|m\rangle$ for any given set of scattering angles. The first term on the right side of Equation (22.11) is the unperturbed or unscattered plane-wave function (17.27) without its normalization factor.

We next generalize Equation (22.11) by forming an arbitrary stationary-state wave function of the form (21.12), which becomes asymptotically

$$\begin{aligned}\Phi_{\mathbf{k}}(\mathbf{r}) &= \sum_m c_m\phi_{\mathbf{k},m}(\mathbf{r})\\ &\underset{r\to\infty}{\longrightarrow} \sum_m c_m e^{i\mathbf{k}\cdot\mathbf{r}}\chi_m + \frac{e^{ikr}}{r}\sum_{m,m'}\langle m'|\mathsf{f}|m\rangle c_m\chi_{m'}\end{aligned}$$

For the mth component of this wave function we then obtain

$$\begin{aligned}\Phi_{\mathbf{k},m}(\mathbf{r}) &\equiv (\chi_m, \Phi_{\mathbf{k}})\\ &\underset{r\to\infty}{\longrightarrow} c_m e^{i\mathbf{k}\cdot\mathbf{r}} + \frac{e^{ikr}}{r}\sum_{m'}\langle m|\mathsf{f}|m'\rangle c_{m'}\end{aligned} \tag{22.12}$$

From Equation (22.12) we can deduce that the amplitude $c_{m,f}$ for scattering into a final spin state m is given by

$$c_{m,f} = \sum_{m''} \langle m|\mathsf{f}|m''\rangle c_{m''} \tag{22.13}$$

The pure-state density matrix that describes the αth scattered particle is, therefore,

$$\begin{aligned}\langle m|\underline{\rho}_{\alpha,f}|m'\rangle &= \sigma^{-1} \sum_{m'',m'''} \langle m|\mathsf{f}|m''\rangle c_{\alpha,m''}(\langle m'|\mathsf{f}|m'''\rangle c_{\alpha,m'''})^* \\ &= \sigma^{-1} \sum_{m'',m'''} \langle m|\mathsf{f}|m''\rangle\langle m''|\underline{\rho}_{\alpha}|m'''\rangle\langle m'''|\mathsf{f}^{\dagger}|m'\rangle\end{aligned} \tag{22.14}$$

where σ^{-1} is a normalization constant to be determined by Equation (21.10). After substituting into Equation (21.17) we obtain the mixed-state density matrix for scattered particles,

$$\underline{\rho}_f = \underline{\sigma}^{-1}(\mathsf{f}\underline{\rho}\mathsf{f}^{\dagger}) \tag{22.15}$$

where $\underline{\rho}$ is the density matrix for the initial unscattered beam and

$$\sigma(\theta, \phi) = \mathrm{Tr}(\mathsf{f}\underline{\rho}\mathsf{f}^{\dagger}) \tag{22.16}$$

This last result and the first equality of Equation (22.14) show that σ is the differential scattering cross section summed over all final spin states. For spin-½ particles both the initial and final density matrices will be of the form (22.9).

To describe the elastic scattering of two spin-½ particles, the density matrices $\underline{\rho}$ and $\underline{\rho}_f$ must both be direct products of 2 × 2 matrices, that is, they must be 4 × 4 matrices. The scattering matrix f will be a scalar function of the initial and final relative momenta ($\mathbf{k}_i$ and $\mathbf{k}_f$) and the Pauli spin matrices $\underline{\boldsymbol{\sigma}}_1$ and $\underline{\boldsymbol{\sigma}}_2$ for the two particles. Thus

$$\mathsf{f} = \mathsf{f}(\mathbf{k}_f, \mathbf{k}_i, \underline{\boldsymbol{\sigma}}_1, \underline{\boldsymbol{\sigma}}_2) \tag{22.17}$$

where $\mathbf{k}_f = (\mathbf{k}_i, \theta, \phi)$.

Exercise 22.2

(a) For the scattering of two identical spin-½ particles initially in a total spin state (S, M_S), where $S(=0$ or $1)$ is conserved, argue that Equation (22.12) must be replaced by the equation

$$\begin{aligned}\Phi_{\mathbf{k}_i S M_S}(\mathbf{r}) \underset{r\to\infty}{\longrightarrow} & \; c_{SM_S}(e^{i\mathbf{k}_i\cdot\mathbf{r}} + (-1)^S e^{-i\mathbf{k}_i\cdot\mathbf{r}}) \\ & + \frac{e^{ikr}}{r} \sum_{M'_S} (\langle SM_S|\mathsf{f}(\mathbf{k}_f, \mathbf{k}_i)|SM'_S\rangle \\ & + (-1)^S\langle SM_S|\mathsf{f}(\mathbf{k}_f, -\mathbf{k}_i)|SM'_S\rangle)c_{SM'_S}\end{aligned}$$

(b) Show for this case that the only part of the scattering matrix (22.17) that needs to be considered is that part satisfying the identity

$$\mathsf{f}(\mathbf{k}_f, \mathbf{k}_i, \underline{\boldsymbol{\sigma}}_1, \underline{\boldsymbol{\sigma}}_2) = -\mathsf{f}(\mathbf{k}_f, -\mathbf{k}_i, \underline{\boldsymbol{\sigma}}_2, \underline{\boldsymbol{\sigma}}_1) \tag{22.18}$$

If identity (22.18) is assumed, then in Equations (22.15) and (22.16) we must make the replacements $\mathsf{f} \to 2\mathsf{f}$.

An operational approach can be used in the application of Equations (22.15) and (22.18) to the analysis of identical-particle elastic-scattering measurements. Inasmuch as the scattering matrix must be a scalar, its most general dependence on the vectors $\mathbf{k}_i$, $\mathbf{k}_f$, $\underline{\boldsymbol{\sigma}}_1$, and $\underline{\boldsymbol{\sigma}}_2$, subject to identity (22.18), can be parameterized rather easily. The analysis of polarization and other measurements, using Equations (22.15) and (22.16), can then be calculated in terms of these unknown and complex energy-dependent parameters. The procedure has been applied to two-nucleon elastic scattering, with the objective of determining experimentally the complete scattering matrix so that it may be compared eventually with theoretical calculations starting from two-nucleon forces.[7]

23 Time Evolution of Statistical Matrix; Liouville's Theorem of Classical Statistical Mechanics

We first examine the time evolution of a quantum-mechanical system in either a pure or mixed state. That is to say, we discuss the equation of motion satisfied by the statistical matrix $\underline{\rho}$, because this equation determines the time evolution of the mixed-state ensemble.

Next we consider the description of systems using classical mechanics and derive Liouville's theorem, which governs the time evolution of the distribution function for an ensemble of classical systems in phase space. The reduction of the quantum-mechanical equation of motion for $\underline{\rho}$ to Liouville's theorem is discussed, and the section ends with a rigorous derivation of the Boltzmann transport equation.

In our analysis of the equation of motion for the statistical matrix $\underline{\rho}$ of Equation (21.3) we shall use the Schrödinger picture, in which the state vectors are time-dependent and physical operators carry only explicit time dependence, if any. Then the equation of motion for the state vector $|j\rangle$ is the Schrödinger equation

$$i\hbar \frac{d}{dt}|j\rangle = \mathsf{H}|j\rangle \tag{23.1}$$

Assuming that the incomplete-information probabilities w_j are constants, determined for the ensemble at $t = 0$, we derive for the time rate of change of $\underset{\sim}{\rho}$ the expression

$$i\hbar\frac{d\underset{\sim}{\rho}}{dt} = \sum_j w_j\left\{\left(i\hbar\frac{d}{dt}|j\rangle\right)\langle j| + |j\rangle\left(i\hbar\frac{d}{dt}\langle j|\right)\right\}$$

$$i\hbar\frac{d\underset{\sim}{\rho}}{dt} = \sum_j w_j(\mathsf{H}\mathsf{P}_j - \mathsf{P}_j\mathsf{H}) = [\mathsf{H}, \underset{\sim}{\rho}] \qquad \textbf{(23.2)}$$

This equation is the quantum mechanical form of *Liouville's theorem*, discussed below. It is a fundamental equation of quantum statistical mechanics, and it gives the important result for systems in equilibrium, or in steady-state transport, that is, when $d\underset{\sim}{\rho}/dt = 0$, that

$$[\mathsf{H}, \underset{\sim}{\rho}] = 0 \qquad \text{(equilibrium)} \qquad (23.3)$$

Thus the equilibrium statistical matrix commutes with the Hamiltonian.

If the Hamiltonian is time-independent, then the solution to the differential equation (23.2) is

$$\underset{\sim}{\rho}(t) = \exp(-i\mathsf{H}t/\hbar)\underset{\sim}{\rho}(0)\exp(i\mathsf{H}t/\hbar) \qquad (23.4)$$

Equation (23.4) is important in the quantum theory of measurement; it gives the statistical matrix to be used when there is a time lapse between successive measurements. It is also quite useful in treatments of nonequilibrium phenomena.

Equation (23.2) can be used to derive the equation of motion for an averaged physical operator $\langle\mathsf{A}\rangle$, Equation (21.9). One finds

$$\begin{aligned} i\hbar\frac{d}{dt}\langle\mathsf{A}\rangle &= i\hbar\,\mathrm{Tr}\left(\frac{d\underset{\sim}{\rho}}{dt}\cdot\mathsf{A}\right) + i\hbar\,\mathrm{Tr}\left(\underset{\sim}{\rho}\frac{\partial\mathsf{A}}{\partial t}\right) \\ &= \mathrm{Tr}([\mathsf{H}, \underset{\sim}{\rho}]\mathsf{A}) + i\hbar\,\mathrm{Tr}\left(\underset{\sim}{\rho}\frac{\partial\mathsf{A}}{\partial t}\right) \\ &= \mathrm{Tr}\left[\underset{\sim}{\rho}\left([\mathsf{A}, \mathsf{H}] + i\hbar\frac{\partial\mathsf{A}}{\partial t}\right)\right] \end{aligned}$$

$$i\hbar\frac{d}{dt}\langle\mathsf{A}\rangle = i\hbar\left\langle\frac{\partial\mathsf{A}}{\partial t}\right\rangle + \langle[\mathsf{A}, \mathsf{H}]\rangle \qquad \textbf{(23.5)}$$

Note that Equations (23.2) and (23.5) hold for both pure-state and mixed-state ensembles. Equation (23.5) gives, in fact, the general time rate of change of $\langle\mathsf{A}\rangle$ in the Schrödinger picture.[8]

Exercise 23.1 Apply Equation (23.5) to calculate the time rate of change of the polarization vector **P**, Equation (22.6), for a beam of spin-½ particles with charge -e and passing through a magnetic field $\mathcal{H}$. The spin Hamiltonian is

$$\mathsf{H} = -\underline{\boldsymbol{\mu}} \cdot \mathcal{H} = \frac{1}{2} g \left(\frac{e\hbar}{2mc} \right) (\underline{\boldsymbol{\sigma}} \cdot \mathcal{H})$$

where g is the Landé g factor for the particles. Use angular-momentum commutation relations to show that

$$\frac{\partial \mathbf{P}}{\partial t} = \frac{1}{2} g \left(\frac{e}{mc} \right) (\mathcal{H} \times \mathbf{P}) \tag{23.6}$$

in agreement with a generalization of Ehrenfest's theorem or the correspondence principle of quantum mechanics.[9]

Classical Statistical Mechanics

It is useful to compare the statistical matrix of quantum statistics with the corresponding distribution function of classical statistical mechanics. In the classical case we describe a system by specifying particle coordinates $\mathbf{q}$ and momenta $\mathbf{p}$, instead of by using states $|i\rangle$. We also use the concept of *phase space* in classical mechanics, and we distinguish between the phase space (called μ space) for a single particle and the phase space (called Γ space) for a whole system of particles.

Consider a particle with s degrees of freedom. Its condition classically is determined completely by the values of a set of s suitably chosen quantities $q_i(i = 1, 2, \ldots, s)$ and the values $\dot{q}_i$ of their time derivatives. The q_i are called *generalized coordinates,* and their time derivatives are called *generalized velocities*. For example, a simple helium atom for which the internal degrees of freedom can be neglected has $s = 3$. A simple diatomic molecule, such as H_2, for which the electronic and spin degrees of freedom can be neglected and for which the interatomic distance is fixed, has $s = 5$. In this second example three of the degrees of freedom are translational, and two are rotational.

It can be shown that the kinetic energy K of a simple particle can always be written as a quadratic function of the $\dot{q}_i$:

$$K = K(q_i, \dot{q}_i) = \sum_{m,n=1}^{s} a_{mn}(q_i)\dot{q}_m\dot{q}_n \tag{23.7}$$

unless the transformation equations that define the q_i depend explicitly on time.[10] Moreover, we assume that the potential energy V is conservative. Then *generalized momenta* p_i are introduced by the definitions

$$p_i = \frac{\partial L(q_j, \dot{q}_j, t)}{\partial \dot{q}_i} \tag{23.8}$$

where $L = K - V$ is the Lagrangian. The Hamiltonian H is introduced by the definition

$$H = \sum_{i=1}^{s} \dot{q}_i p_i - L(q_j, \dot{q}_j, t) \tag{23.9}$$

which reduces to $H = K + V$ under the conditions that the kinetic energy is a quadratic function (23.7) and that the potential energy is velocity independent. The Hamiltonian can also be identified as the total energy when there are electromagnetic interactions.

The equations of motion for a particle in classical mechanics can be chosen to be *Hamilton's equations,* which are differential equations for the generalized coordinates and momenta given by[11]

$$\begin{aligned} \dot{q}_i &= \frac{\partial H}{\partial p_i} \\ \dot{p}_i &= -\frac{\partial H}{\partial q_i} \end{aligned} \tag{23.10}$$

A further equation is $(\partial L/\partial t) = -(\partial H/\partial t)$. It is the generalized coordinates and momenta that are most useful in classical statistical mechanics, rather than generalized coordinates and velocities, primarily because of the simple form Liouville's theorem takes.[12]

The *phase* of a molecule or particle is defined to be a set of values of its s q_i and s p_i coordinates. The behavior of a single molecule can then be imagined as an orbit, or trajectory, in the $2s$-dimensional space of coordinates q_i and p_i. This phase space is called μ space for single molecules (μ for molecule). A volume element in μ space is denoted by $d\omega$, with

$$d\omega = \prod_{i=1}^{s} dq_i \, dp_i \tag{23.11}$$

A system of N molecules, each with s degrees of freedom, can be described classically by using Ns generalized coordinates q_i and Ns generalized momenta p_i. Hamilton's equations (23.10) continue to be valid for the large system, with an N-body Hamiltonian $H^{(N)}$, which we assume, for the present, to be time-independent. The $2Ns$ coordinates need not be the coordinates and momenta of the individual particles, but as in the normal-modes description of a set of coupled oscillators, they may be functional combinations, also called collective coordinates, of the individual particle coordinates. We define the phase space for an N-particle system to be the $2Ns$-dimensional coordinate system in which each microscopic state of the system is represented by a point. This space is called Γ *space* (Γ for gas, although the notion applies to liquids and solids as well) and a volume element in Γ space is denoted by $d\Gamma$, with

$$d\Gamma = \prod_{i=1}^{Ns} dq_i \, dp_i \tag{23.12}$$

The task of solving a set of $2Ns$ coupled differential equations (23.10) in classical mechanics corresponds in quantum mechanics to finding the exact eigenstates $|i\rangle$ of a many-body system, using the methods of Chapter IV, say. We hardly know how to solve these problems definitively for $N = 3$, to say nothing of $N \sim$ Avogadro's number. It would be a hopeless task and, more important, it would be a meaningless undertaking, if only because of the lack of knowledge of $2Ns$ "initial values" for the $2Ns$ coordinates. In fact, the macroscopic properties of systems that are accessible to measurement depend not on values of the coordinates at some specified time, but rather on suitably time-averaged values of these coordinates. Moreover, because of the lack of knowledge of initial values of the coordinates, we may say that any one system is equally well described by the average properties (at a given time, say) of a large collection of identical systems—an ensemble—as it is by the time average of its own trajectory in Γ space. That these two averages are indeed "equal" is the *ergodic hypothesis* of statistical mechanics, to be discussed further in Section 25.

Thus we consider an ensemble of M identical systems, with M corresponding *representative points* in Γ space. The probability dP that the $2Ns$ coordinates of a system lie between the values q_i, p_i, and $(q_i + dq_i)$, $(p_i + dp_i)$ is given in terms of an ensemble probability density $\rho(q_i \cdots q_{Ns}; p_1 \cdots p_{Ns}; t) = \rho(q_i, p_i, t)$ by [13]

$$dP = \rho(q_i, p_i, t) \, d\Gamma \tag{23.13}$$

with a normalization of total probability to unity,

$$\int dP = \int \rho(q_i, p_i, t) \, d\Gamma = 1 \tag{23.14}$$

For a system of N identical particles the distribution function ρ must be symmetric under interchange of particle labels, and for a system at equilibrium it is independent of t. The distribution function ρ is the classical analogue of the statistical matrix $\underline{\rho}$ of quantum mechanics.

Liouville's Theorem

The central problem of classical statistical mechanics is to determine the distribution function $\rho(q_i, p_i, t)$, and for this purpose Liouville's theorem plays a guiding role. A mathematical statement of Liouville's theorem is

$$\frac{d\rho}{dt} = \frac{\partial \rho}{\partial t} + \sum_{i=1}^{Ns} \left(\frac{\partial \rho}{\partial q_i} \dot{q}_i + \frac{\partial \rho}{\partial p_i} \dot{p}_i \right) = 0 \tag{23.15}$$

Its geometrical interpretation is that if we follow the trajectory of a representative point in Γ space, then we find that the density of representative points in its neighborhood is constant. (See Figure V.1). Hence the distribution of representative points of the ensemble moves in Γ space like an incompressible fluid.[14]

Although the theorems and methods of classical statistical mechanics should follow in appropriate limits from corresponding theorems of quantum statistics, and that is the attitude we must adopt ultimately, it is nevertheless of value, if only conceptually, to demonstrate the classical results entirely within the classical framework. Thus we now give a classical proof of Liouville's theorem, and then we discuss its relation to, and how it can be deduced from, the quantum-mechanical equation of motion (23.2) for $\underline{\rho}$.

We assume that the distribution function ρ can be treated as a continuous function of its $2Ns$ coordinates and prove Liouville's theorem, nonrigorously therefore, from this hypothesis. We observe first that the total number of systems in the ensemble is conserved. Correspondingly, the number of representative points per second leaving any "volume" V in Γ space must

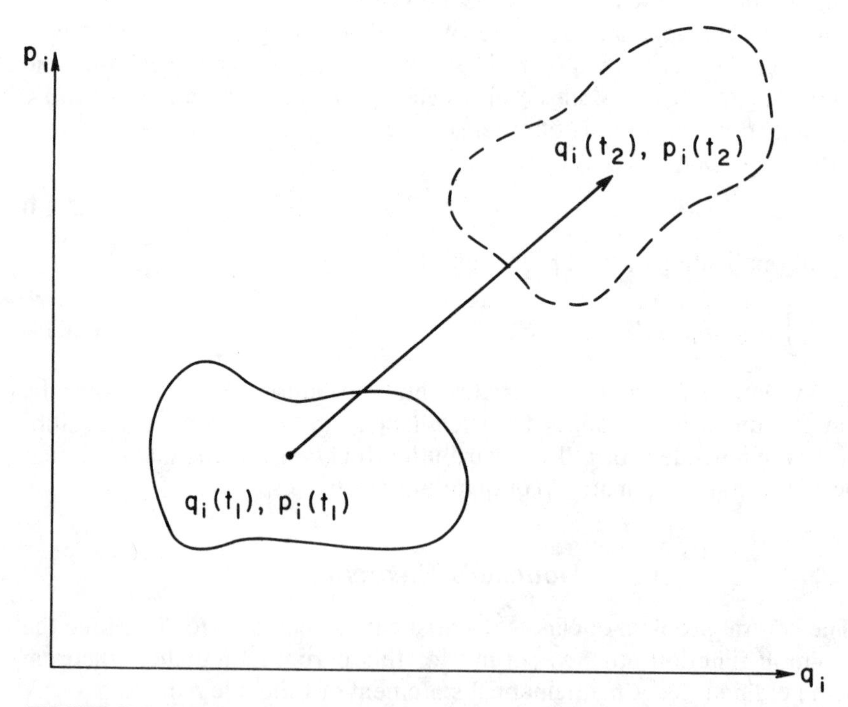

FIGURE V.1 Motion of a collection of representative points in phase space from time t_1 to time t_2. Liouville's theorem states that the phase-space volume of this collection remains constant in time.

be equal to the rate of decrease of the number of representative points within the volume V. If S is the "surface area" of the volume V and $\mathbf{v}$ is a $2Ns$-dimensional velocity vector,

$$\mathbf{v} = (\dot{q}_1, \dot{q}_2, \ldots, \dot{q}_{Ns}; \dot{p}_1, \dot{p}_2, \ldots, \dot{p}_{Ns}) \tag{23.16}$$

then the continuity equation that expresses this equality is

$$-\frac{d}{dt}\int_V \rho(q_i, p_i, t)\, d\Gamma = \int_S dS(\hat{\mathbf{n}} \cdot \mathbf{v})\rho(q_i, p_i, t) \tag{23.17}$$

where $\hat{\mathbf{n}}$ is the outward normal to the surface element dS. With the help of the divergence theorem in $2Ns$-dimensional space, the continuity equation (23.17) can be converted to the form

$$\int_V d\Gamma\left[\frac{\partial \rho}{\partial t} + \nabla \cdot (\mathbf{v}\rho)\right] = 0$$

where ∇ is the $2Ns$-dimensional gradient operator

$$\nabla \equiv \left(\frac{\partial}{\partial q_1}, \frac{\partial}{\partial q_2}, \ldots, \frac{\partial}{\partial q_{Ns}}; \frac{\partial}{\partial p_1}, \frac{\partial}{\partial p_2}, \ldots, \frac{\partial}{\partial p_{Ns}}\right)$$

Since V is an arbitrary volume in Γ space, the integrand of this last equation must vanish identically, and therefore we obtain the differential form of the continuity equation,

$$-\frac{\partial \rho}{\partial t} = \nabla \cdot (\mathbf{v}\rho) = \sum_{i=1}^{Ns}\left[\frac{\partial}{\partial q_i}(\dot{q}_i\rho) + \frac{\partial}{\partial p_i}(\dot{p}_i\rho)\right]$$

Liouville's theorem now follows from this last expression with the aid of Hamilton's equations (23.10).

Exercise 23.2 Complete the above proof of Liouville's theorem (23.15) by verifying the incompressibility condition $\nabla \cdot \mathbf{v} = 0$.

Liouville's theorem can also be expressed in terms of a Poisson bracket, which is defined for functions $A(q_i, p_i)$ and $B(q_i, p_i)$ by the relation

$$\{A, B\} \equiv \sum_{i=1}^{Ns}\left(\frac{\partial A}{\partial q_i}\frac{\partial B}{\partial p_i} - \frac{\partial A}{\partial p_i}\frac{\partial B}{\partial q_i}\right) \tag{23.18}$$

The substitution of Equations (23.10) into Equation (23.15) gives Liouville's theorem in the form

$$\frac{d\rho}{dt} = \frac{\partial \rho}{\partial t} + \{\rho, H^{(N)}\} = 0 \tag{23.19}$$

The relation of this form of Liouville's theorem to the quantum-mechanical equation of motion for $\underline{\rho}$, Equation (23.2), can be understood as follows. One prescription for finding the quantum analogue of a classical equation of motion[15] is that a Poisson bracket goes into a commutator bracket according to

$$\{A, B\} \rightarrow \frac{1}{i\hbar} [\mathsf{A}, \mathsf{B}]$$

If we observe that a representative point in Γ space becomes a state with no implicit time dependence in quantum mechanics, then we also have the correspondence

$$\frac{\partial \rho}{\partial t} \rightarrow \frac{d\underline{\rho}}{dt}$$

With these two prescriptions for transforming Liouville's theorem from classical mechanics to quantum mechanics we immediately obtain Equation (23.2), derived earlier. It should be mentioned, of course, that actually it is a cell of size h^{Ns} in Γ space, and not a point, that is analogous to one quantum-mechanical state, but this clarification does not affect the above argument.

The quantum-mechanical analogue of the classical distribution function ρ, which depends on Ns sets of coordinates (q_i, p_i), is not unique. Consider the simple case $s = 3$ for which the particles have no integral degrees of freedom. Then the statistical matrix $\underline{\rho}$ can be expressed in the position representation in terms of N sets of vector coordinates as $\langle \mathbf{r}_1 \cdots \mathbf{r}_N | \underline{\rho} | \mathbf{r}'_1 \cdots \mathbf{r}'_N \rangle$ or in the momentum representation as $\langle \mathbf{p}_1 \cdots \mathbf{p}_N | \underline{\rho} | \mathbf{p}'_1 \cdots \mathbf{p}'_N \rangle$, but there is no unique prescription for transforming $\underline{\rho}$ to the position-momentum representation of classical statistical mechanics. Indeed, there should be a difficulty with such a prescription, for the Heisenberg uncertainty relation forbids any quantum-mechanical distribution function that would give all N sets of coordinates $(\mathbf{r}_i, \mathbf{p}_i)$ exactly and simultaneously.

One popular and particularly simple quantum-mechanical function of the $3N$ coordinate sets (q_i, p_i) can be defined in terms of the position-space matrix elements $\langle \mathbf{r}_1 \ldots \mathbf{r}_N | \underline{\rho} | \mathbf{r}'_1 \ldots \mathbf{r}'_N \rangle$. This is the *Wigner distribution function* f_w, which is defined and discussed in Appendix E. There it is shown that $f_w(q_i, p_i, t)$ has physical properties one might reasonably expect of such a distribution function; moreover, it is demonstrated that the equation of motion which f_w satisfies reduces in the limit $h \rightarrow 0$ to Liouville's theorem.

Boltzmann Transport Equation

In Section 11 we introduced the one- and two-particle probability densities in phase space, $P_1(\mathbf{q}, \mathbf{p}, t)$ and $P_2(q_1, q_2; p_1, p_2; t)$. We also gave an elementary derivation of the Boltzmann transport equation (11.27), which is an equation of motion for $P_1(\mathbf{q}, \mathbf{p}, t)$. We now provide a rigorous basis for this equation by deducing it from Liouville's Equation (23.19).

We initiate the development by defining m-particle *reduced distribution functions* $P_m(q_1 \ldots q_m; p_1 \ldots p_m; t)$, which are phase-space probability densities for the coordinates and momenta of any set of m particles in a system of N ($\geqq m$) identical particles.[16] We define the P_m in terms of the classical distribution function ρ by

$$P_m(q_1 \ldots q_m; p_1 \ldots p_m; t) \equiv \frac{N!}{(N-m)!}\int \prod_{\alpha=m+1}^{N} (d\omega_\alpha)\rho(q_1 \ldots q_N; p_1 \ldots p_N; t) \qquad \textbf{(23.20)}$$

where the $d\omega_\alpha$ are μ-space volume elements (23.11). The factor $N!/(N - m)!$ is the number of ways of selecting the set of m coordinates (and momenta) in P_m from the set of N coordinates in $\rho(q_i, p_i, t)$, which is itself a symmetric function of its coordinates. According to Equation (23.14), the normalization of the reduced distribution function $P_m(q_i, p_i, t)$ is $N!/(N - m)!$, in agreement with our earlier definitions (11.1) and (11.2) for the cases $m = 1$ and $m = 2$ respectively.

We now derive the Boltzmann transport equation. We define $d\Gamma' \equiv \Pi_{\alpha=2}^{N} d\omega_\alpha$ and use definition (23.20) for $P_1(q_1, p_1, t)$. Then on multiplying Liouville's Equation (23.19) by $Nd\Gamma'$ and integrating over all (and not only the allowed region)[17] of the $(N - 1)$-particle phase space we obtain

$$\frac{\partial}{\partial t}P_1(q_1, p_1, t) + N\int d\Gamma' \{\rho, H^{(N)}\} = 0 \qquad (23.21)$$

Now, according to Equation (23.18) the Poisson bracket of ρ and $H^{(N)}$ can be written as

$$\{\rho, H^{(N)}\} = \left[\frac{\partial\rho}{\partial q_1}\cdot\frac{\partial H^{(N)}}{\partial p_1} - \frac{\partial\rho}{\partial p_1}\cdot\frac{\partial H^{(N)}}{\partial q_1}\right] + \sum_{i=2}^{N}\left[\frac{\partial}{\partial q_i}\cdot\left(\rho\frac{\partial H^{(N)}}{\partial p_i}\right) - \frac{\partial}{\partial p_i}\cdot\left(\rho\frac{\partial H^{(N)}}{\partial q_i}\right)\right]$$

where $\partial/\partial q_i$ and $\partial/\partial p_i$ are gradient operators and we intend to treat the two parts on the right side of this expression differently. In fact, when the second part is substituted into Equation (23.21), then every term in this part can be integrated explicitly to give vanishing surface integrals,

$$\sum_{i=2}^{N}\left\{\int\frac{d\Gamma'}{dq_i}\left(\rho\frac{\partial H^{(N)}}{\partial p_i}\right)\bigg|_{-\infty}^{+\infty} - \int\frac{d\Gamma'}{dp_i}\left(\rho\frac{\partial H^{(N)}}{\partial q_i}\right)\bigg|_{-\infty}^{+\infty}\right\} = 0$$

since $\rho = 0$ at the integration limits.[17]

Remaining is the first term of the Poisson bracket for which the only part of $H^{(N)}$ needed is

$$H_0^{(1)} + \sum_{i=2}^{N} V_{1i}$$

where the one-particle Hamiltonian for particle 1 is given by $H_0^{(1)} = K^{(1)} + U^{(1)}$. Here $K^{(1)}$ is a one-particle kinetic energy, and $U^{(1)}$ is a one-particle potential energy due to external forces. Thus Equation (23.21) becomes

$$\frac{\partial}{\partial t}P_1(q_1, p_1, t) + \{P_1, H_0^{(1)}\}_1 = -N\sum_{i=2}^{N}\int d\Gamma'\{\rho, V_{1i}\}_1 \tag{23.22}$$

where the subscript 1 on the Poisson bracket indicates that the Poisson bracket is to be computed using only the canonical coordinates of particle 1. If now we consider only velocity-independent external forces and set $q_1 \rightarrow \mathbf{r}_1$, then the Poisson bracket of P_1 and $H_0^{(1)}$ becomes

$$\{P_1, H_0^{(1)}\}_1 = \frac{\partial P_1(q_1, p_1, t)}{\partial q_1}\cdot\frac{\partial H_0^{(1)}}{\partial p_1} - \frac{\partial P_1(q_1, p_1, t)}{\partial p_1}\cdot\frac{\partial H_0^{(1)}}{\partial q_1} \tag{23.23}$$

$$\rightarrow \mathbf{v}_1\cdot\nabla_{r_1}P_1(\mathbf{r}_1, \mathbf{p}_1, t) + \mathbf{F}_{\text{ext}}\cdot\nabla_{p_1}P_1(\mathbf{r}, \mathbf{p}, t) \qquad \text{(velocity-independent external forces)}$$

In the second line of this last equation we have used the definition of the external force $\mathbf{F}_{\text{ext}}$ acting on a particle

$$\mathbf{F}_{\text{ext}}(\mathbf{r}) = -\nabla_r U^{(1)}(\mathbf{r}) \tag{23.24}$$

and the velocity expression $\mathbf{v} = \nabla_p K^{(1)} = \mathbf{p}/m$.

The right side of Equation (23.22) can be simplified by noting that the distribution function ρ is symmetric in its N sets of coordinates (q_i, p_i). Therefore, every term in the sum gives the same result as the $i = 2$ term, and we obtain, on setting $q_i \rightarrow \mathbf{r}_i$ and using Equation (23.20) with $m = 2$,

$$\frac{\partial}{\partial t}P_1(\mathbf{r}_1, \mathbf{p}_1, t) + \{P_1, H_0^{(1)}\}_1 = -\int d\omega_2\{P_2, V_{12}\}_1 \tag{23.25}$$

It can be seen that the two-particle distribution function enters into the equation of motion for the one-particle distribution function because of particle interactions. In a similar manner the time development of the m-particle distribution function P_m is influenced by $P_{m+1}(q_i, p_i, t)$, so that a hierarchy of N coupled differential equations of the type (23.25) can be derived. The complete set of equations is called the Bogoliubov–Born–Green–Kirkwood–Yvon (BBGKY) hierarchy of equations, with the mth equation also being equivalent to Liouville's theorem (23.15) for m particles; that is, the mth equation can be written as $dP_m/dt = 0$. See Problem V.6.

We wish now to rewrite the integral on the right side of Equation (23.25) as a *collision integral* (see Section 11). To do this we first use definition (23.18) of the Poisson bracket to get, for velocity-independent internal forces,

$$-\int d\omega_2\{P_2, V_{12}\}_1 = \int d\omega_2 \frac{\partial}{\partial \mathbf{p}_1} P_2(\mathbf{r}_1, \mathbf{r}_2; \mathbf{p}_1, \mathbf{p}_2; t) \cdot \frac{\partial V_{12}}{\partial \mathbf{r}_1}$$

$$= -\int d\omega_2 \frac{\partial}{\partial \mathbf{p}_1} P_2(\mathbf{r}_1, \mathbf{r}_2; \mathbf{p}_1, \mathbf{p}_2; t) \cdot \left(\frac{d\mathbf{p}_1}{dt}\right)_{\text{due to 2}}$$

In the second equality we have used Newton's second law for the force between particles 1 and 2. We recall next that the distribution function P_2 gives the probability for finding any of the other $(N - 1)$ particles in the μ-space element $d\omega_2$ when particle 1 is in the μ-space element $d\omega_1$. Therefore the last integral above yields the time rate at which particle 1 in $d\omega_1$ will receive a momentum transfer $d\mathbf{p}_1$ in time dt due to interaction with any of the other $(N - 1)$ particles in the system; in other words, the right side of Equation (23.25) is a collision integral. We observe, finally, [see Equation (23.17)] that any contribution to the outward flux of a distribution function, that is, of its representative points, through a surface bounding some volume in phase space is minus the time derivative of that distribution function integrated over the volume. We may conclude that

$$-\int d\omega_2\{P_2, V_{12}\}_1 = \left[\frac{\partial}{\partial t} P_1(\mathbf{r}_1, \mathbf{p}_1, t)\right]_{\text{coll}} \tag{23.26}$$

Substitution of this last result and Equation (23.23) into Equation (23.25) yields the famous *Boltzmann equation* of kinetic theory, derived earlier and in less rigorous fashion in Section 11.

$$\left[\frac{\partial}{\partial t} + \mathbf{v} \cdot \nabla_r + \mathbf{F}_{\text{ext}} \cdot \nabla_p\right] P_1(\mathbf{r}, \mathbf{p}, t) = \left(\frac{\partial P_1}{\partial t}\right)_{\text{coll}} \tag{23.27}$$

When the external force arises from applied external electric and magnetic fields, then this kinetic equation must be generalized to account for velocity dependence in both $U^{(1)}$ and V_{12}.[18]

24 Reduced Distribution Functions

Attention is given to the distribution functions that characterize the behavior of only one or two particles in a large system. These "reduced" distribution functions are examined from both classical and quantum-mechanical viewpoints. For the quantum case, reduced-density matrices are also introduced, and it is shown that their diagonal matrix elements are reduced distribution functions.

We mentioned at the beginning of Section 7 that one of the goals of statistical mechanics is to provide a microscopic basis for the laws of thermodynamics. To this end we have prepared ourselves for thermodynamic calculations in statistical mechanics with the developments of Section 7 and elsewhere in Chapter II. But even for systems in thermodynamic equilibrium, to which we have so far restricted much of our discussion, there exists a category of macroscopic phenomena that are accessible to experiment yet that are not describable by classical thermodynamics. To describe these phenomena we need physical quantities other than the thermodynamic variables used to characterize bulk properties of macroscopic systems. We need the physical quantities called *reduced distribution functions,* or often merely *distribution functions,* which characterize in some detail correlation effects among the particles in a system. It is important to emphasize that reduced distribution functions are measurable, to a certain extent, say, by scattering experiments. Moreover, they can be interpreted theoretically only on a microscopic basis. Because they are quasi microscopic, reduced distribution functions are more basic physical quantities for characterizing a macroscopic system than are the classical thermodynamic variables.

We have already given a general definition, Equation (23.20), of an m-particle reduced distribution function P_m for classical systems. It follows from this definition that the relation between P_{m+1} and P_m involves one μ-space integration,

$$\int d\omega_{m+1} P_{m+1}(q_1 \cdots q_{m+1};\, p_1 \cdots p_{m+1};\, t) = (N - m)P_m(q_1 \cdots q_m;\, p_1 \cdots p_m;\, t) \tag{24.1}$$

The factor $(N - m)$ corresponds to the number of ways of selecting the particle identified as number $(m + 1)$ after m have already been identified.

Our development of classical kinetic theory in Chapter III was introduced in Section 11 by considering the one-particle reduced distribution function P_1. A dynamical dependence of P_1 on P_2 was established by consideration of two-particle collisions and the collision integral. In the present section we investigate the formal definition of these two distribution functions in some detail.

We begin by introducing the N-particle classical density functions $n(\mathbf{r})$ and $n(\mathbf{p})$ with the definitions

$$\begin{aligned} n(\mathbf{r}) &\equiv \sum_{\alpha=1}^{N} \delta^{(3)}(\mathbf{r} - \mathbf{r}_\alpha) \\ n(\mathbf{p}) &\equiv \sum_{\alpha=1}^{N} \delta^{(3)}(\mathbf{p} - \mathbf{p}_\alpha) \end{aligned} \tag{24.2}$$

We can understand these definitions by considering first the case $N = 1$ and observing that the one-particle density functions are normalized correctly,

$$\int d^3\,\mathbf{r}n(\mathbf{r})\Big|_{N=1} = \int d^3\,\mathbf{p}n(\mathbf{p})\Big|_{N=1} = 1$$

Note that the integral over position space, for example, is unity as long as the volume of integration includes the location of the one particle. Finally, to get the average density of particles at position $\mathbf{r}$, or with momentum $\mathbf{p}$, we must sum N one-particle density functions over all the particles.[19]

For classical systems we are interested not only in the density functions (24.2), but also in a density function that refers to both the generalized coordinates $\mathbf{r}$ and $\mathbf{p}$ simultaneously. Thus we consider the N-particle density function $n(\mathbf{r}, \mathbf{p})$ defined by

$$n(\mathbf{r}, \mathbf{p}) \equiv \sum_{\alpha=1}^{N^*} \delta^{(3)}(\mathbf{r} - \mathbf{r}_\alpha)\, \delta^{(3)}(\mathbf{p} - \mathbf{p}_\alpha) \tag{24.3}$$

Furthermore, we define the average value $\langle A\rangle$ of a physical quantity $A(\mathbf{r}_1 \dots \mathbf{r}_N;\ \mathbf{p}_1 \dots \mathbf{p}_N, t)$ in terms of the classical distribution function (23.13) by the relation

$$\langle A\rangle \equiv \int d\Gamma A(\mathbf{r}_i, \mathbf{p}_i, t)\rho(\mathbf{r}_i, \mathbf{p}_i, t) \tag{24.4}$$

With these last two definitions in mind, one can write, the classical reduced distribution functions $P_1(\mathbf{r}, \mathbf{p}, t)$ and $P_2(\mathbf{r}, \mathbf{r}'; \mathbf{p}, \mathbf{p}'; t)$ as the average values

$$\begin{aligned} P_1(\mathbf{r}, \mathbf{p}, t) &= \langle n(\mathbf{r}, \mathbf{p})\rangle \\ P_2(\mathbf{r}, \mathbf{r}'; \mathbf{p}, \mathbf{p}'; t) &= \langle n(\mathbf{r}, \mathbf{p})n(\mathbf{r}', \mathbf{p}')\rangle \\ &\quad - \delta^{(3)}(\mathbf{r} - \mathbf{r}')\, \delta^{(3)}(\mathbf{p} - \mathbf{p}')\langle n(\mathbf{r}, \mathbf{p})\rangle \end{aligned} \tag{24.5}$$

These two equations clarify the interpretation of P_1 and P_2 as reduced, or average, distribution functions for one and two particles respectively. The role of the second term on the right side of the expression for P_2 is to remove a single-particle density average from the first term.

Exercise 24.1 Verify that Equations (24.5) for P_1 and P_2 are in agreement with the general definition (23.20) of these two quantities.

In analogy with Equations (24.5) we can also introduce one- and two-particle distribution functions of the position and momentum coordinates alone. We use Equations (24.2) to make the definitions.

$$
\begin{aligned}
P(\mathbf{r}, t) &\equiv \langle n(\mathbf{r}) \rangle \\
P(\mathbf{p}, t) &\equiv \langle n(\mathbf{p}) \rangle \\
P(\mathbf{r}, \mathbf{r}', t) &\equiv \langle n(\mathbf{r}) n(\mathbf{r}') \rangle - \delta^{(3)}(\mathbf{r} - \mathbf{r}') \langle n(\mathbf{r}) \rangle \\
P(\mathbf{p}, \mathbf{p}', t) &\equiv \langle n(\mathbf{p}) n(\mathbf{p}') \rangle - \delta^{(3)}(\mathbf{p} - \mathbf{p}') \langle n(\mathbf{p}) \rangle
\end{aligned}
\tag{24.6}
$$

It is then easy to verify, by referring to Equations (24.5), that

$$
\begin{aligned}
P(\mathbf{r}, t) &= \int d^3\mathbf{p} P_1(\mathbf{r}, \mathbf{p}, t) \\
P(\mathbf{r}, \mathbf{r}', t) &= \int d^3\mathbf{p}\, d^3\mathbf{p}' P_2(\mathbf{r}, \mathbf{r}'; \mathbf{p}, \mathbf{p}', t)
\end{aligned}
\tag{24.7}
$$

Similar relations hold for $P(\mathbf{p}, t)$ and $P(\mathbf{p}, \mathbf{p}', t)$.[20]

The two-particle distribution functions $P(\mathbf{r}, \mathbf{r}', t)$ and $P(\mathbf{p}, \mathbf{p}', t)$ are called *pair distribution functions*. They are of particular interest, since they reflect the effect of correlations between the particles in a system, because of interaction effects and also, in the quantum-mechanical case, wave-function symmetrization. For the classical ideal gas, in which such correlation effects are absent, the pair distribution functions are simply products of one-particle distribution functions:

$$
\begin{aligned}
P(\mathbf{r}_1, \mathbf{r}_2, t) &= P(\mathbf{r}_1, t) P(\mathbf{r}_2, t) \\
P(\mathbf{p}_1, \mathbf{p}_2, t) &= P(\mathbf{p}_1, t) P(\mathbf{p}_2, t) \qquad \text{(classical ideal gas)}
\end{aligned}
\tag{24.8}
$$

We have used this stochastic-independence property earlier as Equation (4.3).[21] In definitions (24.6) of the pair distribution functions the subtracted terms are single-particle averages that, in fact, must be included in these definitions if Equations (24.8) are to hold for the ideal-gas limit.

Quantum-Mechanical Considerations

For quantum-statistical calculations the classical average value (24.4) is replaced by the average value $\langle \mathsf{A} \rangle$ of Equation (21.9b), in which the statistical matrix $\underline{\rho}$ is used. Although there is no direct quantum-mechanical analogy to the averages (24.5), unless one employs the Wigner distribution function of Appendix E,[22] the distribution functions of Equations (24.6) have immediate counterparts in quantum statistics.

Consider first the distribution functions in position space. In this case, Equations (24.6) for the local density $P(\mathbf{r}, t)$ and for the pair distribution function $P(\mathbf{r}, \mathbf{r}', t)$ continue to be valid if one merely replaces the density function $n(\mathbf{r})$ by the density operator $\mathsf{n}(\mathbf{r})$, which is defined in the position representation by the first of Equations (24.2). One can show [see Problem

V.3(d)] that in Fock space the operator $\mathsf{n}(\mathbf{r})$ is given in terms of the field operators $\underline{\psi}(\mathbf{r})$ and $\underline{\psi}^\dagger(\mathbf{r})$ by

$$\mathsf{n}(\mathbf{r}) = \underline{\psi}^\dagger(\mathbf{r})\underline{\psi}(\mathbf{r}) \tag{24.9}$$

an expression that is consistent with Equation (17.24) for the total number operator N.

When treating distribution functions in momentum space quantum mechanically, it is convenient to use wave numbers $\mathbf{k}$ instead of momenta $\mathbf{p} = \hbar\mathbf{k}$. We also consider the periodic-volume case in which these wave numbers form a discrete, rather than a continuous, set of quantum numbers. Thus we use the definition for $n(\mathbf{k})$ in momentum space,

$$n(\mathbf{k}) \equiv \sum_{\alpha=1}^{N} \delta_{\mathbf{k},\,\mathbf{k}_\alpha} = \frac{(2\pi\hbar)^3}{\Omega} n(\mathbf{p}) \tag{24.10}$$

where $n(\mathbf{p})$ is defined by the second of Equations (24.2). Here the relation (17.35) between a Kronecker-δ function and a Dirac-δ function has been used. In view of Equation (24.10), it is useful to define the m-particle distribution function of the discrete wave numbers $(\mathbf{k}_1 \ldots \mathbf{k}_m)$ with the density-of-states factors $\Omega/(2\pi\hbar)^3$ removed,

$$P(\mathbf{k}_1, \mathbf{k}_2, \ldots \mathbf{k}_m, t) \equiv \left[\frac{(2\pi\hbar)^3}{\Omega}\right]^m P(\mathbf{p}_1, \mathbf{p}_2, \ldots, \mathbf{p}_m, t) \tag{24.11}$$

To avoid undue proliferation of notation we have not devised new notation for the functions defined by Equations (24.10) and (24.11) in the discrete wave-number case.

Exercise 24.2 Use Equation (17.12) to show that in Fock space the operator $\mathsf{n}(\mathbf{k})$ is given by Equation (17.5), that is, by

$$\mathsf{n}(\mathbf{k}) = \mathsf{a}_\mathbf{k}^\dagger \mathsf{a}_\mathbf{k} \tag{24.12}$$

With the aid of the preceding three equations for the discrete wave-number case we can now readily adopt, for quantum-mechanical use, the classical expressions (24.6) for the momentum distribution $P(\mathbf{k}, t)$ and the pair distribution function $P(\mathbf{k}, \mathbf{k}', t)$. We obtain

$$P(\mathbf{k}, t) \equiv \langle \mathsf{n}(\mathbf{k}) \rangle \tag{24.13}$$

$$P(\mathbf{k}_1, \mathbf{k}_2, t) \equiv \langle \mathsf{n}(\mathbf{k}_1)\mathsf{n}(\mathbf{k}_2) \rangle - \langle \mathsf{n}(\mathbf{k}_1) \rangle\, \delta_{\mathbf{k}_1,\mathbf{k}_2} \tag{24.14}$$

To proceed further we observe that Equation (24.13) can be rewritten, using Equations (21.9b) and (24.12), in the form

$$\begin{aligned} P(\mathbf{k}, t) &= \langle \mathsf{a}_\mathbf{k}^\dagger \mathsf{a}_\mathbf{k} \rangle \\ &= \mathrm{Tr}(\underline{\rho}\mathsf{a}_\mathbf{k}^\dagger \mathsf{a}_\mathbf{k}) = \mathrm{Tr}(\mathsf{a}_\mathbf{k}\underline{\rho}\mathsf{a}_\mathbf{k}^\dagger) \end{aligned}$$

This notation suggests useful generalizations of the reduced distribution functions in quantum statistical mechanics. These generalizations, introduced by Yang,[23] are called *reduced density matrices;* they represent a quantum-mechanical counterpart to the more general distribution functions (24.5) of classical statistical mechanics.[24]

Reduced Density Matrices

The reduced density matrices ($\underline{\rho}_1, \underline{\rho}_2, \ldots$) are particular off-diagonal matrix elements of the statistical matrix $\underline{\rho}$, as defined by the set of equations:

$$\mathrm{Tr}(\underline{\rho}) = 1$$

$$\langle j|\underline{\rho}_1|i\rangle \equiv \mathrm{Tr}(\mathsf{a}_j\underline{\rho}\mathsf{a}_i^\dagger) \tag{24.15}$$

$$\langle kl|\underline{\rho}_2|ij\rangle \equiv \mathrm{Tr}(\mathsf{a}_k\mathsf{a}_l\underline{\rho}\mathsf{a}_j^\dagger\mathsf{a}_i^\dagger) \quad \text{etc.}$$

For large systems in which there is translational invariance (see end of Section 17), momentum conservation applies to all the reduced density matrices in momentum space. In particular, we have

$$\langle \mathbf{k}|\underline{\rho}_1|\mathbf{k}'\rangle = \mathrm{Tr}(\mathsf{a}_{\mathbf{k}}\underline{\rho}\mathsf{a}_{\mathbf{k}}^\dagger)\,\delta_{\mathbf{k},\mathbf{k}'} = P(\mathbf{k})\,\delta_{\mathbf{k},\mathbf{k}'} \tag{24.16}$$

where notation to represent any time dependence in the distribution functions is suppressed for the rest of this section.

Exercise 24.3 Use the commutation relations (17.7) and (17.8), that is

$$\mathsf{a}_i\mathsf{a}_j^\dagger - \varepsilon\mathsf{a}_j^\dagger\mathsf{a}_i = \delta_{ij}\mathsf{1}, \qquad \mathsf{a}_i\mathsf{a}_j - \varepsilon\mathsf{a}_j\mathsf{a}_i = 0, \qquad \text{etc.}$$

to show, for both statistics, that the diagonal elements of $\underline{\rho}_2$ (in momentum space, say) are pair distribution functions:

$$\langle \mathbf{k}_1\mathbf{k}_2|\underline{\rho}_2|\mathbf{k}_1\mathbf{k}_2\rangle = P(\mathbf{k}_1, \mathbf{k}_2) \tag{24.17}$$

It is straightforward to prove that a simple sum over the reduced density matrix $\underline{\rho}_m$ gives essentially the reduced density matrix $\underline{\rho}_{m-1}$ according to the following set of equations:

$$\sum_i \langle i|\underline{\rho}_1|i\rangle_N = \mathsf{N}$$

$$\sum_i \langle ik|\underline{\rho}_2|ij\rangle_N = (\mathsf{N}-1)\langle k|\underline{\rho}_1|j\rangle_N \tag{24.18}$$

$$\sum_i \langle ijk|\underline{\rho}_3|ilm\rangle_N = (\mathsf{N}-2)\langle jk|\underline{\rho}_2|lm\rangle_N$$

where the averages in these equations are computed using an N-particle statistical matrix. The first of Equations (24.18) follows immediately from definition (24.15) and Equation (17.6) for the number operator **N**.

The second of Equations (24.18) can be verified with the aid of either of the commutation relations (17.9). Thus one writes

$$\begin{aligned}\sum_i \langle ik|\underline{\rho}_2|ij\rangle_N &= \sum_i \mathrm{Tr}_{N-2}(\mathsf{a}_i\mathsf{a}_k\underline{\rho}\mathsf{a}_j^\dagger\mathsf{a}_i^\dagger)\\ &= \sum_i \mathrm{Tr}_N(\underline{\rho}\mathsf{a}_j^\dagger\mathsf{a}_i^\dagger\mathsf{a}_i\mathsf{a}_k)\\ &= \sum_i \mathrm{Tr}_N[\underline{\rho}\mathsf{a}_j^\dagger\mathsf{a}_k(\mathsf{a}_i^\dagger\mathsf{a}_i - 1\,\delta_{ik})]\\ &= (\mathrm{N}-1)\langle k|\underline{\rho}_1|j\rangle_N\end{aligned}$$

Since we are considering averages (21.9b) performed using an N-particle statistical matrix, we have proceeded first to the second equality of this last equation. For a similar reason we have written the third equality before summing over the number operators $\mathsf{n}_i = \mathsf{a}_i^\dagger\mathsf{a}_i$ to get the final equality. However, if one is thinking carefully, then the final result follows immediately from the second equality without recourse to Equations (17.9). All the remaining equations in the set (24.18) can now be verified by using the same line of reasoning for each case.

Equations (24.18) correspond to the classical expression (24.1). The use of an N-particle statistical matrix is only necessary, however, to arrive at the quantum-mechanical results. As soon as we have reasoned that Equations (24.18) are correct, then one can average over all N values to obtain expressions for more general variable-N systems. It follows from these equations that the normalizations of the reduced density matrices are

$$\begin{aligned}\sum_i \langle i|\underline{\rho}_1|i\rangle &= \langle \mathrm{N}\rangle\\ \sum_{ij} \langle ij|\underline{\rho}_2|ij\rangle &= \langle \mathrm{N}(\mathrm{N}-1)\rangle \qquad (24.19)\\ \sum_{ijk} \langle ijk|\underline{\rho}_3|ijk\rangle &= \langle \mathrm{N}(\mathrm{N}-1)(\mathrm{N}-2)\rangle \qquad \text{etc.}\end{aligned}$$

Thus the reduced density matrices are normalized in the same way as the classical reduced distribution functions (23.20).

Finally we show that if the reduced density matrix elements have been (or can be) calculated in one representation, say, the momentum representation, then it is a simple matter to derive expressions for reduced distribution functions in any other representation. In particular, we are interested in position-space distribution functions. In this connection we observe that in the (continuous) position representation one usually uses the field operators $\underline{\psi}(\mathbf{r})$ and $\underline{\psi}^\dagger(\mathbf{r})$, defined by Equations (17.22), instead of discrete-space operators $\mathsf{a}_\mathbf{r}$ and $\mathsf{a}_\mathbf{r}^\dagger$.

Transformation between representations is accomplished quantum mechanically by using the appropriate unitary transformation elements, for example, the plane-wave functions $\langle \mathbf{r}|j\rangle$ of Equation (17.27). Thus, on referring to the result of Exercise 24.3 we can derive for $P(\mathbf{r})$ and $P(\mathbf{r}_1, \mathbf{r}_2)$ the expressions:

$$P(\mathbf{r}) = \langle \mathbf{r}|\underline{\rho}_1|\mathbf{r}\rangle = \sum_{i,j} \phi_i(\mathbf{r})\phi_j^\dagger(\mathbf{r})\langle i|\underline{\rho}_1|j\rangle \tag{24.20}$$

$$P(\mathbf{r}_1, \mathbf{r}_2) = \sum_{ijkl} \phi_i(\mathbf{r}_1)\phi_j(\mathbf{r}_2)\phi_k^\dagger(\mathbf{r}_1)\phi_l^\dagger(\mathbf{r}_2)\langle ij|\underline{\rho}_2|kl\rangle \tag{24.21}$$

As advertised, position-space distribution functions can be related to reduced density matrix elements in some other representation.

For large translationally invariant systems we can choose the $\phi_i(\mathbf{r})$ in these last two equations to be the plane-wave states (17.27). We obtain in this case, neglecting spin effects,

$$P(\mathbf{r}) = \Omega^{-1} \sum_{\mathbf{k},\mathbf{k}'} \exp[i(\mathbf{k} - \mathbf{k}') \cdot \mathbf{r}]\langle \mathbf{k}|\underline{\rho}_1|\mathbf{k}'\rangle = n \tag{24.20a}$$

$$P(\mathbf{r}_1, \mathbf{r}_2) = \Omega^{-2} \sum_{\mathbf{k}_1\mathbf{k}_2,\mathbf{k}'_1\mathbf{k}'_2} \exp[i(\mathbf{k}_1 - \mathbf{k}_1') \cdot (\mathbf{r}_1 - \mathbf{r}_2)]\langle \mathbf{k}_1\mathbf{k}_2|\underline{\rho}_2|\mathbf{k}_1'\mathbf{k}_2'\rangle \tag{24.21a}$$

To derive the second equality of Equation (24.20a) we have used Equation (24.16) and the first of Equations (24.19). Also, conservation of momentum for the matrix elements of $\underline{\rho}_2$ has been used to simplify the exponential factor in Equation (24.21a).

In applications, only the reduced distribution functions for one, two, and sometimes three particles are normally considered. Because the pair-distribution function $P(\mathbf{r}_1, \mathbf{r}_2)$ provides a measure of particle correlations in a system, it can be used to characterize the *long-range order* (Section 49) of crystalline structures. In his study of reduced density matrices Yang[23] discussed the concept that quantum fluids may exhibit *off-diagonal long-range order* (ODLRO). For such systems not only are the reduced distribution functions relevant physical quantities, but also, certain off-diagonal matrix elements of reduced density matrices can be physically significant. In Chapter X we investigate properties of one- and two-particle distribution functions in considerable detail.

25 Macroscopic Irreversibility

The microscopic basis of the second law of thermodynamics is investigated and shown to lead to deeper questions concerning the foundations of statistical mechanics. Our discussion deals mainly with classical statistical mechanics, although much of it is general, or with analogies in the quantum-mechanical case. Topics included are Boltzmann's H theorem, the relevance of time-reversal invariance for macroscopic irreversibility, ergodic theory, and coarse graining. Of necessity, supportive details are referred to in other readings via footnotes.

The second law of thermodynamics, or law of increasing entropy, has already been provided with a statistical basis by our definition, with Equation (4.23), of statistical entropy. But though we have made it plausible that a maximum-entropy principle must govern macroscopic phenomena in nature, such that the condition of thermodynamic equilibrium in a system is associated with a maximum of the entropy (Section 6), we have not, in fact, demonstrated the increasing tendency of the entropy. The definitive demonstration is an elusive goal, and though many results of important physical significance can be cited, important links in the complete logical chain are still missing. It can be fairly stated, however, that by now a qualitative understanding of the irreversibility of macroscopic processes has been achieved. In the present section we attempt to convey this understanding without reference to the specific relaxation processes and times that govern every real system. Then, in Chapter IX we return to a further study of irreversibility and to exploration of interesting questions associated with relaxation and the dissipation of order in irreversible processes.

The origins of entropy increase can be studied at many levels. Most remotely, it is possible to associate with our entire universe an entropy whose increase follows from energy conservation in the general relativistic description of an expanding universe.[25] Although it may be that the increase of this gravitational entropy provides us ultimately with a local arrow of time, it cannot be that the understanding of entropy increase for phenomena in terrestrial laboratories is related intimately to cosmology. We note, however, that between these two extremes are the details of astrophysical evolutionary processes, alluded to in Section 6, whose understanding can be enhanced by the study of terrestrial phenomena. With these remarks for perspective, we now restrict our attention to processes that take place in the laboratory.

A microscopic understanding of irreversibility in nature must rely ultimately on our knowledge of particle interactions and, in particular, on our models of collision processes. Thus from the extreme scale of the preceding paragraph we return to the detailed analysis of Section 11. Our interest, once again, is in the collision integral of the Boltzmann transport equation (11.27), or (23.27), and our focus is on the approximate form (11.14) of the collision integral. Not only was this the original approach historically, by Boltzmann, toward the subject of irreversibility, but also pedagogically it is convenient to begin with these considerations before pursuing more profound questions.

We recall Figure III.1, in which two particles with initial momenta $\mathbf{p}_1$ and $\mathbf{p}_2$ collide elastically to give a final state with momenta $\mathbf{p}_1'$ and $\mathbf{p}_2'$, for a given direction $\hat{\mathbf{q}}$ as defined by Equation (11.4) (now, with $\mathbf{k}_i \to \mathbf{p}_i$). Then, for a dilute gas it was argued that the collision integral at the point $\mathbf{r}$ and time t could be written in terms of this scattering process as Equation (11.14), which together with Equation (11.15) becomes

$$\left[\frac{\partial}{\partial t}P_1(\mathbf{r}, \mathbf{p}_1, t)\right]_{\text{coll}} \cong \int d^3\mathbf{p}_2\, d\omega_q R(\mathbf{p}_1, \mathbf{p}_2, \hat{\mathbf{q}}) \times [P_1(\mathbf{r}, \mathbf{p}_1', t)P_1(\mathbf{r}, \mathbf{p}_2', t) - P_1(\mathbf{r}, \mathbf{p}_1, t)P_1(\mathbf{r}, \mathbf{p}_2, t)] \tag{25.1}$$

where the transition probability $R(\mathbf{p}_1, \mathbf{p}_2, \hat{\mathbf{q}})$ is defined by Equation (11.11). Crucial to the derivation of Equation (25.1) was the *Stosszahlansatz* (11.13), which is equivalent to an assumption of stochastic independence at all times for the two colliding particles. It was stressed that this assumption can only be valid approximately for real gases, even when application is restricted to microscopically large phase-space volume elements [cf. inequality (11.2)]. We note, in passing, that an arrow of time is implicitly included in Equation (25.1) because of our identification, with Figure III.1, of initial and final scattering states.

We have remarked that the *Stosszahlansatz* is not valid when the two colliding particles are localized to too fine a region of phase space, for then their spatial correlations become important. One can also consider the washing-out effect on spatial correlations of a time average, say, over a macroscopically small time interval, as in Equation (15.3). Our interest now is in just this case, in which we use the Boltzmann transport equation on a macroscopic time scale to deduce an expression for the time evolution of the entropy.[26] In such an application the assumption of stochastic independence for two particles that collide as in Figure III.1 is an excellent approximation, and so then is expression (25.1) for the collision integral of a dilute gas. One of the keys to our present understanding of macroscopic irreversibility is the realization that macroscopic observations and descriptions are associated with microscopically large phase-space and time intervals. For this reason our qualifying remarks concerning Equation (25.1) are quite relevant.

We assume now that there are no external forces acting on the dilute gas under consideration and that any position dependence of the one-particle distribution function P_1 can be neglected.[27] The Boltzmann transport equation (23.27) and Equation (25.1) taken together then give

$$\frac{\partial}{\partial t}P_1(\mathbf{p}_1, t) = \int d^3\mathbf{p}_2\, d\omega_q R(\mathbf{p}_1, \mathbf{p}_2, \hat{\mathbf{q}})[P_1(\mathbf{p}_1', t)P_1(\mathbf{p}_2', t) - P_1(\mathbf{p}_1, t)P_1(\mathbf{p}_2, t)] \tag{25.2}$$

Boltzmann's H Theorem

In 1872 Ludwig Boltzmann introduced his famous *H* function $\mathcal{H}(t)$, which is essentially the negative of the statistical entropy. The *H* theorem is a statement that

$$\frac{d\mathcal{H}(t)}{dt} \leqq 0 \tag{25.3}$$

for all closed systems; its definitive proof using quantum statistical mechanics would be equivalent to a proof of the increasing tendency of statistical entropy $S(t)$, that is, of the second law of thermodynamics.

Let us consider, then, the statistical-entropy expression (4.25), which for a gas can be written as[28]

$$S(t) = -N\kappa\left(\sum_i Z_i P_i \ln P_i + \ln N - 1\right) \tag{25.4}$$

where for the cell sizes $Z_i = \delta^3\mathbf{n}_i$ in phase space (μ space) we must use the expression

$$Z_i = \frac{\delta\omega_i}{(2\pi\hbar)^3} = \frac{\delta^3\mathbf{r}_i\,\delta^3\mathbf{p}_i}{(2\pi\hbar)^3}$$

The probabilities P_i are the single-particle probabilities of Equation (4.21), normalized according to

$$\sum_i Z_i P_i = 1$$

so that the relation between P_i and the single-particle distribution function $P_1(\mathbf{r}_i, \mathbf{p}_i, t)$, whose normalization is given by Equation (11.1), is

$$P_i = \frac{(2\pi\hbar)^3}{N} P_1(\mathbf{r}_i, \mathbf{p}_i, t) \tag{25.5}$$

One can show, therefore, that Boltzmann's *H* function, defined for a classical gas by

$$\mathcal{H}(t) \equiv \int d\omega P_1(\mathbf{r}, \mathbf{p}, t) \ln P_1(\mathbf{r}, \mathbf{p}, t) \tag{25.6}$$

is related to the entropy of Equation (25.4) by the simple expression

$$\mathcal{H}(t) = -\kappa^{-1} \lim_{\Omega\to\infty} S(t) + N[1 - \ln(2\pi\hbar)^3] \tag{25.7}$$

Exercise 25.1 Verify Equation (25.7).

We now give a proof of Boltzmann's H theorem (25.3) for a dilute gas in which there are no external forces, and any position dependence of P_1 can be neglected. We differentiate Equation (25.6) with respect to time and use Equation (25.2) to obtain

$$\frac{d\mathcal{H}(t)}{dt} = \Omega \int \mathbf{d}^3\mathbf{p}_1\, d^3\mathbf{p}_2\, d\omega_q R(\mathbf{p}_1, \mathbf{p}_2, \hat{\mathbf{q}})[\ln P_1(\mathbf{p}_1, t) + 1]$$

$$\times [P_1(\mathbf{p}_1', t)P_1(\mathbf{p}_2', t) - P_1(\mathbf{p}_1, t)P_1(\mathbf{p}_2, t)]$$

$$= \frac{\Omega}{2} \int d^3\mathbf{p}_1\, d^3\mathbf{p}_2\, d\omega_q R(\mathbf{p}_1, \mathbf{p}_2, \hat{\mathbf{q}})[P_1(\mathbf{p}_1', t)P_1(\mathbf{p}_2', t)$$

$$- P_1(\mathbf{p}_1, t)P_1(\mathbf{p}_2, t)] \cdot [2 + \ln(P_1(\mathbf{p}_1, t)P_1(\mathbf{p}_2, t)]$$

$$= \frac{\Omega}{2} \int d^3\mathbf{p}_1'\, d^3\mathbf{p}_2'\, d\omega_{q'} R(\mathbf{p}_1', \mathbf{p}_2', \hat{\mathbf{q}}')[P_1(\mathbf{p}_1, t)P_1(\mathbf{p}_2, t)$$

$$- P_1(\mathbf{p}_1', t)P_1(\mathbf{p}_2', t)] \cdot [2 + \ln(P_1(\mathbf{p}_1', t)P_1(\mathbf{p}_2', t))]$$

To derive the second equality of this equation we have relabeled particles 1 and 2, and to derive the third line we have relabeled what we mean by initial and final states. But now in the last line we can replace $d^3\mathbf{p}_1'\, d^3\mathbf{p}_2'\, d\omega_q'$ by $d^3\mathbf{p}_1$ $d^3\mathbf{p}_2\, d\omega_q$, because the Jacobian of the transformation from primed to unprimed variables is unity.[29] Moreover, the positive transition probability factor R for elastic collisions is invariant under time reversal, as well as spatial rotations and reflections, so that[30]

$$R(\mathbf{p}_1', \mathbf{p}_2', \hat{\mathbf{q}}') = R(\mathbf{p}_1, \mathbf{p}_2, \hat{\mathbf{q}})$$

We may, therefore, finally write the preceding equation as

$$\frac{d\mathcal{H}(t)}{dt} = \frac{\Omega}{4} \int d^3\mathbf{p}_1\, d^3\mathbf{p}_2\, d\omega_q R(\mathbf{p}_1, \mathbf{p}_2, \hat{\mathbf{q}})[P_1(\mathbf{p}_1', t)P_1(\mathbf{p}_2', t)$$

$$- P_1(p_1, t)P_1(p_2, t)] \times [\ln(P_1(\mathbf{p}_1, t)P_1(\mathbf{p}_2, t))$$

$$- \ln(P_1(\mathbf{p}_1', t)P_1(\mathbf{p}_2', t))]$$

$$\leqq 0 \tag{25.8}$$

The inequality of this last equation follows, because the integrand can never be positive. Boltzmann's H theorem for a dilute gas is therefore proved, and the H function always decreases or remains constant with time.

Exercise 25.2 Use the fact that $(x - 1)\ln x \geqq 0$ for all $x \geqq 1$ to prove the inequality of Equation (25.8).

The above proof of Boltzmann's H theorem is inadequate as a general proof, primarily because Equation (25.1) for the collision integral is not a sufficiently general expression. Position correlations between particles 1 and 2 have been omitted, as has the influence of multiple collisions included implicitly in the exact collision integral (23.26).[31] Nevertheless, in view of our qualifying remarks below Equation (25.1), one might conjecture that these cited omissions would, if included, provide only small corrections to the above proof for dilute gases.

A view that has evolved about the time development of the H function is that if one were able to calculate its dependence on a microscopic time scale, but retaining still application to finite volume elements in phase space, then one would discover a fluctuating behavior such as is suggested in Figure V.2. Thus, as we have already emphasized, the result achieved by Equation (25.8) must refer to a macroscopic time scale, and this result is indicated by the dashed curve in Figure V.2. By no means has the first of the above assertions been proved, nor without explicit calculation does Equation (25.8) provide an explicit relaxation time for the approach to equilibrium.[32] Theorem (25.8) does provide, however, an initial insight into the nature of macroscopic irreversibility. We now examine Boltzmann's H theorem from another viewpoint by subjecting it to the requirements of microscopic reversibility.

Microscopic Reversibility

"It should not be forgotten, when our ensembles are chosen to illustrate the probabilities of events in the real world, that while the probabilities of subsequent events may often be determined from those of prior events, it is rarely the case that probabilities of prior events can be determined from

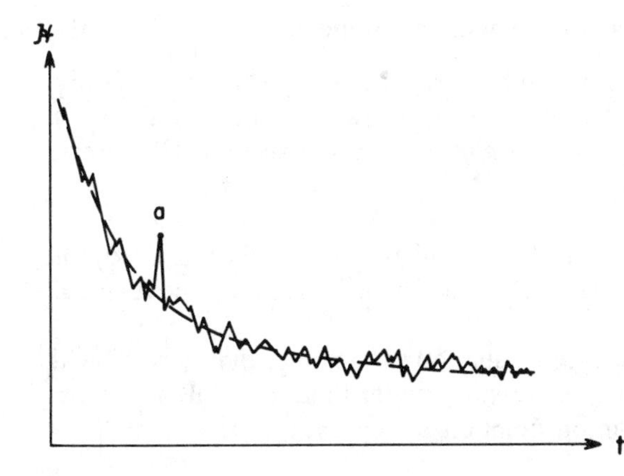

FIGURE V.2 Boltzmann's H function sketched as a function of time. The solid curve is the dependence that would be expected on a microscopic time scale. The corresponding average behavior on a macroscopic time scale is the dashed curve. Point a is a highly improbable fluctuation.

those of subsequent events, for we are rarely justified in excluding the consideration of the anteceding probability of the prior events"—J. W. Gibbs (1901).[33]

It is necessary to reconcile the causality of events implied by physical measurements, as emphasized by the above quotation, with the symmetry in time of physical processes in the microscopic world. Our objective here is to show that macroscopic irreversibility is not inconsistent with microscopic reversibility.

It is generally believed that the laws of physics are invariant under *time reversal,* although there is currently some question whether this property is true for the weak interactions of neutral K mesons, a problem that in any event has little relevance for our present discussion. What time-reversal invariance implies for particle motions is that the kinematics of a particle moving backward in time, with reversed velocities, could not be distinguished by an observer (if he could make such an observation) from the kinematics he does observe for a particle. In other words, from the microscopic point of view nature seems to be inherently reversible.

Mathematically we can characterize time reversal in classical mechanics as follows: If the position and momentum of a particle are given by $\mathbf{r}(t)$ and $\mathbf{p}(t)$, then the time-reversed position and momentum $\mathbf{r}_T(t)$ and $\mathbf{p}_T(t)$ are defined by

$$\begin{aligned} \mathbf{r}_T(t) &= \mathbf{r}(-t) \\ \mathbf{p}_T(t) &= -\mathbf{p}(-t) \end{aligned} \tag{25.9}$$

For a function F of the particle coordinates the time-reversed function F_T is defined by

$$F_T(\mathbf{r}, \mathbf{p}, t) \equiv F(\mathbf{r}_T, \mathbf{p}_T, -t) = F(\mathbf{r}, -\mathbf{p}, -t) \tag{25.10}$$

and time-reversal invariance means that $F_T = F$, that is

$$F(\mathbf{r}, -\mathbf{p}, -t) = F(\mathbf{r}, \mathbf{p}, t) \qquad \text{(time-reversal invariance)} \tag{25.11}$$

The assumption that nature is invariant under time reversal is clearly equivalent to the assumption that the Hamiltonian H satisfies Equation (25.11), since, in principle, all particle motions can be determined from Hamilton's equations (23.10).

Exercise 25.3 Assuming that the Hamiltonian is time-reversal invariant, show that Hamilton's equations (23.10) are invariant under time reversal.

It is easy to verify that Liouville's equation (23.15), which is derived with the aid of Hamilton's equations, is invariant under time reversal, and therefore the time-reversed distribution function ρ_T represents the ensemble associated with a given system just as adequately as does ρ. Moreover, since the representative points in ρ are associated with a random set of initial values for a typical system, we may conclude that the motion of the representative points is invariant under time reversal. It would seem that macroscopic irreversibility can never be derived from Liouville's theorem [or, equivalently, from Boltzmann's equation (23.27)], and that the flaw in the preceding derivation of Boltzmann's H theorem is more basic than the inadequacies of the collision integral (25.1), as stated earlier. Thus if we assume that $P_1(\mathbf{r}, \mathbf{p}, t)$ is invariant under time reversal, the approximate collision integral (25.1) does not change sign when the time is reversed, as does the exact collision integral (23.26). In other words, Equation (25.1) appears not to be time-reversal invariant, as it should be.

Two Historical Paradoxes

Historically, the above reversibility argument, phrased somewhat differently, is referred to as Loschmidt's reversibility paradox (1876). It is now understood that the reversibility argument does not present a paradox. To begin with, and as already mentioned below Equation (25.1), the causal relationship between initial and final scattering states, with its implicit assumption of an arrow (or direction) of time, was used in the derivation in Section 11 of the approximate collision integral. Therefore, when time is reversed in Equation (25.1) we must also interchange the labeling of initial and final states and then, lo and behold, the approximate collision integral does, indeed, change sign under time reversal.

Resolution of the reversibility paradox is now given, with reference to statistical entropy, in more general terms. We consider an initial nonequilibrium condition at time $t = 0$, a condition that could arise either by external intervention in a system or spontaneously in a closed system. The situation is represented in Figure V.3(a) by a decreasing entropy S for $t < 0$. The most probable evolution of the entropy for $t > 0$ is to increase, following the mandate of the second law of thermodynamics, and this behavior of S is also shown in Figure V.3(a). Clearly, when entropy is viewed statistically, then either microscopically or macroscopically the above process is time reversible. See Figure V.3(b).

It is not necessary that the statistical entropy in the given situation increase after $t = 0$, especially when the time scale is considered microscopically, although eventually S is expected to increase. Thus a second possibility for the time dependence of S is depicted in Figure V.4(a); the corresponding time-reversed case, which could also occur, is shown as Figure V.4(b). We may conclude, then, that time-reversed behavior for a system can occur, consistent with the microscopic laws of physics, although it

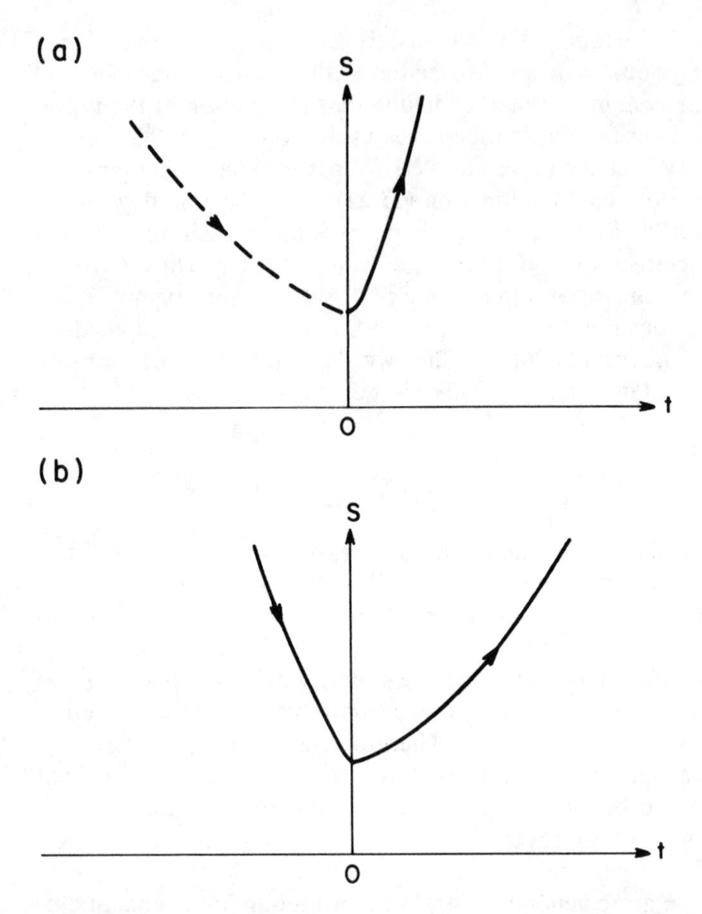

FIGURE V.3 (a) A macroscopic system is brought to a nonequilibrium state at $t = 0$ either by external intervention or spontaneously (dashed line), and subsequently the entropy increases. The time scale could be either microscopic or macroscopic. (b) The time-reversed situation of (a), when it occurs spontaneously, is consistent with both the microscopic laws of physics and with the general statistical description of macroscopic systems.

must be emphasized that behavior such as in Figure V.4 can only be expected on a microscopic time scale. In fact, continually fluctuating values of S are manifested characteristically by observed fluctuation phenomena. *The irreversibility we observe in nature is a consequence of given initial conditions.* Although we may still wish "to explain" irreversible processes in microscopic terms, asymmetry between past and future is not inherent in any general statistical description of nature.

A second paradox, due to Zermelo (1896), uses a theorem of Poincaré on

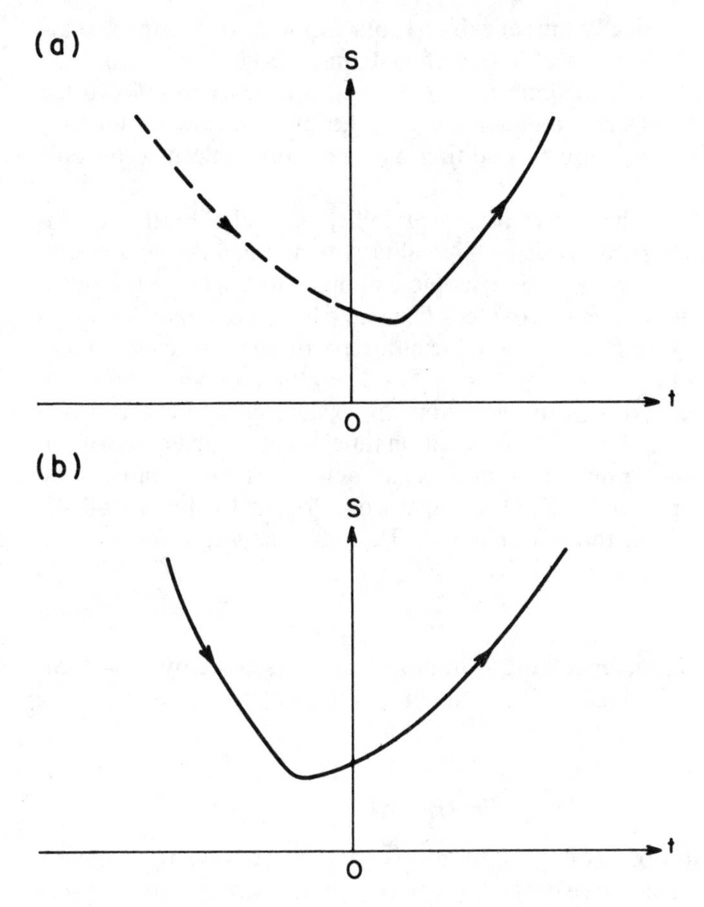

FIGURE V.4 (a) A macroscopic system is brought to a nonequilibrium state at $t = 0$, as in Figure V.3(a), and subsequently the entropy decreases before increasing again. In this case the time scale must be microscopic for the entropy to decrease spontaneously after $t = 0$. (b) The time-reversed situation of (a), when it occurs spontaneously, is consistent with both the microscopic laws of physics and the general statistical description of macroscopic systems.

the time development of a system.[34] The essential point of this paradox, the recurrence paradox, is that given a sufficiently long, but finite, time any classical system is quasi periodic and will retrace its trajectory in Γ space to within prescribed limits (Δq_i, Δp_i). Here we must inquire about the length of relevant recurrence times. Certainly the recurrence of minor fluctuations from the equilibrium condition will be determined by the short, and often measurable, relaxation times of the system. But the recurrence time associated with a macroscopic nonequilibrium state could be many ages of the universe, that is, physically unrealizable. From this viewpoint, then, macroscopic irreversibility is not a matter of violating the laws of mechanics, but rather it is the result of finite observation times. Thus to resolve the recurrence paradox we stress the identification the law of increasing entropy makes between the equilibrium condition and most-probable macroscopic states.

Our clarification of the nature of irreversibility, provided by the resolution of the recurrence paradox, is again qualitative and not a detailed microscopic explanation. However, microscopic examples to illustrate the probability concepts can be easily provided. Consider the free expansion of an almost-ideal gas (Figure II.4) in which the entire partition between two identical chambers is removed at time $t = 0$. Let N be the number of atoms in the gas and assume that measurements on the system could be performed every Δt seconds, such that $\Delta t >>$ (relaxation time to approach equilibrium). We are interested in the number n of measurements that would have to be made so that an expectation equal to unity would occur for finding all the atoms gathered again in the left chamber. This number n is given by [see Problem V.8(a)]

$$n = 2^N \tag{25.12}$$

An estimate of the recurrence time T_r in this example is given by $T_r = n\,\Delta t$, and Table V.1 shows values of T_r as a function of N. For comparison the age of the universe is $\sim 10^{10}$ yr $= 3 \times 10^{17}$ sec.

Ergodic Theory

In our discussion of irreversibility in this section we have used mostly the notion of a representative ensemble whose averages are intended to represent the average behavior of the system being investigated or observed. We have also considered aspects of the dynamical description of a single system. We now probe deeper into the foundations of statistical mechanics and ask for reasons to justify a statement (the ergodic hypothesis) that the macroscopic properties of a closed system, averaged in time, are identical to those of a suitably chosen representative ensemble. *Ergodic theory* seeks to

Table V.1 Recurrence Time T_r Versus N for the Example of Equation (25.12). Δt = Time Interval Between Successive Measurements on the System.

N	5	10		10^2	10^4	10^{21}
$\frac{T_r}{\Delta t}$	32	1024	↑ age of universe	10^{30}	10^{3000}	$10^{3 \times 10^{20}}$

clarify the conditions under which this statement might be true and then attempts to prove it.

The importance of ergodic theory is that though one measures time-averaged physical quantities of a system, ensemble averages are much more easily computed. A fundamental question that can be answered only within the framework of ergodic theory is: What probability weightings are to be given, classically, say, to various regions of phase space? We have assumed previously that equal volumes in phase space must have equal weightings,[35] an assumption that conveniently makes these weightings invariant under canonical transformation.[12] This assumption of equal *a priori* probabilities, also known as Laplace's principle of insufficient reason, can be given some justification by ergodic theory.[36]

We give a definition. A system is *ergodic* if the orbit of its representative points in Γ space, taken from time $t = -\infty$ to $t = +\infty$, passes through all points of an energy surface. Clearly, all orbits of an ergodic system are equivalent, differing only in their starting points, and therefore we could expect for an ergodic system that an ensemble average would be the same as a long-time average for the single system. Here we must stress that whereas ergodic theory, both classically and quantum mechanically, deals only with a system in a definite energy state (or a small spread of energy states), there is, in fact, no inherent conflict between this restriction and the considerations presented at the beginning of Section 21. Thus, although ergodic theory deals only with the justification of constant-energy (microcanonical) ensembles, the more physically realistic canonical ensemble can be justified, in turn, by considering such an ensemble to be a large collection of subsystems within some huge constant-energy system (see Section 26).

Versions of the classical ergodic theorem, which states that the infinite time average of a dynamical quantity for a system is equal to a phase space, or microcanonical ensemble, average, were first proved by G. D. Birkhoff (1931) and J. von Neumann (1932).[36] These proofs used the formalism of measure theory and included the important requirement that an energy surface in Γ space must be *metrically transitive*. A metrically transitive set is one that cannot be separated into two invariant subsets of nonzero measure. Unfortunately, no procedure for establishing whether the energy surfaces of real systems are metrically decomposable was provided by the early ergodic theorems, and therefore other approaches to this subject were often followed. Recently, however, it has been demonstrated by Sinai (1963) that the energy hypersurfaces for a finite gas of hard-sphere atoms are metrically transitive.[37]

In quantum-mechanical versions of the ergodic theorem (von Neumann, 1929),[36] the assumption of metrical transitivity becomes the quite restrictive assumption that the system Hamiltonian have a nondegenerate energy spectrum, which is generally not true. The fact that stationary-state probabilities remain unchanged in time implies that a simple time average can never equal a microcanonical average for systems with energy degeneracy, and this is the reason why energy nondegeneracy is required in the basic quantum ergodic theorems.

In view of the above difficulty with the quantum ergodic theory, von Neumann readily realized the necessity for considering a spread in energy states even for a closed system, a point we emphasized earlier with different motivation at the beginning of Section 21. Thus either by this argument or from a physical viewpoint we are forced to consider the *coarse graining* of macroscopic systems, in which a measurable dynamical quantity A can be specified only to within a limit δA (consistent with the Heisenberg uncertainty principle for conjugate variables, of course). Contrasted with coarse graining is the *fine-grained* treatment of systems, outlined above, in which it is assumed that every dynamical variable A is specified to a limiting accuracy $\delta A = 0$. We only discuss coarse graining from a classical viewpoint and do not review its applications to ergodic theory.

Coarse-Grained Distributions

With the concept of coarse graining one is forced, even for the time average of a single system, to introduce probability notions into physical descriptions from the very beginning. To illustrate the concept we consider again a dilute gas of N molecules distributed into equal-sized cells Z_μ in μ space, similar to those defined above Equation (4.18). Corresponding to each distribution of the N particles into N μ-space cells will be a region, or volume in Γ space, which is labeled by a $6N$-dimensional volume element Z and called a Z star. Coarse graining in the microscopic description is achieved by choosing the Z_μ to be microscopically large [see inequality (11.2)], consistent with experimental limitations on any possible measurement of the Maxwell–Boltzmann distribution. We let n_i be the number of particles in the ith μ-space cell. Then, if we use Equation (4.21), the entropy expression (4.25) can be written as

$$\begin{aligned}-S/K &= \sum_i n_i \ln n_i - \ln N! + N(1 - \ln Z_\mu) \\ &= \sum_i n_i \ln n_i + \text{constant}\end{aligned}$$

and therefore we may investigate irreversibility by studying the time evolution in Γ space of the coarse-grained H function defined by

$$\mathcal{H}_Z(t) \equiv \sum_i n_i \ln n_i \tag{25.13}$$

Note that permutations among the N particles of the system do not change the numbers n_i, and all possible permutations define the extent in Γ space of any one Z star.

With the introduction of Z star regions in Γ space we adopt the nomenclature that the distribution function or occupation number n_i is a *coarse-grained* quantity.[38] Certainly, a given system can never be specified more precisely than by saying that its representative point lies in a given Z star. A program for proving the H theorem using definition (25.13) for $\mathcal{H}_Z(t)$ was formulated in 1912 by P. and T. Ehrenfest,[39] but this program has never been successfully carried through to completion.

Equation (25.13) shows one method for generalizing Boltzmann's H function (25.6). Another possible generalization is to introduce a (Gibbs) H function for the ensemble distribution function ρ. The corresponding H theorem in ensemble theory is outlined in Problem V.4, in which attention is also given to the importance of using a coarse-grained distribution function. Generalizations of the H function suitable for quantum-mechanical discussion have been studied extensively.[40]

We have advanced statistical and measurability arguments to favor the use of coarse-grained distribution functions in statistical mechanics. We have also attempted to show how coarse-grained distributions are essential for an understanding of macroscopic irreversibility (see Figure V.2). Further appreciation of this last statement occurs if we think about mixing two colors of paint, say, red and yellow. As we stir the paint toward its "equilibrium condition" the colors mix until on the coarse-grained level we see orange. But with a microscope we could imagine at the fine-grained level, assuming no chemical reactions, that we would always be able to distinguish streaks of red and yellow paint. In statistical mechanics it is the coarse-grained behavior of a macroscopic system that is of physical interest to us, fortunately, and not the fine-grained unmeasurable behavior. Indeed, macroscopic irreversibility would not be readily apparent for systems described and observed at the fine-grained level.

Irreversibility, the law of increasing entropy—has it been explained? In this section we have only been able to survey various topics that touch this explanation and to indicate where gaps lie. But more profound questions can now be delineated, and in Chapter IX on irreversible processes we present further "derivations" of macroscopic irreversibility. In the intervening chapters we concentrate on some of the many beautiful results of statistical mechanics that have been achieved by using ensemble theory.

By way of providing transition to these applications of ensemble theory, we observe, finally, that the maximum entropy principle, subject to physical constraints, is equivalent for a classical system to maximizing the volume in Γ space occupied by the representative points of an ensemble. The connection is established by noting first that the maximum entropy condition corresponds to a system in thermodynamic equilibrium with a time-independent distribution function ρ. Furthermore, maximizing the entropy corresponds to maximizing missing information about the system, which corresponds, in turn, to maximizing the number of ways of distributing the representative points of the ensemble into a uniform coarse-grained mesh over the allowed regions of Γ space. This condition will be achieved if the occupied volume of Γ space is maximized with uniform occupation (ρ = constant). It will be seen that one of the important goals of ergodic theory and studies of irreversibility, as we have outlined these subjects in this section, is to provide rigorous foundations for the above argument.[41,42]

Bibliography

Texts

R. Jancel, *Foundations of Classical and Quantum Statistical Mechanics* (Pergamon, 1969), translated by W. E. Jones. A definitive exposition of the very difficult and many faceted foundations of statistical mechanics.

L. D. Landau and E. M. Lifshitz, *Statistical Physics* (Addison-Wesley, 1958), translated by E. Peierls and R. F. Peierls, Chapter I. Includes good physical discussions of the statistical matrix of quantum statistical mechanics and the corresponding distribution function of classical statistical mechanics.

O. Penrose, *Foundations of Statistical Mechanics* (Pergamon, 1970). A well-written, nontraditional presentation of the foundations of statistical mechanics, which appeals for one of its axioms to a "Markovian postulate" in the interpretation of measurements.

Review Articles

N. N. Bogoliubov, "Problems of a Dynamical Theory in Statistical Physics" in *Studies in Statistical Mechanics,* Vol. I, edited by J. de Boer and G. E. Uhlenbeck (North-Holland, 1962), translated by E. K. Gora. The original and very readable account of Bogoliubov's introduction of the BBGKY hierarchy and related topics in kinetic theory.

U. Fano, "Description of States in Quantum Mechanics by Density Matrix and Operator Techniques," *Rev. Mod. Phys.* **29** (1957), p. 74.

W. H. Furry, "Some Aspects of the Quantum Theory of Measurement" in *Lectures in Theoretical Physics,* Vol. 8a, edited by W. E. Brittin (University of Colorado, 1966).

D. ter Haar, "Theory and Applications of the Density Matrix," *Rpts. Prog. Phys.* **XXIV** (1964), p. 304 [reprinted in *Many-Body Problems,* edited by S. F. Edwards (Benjamin, 1969)].

Notes

1. Of course, the choices of possible ways for computing the total energy must be over distinct possible ways, so that two ways differing only by an identical-particle permutation are not counted twice.

2. This is the uncertainty relation that is deduced from time-dependent perturbation theory [see L. I. Schiff, *Quantum Mechanics* (McGraw-Hill, 1968), p. 284] as opposed to the time–energy uncertainty relation associated with natural time constants of the system [see A. Messiah, *Quantum Mechanics,* Vol. I (Wiley, 1966), Chapter VIII, Section 13].

3. A detailed development of the following discussion is given in Chapter 5 of *The Feynman Lectures on Physics,* Vol. III (Addison-Wesley, 1965).

4. The question of orthogonality of the states $|i\rangle$, and some associated ambiguities, is discussed in Section I.10 of the article by W. H. Furry, *op. cit.*

5. D. ter Haar, *op. cit.*, p. 342.

6. The scattering amplitude $\langle m'|\mathsf{f}|m\rangle$ is a simple generalization of the amplitude $f(\theta, \phi)$ that characterizes the elastic scattering of spinless particles. It is directly related to the transition, or reaction, matrix. See, for example, L. I. Schiff, *loc. cit.*, Equations (18.10) and (37.21).

7. See L. Wolfenstein, "Polarization of Fast Nucleons," *Ann. Rev. Nucl. Sci.* **6** (1956), p. 43, and C. R. Schumaker and H. A. Bethe, *Phys. Rev.* **121** (1961), p. 1534.

8. L. I. Schiff, *loc. cit.*, Equation (24.4).

9. *Ibid.*, pp. 30 and 384. See also H. Goldstein, *Classical Mechanics* (Addison-Wesley, 1980), Equation (5-103).

10. H. Goldstein, *loc. cit.*, p. 25.

11. *Ibid.*, Section 8-1.

12. The invariance of phase-space volume elements [see Equations (23.11) and (23.12)] under canonical transformation is a more fundamental reason why in classical mechanics we use generalized momenta instead of generalized velocities. The theorem that establishes this invariance is known as the integral invariants of Poincare. *Ibid.*, Section 9-4.

13. The concept of a continuous probability distribution is clarified at the beginning of Section 3.

14. Section 9-8 of H. Goldstein, *loc. cit.*, includes a good discussion of Liouville's theorem. The analogy with an ordinary classical fluid can be understood by comparing the continuity equation for mass,

$$\frac{\partial \rho}{\partial t} + \nabla \cdot (\rho \mathbf{v}) = 0$$

with the total-time-rate-of-change equation

$$\frac{d\rho}{dt} = \frac{\partial \rho}{\partial t} + (\mathbf{v} \cdot \nabla)\rho$$

For an incompressible fluid $d\rho/dt = 0$, which implies that $\nabla \cdot \mathbf{v} = 0$. That $\nabla \cdot \mathbf{v} = 0$ for the motion of representative points in Γ space, where $\mathbf{v}$ is the vector (23.16), follows from Hamilton's equations. See Exercise 23.2.

15. L. I. Schiff, *loc. cit.*, p. 175.

16. Here, the q_i and p_i each represents s variables.

17. The classical distribution function ρ is restricted, for each particle, to the region of μ space that corresponds in position space to the boundaries of the system and in momentum space to mc, where c is the speed of light. Therefore, when we integrate over all of phase space, then ρ is zero at the integration limits. There can be other phase-space restrictions, such as those arising from identical-particle considerations and the periodic nature of angular coordinates.

18. Extensive review of the transport equations that apply to various physical situations in which electric and magnetic fields occur is given in T. Y. Wu, *Kinetic Equations of Gases and Plasmas* (Addison-Wesley, 1966).

19. For the remainder of this section we shall, for simplicity, use the notation $\mathbf{r}_\alpha$ for coordinates, instead of q_α, and neglect any internal degrees of freedom of the particles.

20. Equations (24.7) were encountered previously as Equations (11.18) and (12.13).

21. See also the discussion of the *stosszahlansatz,* Equation (11.13).

22. In this connection, see W. E. Brittin and W. R. Chappell, *Rev. Mod. Phys.* **34** (1962), p. 620. The relation between average values defined quantum mechanically and classically is made explicit with Equations (E.5) and (E.6) in Appendix E by using the Wigner distribution function f_w, where in the limit $\hbar \rightarrow 0, f_w \rightarrow \rho$.

23. C. N. Yang, *Rev. Mod. Phys.* **34** (1962), p. 694.

24. This correspondence becomes more clear when one refers, for the quantum-mechanical case, to definition (E.1) of the Wigner distribution function in Appendix E.

25. For a brief discussion and review of cosmological questions relating to entropy see the article by B. O. Gal-or in *A Critical Review of Thermodynamics,* edited by S. A. Rice, B. O. Gal-or, and A. J. Brainard (Mono Book Corp., Baltimore, 1970).

26. Recall the observation in the introduction to Section 15 that the approximate Boltzmann equation, in which Equation (25.1) is used for the collision integral, cannot be justified for a time interval $t \lesssim \tau_{\text{coll}}$.

27. A derivation of Boltzmann's H theorem that takes into account position dependence of P_1, velocity-independent external forces, and surface effects is given in Section 4.5 of the text by J. H. Ferziger and H. G. Kaper, *Mathematical Theory of Transport Processes in Gases* (North-Holland, 1972).

28. In the entropy expression (25.4) we have included the Gibbs-paradox factor $\kappa \ln(N!)^{-1} \cong -N\kappa(\ln N - 1)$. See the discussion at the end of Section 4.

29. See footnote 5 in Chapter III.

30. See Equation (D.6) of Appendix D.

31. At present Boltzmann's H theorem has not been proved using the exact collision integral. The approximate collision integral (25.1) is deduced from the exact collision integral (23.26) and correction terms are derived in *Kinetic Theory of Dense Gases* by E. G. D. Cohen; see, in particular, pp. 301–306. This article appears in *Lectures in Theoretical Physics,* Vol. IX C, edited by W. E. Brittin (Gordon and Breach, 1967). The method used was outlined originally by N. N. Bogoliubov, *op. cit.*

32. Without careful clarification our treatment of Boltzmann's H theorem can be quite misleading. Thus if it is assumed that at thermodynamic equilibrium the distribution function $P_1(\mathbf{r}, \mathbf{p}, t)$ becomes essentially the Maxwell–Boltzmann distribution of kinetic energies, then one can easily prove that Boltzmann's minimum H function cannot be identified as being proportional to the thermodynamic entropy except in the limiting case of a classical ideal gas [see E. T. Jaynes, *Am. J. Phys.* **33** (1965), p. 391]. Moreover, with the same assumption for $P_1(\mathbf{r}, \mathbf{p}, t)$ it can also be shown that Boltzmann's H function can even increase for some real gases as they approach equilibrium [see E. T. Jaynes, *Phys. Rev.* **A4** (1971), p. 747]. These difficulties are resolved most readily by employing instead of Boltzmann's H function the Gibbs H function of Problem V.4. The difficulties are also resolved when one appreciates that the single-particle energies in the Maxwell–Boltzmann distribution must, in general, include an internal potential-energy contribution (see footnote 19 in Chapter I).

33. J. W. Gibbs, *Elementary Principles in Statistical Mechanics* (Dover, 1960), p. 150.

34. R. Jancel, *op. cit.*, pp. 157–163.

35. See, for example, Equation (23.13) and the discussions below Equations (4.18) and (4.21).

36. Ergodic theory is treated in detail, with emphasis on the physical point of view, in the text by R. Jancel, *loc. cit.* A more mathematically oriented but equally readable account is provided by I. E. Farquhar, *Ergodic Theory in Statistical Mechanics* (Interscience, 1964). A contemporary review of the status of ergodic theory and its relation to mathematical investigations of macroscopic irreversibility is the article "Modern Ergodic Theory" by J. L. Lebowitz and O. Penrose in *Phys. Today* **26** (Feb. 1973), p. 23.

37. Discussion of Sinai's work is given by A. S. Wightman in "Statistical Mechanics and Ergodic Theory, An Expository Lecture," which is included in a collection of articles under the title: *Statistical Mechanics at the Turn of the Decade,* edited by E. G. D. Cohen (Marcel Dekker, 1971).

38. From the outset in our discussions of kinetic and ensemble theory we have adopted the attitude that coarse graining is necessary for the statistical description of macroscopic systems [see, e.g., inequality (11.2) and its discussion] so that the present discussion serves only to clarify and justify this procedure further.

39. P. and T. Ehrenfest, *The Conceptual Foundations of the Statistical Approach in Mechanics,* translated by M. J. Moravcsik (Cornell University Press, 1959).

40. R. Jancel, *loc. cit.,* Chapter VI. See also Equation (45.11) and its discussion.

41. Understanding of irreversible behavior in a macroscopic system and, in particular, an appreciation of important ingredients required for a mathematical description of the approach to equilibrium can be gained from the study of model systems. One such system is the Ehrenfests' urn model, which we discuss in some detail in Section 45. Another example is Kac's ring model, which is treated in G. H. Wannier, *Statistical Physics* (Wiley, 1966), pp. 398–404.

42. We must mention here also that important progress on the foundations of classical statistical mechanics has recently been made by Lanford, who has demonstrated rigorously that in the thermodynamic limit all observables are essentially determined for a system by its density and energy and, moreover, that the entropy is a

concave function of the thermodynamic variables on which it depends. See O. E. Lanford, "Entropy and Equilibrium States in Classical Statistical Mechanics" in *Statistical Mechanics and Mathematical Problems* (Springer-Verlag, 1973), pp. 1–113.

43. F. W. Sears, *Thermodynamics* (Addison-Wesley, 1953), Section 17.1.

Problems

V.1. (a) What is the direction and magnitude of the polarization vector for a beam of electrons that is characterized by the following density matrix?

$$\underline{\rho} = \frac{1}{2}\begin{pmatrix} 1 & e^{-i\pi/4} \\ e^{i\pi/4} & 1 \end{pmatrix}$$

(b) Write down the density matrix for a partially polarized beam of electrons with polarization $\mathbf{P} = P\hat{\mathbf{i}}_z$ and $P < 1$. Write down expressions that relate P to the fractions $w_\pm$ of electrons with polarizations in the $\pm z$ directions, and express $\underline{\rho}$ in terms of the $w_\pm$.

(c) The Hamiltonian H of an electron in a magnetic field $\mathcal{H} = \hat{\mathbf{i}}_z\mathcal{H}$ is given by

$$H = \mu_\beta \underline{\boldsymbol{\sigma}} \cdot \mathcal{H}$$

where the vector $\underline{\boldsymbol{\sigma}}$ is composed of the Pauli spin matrices. Write down the 2×2 statistical matrix $\underline{\rho}$, using probabilities $w(E) \propto e^{-\beta E}$, and evaluate the polarization P_z for this case.

V.2. Derive the Boltzmann transport equation and corresponding Maxwell–Boltzmann distribution function for the general internal motion of a gas that is subject to velocity-independent external forces and is moving with uniform translation $\mathbf{u}$ and rotation $\boldsymbol{\Omega}$. Use the method of Section 23. See also Problem III.1.

V.3. (a) Determine the relation between the two-particle reduced distribution function $P(\mathbf{k}_1, \mathbf{k}_2)$ of Equation (24.14) and the fluctuation in particle number,

$$\langle \Delta n(\mathbf{k}_1)\, \Delta n(\mathbf{k}_2)\rangle \equiv \langle (n(\mathbf{k}_1) - \langle n(\mathbf{k}_1)\rangle) \cdot (n(\mathbf{k}_2) - \langle n(\mathbf{k}_2)\rangle)\rangle$$

$$\xrightarrow[\mathbf{k}_1 = \mathbf{k}_2]{} \langle [\Delta n(\mathbf{k}_1)]^2\rangle$$

(b) What is $P(\mathbf{k}, \mathbf{k})$ for a Fermi system? Neglect spin considerations and use a physical argument. Show, therefore, for a Fermi system that

$$\langle [\Delta n(\mathbf{k})]^2\rangle = \langle n(\mathbf{k})\rangle[1 - \langle n(\mathbf{k})\rangle]$$

(c) Use the result of part (b) to determine $\langle[\Delta n(\mathbf{k})]^2\rangle$ for the classical ideal gas, for which $\langle n(\mathrm{k})\rangle << 1$. Does this suggest an expression for $P(\mathbf{k}_1, \mathbf{k}_2)$, in this case, that agrees with Equation (24.8)?

(d) We can use the density function $n(\mathbf{r})$, as given by Equation (24.2), in quantum statistical calculations if we use the position representation. Show that in Fock space $n(\mathbf{r})$ is given by Equation (24.9), where the field operators $\underline{\psi}(\mathbf{r})$ and $\underline{\psi}^\dagger(\mathbf{r})$ are defined by Equations (17.22). [*Suggestion:* Use Equation (17.16).]

V.4. Examine the H theorem for ensembles in classical statistical mechanics by considering the distribution function ρ at times t_1 and t_2, with $t_2 > t_1$. Both the cases of a fine-grained distribution function $\rho_f(t)$ and a coarse-grained distribution function $\rho_c(t)$ are to be analyzed, where for the ith cell in Γ space, with volume V_i, ρ_c is defined by

$$\rho_c(V_i, t) = V_i^{-1} \int_{V_i} d\Gamma \rho_f$$

Thus at any time t, ρ_c is taken to be constant over each small Γ-space volume element V_i. See Figure V.5.

(a) Consider the Gibbs H function defined by

$$\mathcal{H}(t) \equiv \int d\Gamma \rho (\ln \rho)$$

and variations $\delta\rho$ in the distribution function, subject to the constraints that the average energy $\langle E\rangle$ is known and that $\int \rho \, d\Gamma = 1$. Show that the Gibbs H functions will be minimal with respect to functional variations $\delta\rho$ for distribution functions of the form $\rho = \exp(\lambda - \beta E)$.

(b) Prove that the approach to equilibrium cannot be demonstrated using a fine-grained ensemble, by showing that

$$\frac{d\mathcal{H}_f(t)}{dt} = 0$$

for all ensembles. (*Suggestion:* Use Liouville's theorem.)

(c) Choose $\rho_f(t_1) = \rho_c(t_1)$, consistent with known information at some time t_1 when the ensemble is "prepared"; that is, let $\rho_f(t_1)$ be represented by Figure V.5(b), say, and show that

$$\mathcal{H}_c(t_1) \geqq \mathcal{H}_c(t_2)$$

[*Suggestion:* Write $\rho_f(t_2) = \rho_c(t_2)e^{\Delta}$ and use the fact that $\Delta e^{\Delta} + 1 - e^{\Delta} > 0$ for all $\Delta \neq 0$.]

(d) Discuss results (a), (b), and (c).

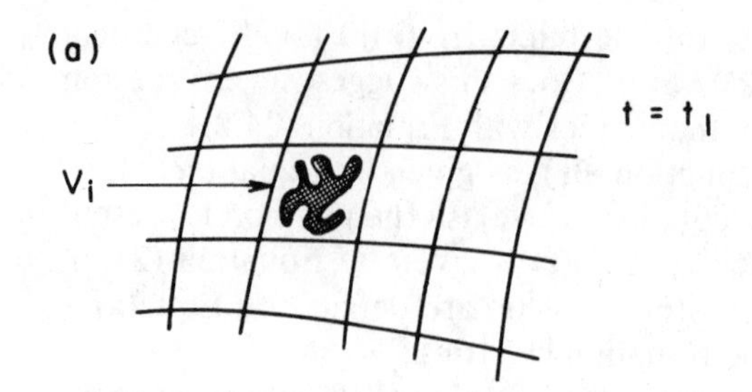

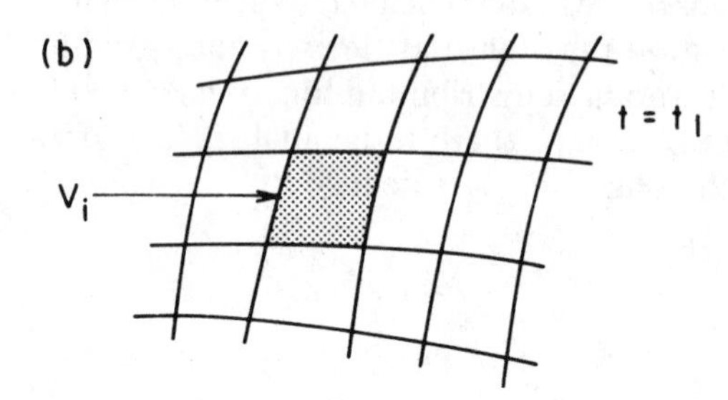

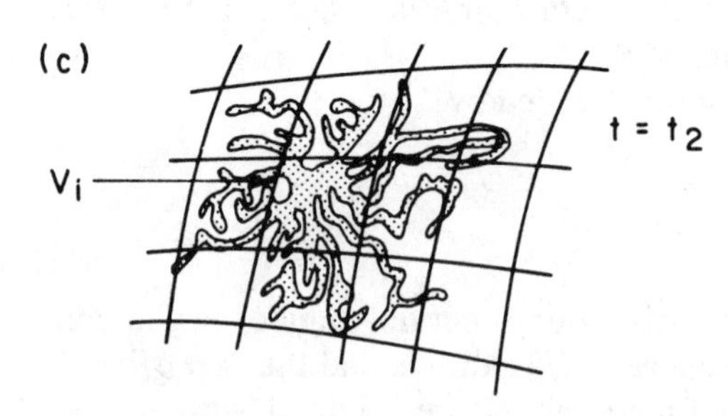

FIGURE V.5 (a) Schematic illustration of a fine-grained distribution function that is limited to occupation of a single cell V_i in phase space at time t_1. (b) The corresponding coarse-grained distribution function at time t_1. (c) The fine-grained distribution function at a later time t_2.

V.5. *Magnetic polarization of thermal neutrons*
Thermal neutrons that pass through magnetized iron are polarized by coherent dipole–dipole interactions with atomic electrons. This process can be described by using the density-matrix formalism as follows:

(a) As the neutron beam proceeds from distance x to $x + dx$ inside the iron, its density matrix changes according to Equation (22.15):

$$\underline{\rho}(x + dx) = (1 + d\mathsf{f})\underline{\rho}(x)(1 + d\mathsf{f})^{\dagger}$$

which can be rewritten in the form

$$\frac{d\underline{\rho}(x)}{dx} = \mathsf{S}\underline{\rho}(x) + \underline{\rho}(x)\mathsf{S}^{\dagger}$$

where $\mathsf{S} \equiv d\mathsf{f}/dx$. Let $\underline{\rho}(x) = C[1 + \mathbf{P}(x) \cdot \underline{\boldsymbol{\sigma}}]$ and assume that the most general form of S is given by $\mathsf{S} = \mathsf{S}^{\dagger} = (1/2\lambda)[1 + \mathcal{M} \cdot \underline{\boldsymbol{\sigma}}]$, where λ is a polarizing length. Show that

$$\frac{d\underline{\rho}}{dx} = \frac{1}{\lambda}[1 + \mathcal{M} \cdot \underline{\boldsymbol{\sigma}}]\underline{\rho}(x) + \frac{\iota C}{\lambda}(\mathbf{P} \times \mathcal{M}) \cdot \underline{\boldsymbol{\sigma}}$$

(b) Assume that the neutron polarization vector **P** is parallel to the iron polarizing vector $\mathcal{M}$, and show that the solution to the equation derived in part (a) is

$$\underline{\rho}(x) = e^{x/\lambda}\left[\cosh\left(\frac{\mathcal{M}x}{\lambda}\right)1 + \hat{\mathcal{M}} \cdot \underline{\boldsymbol{\sigma}} \sinh\left(\frac{\mathcal{M}x}{\lambda}\right)\right]\underline{\rho}(0)$$

(c) Show for initially unpolarized beams that

$$\mathbf{P}(x) = \hat{\mathcal{M}} \tanh\left(\frac{\mathcal{M}x}{\lambda}\right)$$

V.6. (a) Starting from Liouville's equation, derive the equation of motion for the m-particle reduced distribution function $P_m(\mathbf{r}_i, \mathbf{p}_i, t)$. This result gives the BBGKY hierarchy of coupled differential equations for the P_m, Equation (23.25) being the $m = 1$ equation. It is important to distinguish, in the mth equation, between particle interactions among the m identified particles and all other particle interactions.

(b) Show that your result in part (a) is consistent, by integrating the equation for P_m over the μ-space volume element $d\omega_m$ to get the equation for P_{m-1}.

(c) Show that a Liouville's equation is satisfied by each of the P_m; that is, verify that $(dP_m/dt) = 0$. [*Suggestion:* Begin with an equation similar to the first equality of Equation (23.15).]

V.7. Use the result of Problem V.6(a) for the case $m = 2$ to derive the result, for velocity-independent potentials,

$$\frac{\partial \phi}{\partial t} = \frac{1}{2}\langle \delta^{(3)}(\mathbf{r}_1 - \mathbf{r})\left[\frac{\mathbf{p}_1}{m} \cdot \nabla_{r_1} V(\mathbf{r}_{21}) + \frac{\mathbf{p}_2}{m} \cdot \nabla_{r_2} V(\mathbf{r}_{21})\right]\rangle_2$$

$$- \frac{1}{2}\nabla \cdot \langle \delta^{(3)}(\mathbf{r}_1 - \mathbf{r})\frac{\mathbf{p}_1}{m} V(\mathbf{r}_{21})\rangle_2$$

where the average interparticle potential energy $\phi(\mathbf{r}, t)$ is defined by Equation (12.15) and

$$\langle A \rangle_2 = \int d\omega_1\, d\omega_2 A P_2(\mathbf{r}_1, \mathbf{r}_2; \mathbf{p}_1, \mathbf{p}_2; t)$$

V.8. *Spontaneous fluctuations and recurrence times*

(a) Consider the free expansion of an ideal gas as defined above Equation (25.12). Show that during n measurements the probability $W_N(n)$ of finding all N atoms gathered at least once in the original chamber is given by

$$W_N(n) = 1 - \left(1 - \frac{1}{2^N}\right)^n$$

Contrast this result with the expectation $n/2^N$ that the filling of the original chamber with all N atoms will recur after n measurements. Note that this second quantity can be >1.

(b) A room of volume $3 \times 3 \times 3$ cubic meters is at standard conditions (atmospheric pressure and 273K). Estimate the probability that at any instant of time a given 1 cm^3 volume within this room becomes totally devoid of air due to spontaneous fluctuations. What result do you estimate for a 1000 A^3 volume? Estimate the recurrence time for the first case, using a time scale based on the collision time $\tau \sim 10^{-9}$ sec, and express your answer in ages of the universe. [*Suggestion:* Consider an approach similar to that used for solving Problem I.3. Note that the result of Problem II.7(d) can be applied to an analysis of density fluctuations in the atmosphere,[43] but not to this problem.

CHAPTER VI

MACROCANONICAL ENSEMBLES

In this chapter and the following two we are concerned with the description of macroscopic systems in thermodynamic equilibrium. We describe any given system by the average over a time-independent ensemble of identically prepared systems. In the present chapter we treat systems in contact with uniform temperature baths and with constant particle number N. The ensembles that correspond to this physical situation are macrocanonical ensembles, and in Sections 26 and 27 the formalism of macrocanonical ensembles is developed generally. The formal reduction of such descriptions from quantum statistical mechanics to the classical limit is deduced in Section 28. Applications to dilute diatomic gases and to simple crystals are given in Sections 29 and 30.

26 Ensemble Theory, a General Discussion

Our discussion focuses initially on the microcanonical ensemble, one that represents a system with constant total energy. The development is related to topics covered in previous chapters, and then the microcanonical ensemble is applied to a very large uniform system with constant energy. It is shown that a small macroscopic subsystem of such a system behaves similarly to a system in contact with a temperature bath. By this device the macrocanonical ensemble, to describe a system in equilibrium with a temperature bath, is introduced. The intention in this section is not to provide a careful derivation of the equations and methods of the macrocanonical ensemble—a goal that is set for the following section—but to provide a general conceptual perspective of ensemble theory. The section closes with a brief discussion of fluctuations.

For systems in thermodynamic equilibrium we must use a *stationary ensemble,* that is, one constant in time. Then according to Equation (23.3) the statistical matrix $\underline{\rho}$ of quantum statistical mechanics commutes with the many-body Hamiltonian; according to Equation (23.19) the distribution function ρ of classical statistical mechanics is a function of the energy only.

We may use concepts of classical statistical mechanics to elaborate further on stationary ensembles. According to Liouville's theorem, and as we have discussed at the end of Section 25, the distribution function ρ for a stationary ensemble will be one that is uniform, or constant, over the allowed region of Γ space. That stationary ensembles are indeed appropriate for describing real systems in thermodynamic equilibrium, however, requires an assumption; namely, we must assume that the *a priori* probability for finding a certain number of representative points in a volume element $\Delta\Gamma_1$, in the occupied region of Γ space, is proportional to $\Delta\Gamma_1$.[1] Then for systems in equilibrium we would expect that the number of representative points in equal volumes of phase space ($d\Gamma_1 = d\Gamma_2$) would be the same,

$$\rho_1(q_i, p_i)\, d\Gamma_1 = \rho_2(q_i, p_i)\, d\Gamma_2 \tag{26.1}$$

which implies that $\rho_1 = \rho_2$. Thus we expect that ρ = constant in the occupied region of Γ space, and this physical requirement for an equilibrium ensemble is in accord with the description provided by Liouville's theorem for stationary ensembles.

Microcanonical Ensembles

In the introduction to Section 21 we argued in considerable detail that real macroscopic systems in nature cannot be assumed to have a definite energy. On the other hand, attempts to establish the foundations of statistical mechanics[1] often make use of the closed-system model for which the energy is constant. It is not necessary to show our prejudice either for more realistic descriptions of nature or for a more rigorously satisfying exploration of statistical mechanics, because these two apparently opposing positions can be reconciled. We have only to introduce the concept that any real macroscopic system can be considered to be a small subsystem of a huge system. That this huge system is assumed to be a closed system, suitably defined for studies of the foundations of statistical mechanics, is unimportant for the physical description of the real system imbedded in it. The natural energy fluctuations that occur for subsystems of the huge system would be affected only insignificantly by any energy fluctuations of the huge system.[2]

To relate our present development of ensemble theory to previous discussions in which closed systems were treated, we begin by considering systems with constant total energy E. An ensemble of identical systems, each with the same total energy, is called a *microcanonical ensemble*. Not only can we calculate average values for a system that is represented by a microcanonical ensemble, but we also may treat improbable situations; that is, we can calculate the average fluctuations to be expected for physical measurements (of quantities other than N and E).

The discussion of the maximum entropy principle in connection with Equation (4.26) suggests that the statistical matrix or classical distribution function of statistical mechanics should contain all given information about a system. For example, if nothing were known about a system, then the appropriate distribution function in classical statistical mechanics would be

$$\rho(q_i, p_i) = \text{constant over all } \Gamma \text{ space}$$

The ensemble corresponding to such a distribution function is called a *uniform ensemble*. Of course, there is no measurement on a system that does not imply knowledge of its energy. Certainly all thermodynamic functions, for example, heat capacity and temperature, are related to the energy. For a closed classical system the relevant distribution function is that corresponding to a definite total energy E_0.

$$\rho(q_i, p_i) = A\,\delta(E - E_0) \tag{26.2}$$

where A is a constant. This distribution, constant over an energy surface in Γ space, is the microcanonical ensemble distribution function of classical statistical mechanics. If there are other measured constants of the motion besides the energy, then the distribution function (26.2) must be restricted further on the energy surface in Γ space to the region defined by these constants.[3]

What other constants of the motion besides the energy can be measured? Certainly, there are the six constants that comprise the total linear momentum and the total angular momentum of a system, as discussed previously in Section 6. Apart from the scalar invariants, such as total charge and particle number (to be discussed further in Chapter VIII), these are the only "additive" constants of motion for a uniform, or isotropic, system.[4] Therefore, for uniform systems at rest, and in accordance with our discussion at the beginning of this section, we need consider only ensembles with a functional dependence on the energy of the system.

In the foregoing discussion we considered the model in which a real system of interest is imbedded in some huge constant-energy system, a model to which we return again when we discuss macrocanonical ensembles. For the present we discuss an alternative, a closed-system model in which coarse graining is introduced (see discussion near the end of Section 25). In this representation of an actual closed system the energy surface in Γ space has a finite width δE determined by the considerations detailed in the introduction to Section 21. By this means we can relax the condition that the total system energy E is specified precisely, and the microcanonical ensemble can be applied directly, in principle, to calculations for real systems. Such calculations could be ones that utilize the many-body perturbation theory of quantum mechanics, discussed in Chapter IV, or any of several generalizations of this method.

When coarse graining is introduced into the microcanonical ensemble, then the distribution function (26.2) must be generalized to a constant function with a finite width δE. This distribution function also applies in quantum statistical mechanics, according to Equation (21.15), if it is identified as a collection of diagonal matrix elements of the statistical matrix. The introduction of coarse graining has a further benefit, because it permits one to define the temperature and pressure for closed systems, by Equations (6.3) and (7.8) respectively.

$$T = \left(\frac{\partial E}{\partial S}\right)_\Omega \qquad \mathcal{P} = -\left(\frac{\partial E}{\partial \Omega}\right)_S \tag{26.3}$$

The thermodynamic identity (7.9) then follows, and one can proceed to make thermodynamic, or other, calculations of physical interest.

Here a point must be clarified. The appearance of the entropy as an independent variable in Equations (26.3) presents calculational difficulties. To calculate the entropy we must use an equation such as (4.24) in quantum statistics, with the n_i referring to numbers of systems in the ensemble. One can also use, for either the classical or quantum-mechanical case, the expression

$$S = \kappa \ln[n(E)\, \delta E] \tag{26.4}$$

where $n(E)$ is the density of energy states with energy E in the interval δE (see Problem II.4). For the classical case, $n(E) = \partial\Sigma(E)/\partial E$, where the quantity $\Sigma(E)$ is defined to be the volume of Γ space with total energy $\leqslant E$.

It is possible to carry through a calculation of Equation (26.4) for the ideal gas,[5] but the method seems not to have a simple generalization for real systems. On the other hand, the use of Equations (4.24) and (4.18) to derive the classical ideal gas law, as in Exercise 4.1, with the n_i referring to particle numbers in one system, is an application of the microcanonical ensemble that can be generalized. Thus for a Fermi fluid we were able to use simple generalizations of the ideal-gas entropy expressions, Equations (5.5) and (5.8), to arrive at Equation (19.14). We may conclude that in such cases where we can write down an entropy expression, the microcanonical ensemble in conjunction with, say, perturbation theory can be applied to practical thermodynamic calculations.

Macrocanonical Ensembles

The preceding two paragraphs show that although conceptually attractive, the coarse-grained microcanonical ensemble is quite limited in application. Fortunately, many real systems do conform to our first microcanonical-ensemble model system, in which the real system is assumed to be part of some much larger system. We now show how the rest of this larger system can serve as a temperature bath for the system of interest. Our treatment applies to quantum-mechanical as well as to classical systems; it forms a basis for the careful development that follows in Section 27.

For convenience, but without loss of generality, as we shall see below Equation (26.6), we consider a very large closed system of identical particles divided into many macroscopic subsystems, as in Figure VI.1. Statistically, we may say that these subsystems are all weakly interacting; that is, at equilibrium they are almost independent of each other.[6] Of course, the weak interactions are responsible for the approach to equilibrium, but at equilibrium these interactions can be neglected, just as we can neglect the particle interactions in dilute real gases. If now we also specify that each subsystem has the same volume and number of particles, then we may regard the entire system as an ensemble of identical subsystems. In other words, we may write, classically,

$$\rho = \prod_{\mu} \rho_{\mu} \tag{26.5}$$

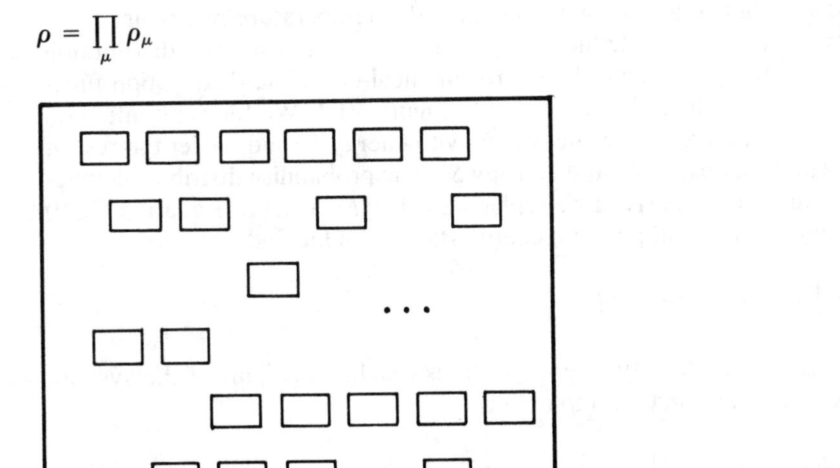

FIGURE VI.1 Macroscopic subsystems in a very large system.

where the subscripts μ refer to the various subsystems. For quantum-mechanical treatments, Equation (26.5) applies, according to Equation (21.15), with $\rho \to w$.

Now, each of the subsystems can be considered to be a distinguishable free particle with energy E_μ. Therefore, in complete analogy with the derivation of the Maxwell–Boltzmann distribution in Exercise 4.1, we may conclude that

$$\rho_\mu = e^{\beta(\Psi - E_\mu)} \tag{26.6}$$

where $e^{\beta\Psi}$ is a normalization constant that characterizes the entire system.[7] At equilibrium the distribution function for an ensemble of subsystems is the product of all the Maxwell–Boltzmann distributions for the subsystems.

Consider next any system in contact with a temperature bath, a common physical situation in thermodynamics. Such a system could be considered to be any one of the subsystems in the above example in so far as its thermodynamic behavior is concerned. Therefore, we may conclude immediately that the distribution function for an ensemble that represents a system in contact with a temperature bath is the same as the Maxwell–Boltzmann distribution function for that system. Gibbs called an ensemble characterized by such a distribution function a canonical ensemble.[8] To distinguish this ensemble from the microcanonical ensemble, we use here the contemporary terminology *macrocanonical ensemble* for ensembles with the distribution function (26.6). Physically, the distinction between a microcanonical and a macrocanonical ensemble is the same as the difference between a closed or isolated system and a system in contact with a temperature reservoir.

It is instructive to deduce the macrocanonical-ensemble distribution function (26.6) by applying the microcanonical-ensemble distribution function (26.2) to the large closed system of Figure VI.1. We focus attention on any one small macroscopic subsystem with energy E, and we let the rest of the system have energy E' and entropy S'. The probability distribution function ρ_E can then be derived by replacing E by $(E' + E)$ in Equation (26.2) and integrating over all possible energy states E'. One finds

$$\rho_E = A \int dE' n(E')\, \delta(E' + E - E_0)$$

where $n(E')$ is the density of energy states for the large part of the system. We next use Equation (26.4) to obtain

$$\rho_E = An(E_0 - E) = A\left(\frac{e^{S'/\kappa}}{\delta E'}\right)_{E' = E_0 - E} \tag{26.7}$$

Since $E \ll E_0$, we may expand S' in a Taylor series about $E' = E_0$ and keep only the first two terms. With the aid of the first of Equations (26.3), we then get for ρ_E the expression

$$\rho_E = \left[\frac{Ae^{S'(E_0)/\kappa}}{\delta E'(E_0)}\right] e^{-E/\kappa T}$$

which is Equation (26.6) apart from the normalization factor.

At the first encounter with ensemble theory one will certainly inquire as to which ensembles are "best." As far as microcanonical and macrocanonical ensembles are concerned, they are both equally good in principle. In general, both could be expected to give the same final predictions for average values of physical quantities. The difference between the physical systems to which they apply, that is, closed systems versus systems in temperature baths, corresponds in thermodynamic calculations only to the difference between the energy functions $E(S, \Omega, N)$ and $\Psi(T, \Omega, N)$. However, because their independent variables include T instead of S, macrocanonical ensembles are much easier to apply in practical calculations. Finally, the underlying difference between the micro- and macrocanonical ensembles is reflected in the different predictions they give for the energy fluctuation ΔE (which is zero, or a given value δE, for microcanonical ensembles).

In the above derivations of the macrocanonical distribution function (26.6), the total particle number N was held fixed in the subsystems. This artificial constraint is removed when one considers the grand canonical ensemble, in which particle exchange between subsystems is allowed in addition to energy exchange. Apart from fluctuations ΔN in total particle number, the predictions for average values calculated by considering a grand canonical ensemble must agree with those calculated by considering a macrocanonical ensemble.[9]

Calculations in statistical mechanics are usually applied in the thermodynamic limit, in which the density remains finite as $(N, \Omega) \to \infty$. This limit is justified physically by, say, the discussion of inequality (11.2), where we indicated the enormous N values associated with ordinary systems. To be sure, surface and other finite-size effects are interesting macroscopic phenomena, and their investigation by thermodynamic methods has been extensively pursued. For reasons of simplicity the microscopic treatment of finite-size effects is not normally considered initially. Occasionally, however, the thermodynamic limit must be taken with extreme care, as with the highly degenerate Bose gas (Section 34). It is convenient, therefore, to use the notation of finite systems, for example, by retaining sums instead of integrals over states up to the stage of numerical evaluation [see Equation (17.29)].

Fluctuations

Some of the interesting questions asked about physical quantities concern the deviations of measured values from their average values. These deviations, or fluctuations, may be appreciable or negligible, depending on the actual situation, but in any event their spread is predictable using the methods of statistical mechanics. Also, if the fluctuations of extensive quantities are small compared with their average values, then we need not exert special care to distinguish average values from most probable values, as we did for molecular speeds with Equations (11.23) and (11.25).

The *fluctuation,* or mean-square deviation, of any physical quantity A is defined quantitatively [see Equation (1.7)] by

$$\begin{aligned}\langle(\Delta A)^2\rangle &\equiv \langle(A - \langle A\rangle)^2\rangle \\ &\equiv \langle A^2\rangle - \langle A\rangle^2 > 0\end{aligned} \tag{26.8}$$

By definition, this quantity is also $(\Delta A)^2_{\text{rms}}$. Let us consider the dimensionless ratio of the root-mean-square (rms) deviation of A to its average value. We wish to show for extensive quantities and $N >> 1$ that ordinarily

$$\frac{(\Delta A)_{\text{rms}}}{\langle A\rangle} \sim \frac{1}{\sqrt{N}} \qquad (N >> 1) \tag{26.9}$$

thereby demonstrating that the fluctuations of extensive quantities are indeed small in the thermodynamic limit.[10]

As defined above Equations (8.14) an extensive quantity is proportional to N at fixed density. Therefore, to prove Equation (26.9) one must show that the fluctuation (26.8) is an extensive quantity when $\langle A\rangle$ itself is. We do not give a general proof of (26.9); rather, we only prove this result for a gas of weakly interacting particles for which one can write

$$A = \sum_{\alpha=1}^{N} a_\alpha$$

where a_α is the quantity A for the αth single particle and $\langle a_\alpha\rangle = (1/N)\langle A\rangle$. Then

$$\langle(\Delta A)^2\rangle = \left\langle\left(\sum_{\alpha=1}^{N} (\Delta a)_\alpha\right)\left(\sum_{\beta=1}^{N} (\Delta a)_\beta\right)\right\rangle$$

But for $N >> 1$,

$$\langle(\Delta a)_\alpha (\Delta a)_\beta\rangle = 0 \qquad \text{when } \alpha \neq \beta$$

since the particle fluctuations are random.[11] Therefore,

$$\langle(\Delta A)^2\rangle = \left\langle\sum_{\alpha=1}^{N} (\Delta a)^2_\alpha\right\rangle \propto N \tag{26.10}$$

as was to be proved; for a dilute gas, the rms deviation of an extensive variable is proportional to $\sqrt{N}$.[12] Examples of the general result (26.9) will be encountered in subsequent sections.

27 Method of the Most Probable Distribution, Thermodynamic Analysis

The most probable distribution of system occupation numbers N_i for a macrocanonical ensemble is determined by maximizing the ensemble entropy with respect to these numbers, subject to constraints of given information. The most probable occupation numbers $\overline{N}_i$ are expressed as system occupation probabilities w_i, and these quantities, in turn, readily determine system average values. The partition function Z is introduced, and prescriptions for the calculation of average values in terms of Z are deduced. The thermodynamics of systems described by macrocanonical ensembles is related with care to the statistical procedures used, and then the section is ended with a short analysis of energy fluctuations. The notation of quantum mechanics is used throughout, with reduction to the classical limit being reserved for the following section.

In quantum statistical mechanics, according to the treatment of Section 21, we must first determine the w_i of Equations (21.2) and (21.15). Then we may form the statistical matrix $\underline{\rho}$ by using Equation (21.3). The method of the most probable distribution that we apply to determine the w_i was introduced previously, in brief form, as Exercise 4.1 for the determination of free-particle occupation numbers in a single system (an ideal gas).

We consider an ensemble of M identical, but distinguishable, weakly interacting systems. This is essentially the model of Figure VI.1, except that now the model has the abstract character that the systems of the ensemble are truly distinguishable; that is, the indistinguishability of particles in two different systems is not a reality, and we make no approximation by its omission. Each system in the ensemble is characterized by the same ordered spectrum of energy levels E_1, E_2, ..., E_n, ..., where we might specify the ordering by requiring that $E_i < E_{i+1}$ for all i. If there are degeneracies, then we let g_i be the degeneracy of the ith state. It is assumed, as in the classical discussion at the beginning of Section 26, that each energy state has an *a priori* probability for occupation that is proportional to g_i. Let N_i be the number of systems in the ensemble with energy E_i. Then the number of possibilities for finding an ensemble with the set of numbers $\{N_1, N_2, ..., N_n, ...\}$ is $W_M(N_i)$ of Equation (4.18), with $C = 1$ and $Z_i \rightarrow g_i$.

$$W_M(N_i) = M! \prod_i \frac{(g_i)^{N_i}}{(N_i!)} \tag{27.1}$$

According to Equation (4.24), the statistical entropy S_{ens} for the ensemble is proportional to $\ln W_M(N_i)$,

$$S_{\text{ens}} \propto \ln W_M(N_i)$$

We can then apply the maximum entropy principle to find the most probable occupation numbers $\overline{N}_i$, corresponding to thermodynamic equilibrium, by maximizing $\ln W_M(N_i)$ with respect to each of the N_i and subject to the constraints of given information

$$\sum_i N_i = M$$
$$\sum_i N_i E_i = E_M \tag{27.2}$$

where E_M is the total energy of the ensemble. According to Lagrange's method of undetermined multipliers, this maximum can be found as the unrestricted maximum of

$$\ln W_M(N_i) + \nu \sum N_i - \beta \sum_i N_i E_i$$

where ν and $-\beta$ are unknown multipliers.

We now require M to be sufficiently large that the significant N_i values are all large numbers. Then these N_i can be treated as continuous variables in the variation procedure for maximizing $\ln W_M(N_i)$. Also, we can then use Stirling's formula (2.6) in the form

$$\ln(N_i!) \cong N_i(\ln N_i - 1) \tag{27.3}$$

which also happens to be valid for the myriad of vanishingly small N_i values.

The maximization of $\ln W_M(N_i)$, using Stirling's formula (27.3), gives

$$0 = \sum_i [\ln g_i) \, \delta N_i - \delta(\ln N_i!) + \nu \, \delta N_i - \beta E_i \, \delta N_i]$$
$$= \sum_i [\ln g_i + \nu - \beta E_i - \ln N_i] \, \delta N_i$$

Then on equating the coefficient of δN_i to zero we find for the most probable value $\overline{N}_i$,[13]

$$\overline{N}_i = g_i \, e^{\nu} \, e^{-\beta E_i} \tag{27.4}$$

To determine the multipliers ν and β we must substitute this expression back into the constraint equations (27.2):

$$M = e^{\nu} \sum_i g_i \, e^{-\beta E_i}$$

$$E_M = e^{\nu} \sum_i g_i E_i \, e^{-\beta E_i}$$

Now, in fact, we are not interested in the number M, which was introduced artificially. The quantities that do interest us are the systems average values $g_i w_i$ and $\langle E \rangle$, defined by

$$g_i w_i \equiv \frac{\overline{N}_i}{M} = \frac{g_i\, e^{-\beta E_i}}{\sum_i g_i\, e^{-\beta E_i}}$$
$$\langle E \rangle \equiv \frac{E_M}{M} = \frac{\sum_i g_i E_i\, e^{-\beta E_i}}{\sum_i g_i\, e^{-\beta E_i}} \tag{27.5}$$

As in Section 21, the quantity w_i is the probability that the system has for occupying any single quantum state with energy E_i.

The unknown multiplier ν does not appear in Equations (27.5), because it is fixed by the value of M that we have chosen to eliminate by considering only system averages. Instead, a new quantity, the denominator on the right sides of Equations (27.5), has appeared. We define this quantity to be the *partition function* Z for the macrocanonical ensemble, because it tells us how to partition, or to weight statistically, the various states i. Henceforth we include the degeneracy factors g_i in the sum over states; that is, each i now refers to a single many-body state. Then

$$Z \equiv \sum_i e^{-\beta E_i} \tag{27.6}$$

Equations (27.5) can be written in terms of Z as

$$w_i = -\beta^{-1}\frac{\partial}{\partial E_i}(\ln Z)$$
$$\langle E \rangle = -\frac{\partial}{\partial \beta}(\ln Z) \tag{27.7}$$

where the system's volume Ω and the system's total particle number N are each to be held constant in each partial differentiation. These last two equations give the calculational prescription (in principle, at least) for determining w_i and $\langle E \rangle$ in quantum statistical mechanics; that is, one must first calculate the partition function and then use Equations (27.7). This procedure can be applied quite generally to the calculation of average values $\langle A \rangle$ if one writes them in the form (21.9a)

$$\langle A \rangle = \sum_i w_i A_i = Z^{-1} \sum_i A_i\, e^{-\beta E_i} \tag{27.8}$$

Of course, one must also know the functional dependence $E(A)$.

Exercise 27.1

(a) Equation (27.4) has been derived for systems at rest. Include the appropriate constraint for systems moving with constant velocity, and show in this case that Equation (27.4) becomes

$$\overline{N}_i = g_i e^{\nu'} \exp[-\beta(E_i - \mathbf{u} \cdot \mathbf{P})]$$

where $\mathbf{P}$ is the momentum of the system.

(b) Argue that $\mathbf{u}$ can be identified as the linear velocity of the moving system. [*Suggestion:* Apply an argument of Galilean, or translational, invariance to Equation (27.4).]

The above development using the maximum entropy principle has identified the macroscopic state of thermodynamic equilibrium for a macrocanonical ensemble. We may argue, therefore, as we did in Section 26, that the system of interest behaves thermodynamically as if it were in contact with a temperature reservoir. We may say that Equations (27.7) contain all of the available information about a simple system with fixed Ω and N in contact with a temperature bath, except that we have not yet determined the Lagrange multiplier β. At first reflection one might think that β is determined by $\langle E \rangle$, but actually we shall now show that β can be identified as $(\kappa T)^{-1}$ and then $\langle E \rangle$ is determined by β. We prove first that β is a universal function of the absolute temperature T.

Consider three different ensembles of systems:

(1) Ensemble A composed of systems (a) with energy levels $E_k^{(a)}$.
(2) Ensemble B composed of systems (b) with energy levels $E_\ell^{(b)}$.
(3) Ensemble $(A + B)$ composed of systems, each of which is the combination of one (a) system and one (b) system, with energy levels E_m.

The component systems (a) and (b) of the third ensemble are assumed to interact only very weakly, so that each energy level E_m can be written as a sum $E_m = E_k^{(a)} + E_\ell^{(b)}$. For the partition function $Z_{(A+B)}$ we then have the following identity:

$$Z_{(A+B)}(\beta) = \sum_m e^{-\beta E_m} = \sum_k \sum_\ell \exp[-\beta(E_k^{(a)} + E_\ell^{(b)})] = Z_A(\beta) \cdot Z_B(\beta),$$

which implies in turn that

$$w_m(\beta) = Z_{(A+B)}^{-1} e^{-\beta E_m} = Z_A^{-1} Z_B^{-1} \exp[-\beta(E_k^{(a)} + E_\ell^{(b)})] = w_k^{(a)}(\beta) \cdot w_\ell^{(b)}(\beta)$$

We now determine the probability $w_k^{(a)'}$ for finding an (a) system with energy $E_k^{(a)}$ in the $(A + B)$ ensemble. This probability is given by summing w_m over all probabilities for finding the (b) system, that is, by summing over all states ℓ.

$$w_k^{(a)'}(\beta) = \sum_\ell w_m = \sum_\ell w_k^{(a)} w_\ell^{(b)} = w_k^{(a)}(\beta)$$

We conclude that if systems (a) and (b) are put into loose contact with each other, then each behaves precisely as it would if put into a heat bath (ensemble) by itself—provided only that the β values are the same. This result can be interpreted only by the conclusion that two equal β values correspond to two equal temperature values, that is, that β is a universal function of the absolute temperature. In the following development we assume only that β is an intensive quantity.

Proof that $\beta = (\kappa T)^{-1}$

According to Equation (27.7) for $\langle E \rangle$, ln Z must be an extensive quantity, a fact that suggests that $F' \equiv \ln Z$ is a thermodynamic energy function (see Section 8). To pursue this suggestion we consider possible variations of F', with equilibrium maintained, due to variations of the system properties. Functional dependence on external parameters such as the total volume Ω occurs implicitly through the energy eigenvalues E_i.[14] Then

$$\begin{aligned} dF' &= \frac{\partial F'}{\partial \beta}\, d\beta + \sum_i \frac{\partial F'}{\partial E_i}\, dE_i \\ &= -\left[\langle E \rangle\, d\beta + \beta \sum_i w_i\, dE_i \right] \end{aligned}$$

Here Equations (27.7) have been used to obtain the second equality. This last equation can be rearranged in the form

$$d(F' + \beta\langle E \rangle) = \beta\left(d\langle E \rangle - \sum_i w_i\, dE_i \right) \tag{27.9}$$

Now the energy levels E_i will vary whenever any of the external parameters λ_μ varies, that is,

$$dE_i = \sum_\mu \left(\frac{\partial E_i}{\partial \lambda_\mu} \right) d\lambda_\mu \tag{27.10}$$

The coefficient $-(\partial E_i/\partial \lambda_\mu)$ can be interpreted as a generalized force $\mathscr{F}_{\mu,i}$ exerted by the system when in its ith state and when λ_μ is varied.

$$\mathscr{F}_{\mu,i} \equiv \frac{\partial E_i}{\partial \lambda_\mu} \tag{27.11}$$

Therefore the *average work* $đW$ done by the system when the external parameters λ_μ are changed by $d\lambda_\mu$ is[15]

$$đW = \sum_i w_i\, đW_i = \sum_i w_i \sum_\mu \mathscr{F}_{\mu,i}\, d\lambda_\mu = -\sum_i w_i\, dE_i \tag{27.12}$$

Equation (27.9) is expressed entirely in terms of average values and may therefore be interpreted thermodynamically. Thus we can use result (27.12) and the first law of thermodynamics (7.11) to identify the right side of Equation (27.9) as $\beta đQ$. But for variations at equilibrium $đQ = Td\langle S\rangle$, according to Equation (7.12). Altogether then Equation (27.9) becomes

$$d(F' + \beta\langle E\rangle) = \beta\, đQ = (\beta T)\, d\langle S\rangle \qquad (27.13)$$

We may observe from the first equality that β is the integrating factor for $đQ$ since the quantity on the left side is an exact differential, and this fact implies that $\beta \propto (1/T)$. Alternatively, we shall argue from the second equality that βT is a universal constant.[16]

Let $G \equiv F' + \beta\langle E\rangle$, so that Equation (27.13) becomes

$$(\beta T)^{-1}\, dG = dS$$

where temporarily we shall write $\langle S\rangle$ as S. Since dS is an exact differential, it must therefore be so that $(\beta T)^{-1} = \phi(G)$. Integration of the equation $\phi(G)dG = dS$ must then yield an expression for G in terms of S, which we write as

$$G = F' + \beta\langle E\rangle = \chi(S) \qquad (27.14)$$

Now G is an extensive quantity, which means that when we apply Equation (27.14) to two systems a and b and to a weakly interacting combined system $(a + b)$, then we must have

$$\begin{aligned}\chi_a(S_a) + \chi_b(S_b) &= \chi_{a+b}(S_{a+b}) \\ &= \chi_{a+b}(S_a + S_b)\end{aligned}$$

where the entropy is also an additive, or extensive, quantity. But the systems a and b are two arbitrary systems, and therefore differentiation of this last result with respect to entropy dependence gives

$$\chi_a'(S_a) = \chi_{a+b}'(S_a + S_b) = \chi_b'(S_b) = \frac{1}{\kappa} \qquad (27.15)$$

where κ is a universal constant. We may conclude that

$$\frac{1}{\kappa} = \chi'(S) = \frac{dG}{dS} = \phi^{-1}(G) = \beta T \qquad (27.16)$$

In other words, $\beta = (\kappa T)^{-1}$. The identification of κ as Boltzmann's constant is made in the next section, following Equation (28.20), after proving the equipartition theorem of classical statistical mechanics. (Note that κ cannot depend on the choice of system.)

Thermodynamic Analysis

We now identify $F' = \ln Z$ by using Equation (27.13), which may be integrated to give

$$F' = \frac{1}{\kappa}\langle S\rangle - \beta\langle E\rangle + \text{constant}$$
$$= -\beta(\langle E\rangle - T\langle S\rangle)$$

where in the second equality we have incorporated the integration constant into the average system entropy $\langle S\rangle$ (see Exercise 27.2). On referring to Equation (8.4) we see that the right side of this expression is $-\beta\Psi$, where Ψ is the Helmholtz free energy. Thus

$$F' = \ln Z = -\beta\Psi \tag{27.17}$$

and Equation (27.7) for $\langle E\rangle$ agrees with Equation (8.7). Moreover, according to Equation (21.15) the matrix elements of the statistical matrix in the energy representation are the w_i, which we can now write, using Equations (27.5), (27.6), and (27.17), as

$$\rho_{ii} = w_i = e^{\beta(\Psi - E_i)} \tag{27.18}$$

in agreement with Equation (26.6).

We consider next the pressure $\mathcal{P}$. According to Equation (7.8),

$$\begin{aligned}\mathcal{P} &= -\left\langle\frac{\partial H}{\partial\Omega}\right\rangle = -\sum_i w_i\left(\frac{\partial E_i}{\partial\Omega}\right)\\ &= \beta^{-1}\sum_i\left(\frac{\partial \ln Z}{\partial E_i}\right)_T\frac{\partial E_i}{\partial\Omega} = \beta^{-1}\left(\frac{\partial \ln Z}{\partial\Omega}\right)_T\\ &= -\left(\frac{\partial\Psi}{\partial\Omega}\right)_T\end{aligned} \tag{27.19}$$

The second line of this equation is obtained by using the first of Equations (27.7), and the third line follows after substituting Equation (27.17). The last equality agrees, as it must, with the second of Equations (8.6).

From the relation $\Psi = \langle E\rangle - T\langle S\rangle$, and with the aid of Equations (27.7) and (27.17), one may deduce that the average entropy $\langle S\rangle$ is given by

$$\begin{aligned}\langle S\rangle &= \frac{1}{T}\left(\frac{\partial(\beta\Psi)}{\partial\beta} - \Psi\right)\\ &= (\kappa T^2)^{-1}\left(\frac{\partial\Psi}{\partial\beta}\right)_\Omega\end{aligned} \tag{27.20}$$

in agreement with the first of Equations (8.6). To complete our reasoning process, we show in the following exercise that this thermodynamically defined average entropy is also the average statistical entropy.

Exercise 27.2 Show by using $T\langle S\rangle = (\langle E\rangle - \Psi)$ and Equation (27.18) that the average entropy expression can be written as

$$\langle S\rangle = -\kappa \sum_i w_i \ln w_i \tag{27.21}$$

in agreement with the right side of Equation (4.25) (with $Z_i = 1$). But we have already shown in Section 4 that Equation (4.25) follows from Equation (4.24) when the number of systems in the ensemble is >>> 1. Therefore, Equation (27.21) agrees with the fundamental statistical-entropy expression for the ensemble,

$$M\langle S\rangle = \kappa \ln W_M(\overline{N}_i) \tag{27.22}$$

We may conclude that in the above development it was consistent to absorb the integration constant for F' into the entropy expression. Equation (27.21) also shows explicitly that the quantum-statistical form of the Gibbs H function (see Problem V.4) equals $-\kappa^{-1}\langle S\rangle$ at thermodynamic equilibrium.

With the preceding thermodynamic analysis we have shown, for systems in thermodynamic equilibrium, that thermodynamic quantities can be expressed as partial derivatives of Ψ in agreement with our earlier conclusion in connection with Equation (27.8). In other words, $Z = e^{-\beta\Psi}$ is the only quantity that needs to be calculated from first principles in equilibrium statistical mechanics. Now the statistical matrix $\underset{\sim}{\rho}$ has the property that it contains all known information about a system [see, e.g., Equation (21.9)], and therefore we should expect to find an intimate relationship between $\underset{\sim}{\rho}$ and Z.

To demonstrate the relationship between $\underset{\sim}{\rho}$ and Z we observe first, by referring to Equations (21.3) and (27.18), that for a macrocanonical ensemble $\underset{\sim}{\rho}$ is given by the operator equation

$$\underset{\sim}{\rho} = e^{\beta\Psi} e^{-\beta \mathsf{H}} \tag{27.23}$$

where H is the many-body, or system, Hamiltonian. See also Equation (23.3). The matrix elements of this operator in the energy representation are indeed the w_i of Equation (27.18), as required by Equation (21.15). Moreover, Equations (27.6) and (27.17) can be used to show that $\mathrm{Tr}(\underset{\sim}{\rho}) = 1$ as required by Equation (21.10). Thus Equation (27.6) for Z can also be written as

$$Z = \mathrm{Tr}(e^{-\beta \mathsf{H}}) \tag{27.24}$$

which shows that Z can be calculated in any representation, and not only in the energy representation. We may conclude finally that Z is the trace of an unnormalized statistical matrix.

Energy Fluctuations

We have indicated below Equation (26.7) that apart from energy fluctuations the results of calculations with micro- and macrocanonical ensembles can be expected to agree. Moreover, we have suggested with Equation (26.9) that for very large N values the rms deviations of extensive quantities, for example, the total energy, will ordinarily be extremely small. It is the large N values of macroscopic systems that lead to agreement between micro- and macrocanonical ensemble calculations. In fact, the two ensembles are essentially the same when $N >>> 1$, since to good approximation most systems in a macrocanonical ensemble have the same internal energy E. We use the result of the following exercise to show this fact explicitly.

Exercise 27.3 Show for a macrocanonical ensemble that the energy fluctuations can be calculated by using the formula

$$\langle(\Delta E)^2\rangle = \frac{\partial^2}{\partial\beta^2}(\ln Z) = -\frac{\partial}{\partial\beta}\langle E\rangle \tag{27.25}$$

where the second equality follows from Equation (27.7) for $\langle E\rangle$.

In Equation (27.25) the partial derivative is to be computed at constant Ω and N. Therefore, on referring to Equation (8.25) we see that

$$\langle(\Delta E)^2\rangle = \kappa T^2\left(\frac{\partial\langle E\rangle}{\partial T}\right)_{\Omega,N} = \kappa T^2 C_V \tag{27.26}$$

Clearly, the energy fluctuation is an extensive quantity, in agreement with expression (26.10). The system energy satisfies Equation (26.9), and its rms deviations are relatively small for large N.[17]

It is not always true that the energy fluctuations in a system are small. Consider, for example, an enclosed gas in a variable temperature bath at constant pressure (instead of constant volume), and let the temperature be varied until the gas starts to condense (Figure VI.2). At the condensation temperature many different energy states are possible corresponding to different vapor–liquid ratios. The energy fluctuations at the condensation temperature are enormous. It is known that one must consider the thermodynamic limit of the partition function to describe this situation quantitatively, but a definitive treatment of phase transitions in statistical mechanics has not yet been formulated.[18]

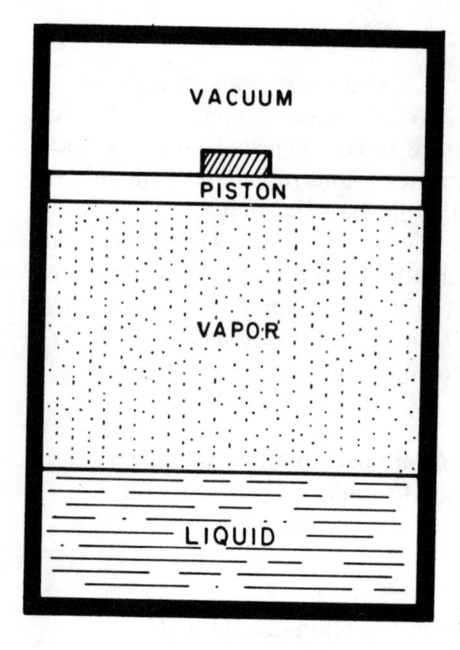

FIGURE VI.2 Liquid–vapor phase change. The frictionless piston holds the vapor pressure constant.

28 Classical Systems

The classical limit of the macrocanonical-ensemble partition function for a translationally invariant system is deduced explicitly by making approximations, that neglect: (1) quantum dynamical effects and (2) considerations of particle indistinguishability (exchange effects). Following this development the classical partition function is used to derive the famous equipartition-of-energy theorem of classical statistical mechanics. The calculation of the partition function for a dilute gas of molecules is then outlined for the case of Boltzmann statistics, in which translational motions are treated classically, whereas molecular internal degrees of freedom may still be analyzed quantum mechanically.

When the temperature of a system is sufficiently high, the partition function for a large translationally invariant system can be written classically as

$$Z = \frac{1}{N!(2\pi\hbar)^{3N}} \int \prod_{\alpha=1}^{N} d^3\mathbf{r}_\alpha \, d^3\mathbf{p}_\alpha \exp[-\beta E(\mathbf{r}_\alpha, \mathbf{p}_\alpha)] \qquad \textbf{(28.1)}$$

To understand this expression qualitatively, we notice first that the trace in Equation (27.24) can be evaluated by using a complete set of single-particle product states. Now each single-particle state corresponds to a μ-space volume $h^3 = (2\pi\hbar)^3$, so that we must divide by N of these factors when we convert the sum over all states to an integral over all Γ space [see also Equation (17.29)]. The factor $(N!)^{-1}$ occurs because of particle indistinguishability, even in the classical limit, as we discussed earlier at the end of Section 4. If the molecules of the system have internal degrees of freedom, then in the classical limit Equation (28.1) must include additional factors

$$\left[h^{-N} \prod_{\alpha=1}^{\dot{N}} dq_\alpha \, dp_\alpha\right]$$

For reasons of simplicity, however, we do not include this possibility in the derivation that follows.

To prove Equation (28.1) we first derive several useful and rigorously correct forms of Equation (27.24) for Z, by using single-particle product states of the type (16.2) to evaluate the trace. Thus we may write

$$\begin{aligned} Z &= \sum_{\{n_i\}} \langle n_1 n_2 \cdots n_i \cdots | e^{-\beta \mathsf{H}} | \mathrm{n}_1 \mathrm{n}_2 \cdots \mathrm{n}_i \cdots \rangle \\ &= \sum_{\mu_1 \leqq \mu_2 \leqq \cdots \leqq \mu_N} \langle \mathrm{n}_1 \mathrm{n}_2 \cdots \mathrm{n}_i \cdots | e^{-\beta \mathsf{H}} | \mathrm{n}_1 \mathrm{n}_2 \cdots \mathrm{n}_i \cdots \rangle \end{aligned} \tag{28.2}$$

where the summation in the first equality is over all possible sets $\{n_i\}$ of n_i values, such that $\Sigma_i n_i = N$. The n_i denote the number of particles in the ith single-particle state, whereas the $\mu_i = \mu_i^{(i)}$ in the second summation refer to the state of the ith particle (with superscripts omitted for convenience). In both equalities we assume that the denumerably infinite set of single-particle states are ordered as described below Equation (16.2). Note that each different N-particle state must be counted only once in the sum over states, which is why we have included the ordering restriction (for identical particles) in the second sum, or trace, over all states.

Now, if we wish to drop the cumbersome ordering restriction in the sum over states, then we can do so provided we divide by the number (16.3) of ways of distributing N particles into the state $\{n_1 n_2 \cdots n_i \cdots\}$. On substituting Equations (16.2), (16.3), and (16.5) we obtain for Z the expression

$$Z = \frac{1}{(N!)^2} \sum_{\mu_1 \cdots \mu_N} \sum_{\mathcal{P}, \mathcal{P}'} \varepsilon^{P+P'} \langle \mu_1^{(1)} \mu_2^{(2)} \cdots \mu_N^{(N)} | \mathrm{U}_P^\dagger \, e^{-\beta \mathsf{H}} \mathrm{U}_{P'} | \mu_1^{(1)} \mu_2^{(2)} \cdots \mu_N^{(N)} \rangle$$

where U_P and $\mathrm{U}_{P'}$ are unitary permutation operators that operate on the superscript (particle) labels, as defined above Equations (16.1). The state vectors $|\mu_1^{(1)} \mu_2^{(2)} \cdots \mu_N^{(N)}\rangle$ are unsymmetrized single-particle product states.

We next observe that $e^{-\beta \mathsf{H}}$ is a symmetric operator in the coordinates of the N particles, since H itself is. Therefore, we can relabel all the $\mathcal{P}$ permutations to the unity permutation, by making corresponding readjustments in the definition of the $\mathscr{P}'$ permutations (as in Exercise 16.1). This process gives $N!$ identical terms, with Z simplifying to

$$Z = \frac{1}{N!} \sum_{\mu_1 \cdots \mu_N} \sum_{\mathscr{P}'} \varepsilon^{P'} \langle \mu_1^{(1)} \mu_2^{(2)} \cdots \mu_N^{(N)} | e^{-\beta \mathsf{H}} U_{P'} | \mu_1^{(1)} \mu_2^{(2)} \cdots \mu_N^{(N)} \rangle$$

$$Z = \frac{1}{N!} \sum_{\mu_1 \cdots \mu_N} \sum_{\mathscr{P}'} \varepsilon^{P'} \langle \mu_1 \mu_2 \cdots \mu_N | e^{-\beta \mathsf{H}} U_{P'} | \mu_1' \mu_2' \cdots \mu_N' \rangle_{\mu_i' = \mu_i} \tag{28.3}$$

In the second equality we have used the fact that we can permute state labels instead of particle labels without changing final results. If, moreover, we agree always to locate the ith particle's state label at the ith location in a bra or ket, then the superscript particle labels can be dropped, as we have done in the second equality.[19] The primes on the states in the ket vector have been introduced to facilitate notation in equations that follow.

With the forms (28.3) for Z we see the appearance, as a consequence of quantum statistics, of the factor $(N!)^{-1}$ in Equation (28.1). In the special case of large translationally invariant systems, for which the μ_i are plane-wave states with no internal degrees of freedom, the second equality of (28.3) becomes

$$Z = \frac{1}{N!} \sum_{\mathbf{k}_1 \cdots \mathbf{k}_N} \sum_{\mathscr{P}'} \varepsilon^{P'} \langle \mathbf{k}_1 \cdots \mathbf{k}_N | e^{-\beta \mathsf{H}} \mathsf{U}_{P'} | \mathbf{k}_1' \cdots \mathbf{k}_N' \rangle_{k_i' = k_i} \tag{28.4}$$

which is still exact.

The Classical Limit

Two approximations are now necessary to reduce Equation (28.4) to the classical form (28.1). For the first approximation we write

$$e^{-\beta \mathsf{H}} \cong e^{-\beta \mathsf{K}} e^{-\beta \mathsf{V}} \tag{28.5}$$

where K and V are the kinetic-energy and potential-energy operators of the system. This approximation is equivalent to assuming that K and V commute, which they do not. In fact, the infinite collection of correction factors that should be included on the right side of (28.5) involves commutators of K and V.[20] These commutators can be neglected if

$$\lambda_T \ll r_o \tag{28.6}$$

where r_o is of the order of the range of the intermolecular forces and λ_T is the thermal wavelength (4.2).[21] If approximation (28.5) is used in Equation (28.4), then we obtain, with $\mathbf{p}_i = \hbar \mathbf{k}_i$,

$$Z \cong \frac{1}{N!} \sum_{\mathbf{p}_1 \cdots \mathbf{p}_N} \exp\left(-\frac{\beta}{2m} \sum_{\alpha=1}^{N} p_\alpha^2\right)$$
$$\times \sum_{\mathscr{P}'} \varepsilon^{P'} \langle \mathbf{p}_1 \cdots \mathbf{p}_N | e^{-\beta \mathrm{V}} \mathrm{U}_{P'} | \mathbf{p}_1' \cdots \mathbf{p}_N' \rangle_{\mathbf{p}_\alpha' = \mathbf{p}_\alpha}$$
$$= \frac{1}{N!} \frac{1}{(2\pi\hbar)^{3N}} \int \left[\prod_{\alpha=1}^{N} d^3\mathbf{p}_\alpha \, d^3\mathbf{r}_\alpha\right] e^{-\beta K} e^{-\beta V}$$
$$\times \sum_{\mathscr{P}'} \varepsilon^{P'} \left[U_{P'} \prod_{\alpha=1}^{N} \exp(i(\mathbf{p}_\alpha' - \mathbf{p}_\alpha) \cdot r_\alpha/\hbar)\right]_{\mathbf{p}_\alpha' = \mathbf{p}_\alpha} \tag{28.7}$$

To derive the second equality in this equation we have used Equation (17.27), without spin considerations, to write the matrix elements of $e^{-\beta \mathrm{V}}$ explicitly in the position representation and have taken the infinite-volume limit with the aid of Equation (17.29). We have also used the classical expression for K,

$$K = \sum_{\alpha=1}^{N} \left(\frac{p_\alpha^2}{2m}\right)$$

The second approximation we make is to drop all of the "exchange" terms in Equation (28.7), keeping only the identity permutation. Equation (28.7) then reduces to Equation (28.1). With regard to the neglected sum over all permutations in Equation (28.7), the partial sum that consists of all binary interchanges is

$$\varepsilon \sum_{\alpha<\beta} \exp[i(\mathbf{p}_\beta - \mathbf{p}_\alpha) \cdot (\mathbf{r}_\alpha - \mathbf{r}_\beta)/\hbar]$$

To compare this quantity, in position space, with the identity permutation we define

$$f(\mathbf{r}_\alpha - \mathbf{r}_\beta) \equiv \frac{\int d^3\mathbf{p} \exp(-\beta p^2/2m) \exp[i\mathbf{p} \cdot (\mathbf{r}_\beta - \mathbf{r}_\alpha)/\hbar]}{\int d^3\mathbf{p} \exp(-\beta p^2/2m)}$$
$$= f(\mathbf{r}_\beta - \mathbf{r}_\alpha) \tag{28.8}$$

If we keep only the correction terms due to binary interchanges, then Equation (28.7) can be written as

$$Z \cong \frac{1}{N!(2\pi\hbar)^{3N}} \int \left[\prod_{\alpha=1}^{N} d^3\mathbf{r}_\alpha \, d^3\mathbf{p}_\alpha\right] e^{-\beta(K+V)} \left[1 + \varepsilon \sum_{\alpha<\beta} f^2(\mathbf{r}_\alpha - \mathbf{r}_\beta)\right] \tag{28.9}$$

Exercise 28.1

(a) Show that

$$f(\mathbf{r} - \mathbf{r}') = \exp\left[-\left(\frac{\pi}{\lambda_T^2}\right)(\mathbf{r} - \mathbf{r}')^2 \right] \tag{28.10}$$

Thus $f(\mathbf{r} - \mathbf{r}')$ vanishes in the high-temperature limit or, more precisely, the exchange terms in Equation (28.9) can be neglected in the temperature-density region where

$$\lambda_T^2 \ll \langle (\mathbf{r}_i - \mathbf{r}_j)^2 \rangle$$

With this argument, we arrive again at the distinguishability condition (16.6),

$$\lambda_T \ll \ell \tag{28.11}$$

where $\ell = n^{-1/3}$ is the average interparticle spacing and $n = N/\Omega$ is the particle density. Conditions (28.6) and (28.11) together define the region of applicability of classical statistical mechanics.

(b) Verify that

$$\langle \mathbf{r} | e^{-\beta K} | \mathbf{r}' \rangle = \lambda_T^{-3} f(\mathbf{r} - \mathbf{r}') \xrightarrow[\beta \to 0]{} \delta^{(3)}(\mathbf{r} - \mathbf{r}') \tag{28.12}$$

This instructive result shows that the operator $e^{-\beta K}$ is not diagonal in the position representation. See also Equation (17.14b).

The function $[1 + \varepsilon \sum_{\alpha<\beta} f^2(\mathbf{r}_\alpha - \mathbf{r}_\beta)]$ in Equation (28.9) can be written as an effective potential-energy factor $\exp(-\beta \tilde{V})$, where

$$\tilde{V} = -\kappa T \ln\left[1 + \varepsilon \sum_{\alpha<\beta} f^2(\mathbf{r}_\alpha - \mathbf{r}_\beta) \right] \cong \sum_{\alpha<\beta} \tilde{V}(\mathbf{r}_\alpha - \mathbf{r}_\beta)$$

$$\tilde{V}(\mathbf{r}) = -\varepsilon \kappa T f^2(\mathbf{r})$$

Thus in the first approximation, when $n\lambda_T^3 \ll 1$, the effect of particle statistics in classical problems is to introduce a "statistical potential energy" $\tilde{V}(\mathbf{r})$, which is attractive for bosons and repulsive for fermions. See Figure VI.3. It must be emphasized, however, that this temperature-dependent potential energy originates solely from the symmetry properties of the many-body wave function and cannot be regarded as a true interparticle potential.

Equipartition Theorem

We consider a classical system of identical molecules, each having s degrees of freedom. The equipartition theorem we investigate now has an implication for such a system, because it gives expressions for the average

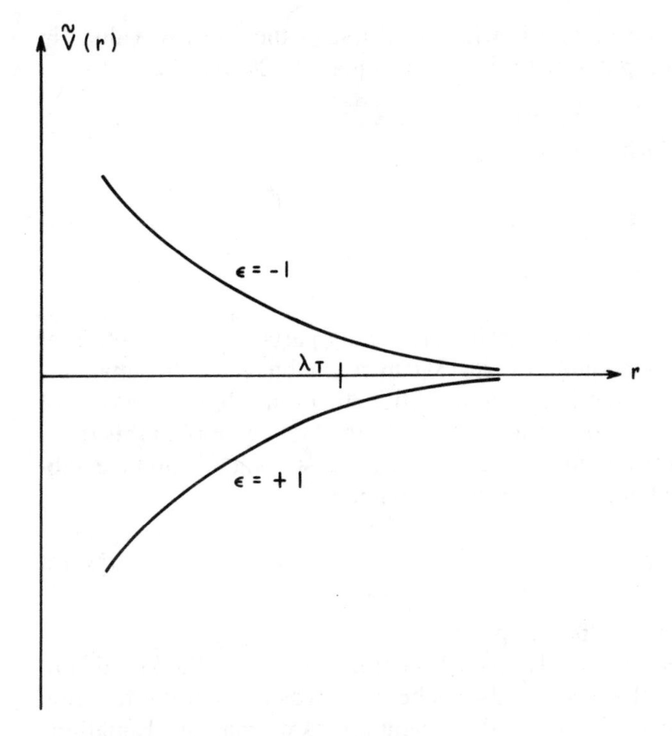

FIGURE VI.3 The statistical potential energy (schematic) between two particles in a dilute gas. For *FD* particles (fermions) $\varepsilon = -1$, and for *BE* particles (bosons) $\varepsilon = +1$.

internal energy and heat capacity. However, the experimental fact that not all degrees of freedom can be thermally excited at ordinary temperatures is in direct violation of this classical result; historically, this contradiction provided a tremendous theoretical challenge to classical statistical mechanics. The resolution of the contradiction using quantum statistics (Sections 29 and 30) was one of the initial triumphs of quantum theory. We consider first the generalized equipartition theorem.

The *generalized equipartition theorem* states that for any canonical coordinates q_i and momenta p_i,

$$\left\langle q_i \frac{\partial H}{\partial q_j} \right\rangle = \left\langle p_i \frac{\partial H}{\partial p_j} \right\rangle = \delta_{ij} \cdot \kappa T \tag{28.13}$$

To prove this theorem we use Equation (28.1) in its most general form

$$Z = e^{-\beta\Psi} = \frac{1}{N!} \int \frac{d\Gamma}{(2\pi\hbar)^{Ns}} \exp[-\beta H(q_i, p_i)] \tag{28.14}$$

where $d\Gamma$ is given by Equation (23.12). The first of the average values in Equation (28.13) is computed with the aid of Equation (26.6) to be

$$\left\langle q_i \frac{\partial H}{\partial q_j} \right\rangle = \frac{1}{Z \cdot N!} \int \frac{d\Gamma}{(2\pi\hbar)^{Ns}} \left(q_i \frac{\partial H}{\partial q_j} \right) e^{-\beta H}$$

$$= -\frac{\kappa T}{Z \cdot N!} \int \frac{d\Gamma}{(2\pi\hbar)^{Ns}} q_i \frac{\partial}{\partial q_j} (e^{-\beta H})$$

$$= \delta_{ij} \cdot \kappa T$$

where the last equality follows after integration by parts and substitution of Equation (28.14). Both inequalities (28.13) can be established in this manner, thereby proving the generalized equipartition theorem. If the second of Hamilton's equations (23.10) is substituted into the first equality of this theorem, with $i = j$, then the *virial theorem* of classical statistics[22] can be established for translational degrees of freedom, namely,

$$-\frac{1}{3} \sum_{\alpha=1}^{N} \langle \mathbf{r} \cdot \dot{\mathbf{p}} \rangle_\alpha = N\kappa T \tag{28.15}$$

where the label α refers to the αth particle.

We have indicated previously, with Equation (23.7), that the kinetic energy K in classical mechanics can always be written as a quadratic function of the $\dot{q}_i$. In the absence of velocity-dependent forces we may use Equations (23.7) and (23.8) to write for the average of K,

$$\langle K \rangle = \frac{1}{2} \sum_{i=1}^{Ns} \langle p_i \dot{q}_i \rangle = \left(\frac{Ns}{2} \right) \kappa T \tag{28.16}$$

Here we have employed the first of Hamilton's equations (23.10) and the generalized equipartition theorem to derive the second equality.

Consider next the case in which the potential energy can be written as a quadratic function of the generalized coordinates,

$$V = \sum_{k,l=1}^{Nt} b_{kl} q_k q_l = \frac{1}{2} \sum_{i=1}^{Nt} \frac{\partial V}{\partial q_i} q_i \tag{28.17}$$

where we have assumed that only $t \leqq s$ of the s molecular coordinates are associated with interaction energies. For example, in a dilute gas we will have $t < s$ when intermolecular interactions can be neglected and, say, t internal coordinates are associated with internal interaction energies. If, further, these t coordinates q_i do not appear in the kinetic-energy function, then we can derive in analogy with Equation (28.16) the result

$$\langle V \rangle = \left(\frac{Nt}{2} \right) \kappa T \tag{28.18}$$

On combining Equations (28.16) and (28.18) we obtain the law of equipartition of energy,

$$\langle E\rangle = \langle K\rangle + \langle V\rangle = N(s + t)\left(\frac{1}{2}\kappa T\right) \tag{28.19}$$

In words, this law states that in a classical system each degree of freedom that contributes to the kinetic energy and each degree of freedom that contributes quadratically to the potential energy, in the form (28.17), is associated with a term $\frac{1}{2}\kappa T$ in the average total energy. The corresponding expression for the heat capacity C_V in a classical system is

$$C_V = \frac{1}{2}N(s + t)\kappa \tag{28.20}$$

which for the classical ideal gas, with $t = 0$ and $s = 3$, reduces to $C_V = \frac{3}{2}N\kappa$. As indicated in Exercise 8.4, this result is one way to establish that κ is indeed Boltzmann's constant. The classical prediction (28.20) is only realized experimentally at very high temperatures, and in the next two sections we use the methods of quantum statistics to show explicitly why this is so for two special cases.

Dilute Gas of Molecules

At the end of Section 16 we defined the case of Boltzmann statistics, to be used in conjunction with classical considerations when the identical particles of a system may be regarded as distinguishable. According to the development in the present section, the applicability of Boltzmann statistics is governed not only by inequality (16.6), or (28.11), but also by (28.6). For a dilute gas of molecules these two conditions may then be combined to give $\lambda_T \ll r_o \ll l$, where now inequality (4.1) has also been included.

We must emphasize that although with Boltzmann statistics the translational motions of a system are treated classically, we are still free to treat any internal molecular dynamics by either classical statistics or quantum statistics. Thus the case of Boltzmann statistics is not completely equivalent to the applications of classical statistical mechanics to a system, using Equation (28.1). We now illustrate this distinction by treating the internal structure of the molecules in a dilute gas quantum mechanically.

For a dilute gas of weakly interacting, that is, essentially noninteracting, molecules the Hamiltonian H can be written as

$$\mathsf{H} = \sum_{\alpha=1}^{N} \mathsf{H}_\alpha$$

For the case of Boltzmann statistics, when $N >>> 1$, Equation (28.3) then becomes

$$
\begin{aligned}
Z_{\text{Boltz}} &= \frac{1}{N!} \sum_{\mu_1 \cdots \mu_N} \langle \mu_1 \cdots \mu_N | \prod_{\alpha=1}^{N} e^{-\beta \mathsf{H}_\alpha} | \mu_1 \cdots \mu_N \rangle \\
&= \frac{1}{N!} \sum_{\mu_1 \cdots \mu_N} \prod_{\alpha=1}^{N} \langle \mu_\alpha | e^{-\beta \mathsf{H}_\alpha} | \mu_\alpha \rangle \\
&\cong \left(\frac{e}{N}\right)^N \prod_{\alpha=1}^{N} \sum_{\mu_\alpha} \langle \mu_\alpha | e^{-\beta \mathsf{H}_\alpha} | \mu_\alpha \rangle \\
&= \sum_{\alpha=1}^{N} Z_\alpha
\end{aligned}
\tag{28.21}
$$

where the single-particle partition function Z_α is defined by

$$
Z_\alpha = \frac{e}{N} \sum_{\mu_\alpha} \langle \mu_\alpha | e^{-\beta \mathsf{H}} | \mu_\alpha \rangle \tag{28.22}
$$

To derive the third equality of Equation (28.21) we have interchanged the sum over all states with the product over all molecules and used Stirling's formula (2.6) for $N!$.[23]

We next write the single-particle Hamiltonian operator as

$$
\mathsf{H} = \frac{p^2}{2m} \mathsf{1} + \mathsf{H}_{\text{int}}
$$

where H_{int} is the Hamiltonian that describes the internal structure of a single molecule. Then the single-particle partition function Z_α takes the form, because the sum over all internal states occurs for each $\mathbf{p}$ value,

$$
Z_\alpha = Z_{\text{tr}} \cdot Z_{\text{int}} \tag{28.23}
$$

Here the (classical) partition function Z_{tr} for the three translational degrees of freedom is given by

$$
\begin{aligned}
Z_{\text{tr}} &= \frac{e}{N} \frac{1}{(2\pi\hbar)^3} \int d^3\mathbf{r}\, d^3\mathbf{p} \exp[-\beta(p^2/2m)] \\
&= \frac{e}{n\lambda_T^3} \qquad (n = N/\Omega)
\end{aligned}
\tag{28.24}
$$

and the partition function Z_{int} for the internal degrees of freedom is

$$
Z_{\text{int}} = \sum_{\mu_{\text{int}}} \langle \mu_{\text{int}} | e^{-\beta \mathsf{H}_{\text{int}}} | \mu_{\text{int}} \rangle \tag{28.25}
$$

With the aid of Equations (28.21), (28.23), and (28.24), we now have for the dilute gas of molecules

$$\ln Z = -\beta\Psi = N(\ln Z_{tr} + \ln Z_{int})$$

$$\ln Z = N\left[\ln\left(\frac{e}{n\lambda_T^3}\right) + \ln Z_{int}\right] \tag{28.26}$$

Exercise 28.2 Derive the ideal-gas equation of state, $\mathcal{P}\Omega = N\kappa T$, by using Equations (27.19) and (28.26). This result shows, as is well known, that the classical ideal-gas law applies to molecules with internal degrees of freedom provided only that intermolecular forces can be neglected.

29 Diatomic Gases

In this section we investigate the calculation of the internal partition function (28.25) for a gas of diatomic molecules, considering first the example of nonidentical atoms within each molecule. For this case contributions due to both vibrational and rotational excitations are written down explicitly, but no effect due to nuclear spin occurs at ordinary temperatures. Both high- and low-temperature limits are studied. The effect of nuclear identity within diatomic molecules is to couple nuclear spin states to molecular rotational states, and the section ends with a general analysis of this effect including application to molecular hydrogen.

Before plunging into the calculations of this section we should note that a large proportion of the chemically stable diatomic molecules have a single lowest electronic level, which carries no angular momentum, and no other electronic level with energy low enough to become appreciably excited below several thousand degrees. We therefore neglect electronic excitations in this section.

Diatomic molecules are characterized by both rotational and vibrational degrees of freedom. It is easily argued[24] that the rotational and vibrational energies for diatomic molecules satisfy the inequalities

$$E_{rot} \ll E_{vib} \ll E_{elec} \tag{29.1}$$

where E_{elec} is of the order of the electronic excitation energy of one of the atoms. As a consequence of (29.1), the molecular Hamiltonian H_{int} can be written in terms of the internuclear distance R, approximately, as

$$\mathsf{H}_{int} = -\frac{\hbar^2}{2\mu}\nabla^2 + U(R) \tag{29.2}$$

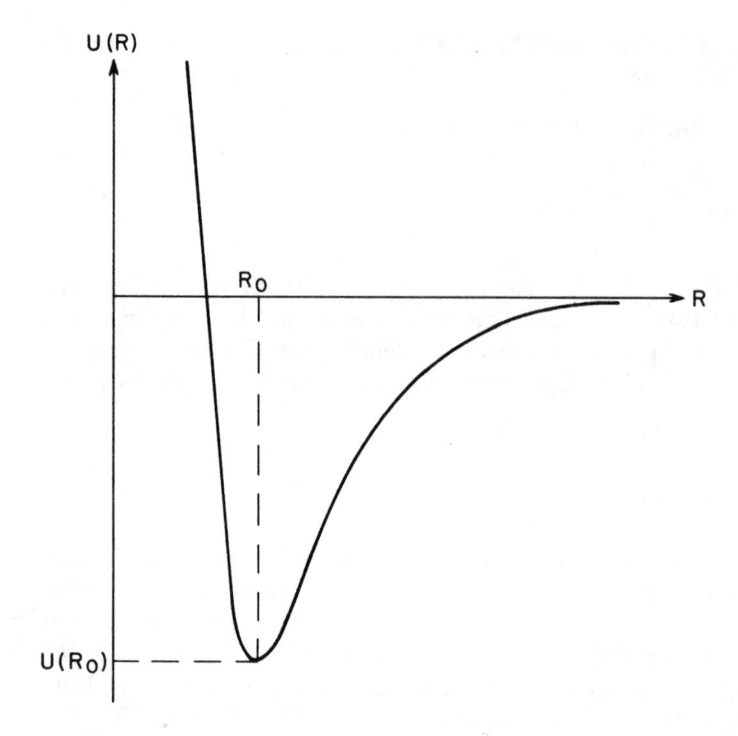

FIGURE VI.4 The form of the internuclear potential energy $U(r)$, with a minimum occurring at $R_o \sim 1$ Å. Typical empirical expressions for $U(R)$ are the Morse potential and the Lennard–Jones potential.[25]

where μ is the reduced mass for the two nuclei and $U(R)$ is an internuclear potential energy. In principle, this potential energy is determined by the electronic states of the atoms, but in practice it is usually fixed empirically (Figure VI.4). The form (29.2) for H_{int} is called the Born–Oppenheimer approximation.

The wave equation formed by using the Hamiltonian (29.2) is separable in spherical coordinates, and the angular solution involves integral quantum numbers K corresponding to molecular rotation with angular momentum squared equal to $K(K + 1)\hbar^2$. The radial equation, for a particular value of K, has an effective potential $W(R)$,

$$W(R) \equiv U(R) + \frac{\hbar^2 K(K + 1)}{2\mu R^2}$$

This potential energy may be expanded about its minimum at $R = R_1$ to give a quadratic, or harmonic oscillator, potential

$$W(R) = W_o + \frac{1}{2}k_o(R - R_1)^2 + \text{(higher-order terms)}$$

$$W_o = -U_o + \frac{\hbar^2 K(K+1)}{2\mu R_o^2} + \cdots$$

where $U(R_o) \equiv -U_o$ and in leading approximation $R_1 \cong R_o$.

To a lowest order of approximation, the energy eigenvalues for a diatomic molecule involve only pure rotational and vibrational modes, without coupling effects, and one finds

$$E_{\text{int}} = -U_o + \frac{\hbar^2 K(K+1)}{2\mu R_o^2} + \hbar\left(\frac{k_o}{\mu}\right)^{1/2}\left(v + \frac{1}{2}\right) + \cdots \tag{29.3}$$

where the vibrational quantum number v takes on positive integral values including zero. Higher-order correction terms to (29.3) can be neglected for temperatures such that $\kappa T < \Theta_v$,[26] where $\kappa^{-1}\Theta_v$, a temperature characteristic of vibrational energies, is defined by

$$\Theta_v \equiv \hbar\left(\frac{k_o}{\mu}\right)^{1/2} \tag{29.4}$$

We also define a temperature $\kappa^{-1}\Theta_r$, which is characteristic of rotational energies, by

$$\Theta_r \equiv \frac{\hbar^2}{2\mu R_o^2} \tag{29.5}$$

Typical empirical values of Θ_r and Θ_v are given in Table VI.1, and these agree with the estimate (29.1).

Table VI.1 Rotational and Vibrational Temperatures for Three Diatomic Molecules*

Molecule	$\kappa^{-1}\Theta_r$	$\kappa^{-1}\Theta_v$
H_2	85.35°	5983°
O_2	2.068°	2239°
I_2	0.0537°	306.9°

*J. E. Mayer and M. G. Mayer, *Statistical Mechanics* (Wiley, 1940), Table AXII. The Mayer-and-Mayer numbers have been revised by using data from the tables of K. P. Huber and G. Herzberg, *Molecular Spectra and Molecular Structure. IV. Constants of Diatomic Molecules* (Van Nostrand Reinhold, 1979).

Calculation of Z_{int}

We now calculate the partition function Z_{int} for molecules with nonidentical atoms using Equation (28.25) and the lowest approximation (29.3) to the internal energy. The internal states are enumerated by summing over all the quantum numbers K and v, so that we may write Z_{int} as

$$Z_{\text{int}} = \sum_{K,v=0}^{\infty} e^{-\beta E_{\text{int}}} = \exp\left[-\beta\left(-U_o + \frac{1}{2}\Theta_v\right)\right] \cdot Z_{\text{rot}} \cdot Z_{\text{vib}} \tag{29.6}$$

where, on using definitions (29.4) and (29.5),

$$Z_{\text{rot}} \equiv g_N \sum_{K=0}^{\infty} (2K+1)\exp[-\beta\Theta_r K(K+1)]$$
$$Z_{\text{vib}} \equiv \sum_{v=0}^{\infty} e^{-\beta\Theta_v \cdot v} \tag{29.7}$$

The factor $(2K + 1)$ in Z_{rot} is the degeneracy, in the absence of a magnetic field, of the rotational levels, and g_N is a degeneracy factor due to nuclear spins.

Let us first evaluate Z_{rot} in the classical limit where $\beta\Theta_r \ll 1$. For this case it is necessary to apply the Euler–Maclaurin summation formula[27] in the form

$$\sum_{n=\alpha}^{\infty} f(n) = \int_a^{\infty} f(n)\,dn + \frac{1}{2}f(a) - \frac{1}{12}f^{(1)}(a) + \frac{1}{720}f^{(3)}(a) + \cdots \tag{29.8}$$

where $f^{(n)}(a)$ is the nth derivative of $f(n)$ evaluated at $n = a$ and we have assumed that $f^{(n)}(\infty) = 0$. Inasmuch as this series (29.8) replaces an infinite sum by an integral with correction terms, it can only be valid where the integral is a good first approximation to the sum. Thus use of this summation formula gives the following expression for Z_{rot}, valid only in the high-temperature, or classical, region.

$$g_N^{-1}Z_{\text{rot}} = (\beta\Theta_r)^{-1}\left[1 + \frac{1}{3}(\beta\Theta_r) + O(\beta\Theta_r)^2\right], \qquad \kappa T \gg \Theta_r \tag{29.9a}$$

We note that the terms in this result are not in one-to-one correspondence with those in Equation (29.8), because in this application each derivative $f^{(n)}(a)$ in the summation formula yields several powers of $(\beta\Theta_r)$.

In the low-temperature region, where $\beta\Theta_r \gg 1$, we obtain directly from the first of Equations (29.7)

$$g_N^{-1}Z_{\text{rot}} = 1 + 3(e^{-\beta\Theta_r})^2 + 5(e^{-\beta\Theta_r})^6 + \cdots, \qquad \kappa T \ll \Theta_r \tag{29.9b}$$

Here we notice that the quantum-mechanical region, where $\kappa T \ll \Theta_r$, is characterized by small probability for excitation of rotational levels. This feature of quantum mechanics, that the level spacing can be greater than κT at low temperatures, explains the lack of agreement between low-temperature observations and the prediction (28.20) of the equipartition theorem. See also Equation (29.13).

Exercise 29.1 Show that the classical "approximation" (28.24) is valid for translational modes in a cubical box of edge length L provided that

$$\Theta_{tr}(n) \equiv \frac{\hbar^2 n^2}{mL^2} \ll \kappa T$$

where n is a quantum number for free-particle states in the box. On choosing $n \sim 10$ and $L \sim 1$ cm, one finds for H_2 molecules that $\kappa^{-1}\Theta_{tr} \sim 10^{-12}K$. Equation (28.24) is indeed correct for a dilute gas whenever the Boltzmann-statistics limit $\lambda_T \ll (\ell, r_o)$ applies.

The partition function Z_{vib} is readily calculated from the second of Equations (29.7) to be

$$\begin{aligned} Z_{vib} &= (1 - e^{-\beta\Theta_v})^{-1} \\ &= 1 + e^{-\beta\Theta_v} + (e^{-\beta\Theta_v})^2 + \cdots \qquad \kappa T \ll \Theta_v \end{aligned} \tag{29.10}$$

We see explicitly from Equations (29.6), (29.9), and (29.10) that the internal partition function makes no contribution to the equation of state, as calculated in Exercise 28.2, because Z_{int} is independent of Ω.

We may now calculate the average energy for a dilute gas of diatomic molecules with nonidentical atoms, by using Equations (27.7), (28.26), (29.6), (29.9), and (29.10). One finds, whenever Boltzmann statistics apply,

$$\frac{\langle E \rangle}{N} = -\frac{\partial}{\partial\beta}\ln\left(\frac{e}{n\lambda_T^3}\right) - \frac{\partial}{\partial\beta}\ln Z_{rot} - \frac{\partial}{\partial\beta}\ln Z_{vib} - U_o + \frac{1}{2}\Theta_v$$

$$\frac{\langle E \rangle}{N} = -U_o + \frac{3}{2}\kappa T + \Theta_v\left[\frac{1}{2} + \frac{1}{e^{\beta\Theta_v} - 1}\right] + E_{rot} \tag{29.11}$$

where

$$\begin{aligned} E_{rot} &= \kappa T\left[1 - \frac{1}{3}(\beta\Theta_r) + O(\beta\Theta_r)^2\right] \qquad \text{if } \Theta_v \gg \kappa T \gg \Theta_r \\ &= 6\Theta_r e^{-2\beta\Theta_r}[1 - 3\,e^{-2\beta\Theta_r} + \cdots] \qquad \text{if } \kappa T \ll \Theta_r \end{aligned} \tag{29.12}$$

As one would expect, the nuclear-spin degeneracy factor g_N does not contribute to the rotational energy E_{rot}. The quantity $-U_o$, which is independent of temperature, is the (electronic) energy of the diatomic molecule at the minimum of the internuclear potential well.

As mentioned below Equation (29.3), correction terms to Equation (29.11) occur when $\kappa T \sim \Theta_v$. If, however, Equation (29.11) is extended arbitrarily into the high-temperature region where $\kappa T \gg \Theta_v \gg \Theta_r$, then the energy per particle becomes

$$\frac{\langle E\rangle}{N} = \frac{7}{2}\kappa T - \left(U_o + \frac{1}{3}\Theta_r\right) + \kappa T\, O(\beta\Theta_v)^2 \cong \frac{7}{2}\kappa T - U_o \tag{29.11a}$$

In this high-temperature limit, the quantum-mechanical zero-point energy $\frac{1}{2}\Theta_v$ does not appear, because of cancellations when the bracket term in the second line of Equation (29.11) is expanded in powers of $\beta\Theta_v$. The result (29.11a) agrees with the law of equipartition of energy (28.19), as one can see by setting $s = 6$ and $t = 1$ for the translation, rotational, and vibrational degrees of freedom that occur for a diatomic molecule. Such a molecule has only two rotational degrees of freedom, because there is no moment of inertia about the internuclear axis.

On using Equations (8.25) and (29.11), the constant-volume heat capacity per particle for a dilute gas of diatomic molecules with nonidentical atoms is found to be

$$\begin{aligned}(N\kappa)^{-1}C_V &= \frac{3}{2} + (\beta\Theta_v)^2 e^{-\beta\Theta_v}(1 - e^{-\beta\Theta_v})^{-2} \\ &\quad + \begin{cases} 1 + O(\beta\Theta_r)^2 & \text{if } \Theta_v \gg \kappa T \gg \Theta_r \\ 12(\beta\Theta_r)^2 e^{-2\beta\Theta_r}(1 - 6e^{-2\beta\Theta_r} + \cdots) & \text{if } \kappa T \ll \Theta_r \end{cases}\end{aligned} \tag{29.13}$$

Figure VI.5 shows a plot of this result for hydrogen gas, and it shows how the classical result (28.20) is approached as the temperature is increased.

Effect of Nuclear Identity

The result (29.13) holds for diatomic molecules in which the two nuclei are different. Because of the possible orientations of the two nuclear spins (S_1 and S_2), the degeneracy factor g_N in the first of Equations (29.7), which has no effect on either the average total energy or the heat capacity, is given by

$$g_N = (2S_1 + 1)(2S_2 + 1)$$

We have assumed that the splitting δE of this degeneracy due to internal and/or external electromagnetic fields is negligible ($\kappa T \gg \delta E$).

Consider the case in which the two nuclei in the diatomic molecule are identical. Then the total nuclear wave function must be symmetric with re-

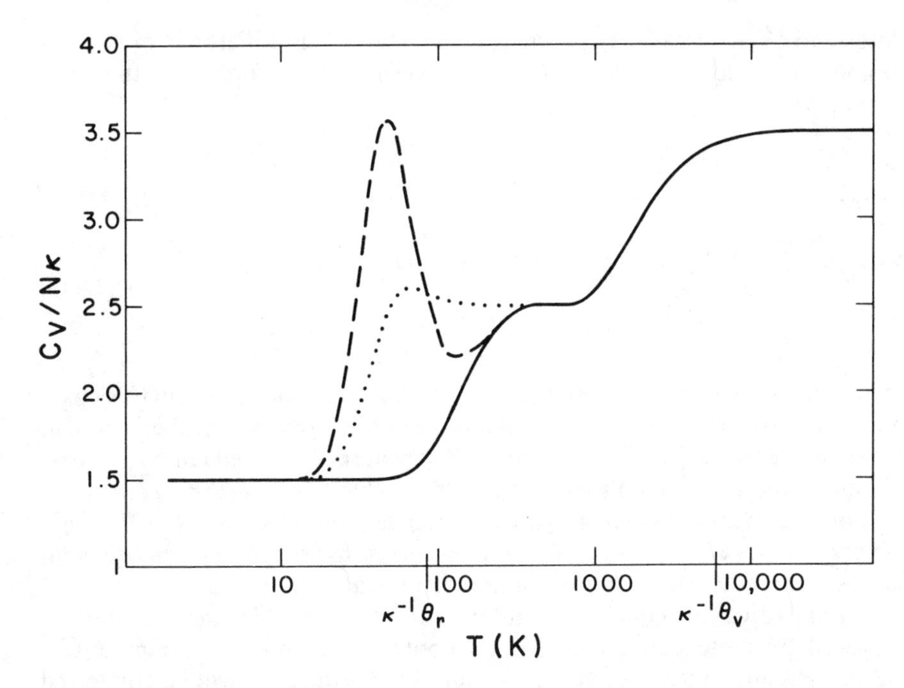

FIGURE VI.5 Specific heat capacity at constant volume for a dilute gas of H_2 molecules. The dotted curve, which is not observed, corresponds to Equation (29.13). The solid-line curve is for Equation (29.19) evaluated using the value $r = 3$. When Equation (29.18) for r is substituted into Equation (29.19), then the dashed curve with the sharp peak at $\sim(1/2)\kappa^{-1}\Theta_r$ results. Not included are correction terms that become important for $\kappa T \gtrsim \Theta_v$.[26]

spect to an interchange of all coordinates (space and spin) for nuclei with zero or integer spin (bosons), or antisymmetric for nuclei with half-odd-integer spin (fermions). Since the vibrational wave function is symmetric, we can satisfy this requirement for each K value by proper choice of the nuclear spin wave functions as shown in Table VI.2.[28]

For two nuclei, each with spin S, there are $S(2S + 1)$ antisymmetric spin

Table VI.2 Allowed Nuclear Spin States for Diatomic Molecules with Identical Nuclei

Rotational States	Nuclei	
	Fermions	Bosons
Even K	Antisymmetric spin states	Symmetric spin states
Odd K	Symmetric spin states	Antisymmetric spin states

states and $(S + 1)(2S + 1)$ symmetric spin states.[29] For diatomic molecules composed of identical Fermi (F) nuclei we must therefore write the rotational partition function as

$$Z_{\rm rot}^{(F)} = S(2S + 1)Z_{\rm even} + (S + 1)(2S + 1)Z_{\rm odd} \tag{29.14}$$

where

$$\begin{aligned} Z_{\rm even} &\equiv \sum_{K=0,2,\ldots} (2K + 1) \exp[-\beta\Theta_r K(K + 1)] \\ Z_{\rm odd} &\equiv \sum_{K=1,3,\ldots} (2K + 1) \exp[-\beta\Theta_r K(K + 1)] \end{aligned} \tag{29.15}$$

The use of an overall factor g_N in $Z_{\rm rot}^{(F)}$, as in (29.7), is not correct in this case. We note, however, for the high-temperature limit $\beta\Theta_r \ll 1$, that one can prove the result, $Z_{\rm odd} = Z_{\rm even} = \frac{1}{2}g_N^{-1}Z_{\rm rot}$, where $Z_{\rm rot}$ is defined by the first of Equations (29.7). See Problem VI.3. In this limit, $Z_{\rm rot}^{(F)} = \frac{1}{2}Z_{\rm rot}$.

Consider next the case of hydrogen (H_2) gas, for which $S = \frac{1}{2}$. For high temperatures ($\kappa T \gg \Theta_r$), the ratio of ortho-hydrogen (symmetric spin states) to para-hydrogen (antisymmetric spin states) will be 3 : 1, because a purely statistical, or equal, distribution into the four different spin states is achieved by molecular collisions. By contrast, at very low temperatures ($\kappa T \ll \Theta_r$) and in thermodynamic equilibrium, all the gas will be converted to para-hydrogen, which has the energetically lowest rotational state. Equilibrium at low temperatures is not easily achieved experimentally, however, because the conversion of ortho-hydrogen to para-hydrogen proceeds rapidly only in the presence of a catalyst.

With the aid of Equation (29.14) and the first equality of Equation (29.11) one can derive a general expression for the rotational energy of H_2 gas. The result is

$$E_{\rm rot}^{(H)} = (1 + r)^{-1}[E_{\rm para} + \Gamma E_{\rm ortho}] \tag{29.16}$$

where

$$\begin{aligned} E_{\rm para} &= -N\frac{\partial(\ln Z_{\rm even})}{\partial\beta} \\ E_{\rm ortho} &= -N\frac{\partial(\ln Z_{\rm odd})}{\partial\beta} \end{aligned} \tag{29.17}$$

The quantity r is defined to be the ratio of the number of ortho-hydrogen molecules to the number of para-hydrogen molecules, which for the equilibrium situation is given by

$$r = \frac{3Z_{\rm odd}}{Z_{\rm even}} = \begin{cases} 3 & \text{if } \kappa T \gg \Theta_r \\ 0 & \text{if } \kappa T \ll \Theta_r \end{cases} \tag{29.18}$$

One can argue independently that Equation (29.18) for r is correct by using the first of Equations (27.5).

An expression for the constant-volume heat capacity of H_2 gas can be derived from Equation (29.16). One finds

$$(N\kappa)^{-1}C_V = \frac{3}{2} + (\beta\Theta_V)^2 e^{-\beta\Theta_V}(1 - e^{-\beta\Theta_V})^{-2}$$
$$+ (1 + r)^{-1}\left[\left(\frac{C_V}{N\kappa}\right)_{\text{para}} + r\left(\frac{C_V}{N\kappa}\right)_{\text{ortho}}\right] \tag{29.19}$$
$$+ (1 + r)^{-2}\left[\left(\frac{E}{N}\right)_{\text{ortho}} - \left(\frac{E}{N}\right)_{\text{para}}\right]\frac{dr}{d(\kappa T)}$$

The calculated curve for C_V, using Equations (29.18) and (29.19), is shown as a dashed line in Figure VI.5. The sharp peak corresponds to the onset of conversion of para-hydrogen to ortho-hydrogen when the temperature is raised to the point where occupation of the $K = 1$ rotational state can occur appreciably (in the presence of a catalyst). This peak is due to the last term in Equation (29.19). For the more general experimental situation in which complete equilibrium is not achieved r must be treated as an empirical parameter in Equation (29.19). When the temperature of ordinary H_2 gas is lowered in the absence of a catalyst, then the high-temperature spin-state ratio $r = 3$ remains constant. The solid-line curve in Figure VI.5 corresponds to this more common experimental situation in which ortho-hydrogen and para-hydrogen can be treated as two different, noninteracting gases.[30]

30 Heat Capacity of Simple Crystals

The simplest crystalline structure, with atoms at roughly equal spacings throughout, is considered. Quantum-mechanical energy levels corresponding to noninteracting normal coordinates are used to write down the partition function for such crystals. The average energy and heat capacity are studied in both the classical and quantum-mechanical limits. For the latter case calculations are performed using the Debye model to represent quantized normal-mode, or phonon, states.

We consider a stable crystal with N atoms (or ions) such that the αth atom with coordinate $\mathbf{r}_\alpha$, has a potential-energy function

$$U(\mathbf{r}_\alpha) \equiv \sum_{\substack{\beta=1 \\ (\beta\neq\alpha)}}^{N} V(\mathbf{r}_\alpha - \mathbf{r}_\beta) \tag{30.1}$$

To denote the equilibrium position we define $\mathbf{r}_\alpha \equiv \mathbf{R}_\alpha$, which implies $\nabla U(\mathbf{R}_\alpha) = 0$, so that for small displacements from equilibrium the total potential energy V can be written as a series expansion.

$$\begin{aligned} V &= \frac{1}{2}\sum_{\alpha=1}^{N} U(\mathbf{r}_\alpha) \\ &= \frac{1}{2}\sum_{\alpha=1}^{N} U(\mathbf{R}_\alpha) + \frac{1}{4}\sum_{\substack{\alpha,\beta=1 \\ (\alpha\neq\beta)}}^{N} \left[\frac{\partial^2}{\partial \mathbf{R}_\alpha\, \partial \mathbf{R}_\beta} V(\mathbf{R}_\alpha - \mathbf{R}_\beta)\right]\cdot \boldsymbol{\xi}_\alpha \boldsymbol{\xi}_\beta \\ &\quad + \frac{1}{4}\sum_{\alpha=1}^{N}\sum_{i,j=i}^{3}\left[\frac{\partial^2}{\partial R_{\alpha,i}\, \partial R_{\alpha,j}} U(\mathbf{R}_\alpha)\right]\xi_{\alpha,i}\xi_{\alpha,j} + \cdots \end{aligned} \tag{30.2}$$

The vector $\boldsymbol{\xi}_\alpha \equiv \mathbf{r}_\alpha - \mathbf{R}_\alpha$ is the displacement of the αth atom from its equilibrium position, and the factor of 1/2 in the first equality is required because every pair of particles has only one interaction. On choosing the first term to be zero, thereby defining the zero of energy, we see that the Hamiltonian of the crystal has the form

$$H(\xi, \pi) \cong \sum_{i=1}^{3N} \frac{\pi_i^2}{2m_i} + \frac{1}{2}\sum_{i,j=1}^{3N} a_{ij}\xi_i\xi_j \tag{30.3}$$

where m_i is the mass of the atom having the coordinate ξ_i and corresponding momentum π_i.

To study the thermodynamic properties of a crystal it is useful to transform the Hamiltonian to *normal coordinates* $\{q_i, p_i\}$. Thus there exists a transformation under which the potential-energy function is transformed to a sum of squares,

$$H(q, p) \cong \frac{1}{2}\sum_{i=1}^{3N} (p_i^2 + \omega_i^2 q_i^2) \tag{30.4}$$

with the kinetic energy remaining diagonal.[31] Note, in Equation (30.4), that the normal coordinates and momenta have been normalized so that the masses in the kinetic-energy term are all unity. In the normal form the Hamiltonian is expressed as a sum over $3N$ independent oscillators, and we now want to determine the equilibrium properties of these oscillators. Actually, these oscillators are not independent, for they interact via the neglected third-order terms in the expansion (30.2) of the potential-energy function. It is these "interaction terms" that result in the establishment of thermodynamic equilibrium in a crystal in the same sense that collisions lead to equilibrium in a gas.

A quantum-mechanical calculation of the energy levels of the crystal, using Equation (30.4) with higher-order terms neglected, results in the expression

$$E(n_1, n_2, \ldots, n_i \ldots) = \sum_{i=1}^{3N-6} \left(n_i + \frac{1}{2}\right)\hbar\omega_i \qquad \textbf{(30.5)}$$

where the n_i can be zero or any positive integer value. Since six of the normal coordinates must correspond to translation and rotation of the crystal as a whole, with corresponding zero frequencies, the summation in Equation (30.5) extends only over $(3N - 6)$ coordinates.

Calculation of Heat Capacity

At ordinary temperatures where correction terms to Equation (30.3) can be neglected, we have for the partition function (28.2)

$$\begin{aligned} Z &= \sum_{\{n_i\}} \prod_{i=1}^{3N-6} \exp\left[-\beta\left(n_i + \frac{1}{2}\right)\hbar\omega_i\right] \\ &= \prod_{i=1}^{3(N-2)} \sum_{n_i=0}^{\infty} \exp\left[-\beta\left(n_i + \frac{1}{2}\right)\hbar\omega_i\right] \end{aligned} \qquad (30.6)$$

The summation in the first line is over all possible sets of n_i values, with all possible n_i values allowed for each i, according to Equation (30.5). After interchanging this summation with the product to obtain the second line of Equation (30.6), the resulting sums are easy to perform. We derive for $\ln Z$, when $N >> 2$,

$$\ln Z = -\sum_{i=1}^{3N} \left[\frac{1}{2}\beta\hbar\omega_i + \ln\left(1 - e^{-\beta\hbar\omega_i}\right)\right] \qquad \textbf{(30.7)}$$

In this expression the ω_i may be functions of the crystal volume and other external (mechanical) parameters, but they do not depend on the temperature.

To calculate the average total energy we substitute Equation (30.7) into the second of Equations (27.7) and obtain

$$\langle E \rangle = \sum_{i=1}^{3N} \left[\frac{1}{2}\hbar\omega_i + \hbar\omega_i(e^{\beta\hbar\omega_i} - 1)^{-1}\right] \qquad \textbf{(30.8)}$$

Comparison of this result with Equation (30.5) shows that the average number $\langle n_i \rangle$ of quanta in the ith mode is given by

$$\langle n_i \rangle = [e^{\beta\hbar\omega_i} - 1]^{-1} \qquad \textbf{(30.9)}$$

a result we discuss below in connection with the Debye model. Expansion of Equation (30.8) in the low- and high-temperature limits yields

$$\begin{aligned}\langle E\rangle &= \sum_{i=1}^{3N} \hbar\omega_i\left[\frac{1}{2} + e^{-\beta\hbar\omega_i} + O(e^{-\beta\hbar\omega_i})^2\right] \qquad \text{if } \kappa T \ll h\tilde{\omega} \\ &= 3N\kappa T + \frac{\beta}{12}\sum_{i=1}^{3N}(\hbar\omega_i)^2 + \cdots \qquad \text{if } \kappa T \gg h\tilde{\omega}\end{aligned} \tag{30.10}$$

where $\tilde{\omega}$ is a mean frequency among the ω_i. In the low-temperature limit the zero-point vibrations give the dominant contribution to the crystal energy. In the high-temperature limit the zero-point vibrations are canceled, as they also were in Equation (29.11a). In this limit the law of equipartition of energy (28.19) is satisfied with $s = t = 3$, where according to Equation (30.4) both kinetic and potential energies depend quadratically on their respective coordinates.

For the heat capacity at constant volume we readily derive from Equation (30.8) the expression

$$\begin{aligned}C_V &= \kappa \sum_{i=1}^{3N} (\beta\hbar\omega_i)^2\, e^{-\beta\hbar\omega_i}(1 - e^{-\beta\hbar\omega_i})^{-2} \\ &= \kappa \sum_{i=1}^{3N} (\beta\hbar\omega_i)^2\, e^{-\beta\hbar\omega_i}[1 + O(e^{-\beta\hbar\omega_i})] \qquad \text{if } \kappa T \ll \hbar\tilde{\omega} \\ &= 3N\kappa - \frac{\kappa}{12}\sum_{i=1}^{3N}(\beta\hbar\omega_i)^2 + \cdots \qquad \text{if } \kappa T \gg \hbar\tilde{\omega}\end{aligned} \tag{30.11}$$

The fact that the heat capacity at high temperatures equals $3N\kappa$ is the famous law of Dulong and Petit for monatomic crystals.[32] Moreover, the heat capacity is known to decrease from $3N\kappa$ with decreasing temperature, in agreement with the third equality of Equation (30.11). At very low temperatures the heat capacity is known to vary as T^3, but this result is not obvious from the second equality of Equation (30.11), and it requires the further analysis given below. Actually, the quantity that is measured is C_P, and not C_V. The small difference between these two quantities is given by Equation (9.10).

The Einstein Model

The above results are deceptive for one reason; namely, the normal-mode frequencies ω_i are unknown quantities unless the lengthy diagonalization procedure that ends with Equation (30.4) is carried through explicitly for a given crystal. Fortunately, such calculations can be bypassed in first approximation by introducing suitable crystal models whose validity may be judged by the accuracy with which they produce agreement with experimental results.

For an initial orientation toward understanding the low-temperature behavior of a crystal, we assume that all the $3N$ frequencies are the same. This model was proposed by Einstein in 1907, and it would be a reasonable model if, say, we were allowed to consider each of the N atoms as vibrating independently in the crystal. In this case Equations (30.8) and (30.11), for $\langle E \rangle$ and C_V, would become

$$\langle E \rangle = 3N\hbar\omega\left(\frac{1}{2} + \frac{1}{e^{\beta\hbar\omega} - 1}\right)$$

$$C_V = 3N\kappa(\beta\hbar\omega)^2\, e^{-\beta\hbar\omega}(1 - e^{-\beta\hbar\omega})^{-2}$$

The Einstein model produced, historically, the first good understanding of room-temperature deviations from the law of Dulong and Petit. It fails to give an understanding of the T^3 dependence of C_V at low temperatures.

The Debye Model

In the Debye model of a crystal it is assumed that oscillations of the classical normal modes correspond quantum-mechanically to particles. As in the Landau theory of a Fermi liquid (Section 19) we call these quantized oscillations quasi-particles, because they are not true particles that could exist separately from the crystal. Quasi-particles are, however, "particle-like," with oscillations of the ith normal mode corresponding to quasi-particles of energy $\hbar\omega_i$.

In a crystal we give the special name *phonon* to a quasi-particle. Phonons are assumed to be spinless and to satisfy Bose–Einstein statistics. Thus Equation (5.10) for the occupation numbers in an ideal Bose gas, with $g = 0$, is in agreement with Equation (30.9) for the $\langle n_i \rangle$. As we clarify in the following two chapters, the chemical potential g is zero for phonons, because their total number is not constrained.

The Debye model assumes that the density of phonon states in a crystal can be calculated in the same way as free-particle states are enumerated in a large box to arrive at Equation (17.28). If, further, we consider only monatomic crystals that are almost isotropic in their properties, then we can refer to the number dn of states in a spherical shell of radius k and thickness dk in wave-number space. From Equation (17.28) we obtain

$$dn = \frac{\Omega}{(2\pi)^3} 4\pi k^2 dk \tag{30.12}$$

We can convert Equation (30.12) to a number of states in the frequency interval $d\omega$ if we know the energy-momentum relation for phonons. We assume, as for light quanta,

$$\omega = kc \tag{30.13}$$

where c is the velocity of phonons in the crystal. [In general, $c = c(k)$, but here we assume $c =$ constant.] On combining Equations (30.12) and (30.13) we derive for the density of phonon states in a crystal,

$$\rho(\omega) \equiv \frac{dn}{d\omega} = \frac{\Omega}{2\pi^2} \cdot \frac{\omega^2}{c^3}$$

Finally, we improve this last expression by observing that for each wave number k there can be two transverse modes, with velocities c_T, and one longitudinal mode, with velocity c_L. Then the density of states in the simplest version of the Debye model is given by

$$\rho(\omega) = \frac{\Omega}{2\pi^2}\,\omega^2\left(\frac{2}{c_T^3} + \frac{1}{c_L^3}\right) \tag{30.14}$$

where, in general, c_T and c_L are not equal.

The total number of phonon states in a monatomic crystal is $3(N - 2) \cong 3N$. Therefore in the Debye model there must be a maximum frequency ω_m, which is determined by the equation

$$3N = \int_0^{\omega_m} d\omega\,\rho(\omega) = \frac{\Omega}{6\pi^2}\left(\frac{2}{c_T^3} + \frac{1}{c_L^3}\right)\omega_m^3 \tag{30.15}$$

This result can be substituted back into Equation (30.14) to give the simpler form for $\rho(\omega)$,

$$\rho(\omega) = 9N\frac{\omega^2}{\omega_m^3} \tag{30.16}$$

Note that Equation (30.15) is not an equation for the total number of phonons; rather, it is an equation that enumerates phonon states.

The cutoff frequency ω_m corresponds to a minimum wavelength λ_m, defined for each velocity mode by

$$\lambda_m \equiv \frac{1}{k_m} = \frac{c}{\omega_m} \tag{30.17}$$

On physical grounds we can expect that there should be a minimum wavelength $\lambda_m \sim a$, where a is the interatomic spacing in the crystal, because acoustic waves with $\lambda < a$ certainly cannot propagate in a crystal. (The so-called optical modes can exist with $\lambda < a$, but these do not occur for simple crystals.) We therefore arrive at the estimate $c \sim \omega_m a$ for the phonon velocities c_T and c_L, where $c_T \sim c_L$.

We now substitute Equation (30.16) into Equation (30.8) to derive the Debye-model expression for $\langle E\rangle$:

$$\begin{aligned}\langle E\rangle &= \int_0^{\omega_m} d\omega\, \rho(\omega)(\hbar\omega)\left[\frac{1}{2} + (e^{\beta\hbar\omega} - 1)^{-1}\right] \\ &= \frac{9N\kappa T}{u^3}\int_0^u x^3\, dx\left(\frac{1}{2} + \frac{1}{e^x - 1}\right)\end{aligned} \tag{30.18}$$

where $u = \beta\hbar\omega_m = T_D/T$ and the *Debye temperature* T_D is defined by

$$T_D \equiv \frac{\hbar\omega_m}{\kappa} \tag{30.19}$$

For many substances $T_D \sim 200K$, which implies that $\hbar\omega_m \sim 1/60$ eV.

Exercise 30.1

(a) Use Equations (30.17) and (30.19), with $\lambda_m \sim 1$ Å, to estimate typical phonon velocities as

$$c \sim 3 \times 10^5 \text{ cm/sec}$$

This velocity is close to that measured for elastic waves in solids, which suggests correctly that phonons are intimately connected with sound propagation.

(b) With other crystal parameters held constant, argue that

$$T_D \propto a^{-1} \qquad (a = \text{lattice spacing})$$

$$T_D \propto M^{-1/2} \qquad (M = \text{atomic weight})$$

We next introduce a quantity $D(u)$, called the *Debye function*, by the equation

$$\begin{aligned}D(u) &\equiv \frac{3}{u^3}\int_0^u x^3\, dx\left(\frac{1}{e^x - 1}\right) \\ &\xrightarrow[u \ll 1]{} 1 - \frac{3}{8}u + \frac{1}{20}u^2 + O(u^3) \\ &\xrightarrow[u \gg 1]{} \frac{\pi^4}{5u^3} - 3\left[1 + O\left(\frac{1}{u}\right)\right]e^{-u} + O(e^{-2u})\end{aligned} \tag{30.20}$$

where the third line can be derived by writing $D(u)$ as

$$\frac{u^3}{3}D(u) = \lim_{\varepsilon\to 0+}\int_\varepsilon^\infty x^3\,dx\left(\frac{e^{-x}}{1-e^{-x}}\right) - \int_u^\infty x^3\,dx\left(\frac{e^{-x}}{1-e^{-x}}\right)$$
$$= \lim_{\varepsilon\to 0+}\sum_{n=1}^{\infty}\left[\int_\varepsilon^\infty x^3\,dx\,e^{-nx} - \int_u^\infty x^3\,dx\,e^{-nx}\right]$$

After integration, the first term in this last expression is well defined in the limit $\varepsilon \to 0+$. The resulting sum is $6 \cdot \zeta(4) = \pi^4/15$, where $\zeta(n)$ is the Riemann zeta function. The Debye function is plotted in Figure VI.6 as a function of $u^{-1} = T/T_D$.

In terms of the Debye function, the energy expression (30.18) becomes

$$\frac{\langle E\rangle}{N} = \frac{9}{8}\kappa T_D + 3\kappa T\cdot D\left(\frac{T_D}{T}\right)$$
$$\xrightarrow[T\ll T_D]{} \frac{9}{8}\kappa T_D\left[1 + \frac{8}{15}\pi^4\left(\frac{T}{T_D}\right)^4 + O\left(\frac{T}{T_D}e^{-T_D/T}\right)\right] \tag{30.21}$$

in which the first term is the zero-point energy. The T^4 dependence of the energy in the low-temperature limit is characteristic for a gas of massless bosons. Another example is the Stefan–Boltzmann law (35.8) for blackbody radiation energy.

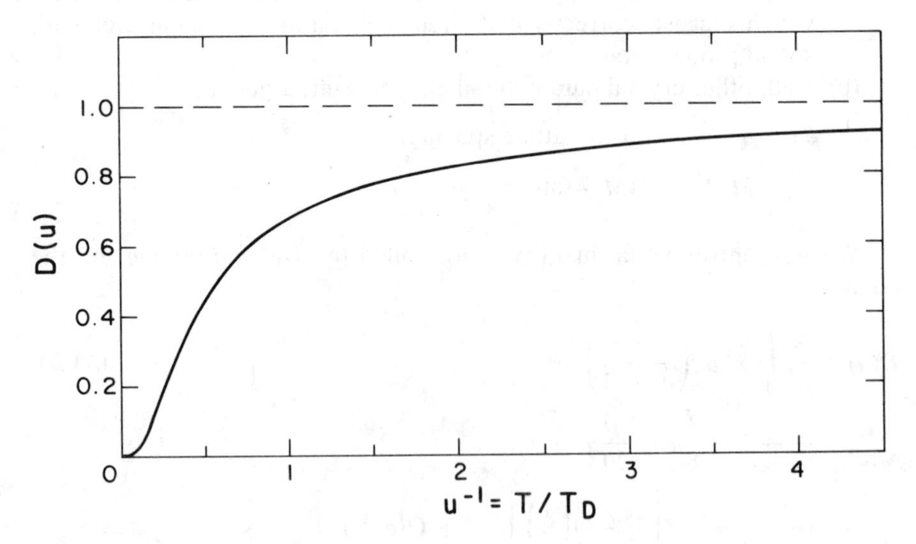

FIGURE VI.6 The Debye function.[33]

Exercise 30.2 Show that the heat capacity expression (30.11) can be written, using the Debye model, as

$$\frac{C_V}{3N\kappa} = \left[D(u) - u\frac{d}{du}D(u)\right]_{u=T_D/T}$$

$$\frac{C_V}{3N\kappa} = 4D\left(\frac{T_D}{T}\right) - \frac{3T_D/T}{e^{T_D/T} - 1} \tag{30.22}$$

Figure VI.7 shows this result graphically.

By using the last line of Equation (30.20) the low-temperature limit of Equation (30.22) can be deduced. One readily finds the famous result of the Debye theory,

$$C_V = \frac{12\pi^4}{5}N\kappa\left[\left(\frac{T}{T_D}\right)^3 + O\left(\frac{T_D}{T}e^{-T_D/T}\right)\right] \tag{30.23}$$

which agrees with observations on a wide variety of monatomic crystals. However, a detailed examination of measurements, when Equation (30.22) is fitted to the data, reveals ~10% deviations in T_D as the temperature is varied. For simple crystals such deviations have been explained by calculations of $\rho(\omega)$, based on microscopic properties. Thus NaCl, which is not even monatomic, was investigated by Kellerman,[34] who found that the ω^2 dependence in Equation (30.16) represents only a rough average of the true

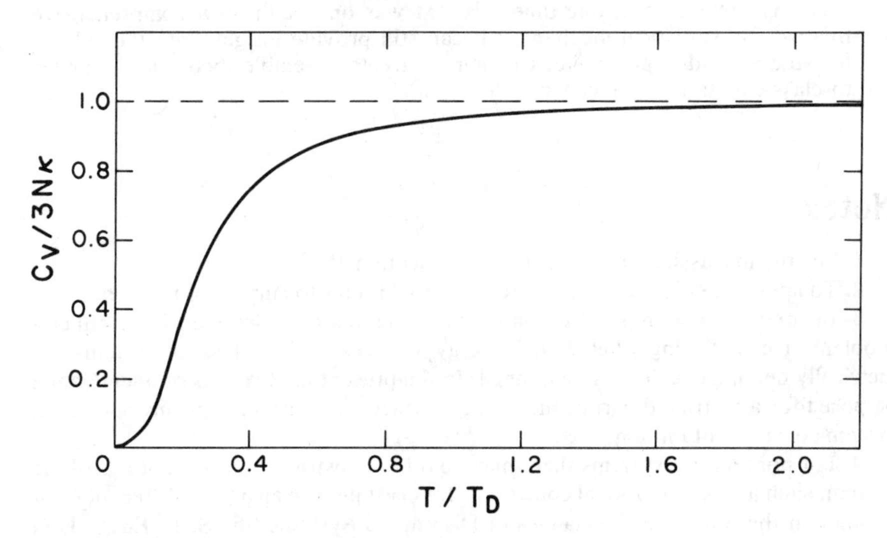

FIGURE VI.7 Plot of the specific heat capacity of a monatomic crystal according to the Debye model.

frequency variation of $\rho(\omega)$. His theoretical calculations of an effective T_D to be used in Equation (30.22) are in excellent agreement with experiment.

Bibliography

D. ter Haar, *Elements of Thermostatics* (Holt, Rinehart and Winston, 1966), Chapters 5 and 6. Excellent discussion is given of ensemble theory, both in classical and quantum physics. The theory of diatomic gases is covered succinctly in Sections 2.4, 3.4, and 3.5.

K. Huang, *Statistical Mechanics* (Wiley, 1967), Chapter 7. A fine treatment of ensemble theory in classical statistical mechanics. The classical limit of the macrocanonical partition function of quantum statistical mechanics is derived in Section 10.2. The Debye model of crystals is treated in Section 12.2.

L. D. Landau and E. M. Lifshitz, *Statistical Physics* (Addison-Wesley, 1958), translated by E. Peierls and R. F. Peierls. Many excellent sections of this text have been used as background material for our Chapter VI. In particular, the reader is referred to Sections 4 and 28 and to relevant sections in Chapters IV and VI.

J. E. Mayer and M. G. Mayer, *Statistical Mechanics* (Wiley, 1940), Chapters 7 and 11. Application of macrocanonical ensemble theory to diatomic gases and to crystals is given with considerable detail and discussion.

E. Schrödinger, *Statistical Thermodynamics* (Cambridge, 1967), Chapter II. A succinct and extremely readable presentation of the method of the most probable distribution.

R. C. Tolman, *The Principles of Statistical Mechanics* (Oxford University Press, 1962), Chapter III. At one time this text was one of the few comprehensive treatises on statistical mechanics; it can still provide insights into the subject for student and expert alike. Chapter III treats ensemble theory as it applies to classical statistical mechanics.

Notes

1. See the discussion of ergodic theory in Section 25.

2. To appreciate this remark, the reader should refer to part (b) of Problem II.7.

3. In connection with our discussion of ergodic theory in Section 25, the difficult problem of establishing whether the energy surfaces of real classical systems are metrically decomposable was mentioned. In the present context this problem would be posed for a restricted part of the energy surface as determined by any additional system constants of motion.

4. For nonuniform systems there may be other constraints on the motions of the system, such as configurational constraints for crystals. See also "The Effect of Constraints on the Statistical Mechanics of Disordered Systems" by S. F. Edwards in *Proceedings of the International Conference on Statistical Mechanics*, Kyoto, 1968 (*J. Phys. Soc. Japan* **26**, supp.).

5. K. Huang, *op. cit.*, Section 7.5.

6. Interactions between subsystems occur, essentially, only between molecules near common surfaces. The interaction energies can therefore be classified as surface energies, which are negligible for sufficiently large macroscopic subsystems. When particle indistinguishability between subsystems is important, then the present treatment must be generalized conceptually to the ensemble treatment of Section 27.

7. In Section 27 we show that Ψ is the Helmholtz free energy of any subsystem.

8. J. W. Gibbs, *Elementary Principles in Statistical Mechanics* (Dover, 1960), Chapter IV.

9. The grand canonical ensemble corresponds in thermodynamic calculations to use of the energy function $\Omega f(T, \Omega, g)$ of Equation (8.22).

10. Exceptions to this general conclusion do occur, as is discussed at the end of Section 27.

11. When particle interactions or exchange effects due to particle indistinguishability are significant, then one finds that

$$\left\langle \sum_{\substack{\alpha,\beta=1 \\ (\alpha\neq\beta)}}^{N} (\Delta a)_\alpha (\Delta a)_\beta \right\rangle = O(N)$$

a result that does not alter conclusion (26.10). See, for example, Problem V.3(a) and Equation (47.21).

12. This result is familiar in probability theory. See, for example, Equations (1.8) and (3.15).

13. In Appendix A the Darwin–Fowler method of mean values is used to show, when $M \ggg 1$, that ensemble average occupation numbers $\langle N_i \rangle$ are equal to most probable occupation numbers $\overline{N}_i$. It then follows, as is shown explicitly in Appendix A, that system average values are independent of which of these two methods has been applied to the ensemble analysis.

14. Recall the discussion in the first part of Section 7, through Equation (7.5).

15. Note that we must consider here infinitesimal work done in a reversible isothermal process, rather than in the adiabatic process of Section 7, because of use of the independent variable T. See the discussion below Equation (7.5).

16. The following proof is taken from E. Schrödinger, *op. cit.* pp. 12 and 13.

17. Further discussion of energy fluctuations is given in Section 8.2 of the text by Huang.

18. See K. Huang, *loc. cit.*, Chapter 15, and also *Phase Transitions and Critical Phenomena Volume 1, Exact Results,* edited by C. Domb and M. S. Green (Academic Press, 1972).

19. Recall that adherence to the particle-state ordering convention of Section 16 was required for the development at the beginning of Section 17. The use of the Fock-space notation of Section 17 would be quite inappropriate for the present classical considerations.

20. A complete expression that gives these correction factors is due to Hausdorff. With $\mathsf{A} = \beta\mathsf{K}$ and $\mathsf{B} = -\beta\mathsf{H}$, Hausdorff's theorem takes the form

$$e^{\mathsf{A}} e^{\mathsf{B}} = e^{\mathsf{A}+\mathsf{B}+\mathsf{C}}$$

where

$$C = \frac{1}{2}[A, B] + \frac{1}{12}([[A, B], B] + [[B, A], A]) + \cdots$$

The original derivation of this theorem is in F. Hausdorff, "Akademieder Wissenschaften," Leipzig, *Mathematisch-Physiche Kasse, Berichte* **58** (1906), p. 19. A contemporary derivation is provided by R. M. Wilcox, *J. Math. Phys.* **8** (1967), p. 962.

21. The first term in the expression for C (footnote 20) can be written as

$$-\frac{1}{2}\beta^2[K, V] = \frac{1}{8\pi}\lambda_T^2\beta \cdot 1 \sum_{\alpha=1}^{N} [\nabla_\alpha^2 V + 2(\nabla_\alpha V \cdot \nabla_\alpha)]$$

It can then be argued that, when this expression is inserted properly into Equation (28.4) and the matrix element evaluated by integrating over position space, effectively this commutator is $\sim\beta V(\lambda_T/r_o)$. See K. Huang, *loc. cit.*, pp. 218–219.

22. The term *virial* was introduced by Clausius in 1870, essentially for the quantity $-\Sigma_{\alpha=1}^{N}\langle \mathbf{r} \cdot \dot{\mathbf{p}}\rangle_\alpha$, which he used in calculations of the pressure of a gas. The virial theorem was then used in kinetic-theory calculations of the equation of state for an imperfect gas. One such result is the virial expansion of the equation of state (Section 38). See also H. Goldstein, *Classical Mechanics* (Addison-Wesley, 1980), Section 3-4.

23. Inasmuch as the factor $(2\pi N)^{-1/2}$ can be written as $[\exp(-(1/2N)\ln 2\pi N)]^N$, we may neglect its contribution to Z_α.

24. L. I. Schiff, *Quantum Mechanics*, 3rd ed. (McGraw-Hill, 1968), Sec. 49. Correction terms to Equation (29.3) are derived in this reference.

25. The Morse *internuclear* potential, for the short-range region $R \sim R_o$, has the form

$$U(r) = U_o[\exp(-2a(R - R_o)) - 2\exp(-a(R - R_o))] \qquad (U_o > 0, a > 0)$$

The Lennard–Jones *intermolecular* potential, more suitable when the full range of R values must be considered, has the form

$$U(R) = \frac{A}{R^n} - \frac{B}{R^m} \qquad (n > m, A > 0, B > 0)$$

with values $m = 6$ and $n = 12$ usually cited. In comparison with the internuclear potential of Figure VI.4, the intermolecular potential generally involves a narrower and weaker attractive part relative to a pronounced hard-core region. Abundant discussion of intermolecular potentials, both from the empirical viewpoint and with relation to dynamical calculations from first principles, can be found in J. O. Hirschfelder, C. F. Curtiss, and R. B. Bird, *Intermolecular Theory of Gases and Liquids* (Wiley, 1954), Chapters 1 (Section 3), 13, and 14. A more recent reference with emphasis on theoretical calculations is H. Margenau and N. R. Kestner, *Theory of Intermolecular Forces* (Pergamon, 1971).

26. J. E. Mayer and M. G. Mayer, *op. cit.*, Section 7d.

27. *Ibid.*, Appendix AIII. The coefficient of $f^{(2k-1)}(a)$ in this formula is $(-1)^k B_k/(2k)!$, where the B_k are *Bernoulli numbers:* $B_1 = 1/6$, $B_2 = 1/30$, $B_3 = 1/42$, ... These coefficients can be determined by application of the Euler–Maclaurin sum-

mation formula to

$$(1 - e^{-a})^{-1} = \sum_{n=0}^{\infty} e^{-an}$$

because the resulting series expansion in powers of a can be calculated independently from the closed expression. The reader is warned that several inequivalent definitions of the Bernouli numbers exist in the literature. Also, the Euler–Maclaurin summation formula is, in general, an asymptotic expansion. Clarification of this point and a rigorous discussion of the formula are given by H. Jeffries and B. J. Jeffries, *Methods of Mathematical Physics* (Cambridge, 1972), pp. 278–283, 498–499.

28. The parity $(-1)^K$ of the rotational wave function is determined by inversion of the spatial coordinates. For a diatomic molecule, this inversion process is equivalent to interchange of the two atoms. Therefore, assuming that the electronic states are properly symmetrized, the spatial symmetry of the nuclear wave function is equal to its rotational parity $(-1)^K$.

29. L. I. Schiff, *loc. cit.*, p. 374.

30. An interesting account of the discovery of this explanation for the behavior of the heat capacity of H_2 gas is contained in an article by D. M. Dennison, *Am. J. Phys.* **42** (1974), p. 1051.

31. H. Goldstein, *loc. cit.*, Sections 6.1–6.3.

32. Molecular crystals, or crystals containing ionic groups, are more complex and must be analyzed with special considerations.

33. Tabulated values are given by M. Abramowitz and I. A. Stegun, *Handbook of Mathematical Functions* (Dover, 1965), p. 998.

34. E. W. Kellerman, *Phil. Trans.* **A238** (1940), p. 513; *Proc. Roy. Soc.* (*London*) **A178** (1941), p. 17.

35. A comprehensive review of "Brownian Motion as a Natural Limit to all Measuring Processes" has been given by R. Bowling Barnes and S. Silverman in *Rev. Mod. Phys.* **6** (1934), p. 162.

36. The more correct derivation of Van der Waals' equation in Section 39 yields $v_a = 2\pi a^3/3$. Note that a is the diameter of a hard sphere, because r is an interparticle distance.

Problems

VI.1. Consider a triangular die with probabilities P_1, P_2, and P_3 for throwing faces 1, 2, and 3. Suppose that the average of a large number of throws is known to be α, that is,

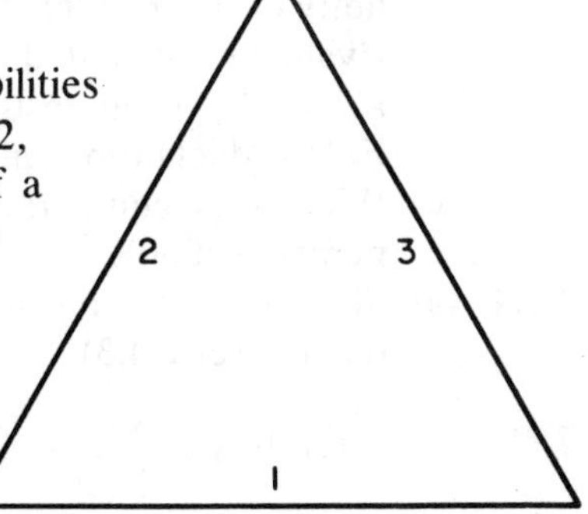

$$\sum_{n=1}^{3} nP_n = \alpha \qquad (1 \leqq \alpha \leqq 3)$$

Find the most probable P_n values

by maximizing the entropy

$$S = -\kappa \sum_n P_n(\ln P_n)$$

and plot the results for $\overline{P}_n$ and $\overline{S}$ as a function of α. What are the P_n values if α is unknown?

VI.2. *Langevin's classical theory of paramagnetism*
Consider a system of N atoms each of which has an intrinsic magnetic moment μ. The Hamiltonian in the presence of an external magnetic field $\mathcal{H}$ is

$$H = H_o(p_i, q_i) - \mu\mathcal{H} \sum_{i=1}^{N} \cos \Theta_i$$

where $H_o(p_i, q_i)$ is the Hamiltonian of the system in the absence of an external magnetic field and Θ_i is the angle between $\mathcal{H}$ and the magnetic moment of the ith atom. Show using classical statistical mechanics that:

(a) The induced magnetic moment $\langle M \rangle$ along the direction of $\mathcal{H}$ is

$$\langle M \rangle = N\mu\left(\coth x - \frac{1}{x}\right)$$

where

$$x \equiv \mu\mathcal{H}/\kappa T$$

(b) The differential magnetic susceptibility per atom is

$$\chi \equiv \frac{1}{N}\frac{d\langle M \rangle}{d\mathcal{H}} = \frac{\mu^2}{\kappa T}\left(\frac{1}{x^2} - \operatorname{csch}^2 x\right)$$

(c) At high temperatures χ satisfies Curie's law, namely, $\chi \propto T^{-1}$. Find the proportionality constant, which is called Curie's constant.

VI.3. (a) Show that application of the Euler–Maclaurin summation formula leads to the result $Z_{\text{even}} = Z_{\text{odd}}$ *identically,* where Z_{even} and Z_{odd} are defined by Equations (29.15). What is the criterion for the validity of this result?

(b) Evaluate the first three terms of Z_{even} in powers of $\beta\Theta_r$ when $\kappa T >> \Theta_r$. (This result requires use of the first four terms of the Euler–Maclaurin summation formula.)

(c) What is the equilibrium ratio of ortho- to para-hydrogen at a temperature of 30K?

VI.4. (a) Show that the entropy of a dilute diatomic gas can be written as (cf. Exercise 4.3):

$$(N\kappa)^{-1}\langle S \rangle = \frac{5}{2} - \ln(n\lambda_T^3) + S_{\text{int}}$$

where the internal entropy per particle κS_{int} is given by

$$S_{\text{int}} = -\beta^2 \frac{\partial}{\partial \beta}(\beta^{-1} \ln Z_{\text{int}})$$

(b) Evaluate S_{int} for both high- and low-temperature limits when the molecules have nonidentical atoms.

(c) Repeat calculation (b) for H_2 gas. Compare the two results both when $\kappa T >> \Theta_r$ and when $\kappa T << \Theta_r$ and explain any differences.

VI.5. *Fluctuations* can be studied from several points of view, as is shown by Problem V.8 and the following two short examples.

(a) Derive Equation (27.26) by using the result of Problem II.7(d). [*Suggestion:* Write

$$\Delta E = \left(\frac{\partial E}{\partial \Omega}\right)_T \Delta\Omega + \left(\frac{\partial E}{\partial T}\right)_\Omega \Delta T.\Bigg]$$

(b) A very sensitive spring balance with spring constant α consists of a quartz spring suspended from a fixed support. The balance is at a temperature T in a location where the acceleration due to gravity is g. Estimate the (minimum) error $(\Delta m)_{\text{rms}}$ of mass measurements when the fluctuations Δy_{rms} are dominated by random molecular bombardments on the mass m. Neglect the mass of the spring, and express your result in terms of m, T, g, and the natural period T_o of the system. Discuss your answer quantitatively for T = 300K and T_o = 10 sec.[35]

VI.6. Consider a monatomic crystal that consists of N atoms. These may be situated in two kinds of position, as shown:

```
                              0   0   0   0
                                x   x   x
0 = Normal position           0   0   0   0
x = Interstitial position       x   x   x
                              0   0   0   0
                                x   x   x
                              0   0   0   0
```

Suppose that there is an equal number N of both kinds of positions, but that the energy of an atom at an interstitial position is larger by an amount ε than that of an atom at a normal position. Then at $T = 0$ all atoms will be in normal positions. Show that at temperature T the most probable number $\bar{n}$ of atoms at interstitial sites, when $\bar{n} >> 1$, is given by

$$\bar{n} = N(e^{\varepsilon/2\kappa T} + 1)^{-1} \xrightarrow[\varepsilon >> \kappa T]{} N e^{-\varepsilon/2\kappa T}$$

[*Suggestion:* Derive an expression for the Helmholtz free energy $\Psi = E - TS$ and apply the minimum condition (9.14).]

VI.7. The energy levels of a one-dimensional anharmonic oscillator can be written in first approximation as

$$E = \Theta_v\left(v + \frac{1}{2}\right)\left[1 - x\left(v + \frac{1}{2}\right)\right] \qquad (v = 0, 1, 2, \ldots)$$

where the notation of Equations (29.3) and (29.4) has been used. The parameter x, assumed to be $\ll 1$, represents the degree of anharmonicity.

(a) Show that the partition function Z for such an oscillator can be written, to first order in x, as

$$Z = \left(1 + xu\frac{\partial^2}{\partial u^2}\right)\left[e^{-u/2}\sum_{v=0}^{\infty}(e^{-u})^v\right]$$

where $u = \beta\Theta_v$.

(b) Use the result of part (a) to calculate the heat capacity C_V to fourth order in u, for $u \ll 1$, and show that

$$\kappa^{-1}C_V = \left(1 - \frac{1}{12}u^2 + \frac{1}{240}u^4\right) + 4x(u^{-1} + \frac{1}{80}u^3) + \cdots$$

VI.8. *Generalizations of the Debye model*

(a) Consider the dispersion relation $\omega = Ak^s$ for wave motions in a solid, where the linear case $s = 1$ is given by Equation (30.13). Show that the contribution of such motions to the heat capacity C_V at low temperatures is $\propto T^{3/s}$. (The case $s = 2$ applies to spin waves propagating in a ferromagnetic system.)

(b) A simple generalization of the Debye model for monatomic crystals is to replace Equation (30.15) by a similar equation for each propagation mode $\alpha = (T, L)$, that is, by

$$N = \frac{\Omega}{6\pi^2}\left(\frac{\omega_m}{c}\right)^3_\alpha$$

Show that Equation (30.22) for the heat capacity of the crystal is generalized in this case to

$$\frac{C_V}{3N\kappa} = \frac{2}{3}c_{V,T} + \frac{1}{3}c_{V,L}$$

where $c_{V,\alpha}$ is given by the right side of Equation (30.22) with

$$T_D \rightarrow T_{D,\alpha} = \kappa^{-1}\hbar\omega_{m,\alpha}$$

Note that the cutoff wavelengths (30.17) must be equal for the two modes in this generalization.

VI.9. *Van der Waals' equation of state*

(a) In this problem we shall deduce Van der Waals' equation of state (39.5) for a dilute gas by generalizing the single-particle partition function with the aid of a so-called "mean-field" approximation. Thus, the classical energy $E(r_\alpha, p_\alpha)$ is first written as a sum over single-particle energies by averaging the potential energy seen by particle α as follows:

$$E = E(r_\alpha, p_\alpha) = \sum_{\alpha=1}^{N} \frac{p_\alpha^2}{2m} + \tfrac{1}{2} \sum_{\substack{\alpha,\beta=1 \\ \alpha\neq\beta}}^{N} V(r_{\alpha\beta})$$

$$\cong \sum_{\alpha=1}^{N} \left[\frac{p_\alpha^2}{2m} + \frac{N-1}{2\Omega} \int d^3\mathbf{r}_\beta V(r_{\alpha\beta}) \right]$$

Argue that in this approximation the single-particle partition function $Z_1 = Z_{tr}$ of Equation (28.24) becomes

$$Z_1 = \left(\frac{e}{n\lambda_T^3}\right) \int \frac{d^3\mathbf{r}_1}{\Omega} \exp\left[-\frac{n\beta}{2} \int d^3\mathbf{r}_{12} V(r_{12}) \right]$$

where $n = N/\Omega$ is the particle density and $\int d^3\mathbf{r}_1 = \Omega$.

(b) Consider the particular model for particle interactions in which there is an attractive square well outside of a repulsive core, that is,

$$V(r) = \begin{cases} \infty & r < a \\ -\varepsilon & a \leqq r \leqq b \\ 0 & r > b \end{cases}$$

Due to the singular character of the hard-core repulsion V_{HC}, Z_1 must be modified. Show that in the mean-field approximation of part (a), a suitable modification is

$$Z_1 \cong \left(\frac{e}{n\lambda_T^3}\right) \left[\int \frac{d^3\mathbf{r}_1}{\Omega} \prod_{j=2}^{N-1} \exp\left(-\frac{\beta}{2} V_{HC}(r_{1j}) \right) \right] \exp\left[-\frac{n\beta}{2} \int_a^b d^3\mathbf{r} V(r) \right]$$

$$= \left(\frac{e}{n\lambda_T^3}\right) \left[1 - \frac{4\pi}{3} n a^3 \right] e^{n\tilde{a}\beta}$$

where $\tilde{a} = (2\pi/3)\,\varepsilon(b^3 - a^3)$ and the effect of the hard-core factors in the integrand of $\int d^3\mathbf{r}_1$ is to exclude from Ω the volume of $(N - 1) \cong N$ hard-sphere volumes.

(c) Make the dilute-gas and weak-attraction approximations, ($nb^3 \sim na^3 << 1$, $\beta\varepsilon << 1$) and derive Van der Waals' equation of state,

$$\left(\mathscr{P} + \frac{\tilde{a}}{v^2}\right)(v - v_a) = \kappa T$$

where $v = n^{-1}$ and $v_a = (4\pi/3)a^3$.[36]

CHAPTER VII

DEGENERATE GASES

A degenerate gas is one for which quantum-mechanical effects are important; it is a quantum fluid. Because kinetic energies have a predominant role in a gas, the primary criterion for a gas to exhibit quantum effects is that its thermal wavelength be greater than its interparticle spacing; that is, we must have $n\lambda_T^3 \gtrsim 1$. In this chapter we consider only ideal quantum gases and deduce first, in Section 31, general expressions for both Fermi and Bose statistics. Then in following sections detailed analyses are given of the low-temperature region where quantum effects are manifest. For the Fermi case the Fermi sea of occupied states emerges as a dominant characteristic property in the low-temperature limit, whereas for the Bose case Bose–Einstein condensation into a single quantum state is exhibited at very low temperatures. Finally, in Section 35 the special case of blackbody radiation is investigated. For all cases low-temperature thermodynamics is explored in detail.

31 Partition Function for Ideal Gases

In this section we begin by calculating the partition function for the quantum-statistical case of the *degenerate ideal gas,*[1] that is, one for which the distinguishability condition (16.6) does not hold. We distinguish between the two cases of Fermi–Dirac (FD) statistics, in which the symmetrization factor $\varepsilon = -1$, and Bose–Einstein (BE) statistics, with $\varepsilon = +1$. Results obtained in Section 5 are rederived and elaborated further. General thermodynamic expressions for quantum ideal gases are written down, and these are applied in the near-classical limit of weak degeneracy.

In Section 28 we showed how to derive the limit of classical statistics from the macrocanonical-ensemble partition function of quantum statistics.

We then deduced the ideal-gas law of classical mechanics, at the end of that section, using the special case of Boltzmann statistics for which internal molecular motions can be treated quantum mechanically. When the ideal gas is degenerate, then this derivation requires fresh consideration before it can be applied to quantum-mechanical cases.

For comparison with the Boltzmann-statistics derivation of Section 28, it is instructive to start our calculation of the ideal-gas partition function for large systems with Equation (28.4). With each plane-wave-state label $\mathbf{k}$ we now also include a label m, for particles with spin, to denote the spin quantum number in the azimuthal direction. We use the condensed notation k to refer to the four state labels $(\mathbf{k}, m)$.

For free-particle systems the N-particle Hamiltonian H is simply a sum over N kinetic-energy operators,

$$\mathsf{H} = \sum_{\alpha=1}^{N} \mathsf{K}_\alpha$$

where for a large system the single-particle eigenvalues are $\omega(k) = \hbar^2 k^2/2m$. Moreover, for a single particle the momentum-space matrix elements of $e^{-\beta\mathsf{K}}$ are

$$\langle k|e^{-\beta\mathsf{K}}|k'\rangle = e^{-\beta\omega(k)}\, \delta_{\mathbf{k},\mathbf{k}'}\, \delta_{m,m'} \tag{31.1}$$

We may, therefore, write Equation (28.4) for Z as

$$Z = \frac{1}{N!} \sum_{k_1 \ldots k_N} \left[\prod_i (n_i!) \exp[-\beta n_i \omega(k_i)] \right] \tag{31.2}$$

where the product is over all single-particle states i. For a Fermi gas the single-particle occupation numbers n_i can be 0 or 1 only; that is, the exchange terms in Equation (28.4) do not contribute for a Fermi gas. Nor do the exchange terms contribute in the case of Boltzmann statistics, for which all $n_i = 1$ for the N occupied states, and this then is the source of the distinction between the present analysis and that given at the end of Section 28. Of course, on the average $\overline{n}_i << 1$ when Boltzmann statistics applies.

Instead of summing over all k_i values $(i = 1, 2, \ldots, N)$ in Equation (31.2), we can sum over all different sets of n_i values. On recalling the combinatorial factor (16.3), we obtain for Z the simple expression

$$Z = \sum_{\{n_i\}} \prod_i [\exp(-\beta\omega(k_i))]^{n_i} \tag{31.3}$$

in which the summation over all sets of n_i values is subject to the restriction $\Sigma_i\, n_i = N$. This formula also follows, of course, directly from the first equality of Equation (28.2). Note that there is no constraint on the partition function (31.3) that the total energy $E = \Sigma_i\, n_i \omega_i$ be fixed, because the gas is assumed to be in a temperature bath characterized by a definite temperature T, rather than energy E.

Use of the Darwin–Fowler Method

To evaluate the sum over all sets $\{n_i\}$ in Equation (31.3) subject to the constraint of fixed N, we appeal to the Darwin–Fowler method of mean values, which is outlined in Appendix A. This method achieves the summation without requiring that the n_i be large, so that Stirling's formula (2.6) is not required, as it was, say, in the derivation of Exercise 5.2.[2] The results we obtain can also be derived by using the formalism of the grand canonical ensemble (Chapter VIII), in which neither E values nor N values are fixed.

To proceed within the present context, we now introduce a function $u(\mathfrak{z})$ by

$$u(\mathfrak{z}) \equiv \sum_{\{n_i\}} \prod_i [\mathfrak{z} \exp(-\beta\omega(k_i))]^{n_i} \tag{31.4}$$

in which there is no restriction on the summation over sets of n_i values. If we allow $\mathfrak{z}$ to be a complex variable, then the partition function Z can be derived from $u(\mathfrak{z})$ by the following prescription:

$$\begin{aligned} Z &= \text{coefficient of } \mathfrak{z}^N \text{ in } u(\mathfrak{z}) \\ &= \frac{1}{2\pi i} \oint d\mathfrak{z} \frac{u(\mathfrak{z})}{\mathfrak{z}^{N+1}} \end{aligned} \tag{31.5}$$

the contour being any closed contour around the origin within the radius of convergence of the integrand. We can evaluate this contour integral after we investigate the analytic properties of the integrand

$$I(\mathfrak{z}) \equiv e^{\nu(\mathfrak{z})} = \frac{u(\mathfrak{z})}{\mathfrak{z}^{N+1}} \tag{31.6}$$

The exponential form is introduced in (31.6), as with Equation (2.5), because of the very rapid variation of $I(\mathfrak{z})$ with $\mathfrak{z}$ when $N >> 1$.

Inasmuch as there is no restriction on the summation over sets of n_i values in Equation (31.4), we may interchange this sum over sets with the product over all i to get for $u(\mathfrak{z})$ the expression

$$u(\mathfrak{z}) = \prod_i \sum_{n_i=0}^{n_{\max}} [\mathfrak{z} \exp(-\beta\omega(k_i))]^{n_i} \tag{31.7}$$

where $n_{\max} = 1$ for a Fermi gas ($\varepsilon = -1$) and $n_{\max} = N$, that is, ∞, for a Bose gas ($\varepsilon = +1$). It is then easy to show that both statistics are included by the formula

$$u(\mathfrak{z}) = \prod_i [1 - \varepsilon\mathfrak{z} \exp(-\beta\omega(k_i))]^{-\varepsilon} \tag{31.8}$$

In Equation (31.6) both the functions $u(\mathfrak{z})$ and $\mathfrak{z}^{N+1}$ increase monotonically with $\mathfrak{z}$, for positive real $\mathfrak{z}$, so that within the region where it is continuously analytic integrand $I(\mathfrak{z})$ has one minimum value determined by the vanishing of its first derivative

$$\frac{dI(\mathfrak{z})}{d\mathfrak{z}} = I(\mathfrak{z})v'(\mathfrak{z})$$

where

$$v'(\mathfrak{z}) = \frac{u'(\mathfrak{z})}{u(\mathfrak{z})} - \frac{N+1}{\mathfrak{z}} \tag{31.9}$$

The minimum value $\mathfrak{z} = \mathfrak{z}_o$ is identified by the condition $v'(\mathfrak{z}_o) = 0$, which with the aid of Equation (31.8) becomes, when $N + 1 \cong N >> 1$,

$$N = \sum_i \left[\frac{\mathfrak{z}_o\, e^{-\beta\omega_i}}{1 - \varepsilon\mathfrak{z}_o\, e^{-\beta\omega_i}} \right] \tag{31.10}$$

Within the region where $I(\mathfrak{z})$ is continuously analytic, that is, for $\mathfrak{z} < \mathfrak{z}_s$, there exists only one positive real solution $\mathfrak{z}_o$ to Equation (31.10).[3] That $I(\mathfrak{z})$ has only one minimum value along the positive real axis for $\mathfrak{z} < \mathfrak{z}_s$ is demonstrated explicitly by the following exercise.

Exercise 31.1

(a) Show that the extremum point $\mathfrak{z}_o$ is indeed a minimum of $I(\mathfrak{z})$ for positive real $\mathfrak{z}$ values by deriving

$$\frac{d^2I(\mathfrak{z})}{d\mathfrak{z}^2} = I(\mathfrak{z})\{[v''(\mathfrak{z}) + \mathfrak{z}^{-1}v'(\mathfrak{z})] + v'(\mathfrak{z})[v'(\mathfrak{z}) - \mathfrak{z}^{-1}]\}$$

and showing that

$$\begin{aligned}[v''(\mathfrak{z}) + \mathfrak{z}^{-1}v'(\mathfrak{z})] &= \mathfrak{z}^{-1}\sum_i \frac{e^{-\beta\omega_i}}{[1 - \varepsilon\mathfrak{z}\, e^{-\beta\omega_i}]^2} \\ &= O(N) > 0\end{aligned} \tag{31.11}$$

Inasmuch as the first derivative changes very rapidly when $N >> 1$, the point $\mathfrak{z} = \mathfrak{z}_o$ where $v'(\mathfrak{z}_o) = 0$ is at an extremely sharp minimum.

(b) Argue that $d^2I(\mathfrak{z})/d\mathfrak{z}^2 > 0$ for all $\mathfrak{z} < \mathfrak{z}_s$. Note that the second term of $d^2I/d\mathfrak{z}^2$ is negative only for an extremely small $\mathfrak{z}$ interval above the value $\mathfrak{z}_o$.

In Appendix B the contour integral (31.5) is evaluated by the method of steepest descents. The method applies when the integrand, expressed as in (31.6), has a minimum for positive real $\mathfrak{z}$ values; when this minimum is extremely sharp the result (B.10) is very accurate. After completing the argument at the end of Appendix B for the present application of the method, we obtain Equation (B.10) for Z:

$$\ln Z = v(\mathfrak{z}_o) - \frac{1}{2}\ln[2\pi v''(\mathfrak{z}_o)]$$

$$= \ln u(\mathfrak{z}_o) - (N+1)\ln \mathfrak{z}_o - \frac{1}{2}\ln[2\pi v''(\mathfrak{z}_o)] \tag{31.12}$$

$$\cong \ln u(\mathfrak{z}_o) - N\ln \mathfrak{z}_o$$

The logarithm of $v''(\mathfrak{z}_o)$ has been neglected in the third line of this last equation, because $v''(\mathfrak{z}_o)$ is proportional to N for large N and $N^{-1}\ln N \to 0$ in the limit $N \to \infty$.

Interpretation of Results

We first identify the pressure $\mathcal{P}$ by using Equation (27.19). Now Equation (31.8) can be used to evaluate $\ln u(\mathfrak{z}_o)$ and the result is a sum over single-particle states, which sum is proportional to the volume in the thermodynamic limit [see Equation (17.29)]. We may therefore derive from Equation (31.12) the expression

$$\mathcal{P} = (\beta\Omega)^{-1}\ln u(\mathfrak{z}_o)$$

$$\mathcal{P} = \frac{-\varepsilon}{\beta\Omega}\sum_i \ln(1 - \varepsilon\mathfrak{z}_o e^{-\beta\omega_i}) \tag{31.13}$$

Equations (31.12) and (31.13), together with $\ln Z = -\beta\Psi$, then yield

$$\beta^{-1}N\ln \mathfrak{z}_o = \Psi + \mathcal{P}\Omega = Ng \tag{31.14}$$

where the second equality follows from Equations (8.9) and (8.17) for the thermodynamic potential. Therefore the parameter $\mathfrak{z}_o$ is related to the chemical potential g by

$$\mathfrak{z}_o = e^{\beta g} \tag{31.15}$$

We next determine the average number $\langle n_i \rangle$ of particles in the state k_i by substituting Equations (27.17) and (27.18) for w_i into Equation (21.9a). We find, with the aid of Equation (31.3),

$$\langle n_i \rangle = Z^{-1}\sum_{\{n_j\}} n_i \prod_j [e^{-\beta\omega_j}]^{n_j}$$

$$= -\beta^{-1}\frac{\partial}{\partial\omega_i}\ln Z$$

$$\langle n_i \rangle = [e^{\beta(\omega_i - g)} - \varepsilon]^{-1} \tag{31.16}$$

Here, Equations (31.8), (31.12), and (31.15) have all been used to derive the third equality. This result agrees with Equation (5.10), providing at the same time an identification of the quantity g in the previous derivation.

The average value for the total energy of an ideal gas can be derived by substituting Equation (31.12) into the second of Equations (27.7). One finds with the aid of Equations (31.8) and (31.16) the obvious result

$$\langle E\rangle = \sum_i \langle n_i\rangle\omega_i \tag{31.17}$$

In the thermodynamic limit (17.29) this expression can be integrated by parts to give the relation between the pressure (31.13) and the energy,

$$\mathcal{P}\Omega = \frac{2}{3}\langle E\rangle \tag{31.18}$$

The factor of 2 derives here from the use of the quadratic energy-momentum relation, $\omega(k) = \hbar^2k^2/2m$. Equation (31.18), implied by previous results only for the classical ideal gas,[4] is now seen to hold quite generally for an ideal gas of nonrelativistic particles.

We can also deduce an expression for the average entropy by using $\Psi = \langle E\rangle - T\langle S\rangle = -\beta^{-1}\ln Z$. Thus we find

$$\begin{aligned}\frac{\langle S\rangle}{\kappa} &= \beta\langle E\rangle + \ln Z\\ &= -\sum_i \beta(g - \omega_i)\langle n_i\rangle + \varepsilon\sum_i \ln(1 + \varepsilon\langle n_i\rangle)\end{aligned}$$

Exercise 31.2 Show that the preceding expression for the average value of the entropy of an ideal gas can be written as

$$\frac{1}{\kappa}\langle S\rangle = -\sum_i [\langle n_i\rangle\ln\langle n_i\rangle - \varepsilon(1 + \varepsilon\langle n_i\rangle)\ln(1 + \varepsilon\langle n_i\rangle)] \tag{31.19}$$

in agreement with the previously derived (and more general!) results (5.5) and (5.7).

The equation of state for an ideal gas is obtained, in principle, by eliminating $\mathfrak{z}_o$ (or g) between Equations (31.10) and (31.13). In view of Equation (31.18), we may consider the energy expression (31.17) instead of the pressure expression. Thus we are led to focus attention on the total energy $\langle E\rangle$ and the density n, which can be rewritten in the thermodynamic limit, with S = particle spin, as

$$\frac{\langle E\rangle}{\Omega} = \frac{(2S + 1)}{2\pi^2}\left(\frac{\hbar^2}{2m}\right)\int_0^\infty k^4\,dk\langle n(k)\rangle \tag{31.20a}$$

$$n = \frac{(2S+1)}{2\pi^2}\int_0^\infty k^2\,dk\langle n(k)\rangle \tag{31.20b}$$

To proceed further we must calculate the integrals in these two expressions for various cases of interest. The detailed analyses we give of Fermi and Bose ideal gases at low temperatures, in Sections 32 and 34, rely heavily on these two expressions.

Exercise 31.3 Quantum-mechanical corrections to classical ideal-gas results, the case of weak degeneracy discussed below, can be derived from power-series expansions of the integrals in Equations (31.20). Use Equations (31.16) and (2.4) to show for integral m values that

$$\begin{aligned}\int_0^\infty k^m\,dk\langle n(k)\rangle &= \varepsilon\sum_{s=1}^\infty(\varepsilon z_o)^s\int_0^\infty k^m\,dk\exp\left(-\frac{s\lambda_T^2}{4\pi}k^2\right)\\ &= \frac{1}{2}\left(\frac{4\pi}{\lambda_T^2}\right)^{1/2(m+1)}\mathcal{N}_m f^{(\varepsilon)}_{(1/2)(m+1)}(z_o)\end{aligned} \tag{31.21}$$

where $\mathcal{N}_m = \sqrt{\pi}2^{-m/2}(m-1)!!$ if m = even, $\mathcal{N}_m = ([m/2]-[1/2])!$ if m = odd, and

$$f^{(\varepsilon)}_\ell(z_o) \equiv \varepsilon\sum_{s=1}^\infty\frac{(\varepsilon z_o)^s}{s^\ell} \tag{31.22}$$

The parameter z_o equals $e^{\beta g}$, and λ_T is the thermal wavelength (4.2). The functions defined by Equation (31.22) play an important role in the study of degenerate gases.

Substitution of Equation (31.21) into Equations (31.20), for the energy and density of an ideal gas, yields the general power-series expansions

$$\begin{aligned}\frac{\langle E\rangle}{\Omega} &= \frac{3}{2}\kappa T\left(\frac{2S+1}{\lambda_T^3}f^{(\varepsilon)}_{5/2}(z_o)\right]\\ n &= \frac{2S+1}{\lambda_T^3}f^{(\varepsilon)}_{3/2}(z_o)\end{aligned} \tag{31.23}$$

Where these power-series expansions fail to converge, and this can be the case at very low temperatures, then we must return to Equations (31.20) for their analytic continuation. Moreover, we show in Section 34 that the density equation must be modified for an ideal Bose gas below its Bose–Einstein condensation temperature.

Weak Degeneracy

The case of weak degeneracy merges with the Boltzmann statistics limit (16.6) in which $n\lambda_T^3 \ll 1$. If we substitute the expansion (31.22) for $\ell = 3/2$ into the density expression (31.23), then we find in this case

$$\frac{n\lambda_T^3}{2S+1} = \mathfrak{z}_o + \frac{\varepsilon\mathfrak{z}_o^2}{2^{3/2}} + \frac{\mathfrak{z}_o^3}{3^{3/2}} + O(\mathfrak{z}_o^4) \ll 1 \tag{31.24}$$

This equation is consistent with the assumption $\mathfrak{z}_o \ll 1$. Thus in the classical, or Boltzmann-statistics, limit we have

$$\mathfrak{z}_o = e^{\beta g} = \frac{n\lambda_T^3}{2S+1} \ll 1 \qquad \text{(Boltzmann statistics)} \tag{31.25}$$

If we insert this last result into Equation (31.16), then we find that the quantum-mechanical expression for $\langle n_i \rangle$ reduces correctly in the classical limit (with $S = 0$) to Equation (4.17). Moreover, the assumption $\mathfrak{z}_o \ll 1$ also gives the correct classical limit for the total energy, for on dividing the two equations (31.23) we find

$$\begin{aligned} \frac{\langle E \rangle}{N} &= \frac{3}{2}\kappa T \frac{f_{5/2}^{(\varepsilon)}(\mathfrak{z}_o)}{f_{3/2}^{(\varepsilon)}(\mathfrak{z}_o)} \\ &= \frac{3}{2}\kappa T \left\{ 1 - \frac{\varepsilon\mathfrak{z}_o}{2^{5/2}} + O(\mathfrak{z}_o^2) \right\} \\ &= \frac{3}{2}\kappa T \left\{ 1 - \frac{\varepsilon}{2^{5/2}}\left(\frac{n\lambda_T^3}{2S+1}\right) + O(n\lambda_T^3)^2 \right\} \end{aligned} \tag{31.26}$$

The final equality of this equation gives the leading correction, due to quantum statistics, to the classical expression (28.19) for the energy of an ideal gas (see also the discussion following Exercise 28.1). Further correction terms in the weak-degeneracy case can be calculated by straightforward iteration. The following exercise gives one method for systematizing the iteration procedure.

Exercise 31.4 We wish to express

$$y = f_{5/2}^{(\varepsilon)}(\mathfrak{z}_o)$$

as a power-series expansion in the variable

$$x = \frac{n\lambda_T^3}{2S+1} = f_{3/2}^{(\varepsilon)}(\mathfrak{z}_o)$$

that is, we wish to find the coefficients a_n in the expansion

$$y = \sum_{n=0}^{\infty} a_n x^n$$

Use the formula

$$a_n = \frac{1}{n!}\left(\frac{d^n y}{dx^n}\right)_{x=0} = \frac{1}{n!}\left\{\left(\frac{dx}{d\mathfrak{z}_o}\right)^{-1}\frac{d}{d\mathfrak{z}_o}\left(\frac{d^{n-1}y}{dx^{n-1}}\right)\right\}_{\mathfrak{z}_o=0}$$

to show:

$$a_0 = 0, \qquad a_1 = 1, \qquad a_2 = -\frac{\varepsilon}{4\sqrt{2}} = -(0.1768)\varepsilon$$

$$a_3 = -\left(\frac{2}{9\sqrt{3}} - \frac{1}{8}\right) = -0.00330, \qquad \text{etc.}$$

Note that these are the coefficients that appear in the final equality of Equation (31.26).
[*Suggestion:* Use

$$\mathfrak{z}_o\frac{d}{d\mathfrak{z}_o}f_\ell^{(\varepsilon)}(\mathfrak{z}_o) = f_{\ell-1}^{(\varepsilon)}(\mathfrak{z}_o)$$

here.]

32 Ideal Fermi Gas

The ideal gas of Fermi particles at very low temperatures is studied. In this limit, $n\lambda_T^3 >> 1$ and the fermions are highly degenerate. It is shown that the distribution function for the single-particle occupation numbers at $T = 0$ corresponds to the Fermi sea. Finite-temperature thermodynamics is associated with thermal excitations at the top of the Fermi sea, and explicit expressions are written down for this limiting case. Physical application of the mathematical deductions is reserved for the following section.

The very degenerate Fermi gas is characterized by $\mathfrak{z}_o > 1$, and we must therefore use Equations (31.20), instead of (31.23), to calculate the energy and the density. According to Equation (31.15) $\mathfrak{z}_o > 1$ is equivalent to $g > 0$, and consequently the mean occupation numbers (31.16), which we now call $\nu(k)$, become at $T = 0$

$$\nu(k) \equiv \langle n(k) \rangle = \frac{e^{\beta(g-\omega_k)}}{1 - \varepsilon\, e^{\beta(g-\omega_k)}} \underset{\substack{\varepsilon=-1 \\ \beta\to\infty}}{\longrightarrow} \begin{cases} 1 & \text{if } \omega_k < g_0 \\ 0 & \text{if } \omega_k > g_0 \end{cases} \tag{32.1}$$

This result is equivalent to a Fermi sea of occupation numbers (see Figure IV.3), and it is in accord with the requirement of the Pauli principle that each single-particle state can be occupied by only one fermion. Therefore the chemical potential at zero temperature is indeed positive; it must be identified with the Fermi energy E_F,

$$g \underset{T\to 0}{\longrightarrow} g_0 \equiv E_F = \frac{\hbar^2 k_F^2}{2m} \tag{32.2}$$

where the Fermi momentum $\hbar k_F$ is defined by using Equation (19.2) for the particle density.

$$n = \ell^{-3} = \frac{(2S+1)}{6\pi^2} k_F^3 \tag{32.3}$$

Although the integral (19.2) that gives this expression is meaningful only at $T = 0$, we nevertheless use Equation (32.3) to define k_F in terms of the density when $T \neq 0$.

To complete our introductory discussion of zero-temperature results, we consider Equation (19.4) for the energy per particle of an ideal Fermi gas. This well-known result can be deduced by using thermodynamic considerations as follows. Thus Equations (8.9) and (8.17) can be combined to give an expression for the entropy,

$$T\langle S \rangle = \langle E \rangle + \mathcal{P}\Omega - Ng \tag{32.4}$$

which vanishes at $T = 0$. On substituting Equation (31.18) into this expression and using $g_0 = E_F$, we obtain Equation (19.4),

$$\frac{\langle E \rangle}{N} \underset{T\to 0}{\longrightarrow} \frac{3}{5} E_F \tag{32.5}$$

Mathematical Procedure

A graph of Equation (32.1) for $\nu(k)$, near $T = 0$ and when $\varepsilon = -1$, is shown as Figure VII.1. It can be seen from this figure that when $T \neq 0$, $\nu(k)$ differs from a step function only in a region of width κT about $\omega(k) = g$. The (negative) derivative of $\nu(k)$ is therefore a very sharply peaked function for small κT, as is shown in Figure VII.2. We can verify analytically, in fact, that this derivative becomes a δ function at $T = 0$. Thus

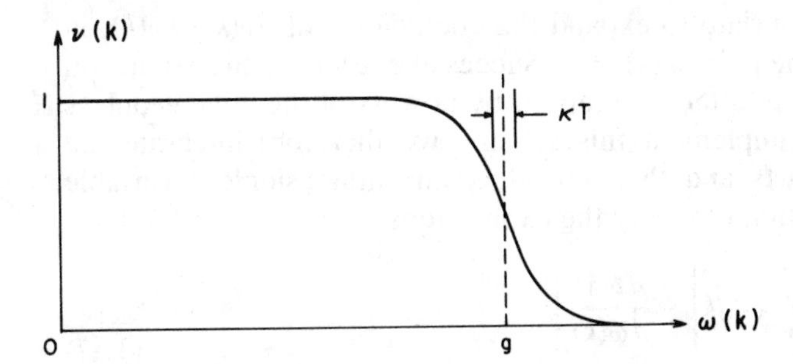

FIGURE VII.1 The average occupation number in an ideal Fermi gas near $T = 0$ ($\beta g = 20$).

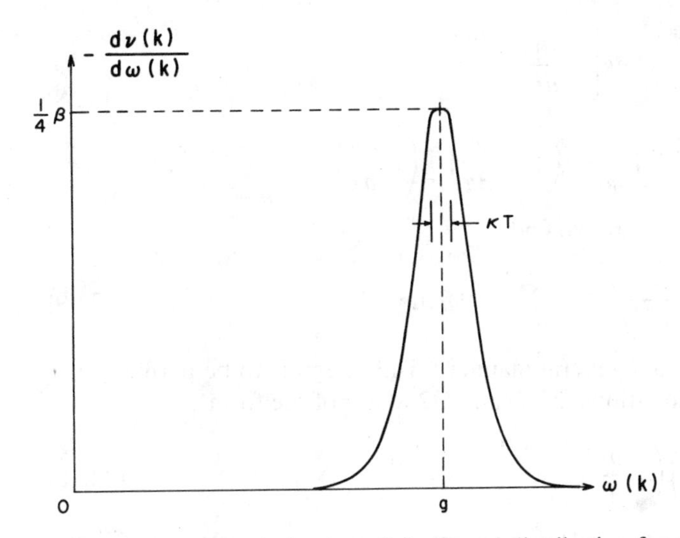

FIGURE VII.2 The derivative of the Fermi distribution function near $T = 0$ ($\beta g = 20$).

$$-\frac{d\nu(k)}{d\omega(k)} = \frac{\beta e^{\beta(\omega_k - g)}}{[e^{\beta(\omega_k - g)} - \varepsilon]^2} = \beta\nu(k)[1 + \varepsilon\nu(k)] \qquad (32.6)$$

$$\xrightarrow[\substack{\varepsilon = -1 \\ \beta \to \infty}]{} \delta(\omega_k - g_0)$$

The constant of proportionality of the δ function is unity, as can be verified by integrating both sides of Equation (32.6) over a small region about the point $\omega(k) = g_0$.

The result (32.6) suggests a systematic procedure for evaluating the integrals in Equations (31.20) for an ideal Fermi gas near $T = 0$. After expressing these integrals in terms of the derivative of $\nu(k) = \langle n(k) \rangle$, instead of $\nu(k)$

itself, it becomes appropriate to expand the coefficients of $d\nu(k)/d\omega(k)$, in the integrands, about the point $\omega(k) = g$. Successive terms in this expansion are expected to decrease, after integration, by powers of the dimensionless parameter $(\kappa T/g)$. To implement this scheme we therefore integrate the integrals (31.20) by parts and then introduce the dimensionless variable $x = \beta[\omega(k) - g]$. We obtain thereby the expressions

$$\begin{aligned}\frac{\langle E\rangle}{\Omega} &= \frac{(2S+1)}{20\pi^2}\left(\frac{\hbar^2}{m}\right)^2 \int_0^\infty k^6\,dk\left[-\frac{d\nu(k)}{d\omega(k)}\right] \\ &= \frac{(2S+1)}{10\pi^2}\left(\frac{2mg}{\hbar^2}\right)^{3/2} g\int_{-\beta g}^\infty dx\left(1+\frac{x}{\beta g}\right)^{5/2}\left(-\frac{d\nu}{dx}\right)\end{aligned} \tag{32.7}$$

$$\begin{aligned}n &= \frac{(2S+1)}{6\pi^2}\left(\frac{\hbar^2}{m}\right)\int_0^\infty k^4\,dk\left[-\frac{d\nu(k)}{d\omega(k)}\right] \\ &= \frac{(2S+1)}{6\pi^2}\left(\frac{2mg}{\hbar^2}\right)^{3/2}\int_{-\beta g}^\infty dx\left(1+\frac{x}{\beta g}\right)^{3/2}\left(-\frac{d\nu}{dx}\right)\end{aligned} \tag{32.8}$$

where from Equation (32.6) we find

$$-\frac{d\nu}{dx} = \frac{1}{(1+e^x)(1+e^{-x})} = -\sum_{m=1}^\infty (-1)^m m e^{-mx} \tag{32.9}$$

We proceed now in a general manner.[5] The integrals to be performed in the second lines of Equations (32.7) and (32.8) are of the form

$$I(F) = \int_{-\beta g}^\infty dx F\left(\frac{x}{\beta g}\right)\left(-\frac{d\nu}{dx}\right) \tag{32.10}$$

where we are interested in the low-temperature region $\beta g \gg 1$, or $\kappa T \ll g$. We first replace the upper integration limit with βg, thereby discarding an exponentially small term. An expansion of $F(x)$ in a Taylor series about $x = 0$ then yields the convergent low-temperature expansion

$$F\left(\frac{x}{\beta g}\right) = F(0) + \sum_{n=1}^\infty \frac{1}{n!}\left(\frac{\kappa T}{g}\right)^n F^{(n)}(0)x^n \tag{32.11}$$

On substituting this series into Equation (32.10) we obtain, since $(-d\nu/dx)$ is an even function of x,

$$\begin{aligned}I(F) &\cong F(0) + 2\sum_{n=2,4,\ldots}^\infty \frac{1}{n!}\left(\frac{\kappa T}{g}\right)^n F^{(n)}(0)\int_0^{\beta g} dx\, x^n\left(-\frac{d\nu}{dx}\right) \\ &\cong F(0) + 2\sum_{\ell=1}^\infty \left(\frac{\kappa T}{g}\right)^{2\ell} F^{(2\ell)}(0) f_{2\ell}^{(-)}(1)\end{aligned} \tag{32.12}$$

The second equality is derived by replacing the upper integration limit of the integrals by ∞ and then substituting the series expansion (32.9), integrating term by term, and using definition (31.22).[6]

Now $f_\ell^{(-)}(1)$ is related simply to $f_\ell^{(+)}(1) = \zeta(\ell)$, where $\zeta(\ell)$ is the *Riemann zeta function*. Moreover, $\zeta(2\ell)$ can be written as a numerical factor times the *Bernoulli number* B_ℓ.[7] Thus

$$
\begin{aligned}
f_{2\ell}^{(-)}(1) &= (1 - 2^{1-2\ell})f_{2\ell}^{(+)}(1) = \left(1 - \frac{1}{2^{2\ell-1}}\right)\zeta(2\ell) \\
&= (2^{2\ell-1} - 1)\frac{\pi^{2\ell}}{(2\ell)!}B_\ell
\end{aligned}
\tag{32.13}
$$

where the lowest three Bernoulli numbers are $B_1 = (1/6)$, $B_2 = (1/30)$, and $B_3 = (1/42)$. The lowest three coefficients $f_{2\ell}^{(-)}(1)$ in Equation (32.12) are then

$$
\begin{aligned}
f_{2\ell}^{(-)}(1) &= \frac{\pi^2}{12} && \text{for } \ell = 1 \\
&= \frac{7\pi^4}{720} && \text{for } \ell = 2 \\
&= \frac{31 \cdot \pi^6}{42 \cdot 720} && \text{for } \ell = 3
\end{aligned}
\tag{32.14}
$$

Exercise 32.1 Consider the application of Equation (32.12) with $\kappa T \ll g$ to the density integral (32.8), for which $F(y) = (1 + y)^{3/2}$.

(a) Estimate the error when ∞ is replaced by βg in the upper integration limit of Equation (32.10).

(b) Estimate the remainder $R_2(F)$ of the series (32.12) when only the $\ell = 1$ and $\ell = 2$ terms are kept, by writing

$$
\begin{aligned}
R_2(F) &\equiv I(F) - F(0) - 2f_2^{(-)}(1)F^{(2)}(0)\left(\frac{\kappa T}{g}\right)^2 \\
&\quad - 2f_4^{(-)}(1)F^{(4)}(0)\left(\frac{\kappa T}{g}\right)^4 \\
&\cong 2\int_0^{\beta g} dx\left[\frac{1}{2}F(y) + \frac{1}{2}F(-y) - F(0) - \frac{1}{2}y^2F^{(2)}(0) - \frac{1}{24}y^4F^{(4)}(0)\right] \\
&\quad \times \left(-\frac{dv}{dx}\right)
\end{aligned}
$$

where $y = (\kappa T/g)x$. Is your answer consistent with the $\ell = 3$ term of Equation (32.12)? (*Suggestion:* Argue first that the square-brackets factor in the integrand is a positive definite quantity times y^6 with a maximum value occurring at $y = 1$.)

Low-Temperature Thermodynamics

We can now evaluate the integrals in Equations (32.7) and (32.8) in the low-temperature limit. For the density we can apply definition (32.10), along with Equations (32.2) and (32.3), and write

$$\begin{aligned} n \equiv \frac{(2S+1)}{6\pi^2}\left(\frac{2mg_0}{\hbar^2}\right)^{3/2} &= \frac{(2S+1)}{6\pi^2}\left(\frac{2mg}{\hbar^2}\right)^{3/2} I\left[\left(1+\frac{x}{\beta g}\right)^{3/2}\right] \\ &= \frac{(2S+1)}{6\pi^2}\left(\frac{2mg}{\hbar^2}\right)^{3/2}\left\{1+\frac{\pi^2}{8}\left(\frac{\kappa T}{g}\right)^2+\frac{7\pi^4}{640}\left(\frac{\kappa T}{g}\right)^4+\mathrm{O}\left(\frac{\kappa T}{g}\right)^6\right\} \end{aligned} \tag{32.15}$$

where Equations (32.12) and (32.14) have been used to derive the second equality. We next divide Equation (32.7) by Equation (32.8) and obtain similarly for the energy per particle,

$$\begin{aligned} \frac{\langle E\rangle}{N} &= \frac{3}{5}g\frac{I\left[\left(1+\frac{x}{\beta g}\right)^{5/2}\right]}{I\left[\left(1+\frac{x}{\beta g}\right)^{3/2}\right]} \\ &= \frac{3}{5}g\left\{1+\frac{\pi^2}{2}\left(\frac{\kappa T}{g}\right)^2-\frac{11\cdot\pi^4}{120}\left(\frac{\kappa T}{g}\right)^4+\mathrm{O}\left(\frac{\kappa T}{g}\right)^6\right\} \end{aligned} \tag{32.16}$$

The energy expression (32.16) is expressed in terms of the independent variables T and g. We can, however, utilize the density relation (32.15) to express the energy in terms of the variables T and n, or equivalently, T and $g_0 = E_F$. Thus the first and last equalities of Equation (32.15) can be combined to give

$$\begin{aligned} \frac{g}{g_0} &= \left[1-\frac{\pi^2}{12}\left(\frac{\kappa T}{g}\right)^2+\frac{\pi^4}{720}\left(\frac{\kappa T}{g}\right)^4+\mathrm{O}\left(\frac{\kappa T}{g}\right)^6\right] \\ \frac{g}{g_0} &= \left[1-\frac{\pi^2}{12}\left(\frac{\kappa T}{g_0}\right)^2-\frac{\pi^4}{80}\left(\frac{\kappa T}{g_0}\right)^4+\mathrm{O}\left(\frac{\kappa T}{g_0}\right)^6\right] \end{aligned} \tag{32.17}$$

Here the second line is derived from the first by simple iteration. Substitution of this expression into Equation (32.16) then yields after some manipulations,

$$\frac{\langle E\rangle}{N} = \frac{3}{5}E_F\left[1+\frac{5}{12}\pi^2\left(\frac{\kappa T}{E_F}\right)^2-\frac{\pi^4}{16}\left(\frac{\kappa T}{E_F}\right)^4+\mathrm{O}\left(\frac{\kappa T}{E_F}\right)^6\right] \tag{32.18}$$

Of course, in the zero-temperature limit this result reduces to Equation (32.5). Notice that the Fermi energy E_F is related directly to the density n by Equations (32.2) and (32.3).

We deduce finally an expression for the entropy per particle in an ideal Fermi gas at equilibrium by using Equation (32.4) together with Equation (31.18). We obtain

$$\frac{\langle S\rangle}{N\kappa} = \beta\left(\frac{5}{3}\frac{\langle E\rangle}{N} - g\right) = \frac{\pi^2}{2}\left(\frac{\kappa T}{E_F}\right)\left[1 - \frac{\pi^2}{10}\left(\frac{\kappa T}{E_F}\right)^2 + \mathrm{O}\left(\frac{\kappa T}{E_F}\right)^4\right] \tag{32.19}$$

where Equations (32.17) and (32.18) have been used to derive the second equality. The heat capacity at constant volume then follows from either of the last two equations [see Equations (8.24) and (8.25)] and is given by

$$(N\kappa)^{-1}C_V = \frac{\pi^2}{2}\left(\frac{\kappa T}{E_F}\right)\left[1 - \frac{3\pi^2}{10}\left(\frac{\kappa T}{E_F}\right)^2 + \mathrm{O}\left(\frac{\kappa T}{E_F}\right)^4\right] \tag{32.20}$$

We note (with $S = ½$) that the parameter $(\kappa T/E_F) = (64/9\pi)^{1/3}(\ell/\lambda_T)^2 \sim (\ell/\lambda_T)^2$ is very small, where ℓ is the interparticle spacing, because $\ell \ll \lambda_T$ for a very degenerate Fermi gas. The result (32.20) is discussed further below Equation (33.3) in the next section.

33 Free-Electron Theory of Metals

Results and concepts from the preceding section on the low-temperature ideal Fermi gas are applied to the valence electrons in a metal. The validity of this model application is discussed, and then three physical properties of the model are analyzed; namely, heat capacity, thermionic emission, and Pauli paramagnetism.

We can understand a number of important physical properties of metals, particularly the simple metals, in terms of a free-electron model. According to this model the most weakly bound, or valence, electrons of the constituent atoms move about freely through the volume of the metal. Thus the valence electrons of the atoms become the conductors of electricity and are called *conduction electrons*. The forces between the conduction electrons and the ion cores are neglected in the free electron approximation. Calculations, due originally to Sommerfeld, proceed as if the conduction electrons were free to move everywhere in the metal.

To be sure, a conduction electron in a metal is not free, but is subject to strong and rapidly varying forces. In this connection the Coulomb repulsion between conduction electrons is more easily dismissed. It is of long range and slowly varying, and therefore the potential on any one electron does not depend strongly on the instantaneous positions of the others. In first approximation one may replace the interelectron potentials by average single-particle potentials. On the other hand, the attractive ionic forces vary greatly over lengths that are ~ 1 Å. The presence of strong ion potentials would seem to be inconsistent with a simple free-electron model. But the conduction electrons all have wavelengths that are $\gtrsim 1$ Å (see Exercise 33.1), and therefore they can experience only average forces due to ionic attractions. We conclude that for both of the electrostatic electron forces we may plausibly consider using an average potential $(-V)$, which is attractive overall and, in first approximation, constant.

A theorem due to F. Bloch states that the single-particle eigenfunctions of the wave equation for a periodic potential are of the form[8]

$$\Psi_{\mathbf{k}}(\mathbf{r}) = e^{i\mathbf{k}\cdot\mathbf{r}}u_{\mathbf{k}}(\mathbf{r})$$

where $u_{\mathbf{k}}(\mathbf{r})$ has a periodicity determined by the potential and $\mathbf{k}$ is a wave number. The associated single-particle energies $\omega'(\mathbf{k})$ involve potential-energy contributions in addition to the kinetic energy $\hbar^2k^2/2m$. When applied to the conduction electrons in a metal, it is the Bloch theorem that underlies the success of the free-electron approximation in which we argue, as above, that the function $u_{\mathbf{k}}(\mathbf{r})$ can be approximated by a constant and the energy $\omega'(k)$ replaced by only its kinetic-energy part.

Even in metals for which the free-electron model works well, the actual charge distribution of the conduction electrons is known to reflect the strong electrostatic potential of the ion cores. The success of the free-electron model, then, stems from the fact that many physical properties are due primarily to the kinetic energies of the electrons, despite these lattice variations. Of course, there are many important effects due to lattice structure. For example, the arrangement of ions determines the density of states for the electrons, and this consideration leads to band structure, or forbidden zones, for electron occupation. When forbidden zones become important, as for insulators, then the density of states cannot be calculated, even approximately, as it is for metals.

Exercise 33.1 Consider copper metal with a density $\rho = 8.9 \text{ gm} \cdot \text{cm}^{-3}$ and $n_v \cong 1$ valence, or conduction, electron per atom.

(a) Show that the wave number k_F for copper is

$$k_F = (1.36 \times 10^8)n_v^{1/3}\text{cm}^{-1} \sim \ell^{-1} \tag{33.1a}$$

The smallest free-electron wavelengths in copper are of the order of the reciprocal of this number, that is, $\gtrsim 1$ Å.

(b) Show that the Fermi energy E_F for copper is

$$E_F \cong 7n_v^{2/3}\ \text{eV} \qquad (33.1b)$$
$$= (8.14 \times 10^4\ \text{deg})n_v^{2/3}\kappa$$

where κ is Boltzmann's constant. For $T = 300\text{K}$

$$\frac{\kappa T}{E_F} \sim \frac{1}{400} \ll 1$$

Thus even at ordinary temperatures the conduction electrons in copper are very degenerate.

(c) Show that the pressure of the electron gas is $\mathcal{P} \sim 10^6$ atm.

On the basis of Exercise 33.1 and the preceding discussion we may think of the electron gas in a metal as is shown in Figure VII.3. The quantity ϕ is called the *work function*. It is the energy that must be supplied to an electron to remove it from the top of the Fermi sea, that is

$$\phi = V - g \qquad (33.2)$$

where the chemical potential g equals E_F at $T = 0\text{K}$.[9] We now consider three simple applications of the free-electron theory of metals.[10]

Heat Capacity

According to Equations (30.20) and (30.22) the lattice, or phonon, contribution to the heat capacity of a metal, using the Debye theory, is

$$(C_V)_{\text{lattice}} = 3N\kappa\left[1 - \frac{1}{20}\left(\frac{T_D}{T}\right)^2 + O\left(\frac{T_D}{T}\right)^3\right] \quad \text{for } T \gg T_D \qquad (33.3)$$
$$= \frac{12}{5}\pi^4 N\kappa\left(\frac{T}{T_D}\right)^3 \quad \text{for } T \ll T_D$$

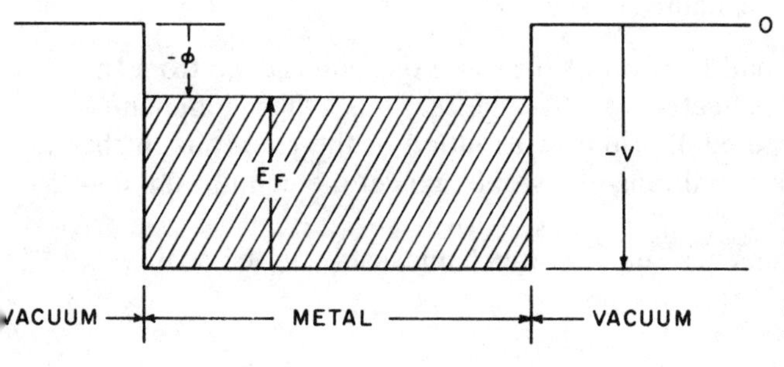

FIGURE VII.3 The Fermi sea of electrons at $T = 0\text{K}$ in a metal. Indicated are the work function ϕ and the average potential $-V$ for the electrons.

where $T_D \sim 200$K for many metals. This result agrees with measurements of C_V for metals, and it is very puzzling if one thinks classically. The reason is that the equipartition theorem, Equations (28.19) and (28.20), requires a contribution to the heat capacity from the electron degrees of freedom. This apparent discrepancy is readily resolved by the quantum-mechanical free-electron theory, for on referring to Figure VII.1, which shows the free-electron distribution function $\nu(\kappa)$, we see that because of the Pauli exclusion principle it is not possible to excite an electron by adding heat energy κT if its energy $\omega(\kappa)$ is $\ll g$. In fact, the fraction of electrons that can be excited is $\sim \kappa T/E_F$. Thus when heat is added to the metal the increase in energy per electron is $\sim(\kappa T)^2/E_F$, and therefore

$$(C_V)_{\text{electrons}} \sim N_e \kappa \left(\frac{\kappa T}{E_F}\right) \ll N_e \kappa \qquad (N_e = N n_v)$$

in agreement with Equation (32.20).

We have indicated that at ordinary temperatures the heat capacity of the free electrons in a metal is negligible in comparison with the heat capacity of the lattice. At very low temperatures, however, the specific heat capacity per atom has a measurable free-electron contribution, that is,

$$\begin{aligned}(N\kappa)^{-1}C_V &= \frac{\pi^2}{2} n_v \kappa \left(\frac{\kappa T}{E_F}\right) + \frac{12}{5}\pi^4 \left(\frac{T}{T_D}\right)^3 \\ &= \gamma T + BT^3\end{aligned} \tag{33.4}$$

where γ and B are constants characteristic of the metal. Measured values of γ differ from those predicted via Equation (33.4) by an "effective mass" correction.[11]

Thermionic Emission

We derive next the Richardson–Dushman equation for thermionic emission. The momentum distribution $\nu(k)$ for the electrons in a metal has a tail (Figure VII.1) that has the same variation with momentum as the Maxwell–Boltzmann distribution. Thus from Equation (32.1) we see that

$$\nu(k) \to e^{\beta(g-\omega_k)} \qquad \text{for } \omega_k > g \text{ and } \beta g \gg 1$$

This means that a very small fraction of the electrons can escape from the metal, especially when it is heated. We assume that this "thermionic emission" does not destroy the equilibrium distribution of the electrons in the metal. Under steady-state conditions one would expect the equilibrium distribution to be maintained.

Now the flux of electrons per quantum-mechanical state in the positive x direction, and striking a (y, z) surface is

$$\frac{1}{\Omega} \cdot \frac{p_x}{m} \nu(k)$$

for electrons with energy ω_k and x momentum p_x. Those with momentum p_x such that $(p_x^2/2m) > V$ (see Figure VII.3) can escape. The emission current density j_x is therefore

$$j_x = -\frac{2e}{h^3} \int_{-\infty}^{+\infty} dp_y \, dp_z \int_{(2mV)^{1/2}}^{\infty} dp_x \left(\frac{p_x}{m}\right) \left[\frac{e^{\beta(g-\omega_k)}}{1 + e^{\beta(g-\omega_k)}}\right] \tag{33.5}$$

where we have used Equation (32.1), $e > 0$, and $\omega_k = (2m)^{-1}(p_x^2 + p_y^2 + p_z^2)$. The p_x integration can be done exactly, and on making use of Equation (33.2) we obtain

$$j_x = -\frac{2e}{h^3} \kappa T \int_{-\infty}^{\infty} dp_y \, dp_z \ln\left\{1 + \exp\left(-\beta\left[\phi + \frac{1}{2m}\left(p_y^2 + p_z^2\right)\right]\right)\right\}$$

$$\cong -\frac{2e}{h^3} \kappa T \, e^{-\phi/\kappa T} \left[\int_{-\infty}^{+\infty} dp \, e^{-\beta p^2/2m}\right]^2$$

where at ordinary temperatures $\kappa T << \phi$ since ϕ is typically a few electron volts in magnitude. The Richardson–Dushmann equation now follows directly from this last expression, after using the second of Equations (2.4), as

$$j_x = -AT^2 \, e^{-\phi/\kappa T} \tag{33.6}$$

where $A = 4\pi e\kappa^2 mh^{-3}$. This expression agrees well with experiment.[12]

Pauli Paramagnetism

In the presence of an external magnetic field the free electrons in a metal are subject to two influences. Their quantized orbital motions around the direction of the field results in a diamagnetic effect called *Landau diamagnetism*.[13] The tendency of their spins to line up parallel to the field direction is a paramagnetic effect called *Pauli paramagnetism*. We discuss only the latter effect here.

In an external magnetic field $\mathcal{H}$ the energy of a free electron becomes

$$\omega_k' = \omega_k \pm \mu_B \mathcal{H}$$

where the plus sign occurs for electrons with spin "up" (that is, $\mathbf{S} \parallel \mathcal{H}$) and the minus sign occurs for electrons with spin "down." The quantity $\mu_B = (e\hbar/2mc) > 0$ is the Bohr magneton. The total magnetization M of the system due to the electron spins is given by the sum over all momentum states of the individual particle magnetic moments $\mu_z = \mp\mu_B$. In the infinite-volume limit we obtain

$$\frac{M}{\Omega} = -\frac{\mu_B}{(2\pi)^3}\int d^3\mathbf{k}\{\nu[\omega(k) + \mu_B\mathcal{H}] - \nu[\omega(k) - \mu_B\mathcal{H}]\} \tag{33.7}$$

where the chemical potentials for spin-up and spin-down electrons must be equal at equilibrium.[14]

We now consider the weak-field limit $\mu_B\mathcal{H} << E_F = \hbar^2k_F^2/2m$. Then on expanding the integrand in Equation (33.7) about the value of the chemical potential when $\mathcal{H} = 0$, we obtain in lowest approximation

$$\frac{M}{\Omega} \cong -\frac{\mu_B}{2\pi^2}\int_0^\infty k^2\,dk\left[\frac{d\nu(k)}{d\omega(k)}\right]_{\mathcal{H}=0}(2\mu_B\mathcal{H})$$

The magnetic susceptibility per unit volume, χ, is the coefficient of $\mathcal{H}$ on the right side of this equation. We obtain for χ the result

$$\begin{aligned}\chi &= \frac{\mu_B^2}{2\pi^2}\left(\frac{\hbar^2}{2m}\right)^{-1}\int_0^\infty k\,d\omega_k\left[-\frac{d\nu(\kappa)}{d\omega(\kappa)}\right]_{\mathcal{H}=0} \\ &= \frac{3}{2}\left(\frac{n\mu_B^2}{E_F}\right)\left(\frac{g}{g_o}\right)^{1/2}\int_{-\beta g}^\infty dx\left(1 + \frac{x}{\beta g}\right)^{1/2}\left(-\frac{d\nu}{dx}\right)\end{aligned} \tag{33.8}$$

where $g_0 = E_F$, $n = \kappa_F^3/3\pi^2$, and $x = \beta[\omega(k) - g]$. The derivative $(-d\nu/dx)$ is given by Equation (32.9).

We can use Equations (32.10), (32.12), and (32.14) to evaluate the integral in Equation (33.8). The result for the Pauli spin susceptibility is

$$\chi = \frac{3}{2}\left(\frac{n\mu_B^2}{E_F}\right)\left(\frac{g}{g_0}\right)^{1/2}\left\{1 - \frac{\pi^2}{24}\left(\frac{\kappa T}{g}\right)^2 - \frac{7\pi^4}{384}\left(\frac{\kappa T}{g}\right)^4 + O\left(\frac{\kappa T}{g}\right)^6\right\}$$

$$\chi = \frac{3}{2}\left(\frac{n\mu_B^2}{E_F}\right)\left\{1 - \frac{\pi^2}{12}\left(\frac{\kappa T}{E_F}\right)^2 - \frac{11\pi^4}{360}\left(\frac{\kappa T}{E_F}\right)^4 + O\left(\frac{\kappa T}{E_F}\right)^6\right\} \tag{33.9}$$

where Equation (32.17) has been used to obtain the second equality.

Exercise 33.2 Derive both equalities of Equation (33.9).

One can understand Pauli paramagnetism at $T = 0$ in the following simple way. On referring to Figure VII.4 we see that the fraction of (spin-down) electrons with uncanceled magnetic moments is $\sim\mu_B\mathcal{H}/E_F$. The total magnetization of the system is, therefore, $M \sim N_e\mu_B(\mu_B\mathcal{H}/E_F)$, in agreement with Equation (33.9).

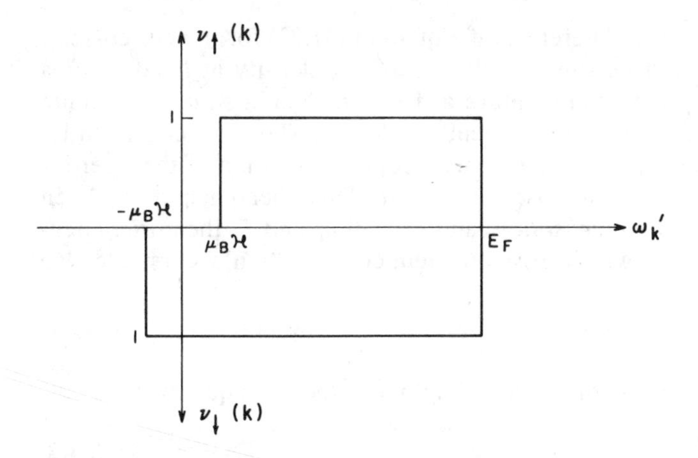

FIGURE VII.4 The average occupation numbers at $T = 0$ for free spin-up ($\uparrow$) and spin-down ($\downarrow$) electrons in a magnetic field $\mathcal{H}$, as a function of their energies $\omega_k' = \omega_k \pm \mu_B \mathcal{H}$.

34 Ideal Bose Gas

Analysis is given of the ideal Bose gas at very low temperatures. It is shown that Bose–Einstein condensation, which is a macroscopic occupation of the zero-momentum single-particle state, occurs when $n\lambda_T^3 \gtrsim 1$. Details of the thermodynamics above and below the critical temperature T_C for this effect are derived. The analogy of Bose–Einstein condensation with an ordinary vapor–liquid phase transition is demonstrated. Finally, the concepts developed in this section are applied to a brief discussion of liquid helium four.

For the ideal Bose gas we set $\varepsilon = +1$ in the equations of Section 31. Also, to simplify equations in this section, the spin S is set equal to zero. We begin by examining the density equation (31.23), which, together with Equation (31.22), may be written as

$$n\lambda_T^3 = f_{3/2}^{(+)}(\mathfrak{z}_o) = \mathfrak{z}_o + \frac{\mathfrak{z}_o^2}{2^{3/2}} + \frac{\mathfrak{z}_o^3}{3^{3/2}} + O(\mathfrak{z}_o^4)$$

We see from this expression that as the density increases, or as the temperature decreases, the parameter $\mathfrak{z}_o = e^{\beta g}$ increases to the value unity. Moreover, according to Equation (31.16) values of $\mathfrak{z}_o > 1$ would be associated with negative occupation numbers for the ideal Bose gas, and therefore such values of $\mathfrak{z}_o$ are unphysical. Now, in fact, the largest value of $f_{3/2}^{(+)}(\mathfrak{z}_o)$ for

$\mathfrak{z}_o \leqslant 1$ occurs at $\mathfrak{z}_o = 1$. Therefore, if Equation (31.23) for n were correct, then there would be a maximum value n_C of the density at fixed T, or a minimum value T_C of the temperature at fixed n. Such a situation is quite unphysical, and we resolve this difficulty below by showing, consistently, for $n > n_C$ or $T < T_C$, that there is macroscopic occupation of the "zero"-momentum state in an ideal Bose gas at rest. This phenomenon has been labeled *Bose–Einstein condensation,* and we call n_C and T_C the *critical* density and temperature at which Bose-Einstein condensation occurs. We also define for this section

$$\alpha \equiv -\beta g = -\ln \mathfrak{z}_o \tag{34.1}$$

The critical point is characterized by either of the two equations

$$\begin{aligned} n_C\lambda_T^3 &= f_{3/2}^{(+)}(1) = 2.612 \\ n\lambda_C^3 &= 2.612 \end{aligned} \tag{34.2}$$

where λ_C is the thermal wavelength (4.2) evaluated at the critical temperature. "Above" the critical point, that is, for $n < n_C$ or $T > T_C$, Equation (31.23) for the density continues to be valid, and we may use Equations (34.1) and (34.2) to rewrite this expression in the alternative forms

$$\begin{aligned} \frac{n}{n_C} &= \frac{\Omega_C}{\Omega} = \frac{f_{3/2}^{(+)}(e^{-\alpha})}{2.612} \qquad \text{for } n \leqslant n_C \text{ or } \Omega \geqslant \Omega_C \\ \left(\frac{T_C}{T}\right)^{3/2} &= \frac{f_{3/2}^{(+)}(e^{-\alpha})}{2.612} \qquad \text{for } T \geqslant T_C \end{aligned} \tag{34.3}$$

For a fixed number of particles the critical density n_C is equivalent to a critical volume Ω_C. Note that Bose–Einstein condensation occurs in the region of high degeneracy where $\lambda_T \gtrsim \ell$, the interparticle spacing.

To investigate the region of Bose–Einstein condensation with some care, it is necessary to know the properties of the functions $f_\ell^{(+)}(e^{-\alpha})$ of Equation (31.22) for α positive and small, and $\ell = ½, 3⁄2$, and $5⁄2$. Asymptotic expansions of the original defining integrals (31.21) for the $f_\ell^{(+)}(e^{-\alpha})$ yield the following expressions[15]:

$$\begin{aligned} f_{1/2}^{(+)}(e^{-\alpha}) &= \left(\frac{\pi}{\alpha}\right)^{1/2} - 1.460 + (0.208)\alpha - (0.0128)\alpha^2 + O(\alpha^3) \\ f_{3/2}^{(+)}(e^{-\alpha}) &= 2.612 - 2(\pi\alpha)^{1/2} + (1.460)\alpha - (0.104)\alpha^2 + O(\alpha^3) \\ f_{5/2}^{(+)}(e^{-\alpha}) &= 1.342 - (2.612)\alpha + \frac{4}{3}(\pi\alpha^3)^{1/2} - (0.730)\alpha^2 + O(\alpha^3) \end{aligned} \tag{34.4}$$

These functions can also be plotted for all α values, as in Figure VII.5. We note from Equation (31.22) that

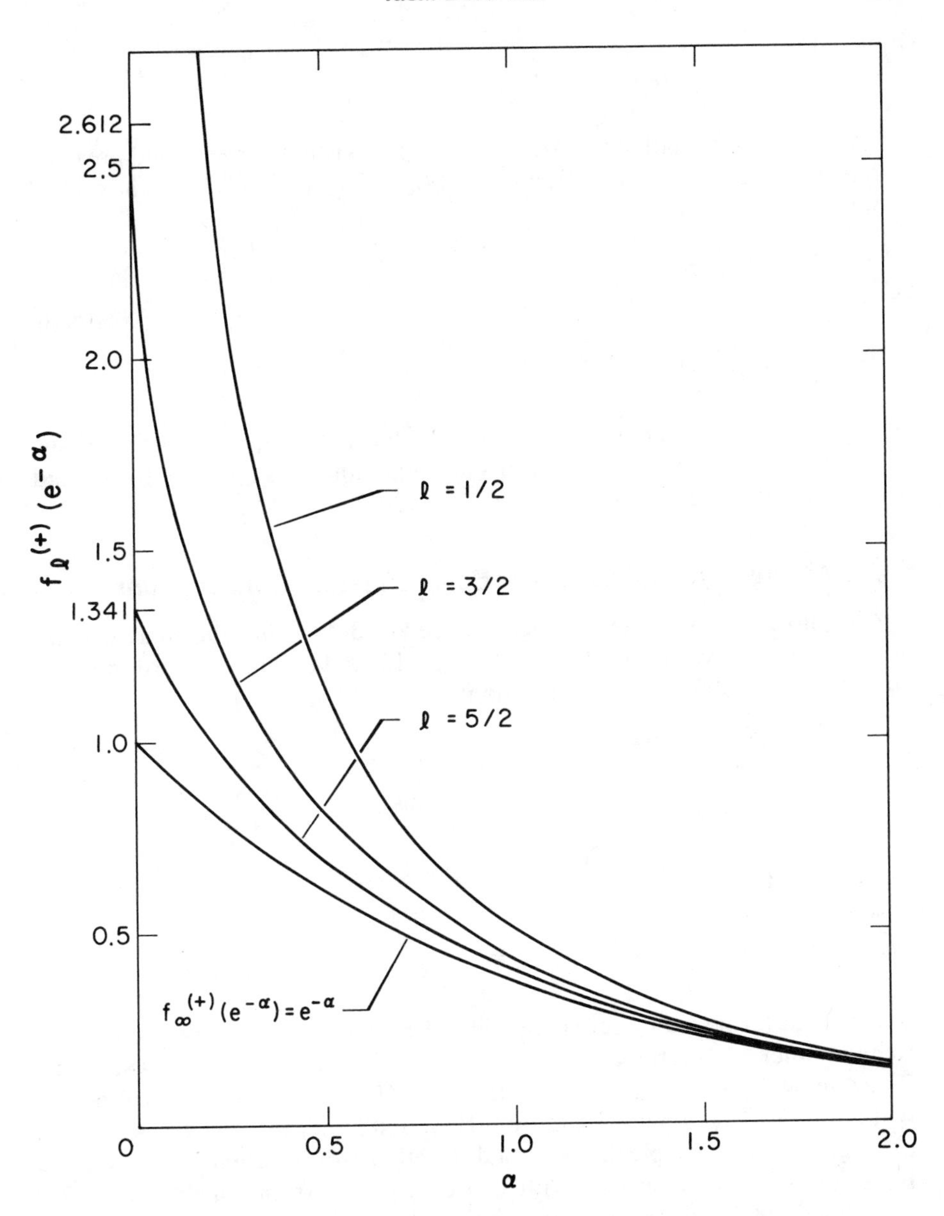

FIGURE VII.5 The functions $f_\ell^{(+)}(e^{-\alpha})$ plotted as a function of α for $\ell = 1/2, 3/2$, and $5/2$. Note that for $\alpha \geqslant 2$ these functions are close to the large-α limit $e^{-\alpha}$, also shown.

$$\frac{df_\ell^{(+)}(e^{-\alpha})}{d\alpha} = -f_{\ell-1}^{(+)}(e^{-\alpha}) \tag{34.5}$$

We can use Equation (34.4) for $f_{5/2}^{(+)}(\mathbf{e}^{-\alpha})$ to write the energy and pressure expressions, Equations (31.23) and (31.18), which are valid above the critical point, as

$$\begin{aligned}\frac{\langle E\rangle}{N} &= \frac{3}{2}\cdot\frac{\mathcal{P}}{n} = \frac{3}{2}\frac{\kappa T}{n\lambda_T^3} f_{5/2}^{(+)}(e^{-\alpha}) \\ &\xrightarrow[T\to T_C^{(+)}]{} \frac{3}{2}\frac{\kappa T_C}{n\lambda_C^3}\cdot(1.342)\end{aligned} \tag{34.6}$$

We shall see that whereas Equation (31.26) for the energy is incorrect below the critical point, the first line of this last equation does continue to be valid below the critical point, with $\alpha = 0$.

Mathematical Analysis of Bose–Einstein Condensation

We now demonstrate the occurrence of Bose–Einstein condensation with some care. We return to Equation (31.10) and exhibit the zero-momentum term separately in this expression:

$$\begin{aligned}N &= \langle n_0\rangle + \sum_{i\neq 0}[e^{\beta\omega_i+\alpha} - 1]^{-1} \\ &= \langle n_0\rangle + \frac{\Omega}{\lambda_T^3} f_{3/2}^{(+)}(e^{-\alpha})\end{aligned} \tag{34.7}$$

where

$$\langle n_0\rangle \equiv (e^\alpha - 1)^{-1} \xrightarrow[T\to T_C^{(+)}]{} \frac{1}{\alpha} \tag{34.8}$$

The second term in the second equality of Equation (34.7) is derived in the same manner as the right side of the density expression (31.23). (See also Exercise 34.1.) If we assume now that $\alpha \sim (1/N)$ below the critical point, that is, for $T < T_C$, then we find from Equation (34.8) that the zero-momentum state is macroscopically occupied. In other words, when $T < T_C$ we get Bose–Einstein condensation with $\langle n_0\rangle$ determined from Equations (34.7), (34.8), and (34.2), for fixed density, by

$$\begin{aligned}\langle n_0\rangle &= \frac{1}{\alpha} = N\left[1 - \frac{f_{3/2}^{(+)}(1)}{n\lambda_T^3}\right] \\ &= N\left[1 - \left(\frac{T}{T_C}\right)^{3/2}\right]\end{aligned} \qquad (T < T_C) \tag{34.9}$$

The important step leading to this result was the return to Equation (31.10), which, before taking the thermodynamic limit, is a discrete summation over occupation numbers $\langle n_i \rangle$.[16] According to Equation (34.9) the fractional occupation $\langle n_0 \rangle / N$ of the zero-momentum state increases monotonically from 0 to 1 as the temperature decreases from T_C to 0 at fixed density.

Exercise 34.1 Demonstrate that it is only the zero-momentum state that is macroscopically occupied, in a large system, when $\alpha \sim (1/N)$. Start from Equation (31.16), which for the lowest energy states, such that $\beta\omega_i << 1$, can be written as

$$\langle n_i \rangle \cong (\beta\omega_i + \alpha)^{-1}$$

Then use the quantization of single-particle states in a box to show that

$$\langle n_i \rangle \sim \left(\frac{N}{n\lambda_T^3}\right)^{2/3} \qquad \text{for } i \neq 0 \text{ and } T < T_C$$

Finally, use this result to justify the second equality of Equation (34.7).

For very large, but finite, N and Ω the second equality of Equation (34.7) can be written as

$$N = (e^{\alpha} - 1)^{-1} + N\left(\frac{T}{T_C}\right)^{3/2} \frac{f_{3/2}^{(+)}(e^{-\alpha})}{2.612} \tag{34.10}$$

where we have used the second of Equations (34.2). This equation can be solved for small α in order to examine the behavior of $\langle n_0 \rangle = \alpha^{-1}$ as $T \to T_C^{(+)}$. The result for temperature intervals

$$0 < (T - T_C) << N^{-1/3} T_C$$

is an approximate expression for $\langle n_0 \rangle$,

$$\langle n_0 \rangle = \frac{1}{\alpha} = (NC)^{2/3}\left[1 - \left(\frac{T - T_C}{T_C}\right)\left(\frac{N}{C^2}\right)^{1/3}\right] \sim N^{2/3} \tag{34.11}$$

where $C = 1.36$. See Problem VII.5. Equation (34.11) shows that at $T = T_C^{(+)}$, $\langle n_0 \rangle / N \to 0$ as $N \to \infty$ and therefore that the transition is smooth from $\langle n_0 \rangle = 0$ when $T \geqslant T_C$ to the expression (34.9) when $T < T_C$. In the thermodynamic limit, however, the function $\langle n_0 \rangle / N$ consists of two analytically different branches, namely, for fixed density,

$$\begin{aligned} \lim_{N \to \infty} \left(\frac{\langle n_0 \rangle}{N}\right) &= 0 \qquad \text{for } T \geqslant T_C \\ &= \left[1 - \left(\frac{T}{T_C}\right)^{3/2}\right] \qquad \text{for } T < T_C \end{aligned} \tag{34.12}$$

Thermodynamics of Ideal Bose Gas

(1) *The temperature region* $T >> T_C$: In the region above the critical point the thermodynamics of the ideal Bose gas is determined, essentially, by eliminating α between Equations (34.6) and (34.3). We have already indicated how to accomplish this objective in the weak-degeneracy limit, with the derivation of Equation (31.26). Thus a convergent power-series expansion of $f_{5/2}^{(+)}(e^{-\alpha})$ when $T >> T_C$ (see also Exercise 31.4) is

$$f_{5/2}^{(+)}(e^{-\alpha}) = x - (0.1768)x^2 - (0.0033)x^3 + O(x^4) \qquad (T >> T_C) \tag{34.13}$$

$$x \equiv f_{3/2}^{(+)}(e^{-\alpha}) = n\lambda_T^3 = (2.612)\left(\frac{T_C}{T}\right)^{3/2}$$

Substitution of this expansion into Equation (34.6) yields for $\langle E\rangle/N$ and the heat capacity $C_V = (\partial E/\partial T)_\Omega$ the expressions

$$\frac{\langle E\rangle}{N} = \frac{3}{2}\frac{\mathcal{P}}{n} = \frac{3}{2}\kappa T\left\{1 - (0.4618)\left(\frac{T_C}{T}\right)^{3/2} - (0.0226)\left(\frac{T_C}{T}\right)^3 + O\left(\frac{T_C}{T}\right)^{9/2}\right\}$$

$$\frac{C_V}{N\kappa} = \frac{3}{2}\left\{1 + (0.231)\left(\frac{T_C}{T}\right)^{3/2} + (0.045)\left(\frac{T_C}{T}\right)^3 + O\left(\frac{T_C}{T}\right)^{9/2}\right\} \tag{34.14}$$

which are useful only when $T >> T_C$.

(2) *The temperature region* $0 \leq (T - T_C) << T_C$: Close to, but above, the critical temperature, where $0 \leq (T - T_C) << T_C$, one can show for fixed density that the energy-per-particle and specific-heat capacity expressions can be expanded as

$$\frac{\langle E\rangle}{N} = \frac{3}{2}\frac{\mathcal{P}}{n} = (0.770)\kappa T_C\left\{1 + \frac{5}{2}\left(\frac{T - T_C}{T_C}\right) - \frac{1}{2}\left(\frac{T - T_C}{T_C}\right)^2 + O\left(\frac{T - T_C}{T_C}\right)^3\right\}$$

$$\frac{C_V}{N\kappa} = (1.926)\left\{1 - 0.4\left(\frac{T - T_C}{T_C}\right) + O\left(\frac{T - T_C}{T_C}\right)^2\right\} \tag{34.15}$$

The derivation of the energy expression is outlined in Problem VII.4(a). The specific-heat expression can be used in part (b) of this problem to show that $(\partial C_V/dT)$ is discontinuous at $T = T_C$ (see also Figure VII.6).

(3) *The temperature region* $T < T_C$: It should be clear that the zero-momentum state and its macroscopic occupation affect neither the total energy nor the pressure. For example, the contribution of the zero-momentum state to the pressure expression (31.13), with $z_o = e^{-\alpha}$, is given by

$$(n\kappa T)\lim_{N\to\infty}\left(-\frac{1}{N}\ln\alpha\right)$$

which vanishes when Equation (34.9) is inserted for α. We may conclude also that Equation (31.18) is valid below the critical point. Thus Equation (34.6), together with the second of Equations (34.2), for constant density, yields the results

$$\frac{\langle E\rangle}{N} = \frac{3}{2}\frac{\mathcal{P}}{n} = (0.770)\left(\frac{T}{T_C}\right)^{3/2}\kappa T \qquad (T < T_C) \qquad \textbf{(34.16)}$$
$$\frac{C_V}{N} = (1.926)\kappa\left(\frac{T}{T_C}\right)^{3/2}$$

With the aid of Equation (34.9) this last energy expression can also be written in the form

$$\frac{\langle E\rangle}{N - \langle n_o\rangle} = (0.770)\kappa T$$

which shows, quite reasonably, that the average energy per nonzero-momentum particle is a simple multiple of κT. In Figure VII.6 we have plotted the constant-volume heat capacity of the ideal Bose gas as a function of temperature for all T.

(4) *Equation of state:* The second equality of Equation (34.6) shows that the pressure is independent of the density below the critical point. Thus, at a given temperature there is a critical density or volume, determined by the

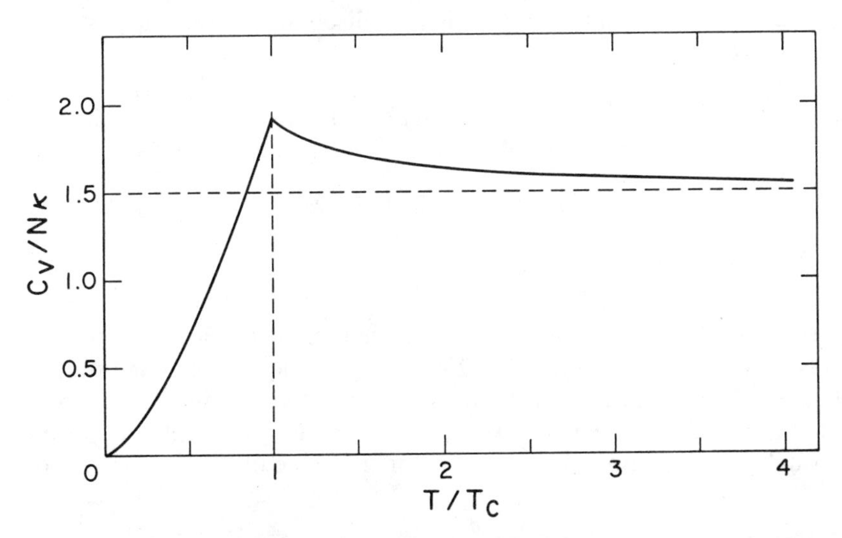

FIGURE VII.6 Specific heat capacity for the ideal Bose gas as a function of temperature. [See Equations (34.14) to (34.16).]

first of Equations (34.2), below which the pressure is constant. The interpretation of this result is that as the density is increased along an isotherm below the critical point, condensation into the zero-momentum state occurs instead of pressure increase. We pursue this interpretation further below, by showing that this condensation process is analogous to the vapor–liquid condensation process that occurs for a real gas.

Explicit expressions for the isotherms of the ideal Bose gas are obtained by using Equations (34.6), (34.14), and (34.15) in conjunction with Equations (34.3). One finds

$$\mathcal{P} = n\kappa T\left\{1 - (0.4618)\left(\frac{n}{n_C}\right) - (0.0226)\left(\frac{n}{n_C}\right)^2 + \mathrm{O}\left(\frac{n}{n_C}\right)^3\right\} \qquad \text{for } n \ll n_C$$

$$= (1.342)\frac{\kappa T}{\lambda_T^3}\left\{1 - \frac{19}{18}\left(\frac{n_C - n}{n_C}\right)^2 + \mathrm{O}\left(\frac{n_C - n}{n_C}\right)^3\right\} \qquad \text{for } 0 \leqslant (n_C - n) \ll n_C \tag{34.17}$$

$$= (1.342)\frac{\kappa T}{\lambda_T^3} = (0.975)\left(\frac{2\pi\hbar^2}{m}\right)n_C^{5/3} \qquad \text{for } n > n_C$$

where $(0.975) = (1.342)(2.612)^{-5/3}$. This result is sketched in Figure VII.7. For $n > n_C$ the isotherms show graphically the Bose–Einstein condensation phenomenon of Equation (34.9) in which, as discussed above, the gas pressure remains constant.

Exercise 34.2 Derive the second equality of Equation (34.17) from the first of Equations (34.15), by introducing the small parameter $y \equiv 1 - (n/n_C)$ and using

$$\frac{T_C}{N} = \left(\frac{n}{n_C}\right)^{2/3} = (1 - y)^{2/3}$$

In the high-density region of Bose–Einstein condensation the pressure depends only on the temperature, as for the liquid–vapor line of an ordinary substance.[17] The analogy suggests that Bose–Einstein condensation can be regarded as a transition between two phases. In fact, differentiation of the third equality in Equation (34.17) yields for the slope of the $\mathcal{P}$–T transition line,

$$\frac{d\mathcal{P}}{dT} = \frac{5}{2}(1.342)\frac{\kappa}{\lambda_T^3} = \frac{n_C\ell}{T} \tag{34.18}$$

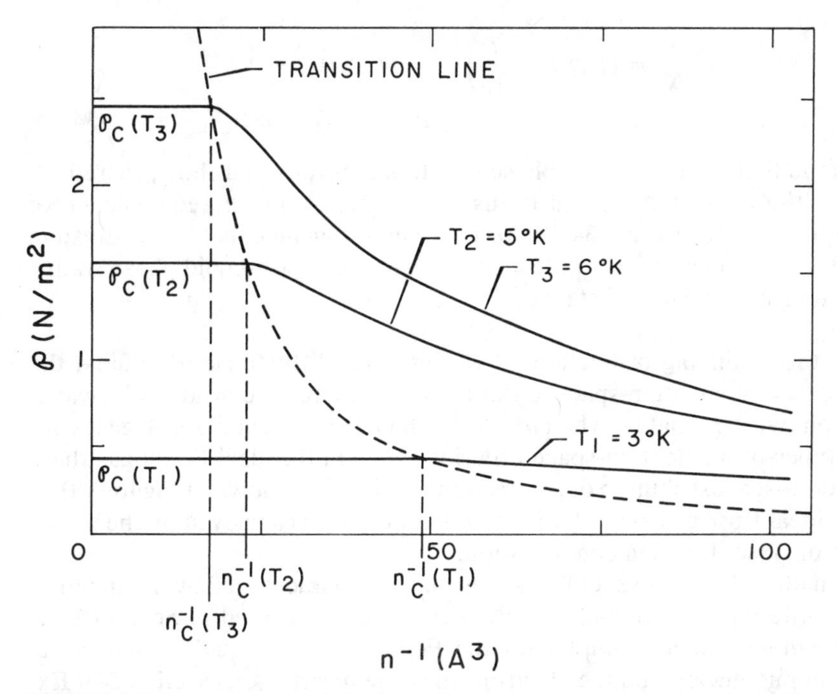

FIGURE VII.7 Isotherms of the ideal Bose gas plotted as a function of the volume per particle n^{-1} [see Equation (34.17)]. The critical density $n_C(T)$ is determined by the first of Equations (34.2), and the pressure $\mathcal{P}_C(T)$ by the third equality of (34.17).

where $n_C(T)$ is defined by the first of Equations (34.2) and

$$\ell \equiv \frac{5}{2}\left(\frac{1.342}{2.612}\right)\kappa T = (1.287)\kappa T \tag{34.19}$$

On comparing Equation (34.18) with the Classius–Clapeyron equation (10.12), we see that the quantity ℓ can be interpreted as a latent heat per particle in the phase transition. Here we take the volume per particle of the nonzero-momentum particles (gas phase) to be n_C^{-1} and the volume per particle of the zero-momentum particles (condensed phase) to be zero at the critical point. Permission to do this occurs via Equation (34.7), which shows that at any temperature and density $n \gtrsim n_C$ the number of particles in the gas phase is the maximum number $(n_C\Omega)$ allowed by the condition $n_C\lambda_T^3 = 2.612$. Of course, below the critical point both phases fill the entire volume Ω.

Exercise 34.3 Since the chemical potential is zero below the critical point we may use the first lines of Equations (32.19) and (34.16) to write for the entropy per particle in an ideal Bose gas.

$$\frac{\langle S \rangle}{N} = (1.287)\kappa\left(\frac{T}{T_C}\right)^{3/2} = (1.287)\kappa\left(\frac{n_C}{n}\right)$$
$$\text{for } T \leq T_C \text{ or } n \geq n_C \qquad (34.20)$$

Argue that the condensed phase is without entropy and show, therefore, that the result (34.20) can be used to demonstrate the equivalence of Equations (10.10) and (34.19) for ℓ. We may conclude that Bose–Einstein condensation is indeed quite analogous to the vapor–liquid phase transition in an ordinary substance.

At the beginning of Section 19 we introduced, for quantum fluids, the concept of momentum-space ordering when there is considerable wave-function overlap, that is, when $n\lambda_T^3 \geq 1$. The ideal Bose and Fermi gases are prototypes of momentum-space ordering. Thus in the ideal Fermi gas there is a Fermi-sea distribution of occupation numbers, as shown in Figure VII.1. In the ideal Bose gas the order in momentum space is achieved by the mechanism of Bose–Einstein condensation.

In nature there are several systems that are characterized by momentum-space ordering. The free-electron gas in metals discussed in Section 33 is one example. Other examples for the Fermi case are liquid helium three, large atomic nuclei, and the electrons in white dwarf stars (Section 54). Examples of Bose gases in nature are blackbody radiation (Section 35) and the phonon gas in crystals (Section 30). The only known Bose liquid is liquid helium four.

Liquid Helium Four

It is believed that the phase transition in which liquid helium four transforms from a normal fluid (He I) to a superfluid (He II) is due to Bose–Einstein condensation. See Figure VII.8. For one thing, the superfluidity exhibited by liquid helium three has a quite different character, and it occurs only below 3 mK. Clearly, the two liquid heliums have some significantly different physical properties, despite the chemical similarity of He^3 and He^4 atoms. The most likely source of this difference in physical properties is the particle statistics of the constituent atoms.[18]

Liquid helium four is a rather loosely packed structure with a density of $\rho = 0.146 \text{ gm/cm}^3$ at near-zero pressures. This density corresponds to an interparticle spacing of $\ell = 3.58$ Å, and therefore apart from their repulsive cores, the helium atoms may be regarded, roughly, as moving about freely in a constant attractive central potential. This argument can, in fact, be made quite quantitative.[19] In first approximation we may regard the helium atoms as free bosons in order to gain qualitative insight into the nature of liquid helium. However, to understand the large zero-point energy in liquid helium

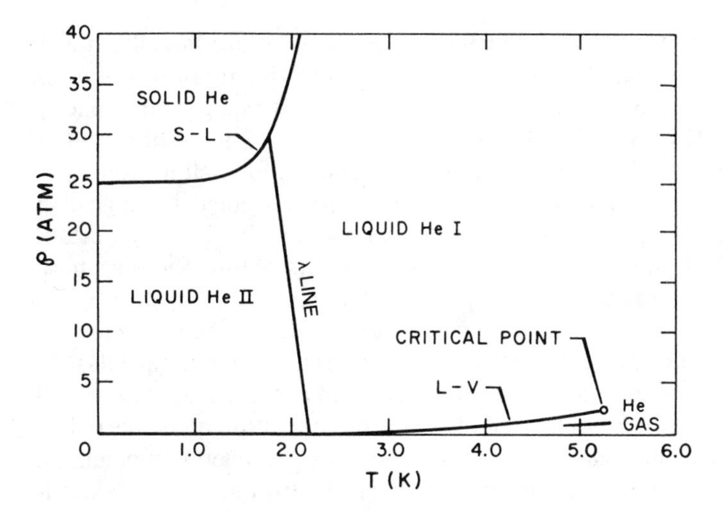

FIGURE VII.8 $\mathcal{P}$–T projection of phase diagram for liquid ^{4}He.

four, we must take into consideration the repulsive cores of the atoms, with diameters $a \sim 2$ Å. We might note, in this connection, that matter composed of any of the other noble elements condenses to the solid state at very low temperatures, because the zero-point energies (which are proportional to m^{-1}) are smaller. Alternatively, matter composed of other light particles, such as hydrogen molecules, solidifies at low temperatures because of stronger interparticle attractions. Indeed, the attractive forces between helium atoms are so weak that even at very low temperatures helium vapor still closely approximates an ideal gas. The quantum-fluid helium is unique in nature.

Encouraging evidence can be given that the λ point in liquid helium four, where He I transforms into He II, at $T_\lambda = 2.18$K for near-zero pressure, is characterized by Bose–Einstein condensation. If we calculate the critical temperature from the second of Equations (34.2), using the density of liquid helium four as given above, then we find $T_C = 3.13$K. This number is remarkably close, perhaps coincidentally, to T_λ. Moreover, the thermal wavelength at T_λ is ~6 Å,[20] which is indeed large compared with the interparticle spacing. We may certainly conclude from these numbers that He II is a very degenerate liquid and that Bose–Einstein condensation ought to occur in it, at least to some extent. It is currently believed that the λ point is characterized by the onset of Bose–Einstein condensation, but that at zero temperature the fraction of bosons in the zero-momentum state is only ~10 to 15%. Theoretical calculations suggest that the *superfluid component* of He II is comprised of $(\mathbf{p}, -\mathbf{p})$ boson pairs in addition to the zero-momentum condensate.

The He I–He II transition point for any given pressure is called the λ *point,* because the specific heat capacity curve in the vicinity of this point is shaped like a (reversed) λ. In fact, the ideal Bose gas has a heat capacity near T_C (Figure VII.6), which resembles somewhat that of liquid helium four near T_λ. The numerical values of C_V (or C_P) near T_λ for helium four are, however, much larger than those of C_V near T_C for the ideal Bose gas. In addition, although the precise nature of the phase transition has been established for the ideal Bose gas (see Exercise 34.3), its nature for the phase transition at T_λ in liquid helium four is still not known.[21]

We have shown that the thermodynamics of the ideal Bose gas below its critical point is determined primarily by its nonzero-momentum particles. In liquid helium four the thermodynamics is governed primarily by its *normal-fluid component,* which is comprised of nonzero-momentum quasi-particles. These quasi-particles are particlelike in that they have an energy–momentum relation $\varepsilon(q)$, which has been measured using neutron-scattering experiments.[22] Moreover, at temperatures below 1.6K the quasi-particles may be regarded as noninteracting so that, essentially, knowledge only of $\varepsilon(q)$ and distribution function (32.1), with $\varepsilon = +1$ and $g = 0$, suffices for thermodynamic calculations below 1.6K. A universally accepted phenomenological theory of He II due to Landau[23] stems from this model, in which the normal fluid comprised of quasi-particles intermingles with the superfluid. The superfluid component is responsible for the remarkable flow properties of liquid helium II.

35 Blackbody Radiation

We treat the problem of electromagnetic radiation in a cavity, or *blackbody radiation,* considered here to be an ideal gas of photons. Since we know that the intrinsic angular momentum of a photon is $\hbar$, photons must obey Bose–Einstein statistics. If the radiation is not in a vacuum, but in a material medium, then for the photon gas to be ideal the interaction of the radiation with the medium must be small. This condition is fulfilled for unionized gases over the whole radiation spectrum, except at frequencies near the absorption lines of the medium. In this section the partition function and thermodynamic behavior of blackbody radiation are calculated.

In the study of a perfect gas, one always assumes tacitly that collisions take place between the otherwise independent atoms to bring about the equilibrium distribution. Light quanta, however, are strictly independent and do not collide (apart from their negligible interactions associated with virtual electron–positron pairs). To attain equilibrium for blackbody radiation there must be some matter present in the cavity, for example, the cavity walls or a few minute black dust particles. By the processes of emission and absorption this matter serves as a catalyst for converting photons of all frequencies into others; total energy is conserved in these processes, but generally total photon number is not.

The radiation field in a large cavity can be considered to consist of a denumerably infinite set of electromagnetic oscillators, corresponding to the various quantum states $\mathbf{k}_i$ in a three-dimensional box. The oscillator frequencies, $\nu_i = (2\pi)^{-1} c|\mathbf{k}_i|$, are related to the total energy E by

$$E = \sum_i n_i h\nu_i \tag{35.1}$$

where n_i is the number of oscillator quanta, or photons, with frequency ν_i, and c is the velocity of light. We do not include the zero-point energies of the electromagnetic oscillators, since these are associated with the vacuum state and not with any measurable effects.

With the above introductory remarks in mind, we turn to the calculation of the partition function of the radiation field. This quantity is given by Equation (31.3) with no restriction on the total photon number, and therefore we have simply

$$\begin{aligned} Z &= \sum_{\{n_i\}} \prod_i e^{-\beta n_i h\nu_i} \\ &= \prod_i \sum_{n_i=0}^{\infty} (e^{-\beta h\nu_i})^{n_i} \\ Z &= \prod_i [1 - e^{-\beta h\nu_i}]^{-1} \end{aligned} \tag{35.2}$$

The derivation of the second and third lines in this equation is in complete analogy with the derivation of Equations (31.7) and (31.8) from Equation (31.4), except that here $\mathfrak{z} = 1$. In analogy with the derivation of Equation (31.16), we use Equation (35.2) to find for the average number $\langle n_i \rangle$ of photons in the cavity,

$$\langle n_i \rangle = [e^{\beta h\nu_i} - 1]^{-1} \tag{35.3}$$

It is clear that the above two results differ formally from those of Section 31 for the ideal Bose gas only by the special condition $\mathfrak{z}_o = 1$ for blackbody radiation. Thus the chemical potential g vanishes identically for photons,

$$g = 0 \qquad \text{(for photons)} \tag{35.4}$$

a fact that follows directly from nonconservation of the total photon number.

To proceed, we next need to calculate the density of photon states in the radiation field. Using periodic boundary conditions, we obtain for the three-dimensional density of states in wave-number space,

$$\frac{d^3\mathbf{n}}{d^3\mathbf{k}} = 2\frac{\Omega}{(2\pi)^3}$$

where the factor of 2 (instead of 3) occurs because the free radiation field consists only of transverse waves for which there are two possible independent polarizations. The density of states per unit frequency interval is then

$$\frac{dn}{d\nu} = \Omega\left(\frac{8\pi\nu^2}{c^3}\right) \tag{35.5}$$

Thermodynamics

By combining Equations (35.3) and (35.5) we obtain the number of photons in a cavity in the frequency interval $d\nu$,

$$\langle n_\nu \rangle \, dn = \Omega[e^{\beta h\nu} - 1]^{-1}\frac{8\pi\nu^2}{c^3}\, d\nu \tag{35.6}$$

The energy density of blackbody radiation of frequency ν, per unit frequency interval, is then

$$\begin{aligned}\Omega^{-1}\frac{d\langle E(\nu)\rangle}{d\nu} &= \frac{8\pi h\nu^3}{c^3}\left[\frac{e^{-\beta h\nu}}{1 - e^{-\beta h\nu}}\right] \\ &= \frac{8\pi h\nu^3}{c^3}e^{-\beta h\nu} \qquad (\text{for } h\nu >> \kappa T) \\ &= \frac{8\pi\nu^2}{c^3}\kappa T \qquad (\text{for } h\nu << \kappa T)\end{aligned} \tag{35.7}$$

The first equality of this last result is *Planck's radiation law,* first deduced by Planck in 1900 in an attempt to derive a general formula that encompassed both *Wien's empirical law* (second line above) and various known classical results.[24] The third line above, due to Lord Rayleigh and Sir James Jeans, was derived subsequently to Planck's original work. Of course, Planck's constant h does not appear in this classical expression, which equals simply the density-of-states factor (35.5) times the thermal energy κT per state.

Exercise 35.1 Express Planck's radiation law as an energy density per unit wavelength interval by using

$$\Omega^{-1}\frac{d\langle E(\nu)\rangle}{d\nu}\, d\nu = -\varepsilon(\lambda)\, d\lambda$$

Then show that the maximum value of $\varepsilon(\lambda)$ occurs at a wavelength λ_{max} given by $\beta hc\lambda_{max}^{-1} = 4.9651$, or

$$\lambda_{max}T = 2897.8\ \mu\text{K} \qquad (1\mu = 10^{-6}\ \text{m})$$

This result that $\lambda_{max} \propto T^{-1}$ is called *Wien's displacement law,* and it was derived originally by Wien using arguments of classical physics.[25]

The total energy density of electromagnetic radiation in a cavity is determined by integrating the first equality of Equation (35.7) over all frequencies. We derive in this manner,

$$\frac{\langle E\rangle}{\Omega} = \frac{8\pi h}{c^3}\int_0^\infty \nu^3\, d\nu\left[\frac{e^{-\beta h\nu}}{1 - e^{-\beta h\nu}}\right]$$

$$= \frac{48\pi}{(hc)^3}(\kappa T)^4 \cdot \sum_{n=1}^{\infty}\frac{1}{n^4}$$

$$\frac{\langle E\rangle}{\Omega} = \frac{8\pi^5}{15}\frac{(\kappa T)^4}{(hc)^3} \qquad \mathbf{(35.8)}$$

which is the famous *Stefan–Boltzmann law* for the temperature dependence of radiation energy in a cavity. The second equality of Equation (35.8) is derived in analogy with Equation (31.21), where the summation term is the Riemann zeta function $\zeta(4)$. The last line then follows by using $\zeta(4) = \pi^4/90$ [see discussion in connection with Equation (32.13)]. The cavity radiancy $\mathcal{R}$ is $c/4$ times this result (cf. Exercise 11.2),

$$\mathcal{R} = \frac{1}{4}c\frac{\langle E\rangle}{\Omega} = \sigma T^4 \qquad (35.9)$$

where

$$\sigma = \frac{2\pi^5\kappa^4}{15h^3c^2} = 5.67 \times 10^{-8} W/m^2K^4$$

is the Stefan–Boltzmann constant.

According to Equation (30.21), the energy density of phonons in a crystal has a T^4 dependence at very low temperatures. The reason that this result agrees with Equation (35.8) for photons in a cavity is that both gases have the same (linear) energy-momentum relation for a single low-momentum particle.[26] In contrast with this T^4 dependence for Bose gases in which the particles are massless, we have derived for ideal gases of particles with mass the low-temperature energy expressions (28.19), (32.18), and (34.16), according to whether the particles obey classical, Fermi–Dirac or Bose–Einstein statistics. See also Problems VI.8(a) and VII.6.

A further thermodynamic distinction between ideal gases of particles with and without mass, that is, with different energy-momentum relations, is found upon calculating the *radiation pressure* of blackbody radiation. Thus from Equations (31.13) and (35.5) we obtain for the radiation-pressure expression, when $\mathfrak{z}_o = 1$ and $\omega_i = h\nu_i$,

$$\mathcal{P} = -(\beta\Omega)^{-1}\int_0^\infty d\nu\left(\Omega\frac{8\pi\nu^2}{c^3}\right)\ln[1 - e^{-\beta h\nu}]$$

$$= \frac{8\pi h}{3c^3}\int_0^\infty \nu^3\, d\nu\left[\frac{e^{-\beta h\nu}}{1 - e^{-\beta h\nu}}\right]$$

$$\mathcal{P} = \frac{1}{3}\frac{\langle E\rangle}{\Omega} \tag{35.10}$$

The second equality of this equation is derived from the first after integrating by parts, and the third follows by comparison with Equation (35.8). Because of the linear energy-momentum relation for photons, we find that the pressure is one-third the kinetic-energy density, instead of two-thirds as in Equation (31.18).

From Equation (35.8) we can derive for the constant-volume heat capacity of blackbody radiation,

$$(\Omega\kappa)^{-1}C_V = \frac{32\pi^5}{15}\left(\frac{\kappa T}{hc}\right)^3 \tag{35.11}$$

The result $\mathcal{P}\Omega = \frac{1}{3}\langle E\rangle$ of Equation (35.10) and the T^4-temperature dependence in Equation (35.8) can also be deduced using arguments of classical physics.[27]

We may, finally, derive an expression for the density of photons in a cavity by integrating Equation (35.6) over all frequencies. We find the result

$$n = \frac{8\pi}{c^3}\int_0^\infty \nu^2\, d\nu\left[\frac{e^{-\beta h\nu}}{1 - e^{-\beta h\nu}}\right]$$

$$= 16\pi\left(\frac{\kappa T}{hc}\right)^3\zeta(3)$$

$$n = (0.244)\left(\frac{\kappa T}{\hbar c}\right)^3 \tag{35.12}$$

where we have used $\hbar$ instead of h in the final equality and $\zeta(3) = 1.202$. In agreement with one of our starting premises in this section the total photon number in a cavity is indeed a variable quantity.[28]

Exercise 35.2 Consider a cavity at 300K and calculate the following numbers for the blackbody radiation:

$$\frac{\langle E \rangle}{\Omega} = 6 \times 10^{-5} \text{ erg/cm}^3$$

$$\mathscr{P} = 2 \times 10^{-5} \text{ dynes/cm}^2$$

$$n = 5.5 \times 10^8 \text{ photons/cm}^3$$

$$\lambda_{av} = 1.8 \times 10^5 \text{ Å}$$

where λ_{av} is defined by $(\langle E \rangle / \langle N \rangle) = (hc/\lambda_{av})$. Comment on the magnitudes of these numbers.

Bibliography

K. Huang, *Statistical Mechanics* (Wiley, 1967), Chapters 11 and 12. In which the ideal Fermi and Bose gases are treated in some detail. Applications to several model systems are included.

C. Kittel, *Introduction to Solid State Physics,* 3rd ed. (Wiley, 1968), Chapter 7. A comprehensive discussion of the free-electron gas in metals.

L. D. Landau and E. M. Lifshitz, *Statistical Physics* (Addison-Wesley, 1958), translated by E. Peierls and R. F. Peierls, Chapter V. Analysis of the ideal Fermi and Bose distribution functions is given for various temperature-density regions. Section 60 on blackbody radiation is especially good.

F. London, *Superfluids,* Vol. II, 2nd rev. ed. (Wiley, 1961), Section 7. A complete analytical treatment of the ideal Bose gas and Bose–Einstein condensation.

J. E. Mayer and M. G. Mayer, *Statistical Mechanics* (Wiley, 1940), Chapter 16. A quite complete review, including abundant mathematical detail and model applications, of the properties of degenerate gases.

R. K. Pathria, *Statistical Mechanics* (Pergamon, 1972), Chapters 7 and 8. Detailed expositions and applications are given of the ideal Bose and Fermi gases.

E. Schrödinger, *Statistical Thermodynamics* (Cambridge, 1967), Chapters VII and VIII. Concise analysis of degenerate gases is given with emphasis on important physical concepts in the development.

Notes

1. The terminology degenerate, as it is used in this chapter, refers to the (average) occupation of a unit cell in phase space by one or more particles (see Figures I.7 and I.8); the degeneracy of quantum-mechanical energy levels is a different concept.

2. See above Equations (5.5) and (5.7).

3. Any points $\mathfrak{z}_s$ are defined to be singularities of $I(\mathfrak{z})$. The radius of convergence of the product function $u(\mathfrak{z})$ is $\mathfrak{z}_c = 1$ along the positive real axis. When $u(\mathfrak{z}_c)$ is well defined, as it is for the ideal Fermi gas ($\varepsilon = -1$), then we may have $\mathfrak{z}_c < \mathfrak{z}_o < \mathfrak{z}_s$ (for positive real $\mathfrak{z}$). When $\mathfrak{z}_c = \mathfrak{z}_s$, then we must have $\mathfrak{z}_o < \mathfrak{z}_c$.

4. Equations (12.29) and (12.30) can be combined for a classical ideal gas at equilibrium to give Equation (31.18). Use of the ideal-gas law in conjunction with Equation (28.16) also yields this result.

5. The method that follows, due in essence to A. Sommerfeld (1928), results in the asymptotic series (32.12). We have utilized some ideas of R. Weinstock [*Am. J. Phys.* **37** (1969), p. 1273], who has shown how the derivation of this series can be made mathematically rigorous. Elaboration of the mathematics associated with integrals over Fermi–Dirac and Bose–Einstein distribution functions $\langle n_i \rangle$ is provided by R. N. Hill, *Am. J. Phys.* **38** (1970), p. 1440.

6. Tabulations of numerical results for these integrals can be found in J. McDougal and E. C. Stoner, *Phil. Trans. Roy. Soc.* (*London*) **A237** (1938), p. 67. Integrals involving the Fermi–Dirac distribution function also occur frequently in theories of interacting fermions, such as the real electron gas in metals and liquid He^3. Straightforward application of the Sommerfeld method does not always prove useful for these integrals, and more elaborate techniques must then be devised. See, for example, A. Wasserman, T. J. Buckholtz, and H. E. DeWitt, *J. Math Phys.* **11** (1970), p. 477; A. Ford, F. Mohling, and J. C. Rainwater, *Ann. Phys.* (*N.Y.*) **84** (1974), p. 80, App. C; and articles by M. L. Glasser, such as *J. Math. Phys.* **10** (1969), p. 1105, and *J. Appl. Phys.* **40** (1969), p. 326. See also Section 53 on relativistic statistics.

7. See M. Abramowitz and I. A. Stegun, *Handbook of Mathematical Functions* (Dover, 1965), Chapter 23, and also footnote 27 of Chapter VI.

8. C. Kittel, *op. cit.*, p. 259.

9. Recall the first equality of Equation (8.16), that is, g is the removal energy for a single electron. In the free-electron model, energies are measured from the "bottom" of the Fermi sea. Such energies differ from those measured relative to the outside vacuum by the amount V; relative to the vacuum the chemical potential is

$$g - V = -\phi$$

10. Discussions of other physical properties that can be understood by using this model—such as electrical conductivity, the Wiedemann–Franz ratio, and the de Haas–van Alphen effect—can be found in the reference texts listed for this chapter.

11. See C. Kittel, *loc. cit.*, pp. 180 and 212, for experimental values of the constants T_D and γ. A brief discussion of effective-mass corrections is given by Kittel on page 214.

12. Representative values of A and ϕ are given in C. Kittel, *loc. cit.*, p. 247.

13. K. Huang, *op. cit.*, Section 11.3.

14. Refer to the general argument that results in Equation (10.6).

15. F. London, *op. cit.*, Appendix.

16. The choice $\omega_0 = 0$ for the lowest energy state has no particular significance, because if we shift the energy $\langle n_0 \rangle$ will still become macroscopic, that is, we will have $\langle n_0 \rangle \sim N$, when $g(T, n) \to \omega_0$.

17. See the second of Figures II.6.

18. It is also striking that under pressure both liquid helium three and liquid helium four solidify to crystals with similar physical characteristics. We can understand this result by observing only that particle statistics affects kinetic energy (associated with momentum-space ordering) much more strongly than it does the potential energy (associated with position-space ordering), which dominates matter in its solid form.

19. F. London, *loc. cit.*, pp. 18–35.

20. See Exercise 11.1.

21. In this connection, see P. T. Sikora and F. Mohling, *J. Low Temp. Phys.* **5** (1971), p. 537.

22. See Figure X.4 and associated discussion at the end of Section 48. The quasi-particles of a Bose fluid are the Bose-statistics counterpart of those discussed for a Fermi fluid in Section 19.

23. L. D. Landau, *J. Phys.* (*USSR*) **11** (1947), p. 91.

24. M. Levitt, "The Original Mystery of Blackbody Radiation," *Fusion* (Jan. 1981), p. 45.

25. This derivation is given by G. H. Wannier, *Statistical Physics* (Wiley, 1966), Section 10.2.

26. As we discussed in connection with Equations (30.13) and (30.23) this linearity does not hold, even approximately, for high-momentum phonons, which is why the T^4 dependence for the energy of simple crystals occurs only at low temperatures.

27. K. Huang, *loc. cit.*, pp. 256–257.

28. Section 3.3 in D. ter Haar, *Elements of Thermostatics* (Holt, Rinehart and Winston, 1966), contains a good treatment of Planck's radiation law including the famous Einstein derivation of Equation (35.6), which uses the principle of detailed balancing for the radiation in a cavity.

29. J. Wiebes and H. C. Kramers, *Proceedings of the Tenth International Conference on Low Temperature Physics* (Moscow, 1966), p. 243.

30. T. D. Lee and C. N. Yang, *Phys. Rev.* **112** (1958), p. 1419. Units used in this reference are chosen so that $\hbar = 2m = 1$.

Problems

VII.1. (a) Determine the fluctuations in n_i for a degenerate ideal gas and show that

$$\langle(\Delta n_i)^2\rangle = \langle n_i\rangle(1 + \varepsilon\langle n_i\rangle)$$

Discuss this result for both the Fermi and Bose cases.

(b) Derive the partition function, Equation (28.24), of Boltzmann statistics for free particles using the method of Section 31. Deduce an explicit expression for the chemical potential in this case and plot its behavior qualitatively as a function of temperature.
Answer: $g_{\text{Boltz}} = \kappa T \ln(n\lambda_T^3)$

VII.2. (a) For a Fermi liquid derive the expression (see Exercise 19.3)

$$\frac{1}{\kappa}C_V = \sum_i (\omega_i' - g)\frac{\partial \nu_i'}{\partial(\omega_i' - g)}\left[-\beta(\omega_i' - g) + \frac{\partial(\omega_i' - g)}{\partial(\kappa T)}\right]$$

where $\nu_i' \equiv \overline{n}_i'$ is given by Equation (19.14). The second term in the brackets is small compared with the first at very low temperatures.

(b) Evaluate the result of part (a) for a large system in the low-temperature limit by making the effective-mass approximation

$$\left.\frac{\partial(\omega_k' - g)}{\partial k}\right|_{k=k_F} = \frac{\hbar^2 k_F}{m^*} \qquad \text{where } \omega_k' = g \text{ at } k = k_F$$

Compare your answer with the heat-capacity expression for an ideal Fermi gas.

VII.3. (a) Show that the speed of sound c in a Fermi fluid can be derived by using the equations

$$c^2 = \left(\frac{\partial \mathcal{P}}{\partial \rho}\right)_S = \left(\frac{\gamma k_F}{3\rho}\right)\left(\frac{\partial \mathcal{P}}{\partial k_F}\right)_T$$

where k_F is defined by Equation (32.3),

$$\gamma \equiv \frac{C_P}{C_V} = 1 - \frac{T}{C_V}\frac{\left(\dfrac{\partial S}{\partial \Omega}\right)_T}{\left(\dfrac{\partial \mathcal{P}}{\partial \Omega}\right)_T}, \qquad \left(\frac{\partial S}{\partial \Omega}\right)_T = -\frac{k_F}{3\Omega}\left(\frac{\partial S}{\partial k_F}\right)_T \qquad \text{etc.}$$

(b) For the ideal Fermi gas derive the low-temperature expansions

$$\gamma = 1 + \frac{\pi^2}{3}\left(\frac{\kappa T}{E_F}\right)^2 - \frac{23\pi^4}{180}\left(\frac{\kappa T}{E_F}\right)^4 + \mathrm{O}\left(\frac{\kappa T}{E_F}\right)^6$$

$$c^2 = \frac{2}{3}\left(\frac{E_F}{m}\right)\left\{1 + \frac{5\pi^2}{12}\left(\frac{\kappa T}{E_F}\right)^2 - \frac{\pi^4}{16}\left(\frac{\kappa T}{E_F}\right)^4 + \mathrm{O}\left(\frac{\kappa T}{E_F}\right)^6\right\}$$

VII.4. *Ideal Bose gas problem*

(a) Derive the first of Equations (34.15) from Equations (34.3) and (34.6) by using the method of Exercise 31.4. [*Suggestion:* Define the variables x and y by

$$x \equiv \left(\frac{T - T_C}{T_C}\right) \qquad \text{and} \qquad \frac{\langle E\rangle}{N} = (0.770)\kappa T_C \cdot y$$

where $(0.770) = \tfrac{3}{2}(1.342)(2.612)^{-1}$.]

(b) Show that the derivative of the constant-volume heat capacity is discontinuous at the critical point by the amount

$$\Delta\left(\frac{\partial C_V}{\partial T}\right)_\Omega \equiv \left[\left(\frac{\partial C_V}{\partial T}\right)_\Omega^{(+)} - \left(\frac{\partial C_V}{\partial T}\right)_\Omega^{(-)}\right] = -(3.66)\frac{N\kappa}{T_C}$$

(c) Consider the region above the critical point and use Equation (34.6) to show that

$$T\left(\frac{\partial \alpha}{\partial T}\right)_{\mathscr{P}} = \frac{5}{2} \cdot \frac{f_{5/2}^{(+)}(e^{-\alpha})}{f_{3/2}^{(+)}(e^{-\alpha})}$$

(d) Show that the heat capacity at constant pressure is infinite at the critical point, that is, verify that

$$C_P(T_C^{(+)}) = (1.12)\frac{N\kappa}{\alpha^{1/2}}\bigg|_{T=T_C} = \infty$$

Discuss this result in physical terms.

VII.5. (a) Substitute the second of Equations (34.4) into Equation (34.10) and derive the expression, valid for small α when $T > T_C$,

$$\alpha = (NC)^{-2/3}\left(\frac{T_C}{T}\right)\left(\frac{\alpha}{e^\alpha - 1}\right)\left\{\frac{1 + (e^\alpha - 1)N\left[\left(\frac{T}{T_C}\right)^{3/2} - 1\right]}{1 - \left(\frac{C'}{C}\right)\alpha^{1/2} + O(\alpha^{3/2})}\right\}^{2/3}$$

where $C = (2.612)^{-1} \cdot 2\sqrt{\pi} = 1.36$ and $C' = (2.612)^{-1} \cdot (1.460) = 0.560$.

(b) Consider the region $0 < (T - T_C) << T_C/N^{1/3}$ and use the result of part (a) to derive Equation (34.11) for $\langle n_0 \rangle = \alpha^{-1}$.

VII.6. Some experimental values[29] for the specific heat capacity C_V of liquid ^{4}He are given in the accompanying table. The measurements were performed at a pressure of 0.1 atm.

Temperature, K	C_V, millijoules/g-deg
0.3	0.55
0.4	1.29
0.5	2.53
0.6	4.76
0.7	10.3
0.8	23.1
0.9	51.3
1.0	106.5

(a) Show graphically that below 0.6K the behavior of C_V is characteristic of that of a gas of phonons.

(b) Use these data to calculate the speed of sound in liquid ^{4}He at very low temperatures.

VII.7. The energy eigenvalues of a low-density gas of Bose hard spheres (characterized by $na^3 << 1$, a = hard-sphere diameter) are given for $N >> 1$ by[30]

$$E(\xi, n_+) = [E_0(\xi) + E_{qp}(\xi)]\frac{\hbar^2}{2m}$$

$$\frac{1}{N}E_0(\xi) = 4\pi na\left\{[1 + (1 - \xi)^2] + \frac{128}{15\sqrt{\pi}}(n\xi^5a^3)^{1/2} + O(na^3\ln(na^3))\right\}$$

$$E_{qp}(\xi) = \sum_{\mathbf{p}\neq 0} n_+ \cdot p(p^2 + p_o^2)^{1/2}$$

where $p_o \equiv (16\pi n\xi a)^{1/2}$ and ξ is determined by the two relations

$$\sum_{\mathbf{p}\neq 0} \langle n(\mathbf{p})\rangle = (1 - \xi)N$$

$$\langle n(\mathbf{p})\rangle = \left\{\frac{(1 + 2y^2)}{4y(1 + y^2)^{1/2}}(1 + 2n_+) - \frac{1}{2}\right\} \xrightarrow[y>>1]{} n_+(\mathbf{p})$$

with $y = p/p_o$. From these equations it can be seen that ξ is the fraction of bosons in the zero-momentum state and $n_+(\mathbf{p})$ is the number of quasi-particles (qp) with momentum $\hbar p$ and energy $\varepsilon'(p)$, where in the first approximation

$$\varepsilon'(p) = \frac{\hbar^2}{2m}p(p^2 + p_o^2)^{1/2}$$

We set $\langle n(\mathbf{p})\rangle \cong n_+(\mathbf{p})$ in this problem.

(a) Write down the partition function $Z(\xi)$, for fixed ξ, for this energy-level spectrum.

(b) Evaluate the partition function for $N >> 1$, using the Darwin–Fowler method as in Section 31.

(c) Determine an expression for the average value $\langle n_+(\mathbf{p})\rangle$.

(d) How are the results of parts (a), (b), and (c) changed if the constraint of fixed ξ is removed? Discuss the relative merits of these two procedures.

VII.8. In first approximation one can assume that both the sun and the earth absorb all electromagnetic radiation incident on them. Moreover, the earth has reached a steady state so that its mean temperature T is essentially constant in time, despite the fact that the earth continually absorbs and emits radiation.

Let the surface temperature of the sun be denoted by T_o, its radius by R, the radius of the earth by r, and the mean distance between the sun and the earth by L.

(a) Find an approximate expression for the temperature T of the earth in terms of the astronomical parameters mentioned above.

(b) Calculate this temperature T numerically, using $T_o = 5500\text{K}$, $R = 7 \times 10^{10}$ cm, $r = 6.37 \times 10^8$ cm and $L = 1.5 \times 10^{13}$ cm. Discuss your answer.

CHAPTER VIII

GRAND CANONICAL ENSEMBLES

The *grand canonical ensemble* is the most powerful tool of statistical mechanics. It differs from the macrocanonical ensemble only in the inclusion, as given information, of system constants other than the energy, linear momentum, and angular momentum. In particular, the total particle number N is the most commonly included additional constant. The inclusion of N as additional known information can simplify statistical calculations considerably; it also makes possible, surprisingly it sometimes seems, the calculation of the rms deviation $(\Delta N)_{\text{rms}}$ in a system. A general method for applying the grand canonical ensemble to physical problems is outlined in Section 36. Then it is shown how the grand partition function Z_G, which characterizes this method, can be expanded in terms of cluster integrals. A particularly simple and physical expression for $\Omega^{-1} \ln Z_G$, which is the grand potential f, results. Application of this result to derivation of the virial expansion of the equation of state for a dilute real gas is given, and Van der Waals' equation, which can be related to this expansion, is discussed. Finally, ionized gases are treated to illustrate how the grand canonical ensemble may be applied when further known information, in this case the total charge of the system, must be considered.

36 Grand Partition Function

The grand canonical ensemble is introduced as a method for solving problems in statistical physics. To treat equilibrium situations, the case of interest in this chapter, the method of the most probable distribution is applied. The result provides expressions for system average values, which are shown to be related by simple differentiation to the grand partition function

Z_G. The thermodynamic energy function associated with Z_G is found to be the grand potential f. Thermodynamic identification of the temperature and chemical potential is made, and these quantities are shown to characterize heat and particle baths, respectively; together, these baths give a physical representation of the grand canonical ensemble at equilibrium. The section concludes with an examination of particle-number fluctuations.

We did not introduce the grand canonical ensemble earlier, because it was not necessary to do so in any of our previous applications. It seemed better to solve many problems in an easier conceptual context, thus saving for the application of the grand canonical ensemble problems eminently suitable for this more powerful method. To be sure, all the applications in the preceding two chapters can also be derived using grand canonical ensembles. In fact, when particle-number fluctuations are small, as is usually the case [see Equation (26.9)], then the grand canonical ensemble and the macrocanonical ensemble are equivalent.[1]

Physical systems are usually associated with density fluctuations. In fact, when we introduced the many identical subsystems of a given large system in connection with Figure VI.1, it was really inappropriate to fix the particle numbers of the subsystems in addition to requiring that they all have the same volume. Thus any one subsystem in Figure VI.1 must be considered to be in contact with both a large energy and a large particle reservoir. The energy reservoir is generally referred to as a temperature bath, and we call the particle reservoir an *activity* (or fugacity) *bath*. Alternatively, we may say that the many subsystems of the large system form a grand canonical ensemble. The equivalence of these two descriptions of a system is demonstrated more carefully in the development below Equation (36.10). For large systems in which chemical or nuclear reactions can take place and the particle numbers vary with the activity coefficients (fugacities) of the system even in the absence of external particle exchange, it is essential to use a grand canonical ensemble to calculate equilibrium properties microscopically.

Exercise 36.1 When the particle number is included as a variable for a subsystem of identical particles, as in Figure VI.1, then we can write for the classical probability distribution function $\rho_{E,N}$, the expression

$$\rho_{E,N} = A \int dE'\, dN' n(E', N')\, \delta(E' + E - E_o)\, \delta(N' + N - N_o)$$

where E_o and N_o are the total energy and particle number of the system. The corresponding primed and unprimed quantities refer to a large and small subsystem, respectively, which together form the given system.

Generalize the derivation associated with Equation (26.7) to show that

$$\rho_{E,N} = \left[\frac{A}{\delta E'(E_o, N_o)} \frac{\exp[S'(E_o, N_o)/\kappa]}{\delta N'(E_o, N_o)}\right] \exp\left[-\frac{1}{\kappa T}(E - gN)\right] \propto \mathfrak{z}^N \rho_E$$

where g is the chemical potential of the system and $\delta E'$ and $\delta N'$ are rms deviations for the large subsystem.

The quantity $\mathfrak{z}$ where

$$\mathfrak{z} \equiv e^{\beta g} \tag{36.1}$$

is called the *fugacity* in statistical mechanics. For an ideal gas this is the quantity $\mathfrak{z}_o$ of Equation (31.15).

We now consider a general mathematical procedure for using the grand canonical ensemble to describe a large system in a quantum-mechanical mixed state. We assume that the system has several components; that is, that there are several different kinds of particles such as electrons, photons. and helium atoms in the system. We label the different kinds of particles by $\alpha, \beta, \gamma, \ldots$ and let the respective particle numbers be $N_\alpha, N_\beta, N_\gamma, \cdots$. Then the ith energy level in the system is a function of the set of these numbers, that is, $i = i\{N_\alpha\}$. If further we let M_i be the number of systems in the state i in the grand canonical ensemble, then in analogy with Equation (27.1) we obtain for the number of possibilities $W_M(M_i)$ for finding the set of numbers M_i,

$$W_M(M_i) = M! \prod_i (M_i!)^{-1} \tag{36.2}$$

where the total number of systems in the ensemble is M. For convenience, we set the energy degeneracy g_i of each state i equal to unity.[2]

We consider next the constraints to which the unnormalized probabilities $W_M(M_i)$ are subjected for a grand canonical ensemble. In addition to constraints (27.2) for systems at rest, we also assume that each of the particle numbers $N_{M,\alpha}$ is conserved for the entire ensemble. We thereby obtain for the complete list of constraints on the $W_M(M_i)$ the set of equations,

$$\begin{aligned} \sum_i M_i &= M \\ \sum_i M_i E_i &= E_M \\ \sum_i M_i N_\alpha &= N_{M,\alpha} \qquad (\text{for } \alpha = \alpha, \beta, \gamma, \cdots) \end{aligned} \tag{36.3}$$

where the sum over states i is intended to denote sums over both energy levels and particle numbers $N_\alpha, N_\beta, \cdots$.

Now if α and β particles can transform into one another by a reaction, then $N_{M,\alpha\beta} \equiv (N_{M,\alpha} + N_{M,\beta})$ is the only ensemble constant we can specify for α and β particles. In addition, if γ particles are photons, then we cannot specify $N_{M,\gamma}$. Finally, if there are other ensemble constants such as the total charge, then we must include them in the above list of constraints. Often, however, special conditions such as the ones we have cited can be incorporated into the theory as conditions on the chemical potentials g_α,[3] and therefore there is not much loss in generality in assuming that Equations (36.3) summarize the complete list of constraints for a multicomponent system. When this list does not suffice, then the general method given below for finding the grand partition function can still be used.

Method of the Most Probable Distribution

We have seen that the method of the most probable distribution for determining the most probable occupation numbers $\overline{M}_i$ leads to a simple and correct procedure for calculating the macroscopic properties of a system in thermodynamic equilibrium. Thus for macrocanonical ensembles we show at the end of Appendix A that the more rigorous Darwin–Fowler method of mean values normally gives identical thermodynamic predictions. To be specific, most probable values $\overline{M}_i$ are equal to average values $\langle M_i \rangle$ in the limit $M \to \infty$ and when the fluctuations $\langle (\Delta M_i)^2 \rangle$ are small.[4] For this reason we may, hesitating only in the case when occupation-number fluctuations are large, apply the method of the most probable distribution to the grand canonical ensemble.

As in Section 27 we determine the maximum of $\ln W_M(M_i)$, subject to the constraints (36.3), by using Lagrange's method of undetermined multipliers. We have discussed before that this procedure is equivalent to maximizing the ensemble entropy. Thus we seek the unrestricted maximum of

$$\begin{aligned} \phi_M(M_i) &\equiv \ln W_M(M_i) + \nu \sum_i M_i + \sum_\alpha \lambda_\alpha \sum_i M_i N_\alpha - \beta \sum_i M_i E_i \\ &\cong M(\ln M - 1) + \sum_i M_i \left[1 - \ln M_i + \nu + \sum_\alpha N_\alpha \lambda_\alpha - \beta E_i\right] \end{aligned} \tag{36.4}$$

where ν, λ_α, and $-\beta$, with $\alpha \in \{\alpha, \beta, \gamma \cdots\}$, are the unknown multipliers. To derive the second line of this equation we have used Stirling's formula as is discussed in connection with Equation (27.3). In analogy with the derivation of Equation (27.4) we find for the most probable value $\overline{M}_i$,

$$\overline{M}_i = e^\nu \exp\left[\sum_\alpha N_\alpha \lambda_\alpha - \beta E_i\right]$$

Now this quantity is not as interesting to us as is the probability w_i that a system be in the ith state, with particle numbers N_α, N_β, $\cdots$. Proceeding as with Equations (27.5), we obtain for w_i, for the average system energy $\langle E \rangle$, and for the average number $\langle N_\alpha \rangle$ of α particles in the system,

$$w_i \equiv \frac{\overline{M}_i}{M} = Z_G^{-1} \exp\left[\sum_\alpha N_\alpha \lambda_\alpha - \beta E_i\right]$$

$$\langle E \rangle \equiv \sum_i w_i E_i = Z_G^{-1} \sum_i E_i \exp\left[\sum_\alpha N_\alpha \lambda_\alpha - \beta E_i\right] \tag{36.5}$$

$$\langle N_\alpha \rangle \equiv \sum_i w_i N_\alpha = Z_G^{-1} \sum_i N_\alpha \exp\left[\sum_\alpha N_\alpha \lambda_\alpha - \beta E_i\right] \qquad (\text{for } \alpha = \alpha, \beta, \gamma, \cdots)$$

where the *grand partition function* Z_G is defined by

$$Z_G \equiv \sum_i \exp\left[\sum_\alpha N_\alpha \lambda_\alpha - \beta E_i\right] \tag{36.6}$$

The grand partition function, or rather its logarithm, is of central importance in the remaining development of this section.

We henceforth distinguish the sum over all possible numbers of particles from the sum over all different quantum states for a given set $\{N_\alpha\}$ of these numbers. We also introduce the quantity Ωf to represent the logarithm of the grand partition function. Then Equation (36.6) for Z_G can be rewritten as

$$Z_G \equiv e^{\Omega f} = \sum_{N_\alpha, N_\beta, \ldots = 0}^{\infty} \exp\left(\sum_\alpha N_\alpha \lambda_\alpha \sum_{i = i\{N_\alpha\}}\right) \exp[-\beta E_i(N_\alpha, N_\beta, \ldots)] \tag{36.6a}$$

Here the upper limit in the first sum, over all different sets of particle numbers, should really be a set of extremely large numbers, rather than ∞, for finite systems. We can accommodate this qualification by assuming that no states i occur whenever the particle numbers exceed their finite-system limits.

The set of equations (36.5) can all be written as partial derivatives of (Ωf) with respect to various system parameters. Thus we obtain

$$w_i = -\beta^{-1} \frac{\partial}{\partial E_i}(\Omega f)$$

$$\langle E \rangle = -\frac{\partial(\Omega f)}{\partial \beta} \tag{36.7}$$

$$\langle N_\alpha \rangle = \frac{\partial(\Omega f)}{\partial \lambda_\alpha} \qquad (\text{for } \alpha = \alpha, \beta, \gamma, \cdots)$$

Assuming that $\beta = (\kappa T)^{-1}$, as we show below, the first two of this last set of equations indicates that (Ωf) is an extensive quantity,[5] whereas the third indicates that the λ_α are intensive quantities. We proceed now to the thermodynamic identification of the Lagrange multipliers, β and the λ_α, and show that the quantity f is the grand potential defined by Equation (8.22).

Thermodynamic Identification of Lagrange Multipliers

The quantity $\Omega f = \ln Z_G$ is a function of β, the λ_α, and any external parameters such as the volume Ω. We follow the method of Section 27 and consider a general variation at equilibrium in Ωf due to variations in these quantities. Thus we write

$$\begin{aligned}\delta(\Omega f) &= \frac{\partial(\Omega f)}{\partial\beta}\,\delta\beta + \sum_\alpha \frac{\partial(\Omega f)}{\partial\lambda_\alpha}\,\delta\lambda_\alpha + \sum_{i=i\{N_\alpha\}} \frac{\partial(\Omega f)}{\partial E_i}\,\delta E_i \\ &= -\langle E\rangle\,\delta\beta + \sum_\alpha \langle N_\alpha\rangle\,\delta\lambda_\alpha - \beta\sum_i w_i\,\delta E_i\end{aligned} \tag{36.8}$$

where we have used Equations (36.7) to obtain the second equality. We require next the first law of thermodynamics (7.11) or, rather, the thermodynamic identity in the form (8.15), which takes into account particle-number variations,[6]

$$\delta\langle E\rangle = T\,\delta\langle S\rangle - \delta W + \sum_\alpha g_\alpha\,\delta\langle N_\alpha\rangle \tag{36.9}$$

We can now rearrange the second equality of Equation (36.8), and use Equations (27.12) and (36.9), to obtain the expression

$$\delta\left[\Omega f + \beta\langle E\rangle - \sum_\alpha \langle N_\alpha\rangle\lambda_\alpha\right] = \sum_\alpha (\beta g_\alpha - \lambda_\alpha)\,\delta\langle N_\alpha\rangle + \beta T\,\delta\langle S\rangle \tag{36.10}$$

The following thermodynamic analysis and identification of β and the λ_α stems from this basic equation.

The result (36.10) is true for any arbitrary variations of the system parameters; in particular, it must be valid when the particle numbers are all held fixed, that is, when $\delta\langle N_\alpha\rangle = 0$ for all α. In this case Equation (36.10) reduces to an expression that has the form of Equation (27.13), and we may use the argument terminating with Equation (27.16) that $\beta = (\kappa T)^{-1}$. That κ, here, is Boltzmann's constant is proved conclusively in Section 38 when we derive the ideal-gas laws of Section 31 using the present formalism.

We consider now an expression for the pressure $\mathcal{P}$ in a system. In analogy with the derivation of Equation (27.19) we find the relation

$$\begin{aligned}\mathcal{P} &= \beta^{-1}\frac{\partial(\Omega f)}{\partial\Omega}\bigg|_{T,g_\alpha} \\ &= (\kappa T)f\end{aligned} \tag{36.11}$$

where the second equality holds because Ωf is an extensive quantity whose natural thermodynamic variables are T, Ω, and the g_α.[7] On comparing this result with Equation (8.22) we verify that f is indeed the grand potential.

We show now, self-consistently, that the first term on the right side of Equation (36.10) is zero identically, that is,

$$\lambda_\alpha = \beta g_\alpha \tag{36.12}$$

where g_α is the chemical potential of component α in the system. With this identification, with $\beta = (\kappa T)^{-1}$ and upon using expression (36.11), Equation (36.10) gives

$$\langle S \rangle = \kappa\beta\left[\Omega\mathcal{P} + \langle E \rangle - \sum_\alpha \langle N_\alpha \rangle g_\alpha\right] \tag{36.13}$$

Any arbitrary integration constants must be absorbed into the definition of the entropy in order that it satisfy Nernst's theorem (9.22). Comparison of this equation with Equations (8.9) and (10.1) shows consistently that the thermodynamic potential G is given by

$$G = \sum_\alpha \langle N_\alpha \rangle g_\alpha \tag{36.14}$$

We may conclude that identification (36.12) must be correct.

Since for any representative system the effect of all the other systems in a grand canonical ensemble can be characterized entirely by the parameters β and the g_α, we may conclude that the rest of the grand canonical ensemble is equivalent to temperature and activity baths as discussed at the beginning of this section.

According to Equation (21.15), the matrix elements of the statistical matrix in the energy representation are the w_i, which we may now write, using Equations (36.5), (36.12), and (36.14), as

$$\rho_{ii} = w_i = e^{-\Omega f} e^{\beta(G - E_i)} \tag{36.15}$$

where $i = i\{N_\alpha\}$. The statistical matrix operator therefore is given by the relation

$$\underline{\rho} = e^{-\Omega f} \exp\left[\beta\left(\sum_\alpha \mathsf{N}_\alpha g_\alpha - \mathsf{H}\right)\right] \tag{36.15a}$$

where H is the multicomponent system Hamiltonian and the N_α are particle-number operators. In analogy with Equation (27.24), we may write Equation (36.6a) for the grand partition function as the representation-independent trace,

$$Z_G = e^{\Omega f} = \mathrm{Tr}\left[\exp\left\{\beta\left(\sum_\alpha \mathsf{N}_\alpha g_\alpha - \mathsf{H}\right)\right\}\right] \tag{36.16}$$

These last two equations, taken together, are equivalent to $\mathrm{Tr}(\boldsymbol{\rho}) = 1$, which is Equation (21.10). Inasmuch as $\beta\Psi = -\Omega f + \beta G$, the w_i for a grand canonical ensemble differ from (27.18) for a macrocanonical ensemble only in the choice of independent variables [see Equations (8.19) and (8.20)]. For the macrocanonical ensemble the independent variables are the set $\{T \text{ (or } \beta), \Omega, N_\alpha\}$; for the grand canonical ensemble they are the set $\{T, \Omega, g_\alpha\}$.[8]

Whenever we replace the λ_α in the grand partition function by βg_α, then the additional β dependence introduced must be taken into account in Equation (36.7) for the system energy $\langle E\rangle$. We obtain, in this case, the modified expression

$$\langle E\rangle = -\frac{\partial(\Omega f)}{\partial\beta} + \sum_\alpha \frac{\partial(\Omega f)}{\partial\lambda_\alpha} g_\alpha$$

$$\langle E\rangle = G - \frac{\partial(\Omega f)}{\partial\beta} \tag{36.17}$$

where we have used the last of Equations (36.7) and Equation (36.14) to derive the second equality. For a single-component system this result is equivalent to Equation (9.5), which was derived from purely thermodynamic considerations. After substituting Equations (36.11) and (36.17) into Equation (36.13) we can deduce an alternate expression for the entropy,

$$\begin{aligned}\langle S\rangle &= -\kappa\beta^2\frac{\partial(\beta^{-1}\Omega f)}{\partial\beta} \\ &= \frac{\partial(\kappa T\Omega f)}{\partial T}\end{aligned} \tag{36.18}$$

Finally, when we substitute Equation (36.12) into the last of Equations (36.7), we obtain expressions for the particle densities $n_\alpha = \langle N_\alpha\rangle/\Omega$,

$$n_\alpha = \beta^{-1}\frac{\partial f}{\partial g_\alpha} \qquad (\text{for } \alpha = \alpha, \beta, \gamma, \ldots) \tag{36.19}$$

At the beginning of this section we suggested that grand canonical ensembles and macrocanonical ensembles must be equivalent when particle-number fluctuations are small. To investigate this question further we now calculate $(\Delta N_\alpha)^2_{\mathrm{rms}}$, using Equations (36.15), (36.16), and (36.19) as follows:

$$\begin{aligned}(\Delta N_\alpha)^2_{\mathrm{rms}} &\equiv \langle(\Delta N_\alpha)^2\rangle = \langle N_\alpha^2\rangle - \langle N_\alpha\rangle^2 \\ &= \beta^{-2}\frac{\partial^2}{\partial g_\alpha^2}(\Omega f) \\ &= \kappa T\frac{\partial\langle N_\alpha\rangle}{\partial g_\alpha}\end{aligned} \tag{36.20}$$

where T and Ω are to be held fixed when calculating the partial derivatives.

Exercise 36.2

(a) For a single-component system, use $\mathcal{P} = -(\partial\psi/\partial v)_T$ and the third of Equations (8.15) to prove that

$$\left(\frac{\partial g}{\partial N}\right)_{T,\Omega} = \frac{v^2}{N}\left(\frac{\partial^2\psi}{\partial v^2}\right)_T = -\frac{v^2}{N}\left(\frac{\partial\mathcal{P}}{\partial v}\right)_T$$

where $v = \Omega/N$ and $\psi = \Psi/N$.

(b) Substitute this result into Equation (36.20), use Equation (9.2) for the isothermal compressibility, and derive the result

$$(\Delta N)^2_{\text{rms}} = \frac{\kappa T\Omega}{v^2}\cdot\kappa_T \qquad \textbf{(36.21)}$$

which is equivalent to the answer to Problem II.7(d). We may conclude that the particle-number fluctuations are small, in the sense of Equation (26.9), when

$$v\kappa_T = -\left(\frac{\partial v}{\partial\mathcal{P}}\right)_T$$

is well defined.

Near a phase transition or close to a critical point, where we may have $\kappa_T \to \infty$, Equation (36.21) shows that particle-number (density) fluctuations can become extremely large. In such cases special analyses of the physical problems are required. Physically, density fluctuations at a liquid–vapor phase transition become extremely large, because the system is composed of two phases of different densities so that the number of particles in any given volume can have a whole range of values, depending on the proportions of each phase present. Further discussion of this phase transition is given at the end of Section 39. An analysis of fluctuations near the critical point, which is a limiting case of the liquid–vapor phase transition, is given at the end of Section 49.

37 Cluster-Integral Expansion of Grand Potential

The development of Section 36 shows that $\ln Z_G = \Omega f$, rather than the grand partition function Z_G itself, is an extensive quantity. Moreover, since the grand potential f is directly related to the pressure, by Equation (36.11), and therefore gives directly the equation of state of a substance, we might expect to be able to find a dynamical expression for f instead of Z_G. In fact, such an expression, the cluster-integral expansion of f, is derived from the grand partition function in this section. Unfortunately, explicit expressions for the cluster integrals themselves are neither easier nor harder to compute than the trace of (36.16) that gives Z_G. Although it is possible to deduce a perturbation series for the cluster integrals, in this section we attempt only to clarify how the two effects of particle dynamics and statistics both contribute to general expressions. The Boltzmann-statistics limit of the cluster-integral expansion is also obtained. The section is introduced with a brief outline of explicit results to be achieved.

The grand potential f is expected to be an intensive quantity, as can be seen from Equation (36.11) and other thermodynamic relations. That $\Omega^{-1} \ln Z_G$ is an intensive quantity is not so obvious from Equation (36.16). In fact, a general study of the thermodynamic limit of $\Omega^{-1}\ln Z_G$ using microscopic considerations can become quite involved.[9] For our present purposes we assume merely that this limit exists; that is, we assume for $f \equiv \Omega^{-1}\ln Z_G$ that

$$\lim_{\Omega\to\infty} f(\beta, \Omega, g_\alpha) = f(\beta, g_\alpha) \tag{37.1}$$

where $f(\beta, g_\alpha)$ is a well-defined quantity.

We consider now the so-called cluster-integral expansion of the grand potential, due originally to H. D. Ursell and to J. E. Mayer. The form and method of derivation of the expansion we use were developed by Kahn and Uhlenbeck.[10] The general result for a single-component system is the expression,

$$f(\beta, \Omega, g) = \sum_{N=1}^{\infty} b_N$$

$$f(\beta, \Omega, g) = \Omega^{-1} \sum_{N=1}^{\infty} \frac{1}{N!} \sum_{\mu_1\cdots\mu_N} \exp\left[\beta\left(Ng - \sum_{i=1}^{N} \omega_i\right)\right] U^{(S)}\begin{pmatrix}\mu_1 & \cdots & \mu_N\\ \mu_1 & \cdots & \mu_N\end{pmatrix} \tag{37.2}$$

where the *quantum cluster functions* $U_N^{(S)}$ are defined by the so-called Ursell equations (37.7). The extension of this result to a general multicomponent system is given as Exercise 37.3. In the limit $\lambda_T \ll \ell$, where exchange effects can be neglected, Equation (37.2) reduces to the grand potential (37.17) for Boltzmann statistics.

The grand partition function can be written similarly to Equation (37.2). Thus we shall show that use of the quantum functions $W_N^{(S)}$, instead of the $U_N^{(S)}$, gives Z_G instead of Equation (37.2) for Ωf.

$$Z_G = \sum_{N=0}^{\infty} \frac{1}{N!} \sum_{\mu_1\cdots\mu_N} \exp\left[\beta\left(Ng - \sum_{i=1}^{N} \omega_i\right)\right] W^{(S)}\begin{pmatrix}\mu_1\mu_2 & \cdots & \mu_N\\ \mu_1\mu_2 & \cdots & \mu_N\end{pmatrix} \tag{37.3}$$

The functions $W_N^{(S)}$ are defined by Equations (37.5) and (37.6). Note that the summation over all N values in Equation (37.2) does not include the term $N = 0$; there is no zero-particle cluster.

Equation (37.3) is a particular explicit form for the grand partition function (36.16) of a single-component system. In this representation of Z_G single-particle eigenstates $|\mu\rangle$ of the "unperturbed" Hamiltonian H_o, with eigenvalues ω_i, have been utilized. To demonstrate that Equation (37.3) is a correct expression for Z_G, we observe first, for a single-component system, that the grand partition function Z_G can be written as a sum over all macrocanonical ensemble partition functions $Z(N)$; that is, from Equations (36.16) and (27.24) we obtain

$$Z_G = \sum_{N=0}^{\infty} e^{\beta N g} Z(N)$$

Substitution of Equation (28.3) for $Z(N)$ into this expression then yields immediately the result

$$\begin{aligned} Z_G &= \sum_{N=0}^{\infty} \frac{1}{N!} e^{\beta N g} \sum_{\mu_1 \cdots \mu_N} \sum_{\mathscr{P}'} \varepsilon^{P'} \langle \mu_1 \mu_2 \cdots \mu_N | e^{-\beta \mathsf{H}} U_{P'} | \mu_1' \mu_2' \cdots \mu_N' \rangle_{\mu_i' = \mu_i} \\ &= \sum_{N=0}^{\infty} \frac{1}{N!} e^{\beta N g} \sum_{\mu_1 \cdots \mu_N} \exp\left(-\beta \sum_{i=1}^{N} \omega_i \right) \sum_{\mathscr{P}'} \varepsilon^{P'} \\ &\quad \times \{ U_{P'} \langle \mu_1 \cdots \mu_N | \mathsf{W}(\beta) | \mu_1' \cdots \mu_N' \rangle \}_{\mu_i' = \mu_i} \end{aligned} \tag{37.4}$$

where the operator $\mathsf{W}(\beta)$ is defined by

$$\mathsf{W}(\beta) \equiv e^{\beta \mathsf{H}_o} e^{-\beta \mathsf{H}} \tag{37.5}$$

Note that use of primed variables in the ket $|\mu_1' \cdots \mu_N'\rangle$ enables us to take the permutation operator $U_{P'}$ outside of the matrix element of $\mathsf{W}(\beta)$.

The introduction of the operator $\mathsf{W}(\beta)$ is related somewhat to the use of the interaction picture in quantum mechanics. Thus in the absence of interactions, in which case $\mathsf{H} \rightarrow \mathsf{H}_o$, this operator reduces to the identity operator. Therefore, the matrix elements of $\mathsf{W}(\beta)$ determine the effect of particle interactions on the properties of a system. However, the cluster-integral analysis that follows now does not rely on the introduction of the operator $\mathsf{W}(\beta)$; one may, instead, carry through this analysis starting from the first equality of Equation (37.4). The free-particle exponential factor $\exp(-\beta \sum_{i=1}^{N} \omega_i)$ occurs because of our use of the operator $\mathsf{W}(\beta)$.

To proceed we introduce next a notation that enables one to deal both efficiently and generally with exchange matrix elements. We define the quantum function $W_N^{(S)}$ to be the sum over all permutations $\mathscr{P}'$ that appears in Equation (37.4).

$$\begin{aligned} W^{(S)} \begin{pmatrix} \mu_1 \mu_2 \cdots \mu_N \\ \mu_1' \mu_2' \cdots \mu_N' \end{pmatrix} &\equiv \sum_{\mathscr{P}'} \varepsilon^{P'} U_{P'} W \begin{pmatrix} \mu_1 \mu_2 \cdots \mu_N \\ \mu_1' \mu_2' \cdots \mu_N' \end{pmatrix} \\ &= \sum_{\mathscr{P}'} \varepsilon^{P'} U_{P'} \langle \mu_1 \mu_2 \cdots \mu_N | \mathsf{W}(\beta) | \mu_1' \mu_2' \cdots \mu_N' \rangle \end{aligned} \tag{37.6}$$

The column notation for the states (μ_i, μ_i') has the advantage that it is visually easy to examine the permutation of bottom-row coordinates with respect to top-row coordinates. The superscript (S) on the W functions denotes that symmetrized (or antisymmetrized) matrix elements of $\mathsf{W}(\beta)$ are being considered. According to Equations (16.2) and (16.5) these matrix elements (37.6) are not properly normalized, but normalization has already been treated properly in the derivation of Equation (37.4), or (28.3). Equation (37.3) can now be derived simply by substituting (37.6) into (37.4).

The Ursell Equations

Only the diagonal case, in which $\mu_i' = \mu_i$ in the $W_N^{(S)}$, is required for the grand partition function (37.3). One can readily imagine, for this case, that the general expression for $W_N^{(S)}$ will include terms that factor into two parts with no μ_i values in common. Such factoring is the clustering property of the $W_N^{(S)}$ functions; it occurs, in fact, in all possible ways for the general off-diagonal case as well as for the diagonal case. We discuss this clustering property in a more physical sense below, for the particular example of the position representation $\mu_i = \mathbf{r}_i$. Of course, the states $|\mathbf{r}\rangle$ are not eigenstates of H_o, but the Ursell equations below are valid for any representation that involves a complete set of single-particle states.

We do not demonstrate explicitly that the $W_N^{(S)}$ functions exhibit generally a clustering property.[11] However, we can now account for this property by introducing nonfactorizable quantum cluster functions $U_N^{(S)}$ via the following equations, the Ursell equations. It will be seen that the Nth equation defines $U_N^{(S)}$, and to emphasize that these equations are valid in any single-particle representation we use numerals (1, 2, 3, ...) instead of the particular set of states (μ_1, μ_2, μ_3, $\cdots$). Then

$$
\begin{aligned}
W^{(S)}\begin{pmatrix}1\\1'\end{pmatrix} &= U^{(S)}\begin{pmatrix}1\\1'\end{pmatrix}\\
W^{(S)}\begin{pmatrix}1\,2\\1'2'\end{pmatrix} &= U^{(S)}\begin{pmatrix}1\\1'\end{pmatrix}U^{(S)}\begin{pmatrix}2\\2'\end{pmatrix} + U^{(S)}\begin{pmatrix}1\,2\\1'2'\end{pmatrix}\\
W^{(S)}\begin{pmatrix}1\,2\,3\\1'2'3'\end{pmatrix} &= U^{(S)}\begin{pmatrix}1\\1'\end{pmatrix}U^{(S)}\begin{pmatrix}2\\2'\end{pmatrix}U^{(S)}\begin{pmatrix}3\\3'\end{pmatrix}\\
&\quad + U^{(S)}\begin{pmatrix}1\\1'\end{pmatrix}U^{(S)}\begin{pmatrix}2\,3\\2'3'\end{pmatrix} + U^{(S)}\begin{pmatrix}2\\2'\end{pmatrix}U^{(S)}\begin{pmatrix}1\,3\\1'3'\end{pmatrix}\\
&\quad + U^{(S)}\begin{pmatrix}3\\3'\end{pmatrix}U^{(S)}\begin{pmatrix}1\,2\\1'2'\end{pmatrix} + U^{(S)}\begin{pmatrix}1\,2\,3\\1'2'3'\end{pmatrix}\\
&\;\;\vdots\\
W^{(S)}\begin{pmatrix}1\,2\,\cdots\,N\\1'2'\cdots N'\end{pmatrix} &= \sum U_{n_1}^{(S)}U_{n_2}^{(S)}\cdots U_{n_k}^{(S)}\\
&\;\;\vdots
\end{aligned}
\tag{37.7}
$$

In the Nth of Equations (37.7) the summation on the right side extends:

(a) over all possible partitions $\{n_i\}$ of N as a sum of positive numbers $(n_1, n_2, \ldots, n_k)$;

(b) for each partition, over all possible ways in which N particles can be divided into distinguishable groups consisting, respectively, of $n_1, n_2, \ldots, n_k$ particles.

With reference to rule (b), the cluster functions $U^{(S)}\binom{1\,2}{1'2'}$ and $U^{(S)}\binom{2\,1}{2'1'}$ are to be regarded as belonging to the same groupings of particles 1 and 2. This point is clarified by the following exercise.

Exercise 37.1

(a) Argue using the symmetry of the Hamiltonians H and H_o that

$$\Pi W\begin{pmatrix} 1\,2 & \cdots & N \\ 1'2' & \cdots & N' \end{pmatrix} = W\begin{pmatrix} 1\,2 & \cdots & N \\ 1'2' & \cdots & N' \end{pmatrix} \tag{37.8}$$

where Π is any operator that permutes columns within the W_N function. The unsymmetrized function W_N is defined by the second equality of Equation (37.6).

(b) Use Equations (37.6) and (37.7) to deduce property (37.8) for the functions $W_N^{(S)}$ and $U_N^{(S)}$.

In the application of the Ursell equations to the expansion of the grand partition function (37.3) we only require the diagonal case in which $\mu_i' = \mu_i$. The more general off-diagonal case is required when we consider the permutation problem later in this section. For the diagonal case and in the position representation, we can give a particularly simple interpretation of the Ursell equations. We may argue that the Nth equation corresponds to all possible ways in which N particles can cluster in a box. A physical cluster, in this case, is one in which every particle in the cluster is removed from all

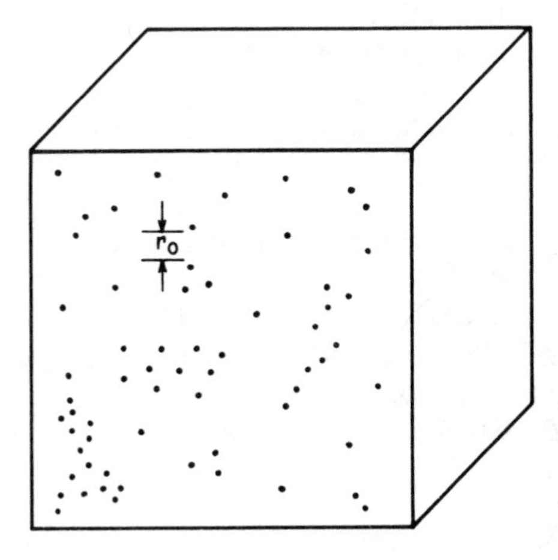

FIGURE VIII.1 Schematic representation of position-space clustering in a system. Fourteen of the particles are not clustered, because they are more than a distance $r_o(>\lambda_T$, in this case) from all other particles; there are two 2-particle clusters, and so on.

other particles in the box by at least a distance D, yet no one particle in the cluster is at a distance greater than D from all the other particles in the cluster. See Figure VIII.1. The distance D is the greater of the lengths r_o and λ_T, where r_o is a measure of the range of particle interactions and λ_T is the thermal wavelength.[12] We may associate the length r_o with classical clustering and λ_T with quantum clustering. That clustering actually occurs microscopically in a system follows from the fact that density fluctuations [see Equation (36.21)] can be relatively large for small volumes.

To sharpen the notion of position-space clustering, as discussed in the preceding paragraph, let us consider a situation in which N particles are divided into two noninteracting groups, with N_1 particles in one group and N_2 particles in the other group, that is, with $|\mathbf{r}_i - \mathbf{r}_j| > r_o$ for all $i \in$ group 1 and $j \in$ group 2. For this case the N-particle Hamiltonian can be written to excellent approximation as $\mathsf{H}_N = \mathsf{H}_{N_1} + \mathsf{H}_{N_2}$, which implies that

$$\exp(-\beta \mathsf{H}_N) = \exp(-\beta \mathsf{H}_{N_1}) \exp(-\beta \mathsf{H}_{N_2}) \qquad \text{(noninteracting groups)}$$

It follows, for this case, that the unsymmetrized W_N function of Boltzmann statistics satisfies the relation

$$W_N = W_{N_1} W_{N_2} \qquad \text{(classical clustering)}$$

When, in addition to the above condition on $|\mathbf{r}_i - \mathbf{r}_j|$, we also have the condition $|\mathbf{r}_i - \mathbf{r}_j| > \lambda_T$ for all $i \in$ group 1 and $j \in$ group 2, then it is possible to show that

$$W_N^{(S)} = W_{N_1}^{(S)} W_{N_2}^{(S)} \qquad \text{(classical and quantum clustering)} \tag{37.9}$$

The proof uses an argument similar to the one given in Section 28 for deducing the classical partition function (28.1) from Equation (28.7).[13]

Exercise 37.2 Construct an induction proof, using Equation (37.9), to prove the following theorem: "A position-space cluster function $U_K^{(S)}$ vanishes if its K coordinates do not all correspond to the same physical cluster of K particles."
(*Suggestion:* First prove this theorem for the case $K = 2$ and then assume that it is true for all $K \leq N$.)

The association of the cluster functions $U_N^{(S)}$ with physical clustering (in position space) is not necessary to derive the cluster-integral expansion of the grand partition function Z_G. Thus, as indicated in footnote 11 and we now emphasize, the Ursell equations (37.7) are valid as a general mathematical property of the $W_N^{(S)}$ defined by Equation (37.6).

We proceed with the expansion of Z_G, which results in Equation (37.2), and define the (Mayer) cluster integrals b_N by

$$b_N(\beta, \Omega, g) \equiv (\Omega \cdot N!)^{-1} \sum_{\mu_1 \cdots \mu_N} \exp\left[\beta\left(Ng - \sum_{i=1}^{N} \omega_i\right)\right] U^{(S)}\begin{pmatrix} \mu_1 \cdots \mu_N \\ \mu_1 \cdots \mu_N \end{pmatrix} \tag{37.10}$$

In connection with this definition it must be stressed that the b_N are volume-independent for large systems, because the $U_N^{(S)}$, by definition, can never be separated into products of smaller cluster functions.[14] In a rough sense, a cluster is like a free particle with internal degrees of freedom so that only one of the sums over states ($\mu_1 \cdots \mu_N$) in b_N can represent translational degrees of freedom for the entire cluster.

We now substitute the Ursell equations (37.7) into Equation (37.3) for Z_G. After relabeling μ_i values and using the invariance of the $U_N^{(S)}$ under column permutations, it can be seen that each of the different arrangements of the particles, as described in rule (b) below Equations (37.7), will give the same contribution to Z_G. We define λ_n to be the number of n-particle clusters in a given partition of N, as described in rule (a) below Equations (37.7). Then the number of different ways in which N particles can be arranged into the partition $\{\lambda_1, \lambda_2, \cdots, \lambda_N\}$ is[15]

$$C_N(\lambda_1, \lambda_2, \ldots, \lambda_N) = \frac{N!}{\prod_{n=1}^{N} (n!)^{\lambda_n} \lambda_n!} \qquad \left(\sum_{n=1}^{N} n\lambda_n = N\right) \tag{37.11}$$

As indicated above, each of these different arrangements gives the same contribution to Z_G so that Equation (37.3) can now be rewritten as

$$Z_G = \sum_{N=0}^{\infty} \frac{1}{N!} \sum_{\substack{\{\lambda_n\} \\ (\Sigma n\lambda_n = N)}} C_N(\lambda_n) \prod_{n=1}^{N} (\Omega n! \cdot b_n)^{\lambda_n} \tag{37.12}$$

where we have used definition (37.10) of the cluster integral b_n.

In analogy with the derivation of Equation (31.7) from Equation (31.4), we write Z_G further, using Equation (37.11), as

$$\begin{aligned} Z_G = e^{\Omega f} &= \sum_{\{\lambda_n\}} \prod_{n=1}^{N} \frac{(\Omega b_n)^{\lambda_n}}{\lambda_n!} \\ &= \prod_{n=1}^{\infty} \sum_{\lambda_n=0}^{\infty} \frac{(\Omega b_n)^{\lambda_n}}{\lambda_n!} \\ &= \prod_{n=1}^{\infty} e^{\Omega b_n} \end{aligned}$$

In the first equality each different partition $\{\lambda_n\}$ is characterized by an N value, given by $\Sigma n\lambda_n = N$, and all possible N values are included in the sum over all partitions. From the third equality it is now straightforward to verify Equation (37.2), which result is the cluster-integral expansion of the grand potential.

Exercise 37.3 Generalize the derivation of Equation (37.2) from Equation (36.16) so that it applies to a multicomponent system, and show that the multicomponent grand potential is given by

$$f(\beta, \Omega, g_\alpha) = \Omega^{-1} \sum_{\substack{N_\alpha, N_\beta, \cdots = 0 \\ (\Sigma N_\alpha \neq 0)}}^{\infty} \prod_\alpha \frac{1}{N_\alpha!} \sum_{\mu_1 \cdots \mu_N}' \exp\left[\beta\left(\sum_\alpha N_\alpha g_\alpha - \sum_{i=1}^{N} \omega_i\right)\right]$$

$$\times\, U^{(S)}\begin{pmatrix} \mu_1 & \cdots & \mu_N \\ \mu_1 & \cdots & \mu_N \end{pmatrix} \qquad \textbf{(37.13)}$$

The prime on the summation over all single-particle states indicates that N_α of the states refer (only) to α particles, N_β of the states refer to β particles, and so on, where $\Sigma_\alpha N_\alpha = N$.
[*Suggestion:* Devise a notation so that the multicomponent grand partition function can be written in the form (37.3).]

The Permutation Problem

The cluster-integral expansion of the grand potential in terms of quantum cluster functions is elegant and physically meaningful, but it would not be very useful if one could not find an explicit procedure for calculating the $U_N^{(S)}$. In fact, the $U_N^{(S)}$ can be identified mathematically by following a procedure indicated in Problem VIII.5 (see also Section 50). Rather than pursue this formal direction, it is instructive to focus attention here on physical aspects that are bypassed by more general procedures. Our intention is to separate the clustering effect due to particle dynamics from that due to particle exchange. In particular, we shall introduce clusters U_N associated entirely with particle dynamics, that is, those of Boltzmann statistics, and we shall show how to express the $U_N^{(S)}$ of quantum statistics entirely in terms of off-diagonal U_K, with $K \leq N$. We refer to this separation of clustering causes as the *permutation problem*.

The cluster functions U_N of Boltzmann statistics are defined in terms of corresponding W_N [see the right sides of Equation (37.6)] by the Ursell equations (37.7) with all superscripts (S) omitted. For the present development we require these equations in their general off-diagonal form. We also introduce a quantity T_N, which is defined in terms of the U_N by the equation

$$T\begin{pmatrix} 1\,2 & \cdots & N \\ 1'2' & \cdots & N' \end{pmatrix} = \varepsilon^N \sum_{\mathcal{P}'} \varepsilon^{P'} U_{P'} U\begin{pmatrix} 1\,2 & \cdots & N \\ 1'2' & \cdots & N' \end{pmatrix} \qquad (37.14)$$

At first sight, one might think that these functions T_N could be identified with the $U_N^{(S)}$, but as we shall show below, the T_N account for only one aspect of the permutation problem. The arbitrary factor ε^N is introduced in definition (37.14) only because it tends to incorporate sign effects in $U_N^{(S)}$ due to exchange. By starting from Equation (37.8), one can show that both the functions U_N and T_N are invariant under the column-permuting operator Π.

Consider now the first equality of Equation (37.6). If one substitutes the quantum Ursell equations into the left side and the Boltzmann–Ursell equations into the right side and compares terms on both sides sequentially for the cases $N = 1, 2$, and 3, then one finds the results:

$$
\begin{aligned}
U^{(S)}\begin{pmatrix}1\\1'\end{pmatrix} &= U\begin{pmatrix}1\\1'\end{pmatrix} = \varepsilon T\begin{pmatrix}1\\1'\end{pmatrix}\\
U^{(S)}\begin{pmatrix}1\,2\\1'2'\end{pmatrix} &= T\begin{pmatrix}1\,2\\1'2'\end{pmatrix} + \varepsilon T\begin{pmatrix}1\\2'\end{pmatrix}T\begin{pmatrix}2\\1'\end{pmatrix}\\
U^{(S)}\begin{pmatrix}1\,2\,3\\1'2'3'\end{pmatrix} &= \varepsilon T\begin{pmatrix}1\,2\,3\\1'2'3'\end{pmatrix} + \varepsilon T\begin{pmatrix}1\\2'\end{pmatrix}T\begin{pmatrix}2\\3'\end{pmatrix}T\begin{pmatrix}3\\1'\end{pmatrix}\\
&\quad + \varepsilon T\begin{pmatrix}1\\3'\end{pmatrix}T\begin{pmatrix}3\\2'\end{pmatrix}T\begin{pmatrix}2\\1'\end{pmatrix} + T\begin{pmatrix}1\\2'\end{pmatrix}T\begin{pmatrix}2\,3\\1'3'\end{pmatrix}\\
&\quad + T\begin{pmatrix}1\\3'\end{pmatrix}T\begin{pmatrix}2\,3\\2'1'\end{pmatrix} + T\begin{pmatrix}2\\1'\end{pmatrix}T\begin{pmatrix}3\,1\\3'2'\end{pmatrix} + T\begin{pmatrix}2\\3'\end{pmatrix}T\begin{pmatrix}3\,1\\2'1'\end{pmatrix}\\
&\quad + T\begin{pmatrix}3\\1'\end{pmatrix}T\begin{pmatrix}1\,2\\3'2'\end{pmatrix} + T\begin{pmatrix}3\\2'\end{pmatrix}T\begin{pmatrix}1\,2\\1'3'\end{pmatrix}
\end{aligned}
\tag{37.15}
$$

These equations solve the permutation problem for $N = 1, 2$, and 3. They show that Boltzmann cluster functions, in the combinations T_K, and for all $K \leqslant N$, should occur in the general expression for $U_N^{(S)}$. Of course, further analysis, which we give only for the easy cases $N = 1$ and $N = 2$ in Section 38, is required to calculate the Boltzmann cluster functions.[16]

Exercise 37.4 Derive Equations (37.15).

On referring to Equations (37.2) and (37.15), one can see that whereas only diagonal quantum cluster-functions $U_N^{(S)}$ are needed for the calculation of the grand potential, the calculation of the $U_N^{(S)}$ themselves (using the method presented here) requires off-diagonal Boltzmann cluster-functions U_K in the combinations T_K. The off-diagonal terms give the effect of exchange for identical-particle systems. By systematizing the derivation of Equations (37.15) one can indeed express the general $U_N^{(S)}$ entirely in terms of T_K with $K = 1, \cdots, N$. The general expression then solves the permutation problem.

We observe now that the T-function products in Equations (37.15) are all "connected." A *connected* product of T functions is defined to be one that cannot be factored (topologically) into two or more separate products when lines are drawn between all the n and corresponding n'. That only connected products of T functions should occur in $U_N^{(S)}$ follows from the fact that the cluster integrals (37.10) are intensive quantities for large systems.

We define a connected product of T functions more precisely by requiring that every such product be associated with a factor $\varepsilon^{N+P'}$.[17] After recalling that the T functions are invariant under column permutations, we can then summarize compactly the three equations (37.15) by the expression

$$U_N^{(S)} = \sum \left[\begin{array}{l} \text{all different connected products} \\ \text{of } T_K \text{ functions, with } K \leqslant N \end{array} \right] \tag{37.16}$$

The proof of this expression for general N is given as Problem VIII.6.

The next logical step in the present development would be to substitute Equation (37.16) into Equation (37.2) and to examine the resulting expression after relabeling and collecting together identical terms. In our specific applications we only need to consider a few simple cases using Equations (37.15), and therefore we do not pursue the general analysis further here. However, after completing the general analysis the grand potential can be written directly in terms of T_K products with the aid of a diagrammatic procedure, and the interested reader is referred to Appendix F for this further development of the theory. The use of diagrams to represent the terms of a general dynamical expansion is characteristic of the many-body problem (see also Section 50) as well as of modern field theories.

The grand potential f_{Boltz} of Boltzmann statistics is also of interest, and it can be deduced from Equation (37.2) simply by neglecting all exchange terms. Since this approximation for f is equivalent to neglecting all exchange terms in Equation (37.6) for $W_N^{(S)}$, we may conclude immediately that f_{Boltz} is given by replacing the $U_N^{(S)}$ in Equation (37.2) by the U_N, that is,

$$f_{\text{Boltz}}(\beta, \Omega, g) = \Omega^{-1} \sum_{N=1}^{\infty} \frac{1}{N!} \sum_{\mu_1 \cdots \mu_N} \exp\left[\beta\left(Ng - \sum_{i=1}^{N} \omega_i\right)\right] U\begin{pmatrix} \mu_1 & \cdots & \mu_N \\ \mu_1 & \cdots & \mu_N \end{pmatrix} \tag{37.17}$$

The classical limit of this expression yields the original Mayer cluster-integral expansion of the equation of state [see Problem VIII.1].

38 Virial Expansion of Equation of State

For a low-density gas corrections to the classical equation of state for an ideal gas are often expressed as an expansion in powers of the density n, which expansion may involve small parameters such as $n\lambda_T^3$, nr_o^3, or other dimensionless quantities. Such a series is called the *virial expansion* of the equation of state.[18] In this section we show generally how the virial expansion may be deduced from the cluster-integral expansion of the grand potential. For an ideal degenerate gas the expansion arises only because of particle identity, and this special case is investigated to illustrate application of the formalism. Explicit general formulas are then derived for the second virial coefficient a_2.

The only dimensionless parameter that enters into the virial expansion of the equation of state for the degenerate ideal gas is $n\lambda_T^3$, where λ_T is the thermal wavelength (4.2). To focus attention generally on some small dimensionless parameter we choose $n\lambda_T^3$ arbitrarily, with an additional factor $(2S + 1)^{-1}$ included,[19] where S is the particle spin, and write the virial expansion for a dilute gas as

$$\frac{\mathcal{P}}{n\kappa T} = \sum_{N=1}^{\infty} a_N \left(\frac{n\lambda_T^3}{2S + 1}\right)^{N-1} \tag{38.1}$$

in which $a_1 = 1$. Our objective now is to deduce this expansion from the cluster-integral expansion of the preceding section.

We wish to derive theoretical expressions for the *virial coefficients* a_N. We assume again the validity of Equation (37.1) and substitute Equations (37.2) and (36.1) into (36.11) to obtain for the pressure of a large single-component system

$$\frac{\mathcal{P}}{n\kappa T} = \frac{2S + 1}{n\lambda_T^3} \sum_{N=1}^{\infty} c_N \mathfrak{z}^N \tag{38.2}$$

where the dimensionless coefficients c_N are defined in terms of the cluster integrals (37.10) by

$$\begin{aligned} c_N(\beta) &\equiv (2S + 1)^{-1}\lambda_T^3 \mathfrak{z}^{-N} \mathrm{b}_N \\ &= \frac{1}{N!}\left(\frac{\lambda_T^3}{\Omega}\right)(2S + 1)^{-1} \sum_{k_1 \cdots k_N} \exp\left(-\beta \sum_{i=1}^{N} \omega_i\right) U^{(S)}\begin{pmatrix} k_1 & \cdots & k_N \\ k_1 & \cdots & k_N \end{pmatrix} \end{aligned} \tag{38.3}$$

In the second equality we have used the fact that for a large volume of gas at rest the single-particle states μ_i become the momentum states $k_i = (\mathbf{k}, m)_i$.

To express Equation (38.2) in the form (38.1), instead of as an expansion in powers of $\mathfrak{z} = e^{\beta g}$, we must use Equation (36.19) in the form

$$\begin{aligned} \left(\frac{n\lambda_T^3}{2S + 1}\right) &= \left(\frac{\lambda_T^3}{2S + 1}\right) \cdot \mathfrak{z}\frac{\partial f}{\partial \mathfrak{z}} \\ &= \sum_{N=1}^{\infty} N c_N \mathfrak{z}^N \end{aligned} \tag{38.4}$$

where $f = \mathcal{P}/\kappa T$. The following exercise now establishes relations between the coefficients a_N and c_N.

Exercise 38.1 Invert the power series (38.4) to obtain the result, assuming $c_1 = 1$,

$$\mathfrak{z} = \left(\frac{n\lambda_T^3}{2S+1}\right) - 2c_2\left(\frac{n\lambda_T^3}{2S+1}\right)^2 + (8c_2^2 - 3c_3)\left(\frac{n\lambda_T^3}{2S+1}\right)^3 + O\left(\frac{n\lambda_T^3}{2S+1}\right)^4 \tag{38.5}$$

Substitute this series into Equation (38.2) and show by comparison with Equation (38.1) that[20]

$$\begin{aligned} a_1 &= 1 \\ a_2 &= -c_2 \\ a_3 &= 4c_2^2 - 2c_3 \\ &\vdots \end{aligned} \tag{38.6}$$

Expressions for the c_N

Explicit expressions for the coefficients, c_1, c_2, and c_3 in Equation (38.2) can be derived by substituting Equations (37.15) into Equation (38.3).[21] To simplify our lowest-order equations, we first determine an expression for $T\binom{k}{k'}$. On referring to Equations (37.5) and (37.6), and the first of Equations (37.7) and (37.15), one readily finds that

$$T\binom{k}{k'} = \varepsilon\, \delta_{k,k'} \tag{38.7}$$

After some relabeling and collecting together of identical terms, one can then establish the following expressions for c_1, c_2, and c_3:

$$\begin{aligned} c_1 &= \frac{\lambda_T^3}{(2S+1)\Omega}\sum_k e^{-\beta\omega_k} \\ c_2 &= \frac{\lambda_T^3}{2(2S+1)\Omega}\left\{\varepsilon\sum_k e^{-2\beta\omega_k} + \sum_{k_1k_2} e^{-\beta(\omega_1+\omega_2)}T\binom{k_1k_2}{k_1k_2}\right\} \\ c_3 &= \frac{\lambda_T^3}{(2S+1)\Omega}\left\{\frac{1}{3}\sum_k e^{-3\beta\omega_k} + \varepsilon\sum_{k_1k_2} e^{-2\beta\omega_1}e^{-\beta\omega_2}T\binom{k_1k_2}{k_1k_2}\right. \\ &\quad \left. + \frac{\varepsilon}{6}\sum_{k_1k_2k_3} e^{-\beta(\omega_1+\omega_2+\omega_3)}T\binom{k_1k_2k_3}{k_1k_2k_3}\right\} \end{aligned} \tag{38.8}$$

Exercise 38.2 Verify Equations (38.8) by substituting Equations (37.15) into Equation (38.3).

The formal procedure for calculating the coefficients c_N can be summarized roughly as follows: One requires explicit calculation of either quantum cluster integrals [Equation (38.3)] or combinations of Boltzmann cluster integrals [Equations (38.8)]. According to Equations (37.5) to (37.7) these integrals are essentially either symmetrized or unsymmetrized N-particle traces of the operator $e^{-\beta \mathsf{H}}$, with certain (factoring) parts removed. Equations (38.14) and (38.15) display this procedure for the case $N = 2$.

The virial expansion (38.1) for an imperfect classical gas applies to a quantum ideal gas when the equation of state for the latter can be written as a power series in $n\lambda_T^3$ or $\mathfrak{z}$. We can exhibit the series explicitly for this case after writing down the grand potential f_o for an ideal gas. Since the grand partition function Z_G is related to the partition function $Z(N)$ for a macrocanonical ensemble by the equation

$$Z_G = e^{\Omega f} = \sum_{N=0}^{\infty} \mathfrak{z}^N Z(N) \tag{38.9}$$

we may deduce the ideal-gas grand potential directly from results in Section 31. Thus the quantity $u(\mathfrak{z})$ defined by Equation (31.4) is the grand partition function for an ideal gas, because, as we have argued in connection with this equation, it is of the form (38.9). Then, according to Equation (31.8) the grand potential f_o is[22]

$$\begin{aligned} f_o &= -\frac{\varepsilon}{\Omega} \sum_k \ln[1 - \varepsilon \mathfrak{z} e^{-\beta\omega_k}] \\ &= \Omega^{-1} \sum_{N=1}^{\infty} \frac{\mathfrak{z}^N}{N} \varepsilon^{N+1} \sum_k e^{-N\beta\omega_k} \end{aligned} \tag{38.10}$$

We now compare Equations (38.2) and (38.10), using $f = \mathcal{P}/\kappa T$, and obtain for the ideal-gas coefficient $c_N^{(0)}$ the expression

$$\begin{aligned} c_N^{(0)} &= \frac{\lambda_T^3}{(2S + 1)\Omega} \left(\frac{\varepsilon^{N+1}}{N} \right) \sum_k e^{-N\beta\omega_k} \\ &\xrightarrow[\Omega\to\infty]{} \frac{\lambda_T^3}{(2\pi)^3} \cdot \frac{\varepsilon^{N+1}}{N} \int d^3\mathbf{k} e^{-N\beta\omega_k} \\ &= \frac{\varepsilon^{N+1}}{N^{5/2}} \end{aligned} \tag{38.11}$$

where the final equality is derived by using the second of Equations (2.4). The first equality agrees with the leading terms of Equations (38.8). When exchange effects are included for an ideal gas it is clear that all the virial coefficients in Equation (38.1) are nonvanishing. We obtain the classical ideal-gas law $\mathcal{P} = n\kappa T$ only in the limit $n\lambda_T^3 \to 0$ when exchange terms can be neglected. We note finally from Equation (38.11) that $c_1 = c_1^{(0)} = 1$, as was assumed in Exercise 38.1.

Second Virial Coefficient

Experimental determination of the virial coefficients is achieved by fitting the $\mathcal{P}\Omega$ isotherms for a gas to the (in principle, infinite) series

$$\frac{\mathcal{P}}{n\kappa T} = 1 + B\left(\frac{n}{N_A}\right) + C\left(\frac{n}{N_A}\right)^2 + \cdots \tag{38.12}$$

where N_A is Avogadro's number, that is, n/N_A is the molar density of the gas. Thus $B(T)$ and $C(T)$ are the experimentally determined second and third virial coefficients respectively. Comparison with Equation (38.1) yields $a_2 = (2S + 1)B(T)/N_A\lambda_T^3$, and so forth. The second virial coefficient is the best known, and sometimes only known, coefficient for a gas. Comparison with theoretical calculations over a wide temperature range (see, e.g., Problem VIII.3.) can yield detailed information about intermolecular potentials. See also footnote 25 in Chapter VI.

On combining Equations (38.6), (38.8), (38.11), and (37.14), we obtain an explicit expression for the second virial coefficient a_2 in terms of the Boltzmann cluster function U_2.

$$\begin{aligned} a_2(\beta) = & -\frac{\varepsilon}{4\sqrt{2}} - \frac{1}{2}\frac{\lambda_T^3\Omega}{(2S+1)(2\pi)^6}\sum_{m_1,m_2}\int d^3\mathbf{k}_1\, d^3\mathbf{k}_2\, e^{-\beta(\omega_1+\omega_2)} \\ & \times\left[U\begin{pmatrix}k_1k_2\\k_1k_2\end{pmatrix} + \varepsilon U\begin{pmatrix}k_1k_2\\k_2k_1\end{pmatrix}\right] \end{aligned} \tag{38.13}$$

Here we have also taken the infinite-volume limit (17.29). The second virial coefficient is determined by the two-particle cluster function, which can, in principle, be calculated for any given interparticle interaction, as we discuss further below. The fitting of a derived result to an experimental curve can then determine unknown parameters of the chosen interaction.

Higher-order coefficients, such as a_3, which depends on c_3 of Equations (38.8), involve many-particle cluster functions, and these are very difficult to calculate. The reason for this difficulty is simply that evaluation of these functions for realistic interactions requires a solution, even if only in some convergent approximate manner, of the N-body dynamical problem. And whether classical or quantum mechanical, such solutions are invariably not known for $N \geqslant 3$.

According to Equations (37.5) and (37.6), and the first two (Boltzmann) Ursell equations (37.7), the two-particle cluster function can be determined as follows:

$$\begin{aligned} U\begin{pmatrix} k_1k_2 \\ k_1'k_2' \end{pmatrix} &= W\begin{pmatrix} k_1k_2 \\ k_1'k_2' \end{pmatrix} - W\begin{pmatrix} k_1 \\ k_1' \end{pmatrix} W\begin{pmatrix} k_2 \\ k_2' \end{pmatrix} \\ &= \langle k_1k_2|\mathsf{W}(\beta)|k_1'k_2'\rangle - \langle k_1|k_1'\rangle\langle k_2|k_2'\rangle \\ &= \langle k_1k_2|\mathsf{U}_2(\beta)|k_1'k_2'\rangle \end{aligned} \tag{38.14}$$

where the two-particle operator $\mathsf{U}_2(\beta)$ is defined by

$$\mathsf{U}_2(\beta) \equiv e^{\beta \mathsf{H}_o} e^{-\beta \mathsf{H}} - 1 \tag{38.15}$$

If we substitute these last results into Equation (38.13), then we can derive an explicit expression for a_2. It is convenient for the manipulations that follow to incorporate the exponential factor $\exp[-\beta(\omega_1 + \omega_2)]$ into the matrix element of $\mathsf{U}_2(\beta)$. We obtain

$$\begin{aligned} a_2 &= -\frac{\varepsilon}{4\sqrt{2}} - \frac{1}{2}\frac{\lambda^3{}_T\Omega}{(2S+1)(2\pi)^6}\sum_{m_1m_2}\int d^3k_1\, d^3k_2\{\langle k_1k_2|e^{-\beta\mathsf{H}_o} - e^{-\beta\mathsf{H}_o}|k_1k_2\rangle \\ &\quad + \varepsilon\langle k_1k_2|e^{-\beta\mathsf{H}} - e^{-\beta\mathsf{H}_o}|k_2k_1\rangle\} \\ &= -\frac{\varepsilon}{4\sqrt{2}} - \frac{1}{2}\cdot\frac{\lambda^3{}_T\Omega}{(2\pi)^6}\int d^3\mathbf{k}_1\, d^3\mathbf{k}_2\{(2S+1)\langle \mathbf{k}_1\mathbf{k}_2|e^{-\beta\mathsf{H}} - e^{-\beta\mathsf{H}_o}|\mathbf{k}_1\mathbf{k}_2\rangle \\ &\quad + \varepsilon\langle \mathbf{k}_1\mathbf{k}_2|e^{-\beta\mathsf{H}} - e^{-\beta\mathsf{H}_o}|\mathbf{k}_2\mathbf{k}_1\rangle\} \\ &= -\frac{\varepsilon}{4\sqrt{2}} - \frac{\sqrt{2}\Omega}{(2\pi)^3}\int d^3\mathbf{k}_{12}\{(2S+1)\langle \mathbf{k}_{12}|(e^{-\beta\mathsf{H}} - e^{-\beta\mathsf{H}_o})_{\text{rel}}|\mathbf{k}_{12}\rangle \\ &\quad + \varepsilon\langle \mathbf{k}_{12}|(e^{-\beta\mathsf{H}} - e^{-\beta\mathsf{H}_o})_{\text{rel}}|\mathbf{k}_{21}\rangle\} \end{aligned} \tag{38.16}$$

In the second equality of this equation we have assumed the special case of spin-independent interactions and performed the sum over spin states, m_1 and m_2. To derive the third equality, we have transformed to center-of-momentum (CM) and relative (rel) coordinates[23] and used the fact that the Hamiltonians H and H_o are separable, for two particles, in these coordinates. Since the CM coordinate $\mathbf{K}_{12}$ is invariant under exchange, that is, $\mathbf{K}_{12} = \mathbf{k}_1 + \mathbf{k}_2 = \mathbf{K}_{21}$, both the direct and exchange integrals in the second equality of Equation (38.16) are associated with the same factor,

$$\begin{aligned} \int d^3\mathbf{K}_{12}\langle \mathbf{K}_{12}|(e^{-\beta\mathsf{H}_o})_{\text{CM}}|\mathbf{K}_{12}\rangle &= \int d^3\mathbf{K}_{12}\exp\left[-\beta\frac{\hbar^2K_{12}^2}{4m}\right] \\ &= 2\sqrt{2}(2\pi)^3\lambda_T^{-3} \end{aligned} \tag{38.17}$$

Exercise 38.3 Verify Equation (38.17) and the third equality in Equation (38.16) by using the transformations to CM($\mathbf{R}_{12}$, $\mathbf{K}_{12}$) and rel($\mathbf{r}_{12}$, $\mathbf{k}_{12}$) coordinates:

$$\mathbf{K}_{12} = (\mathbf{k}_1 + \mathbf{k}_2); \qquad \mathbf{R}_{12} = \frac{1}{2}(\mathbf{r}_1 + \mathbf{r}_2)$$

$$\mathbf{k}_{12} = \frac{1}{2}(\mathbf{k}_1 - \mathbf{k}_2); \qquad \mathbf{r}_{12} = \mathbf{r}_1 - \mathbf{r}_2$$

$$\nabla_1^2 + \nabla_2^2 = \frac{1}{2}\nabla_R^2 + 2\,\nabla_r^2$$

$$\mathbf{k}_1 \cdot \mathbf{r}_1 + \mathbf{k}_2 \cdot \mathbf{r}_2 = \mathbf{K}_{12} \cdot \mathbf{R}_{12} + \mathbf{k}_{12} \cdot \mathbf{r}_{12}$$

The general expression for the second virial coefficient consists of two distinct parts,

$$a_2(\beta) = a_2^{(0)} + a_2^{(V)}(\beta) \tag{38.18}$$

where

$$a_2^{(0)} = -\frac{\varepsilon}{4\sqrt{2}} \tag{38.19}$$

is the free-particle exchange contribution of Equation (38.11), for $N = 2$, and $a_2^{(V)}$ is the contribution including an exchange integral due to the existence of particle interactions,

$$\begin{aligned} a_2^{(V)}(\beta) &= -\frac{\sqrt{2}\Omega}{(2\pi)^3}\int d^3\mathbf{k}_{12}\{(2S+1)\langle\mathbf{k}_{12}|(e^{-\beta\mathsf{H}} - e^{-\beta\mathsf{H}_o})_{\text{rel}}|\mathbf{k}_{12}\rangle \\ &\quad + \varepsilon\langle\mathbf{k}_{12}|(e^{-\beta\mathsf{H}} - e^{-\beta\mathsf{H}_o})_{\text{rel}}|\mathbf{k}_{21}\rangle\} \\ &= -\sqrt{2}\int d^3\mathbf{r}_{12}\{(2S+1)\langle\mathbf{r}_{12}|(e^{-\beta\mathsf{H}} - e^{-\beta\mathsf{H}_o})_{\text{rel}}|\mathbf{r}_{12}\rangle \\ &\quad + \varepsilon\langle\mathbf{r}_{12}|(e^{-\beta\mathsf{H}} - e^{-\beta\mathsf{H}_o})_{\text{rel}}|\mathbf{r}_{21}\rangle\} \end{aligned} \tag{38.20}$$

The first equality of this equation is useful for the derivation of expressions to be applied to low-temperature data analyses. See, for example, Problem VIII.4. The second equality, which can be derived from the first by using the unitary-transformation matrix elements (17.27), is more useful for high-temperature regions of analysis. For realistic interactions in which there is a repulsive core the exchange integral in these equalities decreases exponentially as the temperature increases, that is, for $\lambda_T \lesssim$ (repulsive-core diameter).[24]

Finally, we deduce the classical (or high-temperature) limit of $a_2^{(V)}(\beta)$. In this case we can neglect the exchange integral in Equation (38.20) by using the argument of Section 28, which ends with Equations (28.9) to (28.11).

This development utilizes approximation (28.5), valid when $\lambda_T \ll r_o$. Thus we obtain from Equation (38.20) an expression for the second virial coefficient $a_{2,cl}(\beta)$ of classical statistical mechanics,

$$a_{2,cl}(\beta) = -(2S+1)\sqrt{2}\int d^3\mathbf{r}_{12}\langle\mathbf{r}_{12}|(e^{-\beta H_o})_{\mathrm{rel}}|\mathbf{r}_{12}\rangle[e^{-\beta V(\mathbf{r}_{12})}-1]$$

$$a_{2,cl}(\beta) = -\frac{(2S+1)}{2\lambda_T^3}\int d^3\mathbf{r}[e^{-\beta V(\mathbf{r})}-1] \tag{38.21}$$

Here we have used Equations (28.10) and (28.12) to derive the second equality, with $m \to m/2$ for the reduced mass in the relative-coordinates system. The quantum-mechanical factor $(2S+1)\lambda_T^{-3}$ cancels when this expression for a_2 is substituted into Equation (38.1).

Inasmuch as the integral in Equation (38.21) cuts off at $r \sim r_o$, where r_o is the range of the particle interaction $V(r)$, we may argue that the classical second virial coefficient is $\sim(r_o/\lambda_T)^3$. In the domain of classical physics where $r_o \gg \lambda_T$, we conclude that $a_{2,cl}$ is much greater than the free-particle exchange term $a_2^{(0)}$ of Equation (38.19). Moreover, it can be seen from Equation (38.1) that in the classical region the virial expansion of the equation of state becomes an expansion in powers of the parameter nr_o^3, instead of $n\lambda_T^3$. In general, of course, both lengths r_o and λ_T will occur in the virial expansion.[25]

39 Van der Waals' Equation and the Maxwell Construction

Van der Waals' equation gives the isotherms of an imperfect gas to good accuracy for dilute gases and qualitatively for dense gases. It can be derived for the dilute case, and we do so in this section, from the virial expansion of the equation of state. For a dense gas we show that Van der Waals' equation exhibits a liquid–vapor phase transition provided the so-called Maxwell construction is applied. The significance of the Maxwell construction is investigated generally for a substance described by a grand canonical ensemble.

Consider the function

$$f(\mathbf{r}) \equiv e^{-\beta V(\mathbf{r})} - 1 \tag{39.1}$$

which appears in Equation (38.21). When written in terms of this function, the leading virial coefficient $a_{2,cl}$ gives for the equation of state (38.1) the imperfect-gas law

$$\frac{\mathcal{P}}{n\kappa T} = 1 - \frac{n}{2}\int d^3\mathbf{r}\, f(\mathbf{r}) \tag{39.2}$$

On referring to Equation (39.1), we obtain the physically reasonable result that $\mathcal{P}/n\kappa T < 1$ for attractive interactions and $\mathcal{P}/n\kappa T > 1$ for repulsive interactions.

We investigate first the dependence of $f(\mathbf{r})$ on r.[26] For ordinary gases, the intermolecular potential appears typically as is shown in Figure VI.4. We may idealize this interaction by defining it to consist of a hard-sphere repulsion, with sphere diameter a,[27] and a short-range attractive well. Corresponding to this interaction we obtain the r dependence of $f(r)$ shown in Figure VIII.2. *Van der Waals' equation of state* can now be derived from Equation (39.2) using this sketch of $f(r)$ as a basis. We note in passing, from Equation (39.2) and since

$$|f(r)| \rightarrow |\beta V(r)| << 1 \qquad \text{as } r \rightarrow \infty$$

that a minimum requirement for the virial expansion (38.1) to converge is that $V(r)$ must fall off faster than r^{-3} at large distances. In fact, at large distances $V(r) \propto r^{-6}$ generally for molecular interactions.[28]

Let us divide the integral in Equation (39.2) into two parts, corresponding to the attractive and repulsive regions of Figure VIII.2. We define

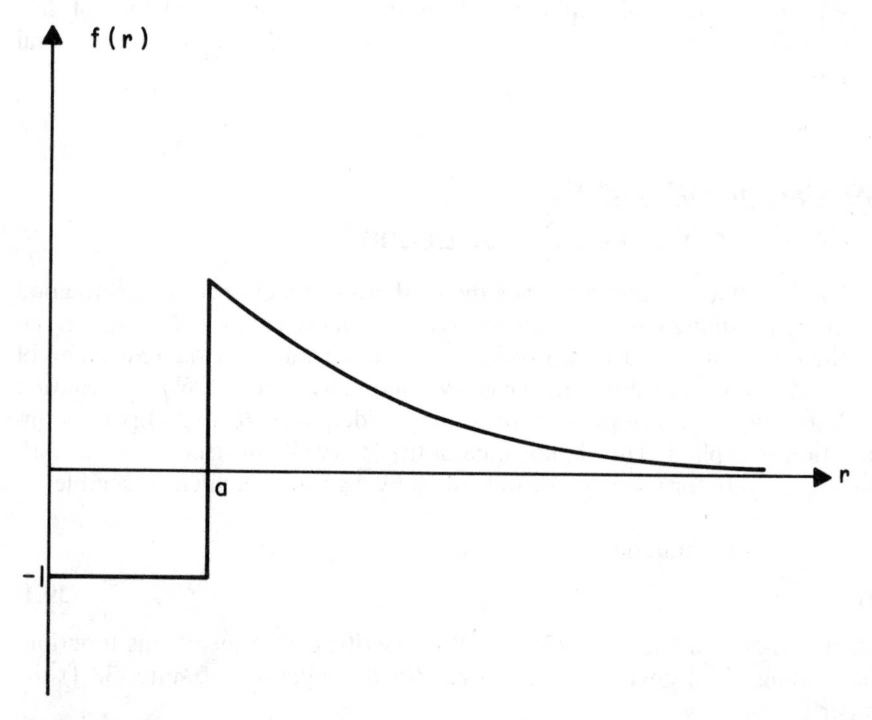

FIGURE VIII.2 The r dependence of the function $f(r)$ for an idealized intermolecular potential $V(r)$.

$$\tilde{a} \equiv \frac{1}{2\beta}\int_a^\infty d^3\mathbf{r}\, f(r) = \frac{2\pi}{\beta}\int_a^\infty r^2\, dr\, f(r)$$
$$b \equiv -\frac{1}{2}\int_0^a d^3\mathbf{r}\, f(r) = -2\pi\int_0^a r^2\, dr\, f(r) \tag{39.3}$$

On assuming that $-\beta V(r) << 1$ for $r \geqslant a$ and that the potential is infinite for $r < a$, these two quantities $\tilde{a}$ and b become

$$\tilde{a} \cong -2\pi\int_a^\infty r^2 dr\, V(r)$$
$$b = \left(\frac{2\pi}{3}\right)a^3 \tag{39.4}$$

The quantity $\tilde{a}$ is a measure of the strength of the attractive interparticle forces, and b, which is four times the volume of a single hard sphere, is a measure of the repulsive forces.

If we use the notation $v = n^{-1}$ for the volume per particle and substitute definitions (39.3) into Equation (39.2), then the imperfect gas law becomes

$$\mathcal{P}v = \kappa T\left(1 + \frac{b}{v}\right) - \frac{\tilde{a}}{v}$$

For low densities such that $b << v$, this equation of state can be written as

$$\mathcal{P}v = \frac{\kappa T}{1 - b/v} - \frac{\tilde{a}}{v}$$

or

$$\left(\mathcal{P} + \frac{\tilde{a}}{v^2}\right)(v - \mathrm{b}) = \kappa \mathrm{T} \tag{39.5}$$

which is Van der Waals' equation of state. In addition to the above derivation using the low-density approximation $na^3 < nr_o^3 << 1$, it has also been shown that Van der Waals' equation follows rigorously for classical systems in the limit of weak infinite-range interactions.[29] Moreover, this equation has been applied empirically to the study of real gases over a wide range of densities. It is therefore instructive to assume its general validity and to examine it further on its own merits. In Figure VIII.3 we show some isotherms corresponding to Van der Waals' equation of state. These resemble qualitatively the isotherms for a real gas, as depicted in Figure II.6.

Below the critical temperature T_C of a Van der Waals' gas there is an unphysical region where $\partial\mathcal{P}/\partial v > 0$.[30] In this region we draw a horizontal line, according to the *Maxwell construction* described below, and interpret

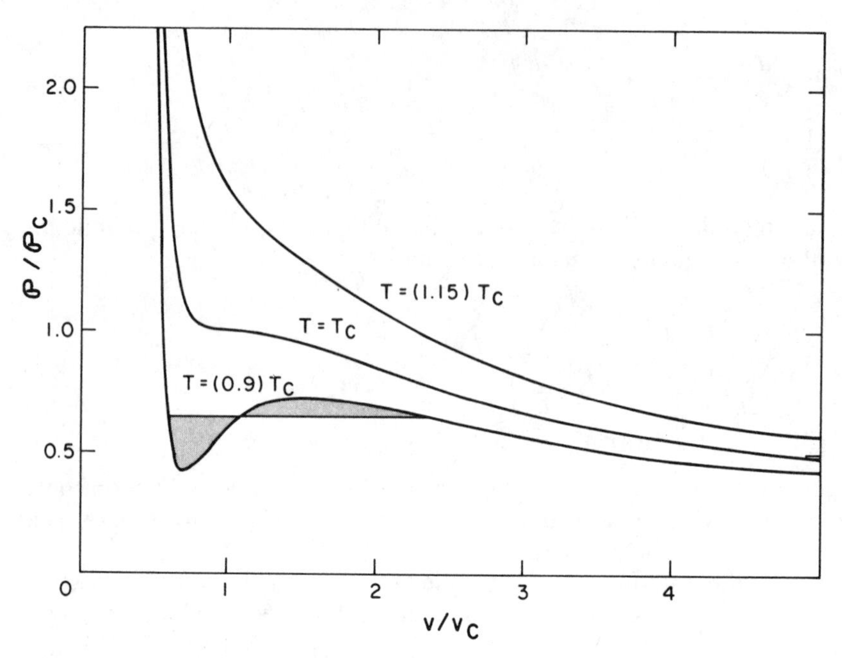

FIGURE VIII.3 Three isotherms of a Van der Waals' gas: for $T > T_C$, $T = T_C$, and $T < T_C$, where the subscript C refers to the critical point.

the constant pressure between v_1 and v_2 as corresponding to a liquid–vapor transition. By this procedure we eliminate the unphysical portions of isotherms where they occur in Van der Waals' equation. Of course, the appearance of vapor–liquid condensation stemming from the imperfect-gas law (39.2) cannot be taken seriously.

The Maxwell Construction

We first describe the Maxwell construction in a general manner and then elaborate on its significance. Consider a change in the Helmholtz free energy per particle, $\psi(T,v)$, at fixed temperature. According to Equation (8.5), the change between two specific volumes v_0 and v is

$$\psi(v) - \psi(v_0) = -\int_{v_0}^{v} dv\,\mathcal{P}(v) \qquad (T = \text{const}) \tag{39.6}$$

In other words, the area under an isotherm in a $\mathcal{P}$–v diagram gives the change in $-\psi(v)$ for fixed T. If we calculate this change for the $\mathcal{P}$–v isotherm in Figure VIII.4(a), starting from some specific volume $v_0 < v_1$, then we obtain the curve shown in Figure VIII.4(b). The dashed straight line 1–4–2

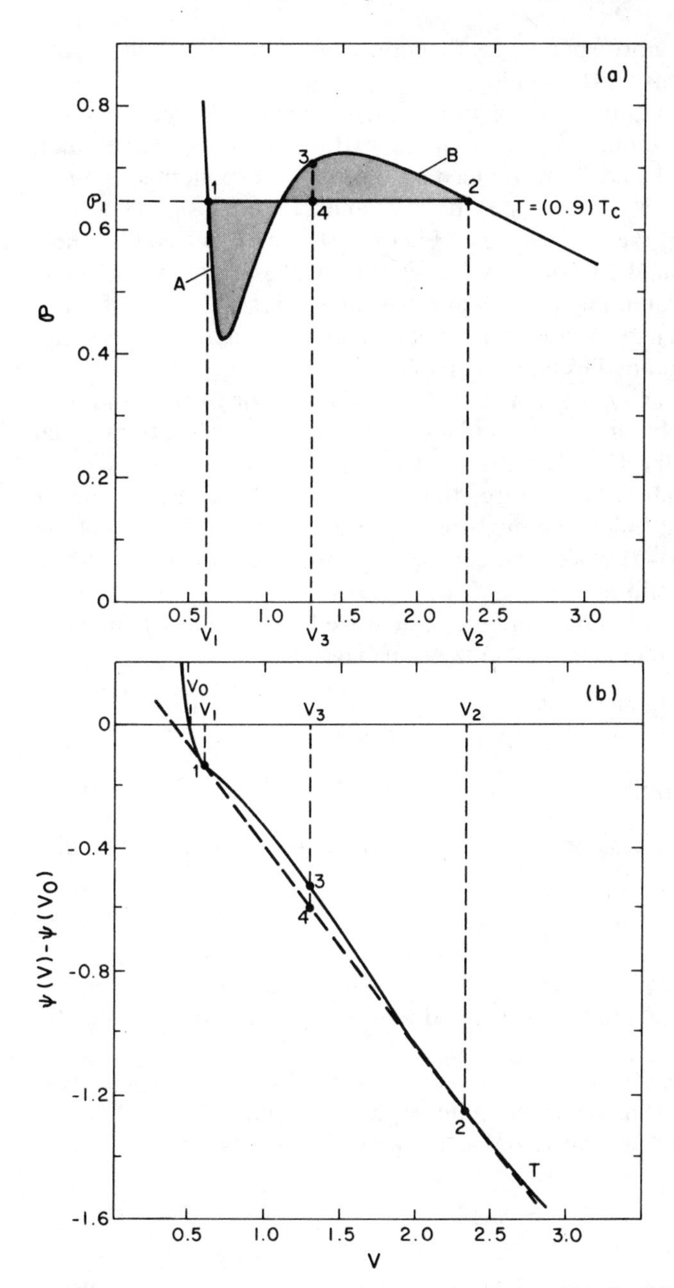

FIGURE VIII.4 The Maxwell construction for a Van der Waals' isotherm (see Figure VIII.3). The dimensions are with v in units of v_C, $\mathcal{P}$ in units of $\mathcal{P}_C$ and ψ in units of $(\mathcal{P}_C v_C)$.

in Figure VIII.4(b) corresponds to the horizontal line 1–4–2 in Figure VIII.4(a), as we discuss further below.

According to Equation (39.6), $\psi(v)$ is a monotonically decreasing function of v. For the ψ–v curve shown in Figure VIII.4(b) there is one common tangent line; points 1 and 2 are defined to be the corresponding common tangent points. Since $\mathcal{P} = -(\partial\psi/\partial v)_T$, points 1 and 2 have the same pressure $\mathcal{P}_1$, as is indicated in the $\mathcal{P}$–v curve of Figure VIII.4(a). Because the thermodynamic states 1 and 2 have the same $\mathcal{P}$ and T values, they can be interpreted as the pure-liquid and pure-vapor end points, respectively, of a liquid–vapor transition line. A condition under which the liquid and vapor can coexist is shown as point 4 in Figure VIII.4(a).

Let us consider next point 3 in the ψ–v diagram. For fixed v and T, a lower value of ψ can be obtained at point 4. Moreover, according to minimal condition (9.14) for the Helmholtz free energy, point 4 must correspond to the equilibrium situation. Therefore, the isotherm in the $\mathcal{P}$–v diagram of Figure VIII.4(a) must follow the mixture-of-phases line 1–4–2 instead of the unphysical curve 1–3–2, where the constant-pressure line is determined by the Maxwell construction.

The Maxwell construction can be characterized more simply. The condition for the common tangent line in the ψ–v diagram is

$$\begin{aligned}\frac{\psi(v_2) - \psi(v_1)}{v_2 - v_1} &= \left(\frac{\partial\psi}{\partial v}\right)_{v=v_1} = \left(\frac{\partial\psi}{\partial v}\right)_{v=v_2} \\ &= -\mathcal{P}(v_1) \quad = -\mathcal{P}(v_2)\end{aligned} \tag{39.7}$$

If this equation is now combined with Equation (39.6), we obtain the result

$$\mathcal{P}_1(v_2 - v_1) = \int_{v_1}^{v_2} \mathcal{P}(v)\, dv \qquad (T = \text{const}) \tag{39.8}$$

In other words, the two dashed portions in Figure VIII.4(a), and in Figure VIII.3, must have equal areas. This condition than determines the specific volumes v_1 and v_2.

According to Equation (36.21) an exact calculation of the grand potential, to be meaningful, must always yield a finite positive number for the fluctuation $(\Delta N)^2_{\text{rms}}$ and, therefore, also for the isothermal compressibility

$$\kappa_T = -\left(v\frac{\partial\mathcal{P}}{\partial v}\right)_T^{-1}$$

But in the region of a liquid-vapor phase transition $\kappa_T = \infty$, with the consequence, as we shall discuss further, that a grand canonical ensemble cannot describe this region. On the other hand, N and Ω are both fixed for a macrocanonical ensemble, so that calculations of the equation of state in the liquid–vapor transition region are possible in this case. When done approximately, such calculations might even yield a region of negative κ_T values where a phase transition occurs. In this case we would be justified in using the Maxwell construction to find the actual constant-pressure part of the isotherms where the phase transition occurs.

That an actual calculation could yield an isotherm with an unphysical portion might not be too surprising in view of the following possible interpretation. It is known that a liquid can be maintained at a pressure lower than its vapor pressure and that supersaturated vapor can occur at pressures higher than its equilibrium vapor pressure. Such metastable thermodynamic states could correspond to the regions A and B, where $\kappa_T > 0$, along the isotherm 1–3–2 in Figure VIII.4(a).

Description of a Liquid–Vapor Phase Transition by a Grand Canonical Ensemble

It can be expected that in the region of a phase transition, and in the thermodynamic limit, the grand partition function will exhibit special analytic behavior as a function of $\mathfrak{z} = e^{\beta g}$.[31] Thus for $T < T_C$ the actual isotherms of a substance have three analytically distinct branches, which we have deduced by the Maxwell construction. Moreover, the Maxwell construction corresponds to analytic solutions for which the thermodynamic states 1 and 2 in Figure VIII.4 have the same $\mathfrak{z}$ values, a result we deduced previously as Equation (10.6).[32]

We show now that a grand canonical ensemble cannot describe the liquid–vapor phase-transition region as a function of v, that is, the region $v_1 < v < v_2$ in Figure VIII.4, although it will give the end points $v = v_1$ and $v = v_2$ corresponding to a Maxwell construction.[33] We first prove two theorems and corollaries:

Theorem 1 Let Ψ be the Helmholtz free energy calculated for given T, Ω, and N using a macrocanonical ensemble and define, with T dependences suppressed,

$$\tilde{f}(v, \mathfrak{z}) \equiv \phi(v) + v^{-1} \ln \mathfrak{z} \tag{39.9}$$

where $v = \Omega/N$, $\mathfrak{z}$ is the fugacity (36.1) and $\phi(v)$ is related to $\Psi(v)$ by

$$\phi(v) = -\beta\Omega^{-1}\Psi(v) \tag{39.10}$$

Then

$$\lim_{\Omega\to\infty} f(\Omega, \mathfrak{z}) = \tilde{f}(\bar{v}, \mathfrak{z}) \tag{39.11}$$

where $f = \Omega^{-1} \ln Z_G$ is the grand potential and $\bar{v}$ is the value of v, for fixed $\mathfrak{z}$, at which $\bar{f}(v, \mathfrak{z})$ is a maximum, that is,

$$\left(\frac{\partial \bar{f}}{\partial v}\right)_{v=\bar{v}} = 0$$
$$\left(\frac{\partial^2 \bar{f}}{\partial v^2}\right)_{v=\bar{v}} \leq 0 \tag{39.12}$$

To prove Theorem 1 we note first, after referring to Equations (38.9), (27.17), (39.9), and (39.10), that the grand partition function Z_G is given by

$$Z_G(\Omega, \mathfrak{z}) = \sum_{N=0}^{\infty} e^{\Omega\bar{f}(v,\mathfrak{z})} \tag{39.13}$$

We now assume that the molecules of the system interact with an intermolecular potential that includes a hard-sphere repulsion with hard-sphere diameter a. If $N_o(\Omega)$ is the maximum number of molecules that can be squeezed into Ω, such that no two molecules are separated by a distance less than a, then there will be a maximum density $n_{\max} = N_o/\Omega$ and we may write

$$N_o = n_{\max}\Omega$$

The infinite series (39.13) will end at this value of N.

Consider next the value $\bar{f}_o(\mathfrak{z})$ at which $\bar{f}(\Omega/N, \mathfrak{z})$ takes its largest value for $0 \leq N \leq N_o$. Then the following inequality must hold:

$$e^{\Omega\bar{f}_o(\mathfrak{z})} \leq Z_G \leq N_o\, e^{\Omega\bar{f}_o(\mathfrak{z})}$$

or

$$e^{\Omega\bar{f}_o(\mathfrak{z})} \leq e^{\Omega f(\Omega,\mathfrak{z})} \leq n_{\max}\Omega\, e^{\Omega\bar{f}_o(\mathfrak{z})}$$

But this inequality must also be true for the logarithms of the terms; that is, it must be so that

$$\bar{f}_o(\mathfrak{z}) \leq f(\Omega, \mathfrak{z}) \leq \bar{f}_o(\mathfrak{z}) + \Omega^{-1} \ln (n_{\max}\Omega)$$

In the limit $\Omega \to \infty$, we obtain Equations (39.11) and (39.12). *Q.E.D.*

Corollary to Theorem 1 If condition (39.12) determines more than one value of $\bar{v}$ for given $\mathfrak{z}$, then the value that gives the largest $\bar{f}(\bar{v}, \mathfrak{z})$ must be chosen. This corollary follows immediately from the proof of Theorem 1. It corresponds to minimal condition (9.14) for the Helmholtz free energy. If we note that the pressure $\mathcal{P}_G$, as calculated for a grand canonical ensemble using (36.11), is given by

$$\mathcal{P}_G(\bar{v}) = \kappa T\bar{f}(\bar{v}, \mathfrak{z}) \tag{39.14}$$

then this corollary also applies to $\mathcal{P}_G$. [We shall discuss below how condition (39.12) can be used to solve for $\mathfrak{z}$ as a function of $\bar{v}$.]

Theorem 2 Let $\mathcal{P}(v)$ be the pressure calculated using a macrocanonical ensemble. Then calculations with a grand canonical ensemble cannot have solutions for v values where $\partial\mathcal{P}/\partial v > 0$.

To prove Theorem 2 we assume first that approximate macrocanonical-ensemble calculations may give solutions for $\mathcal{P}(v)$ with regions where $\partial\mathcal{P}/\partial v > 0$, as in Figure VIII.4(a). (Of course, an exact solution could not exhibit such an unphysical character in the thermodynamic limit.) Now the pressure $\mathcal{P}(v)$ is related to the function $\phi(v)$ of Equation (39.10) by Equation (27.19). One finds

$$\beta\mathcal{P}(v) = \phi(v) + v\frac{\partial\phi}{\partial v} \tag{39.15}$$

This result can be used in conjunction with Equation (39.9) to give the differential equation for $\tilde{f}(v, \mathfrak{z})$,

$$\frac{\partial^2\tilde{f}}{\partial v^2} + \frac{2}{v}\frac{\partial\tilde{f}}{\partial v} = \beta v^{-1}\frac{\partial\mathcal{P}}{\partial v} \tag{39.16}$$

Now, Equation (39.16) must hold for all v values and, in particular, it must hold for $v = \bar{v}$ corresponding to the maximum condition of Theorem 1. Conditions (39.12) then imply that $(\partial\mathcal{P}/\partial v)_{v=\bar{v}} \leqslant 0$ for all values $\bar{v}$ where grand-canonical-ensemble calculations have a solution. *Q.E.D.*

Corollary to Theorem 2 In regions of $\bar{v}$ values where calculations using a grand canonical ensemble have a solution,

$$\mathcal{P}_G(\bar{v}) = \mathcal{P}(\bar{v}) \tag{39.17}$$

Inasmuch as the first of conditions (39.12) determines $\mathfrak{z}$ as a function of $\bar{v}$ by the equation

$$\ln \mathfrak{z} = \left(v^2\frac{\partial\phi}{\partial v}\right)_{v=\bar{v}} \tag{39.18}$$

we may use $\beta\mathcal{P}_G = \tilde{f}(\bar{v})$ together with Equations (39.9) and (39.15) to verify this corollary.

To complete our objective, as stated before Theorem 1, we must now consider the full range of v values corresponding to a liquid–vapor phase transition. Figure VIII.5(a) shows a typical isotherm for a substance in the region of a liquid–vapor phase transition, along with an assumed (dashed and dotted) curve in the transition region corresponding to the result of an approximate macrocanonical-ensemble calculation.[34] The following exercise

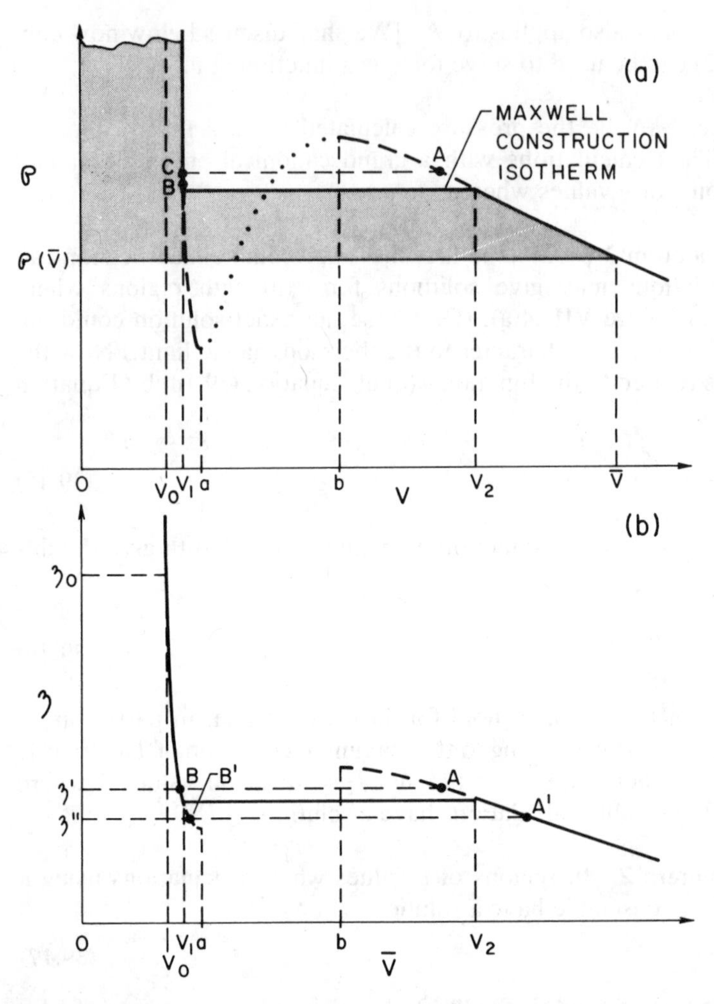

FIGURE VIII.5 (a) An isotherm for a substance in the region of a liquid–vapor phase transition. The dashed and dotted portion corresponds to the assumed result of an approximate calculation using a macrocanonical ensemble.[34] Theorem 2 shows that for such a calculation the region $a < v < b$ cannot be obtained using a grand canonical ensemble. The shaded area, which extends up to $\mathcal{P}(v_0)$, is interpreted in Exercise 39.1. Points A, B, and C are discussed in Exercise 39.2. (b) Plot of $\mathfrak{z}$ versus $\bar{v}$, as determined by Equation (39.19). The dashed-line portions correspond to the approximate curve in Figure (a), for $v_1 < \bar{v} < v_2$, except that in Figure (b) the part for $a < \bar{v} < b$ is excluded because of Theorem 2. The points A, B, A', and B' are discussed above Exercise 39.2.

establishes how such curves may be used graphically to solve for $\mathfrak{z}$ as a function of $\bar{v}$.

Exercise 39.1

(a) Use Equation (39.15) or (39.6) to show that

$$\beta \int_{v_0}^{v} dv' \, \mathcal{P}(v') = [v \, \phi(v) - v_0 \, \phi(v_0)]$$

and then use Equation (39.18) to derive the result

$$-\kappa T \ln(\mathfrak{z}/\mathfrak{z}_0) = \int_{v_0}^{\bar{v}} dv' \, \mathcal{P}(v') - (\bar{v} - v_0)\mathcal{P}(\bar{v}) + v_0[\mathcal{P}(v_0) - \mathcal{P}(\bar{v})] \tag{39.19}$$

where $\mathfrak{z}_0$ is the value of $\mathfrak{z}$ at $\bar{v} = v_0$.

(b) Equation (39.19) gives $\mathfrak{z}$ as a function of $\bar{v}$, with $v_0(>n_{\max}^{-1})$ equal to an arbitrary specific volume. For the real isotherm of Figure VIII.5(a) the terms on the right side of this equation correspond to the shaded area. Use this fact to argue that a graph of $\mathfrak{z}$ versus $\bar{v}$ must appear as in Figure VIII.5(b), with the dashed portions corresponding to those of Figure VIII.5(a). Note that for the real isotherm $\mathfrak{z}$ is constant in the region $v_1 < \bar{v} < v_2$, as it must be.

In Figure VIII.5(b) points A and B correspond to the same value $\mathfrak{z}'$, and points A' and B' correspond to the same value $\mathfrak{z}''$. The corresponding pressure values associated with A and B are assumed (incorrectly) to be located as shown in Figure VIII.5(a). We now complete our description of the liquid–vapor phase transition by a grand canonical ensemble by showing that

$$\mathcal{P}(v_B) > \mathcal{P}(v_A) \tag{39.20}$$

$$\mathcal{P}(v_{A'}) > \mathcal{P}(v_{B'})$$

The proof of these inequalities is outlined in the following exercise. Together with the corollary to Theorem 1, they show that the dashed portions of the isotherm in Figure VIII.5(a), corresponding to the assumed approximate calculation of $\mathcal{P}(v)$ and determined by a Maxwell construction, must be discarded in the determination of $\mathcal{P}_G(\bar{v})$. In other words, Equation (39.17) is only valid for $\bar{v}$ values outside of the interval $\bar{v}_1 < \bar{v} < v_2$. Such a result is understandable intuitively, because the grand partition function $Z_G(\Omega, \mathfrak{z})$ cannot give a unique value $\bar{v}$ for given Ω and $\mathfrak{z}$ in the region of a liquid–vapor phase transition.

Exercise 39.2

(a) Consider point C in Figure VIII.5(a), defined by the equality

$$\mathcal{P}(v_C) = \mathcal{P}(v_A)$$

and assume that $\mathcal{P}(v_B) < \mathcal{P}(v_A)$. Use the fact that the points A and B have a common value $\mathfrak{z}'$ to derive the equation

$$\int_{v_B}^{v_A} dv' \, \mathcal{P}(v') = v_A \, \mathcal{P}(v_A) - v_B \, \mathcal{P}(v_B)$$

for the approximately calculated isotherm in Figure VIII.5(a). Then deduce an area inequality that has the consequence that it contradicts the assumed pressure inequality, thereby verifying the first of inequalities (39.20).

(b) Construct a similar proof to verify the second of inequalities (39.20).

Although the explicit calculation of phase transitions is still an esoteric and largely unexplored branch of theoretical physics, it is clear that we already know, on thermodynamic and other grounds, a great deal about the nature of any analytic solutions.

40 Ionized Gases

In this section we consider two problems related to multicomponent systems of charged particles, or ionized gases. The first is to derive their equilibrium densities, or particle concentrations, assuming that the particles react but are noninteracting. This derivation results for the simplest case in the *Saha equilibrium equation* and, more generally, in the *law of mass action*. The second question we investigate is the effect of density fluctuations, or clustering, on the Coulomb interactions between particles. This study results in the *Debye–Hückel potential* for the electrostatic screened interaction between particles and its consequent effect on the thermodynamics of the system. Electrostatic screening between charged particles, in which their long-range $1/r$ Coulomb potential is reduced exponentially to a finite range, arises because of the presence of charges with both signs in an ionized gas.

We wish to use the classical ideal-gas law $\mathcal{P} = n\kappa T$ as a point of departure for our treatment of ionized gases. For the ideal-gas law, in the sense of Exercise 28.2, to apply even approximately to a system of charged particles it is necessary that their mean Coulomb energy be small compared with their thermal energy. For α particles with charge $z_\alpha e$, we must satisfy the condition

$$n_\alpha^{1/3}(z_\alpha e)^2/\varepsilon << \kappa T$$

where ε is the dielectric constant of the medium. This condition can be rewritten in the form

$$\frac{e^2}{\varepsilon\kappa T}\sum_\alpha z_\alpha^2 n_\alpha << \sum_\alpha n_\alpha^{2/3} \sim n^{2/3}$$

where we have summed over all charged-particle components α and $n = \Sigma_\alpha n_\alpha$ is the total density of charged particles. The inequality can now be expressed more simply as

$$n\lambda_D^3 >> 1 \tag{40.1}$$

where the *Debye length* λ_D is defined by

$$\lambda_D \equiv \left(\frac{4\pi\, e^2}{\varepsilon\kappa T}\sum_\alpha z_\alpha^2 n_\alpha\right)^{-1/2} \tag{40.2}$$

Condition (40.1) is equivalent to $\ell << \lambda_D$, where $\ell \equiv n^{-1/3}$ is the mean spacing of charged particles in the system. If we assume that $r_o = \lambda_D$, then the classical clustering condition of Section 37 (see Figure VIII.1) is readily achieved. However, this physically plausible identification is not associated with an immediate simplification of microscopic theory. The long-range nature of the Coulomb interaction invalidates straightforward application of ordinary perturbation expansions, for example, the virial expansion (38.1), to an ionized gas.[35] The simplest theory that treats the Coulomb interaction correctly, leading thereby to its screening, is the Debye–Hückel theory discussed below.

A further criterion must be introduced before we begin our formal discussion of ionized gases, namely, in order for electrostatic screening to occur, the gas must be electrically neutral, that is, we must have

$$\sum_\alpha z_\alpha n_\alpha = 0 \tag{40.3}$$

We also treat the ionized gas classically in this section by assuming $n_\alpha\lambda_{T,\alpha}^3 << 1$ for all particle components α and neglecting any effects due to particle statistics.

The Saha Equation[36]

The grand partition function for a multicomponent system is given by Equation (36.6a), with Equation (36.12) for the Lagrange multipliers λ_α. Let us consider the particular case of a three-component system, in which particles A and B can combine to form C, and vice versa,

$$A + B \rightleftharpoons C \tag{40.4}$$

If further we assume that these particles are otherwise noninteracting, although they may have internal degrees of freedom, then the grand partition function factors into three parts

$$Z_G = e^{\Omega(f_A + f_B + f_C)} \tag{40.5}$$

where, with $\alpha = \{A, B, C\}$,

$$e^{\Omega f_\alpha} = \sum_{N_\alpha=0}^{\infty} e^{\beta N_\alpha g_\alpha} \sum_{i=i\{N_\alpha\}} e^{-\beta E_i} \tag{40.6}$$

is the grand partition function for a single-component system.

For a multicomponent system in equilibrium and with reacting components, variations of the thermodynamic potential G at constant temperature and pressure must satisfy the minimal condition[37]

$$dG \equiv \sum_\alpha g_\alpha \, dN_\alpha = 0 \tag{40.7}$$

Moreover, if we assume that neutrality condition (40.3) holds, then we must also have

$$\sum_\alpha z_\alpha \, dN_\alpha = 0 \tag{40.8}$$

Our particular problem (40.4) dictates that we choose $2dN_C = -(dN_A + dN_B)$. If further we assume $z_C = 0$ and $z_B = -z_A$, then Equation (40.8) gives $dN_B = dN_A$. Equation (40.7) then becomes, for the present case, $(g_A + g_B - g_C)\, dN_A = 0$, which implies that

$$g_C = g_A + g_B \tag{40.9}$$

This relation between the chemical potentials must be used together with Equations (40.5) and (40.6) to determine the thermodynamic properties of the particular ionized gas in which reaction (40.4) takes place.

Exercise 40.1 In the derivation of Equations (36.5) and (36.6) for a grand canonical ensemble we used the constraint equations (36.3) and discussed below them how other conditions can be incorporated into the grand partition function. An example of a further condition is the chemical equilibrium condition (40.9). Derive Equation (40.9) directly for a grand canonical ensemble by considering, instead of the third of the constraint equations (36.3), the two constraint equations,

$$\sum_i M_i(N_A - N_B) = Q_M \qquad \left(\text{total charge} \times \frac{1}{e}\right)$$

$$\sum_i M_i(N_A + N_B + 2N_C) = N_M \qquad (\text{total number of } A \text{ and } B \text{ particles}).$$

[*Suggestion:* Write your result for $\bar{M}_i$ in the form obtained previously below (36.4).]

According to Equation (38.10), the grand potential for noninteracting particles can be written as

$$\begin{aligned} f_\alpha &= -\frac{\varepsilon_\alpha}{\Omega}\sum_\mu \ln\{1 - \varepsilon_\alpha \exp(\beta[g_\alpha - \omega_\mu^{(\alpha)} - \omega_0^{(\alpha)}])\} \\ &\cong \Omega^{-1}\exp\left(\beta\left[g_\alpha - \omega_0^{(\alpha)}\right]\right)\sum_\mu \exp\left(-\beta\omega_\mu^{(\alpha)}\right) \end{aligned} \tag{40.10}$$

where the sum over all single-particle states μ includes internal states of particle α, with $\omega_\mu^{(\alpha)}$ being measured relative to a zero of energy when particle α is at rest and unexcited. If particle α has a binding energy, then $\omega_0^{(\alpha)} < 0$. In going to the second line of Equation (40.10) we have made the assumption of Boltzmann statistics,

$$\exp(\beta[g_\alpha - \omega_0^{(\alpha)}]) << 1$$

For the present case we shall set $\omega_0^{(A)} = \omega_0^{(B)} = 0$ and $\omega_0^{(C)} = -\gamma^2$, γ^2 being the dissociation energy of particle C.

Equations (36.11) and (40.5) show that the pressure $\mathcal{P}$ is a sum over partial pressures $\mathcal{P}_\alpha$,

$$\mathcal{P} = \sum_\alpha \mathcal{P}_\alpha \tag{40.11}$$

which is the *law of partial pressures* for a multicomponent ideal gas. Moreover, by using Equation (36.19) we can see that each component in the ideal gas obeys the classical ideal-gas law separately, that is,

$$\begin{aligned} n_\alpha = \frac{\mathcal{P}_\alpha}{\kappa T} &= \Omega^{-1}\exp\left(\beta\left[g_\alpha - \omega_0^{(\alpha)}\right]\right)\sum_\mu \exp\left(-\beta\omega_\mu^{(\alpha)}\right) \\ &= \frac{1}{\lambda_{T,\alpha}^3}\exp(\beta[g_\alpha - \omega_0^{(\alpha)}])Z_{\text{int}}^{(\alpha)} \end{aligned} \tag{40.12}$$

where we have also used the calculation of Equation (28.24). The quantity $Z_{\text{int}}^{(\alpha)}$ is the partition function (28.25) for internal degrees of freedom including spin of particle α, with the zero of energy occurring at the lowest state.

If we use Equation (40.12) to form the product $n_A n_B/n_C$ and substitute Equation (40.9), then we can derive the relation

$$\frac{n_A n_B}{n_C} = \lambda_{T,\mu}^{-3} e^{-\gamma^2/\kappa T}\left[\frac{Z_{\text{int}}^{(A)} Z_{\text{int}}^{(B)}}{Z_{\text{int}}^{(C)}}\right] \qquad \textbf{(40.13)}$$

where $\lambda_{T,\mu}$ is the thermal wavelength calculated for a mass $\mu = M_A M_B/M_C$. This result is the Saha equation, relating the equilibrium densities n_A, n_B, and n_C, whose right side is a function of temperature only.

Exercise 40.2 The Saha equation is a special case of the law of mass action, which can be derived by using the general reaction equation (40.7) instead of Equation (40.9). One first writes Equation (40.7) in the form

$$\sum_\alpha \nu_\alpha g_\alpha = 0$$

where the ν_α are (positive or negative) integers, determined by the chemical reaction of interest and whose lowest common denominator is unity. Derive the law of mass action in the form[38]

$$\prod_\alpha (n_\alpha^{\nu_\alpha}) = \prod_\alpha [\lambda_{T,\alpha}^{-3} \exp(-\beta\omega_0^{(\alpha)}) Z_{\text{int}}^{(\alpha)}]^{\nu_\alpha} \qquad \textbf{(40.14)}$$

This law reduces to Equation (40.13) for the special case of reaction (40.4), with $\nu_A = \nu_B = -\nu_C = 1$.

Debye–Hückel Theory

In the simplest version of the Debye–Hückel theory of ionized gases one begins by calculating the average potential $\phi_\alpha(\mathbf{r})$ in the neighborhood of a charged particle α at $\mathbf{r} = 0$. This potential is determined classically from Poisson's equation,

$$\nabla^2\phi_\alpha(\mathbf{r}) = -\frac{4\pi}{\varepsilon}[\rho_\alpha(\mathbf{r}) + z_\alpha e\, \delta^{(3)}(\mathbf{r})] \qquad (40.15)$$

where we use electrostatic units and ε is the dielectric constant, which differs from unity in a polarizable medium such as an electrolytic solution. The quantity $\rho_\alpha(\mathbf{r})$ is the average charge density at $\mathbf{r}$ due to all charges other than the particle α at $\mathbf{r} = 0$. It is related to the two-particle reduced distribution function $P_{\beta\alpha}(\mathbf{r}_1, \mathbf{r}_2)$ by the equation

$$n_\alpha \rho_\alpha(\mathbf{r}) = \sum_\beta (z_\beta e) P_{\beta\alpha}(\mathbf{r}, 0) \qquad (40.16)$$

where the n_α are related in first approximation by Equation (40.14). The two-particle function $P_{\beta\alpha}(\mathbf{r}_1, \mathbf{r}_2)$ is defined by a simple multicomponent generalization of the third of Equations (24.6).

We now assume that $P_{\beta\alpha}(\mathbf{r}, 0)$ is given to good approximation by the expression

$$P_{\beta\alpha}(\mathbf{r}, 0) \cong n_\alpha n_\beta \, e^{-z_\beta e \phi_\alpha(\mathbf{r})/\kappa T} \tag{40.17}$$

which may seem quite reasonable in view of Equation (11.29).[39] Equation (40.17) gives a correction, because of density fluctuations associated with the Coulomb interaction, to the free-particle expression $n_\alpha n_\beta$ for $P_{\beta\alpha}(\mathbf{r}, 0)$. If Equations (40.16) and (40.17) are substituted into Equation (40.15), then we obtain the *Poisson–Boltzmann equation* for $\phi_\alpha(\mathbf{r})$

$$\nabla^2 \phi_\alpha(\mathbf{r}) = -\frac{4\pi e}{\varepsilon}\left[\sum_\beta z_\beta \mathrm{n}_\beta \, e^{-z_\beta e \phi_\alpha(\mathbf{r})/\kappa T} + z_\alpha \delta^{(3)}(\mathbf{r})\right] \tag{40.18}$$

We next use approximation (40.1), which means that on the average the argument of the exponential on the right side of Equation (40.18) is small. In this approximation we may expand the exponential and cancel the leading term by using neutrality condition (40.3). The Poisson–Boltzmann equation then takes the approximate form

$$\left(\nabla^2 - \frac{1}{\lambda_D{}^2}\right)\phi_\alpha(\mathbf{r}) = -\frac{4\pi z_\alpha e}{\varepsilon}\,\delta^{(3)}(\mathbf{r}) \tag{40.19}$$

where the Debye length λ_D is defined by Equation (40.2). The spherically symmetric solution, vanishing as $r \to \infty$, to this equation is

$$\phi_\alpha(r) = \left(\frac{z_\alpha e}{\varepsilon}\right)\frac{e^{-r/\lambda_D}}{r} \tag{40.20}$$

which is the Debye–Hückel potential for point charges. An isolated point charge has an electrostatic potential corresponding to the limit $\lambda_D \to \infty$. In a neutral ionized gas neighboring charged particles arrange themselves on the average so as to reduce, or screen, this $1/r$ potential exponentially.

Thermodynamic Calculations

We now use the Debye–Hückel potential to calculate contributions to various thermodynamic energies. A subscript zero indicates ideal-gas quantities. Thus the change in the average total energy due to Coulomb interaction between the particles can be deduced by using the formula

$$\Omega^{-1}(\langle E\rangle - \langle E\rangle_0) = \frac{1}{2}\sum_\alpha (z_\alpha e n_\alpha)\left[\phi_\alpha(r) - \frac{z_\alpha e}{\varepsilon r}\right]_{r=0}$$

$$= \frac{1}{2}\sum_\alpha \frac{(z_\alpha e)^2}{\varepsilon} n_\alpha\left(-\frac{1}{\lambda_D}\right)$$

$$\Omega^{-1}(\langle E\rangle - \langle E\rangle_0) = -\frac{\kappa T}{8\pi\lambda_D^3} \quad \textbf{(40.21)}$$

On the right side of the first line of this equation we have subtracted the self-potential of particle α at $\mathbf{r} = 0$, leaving the average potential due to clustering of other charges about this one charge. The second and third lines of Equation (40.21) are then derived with the aid of Equations (40.20) and (40.2).

We next calculate the Helmholtz free energy Ψ by integrating Equation (8.7), rewritten as $\langle E\rangle/T^2 = -(\partial/\partial T)(\Psi/T)$. Inasmuch as the isolated point-charge condition corresponds to $\lambda_D \to \infty$, we may conclude that $\Psi \to \Psi_0$ as $T \to \infty$. (At constant volume Debye–Hückel theory should be most accurate in the high-temperature limit.) We then obtain

$$\Psi - \Psi_0 = T\int_T^\infty \frac{dT'}{T'^2}(\langle E\rangle - \langle E\rangle_0)$$

$$= \frac{2}{3}(\langle E\rangle - \langle E\rangle_0) = -\frac{\kappa T\Omega}{12\pi\lambda_D^3} \quad (40.22)$$

We also calculate the correction, arising from Coulomb interaction, to the chemical potential by using the third equality of Equation (8.16),

$$g_\alpha - g_{\alpha,0} = \frac{\partial}{\partial N_\alpha}(\Psi - \Psi_0)\bigg|_{\Omega,T}$$

$$g_\alpha - g_{\alpha,0} = -\frac{1}{2}\frac{(z_\alpha e)^2}{\varepsilon\lambda_D} \quad \textbf{(40.23)}$$

Finally, Debye–Hückel theory gives a correction to the classical ideal-gas law. By using Equations (8.19), (10.1), (40.2), (40.22), and (40.23), we obtain

$$\frac{1}{\kappa T}(\mathcal{P} - \mathcal{P}_0) = f - f_0 = -(24\pi\lambda_D^3)^{-1} \quad (40.24)$$

where $f = \mathcal{P}/\kappa T$ is the grand potential.

We now define the so-called *activity coefficients* γ_α, used in chemistry. These are related to the fugacities $\mathfrak{z}_\alpha$ of Equation (36.1) by

$$\gamma_\alpha = \frac{\mathfrak{z}_\alpha}{\mathfrak{z}_{\alpha,0}} = e^{(g_\alpha - g_{\alpha,0})/\kappa T}$$
$$= \mathfrak{z}_\alpha\left(\frac{2S_\alpha + 1}{n_\alpha \lambda_{T,\alpha}^3}\right) \tag{40.25}$$

where the third equality follows from Equation (38.5). (One usually sets $S_\alpha = 0$ for classical particles.) For a reaction of the type (40.4), with $z_A = -z_B = 1$ and $z_C = 0$, one further defines a mean activity coefficient $\gamma_\pm$ by the expression

$$\gamma_\pm \equiv (\gamma_A \gamma_B)^{1/2}$$
$$= \exp\left[\frac{1}{2\kappa T}(g_A - g_{A,0} + g_B - g_{B,0})\right]$$
$$\longrightarrow \exp\left(-\frac{e^2}{2\kappa T \varepsilon \lambda_D}\right) \qquad \text{(Debye–Hückel theory)} \tag{40.26}$$

where we have used Equations (40.25) and (40.23) to derive the second and third lines of this equation. The third equality is called the *Debye–Hückel limiting law,* and it agrees well with measured values of $\gamma_\pm$ for very dilute solutions of binary salts in water ($\varepsilon = 78.54$). In more concentrated solutions of electrolytes, ion sizes and other effects neglected here become important.

To derive reliable corrections to Debye–Hückel theory, even for the idealized model of point particles, one must refer to the methods of statistical mechanics. The most powerful approach, in which one applies the grand canonical ensemble, yields a fugacity expansion of the grand potential, instead of a density expansion. Moreover, because of the long-range Coulomb interaction, the cluster-integral expansion (37.13) must be rearranged to a convergent form by summation of so-called ring structures.[40] By these methods one obtains to leading order the expression

$$f = \sum_\alpha (2S_\alpha + 1)\mathfrak{z}_\alpha \lambda_{T,\alpha}^{-3} + (12\pi \lambda'^{3}_{D})^{-1} \tag{40.27}$$

where λ'_D differs from the Debye length (40.2) by the (density → fugacity) replacement $n_\alpha \to (2S_\alpha + 1)\mathfrak{z}_\alpha \lambda_{T,\alpha}^{-3}$. Corrections to this result have been derived by many authors.

We conclude our discussion here by showing how the fugacity expansion (40.27) relates to the density expansion (40.24). Thus by using Equation (36.19) one can derive from (40.27) the result

$$n_\alpha = (2S_\alpha + 1)\frac{\mathfrak{z}_\alpha}{\lambda_{T,\alpha}^3}\left\{1 + \frac{(z_\alpha e)^2}{2\kappa T \varepsilon \lambda'_D}\right\} \tag{40.28}$$

When the second term in the brackets is much less than unity, then we may solve this equation for the fugacities z_α by an iteration procedure and thereby derive Equation (40.24) from Equation (40.27). In fact, the second term in the brackets is not necessarily small compared with unity for electrolytic solutions (even at temperatures for which further point-particle model corrections are small). To allow for this case, the γ_α of Equation (40.25) can be derived directly from Equation (40.28) using a graphical procedure, and one finds that $\gamma_\pm$ calculated in this way gives a better correspondence with measured data than does the Debye–Hückel limiting law.[41] See Figure VIII.6.

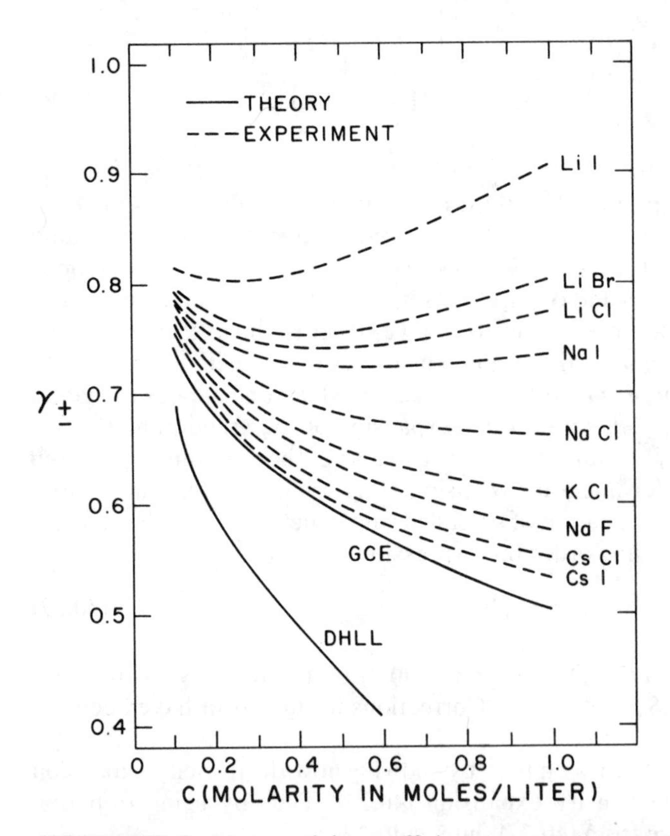

FIGURE VIII.6 Mean activity coefficient versus concentration in molarity units, $c = 10^3 n_\pm / N_o$, where $n_+ = n_-$ and N_o is Avogadro's number. The experimental curves are for aqueous solutions of electrolytes at 25°C. The Debye–Hückel limiting law (DHLL) is from Equation (40.26), and the grand canonical ensemble (GCE) curve is calculated according to the procedure of note 41.

Bibliography

N. Davidson, *Statistical Mechanics* (McGraw-Hill, 1962), Chapter 21. The Debye–Hückel theory of electrolytic solutions presented from a chemist's point of view.

K. Huang, *Statistical Mechanics* (Wiley, 1967), Chapters 8, 14, and 15. Development of both theory and applications of the grand canonical ensemble is rarely encountered in statistical mechanics textbooks. Huang's book is one of the few where a full treatment of this subject is presented. Section 2.3 has a good discussion of the thermodynamics associated with Van der Waals' equation of state.

D. ter Haar, *Elements of Statistical Mechanics* (Reinhart, 1954), Chapters VI, VII, and VIII. The theory of the grand canonical ensemble is outlined with some care. Chapter VIII gives a quite complete treatment of the cluster-integral expansion of the grand potential as it applies to the equation of state for an imperfect gas.

L. D. Landau and E. M. Lifshitz, *Statistical Physics* (Addison-Wesley, 1958), translated by E. Peierls and R. F. Peierls. Further discussion of material covered in our Section 40, on the theory of ionized gases, can be found in Sections 74, 100, and 101 of this text.

J. E. Mayer and M. G. Mayer, *Statistical Mechanics* (Wiley, 1940), Chapters 12 and 13. Chapter 13 contains an older, but still relevant, classical treatment of the cluster-integral expansion of the partition function for imperfect gases. The analysis of Van der Waals' equation in Chapter 12 in quite thorough.

Notes

1. For an explicit demonstration of this equivalence, see K. Huang, *op. cit.*, Sections 8.4–8.6. The proof given by Huang rests on the physical assumption that the isothermal compressibility, $\kappa_T = -\Omega^{-1}\, \partial\Omega/\partial\mathcal{P}$, is a nonnegative quantity [see discussion of inequality (9.20)]. At the end of Section 39 we show that an exact calculation of the grand canonical ensemble would never contradict this assumption.

2. See the discussion preceding Equation (27.6).

3. As is shown below [see Equations (36.4) and (36.12)], the chemical potentials g_α correlate directly with the particle-number constraints. Examples of special conditions on the g_α are Equations (35.4) and (40.9).

4. An example when the fluctuations $\langle(\Delta M_i)^2\rangle$ are large for the grand canonical ensemble occurs at a liquid–vapor phase transition. The description by a grand canonical ensemble of this particular case is analyzed in detail in the second half of Section 39.

5. Refer to the discussion above Equations (8.14).

6. See the discussion at the beginning of Section 10 in which the multicomponent generalization of $g\delta N$ is introduced.

7. See, in this connection, Equations (8.14).

8. In addition to the system volume Ω, the energy levels E_i may depend upon other mechanical parameters. See Equation (27.10).

9. See, for example, K. Huang, *loc. cit.*, Chapter 15. Rigorous proofs of the existence of the thermodynamic limit are given in D. Ruelle, *Statistical Mechanics* (Benjamin, 1969), Chapters 2 and 3, and by J. L. Lebowitz and E. H. Lieb, *Phys. Rev. Lett.* **22** (1969), p. 631.

10. B. Kahn and G. E. Uhlenbeck, *Physica* **5** (1938), p. 399.

11. The clustering property for the position representation is treated below in some detail. More generally, this property can be demonstrated by using a perturbation expansion in powers of the interaction Hamiltonian V, similar to the expansion of the S matrix that is performed in field theory. The nature of this expansion is indicated in Problem VIII.5 (see also Section 50).

12. Recall the discussion of inequality (28.11).

13. This proof of Equation (37.9), due to H. A. Kramers, is outlined in D. ter Haar, *op. cit.*, pp. 184–186.

14. This observation is not quite correct if macroscopic occupation of the zero-momentum state occurs for a Bose system below its BE condensation point. Modifications of the present development must be made for this special case. See T. D. Lee and C. N. Yang, *Phys. Rev.* **117** (1960), p. 897, or the article by F. Mohling in *Symposia on Theoretical Physics,* Vol. 2, edited by A. Ramakrishnan (Plenum Press, 1966).

15. Although we are considering a single-component system of *identical* particles, the sums over single-particle states in Equation (37.3) permit all possible partitions of N particles as defined by rules (a) and (b) below the Ursell equations (37.7).

16. This approach to the identification of a general expression for the $U_N^{(S)}$, which results in a perturbation expansion in terms of two-particle interaction quantities, while cumbersome, has been pursued to completion. See F. Mohling, *Phys. Rev.* **122** (1961), p. 1043.

17. The sign $\varepsilon^{P'}$ is the sign, for Fermi statistics, of the permutation $\mathcal{P}'$ of bottom-row coordinates with respect to top-row coordinates.

18. The origin of this terminology, as it relates to the virial theorem (28.15), is explained in footnote 22 of Chapter VI.

19. The introduction of this extra factor is motivated by Equation (31.26) for the degenerate ideal gas.

20. The procedure of Exercise 31.4 can be applied here. A completely general and elegant method for deriving these results is given by J. Kilpatrick and D. I. Ford, *Am. J. Phys.* **37** (1969), p. 881.

21. A general expression for c_N can be derived by using the method and results of Appendix F.

22. This result is also derived near the end of Appendix F as a special case of the general development outlined there. The interested reader may now wish to pursue the reasoning process left uncompleted in the paragraph preceding Equation (36.11). Note that Equation (38.10) for f_o agrees with pressure expression (31.13) for an ideal gas.

23. See Equations (11.3).

24. Exponential suppression at high temperatures of the exchange term in the second virial coefficient was first demonstrated rigorously by E. H. Lieb, *J. Math. Phys.* **8** (1967), p. 43; see also R. N. Hill, *J. Math. Phys.* **9** (1968), p. 1534. An extension of these proofs to treat the exchange terms of higher-order virial coeffi-

cients has been given by L. W. Bruch; *Prog. Theor. Phys.* **50** (1973), pp. 1537, 1835; see also R. N. Hill, *J. Stat. Phys.* **11** (1974), p. 207.

25. The calculation of the second virial coefficient in powers of (λ_T/a) for high temperatures and a hard-sphere interaction, with a = hard-sphere diameter, has had a particularly interesting history. See J. J. D'Arruda and R. N. Hill, *Phys. Rev.* **A1** (1969), p. 1791, and earlier references cited therein. For a comprehensive review see *The Virial Equation of State* by E. A. Mason and T. H. Spurling (Pergamon, 1969).

26. The function $f(\mathbf{r})$ is the basic two-particle interaction quantity in the Mayer cluster-integral expansion of the equation of state, as derived using classical statistical mechanics. See Problem VIII.1.

27. The distance of closest approach for a hard-sphere repulsion occurs at a distance where $r_{12} = |\mathbf{r}_1 - \mathbf{r}_2| = a$, and therefore a is the hard-sphere diameter.

28. See footnote 25 in Chapter VI. A major exception occurs for ionized gases, which must be treated specially.

29. J. L. Lebowitz and O. Penrose, *J. Math. Phys.* **7** (1966), p. 98. See also J. L. Lebowitz, *Physica* **73** (1974), p. 48.

30. See definition (9.2) of the isothermal compressibility κ_T and the discussion of inequality (9.20).

31. Discussion of the possible nature of this analytic behavior as a function of $\mathfrak{z}$ is given in K. Huang, *loc. cit.,* Sec. 15.2.

32. See also Figure II.7.

33. This development is adapted from K. Huang, *loc. cit.,* Section 8.7.

34. For convenience, we have used the Van der Waals' isotherm of Figure VIII.4(a).

35. See, for example, the condition deduced below Equation (39.2) on the long-range behavior of $V(\mathbf{r})$. Perturbation series in the potential V must be rearranged, with partial-series summations, in order to exhibit Coulomb screening.

36. Our treatment of the Saha equation follows that given by D. ter Haar, *Elements of Thermostatics* (Holt, Rinehart and Winston, 1966), Section 6.9.

37. See Equations (9.15) and (10.1).

38. The law of mass action is developed entirely from the thermodynamic point of view in P. S. Epstein, *Textbook of Thermodynamics* (Wiley, 1947), Section 51.

39. Equation (40.17) is derived in Section 47 for a single-component system with a general interaction potential $V(\mathbf{r})$; see Equations (47.3) and (47.18). When $V(\mathbf{r}) \to 0$, then Equation (40.17) reduces to the ideal-gas limit, which for a single-component system is the first of Equations (24.8).

40. See W. T. Grandy and F. Mohling, *Ann. Phys.* **34** (1965), p. 424, for a detailed analysis of this summation procedure. See also footnote 35.

41. R. W. Jones and F. Mohling, *J. Phys. Chem.* **75** (1971), p. 3790. The GCE curve in Figure VIII.6 includes one important point-particle-model correction term to Equations (40.27) and (40.28), as is explained in this reference.

42. K. Huang, *loc. cit.* Sec. 14.1.

43. J. A. Beattie, J. S. Brierley, and R. J. Barriault, *J. Chem. Phys.* **20** (1952), p. 1615. The number N_A is Avogadro's number.

44. The answer given here has used Levinson's theorem, $\lim_{k\to 0} \delta_L(k) = n_L\pi$, where n_L is the number of bound states with angular momentum L. See L. I. Schiff, *Quantum Mechanics,* 3rd ed. (McGraw-Hill, 1968), Equation (39.36).

45. See, for example, D. ter Haar, *Elements of Statistical Mechnics* (Rinehart, 1954), Sections 8.3 and 8.7.

46. A comprehensive discussion of the first million years in the evolution of the universe according to the "big bang," or fireball, model is given by E. R. Harrison in *Physics Today*, **21** (no. 6) (1968), p. 31. Problem VIII.9 deals with conditions after the first second using this model. The problem can be extended into the slightly earlier relativistic region by using results from Section 53. See also: Steven Weinberg, *The First Three Minutes* (Basic Books, 1977).

Problems

VIII.1. (a) Show for a large fluid system with velocity-independent interactions that in the classical limit the diagonal Boltzmann W_N function can be written in the momentum-independent form

$$W\begin{pmatrix} k_1 \cdots k_N \\ k_1 \cdots k_N \end{pmatrix} = \Omega^{-N} \int d^3\mathbf{r}_1 \cdots d^3\mathbf{r}_N \, e^{-\beta V(\mathbf{r}_1, \cdots \mathbf{r}_N)}$$

(b) Use Equation (37.17) and the result of part (a) to derive the expression for the grand potential of classical statistical mechanics,

$$f = \Omega^{-1} \sum_{N=1}^{\infty} \frac{1}{N!} \left(\frac{\mathfrak{z}}{\lambda_T^3} \right)^N \int d^3\mathbf{r}_1 \cdots d^3\mathbf{r}_N \, U(\mathbf{r}_1, \cdots \mathbf{r}_N)$$

where $U_N(\mathbf{r}_i)$ can be deduced from $e^{-\beta V_N(\mathbf{r}_i)}$ by using classical Ursell equations. Note that the double-row notation of Equations (37.7) loses significance in the classical limit.

(c) Use the result of part (b) to deduce the Mayer cluster-integral expansion of the grand potential in terms of the functions $f(\mathbf{r}_{ij})$ defined by Equation (39.1).[42]

[*Suggestion:* Consider the relation

$$e^{-\beta V_N} = \prod_{\alpha<\beta}^{N} (1 + f_{\alpha\beta})$$

here.]

VIII.2. (a) Show that the cluster integrals b_2 and b_3 of Equation (37.10) can be written in the classical limit of Problem VIII.1 as

$$b_2 = \frac{1}{2} \left(\frac{\mathfrak{z}}{\lambda_T^3} \right)^2 \Omega^{-1} \int d^3\mathbf{r}_1 \, d^3\mathbf{r}_2 \, f_{12}$$

$$b_3 = 2 \left(\frac{\lambda_T^3}{\mathfrak{z}} \right) b_2^2 + \frac{1}{6} \left(\frac{\mathfrak{z}}{\lambda_T^3} \right)^3 \Omega^{-1} \int d^3\mathbf{r}_1 \, d^3\mathbf{r}_2 \, d^3\mathbf{r}_3 \, f_{12} f_{23} f_{31}$$

(b) Derive a general expression for the classical third virial coefficient $a_{3,cl}(\beta)$.

VIII.3. The second virial coefficient, $a_2 \equiv N_A^{-1}\lambda_T^{-3}B(T)$, has been measured as a function of temperature for krypton, and some of the data are given in the accompanying table.[43] Assume that the intermolecular potential $V(r)$ between krypton atoms has the form of a square well outside of an infinite repulsive hard core, as shown. Determine what you consider to be the best choices for the constants a, b, and $\varepsilon \equiv V_o/\kappa$, the first two being measured in angstroms and the second in degrees. Use the classical expression (38.21) for your calculations.

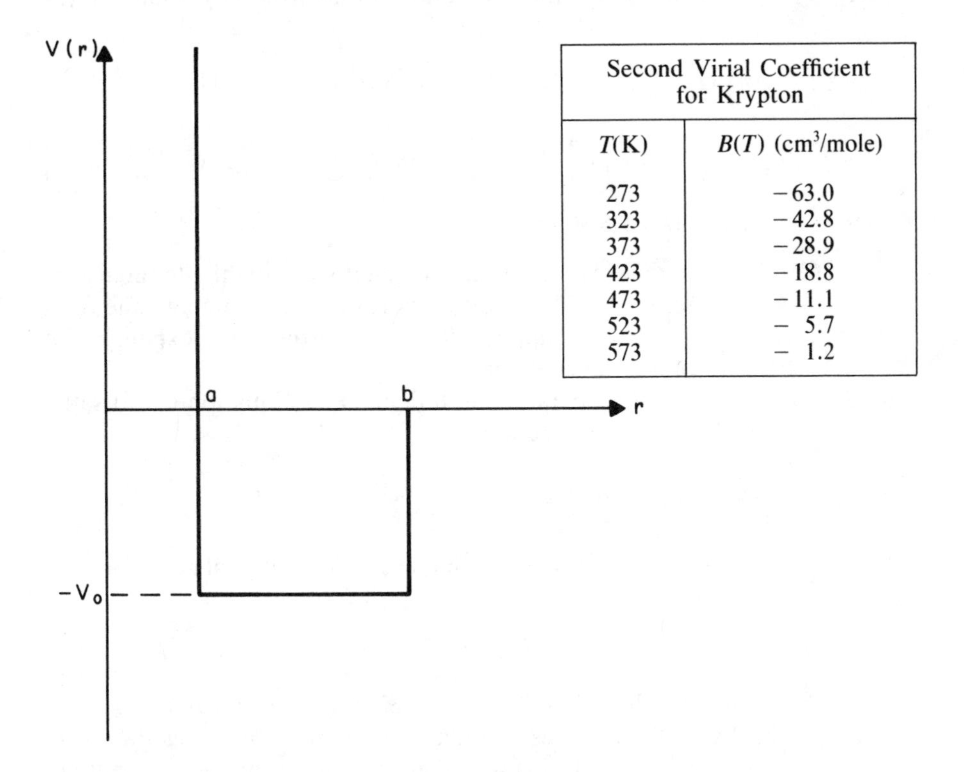

Second Virial Coefficient for Krypton	
T(K)	$B(T)$ (cm³/mole)
273	−63.0
323	−42.8
373	−28.9
423	−18.8
473	−11.1
523	− 5.7
573	− 1.2

VIII.4. (a) Show that the second virial coefficient for identical particles can be written as

$$a_2 = -\frac{\varepsilon}{4\sqrt{2}} - 2\sqrt{2}\,(2S + 1)^{-1}\left(\sum_k e^{-\beta\omega(k)} - \sum_{k_o} e^{-\beta\omega(k_o)}\right)$$

where the sums over quantum numbers k and k_o denumerate all different eigenstates of the two-particle Hamiltonians H_{rel} and $\mathsf{H}_{o,rel}$ respectively.

[*Suggestion:* Recall Exercise 16.1 (a).]

(b) Exhibit a partial-wave expansion of a_2 in the form

$$a_2 = -\frac{\varepsilon}{4\sqrt{2}} + \sum_L a_{2,L}$$

for the case of a spin-independent interaction, but with $S \neq 0$, noting in particular the differences that occur for Fermi–Dirac and Bose–Einstein statistics.

(c) Express $a_{2,L}$ as an integral over the states k, with the density-of-states factor expressed in terms of the phase shift $\delta_L(k)$ for the Lth partial wave. Include contributions from any bound states γ.[44]

Answer: $a_{2,L} = -\sqrt{2}(2L + 1)[2S + 1 + \varepsilon(-1)^L]$

$$\times \left\{\left(\frac{\lambda_T}{\pi}\right)^2 \int_0^\infty k\, dk\, \delta_L(k)\, e^{-\beta\omega(k,L)} + \sum_\gamma [e^{-\beta\omega(\gamma,L)} - 1]\right\}$$

where $\omega(k,L) = \hbar^2 k^2/m$.

(d) Evaluate the S-wave term $a_{2,0}$ explicitly for an infinite repulsive-core interaction. What does this result indicate about the temperature region of applicability of a partial-wave expansion of the second virial coefficient?

VIII.5. (a) Show that the operator $\mathsf{W}(\beta)$ defined by Equation (37.5) satisfies the differential equation

$$\frac{\partial}{\partial\beta}\mathsf{W}(\beta) = -\mathsf{V}(\beta)\mathsf{W}(\beta)$$

where $\mathsf{V}(\beta) \equiv e^{\beta \mathsf{H}_o}\mathsf{V}e^{-\beta \mathsf{H}_o}$, and deduce the integral equation

$$\mathsf{W}(\beta) = 1 - \int_0^\beta dt\, \mathsf{V}(t)\mathsf{W}(t)$$

(b) Show that the $W_N^{(S)}$ of Equation (37.6) can be calculated directly in the number representation of Section 17 by using state vectors of the type (17.21) with their normalization factors omitted.

(c) The quantum cluster function $U_N^{(S)}$, for $N \geqslant 2$, can be expressed entirely in terms of temperature integrals over two-body functions (or matrix elements). A procedure for determining such an expression can be derived from the iterated form of the integral equation for $\mathsf{W}(\beta)$ and using the result of part (b). The operator $\mathsf{V}(\beta)$ can be expressed in the number representation by using Equation (17.19b) and the identities:

$$\mathsf{a}_k(\beta) \equiv e^{\beta\mathsf{H}_o}\mathsf{a}_k e^{-\beta\mathsf{H}_o} = \mathsf{a}_k e^{-\beta\omega_k}$$

$$a_k^\dagger(\beta) \equiv e^{\beta H_o} a_k^\dagger e^{-\beta H_o} = a_k^\dagger e^{\beta\omega_k}$$

Develop this procedure in detail for the case $N = 2$, and relate your result to the second of Equations (37.15) and to Equation (38.14).

[*Suggestion:* Use the factor ε to write both sets of commutation relations (17.7) and (17.8) as a single set. Then use these relations systematically to commute annihilation operators a_k, within the matrix elements of terms in $W(\beta)$, to the position where $a_k|0\rangle = 0$ can be used. For example,
$\langle 0|a_2 a_1 a_{1'}^\dagger a_{2'}^\dagger|0\rangle = \delta_{1,1'}\,\delta_{2,2'} + \varepsilon\,\delta_{2,1'}\,\delta_{1,2'}$

VIII.6. (a) Verify that Equation (37.16) summarizes Equations (37.15) correctly for the cases $N = 1, 2$ and 3.

(b) Prove Equation (37.16). This proof can be accomplished by assuming that Equation (37.16) is correct, substituting it into the Ursell equation for $W_N^{(S)}$, and then verifying that the resulting expression agrees with the right side of Equation (37.6) when Boltzmann Ursell equations are used for the W_N.

VIII.7. (a) Determine the thermodynamic variables (T_C, $\mathcal{P}_C$, and v_C) at the critical point for Van der Waals' equation of state.

(b) Show that Van der Waals' equation, written in terms of the reduced variables

$$T_r \equiv \frac{T}{T_C}, \qquad \mathcal{P}_r \equiv \frac{\mathcal{P}}{\mathcal{P}_C}, \qquad v_r \equiv \frac{v}{v_C}$$

is independent of the interaction constants $\tilde{a}$ and b. This is the law of corresponding states for Van der Waals' equation of state.

(c) Consult other texts and articles, and report in some detail on the law of corresponding states for a few real gases.[45]

VIII.8. Evaluate the equilibrium constant

$$K_P(T) \equiv \frac{\mathcal{P}_N^2}{\mathcal{P}_{N_2}} = \frac{n_N^2}{n_{N_2}}\kappa T$$

for the dissociative nitrogen reaction $N_2 \rightleftharpoons 2N$ at $T = 5000K$ under the following assumptions:

(i) The characteristic temperatures of rotation and vibration for the N_2 molecule are $\kappa^{-1}\theta_r = 2.86K$ and $\kappa^{-1}\theta_v = 3.35 \times 10^3 K$ respectively.

(ii) The dissociation energy, including correction for the zero-point vibrational energy, is $\gamma^2 = 225.0$ kcal/mol

(iii) The electronic ground state of the N_2 molecule has no degeneracy, but that of the N atom has a degeneracy 4 due to electron spin.

(iv) Electronic excitation can be neglected.

[*Suggestions:* Show that Equation (40.14) can be applied with $\nu_1 = 2$ and $\nu_2 = -1$. Recall discussion of Equations (29.14) and (29.15).]

Answer: $\log_{10} K_P = -2.95$, when K_P is expressed in atmospheres.

VIII.9. *Early stage of expanding universe*[46]

(a) Consider an early stage of the universe when its temperature satisfied the condition $\kappa T \leqslant (0.1)mc^2$, where m is the electron mass. Assume that equilibrium is maintained in the production and annihilation of (e^+, e^-) pairs from photons. Derive an expression for the product of densities $n_+ n_-$, and then show that

$$n_\mp = \pm\frac{n_o}{2} + \left[\left(\frac{n_o}{2}\right)^2 + 4\lambda_T^{-6}\, e^{-2mc^2/\kappa T}\right]^{1/2}$$

where n_o is the (approximately) constant density difference $(n_- - n_+)$ during this stage of evolution.

(b) Use the result of part (a) to show that $n_+ \cong n_-$, with little dependence on n_o, when $\kappa T \leqslant (0.1)mc^2$. Give a numerical estimate for the ratio (n_-/n_γ) under these conditions, where n_γ is the equilibrium photon density (35.12).

CHAPTER IX

IRREVERSIBLE PROCESSES

Random events at the microscopic level lead to irreversible phenomena in a macroscopic system. The chapter deals with attempts to establish this connection quantitatively. An important cornerstone is the fluctuation–dissipation theorem, whereby kinetic coefficients that characterize macroscopic dissipation are related to microscopic fluctuations. This theorem is established first for one-dimensional Brownian motion and then for transport processes in whch the condition of local equilibrium holds. For the more general case the Onsager reciprocal relations between various kinetic coefficients are derived. The relaxation of spins in a fluid, from an initial nonequilibrium condition toward a final equilibrium state, is studied using quantum-mechanical procedures. Finally, Section 45 outlines various methods by which we can investigate the approach to equilibrium in a macroscopic system.

41 Fluctuation–Dissipation Theorem for Brownian Motion

Brownian motion is a prototype of the random process, and in this section we investigate again the one-dimensional example of Section 15, with the goal here of understanding better the relationship between fluctuation phenomena and energy dissipation (or, more properly, information dissipation). The important general result of this development is the *fluctuation–dissipation theorem,* which relates short-time correlations of dynamical variables to dissipation constants. The theorem shows that a random, or fluctuating, force can be reinforced briefly so as to produce an observable effect, but the corresponding momentary decrease in entropy is then destroyed, or dissipated, by the same cause that produced it. In the study of nonequilibrium thermodynamics dissipation constants are called *ki-*

netic coefficients (see Section 42). In fact, the fluctuation–dissipation theorem is a cornerstone for the theoretical foundations of irreversible phenomena.

Random events are the mechnism whereby a system arrives irreversibly to its most probable coarse-grained, or macroscopic, state. Thus macroscopic order tends to be dissipated by random events at the microscopic level. A quantitative statement of this last fact, for the case of Brownian motion, is given by Equation (15.3), which relates the averaged noise force $\langle F(t)\rangle$ to the damping of a macroscopic particle, that is, to fluid friction. The fluctuation–dissipation theorem provides an expression for the damping coefficient α, and in this section we deduce this expression explicitly. We are also led to investigate the properties of (classical) time-correlation functions, because the essence of the fluctuation–dissipation theorem is that it provides expressions for dissipation constants in terms of such functions.

We begin by amplifying, in some detail, the discussion below Equation (15.2). We recall that our model of Brownian motion was a macroscopic particle, or subsystem, immersed in a medium through which this particle moved in response to the random microscopic force $F(t)$. We assume now that at any given time t the medium itself is in thermodynamic equilibrium at temperature T. Thus an appreciable energy transfer to or from the particle, by the mechanism of microscopic collisions, is rapidly absorbed by the medium in a macroscopically small time τ that is large compared with the average collision time τ_{coll}. It must be emphasized that $\tau >> \tau_{\text{coll}}$, because only on a macroscopic time scale characterized by such minimum intervals τ is it meaningful to refer to thermodynamic equilibrium of the medium.

At a later time $t + \tau$, the probability distribution function $P(t + \tau)$, that the particle has energy $E_{\text{part}}(t + \tau)$, is related to the corresponding $P(t)$ by

$$P(t + \tau) = P(t)\, e^{\beta \Delta E_{\text{med}}} \tag{41.1}$$

Here $\Delta E_{\text{med}} = -[E_{\text{part}}(t + \tau) - E_{\text{part}}(t)]$ is the energy change of the medium in time τ. This equation is equivalent to Equation (26.6), and it is justified by the derivation, which applies in the present case, associated with Equation (26.7). We note that both the macroscopic particle and medium can be treated as classical systems, although the particles of the medium may themselves behave quantum mechanically.

We now calculate the ensemble average of the random force $F(t')$, for t' in the interval $t \leqslant (t' \equiv t + \tau') \leqslant t + \tau$, assuming that this force depends on the energy state μ of the particle. Thus for a particle initially at rest, say, its subsequent energy state μ will depend directly on the force to which it is subjected. We may therefore write, using Equation (41.1),

$$\langle F(t')\rangle \equiv \sum_{\mu} F_{\mu}(t')P_{\mu}(t + \tau') \cong \sum_{\mu} F_{\mu}(t')P_{\mu}(t)[1 + \beta \Delta E_{\text{med}}]$$

$$= \beta \sum_{\mu} F_{\mu}(t')P_{\mu}(t)\, \Delta E_{\text{med}}$$

$$\langle F(t')\rangle \cong \beta\, \langle F(t')\Delta E_{\text{med}}(t,t')\rangle_0 \qquad \textbf{(41.2)}$$

In the second equality of this equation we have assumed that $|\Delta E_{\text{med}}| << \kappa T$ for any macroscopically small time interval τ', and in the third equality we have assumed that an ensemble average of $F(t')$ at the time t vanishes. Only when the time dependence on τ' of the thermodynamic states of the medium is included, as in the first equality of (41.2), do we get a nonvanishing ensemble average of $F(t')$. To denote an ensemble average over thermodynamic states of the medium at time $\tau' = 0$, we use a subscript zero. Such an average in the final equality of Equation (41.2) is not zero, because, as we show next, this ensemble average involves a quadratic product of the random force F at two closely spaced times.[1]

Now, the energy change ΔE_{med} in the time interval $\tau' = t' - t$ is simply the negative of the work done on the particle by the random force F, that is,

$$\Delta E_{\text{med}}(t,\, t') = -\int_t^{t'} dt''\, u(t'')F(t'') \cong -u[t] \int_t^{t'} dt''\, F(t'')$$

In the second equality of this expression we have assumed that the macroscopic particle velocity $u(t)$ does not vary appreciably over a macroscopically small time interval $t' - t$. On substituting this relation into Equation (41.2) we obtain the result

$$\langle F(t')\rangle = -\beta\langle u(t)\rangle_0 \int_t^{t'} dt''\, \langle F(t')F(t'')\rangle_0 \qquad (41.3)$$

in which a time-varying ensemble average at time t', on the left side, has been related to a fixed-time ensemble average at the earlier time t. We have averaged separately over $u(t)$, because $u(t)$ varies much more slowly than does $F(t')$.

We now substitute Equation (41.3) into Equation (15.2), and transform to the new time variable $s \equiv t'' - t'$, to obtain the important result

$$\begin{aligned} m[\langle u(t+\tau)\rangle - \langle u(t)\rangle] &= F_{\text{ext}}(t)\cdot\tau - \beta\langle u(t)\rangle_0 \\ &\quad \times \int_t^{t+\tau} dt' \int_t^{t'} dt''\, \langle F(t')F(t'')\rangle_0 \\ &= F_{\text{ext}}(t)\cdot\tau - \beta\langle u(t)\rangle_0 \\ &\quad \times \int_t^{t+\tau} dt' \int_{t-t'}^{0} ds\, K(s) \end{aligned} \qquad (41.4)$$

where the *time-correlation function* $K(s)$ of the random force $F(t')$ is defined by

$$K(s) \equiv \langle F(t')F(t' + s)\rangle_0 \qquad \textbf{(41.5)}$$

The time-correlation function cannot depend on the time t at which the equilibrium-ensemble average is performed, and therefore it also cannot depend on the time t'. The fluctuation–dissipation theorem, which we derive heuristically below, is a statement about the double time integral of $K(s)$ in Equation (41.4). The theorem results in a relation between this integral and the damping constant α in Equation (15.3).

Properties of Time-Correlation Function

Correlation functions occur frequently in statistical physics, and they have several properties of general interest. In the following we can consider $F(t)$ to be any random function of t and for convenience write t instead of t' and drop the subscript zero from the averaging brackets. The first property to note is that $K^{1/2}(0)$ is the rms fluctuation ΔF_{rms} when $\langle F\rangle = 0$, that is,

$$K(0) = \langle F^2(t)\rangle = (\Delta F)^2_{\text{rms}} > 0 \qquad (41.6)$$

Thus the correlation function is a generalization of the rms fluctuation. Correlation functions in position and momentum space are closely related to two-particle reduced distribution functions,[2] as we discuss further in Chapter X.

If s becomes sufficiently large, then $F(t)$ and $F(t + s)$ must become uncorrelated, that is, the probability that F assumes a certain value at time $t + s$ must be independent of the value it had at the much earlier time t, and we obtain

$$K(s) \xrightarrow[s\to\infty]{} \langle F(t)\rangle\langle F(t + s)\rangle = 0, \qquad \text{if } \langle F\rangle = 0 \qquad (41.7)$$

In fact, one can show quite generally, for all s, that

$$|K(s)| \leqslant K(0) \qquad (41.8)$$

Exercise 41.1 Verify the inequality (41.8) by expanding the relation

$$\langle [F(t) \pm F(t + s)]^2\rangle \geqslant 0$$

Finally, since for an equilibrium ensemble $K(s)$ is independent of the time t, one can readily show by substituting $t \to t - s$ that

$$K(s) = K(-s) \qquad \textbf{(41.9)}$$

For the particular case of the random force $F(t)$, the decay of $K(s)$, as dic-

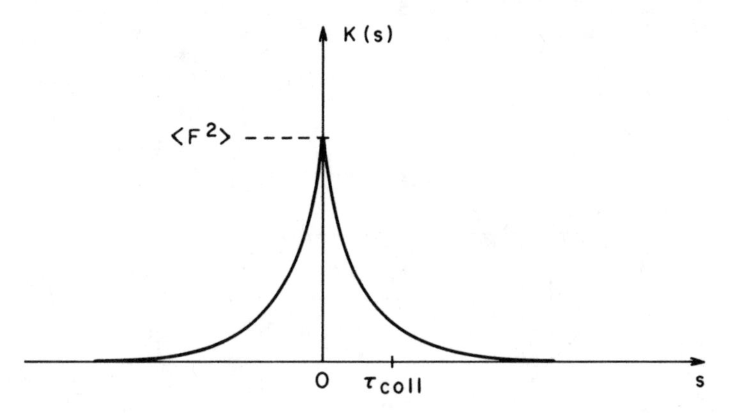

FIGURE IX.1 The correlation function $K(s)$ for the random force $F(t)$ plotted as a function of the time s.

tated by Equations (41.7) and (41.8), must occur in the time scale of a collision time τ_{coll}. On combining this fact with Equation (41.9) we conclude that $K(s)$ has the symmetric shape shown in Figure IX.1.

Fluctuation–Dissipation Theorem

We return now to the evaluation of the double integral over $K(s)$ in Equation (41.4). If we interchange the order of integration over t' and s (see Figure IX.2), then we obtain the result

$$\int_t^{t+\tau} dt' \int_{t-t'}^{0} ds\, K(s) = \int_{-\tau}^{0} ds \int_{t-s}^{t+\tau} dt'\, K(s)$$

$$= \int_{-\tau}^{0} ds(\tau + s)\, K(s) \cong \tau \int_{-\tau}^{0} ds\, K(s)$$

where integration over s in the second equality must give $s \sim \tau_{\text{coll}}$ and, by assumption, $\tau_{\text{coll}} \ll \tau$. Thus this double integral is proportional to τ, as was assumed with Equation (15.3) in the earlier analysis, instead of to τ^2. In fact, we may extend the lower integration limit to $-\infty$ in the final equality of this last result. Then upon using Equation (41.9) we obtain

$$\int_t^{t+\tau} dt' \int_{t-t'}^{0} ds\, K(s) \cong \frac{\tau}{2} \int_{-\infty}^{\infty} ds\, K(s) \tag{41.10}$$

Equation (41.10), when substituted into Equation (41.4), leads to our previous result:

$$m\langle \Delta u(t)\rangle = m[\langle u(t + \tau)\rangle - \langle u(t)\rangle] = [F_{\text{ext}}(t) - \alpha\langle u(t)\rangle]\tau \tag{41.11}$$

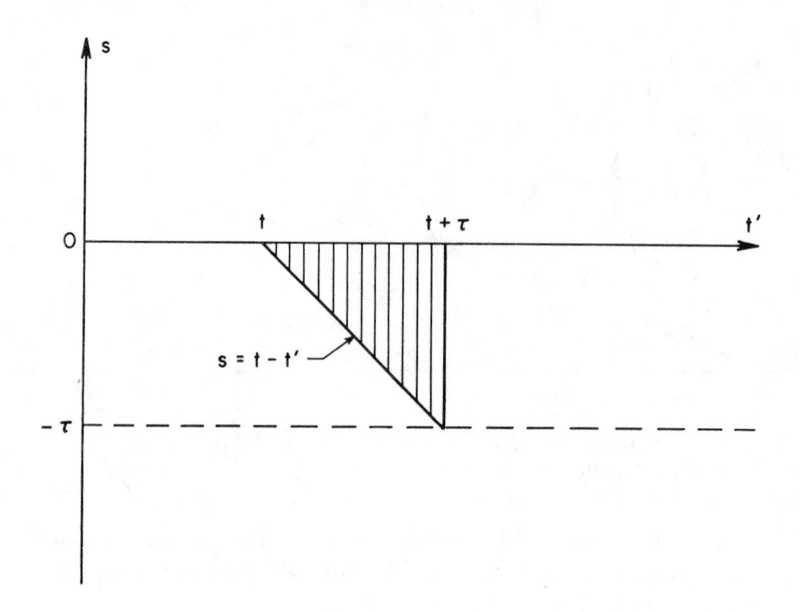

FIGURE IX.2 Domain of integration over $K(s)$ in Equation (41.4).

where now

$$\alpha \equiv \frac{1}{2\kappa T}\int_{-\infty}^{+\infty} ds\, K(s) = \frac{1}{2\kappa T}\int_{-\infty}^{\infty} ds\, \langle F(0)F(s)\rangle_0 \tag{41.12}$$

This expression for the damping constant α is the fluctuation–dissipation theorem for Brownian motion.

It is plausible that there exists a relation between fluctuations and dissipation when the interaction between a subsystem, for example, the macroscopic particle, and its environment is strong. Indeed, the random force $F(t)$ that produces fluctuations is large in this case. But if the interaction with the external system is strong, then the subsystem will also rapidly come to equilibrium if it is not in equilibrium initially. This return to equilibrium is governed by the dissipation constant α. The fluctuation–dissipation theorem relates the two opposing tendencies, fluctuation away from and dissipative return to equilibrium, of common origin and characterized by $K(s)$ and α.

As we have indicated in Section 15, Equation (41.11) is equivalent to Equation (15.4), which in the absence of external forces becomes

$$\frac{d\langle u(t)\rangle}{dt} = -\gamma\langle u(t)\rangle$$

where $\gamma \equiv \alpha/m$. This equation may be integrated, subject to the initial condition $\langle u(0)\rangle \neq 0$, to give

$$\langle u(t)\rangle = \langle u(0)\rangle\, e^{-\gamma t} \tag{41.13}$$

Thus $\langle u(t)\rangle$ approaches its true equilibrium value, zero, with a time constant γ^{-1}. Since we have assumed above, with Equation (41.3), that $\langle u(t)\rangle$ varies slowly with t in a time interval τ, we must have that

$$-\frac{\tau}{\langle u(t)\rangle}\frac{d\langle u(t)\rangle}{dt} = \gamma\tau \ll 1$$

In other words, the validity of our analysis of the one-dimensional Brownian-motion example is subject to the condition

$$\gamma^{-1} \gg \tau \gg \tau_{\text{coll}} \qquad (\gamma = \alpha/m) \tag{41.14}$$

The Brownian-motion data of Problem III.7 give $\gamma^{-1} \sim 10^{-8}$ sec for soot particles in a solution for which we can estimate $\tau_{\text{coll}} \sim 10^{-13}$ sec. Condition (41.14) is therefore easily satisfied in this example.

Exercise 41.2 Consider an electrical conductor with self-inductance L and carrying a current $I(t)$. If $\mathcal{V}(t)$ denotes an applied emf and $V(t)$ the effective fluctuating emf, representing the interaction of a conduction electron with all random microscopic influences on it, then the current $I(t)$ satisfies the equation

$$L\frac{dI(t)}{dt} = \mathcal{V}(t) + V(t)$$

Show that an ensemble average of this equation gives the ordinary circuit equation

$$L\frac{d\langle I\rangle}{dt} = \mathcal{V} - R\langle I\rangle$$

where the electrical resistance R of the conductor is given in terms of the fluctuating voltage $V(t)$ by

$$R = \frac{1}{2\kappa T}\int_{-\infty}^{+\infty} ds\, \langle V(0)V(s)\rangle_0$$

Exercise 15.1 shows that the resistance R can also be expressed in terms of a microscopic damping constant α associated with the average electron drift velocity $\langle u\rangle$.[3]

Equation (41.11) provides a relation between the average velocity change $\langle \Delta u(t)\rangle$ and the damping constant α. We can also find an expression for the rms fluctuations in $u(t)$, starting from the basic equation (15.1). In the absence of external forces we have for the integral of Equation (15.1) over a time interval $\tau >> \tau_{\text{coll}}$,

$$m\,\Delta u(\tau) \equiv m[u(t + \tau) - u(t)] = \int_t^{t+\tau} dt'\, F(t') \tag{41.15}$$

This equation can be squared and then averaged over an ensemble, considered to be at the "fixed" time t, to give

$$\begin{aligned} m^2 \langle [\Delta u(\tau)]^2 \rangle &= \int_t^{t+\tau} dt' \int_t^{t+\tau} dt'' \langle F(t')F(t'')\rangle_0 \\ &= \int_t^{t+\tau} dt' \int_{t-t'}^{t+\tau-t'} ds \langle F(t')F(t' + s)\rangle_0 \\ &= \int_t^{t+\tau} dt' \int_{t-t'}^{t+\tau-t'} ds\, K(s) \end{aligned} \tag{41.16}$$

where we have again introduced the relative time variable $s = t'' - t'$. We do not have to consider a time-varying ensemble, as in Equations (15.2) and (41.2), because the leading contribution due to the noise force $F(t')$ is nonvanishing in the present case. Possible correction terms, which could arise if the quadratic average were computed using a time-varying ensemble, are negligible compared with the double-product term in Equation (41.16) provided only that the approximation $\beta|\Delta E_{\text{med}}| << 1$, used in Equation (41.2), is valid.

If we interchange the order of integration over t' and s in Equation (41.16), as in the derivation of Equation (41.10), then we obtain the result for $\tau >> \tau_{\text{coll}}$,

$$\begin{aligned} \langle [\Delta u(\tau)]^2 \rangle &= \frac{1}{m^2}\left[\int_0^{\tau} ds(\tau - s)K(s) + \int_{-\tau}^{0} ds(\tau + s)K(s)\right] \\ &\cong \frac{\tau}{m^2}\int_{-\infty}^{+\infty} ds\, K(s) \end{aligned}$$

$$\langle [\Delta u(\tau)]^2 \rangle = \frac{\tau}{m^2}(2\kappa T\alpha) = \left(\frac{2\kappa T}{m}\right)\gamma\tau \tag{41.17}$$

Here we have used the fluctuation–dissipation theorem (41.12) to write the third equality, and $\gamma = \alpha/m$. We can explore this result for the case in which the macroscopic particle has a motion due only to microscopic momentum transfers, and not to any initially prepared nonequilibrium situation. Then on making use of the equipartition theorem (28.16) and inequality (41.14), we conclude for Brownian motion in one dimension that the fluctuations $\Delta u(\tau)_{\rm rms}$ are small.

$$\frac{\Delta u(\tau)_{\rm rms}}{\langle u^2\rangle^{1/2}} = \left(\frac{2\kappa T}{m\langle u^2\rangle}\right)^{1/2} (\gamma\tau)^{1/2} = (2\gamma\tau)^{1/2} \ll 1 \tag{41.18}$$

Velocity Correlation Function

It is instructive to calculate the velocity correlation function $K_u(s)$, for the macroscopic particle, as defined by

$$K_u(s) \equiv \langle u(t')u(t' + s)\rangle_0 = \langle u(0)u(s)\rangle_0 \tag{41.19}$$

We consider again the case in which the macroscopic particle has motion due only to microscopic momentum transfers. In view of the long correlation time to be expected for this function,[4] we may use the Langevin equation (15.7) for its calculation. In the absence of external forces we obtain, after integrating over the time interval τ, such that $\gamma^{-1} \gg \tau \gg \tau_{\rm coll}$,

$$u(s + \tau) - u(s) = -\gamma u(s)\,\tau + \frac{1}{m}\int_s^{s+\tau} dt\, F'(t)$$

On multiplying both sides of this expression by $u(0)$ and taking an ensemble average at the "fixed" time $s = 0$, we derive the result

$$\begin{aligned} K_u(s + \tau) - K_u(s) &= -\gamma\tau K_u(s) + \frac{1}{m}\langle u(0)\int_s^{s+\tau} dt\, F'(t)\rangle_0 \\ &= -\gamma\tau K_u(s) \end{aligned}$$

where the second term on the right side of the first equality vanishes, as was explained below Equation (15.6). But for macroscopically infinitesimal τ values this equation is equivalent to

$$\frac{dK_u(s)}{ds} = -\gamma K_u(s) \tag{41.20}$$

The solution to the simple differential equation (41.20) is

$$K_u(s) = K_u(0)\, e^{-\gamma|s|}$$

where we have written $|s|$ in the exponential because of symmetry property (41.9). Since $\langle u^2(0)\rangle = \kappa T/m$ for one-dimensional Brownian motion, we then obtain

$$K_u(s) = \langle u(0)u(s)\rangle_0 = \left(\frac{\kappa T}{m}\right) e^{-\gamma|s|} \tag{41.21}$$

On referring to inequality (41.14) and Figure IX.1, we see that the macroscopic particle velocity is, indeed, correlated over a much longer time interval than is the random force $F(t)$.

Exercise 41.3 Consider the equation

$$x(t) = \int_0^t dt' \, u(t')$$

where $x(t)$ is the position of the macroscopic particle at time t, with the origin chosen so that $x(0) = 0$. Use this equation and result (41.21) to derive the expression (15.12) for $\langle x^2(t) \rangle$.

42 Irreversible Thermodynamics

In the Brownian-motion example of the preceding section the damping force on a macroscopic particle is a consequence of a randomly varying microscopic force; the damping force is proportional to the speed of the particle. To describe transport phenomena we turn the problem around and investigate the average flow rates, or fluxes, that arise as a consequence of random microscopically defined fluxes. The macroscopic fluxes then depend on generalized forces, or *affinities,* such as a temperature difference or gradient, that are associated with a nonequilibrium condition in the system. For small deviations from equilibrium the *linear Markoff process* applies and the macroscopic fluxes are directly proportional to the affinities.

In the present section we study transport processes for the two possibilities of continuously varying affinities, and when the nonequilibrium condition is due to a discontinuity between separate parts of a system, each internally in equilibrium. The thermodynamics of such transport phenomena, called *irreversible thermodynamics,* can often be described by the equations of the linear Markoff process and associated continuity equations for the macroscopic fluxes. Two examples of the former are the equation of heat flow, Equation (13.1), $\mathbf{q} = -K\nabla T$, and Equation (13.2) for the pressure tensor $\mathcal{P}_{ij}$. In these relations the thermal conductivity coefficient K and the viscosity coefficient μ are called kinetic coefficients, and such coefficients represent generalizations of the damping constant α of Equation (15.3).

We use the general notation $L_{\mu\nu}$ for kinetic coefficients, and our central objective in this section is to deduce classically a general form of the fluctuation–dissipation theorem for the $L_{\mu\nu}$. The section is concluded with a derivation of symmetry relations, the *Onsager reciprocal relations,* between various $L_{\mu\nu}$.

To proceed we must develop general notation to describe irreversible processes and associated kinetic coefficients. It should be stressed that we do not demonstrate in this section, as we did in Sections 13 and 14, how to calculate kinetic coefficients explicitly in terms of particle interactions at the microscopic level. Rather, we are interested here only in treating general features of irreversible phenomena, as can be deduced from microscopically based relations such as the fluctuation–dissipation theorem.

The first step of our analysis is to specify a set of n extensive variables, $X_\mu(t)$, which characterize the closed system under investigation. When we are interested in variations within the system, then we shall consider, instead of the X_μ, a corresponding set of $(n - 1)$ intensive variables $x_\mu(t) \equiv \Omega^{-1}X_\mu$, defined for macroscopically small volume elements. Thus the volume is only considered to be a variable in the first of these two cases. For comparison, the Brownian-motion example of Section 41 was characterized by a single variable, the velocity $u(t)$ of the macroscopic particle.

We are interested in situations in which separate parts of a system are internally in equilibrium or in the case of continuous systems where there is local equilibrium.[5] In either case the system can be characterized by local time-dependent values of the temperature, pressure, and other thermodynamic variables. For the case in which there are discrete subsystems, small changes δX_μ in the extensive variables will produce small deviations δS_{sys} in the total entropy of the system. Now, in general we can define a set of intensive parameters F_μ in terms of the local entropy function $S(X_1, X_2, \ldots)$ by the equation

$$F_\mu \equiv \frac{\partial S(X_1, X_2, \ldots X_\mu \ldots)}{\partial X_\mu} \tag{42.1}$$

Reference to the thermodynamic identity, that is, to the first of Equations (8.15), shows that typically the F_μ will be local position- and time-dependent quantities such as T^{-1}, $\mathcal{P}/T$ and $-g/T$.

The F_μ cannot be the generalized forces, or affinities, in a system, because they do not give a direct measure of a nonequilibrium condition. For example, in a system characterized by continuously (and slowly) varying parameters $F_\mu(\mathbf{r}, t)$, the affinities $\mathcal{F}_\mu(\mathbf{r}, t)$ are vector quantities defined to be gradients of the F_μ.

$$\mathcal{F}_\mu(\mathbf{r}, t) \equiv \nabla F_\mu(\mathbf{r}, t) \qquad \text{(continuous system)} \tag{42.2}$$

For the *discrete* case, in which the nonequilibrium condition is associated with a discontinuity between separate parts of a system, each internally in equilibrium, the affinities will generally be differences between the F_μ of the subsystems.

The affinities $\mathcal{F}_\mu$ for the discrete case can be defined more precisely by generalizing Equation (42.1). Thus we write, introducing scalar quantities in this case,

$$\mathcal{F}_\mu \equiv \frac{\partial S_{\text{sys}}}{\partial X_\mu} \qquad \text{(discrete subsystems)} \tag{42.3}$$

where S_{sys} is the sum over the entropies of the various subsystems. Our goal in the first part of this section can now be specified as the determination of expressions for entropy variations when affinities are present. For the discrete case, the task is accomplished readily by substituting the thermodynamic identity for each subsystem i into the sum $\delta S_{\text{sys}} = \Sigma_i \delta S_i$. According to Equation (42.3) the general form of the result will be

$$\delta S_{\text{sys}} = \sum_\mu \mathcal{F}_\mu \, \delta X_\mu \tag{42.4}$$

where both Equations (42.3) and (42.4) must refer to a set of linearly independent extensive variables X_μ. For example, when there are two subsystems with a fixed total volume $\Omega = \Omega_1 + \Omega_2$ then there is only one independent volume variable, either Ω_1 or Ω_2. Also, we may anticipate that the $\mathcal{F}_\mu$ will themselves often depend linearly on some or all of the δX_ν. Thus a generalization of the discussion associated with Equation (7.1) shows that δS_{sys} must be a positive quadratic function of the δX_ν for small deviations from equilibrium.

As an example of the preceding two equations, consider a composite system composed of two subsystems. An extensive parameter has values X and X' in the two subsystems, respectively, and we require that the sum $(X + X')$ be constant. We choose X to be the independent extensive variable for the system and differentiate the system entropy $S_{\text{sys}} = S + S'$ with respect to X to find for the affinity $\mathcal{F}$, the expression

$$\begin{aligned}\mathcal{F} &= \frac{\partial S}{\partial X} + \frac{\partial S'}{\partial X} = \frac{\partial S}{\partial X} - \frac{\partial S'}{\partial X'} \\ &= F - F'\end{aligned}$$

which quantity must vanish at thermodynamic equilibrium.

For definiteness, consider two systems separated by a diathermal (or heat-conducting) wall, and let X be the energy E. Then the affinity is $\mathcal{F} = (1/T) - (1/T')$. No heat flows across the diathermal wall if the difference in inverse temperature vanishes. But a nonzero difference in inverse temperature, acting as a generalized force, drives a flow of heat between the subsystems. Similarly, if X is the volume, then the affinity is $\mathcal{F} = (\mathcal{P}/T) - (\mathcal{P}'/T')$.

Continuous Systems

For continuous systems we use definition (42.1) written as $F_\mu = \partial s/\partial x_\mu$, where $s \equiv S/\Omega$ for macroscopically small volume elements Ω, and obtain the basic thermodynamic identity when there is local equilibrium,

$$ds = \sum_\mu F_\mu \, dx_\mu \tag{42.5}$$

We emphasize here that the local intensive quantity $F_\mu(\mathbf{r}, t)$ must be the same function of the local independent thermodynamic variables $X_\mu(\mathbf{r}, t)$ as it would be at thermodynamic equilibrium.

We next introduce the flux vectors $\langle \mathbf{J}_\mu \rangle$ and $\langle \mathbf{J}_S \rangle$, corresponding to flow of the quantities X_μ and S respectively. We do not apply these considerations to the volume Ω. Moreover, the particle flux vector $\langle \mathbf{J}_N \rangle = n\mathbf{u}$, where $\mathbf{u}$ is the local instantaneous velocity in the system, has a particularly simple expression. In general, however, we cannot give simple macroscopic definitions in terms of x_μ and $\mathbf{u}$ for the flux vectors $\langle \mathbf{J}_\mu \rangle$. For example, the heat-flux vector $\langle \mathbf{J}_Q \rangle = \mathbf{q}_K + \mathbf{q}_V$ is defined classically in terms of microscopic quantities by Equations (12.31) and (12.18), with no corresponding macroscopic analogue as for $\langle \mathbf{J}_N \rangle$. Nevertheless, we can use the concept that $\langle \mathbf{J}_\mu \rangle$ is the amount of X_μ that flows across a unit area, in the direction of $\langle \mathbf{J}_\mu \rangle$, per unit time. Furthermore, we may expect that the thermodynamic identity (42.5) must also hold for fluxes, in the form

$$\langle \mathbf{J}_S \rangle = \sum_\mu F_\mu \langle \mathbf{J}_\mu \rangle \tag{42.6}$$

Exercise 42.1

(a) Neither the heat-flux vector $\langle \mathbf{J}_Q \rangle$ nor the energy-flux vector $\mathbf{Q}$, defined by Equation (12.17), corresponds to the internal-energy-flux vector $\langle \mathbf{J}_E \rangle$. Show that $\mathbf{Q}$ and $\langle \mathbf{J}_Q \rangle$ are related by

$$\mathbf{Q} = \langle \mathbf{J}_Q \rangle + \rho_E \mathbf{u} + \overleftrightarrow{\mathscr{P}} \cdot \mathbf{u}$$

where the energy density ρ_E is defined by the first of Equations (12.16) and the pressure tensor $\mathscr{P}_{ij}$ by Equation (12.29). We shall define $\langle \mathbf{J}_E \rangle$ by the relation

$$\langle \mathbf{J}_E \rangle \equiv \langle \mathbf{J}_Q \rangle + (\rho e)\mathbf{u} \tag{42.7a}$$

where e is the internal energy per unit mass of Equation (12.30). Alternatively, we may write

$$\mathbf{Q} = \langle \mathbf{J}_E \rangle + \left(\frac{1}{2}\rho u^2\right)\mathbf{u} + \overleftrightarrow{\mathscr{P}} \cdot \mathbf{u} \tag{42.7b}$$

(b) Starting from either Equation (12.19) or (12.34), show that

$$\frac{\partial(\rho e)}{\partial t} + \nabla \cdot \langle \mathbf{J}_E \rangle = - \overset{\leftrightarrow}{\mathcal{P}} \cdot \overset{\leftrightarrow}{\Lambda} \tag{42.8}$$

where Λ_{ij} is the velocity-gradient tensor of Equation (12.32). The right side of Equation (42.8), which involves this factor, will be neglected in the following development, in which case this equation becomes a continuity equation for the internal-energy density $x_E = \rho e$.

We expect for the conserved quantities X_μ ($\neq \Omega$) that there will be no sources or sinks in the system. When this is true, a continuity equation (12.8) holds for each of the variables $x_\mu(\mathbf{r}, t)$, that is,

$$\frac{\partial x_\mu}{\partial t} + \nabla \cdot \langle \mathbf{J}_\mu \rangle = 0 \tag{42.9}$$

Of course, for an irreversible process the entropy density s cannot satisfy a continuity equation.

We now compute the local irreversible rate of entropy production $(\partial s/\partial t)_{\text{irr}}$ in a continuous system by identifying this quantity as a source term for the right side of (42.9) when $x_\mu = s$.

$$\left(\frac{\partial s}{\partial t}\right)_{\text{irr}} \equiv \frac{\partial s}{\partial t} + \nabla \cdot \langle \mathbf{J}_S \rangle$$

$$= \sum_\mu F_\mu \frac{\partial x_\mu}{\partial t} + \sum_\mu \nabla \cdot (F_\mu \langle \mathbf{J}_\mu \rangle)$$

$$\left(\frac{\partial s}{\partial t}\right)_{\text{irr}} = \sum_\mu \mathcal{F}_\mu \cdot \langle \mathbf{J}_\mu \rangle \tag{42.10}$$

We have used Equations (42.2), (42.5), (42.6), and (42.9) to derive the second and third equalities of this expression. On integrating the third equality over a macroscopically small time interval τ, we obtain the local (positive) entropy change

$$\begin{aligned} \delta s &= \sum_\mu \int_t^{t+\tau} dt' \, \mathcal{F}_\mu(t') \cdot \langle \mathbf{J}_\mu(t') \rangle \\ &\cong \sum_\mu \mathcal{F}_\mu(t) \cdot \int_t^{t+\tau} dt' \, \langle \mathbf{J}_\mu(t') \rangle \end{aligned} \tag{42.11}$$

As in Equation (41.3) for the Brownian-motion example, the affinities $\mathcal{F}_\mu$ are assumed to vary slowly, that is, not appreciably, during the time interval τ.

Because they are inherently thermodynamic variables, it is not necessary to include ensemble-average brackets with F_μ and $\mathscr{F}_\mu$.

Equation (42.11) is the form that Equation (42.4) takes for continuous systems. We emphasize that whereas an affinity is defined to be the difference between intensive quantities F_μ when there are discrete subsystems, it is the gradient of such a quantity in a continuously varying system.

Linear Processes

Each of the fluxes depends, in general, on all of the affinities that are the basic causes, macroscopically, of transport phenomena. Although it is possible for this dependence to vary according to the entire complicated time evolution, or history, of the affinities, we assume here that the dependence is without memory, or *Markoffian*.[6] Thus in a Markoff process, each local flux depends only on the instantaneous local values of the affinities—a situation that is quite common, but not universal. For example, electrical systems with capacitance and inductance are not Markoffian.

Even for Markoff processes the dependence of the fluxes on the affinities will be generally intricate. But for many experimental situations the affinities are sufficiently small that a linear approximation is completely adequate. For such cases we obtain the simple equations of a *linear Markoff process,*

$$\langle \dot{X}_\mu \rangle \equiv \left\langle \frac{dX_\mu}{dt} \right\rangle = \sum_\nu L_{\mu\nu} \mathscr{F}_\nu \qquad \text{(discrete subsystems)}$$

$$\langle J_\mu \rangle_i = \sum_\nu \sum_{j=1}^{3} L_{\mu\nu}^{(i,j)} \mathscr{F}_{\nu,j} \qquad \text{(continuous system)} \tag{42.12}$$

where for the case of discrete subsystems the fluxes are defined to be the quantities $\langle \dot{X}_\mu \rangle$. The heat-flow equation, $\langle \mathbf{J}_Q \rangle = -K\nabla T$, is a simple example of the second of these equations. In general, the $L_{\mu\nu}$ will be off-diagonal tensors, and the fluxes will be vector quantities with components $i = (1, 2, 3)$. The counterpart of these equations in the Brownian-motion example is Equation (15.3) when it is divided through by τ.

The $L_{\mu\nu}$ are the kinetic coefficients of irreversible thermodynamics, and we give below a statistical-mechanical basis for these coefficients in terms of a general fluctuation–dissipation theorem. In this connection, we must assume that the $X_\mu(\mathbf{r}, t)$ and $\mathbf{J}_\mu(\mathbf{r}, t)$ can all be defined microscopically. For example, the ensemble-averaged particle and internal-energy flux vectors are defined for classical systems by $\langle \mathbf{J}_N \rangle = n\mathbf{u}$ and by Equation (42.7a) for $\langle \mathbf{J}_E \rangle$. These fluxes can also be readily written down for a single system in the ensemble. The rapidly varying particle-flux $\mathbf{J}_N$ is given by

$$\mathbf{J}_N \equiv \frac{1}{\Delta\Omega} \sum_{i \in \Delta\Omega} \mathbf{v}_i$$

where the summation is over all the particles, with velocities $\mathbf{v}_i$, in a macroscopically small volume $\Delta\Omega$. The internal energy density $x_E = \rho e$ in Equation (42.7a) becomes, for a single system, the quantity

$$x_E = \frac{1}{\Delta\Omega} \sum_{i \in \Delta\Omega} \left[\frac{m}{2} (\mathbf{v}_i - \mathbf{u})^2 + \frac{1}{2} \sum_{j \in \Delta\Omega} V_{ij} \right]$$

where $\mathbf{u} \equiv n^{-1}\mathbf{J}_N$, and so forth. On the other hand, the intensive quantities $F_\mu(\mathbf{r}, t)$ and corresponding affinities must always represent slowly varying parameters of an ensemble for the system. With these observations as a basis, we now assume that the quantity δs, derived microscopically, will be given by the expression on the right side of Equation (42.11) with the averaging brackets removed from $\mathbf{J}_\mu(t')$.

To initiate a proof of the general fluctuation–dissipation theorem, we express the quantity $\langle \mathbf{J}_\mu \rangle$ appearing in Equation (42.12), with care, as

$$\langle \mathbf{J}_\mu \rangle \to \overline{\langle \mathbf{J}_\mu(\mathbf{r}, t) \rangle} \equiv \frac{1}{\tau} \int_t^{t+\tau} dt' \, \langle \mathbf{J}_\mu(\mathbf{r}, t') \rangle \tag{42.13}$$

where $\tau >> \tau_{\text{coll}}$ is a macroscopically small time interval. As with Equation (15.3), the important point here is that we must consider a short-time average over $\langle \mathbf{J}_\mu \rangle$, because this flux itself involves an average over a time-varying ensemble. On the one hand, the time variation of $\langle \mathbf{J}_\mu \rangle$ is crucial for an explanation of irreversibility; on the other hand, a short-time average of $\langle \mathbf{J}_\mu \rangle$ is the only meaningful average to consider on a macroscopic time scale. We calculate the right side of Equation (42.13), using the methods of Section 41, by applying it to a system in which there are only small deviations, or fluctuations, from local thermodynamic equilibrium. In so doing, we will be able to relate the kinetic coefficients $L_{\mu\nu}$ to short-time correlations in such a system, and this now is our immediate objective.

Thus we consider continuously varying systems and use the result of Problem II.7(a) to write

$$\begin{aligned} P(t + \tau') &= P(t) \exp\left(\frac{1}{\kappa} \cdot \delta S_{\text{sys}}(t, t') \right) \\ &= P(t) \exp\left(\frac{1}{\kappa} \sum_{\Delta\Omega} \delta s(t, t') \Delta\Omega \right) \\ &= \prod_{\Delta\Omega} P(\mathbf{r}, t + \tau') \end{aligned} \tag{42.14}$$

where $t \leqslant (t' \equiv t + \tau') \leqslant t + \tau$ and the summation is over all small volume elements $\Delta\Omega$ in the system. Clearly, we can identify the local probability

$$
\begin{aligned}
P(\mathbf{r}, t + \tau') &= P(\mathbf{r}, t) \exp\left(\frac{1}{\kappa}\, \delta s(t, t')\Delta\Omega\right) \\
&\cong P(\mathbf{r}, t)\left(1 + \frac{1}{\kappa}\, \delta s(t, t')\Delta\Omega\right)
\end{aligned}
\tag{42.15}
$$

as the probability distribution function for the macroscopically small subsystem of volume $\Delta\Omega$. Correlation with the rest of the system occurs, ultimately, via the affinities $\mathcal{F}_\mu(\mathbf{r}, t)$. The approximation in the second line of Equation (42.15) corresponds to the linear approximations (42.12). Also, remarks given below Equation (41.2) in connection with the analogous Brownian-motion equations are relevant in the present case.

We now substitute the second equality of this last expression into the ith component of Equation (42.13), with $i = 1, 2,$ or 3, so that the time-varying ensemble average of $J_\mu(t')_i$ on the right side can be expressed in terms of a fixed-time ensemble average. On using the notation of Section 41, and with $\mathbf{r}$ dependences suppressed, we obtain

$$
\begin{aligned}
\langle\overline{J_\mu(t)_i}\rangle &= \frac{1}{\tau}\int_t^{t+\tau} dt' \left\langle J_\mu(t')_i \left[1 + \frac{1}{\kappa}\, \delta s(t, t')\Delta\Omega\right]\right\rangle_0 \\
&= \frac{\Delta\Omega}{\kappa\tau}\sum_{j=1}^{3}\int_t^{t+\tau} dt' \sum_\nu \mathcal{F}_\nu(t)_j \\
&\quad \times \left\langle J_\mu(t')_i \int_t^{t'} dt''\, J_\nu(t'')_j\right\rangle_0 \\
&= \frac{\Delta\Omega}{\kappa\tau}\sum_\nu\sum_{j=1}^{3} \mathcal{F}_\nu(t)_j \\
&\quad \times \int_t^{t+\tau} dt' \int_{t-t'}^{0} ds\, \langle J_\mu(t')_i J_\nu(t' + s)_j\rangle_0 \\
&= \frac{1}{\kappa\tau}\sum_\nu\sum_{j=1}^{3} \mathcal{F}_\nu(t)_j \int_{-\tau}^{0} ds\, (\tau + s) K_{\mu\nu}^{(i,j)}(s)
\end{aligned}
\tag{42.16}
$$

where the "cross-correlation" function $K_{\mu\nu}^{(i,j)}(s)$ is defined by

$$
K_{\mu\nu}^{(i,j)}(s) \equiv \Delta\Omega\langle J_\mu(t')_i J_\nu(t' + s)_j\rangle_0 = \Delta\Omega\langle J_\mu(0)_i J_\nu(s)_j\rangle_0 \tag{42.17}
$$

To derive the second equality of Equation (42.16) we have used the microscopic form of Equation (42.11) and the fact that $\langle J_\mu(t')_i\rangle_0 = 0$ at thermodynamic equilibrium. The equilibrium-ensemble average in Equation (42.17) is for a macroscopically small system of volume $\Delta\Omega$. Because it involves microscopic correlations, this average will be independent of $\Delta\Omega$ provided only that $\Delta\Omega >> (\text{mean free path})^3$.[7]

As was argued above Equation (41.10), we can use the fact that $K_{\mu\nu}^{(i,j)}$ decays in a time $\sim\tau_{\text{coll}} << \tau$ to drop the term s in the integrand of the last equality of Equation (42.16). We then obtain the second of Equations (42.12) from this expression by identifying the kinetic coefficients $L_{\mu\nu}^{(i,j)}$ in terms of the cross-correlation function as

$$\begin{aligned} L_{\mu\nu}^{(i,j)} &\equiv \frac{1}{\kappa}\int_{-\infty}^{0} ds\, K_{\mu\nu}^{(i,j)}(s) \\ &= \frac{\Delta\Omega}{\kappa}\int_{-\infty}^{0} ds\, \langle J_\mu(0)_i J_\nu(s)_j\rangle_0 \end{aligned} \qquad \text{(continuous system)} \qquad \textbf{(42.18)}$$

In this definition we have, without loss of accuracy, set $\tau \to \infty$ in the lower integration limit of the integral over ds. Equation (42.18) is a general form of the fluctuation–dissipation theorem for continuous systems.

The important steps in the analyses of these last two sections that invalidate any derivation of irreversibility from microscopic physical laws [see the discussion following Equation (25.11)] are the use of Equations (41.1) and (42.14) for a vanishingly small time interval.[1] In this limit the use of these equations cannot be justified, and therefore we must regard our proof of the fluctuation–dissipation theorem as nonrigorous.

Exercise 42.2 Show for the case of discrete subsystems that the $L_{\mu\nu}$ in the first of Equations (42.12) can be written as

$$L_{\mu\nu} = \frac{1}{\kappa}\int_{-\infty}^{0} ds\, \langle \dot{X}_\mu(0)\dot{X}_\nu(s)\rangle_0 \qquad \text{(discrete subsystems)} \qquad \textbf{(42.18a)}$$

Onsager Reciprocal Relations

The Onsager reciprocal relations can now be deduced from the general fluctuation–dissipation theorem, as expressed by Equations (42.18). First, we observe that the form that property (41.9) takes for the cross-correlation function (42.17) is

$$K_{\mu\nu}^{(i,j)}(s) = K_{\nu\mu}^{(j,i)}(-s) \qquad (42.19)$$

We cannot conclude, therefore, that this function is even in s unless $\mu = \nu$ and $i = j$. On the other hand, a physical argument based upon time-reversal invariance does lead to further interesting results.

If the sign of the time is reversed from s to $-s$, then all particle velocities, or momenta, also reverse their signs. Furthermore, if any external magnetic field $\mathcal{H}$ is present, then it also reverses its sign, since the current producing it must be reversed. But the microscopic equations of motion are invariant under time reversal and therefore we may conclude, using the notation of Equation (25.10), that

$$K_{\mu\nu,T}^{(i,j)}(s, \mathcal{H}) \equiv K_{\mu\nu}^{(i,j)}(-s, -\mathcal{H}) = \pm K_{\mu\nu}^{(i,j)}(s, \mathcal{H}) \tag{42.20}$$

where the sign $\pm$ is determined by the time-reversal properties of the fluxes $J_{\mu,i}$ and $J_{\nu,j}$ (or $\dot{X}_\mu$ and $\dot{X}_\nu$).

If now we combine Equations (42.19) and (42.20), then we obtain the result

$$K_{\mu\nu}^{(i,j)}(s, \mathcal{H}) = \pm K_{\nu\mu}^{(j,i)}(s, -\mathcal{H})$$

Substitution into Equations (42.18) then yields the Onsager relations

$$L_{\mu\nu}^{(i,j)}(\mathcal{H}) = \pm L_{\nu\mu}^{(j,i)}(-\mathcal{H}) \tag{42.21}$$

where the superscripts are to be dropped for the case of discrete subsystems.

43 Applications of Onsager Relations

In this section we study two examples: one in which there are two discrete subsystems and one involving a continuously varying system. These examples illustrate the two cases discussed generally in the preceding section. In the first we imagine that a fluid is enclosed in a container in which there are two compartments connected by a hole. Although particles can pass easily through the hole, it is small enough so that within each compartment the fluid is in thermodynamic equilibrium. In the second example we investigate thermoelectric effects of electrons in wires, in which there is an interplay between thermal and electrical conductivity. In both examples we show how the Onsager reciprocal relations can be used to relate energy flow to particle flow.

We consider first the problem of a fluid enclosed in a container in which there are two constant-volume compartments connected by a small hole (see Figure IX.3). Two distinct processes are considered in this first example:

(a) The *thermomechanical effect* in which both compartments are kept at the same temperature. Associated with any pressure difference there is a net flow of particles and energy, as in the Joule–Thomson process of Exercise 8.1.

(b) The *thermomolecular pressure-difference effect,* in which no fluid passes through the hole when certain pressure and temperature differences are maintained between the compartments by external reservoirs. For this effect there will generally be a net energy flow, and we are interested in the steady-state condition.

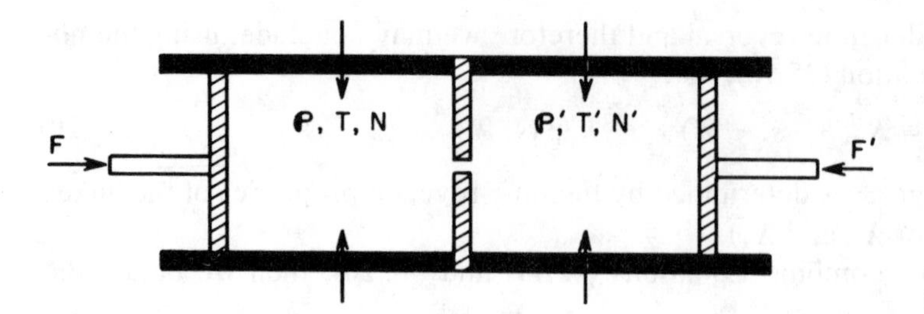

FIGURE IX.3 Two compartments of a container connected by a small hole have their pressures adjusted by external forces applied to pistons, which are then held in place so that the two volumes are constant. The vertical arrows indicate heat flow from (or to) external temperature reservoirs.

Both effects depend on the size of the hole between the compartments, and they can be related quite generally by applying the Onsager relations.

Since the volumes of the compartments are held fixed, the thermodynamic identity (42.4) reduces to

$$\delta S_{\text{sys}} = \left(\frac{\delta E}{T} + \frac{\delta E'}{T'}\right) - \left(\frac{g}{T}\,\delta N + \frac{g'}{T'}\,\delta N'\right)$$

where the unprimed and primed quantities refer to the two different compartments as in Figure IX.3. Now in both of the processes described above, $(E + E')$ and $(N + N')$ can be assumed constant.[8] On choosing E and N to be the independent extensive variables, we obtain for δS_{sys} the expression

$$\delta S_{\text{sys}} = \mathscr{F}_E\,\delta E + \mathscr{F}_N\,\delta N \tag{43.1}$$

where the affinities $\mathscr{F}_E$ and $\mathscr{F}_N$ are given by

$$\begin{aligned} \mathscr{F}_E &= \frac{1}{T} - \frac{1}{T'} \equiv \Delta\left(\frac{1}{T}\right) \\ \mathscr{F}_N &= -\left(\frac{g}{T} - \frac{g'}{T'}\right) \equiv -\Delta\left(\frac{g}{T}\right) \end{aligned} \tag{43.2}$$

The associated linear phenomenological equations (42.12) are

$$\begin{aligned} \frac{dE}{dT} &= L_{11}\mathscr{F}_E + L_{12}\mathscr{F}_N \\ \frac{dN}{dt} &= L_{21}\mathscr{F}_E + L_{22}\mathscr{F}_N \end{aligned} \tag{43.3}$$

where according to the Onsager relation (42.21), $L_{21} = L_{12}$.

We next select more physical "generalized forces" to be ΔT and $\Delta\mathscr{P}$, instead of $\mathscr{F}_E$ and $\mathscr{F}_N$, and we consider only the case of small temperature and pressure differences between the compartments. We also use Equation (8.18), written in the form

$$\Delta\left(\frac{g}{T}\right) = -\left(\frac{w}{T^2}\right)\Delta T + \left(\frac{1}{nT}\right)\Delta\mathcal{P} \tag{43.4}$$

where w is the average enthalpy per particle

$$w = \frac{E}{N} + \frac{\mathcal{P}}{n} \cong \frac{E'}{N'} + \frac{\mathcal{P}'}{n'} \tag{43.5}$$

Equation (43.5) for w refers to either compartment, because ΔT and $\Delta\mathcal{P}$ are both small quantities. Then the phenomenological relations (43.3) become

$$\begin{aligned} \frac{dE}{dt} &= \frac{1}{T^2}(wL_{12} - L_{11})\Delta T - \frac{L_{12}}{nT}\Delta\mathcal{P} \\ \frac{dN}{dt} &= \frac{1}{T^2}(wL_{22} - L_{12})\Delta T - \frac{L_{22}}{nT}\Delta\mathcal{P} \end{aligned} \tag{43.6}$$

We can now apply these expressions to the two processes of interest.

Exercise 43.1 Show that when written in terms of the independent variables T and $\mathcal{P}$, the differential heat function dW becomes

$$dW = C_P\, dT + \Omega(1 - \alpha T)\, d\mathcal{P}$$

where α is the coefficient of thermal expansion (9.1). For small ΔT and $\Delta\mathcal{P}$ this expression gives the difference ΔW in enthalpies between the two compartments. We do not require this expression in the analysis that follows, but it is instructive to examine it in various contexts such as in the ideal-gas limit where $\alpha T = 1$.

For process (a), $\Delta T = 0$ and we find immediately for the ratio of the energy and particle current,

$$\varepsilon_a \equiv \frac{\dot{E}}{\dot{N}} = \frac{L_{12}}{L_{22}} \tag{43.7}$$

The quantity ε_a is the energy per particle passing through the hole in process (a). For process (b), $\dot{N} = 0$ and we obtain for the ratio $\Delta\mathcal{P}/\Delta T$ in the steady-state condition, the expression

$$\frac{\Delta\mathcal{P}}{\Delta T} = \frac{n}{T}\left(w - \frac{L_{12}}{L_{22}}\right) = \frac{n}{T}(w - \varepsilon_a) \tag{43.8}$$

That this result for process (b) can be expressed in terms of ε_a for process (a) is a direct consequence of the Onsager relation $L_{21} = L_{12}$. We note, since generally $w \geqslant \varepsilon$ (see below), that the steady-state condition with $\dot{N} = 0$ occurs when both ΔT and $\Delta \mathcal{P}$ have the same sign.

We now evaluate the quantity ε_a for two limiting cases. In the one limit the hole is large enough so that the pressure of the thermomechanical effect is effective in doing work on the fluid that passes through the hole. This condition is that of the Joule–Thomson process (Exercise 8.1), in which work $\Delta(\mathcal{P}/n)$ per particle is added to the fluid as it passes through the hole.[9] In other words, ε_a is equal to the average heat function, or enthalpy, w in this limit. But this implies for process (b), in which $\dot{N} = 0$, that $\Delta \mathcal{P} = 0$ for large holes and therefore that energy transfer in the system, that is, heat flow, is due to ΔT only. A steady-state condition can be sustained only when there is no pressure difference.

In the opposite limit, the hole is small compared with the mean free path of the particles in the fluid. Every molecule that arrives at the hole passes freely through it. We evaluate ε_a, in this limit, only for a classical ideal gas. According to the result of Exercise 11.2, the average kinetic energy of particles passing through the hole is then $\varepsilon_a = 2\kappa T$. But for an ideal gas we have from Equation (43.5), apart from any internal energies that affect both ε_a and w similarly,

$$w = \frac{3}{2}\kappa T + \kappa T = \frac{5}{2}\kappa T$$

Therefore, we obtain from Equation (43.8) the result, for very small holes and an ideal gas,

$$\frac{\Delta \mathcal{P}}{\Delta T} = \frac{1}{2} n\kappa = \frac{1}{2}\frac{\mathcal{P}}{T} \qquad (\text{when } \dot{N} = 0)$$

We can integrate this result, for the case of small pressure and temperature differences, to find for the pressure–temperature relation between the two compartments when there is steady-state energy flow,

$$\frac{\mathcal{P}}{\sqrt{T}} = \frac{\mathcal{P}'}{\sqrt{T'}} \tag{43.9}$$

An "equilibrium" thermomolecular pressure difference can occur in liquid helium II for two compartments at different temperatures and connected by a fine capillary, or superleak.[10] Flow of superfluid helium through the superleak, as it attempts to establish the appropriate pressure difference for the effect, results in the spectacular *fountain effect*.

Thermoelectric Effects

In our second example we investigate thermoelectric effects of electrons in wires, restricting our discussion to one-dimensional cases only.[11] As in the problem treated above, we first write down the basic equations for energy flow and particle (electron) flow. These are related directly to expressions for heat flow and electric current, so that the corresponding coefficients of *thermal conductivity* and *electrical conductivity* can be introduced. After examining these two cases a third case, the thermocouple, is considered. The *thermoelectric power* for each wire in the thermocouple is defined, and this entity provides a third measurable quantity in terms of which the basic kinetic coefficients may be expressed. Because of the fundamental Onsager relation in this example, all further thermoelectric effects can be expressed in terms of the measurable quantities associated with these first three effects.

The first of Equations (8.15), which is the thermodynamic identity, when written per unit volume for the free electrons in a metal has the form of Equation (42.5)

$$ds = F_E\, d\left(\frac{E}{\Omega}\right) + F_N\, dn$$

where the intensive quantities F_E and F_N are given by

$$F_E = \frac{1}{T} \qquad F_N = -\frac{g}{T} \tag{43.10}$$

Corresponding to Equation (42.6) we obtain the equation of electron fluxes,[12]

$$\mathbf{J}_S = \frac{1}{T}\mathbf{J}_E - \frac{g}{T}\mathbf{J}_N \tag{43.11}$$

As a special case of Equation (42.10) we then get the entropy-production equation

$$\left(\frac{\partial s}{\partial t}\right)_{\text{irr}} = \nabla\left(\frac{1}{T}\right)\cdot\mathbf{J}_E - \nabla\left(\frac{g}{T}\right)\cdot\mathbf{J}_N \tag{43.12}$$

in which the affinities $\mathcal{F}_E$ and $\mathcal{F}_N$ are $\nabla(1/T)$ and $-\nabla(g/T)$, respectively, for the currents $\mathbf{J}_E$ and $\mathbf{J}_N$.

We focus attention on the electric current density $-eJ_N$, for conduction electrons with charge $-e$, and therefore we consider the flux $-\mathbf{J}_N$, with corresponding affinity $+\nabla(g/T)$, instead of $\mathbf{J}_N$. For the present one-dimensional problem the second of the linear phenomenological relations (42.12) then corresponds to the two equations

$$J_E = L'_{11}\,\nabla\left(\frac{1}{T}\right) + L'_{12}\,\nabla\left(\frac{g}{T}\right) \tag{43.13a}$$

$$-J_N = L'_{21}\ \nabla\left(\frac{1}{T}\right) + L'_{22}\,\nabla\left(\frac{g}{T}\right) \tag{43.13b}$$

where (assuming $\mathcal{H} = 0$) the Onsager relation (42.21) specifies that $L'_{21} = L'_{12}$. For convenience, we use the gradient symbol ∇ to represent a one-dimensional partial derivative in the present example.

Although $\mathbf{J}_E$ is a current density for total internal energy, we prefer generally to discuss the current density of heat. In analogy with the equilibrium relation (7.12), $đQ = T\,dS$, we therefore define a heat current density $\mathbf{J}_Q$ by the relation

$$\mathbf{J}_Q \equiv T\mathbf{J}_S = \mathbf{J}_E - g\mathbf{J}_N \tag{43.14}$$

where we have used Equation (43.11). Thus to get $\mathbf{J}_Q$ we subtract from $\mathbf{J}_E$ the energy flow of macroscopic motion.[13] Then the phenomenological relations (43.13) can be rewritten for J_Q and J_N as

$$\begin{aligned} J_Q &= L_{11}\,\nabla\frac{1}{T} + L_{12}\frac{1}{T}\nabla g \\ -J_N &= L_{21}\,\nabla\frac{1}{T} + L_{22}\frac{1}{T}\nabla g \end{aligned} \tag{43.15}$$

Exercise 43.2

(a) Determine the relation between the $L_{\mu\nu}$ of Equation (43.15) and the $L'_{\mu\nu}$ of Equation (43.13), and show using $L'_{21} = L'_{12}$ that

$$L_{21} = L_{12}$$

(b) Show using the continuity equation (42.9) that the entropy-production equation (43.12) can be written for steady-state flow as

$$\begin{aligned} \left(\frac{\partial s}{\partial t}\right)_{\text{irr}} &= \nabla\left(\frac{1}{T}\right)\cdot\mathbf{J}_Q - \frac{1}{T}(\nabla g)\cdot\mathbf{J}_N \\ &= \nabla\left(\frac{1}{T}\right)\cdot\mathbf{J}_Q + \frac{1}{T}\nabla\cdot\mathbf{J}_Q \end{aligned}$$

In this second equality, the first term is due to the flow of heat when there is a temperature gradient and the second term is due to the appearance of heat in the irreversible flow.

Identification of the Kinetic Coefficients

We now introduce the thermal and electrical conductivity coefficients in favor of two of the kinetic coefficients $L_{\mu\nu}$. First we use the result of Problem II.5(a) to write the chemical potential in the presence of an external (electric) potential energy $U(\mathbf{r})$ as

$$g = g_o(T, \mathcal{P}) + U(\mathbf{r}) \tag{43.16}$$

where $g_o\ (T, \mathcal{P})$ is the chemical potential in the absence of any external fields. We assume that this result is also true for nonequilibrium systems in which the condition of local equilibrium holds. Then the *electrical conductivity* σ for conduction electrons can be defined, in the absence of any temperature or pressure gradients, by

$$\sigma \equiv -\frac{eJ_N}{e^{-1}\nabla g} \qquad (\text{for } \nabla T = \nabla \mathcal{P} = 0)$$

$$\sigma = e^2 L_{22}/T \tag{43.17}$$

where the second equality follows from the second of Equations (43.15). Inasmuch as $\nabla g = \nabla U$ in this case, the defining equality for σ corresponds to the ratio of electric current density to electric field, as it should.

The coefficient of thermal conductivity K is defined similarly, for zero electric current, by

$$K = -\frac{J_Q}{\nabla T} \qquad (\text{for } J_N = 0)$$

$$K = \frac{1}{T^2}\left(L_{11} - \frac{L_{12}^2}{L_{22}}\right) \tag{43.18}$$

The second equality of this equation has been derived from the first by using Equations (43.15) together with the Onsager relation $L_{21} = L_{12}$.

Consider next the thermocouple of Figure IX.4 to which a voltmeter at temperature T is connected. We assume that heat can flow through the voltmeter without resistance, but that no electric current can be transmitted. We also assume $T_2 > T_1$. Since $J_N = 0$, we obtain from the second of the kinetic equations (43.15), for either conductor,

$$\nabla g = \frac{L_{12}}{L_{22}} \cdot \frac{\nabla T}{T} \tag{43.19}$$

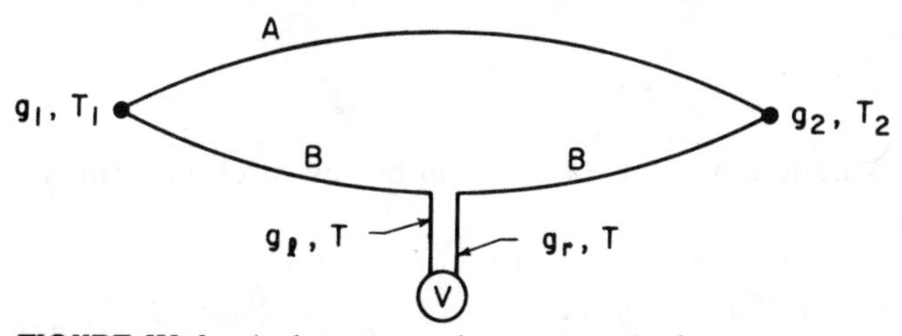

FIGURE IX.4 A thermocouple composed of two materials A and B, connected to an electric-current-blocking voltmeter V.

This equation can be integrated from T_1 to T_2, along the two paths of the thermocouple, to give

$$g_2 - g_1 = \int_{T_1}^{T_2} \frac{dT}{T}\left(\frac{L_{12}}{L_{22}}\right)_A$$

$$= \int_{T_1}^{T_2} \frac{dT}{T}\left(\frac{L_{12}}{L_{22}}\right)_B + (g_r - g_\ell)$$

Inasmuch as there is no temperature difference across the voltmeter a voltage V, defined to be the electromotive force (emf) of the thermocouple, is given simply by

$$V = -\frac{1}{e}(g_r - g_\ell) = \frac{1}{e}\int_{T_1}^{T_2} \frac{dT}{T}\left[\left(\frac{L_{12}}{L_{22}}\right)_B - \left(\frac{L_{12}}{L_{22}}\right)_A\right] \tag{43.20}$$

The production of an emf in a thermocouple under conditions of zero electric current is called the *Seebeck effect*.

The *thermoelectric* power ε_{AB} of the thermocouple is now defined to be the change of the voltage V per unit change in the hot-junction temperature T_2,

$$\varepsilon_{AB}(T) \equiv \frac{1}{eT}\left[\left(\frac{L_{12}}{L_{22}}\right)_B - \left(\frac{L_{12}}{L_{22}}\right)_A\right] = \varepsilon_A - \varepsilon_B \tag{43.21}$$

Here we have used Equation (43.20) and then introduced the absolute thermoelectric power ε of a single medium by the definition [see Equation (43.19)]

$$\varepsilon \equiv -\frac{1}{e}\frac{\partial g}{\partial T} = -\frac{L_{12}}{eTL_{22}} \tag{43.22}$$

Of course, the term power does not have its usual meaning here. We can now replace the three kinetic coefficients L_{11}, L_{12}, and L_{22}, using Equations (43.17), (43.18), and (43.22), by the three physically significant properties of a metal: σ, K, and ε.

Exercise 43.3

(a) Show that the kinetic equations (43.15) can be rewritten in terms of σ, K, and ε as

$$\begin{aligned} J_Q &= -(K + T\sigma\varepsilon^2)\nabla T - \frac{\sigma T\varepsilon}{e}\nabla g \\ -J_N &= \frac{\sigma\varepsilon}{e}\nabla T + \frac{\sigma}{e^2}\nabla g \end{aligned} \tag{43.23}$$

(b) Eliminate ∇g between these two equations to obtain

$$J_Q = -K\nabla T + T\varepsilon e J_N \tag{43.24}$$

Thus the heat flow $(-K\nabla T)$ is augmented by a second term when there is electric current in a metal. According to the relation $J_Q = TJ_S$, this second term is associated with an entropy εe per electron. Therefore, the thermoelectric power ε can also be viewed as the entropy per coulomb transported in electron flow.

Other thermoelectric effects that can be studied with the aid of the above analysis are the *Peltier* and *Thomson effects*. We consider as a final application here the former effect, in which heat is evolved when an electric current flows across a junction of two dissimilar metals A and B kept at constant temperature. Since both g [see Equation (43.16)] and J_N are continuous across the junction, it follows from Equation (43.14) that discontinuities in J_E and J_Q must be equal. Since $\nabla T = 0$ in this case, we obtain from Equation (43.24) for the discontinuity in J_Q,

$$J_{Q,A} - J_{Q,B} = TeJ(\varepsilon_A - \varepsilon_B)$$

The *Peltier coefficient* π_{AB} is now defined as the heat that must be supplied to the junction when unit electric current density eJ_N passes from conductor B to conductor A. Thus,

$$\pi_{AB} \equiv \frac{1}{eJ_N}(J_{Q,A} - J_{Q,B}) = T(\varepsilon_A - \varepsilon_B) \tag{43.25}$$

This relation between the Peltier coefficient and the thermoelectric power $\varepsilon_{AB} = \varepsilon_A - \varepsilon_B$ was first deduced empirically by Kelvin in 1854.

44 Nuclear Spin Relaxation

In the preceding three sections the problems we investigated were all amenable, essentially, to treatment by classical statistical methods. We now deal with a problem in which we must utilize quantum statistical methods, namely, the relaxation of nuclear magnetization, or nuclear spin orientation, in a system from some initially prepared mixed state. Understanding of the intrinsic relaxation times T_1 and T_2 is fundamental to the microscopic explanation of magnetic resonance experiments.

For reasons of simplicity, we present in this section an idealized model of the nuclear spin interactions and consider only the case of relaxation in liquids and gases in which complications due to lattice structure are not present. It is shown that the relaxation times can be expressed in terms of a *spectral density* $J(\omega)$ which is defined to be the Fourier transform of a time-correlation function. For the idealized model considered the relevant correlation function involves the time correlations of a random magnetic field in the fluid.

For the model fluid of this section we shall assume that the Hamiltonian $\hbar\mathsf{H}$ has the form

$$\hbar\mathsf{H} = \hbar(\mathsf{H}_o + \mathsf{H}_\ell + \mathsf{H}') \tag{44.1}$$

where H_o describes the magnetic energy of all the particles α, each with (nuclear) spin angular momentum $\hbar S_\alpha$, in an external magnetic field $\mathcal{H}_o$, $\hbar\mathsf{H}_\ell$ is the Hamiltonian of the entire liquid when $S_\alpha = 0$ and H' characterizes the liquid-spin system interactions. Generally H_o is written in the form

$$\mathsf{H}_o = -\hbar^{-1}\sum_\alpha \boldsymbol{\mu}_\alpha \cdot \mathcal{H}_o = -\sum_\alpha \gamma_\alpha \mathbf{S}_\alpha \cdot \mathcal{H}_o \tag{44.2}$$

where $\gamma_\alpha \equiv \mu_\alpha / S_\alpha \hbar$ is the nuclear gyromagnetic ratio. We also assume that both $\hbar\mathsf{H}_o$ and $\hbar\mathsf{H}_\ell$ are time-independent Hamiltonians.

Consider now the solution (23.4) for the statistical matrix when the Hamiltonian is time independent. This result suggests that we might deal first with H_o and H_ℓ, using such an expression, and then focus attention on H', which is time dependent on a microscopic time scale due to the continually fluctuating locations of particle positions and orientations. We therefore define the statistical matrix $\underline{\rho}_I(t)$ by the equation

$$\underline{\rho}_I(t) \equiv e^{i(\mathsf{H}_o + \mathsf{H}_\ell)t} \underline{\rho}(t)\, e^{-i(\mathsf{H}_o + \mathsf{H}_\ell)t} \tag{44.3}$$

and then we find that the equation of motion for $\underline{\rho}_I(t)$ is similar to Equation (23.2) for $\underline{\rho}(t)$.

$$i\frac{d\underline{\rho}_I(t)}{dt} = [\mathsf{H}_I'(t), \underline{\rho}_I(t)] \tag{44.4}$$

where

$$\mathsf{H}_I'(t) \equiv e^{i(\mathsf{H}_o + \mathsf{H}_\ell)t} \mathsf{H}'(t) e^{-i(\mathsf{H}_o + \mathsf{H}_\ell)t} \tag{44.5}$$

With the preceding three equations we have transformed from the Schrödinger picture to the interaction picture of quantum mechanics.[14]

Exercise 44.1 Prove Equation (44.4), using Equations (23.2) and (44.3).

We now rewrite the differential equation (44.4) as an integral equation,

$$\underline{\rho}_I(t) = \underline{\rho}_I(0) - i\int_0^t dt'\, [\mathsf{H}_I'(t'), \underline{\rho}_I(t')] \tag{44.6}$$

where the first term on the right side represents the statistical matrix $\underline{\rho}_I(0) = \underline{\rho}(0)$ of the initially prepared mixed state at time $t = 0$. Next we consider a macroscopically small time interval t, such that $t >> \tau_c$, where τ_c is an average spin-correlation time for the atomic nuclei in the liquid. Inasmuch as the interaction Hamiltonian is expected to have only a small effect on $\underline{\rho}_I(t)$ over such a time interval, we may expand the right side of Equation (44.6) by iteration to obtain

$$\underline{\rho}_I(t) = \underline{\rho}_I(0) - i\int_0^t dt' \, [\mathsf{H}_I'(t'), \underline{\rho}_I(0)]$$
$$- \int_0^t dt' \int_0^{t'} dt'' \, [\mathsf{H}_I'(t'), [\mathsf{H}_I'(t''), \underline{\rho}_I(0)]] + \cdots$$

But for $t >> \tau_c$, the second term of this expression is expected to be zero for a randomly fluctuating interaction $\mathsf{H}'(t)$. We also neglect terms higher order than the nonvanishing second-order, or third, term on the right side, as in the approximations of Equations (41.2) and (42.15). We then differentiate this approximate expression for $\underline{\rho}_I(t)$ and derive thereby a useful replacement for Equation (44.4),

$$\frac{d\underline{\rho}_I(t)}{dt} = -\int_0^t ds \, [\mathsf{H}_I'(t), [\mathsf{H}_I'(t - s), \underline{\rho}_I(0)]] + \cdots$$

$$\frac{d\underline{\rho}_I(t)}{dt} \cong -\int_0^\infty ds \, [\mathsf{H}_I(t), [\mathsf{H}_I'(t - s), \underline{\rho}_I(0)]] \qquad \textbf{(44.7)}$$

We have made the change of variable $t'' = t - s$ and then used the fact that the product of the two interaction-Hamiltonian operators in this equation must vanish when integrated over intervals of s such that $\Delta s >> \tau_c$. In particular, the only region of s values that can be expected to give a nonvanishing integral in Equation (44.7) is the tiny region $s \lesssim \tau_c$.

We might expect, in zeroth approximation, that the statistical matrix $\underline{\rho}(t)$ can be written as the time-independent direct product $\underline{\rho} \cong \underline{\rho}_\ell(\mathsf{H}_\ell) \cdot \underline{\tilde{\sigma}}_o(\mathsf{H}_o)$, where

$$\underline{\rho}_\ell(\mathsf{H}_\ell) = Z_\ell^{-1} \, e^{-\beta\hbar\mathsf{H}_\ell}$$
$$\underline{\tilde{\sigma}}_o(\mathsf{H}_o) = Z_S^{-1} \, e^{-\beta\hbar\mathsf{H}_o} \qquad (44.8)$$

Here Z_ℓ and Z_S are liquid and N-particle spin-system partition functions respectively.[15] Moreover, the liquid system, because of its much larger heat capacity, should remain in thermodynamic equilibrium on a macroscopic time scale when we consider the effects of the interaction Hamiltonian H'. Only the spin system will be affected appreciably by H', as in the corresponding approximation (41.1) for the Brownian-motion problem. Thus instead of just $\underline{\rho} \cong \underline{\rho}_\ell \tilde{\underline{\sigma}}_o$, we are led to write for $\underline{\rho}(t)$ and $\underline{\rho}_I(t)$ the approximate expressions

$$\underline{\rho}(t) \cong \underline{\rho}_\ell(\mathsf{H}_\ell)[\tilde{\underline{\sigma}}_o(\mathsf{H}_o) + \tilde{\underline{\sigma}}'(t)]$$
$$\underline{\rho}_I(t) \cong \underline{\rho}_\ell[\tilde{\underline{\sigma}}_o + \tilde{\underline{\sigma}}_I'(t)] \tag{44.9}$$

where $\tilde{\underline{\sigma}}_I'(t)$ is related to $\tilde{\underline{\sigma}}'(t)$ by an equation of the type (44.3). It is the matrix $\tilde{\underline{\sigma}}'$ that reflects the influence of the interaction H' on the time evolution of spin polarization in the system.

We shall assume that both H_o and $\tilde{\underline{\sigma}}'(t)$ commute with H_ℓ, which means, effectively, that in spin space alone the statistical matrix $\tilde{\underline{\sigma}}'(t)$ depends on the liquid system only through the temperature parameter β. To exhibit this conclusion explicitly, we make a notational distinction between the matrices $\underline{\sigma}_o$ and $\underline{\sigma}'(t)$ in spin space alone as opposed to the matrices $\tilde{\underline{\sigma}}_o$ and $\tilde{\underline{\sigma}}'(t)$ in the product space of spin and liquid states. In particular, the matrix $\underline{\sigma}_I'(t)$ is related to $\tilde{\underline{\sigma}}_I'(t)$ by the definition

$$\underline{\sigma}_I'(t) \equiv \sum_\mu \langle\mu|\underline{\rho}_\ell|\mu\rangle\langle\mu|\tilde{\underline{\sigma}}_I'(t)|\mu\rangle \tag{44.10}$$

where both $\underline{\rho}_\ell$ and $\tilde{\underline{\sigma}}_I'(t)$ are diagonal in the representation μ. A similar relation gives $\underline{\sigma}_o$ in terms of $\tilde{\underline{\sigma}}_o$, but since $\tilde{\underline{\sigma}}_o(\mathsf{H}_o)$ is a unit matrix in the space of liquid states and $\mathrm{Tr}(\underline{\rho}_\ell) = 1$, there is no significant difference between $\underline{\sigma}_o$ and $\tilde{\underline{\sigma}}_o$. Thus upon using Equation (44.9) and taking the trace of Equation (44.7) over all liquid states μ, we find for the time rate of change of $\underline{\sigma}_I'(t)$,

$$\begin{aligned}
\frac{d\underline{\sigma}_I'(t)}{dt} &= \sum_\mu \langle\mu|\underline{\rho}_\ell|\mu\rangle\left\langle\mu\left|\frac{d\tilde{\underline{\sigma}}_I'(t)}{dt}\right|\mu\right\rangle \\
&\cong \sum_\mu \left\langle\mu\left|\frac{d\underline{\rho}_I(t)}{dt}\right|\mu\right\rangle \\
&\cong -\int_0^\infty ds \sum_\mu \langle\mu|[\mathsf{H}_I'(t), [\mathsf{H}_I'(t-s), (\tilde{\underline{\sigma}}_o + \tilde{\underline{\sigma}}_I'(0))\underline{\rho}_\ell]]|\mu\rangle
\end{aligned} \tag{44.11}$$

Now $(1/N)\langle\beta\hbar\mathsf{H}_o\rangle \sim \mu_\alpha\mathscr{H}_o/\kappa T << 1$ for atomic magnetic moments, even at $T \sim 10$ K and $\mathscr{H}_o = 10^4$ gauss; nuclear magnetic energies are smaller yet by a factor that is $\sim 10^{-3}$. By using this fact, together with some manipulations of the commutator, it is possible to show (Problem IX.2) that Equation (44.11) can be rewritten, to excellent approximation, in the form

$$\frac{d\underline{\sigma}_I'(t)}{dt} = -\int_0^\infty ds \sum_\mu \langle\mu|\underline{\rho}_\ell|\mu\rangle\langle\mu|[\mathsf{H}_I'(t), [\mathsf{H}_I'(t-s), \tilde{\underline{\sigma}}_I'(0)]]|\mu\rangle \tag{44.12}$$

Equation (44.12), which is still quite general, provides the basis for most analyses of spin relaxation in liquids and gases using magnetic resonance experiments. We note that $\underline{\sigma}_I'(t)$ is a matrix in the spin space of all the particles α of the fluid, so that Equation (44.12) can include exchange effects if they are important in any specific application. We are not interested in such effects here, however.

Polarization Vectors

We next use Equation (21.9b) to compute the expectation value of the ith component of the polarization vector $\mathbf{P}(t)$ for the particle spins. For this calculation, we use eigenstates of H_o in the Schrödinger picture and obtain

$$P_i(t) \equiv \frac{1}{N}\sum_\alpha \langle \mathsf{S}_{\alpha,i}\rangle = \frac{1}{N}\sum_\alpha \mathrm{Tr}\left[e^{i\mathsf{H}_o t}[\underline{\sigma}_o + \underline{\sigma}'(t)]\,\mathsf{S}_{\alpha,i}\,e^{-i\mathsf{H}_o t}\right]$$

The exponential factors give the time dependence of the Schrödinger-picture eigenstates, so that in the second equality the trace is to be computed using stationary states. On introducing the notation of the interaction picture, Equations (44.3) and (44.5), we obtain the equivalent expression

$$P_i(t) = \mathrm{Tr}([\underline{\sigma}_o + \underline{\sigma}_I'(t)]\mathsf{S}_{I,i}(t)) \tag{44.13}$$

where now we have assumed that all particles α have the same average value for their spins. The operator S_I is a one-particle spin operator in the product space of all N spins.

Exercise 44.2 We can use Equations (44.2) and (44.5) to write $\mathsf{S}_I(t)$ in the form

$$\mathbf{S}_I(t) = e^{-i\omega \mathsf{S}_z t}\,\mathbf{S}\,e^{i\omega \mathsf{S}_z t} \tag{44.14}$$

where $\omega = \gamma\mathcal{H}_o$ is the Larmor angular-precession frequency and we have assumed $\mathcal{H}_o = \mathcal{H}_o\hat{\mathbf{i}}_z$. This interaction picture operator can be evaluated explicitly with the aid of the operator-expansion theorem

$$e^{\mathsf{A}}\mathsf{B}e^{-\mathsf{A}} = \sum_{n=0}^{\infty}\frac{1}{n!}\mathsf{O}_n$$

where the operators O_n are given by the recursion relation $\mathsf{O}_{n+1} = [\mathsf{A}, \mathsf{O}_n]$ with $\mathsf{O}_0 \equiv \mathsf{B}$.[16] Show, using the commutation relations for the operators S_i, that the three components of $\mathbf{S}_I(t)$ are

$$\begin{aligned}
\mathsf{S}_{I,x}(t) &= \mathsf{S}_x \cos\omega t + \mathsf{S}_y \sin\omega t \\
\mathsf{S}_{I,y}(t) &= -\mathsf{S}_x \sin\omega t + \mathsf{S}_y \cos\omega t \\
\mathsf{S}_{I,z}(t) &= \mathsf{S}_z
\end{aligned} \tag{44.15}$$

Thus, $\mathsf{S}_I(t)$ differs from $\mathbf{S}$ by a right-hand rotation (counterclockwise) about the z-axis through the angle ωt.

If we substitute the results of Exercise 44.2 into Equation (44.13), then we find for $\mathbf{P}(t)$ the three component expressions,

$$P_x(t) = P'_x(t)\cos\omega t + P'_y(t)\sin\omega t,$$

$$P_y(t) = -P'_x(t)\sin\omega t + P'_y(t)\cos\omega t \qquad \textbf{(44.16)}$$

$$P_z(t) = P_{o,z} + P'_z(t)$$

where $P_{o,z}$ is the value of the polarization vector under conditions of thermodynamic equilibrium in the external magnetic field $\mathcal{H}_o\hat{\mathbf{i}}_z$ and the polarization vector $P'_i(t)$ is defined by

$$P'_i(t) \equiv \mathrm{Tr}(\underline{\sigma}'_I(t)\,\mathsf{S}_i) \qquad \textbf{(44.17)}$$

According to Equations (44.16), $\mathbf{P}'(t)$ rotates at the Larmor frequency clockwise about the z-axis relative to $\mathbf{P}(t)$, which is the polarization vector expressed in the laboratory coordinate system. We now consider the calculation of $\mathbf{P}'(t)$.

The most important mechanism responsible for spin relaxation in liquids is the magnetic dipole–dipole interaction. To simplify our calculations here we choose, instead of this interaction, an idealized interaction in which the effect of neighboring spins is represented by a random magnetic field $\mathcal{H}^\alpha(t)$.[17] Thus we choose a Zeeman-coupling model for $\mathsf{H}'(t)$ and write, as with Equation (44.2),

$$\mathsf{H}'(t) = -\sum_\alpha \gamma_\alpha \mathbf{S}_\alpha \cdot \mathcal{H}^{(\alpha)}(t) \qquad \textbf{(44.18)}$$

Now the random magnetic field $\mathcal{H}^{(\alpha)}(t)$ must depend on the locations of all the neighboring atoms in the liquid in such a way that for liquid-state matrix elements in the representation μ, $\langle\mu|\mathcal{H}_I^{(\alpha)}(t)|\mu\rangle = \langle\mu|\mathcal{H}^{(\alpha)}(t)|\mu\rangle \sim 1$ gauss, say, for all α and t. On referring to Equation (44.12), we are led to define a liquid correlation function $K_\ell^{(\alpha)}(s)$, for an isotropic liquid, by the expression

$$K_\ell^{(\alpha)}(s)\,\delta_{j,k} \equiv \sum_\mu \langle\mu|\underline{\rho}_\ell|\mu\rangle\langle\mu|\mathcal{H}_{I,j}^{(\alpha)}(t)\mathcal{H}_{I,k}^{(\alpha)}(t-s)|\mu\rangle \qquad \textbf{(44.19)}$$

$$= \frac{1}{3}\,\delta_{j,k}\sum_{i=1}^{3}\sum_{\mu,\nu}\langle\mu|\underline{\rho}_\ell|\mu\rangle$$

$$\times\; e^{i(E_\mu - E_\nu)s/\hbar}\langle\mu|\mathcal{H}_i^{(\alpha)}(t)|\nu\rangle\langle\nu|\mathcal{H}_i^{(\alpha)}(t-s)|\mu\rangle$$

$$\xrightarrow[s\to 0]{}\; \frac{1}{3}\,\delta_{j,k}\sum_\mu \langle\mu|\underline{\rho}_\ell|\mu\rangle\langle\mu|(\mathcal{H}^{(\alpha)}(t))^2|\mu\rangle$$

The liquid correlation function (44.19) cannot depend on the particular atom α that was selected for its definition, and therefore we may henceforth drop the superscript α from its notation. Moreover, if we observe that $K_\ell(s)$ is nonvanishing only when $s \lesssim \tau_c$, where $\tau_c \sim 10^{-7}$ sec, say,[18] then we can conclude that the liquid states ν that contribute to (44.19) must have energies such that $|E_\nu - E_\mu| < 10^{-20}$ erg $\sim 10^{-8}$ eV. But this energy difference is very much greater than the level spacing of the liquid (or gas) states, yet one would expect that only states ν close (i.e., similar) to the state μ could contribute significantly to $K_\ell(s)$. In other words, for appreciably contributing states ν, the exponential factor in Equation (44.19) may be tiny, and $K_\ell(s)$ will then be a real correlation function that satisfies the classical relation (41.9). In fact, the energy levels of the nuclear spin states in a fluctuating 1-gauss magnetic field ($\mathcal{H}_{\rm rms} \sim 1$ gauss) are spaced by energies about the size of the nuclear Bohr magneton, that is, $\sim 10^{-23}$ erg. By energy conservation, this number must then determine the appreciably contributing states ν in Equation (44.19).

We next observe that on the macroscopic time scale t of (44.12), the matrix $\underline{\tilde{\sigma}}_I'(0)$ cannot be affected significantly, and this implies that $\langle \nu|\underline{\tilde{\sigma}}_I'(0)|\nu\rangle \cong \langle \mu|\underline{\tilde{\sigma}}_I'(0)|\mu\rangle$ for relevant states μ and ν in Equation (44.19). In essence, we may conclude that: (1) $\mathcal{H}^{(\alpha)}$ and $\underline{\sigma}_I'$ can be regarded as commuting quantities; and (2) the matrix $\langle \mu|\underline{\tilde{\sigma}}_I'(0)|\mu\rangle$ varies very slowly over the spectrum of liquid energy states. These two deductions are used to derive the second and third equalities of Equation (44.20).

To determine $P_i'(t)$ we now differentiate Equation (44.17) with respect to t and substitute Equations (44.12) and (44.18) to obtain

$$\frac{dP_i'(t)}{dt} = -\int_0^\infty ds\, \mathrm{Tr}\Big(\sum_\mu \langle \mu|\underline{\rho}_\ell|\mu\rangle \times \langle \mu|[\mathsf{H}_I'(t), [\mathsf{H}_I'(t-s), \underline{\tilde{\sigma}}_I'(0)]]|\mu\rangle\, \mathsf{S}_i\Big)$$

$$\cong -\frac{1}{3}\gamma^2 \sum_{j=1}^{3} \int_0^\infty ds\, \mathrm{Tr}\Big\{\sum_\mu \langle \mu|\underline{\rho}_\ell|\mu\rangle \times \langle \mu|\mathcal{H}_I^{(\alpha)}(t)\cdot \mathcal{H}_I^{(\alpha)}(t-s)|\mu\rangle \times [\mathsf{S}_{I,j}(t), [\mathsf{S}_{I,j}(t-s), \langle \mu|\underline{\tilde{\sigma}}_I'(0)|\mu\rangle]]\, \mathsf{S}_i\Big\}$$

$$\frac{dP_i'(t)}{dt} \cong -\gamma^2 \sum_{j=1}^{3} \int_0^\infty ds\, K_\ell(s) \times \mathrm{Tr}([\mathsf{S}_{I,j}(t), [\mathsf{S}_{I,j}(t-s), \underline{\sigma}_I'(0)]]\, \mathsf{S}_i) \qquad \textbf{(44.20)}$$

Only one of the N terms in (44.18) contributes to the average over S_i in the second equality of Equation (44.20), because with the Zeeman-coupling model the statistical matrix for spins is diagonal in the product space of the N particles. To obtain the third equality of Equation (44.20) we have assumed, as suggested above, that $\langle\mu|\bar{\underline{\sigma}}'(0)|\mu\rangle$ varies slowly over the liquid-state spectrum and can be approximated by the matrix $\underline{\sigma}'(0)$ of Equation (44.10); then we have substituted definition (44.19) for $K_\ell(s)$.

The next step in the analysis is to substitute Equations (44.15) into the final equality of Equation (44.20) and to expand the double commutator bracket. It is somewhat tedious, although straightforward, to simplify the resulting expression to the following form (with $\underline{\sigma}_I'(0) = \underline{\sigma}_I'$).

$$
\begin{aligned}
\frac{dP_i'}{dt} &= -\gamma^2 \int_0^\infty ds\, K_\ell(s) \\
&\quad \times \mathrm{Tr}(\mathsf{S}_i\{[(\mathsf{S}_x^2 + \mathsf{S}_y^2)\underline{\sigma}_I' - 2(\mathsf{S}_x\underline{\sigma}_I'\mathsf{S}_x + \mathsf{S}_y\underline{\sigma}_I'\mathsf{S}_y) \\
&\qquad + \underline{\sigma}_I'(\mathsf{S}_x^2 + \mathsf{S}_y^2)] \cos \omega s \\
&\qquad + i[\underline{\sigma}_I'\mathsf{S}_z - \mathsf{S}_z\underline{\sigma}_I'] \sin \omega s + [\mathsf{S}_z, [\mathsf{S}_z, \underline{\sigma}_I']]\}) \qquad (44.21) \\
&= -\frac{1}{3}\gamma^2\langle\mathcal{H}^2\rangle \int_0^\infty ds\, G_\ell(s) \\
&\quad \times \mathrm{Tr}(\mathsf{S}_i\{([\mathsf{S}_x, [\mathsf{S}_x, \underline{\sigma}_I']] + [\mathsf{S}_y, [\mathsf{S}_y, \underline{\sigma}_I']]) \cos \omega s \\
&\qquad - i[\mathsf{S}_z, \underline{\sigma}_I'] \sin \omega s + [\mathsf{S}_z, [\mathsf{S}_z, \underline{\sigma}_I']]\})
\end{aligned}
$$

To arrive at the second equality we have removed a factor $\frac{1}{3}\langle\mathcal{H}^2\rangle$ from the time-correlation function $K_\ell(s)$ of Equation (44.19). This step results in a dimensionless time-correlation function $G_\ell(s)$, defined by

$$
K_\ell(s) \equiv \frac{1}{3}\langle\mathcal{H}^2\rangle G_\ell(s) \qquad (44.22)
$$

where according to inequality (41.8) $G_\ell(s) \leq G_\ell(0) = 1$.

Exercise 44.3 Verify Equation (44.21).

Fourier Analysis and the Spectral Density

We introduce now the so-called *spectral density* $J(\omega)$ of a fluctuating function [$\mathcal{H}(t)$ in the present case] by the Fourier integrals

$$G(s) \equiv \int_{-\infty}^{+\infty} d\omega\, J(\omega)\, e^{-i\omega s}$$
$$J(\omega) = \frac{1}{2\pi}\int_{-\infty}^{+\infty} ds\, G(s)\, e^{i\omega s} \qquad \mathbf{(44.23)}$$

These integrals are known in the literature as the *Wiener–Khintchine relations*. By using property (41.9) of the correlation function, we may deduce from the second of these relations that $J(\omega)$ is real and an even function, that is,

$$J(\omega) = \frac{1}{2\pi}\int_{-\infty}^{+\infty} ds\, G(s) \cos \omega s$$
$$= \frac{1}{\pi}\int_{0}^{\infty} ds\, G(s) \cos \omega s = J(-\omega) \qquad (44.24)$$

This equation is only valid when the time-correlation function $G(s)$ is treated classically. See Problem IX.3(a). From the first of relations (44.23), we find the normalization condition for the spectral density,

$$\int_{-\infty}^{+\infty} d\omega\, J(\omega) = G(0) = 1 \qquad (44.25)$$

Inasmuch as the correlation function $G(s)$ drops to zero at $s \sim \tau_c$, where τ_c is a microscopic correlation time, we may conclude from Equations (44.23) that the spread of frequencies in $J(\omega)$ is over a range $|\omega| \lesssim \tau_c^{-1}$. We may, in fact, use the condition $G(s) \leqslant G(0) = 1$ to define the correlation time by the expression

$$\tau_c \equiv \int_0^{\infty} ds\, G(s) = \pi J(0) \qquad (44.26)$$

Spin Relaxation Times

We now define a time $T_1(\omega)$, in terms of the liquid spectral density, by

$$\frac{1}{T_1(\omega)} \equiv \frac{2\pi}{3}\gamma^2\langle \mathscr{H}^2\rangle J(\omega) \qquad \mathbf{(44.27)}$$

where $\omega = \gamma\mathscr{H}_o$ for the liquid. We shall see below that $T_1(\omega)$, not to be confused with τ_c of Equation (44.26), is one of two spin relaxation times in the liquid. In analogy with Equation (44.24) we also define a second spectral-density function $\tilde{J}(\omega)$ by the equation

$$\tilde{J}(\omega) \equiv \frac{1}{\pi}\int_0^\infty ds\, G(s) \sin \omega s \tag{44.28}$$

With this notation, Equation (44.21) can be simplified to the form

$$\begin{aligned}
\dot{P}_i'(t) &= \frac{dP_i'}{dt} \\
&= -\frac{1}{2T_1(\omega)} \mathrm{Tr}(\mathsf{S}_i\{[\mathsf{S}_x, [\mathsf{S}_x, \underline{\sigma}_I']] + [\mathsf{S}_y, [\mathsf{S}_y, \underline{\sigma}_I']]\}) \\
&\quad + \frac{i}{2T_1(\omega)} \frac{\tilde{J}(\omega)}{J(\omega)} \mathrm{Tr}(\mathsf{S}_i[\mathsf{S}_z, \underline{\sigma}_I']) - \frac{1}{2T_1(0)} \mathrm{Tr}(\mathsf{S}_i[\mathsf{S}_z, [\mathsf{S}_z, \underline{\sigma}_I']]) \qquad (44.29) \\
&= -\frac{1}{2T_1(\omega)}\left\{\langle[[\mathsf{S}_i, \mathsf{S}_x], \mathsf{S}_x]\rangle' + \langle[[\mathsf{S}_i, \mathsf{S}_y], \mathsf{S}_y]\rangle' - i\frac{\tilde{J}(\omega)}{J(\omega)}\langle[\mathsf{S}_i, \mathsf{S}_z]\rangle'\right\} \\
&\quad - \frac{1}{2T_1(0)}\rangle[[\mathsf{S}_i, \mathsf{S}_z], \mathsf{S}_z]\rangle'
\end{aligned}$$

To derive the third equality of this expression, a number of manipulations involving the cyclic property of the trace are required. We have also used the notation of Equation (21.9b) for the computation of average values, where the prime on the brackets, $\langle\ \rangle'$, indicates use of the statistical matrix $\underline{\sigma}_I'(t)$. We use the matrix $\underline{\sigma}_I'(t)$ instead of $\underline{\sigma}_I'(0)$, because, in the first place, for the approximate expression (44.7) to be valid, we had to consider a macroscopically small time interval t (with, of course, $t >> \tau_c$); for such a time interval $\underline{\sigma}_I'(t) \cong \underline{\sigma}_I'(0)$. Therefore, since there is nothing special about the particular chosen origin of time, Equation (44.29) can be applied at all times t by merely replacing $\underline{\sigma}_I'(0)$ by $\underline{\sigma}_I'(t)$.

The spin relaxation time T_1 is called the *longitudinal relaxation time*, because, as we show next, it is the characteristic time for decay of $P_z'(t)$ to its equilibrium value of zero. Similarly, the characteristic time for decay of $P_x'(t)$ and $P_y'(t)$ to zero is T_2, the *transverse relaxation time*, which is defined by

$$\frac{1}{T_2(\omega)} = \frac{1}{2}\left[\frac{1}{T_1(0)} + \frac{1}{T_1(\omega)}\right] \tag{44.30}$$

To appreciate these definitions, one can evaluate the commutator brackets in the third equality of Equation (44.29) and use definition (44.17) to obtain the three component expressions,

$$\dot{P}_x'(t) = -\frac{1}{T_2(\omega)}P_x'(t) + \frac{1}{2T_1(\omega)}\frac{\tilde{J}(\omega)}{J(\omega)}P_y'(t)$$

$$\dot{P}'_y(t) = -\frac{1}{T_2(\omega)}P'_y(t) - \frac{1}{2T_1(\omega)}\frac{\tilde{J}(\omega)}{J(\omega)}P'_x(t) \tag{44.31}$$

$$\dot{P}'_z(t) = -\frac{1}{T_1(\omega)}P'_z(t)$$

These equations are the final results of this section, and we discuss them below and in the next section.

Exercise 44.4 Define a polarization vector $\mathbf{P}''(t)$ by the replacements in the first two of Equations (44.16): $P_i(t) \to P'_i(t)$, $P'_i(t) \to P''_i(t)$ and $\omega \to \tilde{\omega}$, where the angular frequency $\tilde{\omega}$ is defined by [see Equation (44.27)]

$$\tilde{\omega} \equiv \frac{1}{2T_1(\omega)}\frac{\tilde{J}(\omega)}{J(\omega)} = \frac{\pi}{3}\gamma^2\langle\mathcal{H}^2\rangle\tilde{J}(\omega)$$

We also set $P''_z(t) \equiv P'_z(t)$. Show that the first two of Equations (44.31) can be rewritten for $P''_i(t)$, $i = (x, y)$, simply as

$$\dot{P}''_i(t) = -\frac{1}{T_2(\omega)}P''_i(t) \qquad (\text{for } i = x, y)$$

According to this result and the first two of Equations (44.16), the polarization vector $\mathbf{P}''(t)$ precesses clockwise about the z-axis relative to $P(t)$ with an angular frequency $(\omega + \tilde{\omega})$.

For the "time" approximations we have made in this section to be valid, we must require the condition $T_1(\omega) >> \tau_c$ [see also (41.14)]. On referring to Equations (44.26) and (44.27), we see that this condition is equivalent to the requirement

$$\frac{\tau_c}{T_1(\omega)} \sim \gamma^2\langle\mathcal{H}^2\rangle J(\omega)\tau_c \sim \gamma^2\langle\mathcal{H}^2\rangle\tau_c^2 << 1 \tag{44.32}$$

For a random magnetic field $\langle\mathcal{H}^2\rangle^{1/2} \sim 1$ gauss, this requirement becomes a condition on τ_c, $\tau_c << \gamma^{-1} \sim \hbar/\mu \sim 10^{-4}$ sec. Thus the time correlation of nuclear magnetic moments on a microscopic scale must occur over time intervals $\tau_c << 10^{-4}$ sec, which is a quite reasonable requirement for liquids and gases. The ratio of frequencies $\tilde{\omega}/\omega = \tilde{\omega}/\gamma\mathcal{H}_o$ is in this case very small since

$$\left|\frac{\tilde{\omega}}{\omega}\right| \sim \frac{\gamma\langle\mathcal{H}^2\rangle\tau_c}{\mathcal{H}_0} << 1$$

Therefore we may neglect the terms involving $\tilde{J}(\omega)$ in Equations (44.29) and (44.31).[19]

45 Approach to Equilibrium

In Section 25 we discussed irreversibility, or the approach to equilibrium, for a macroscopic system mostly in qualitative terms. In the present section we give this subject a more quantitative character within the context of a survey of more sophisticated methods for its analysis.

There are, to begin with, several fundamental approaches used in contemporary treatments of the approach to equilibrium of a system. For example, a version of the H theorem, introduced previously in Section 25, follows generally from the *master equation* for the time development of system probabilities. We show for the spin relaxation problem of the preceding section that a master equation is equivalent to the decay equation for the polarization component $P_z(t)$. Then to illustrate another method we give a quantum-mechanical derivation, due to Kubo, of the electrical conductivity tensor $\sigma_{\mu\nu}(\omega)$. The result is closely related to the classical fluctuation–dissipation theorem (42.18).

An important topic of this section is to discuss the microscopic basis of the master equation. We show that the master equation is a consequence of the assumption that successive stages in the time evolution of a macroscopic system can be described as a Markov process. The associated mathematical description is given by the *Chapman–Kolmogorov equation*. Subsequent to this discussion we apply these ideas to an analysis of the Ehrenfests' urn model and to the calculation of recurrence times for the model. The model shows explicitly that the observed condition we call thermodynamic equilibrium is that (macroscopic) condition and its fluctuations that recur most frequently. Finally, Svedberg's experiment on the frequencies with which various numbers of colloidal particles occur in a restricted region of a solution is cited as a further demonstration of this last conclusion.

We begin by continuing the discussion of the nuclear spin relaxation problem of Section 44 and discussing explicitly the approach to equilibrium. Thus on combining the third of Equations (44.16) and (44.31), we obtain the decay equation for the polarization component $P_z(t)$,

$$\frac{dP_z(t)}{dt} = \frac{dP'_z(t)}{dt} = -\frac{1}{T_1}[P_z(t) - P_{o,z}] \tag{45.1}$$

The solution to this equation is

$$P_z(t) = P_{o,z} + [P_z(0) - P_{o,z}]e^{-t/T_1} \tag{45.2}$$

where $P_z(0)$ is the value of the nuclear polarization $P_z(t)$ at $t = 0$. After several relaxation times T_1, the system decays from this initially prepared condition to its equilibrium value $P_{o,z}$. Initial conditions involving transverse components of polarization, as well as physical situations in which there are small, oscillating, transverse magnetic fields, can also be understood by starting from equations of the type of (44.16) and (44.31).[20] Resulting decay equations of the form (45.1) then characterize the general approach to equilibrium of the spin system.

The Master Equation

The approach to equilibrium of a system that starts from some initially prepared nonequilibrium state is governed, in general, by a decay equation for its statistical matrix. In Section 44, we derived such a decay equation [(44.12)] from the quantum-mechanical form of Liouville's theorem. The *master equation* expresses generally the decay rate dw_i/dt, for system probabilities w_i, which one will obtain from such calculations. We give here a brief derivation of the master equation without detailed reference to the underlying microscopic equations of Section 23. But first we discuss the master equation and how it can be applied to various problems of interest.

The master equation for a system, due originally to W. Pauli,[21] is usually written as an equation for the diagonal matrix elements $w_i(t)$, Equation (21.15), of the statistical matrix. The simplest form of this equation is

$$\frac{dw_i}{dt} = \sum_j (W_{ij}w_j - W_{ji}w_i) \qquad \textbf{(45.3)}$$

where W_{ij} is the probability that a transition will occur from the many-body state j to the state i. The right side of this equation is merely a statement that the time rate of change of w_i is determined by subtracting the rate of transitions out of the state i from the rate of transitions into the state i, as was argued similarly in connection with Equation (11.6). When the energy levels of the system of interest are very closely spaced, as they always are for a macroscopic system (but not for nuclear spin states), then the probabilities w_i will be essentially constant over a small energy range. We must then consider the master equation from the "coarse-grained" viewpoint in which $Z_i >> 1$ states i are all associated with the same probability value w_i.[22] This observation is crucial for the derivation of the master equation, as is discussed below. For the present we assume that Equation (45.3) is correct and discuss in some detail the transition probabilities W_{ij}.

The master equation applies generally to a macroscopic system in interaction with its environment and, in particular, to the common situation in which the system of interest is in contact with an external reservoir, for example, the temperature-fugacity bath discussed at the beginning of Section 36. The system of interest could be the nuclear spin system of Section 44. For such situations, we label reservoir states by s and t, and assume further that in any state transitions energy E and particle numbers N_α are all conserved quantities in the complete system-reservoir closed system.[23] Thus we must have for all transitions $(i, s) \to (j, t)$,

$$\begin{aligned} E_i + \tilde{E}_s &= E_j + \tilde{E}_t \\ N_{\alpha,i} + \tilde{N}_{\alpha,s} &= N_{\alpha,j} + \tilde{N}_{\alpha,t} \end{aligned} \tag{45.4}$$

where $\tilde{E}$ and $\tilde{N}_\alpha$ denote reservoir quantities. If next we write $W_{is,jt}$ for the transition probabilities (excluding density-of-states factors) in the large closed system, then the relation of W_{ij} to $W_{is,jt}$ is given by

$$W_{ij} = \sum_{s,t} W_{is,jt}\tilde{w}_t \tag{45.5}$$

where $\tilde{w}_t$ is the statistical matrix for the reservoir. We also assume that the system-reservoir transition probabilities satisfy the property

$$W_{is,jt} = W_{jt,is} \tag{45.6}$$

which is valid provided only that the weak system-reservoir interaction Hamiltonian is Hermitean.[24]

We now introduce the assumption, made throughout this chapter, that the reservoir because of its large size remains always in thermodynamic equilibrium (on a macroscopic time scale). Then $\tilde{w}_t$ will be given by Equation (36.15). We also define the quantity R_{ij} in terms of the W_{ij} by

$$R_{ij} \equiv W_{ij}\, e^{\beta(G_j - E_j)} \tag{45.7}$$

where $G_j \equiv \Sigma_\alpha N_{\alpha,j}\, g_\alpha$ and the thermodynamic quantities β and g_α are reservoir parameters. It follows from this definition and Equations (45.4) to (45.6) that

$$R_{ij} = R_{ji} \tag{45.8}$$

Use of this theorem, together with definition (45.7), enables us to rewrite the master equation (45.3) in the useful form

$$\frac{dw_i}{dt} = \sum_j R_{ij}[w_j(t)\, e^{\beta(E_j - G_j)} - w_i(t)\, e^{\beta(E_i - G_i)}] \tag{45.9}$$

Clearly, when the system is in thermodynamic equilibrium with the reservoir, then the right side of this equation is zero, and the statistical matrix is constant in time.

Exercise 45.1 Prove (45.8) using Equations (45.4) to (45.7).

It is instructive to derive the decay equation (45.1) for particle spins from the master equation. In interacting with a liquid, the particle spins can be considered to be in a temperature bath only, so that we set $G_j \to 0$ in Equation (45.9). Also, we restrict ourselves to the simple case of spin $S = ½$ particles. Now for a single particle $\beta E_\pm = \pm \beta\mu\mathcal{H}_o$ is a small quantity in absolute magnitude for all but the lowest temperatures. With the aid of Equations (21.2) and (22.6) we can therefore derive from Equation (45.9) a single-spin decay equation, using $m_\pm = \pm ½$, as follows:

$$\begin{aligned}\frac{dP_z}{dt} &= 2\sum_i m_i \frac{dw_i}{dt} = -2\sum_{i,j} R_{ij} m_i (w_i\, e^{\beta E_i} - w_j\, e^{\beta E_j}) \\ &\cong -2R_{+-}\,[(w_+ - w_-) - \beta\mu\mathcal{H}_o(w_+ + w_-)] \\ &= -\,2R_{+-}\,[P_z(t) - \beta\mu\mathcal{H}_o]\end{aligned} \tag{45.10}$$

Equation (45.1) now follows from this result upon identification of the transition probability R_{+-} as $(2T_1)^{-1}$ and $\beta\mu\mathcal{H}_o$ as $P_{o,z}$.[25]

We consider next the Gibbs H function of Problem V.4, which we define here quantum mechanically in terms of the system probabilities w_i to be

$$\mathcal{H}(t) \equiv \sum_i w_i \ln w_i \tag{45.11}$$

According to Equation (4.25), $\mathcal{H}(t)$ is the negative of the statistical entropy $(\mathcal{H} = -S/\kappa)$ for the system. See also Exercise 27.2. The tendency of the entropy of isolated systems to increase must correspond therefore to a decreasing tendency of $\mathcal{H}(t)$. The H theorem, $d\mathcal{H}/dt \leq 0$, can, in fact, be derived from the master equation (see Problem IX.4). Its interpretation can be understood by referring to Figure V.2. Thus the master equation, which can be valid only for macroscopically small time intervals t, such that $t >> \tau_c$ where τ_c is a microscopic correlation time, governs the approach to equilibrium of a system from any given nonequilibrium condition. A solution to the master equation, when inserted into Equation (45.11), could give for $\mathcal{H}(t)$ only the dashed line in Figure V.2.

Kubo's Method

The general nonequilibrium condition is associated with dissipative processes that promote the tendency toward thermodynamic equilibrium. It is instructive to depart briefly from our discussion of the master equation to illustrate a general method for treating such processes. In Section 42, we have indeed outlined the theory of a large class of these processes, those which are Markoff and linear [Equations (42.12)], but the development there was entirely classical. Now we wish to indicate how these problems may be treated quantum mechanically. We only illustrate the general method, which is due to Kubo, by a calculation of the electrical conductivity tensor $\sigma_{\mu\nu}(\omega)$. For isotropic systems and in the zero-frequency limit, this tensor becomes the scalar electrical conductivity σ of Equation (43.17).

The essence of Kubo's method is to investigate the response of a system to a weak, time-dependent, oscillating perturbation that is switched on slowly (adiabatically) starting in the distant past. In the final result this perturbing interaction no longer appears, so that its role is only to facilitate calculation of the dissipative coefficient, or *linear response,* to the interaction. [This last concept was also used for the calculation in Section 42 of the kinetic coefficients in Equations (42.12).] Thus we assume the total Hamiltonian to be of the form

$$\mathsf{H}(t) = \mathsf{H}_o + \mathsf{V}\, e^{i(\omega - i\varepsilon)t} \qquad (\varepsilon = 0+) \xrightarrow[t\to -\infty]{} \mathsf{H}_o \tag{45.12}$$

where H_o is the Hamiltonian in the absence of the perturbation V that oscillates with angular frequency ω. The operator V itself is taken to be time independent.

We assume next that in the distant past, when the perturbation V was turned off, the system was at thermodynamic equilibrium and described by the statistical matrix $\underline{\rho}_o$. In the absence of particle-number fluctuations $\underline{\rho}_o$ is given by Equation (27.23) using the Hamiltonian H_o. The statistical $\underline{\rho}(t)$ for more recent times must then be of the form

$$\underline{\rho}(t) = \underline{\rho}_o + \Delta\underline{\rho} \qquad \underline{\rho}_o = Z^{-1}\, e^{-\beta \mathsf{H}_o} \tag{45.13}$$

where according to Equation (23.2) $\Delta\underline{\rho}$ satisfies the equation of motion

$$i\hbar \frac{d}{dt}(\Delta\underline{\rho}) = [\mathsf{H}, \Delta\underline{\rho}] + [\mathsf{H}', \underline{\rho}_o] \tag{45.14}$$

in which $\mathsf{H}' \equiv \mathsf{V}\, e^{i(\omega - i\varepsilon)t}$. On introducing the interaction representation, using Equations (44.3) and (44.5) with $\mathsf{H}_\ell = 0$ and $\mathsf{H}_o \to \mathsf{H}_o/\hbar$, this equation becomes

$$\begin{aligned} i\hbar \frac{d}{dt}(\Delta\underline{\rho}_I) &= [\mathsf{H}'_I, \underline{\rho}_o] + [\mathsf{H}'_I, \Delta\underline{\rho}_I] \\ &\cong [\mathsf{H}'_I, \underline{\rho}_o] \end{aligned}$$

where we have neglected the second commutator in the first equality because the interaction V is weak. As with Equation (44.6) in the preceding section, this differential equation is equivalent to an integral equation for $\Delta\underline{\rho}$.

$$\Delta\underline{\rho} = \exp(-i\mathsf{H}_o t/\hbar)\, \Delta\underline{\rho}_I \exp(i\mathsf{H}_o t/\hbar)$$

$$\Delta \underline{\rho} \cong (i\hbar)^{-1} \int_{-\infty}^{t} ds \exp[i\mathsf{H}_o(s - t)/\hbar]$$

$$\times [\mathsf{V}, \underline{\rho}_o]\, e^{i(\omega - i\varepsilon)s} \exp[-i\mathsf{H}_o(s - t)/\hbar] \qquad \textbf{(45.15)}$$

The relation (45.15) can be used for the calculation of average values of interest. Thus if Equations (45.13) and (45.15) are substituted into Equation (21.9b), then a general average-value expression is obtained.

$$\langle \mathsf{A}(t) \rangle \cong \langle \mathsf{A} \rangle_o + (i\hbar)^{-1}\mathrm{Tr} \int_{-\infty}^{t} ds\, e^{i(\omega - i\varepsilon)s} \exp[i\mathsf{H}_o(s - t)/\hbar][\mathsf{V}, \underline{\rho}_o]$$
$$\times \exp[-i\mathsf{H}_o(s - t)/\hbar]\mathsf{A}(t) \qquad \textbf{(45.16)}$$

where the physical observable A may be time dependent and $\langle \mathsf{A} \rangle_o \equiv \mathrm{Tr}(\underline{\rho}_o \mathsf{A})$.

We now apply Equation (45.16) to the calculation of the average electric-current density, with $\mathsf{A} \to \mathsf{j}_\mu (\mu = 1, 2, 3)$. Here $\mathbf{j} = \Sigma_\alpha q_\alpha \mathbf{J}_\alpha$, where the summation is over all different kinds of charge carriers α with charge-per-particle q_α and particle flux $\mathbf{J}_\alpha$. Considering, for simplicity, only one kind of charge carrier, we next choose V to be an electric-dipole interaction,

$$\mathsf{V} = -q \sum_k (\mathbf{x}_k \cdot \mathscr{E}) \qquad (45.17)$$

where $\mathscr{E}$ is the amplitude of an applied electric field that alternates with angular frequency ω, and the summation is over all charge carriers k at locations $\mathbf{x}_k$. Inasmuch as $\langle \mathsf{j}_\mu \rangle_o = 0$ we then obtain from Equation (45.16), after setting $\omega - i\varepsilon \to \omega$, the expression

$$\langle \mathsf{j}_\mu \rangle = \left(\frac{i}{\hbar}\right) \mathrm{Tr} \int_{-\infty}^{t} ds\, e^{i\omega s} [q \sum_k (\mathbf{x}_k \cdot \mathscr{E}), \underline{\rho}_o]\, \mathsf{j}_{\mu,I}(t - s)$$
$$= \left(\frac{i}{\hbar}\right) \mathrm{Tr} \int_{0}^{\infty} dt'\, e^{i\omega(t - t')} [q \sum_k (\mathbf{x}_k \cdot \mathscr{E}), \underline{\rho}_o]\, \mathsf{j}_{\mu,I}(t') \qquad (45.18)$$

To obtain the first equality we have used the cyclic property of the trace and definition (44.5) of an operator in the interaction representation. In the second equality we have transformed to the variable $t' = t - s$.

The complex electrical conductivity tensor $\sigma_{\mu\nu}(\omega)$ is defined by the relation

$$\langle \mathsf{j}_\mu \rangle = \sum_\nu \sigma_{\mu\nu}(\omega) \mathscr{E}_\nu\, e^{i\omega t} \qquad \textbf{(45.19)}$$

It follows, on comparison with Equation (45.18), that

$$\sigma_{\mu\nu}(\omega) = \left(\frac{i}{\hbar}\right) \mathrm{Tr} \int_0^\infty ds\, e^{-i\omega s}\, [q \sum_k \mathsf{x}_{k\nu}, \underline{\rho}_o]\, \mathrm{j}_{\mu,I}(s) \tag{45.20}$$

where $\mathsf{x}_{k\nu}$ is the νth component of the vector $\mathbf{x}_k$.

Exercise 45.2

(a) Use Ehrenfest's theorem for the time-independent operator B,[26] that is,

$$\langle m|\dot{\mathsf{B}}|n\rangle = \frac{d}{dt}\langle m|\mathsf{B}|n\rangle$$

and the Schrödinger equation (23.1) to verify the operator identity

$$[\mathsf{B}, e^{-\beta \mathsf{H}_o}] = -i\hbar e^{-\beta \mathsf{H}_o} \int_0^\beta d\lambda\, e^{\lambda \mathsf{H}_o}\, \dot{\mathsf{B}} e^{-\lambda \mathsf{H}_o} \tag{45.21}$$

Note that $\dot{\mathsf{B}}$ is the quantum-mechanical operator corresponding to the classical quantity $d\mathsf{B}/dt$. [*Suggestion:* Verify (45.21) by taking matrix elements of both sides in the representation in which H_o is diagonal.]

(b) Argue that if $\mathsf{B} = q \Sigma_k \mathsf{x}_{k\nu}$, then $\dot{\mathsf{B}} = \Delta\Omega \mathrm{j}_\nu$, where $\Delta\Omega$ is the (small) volume occupied by the charge carriers corresponding to the summation over k.

The results of Exercise 45.2 can be substituted into Equation (45.20). This gives the final expression

$$\sigma_{\mu\nu}(\omega) = \Delta\Omega \int_0^\infty ds\, e^{-i\omega s} \int_0^\beta d\lambda\, \langle \mathrm{j}_{\nu,I}(-i\hbar\lambda)\mathrm{j}_{\mu,I}(s)\rangle_o \tag{45.22}$$

Thus Kubo's method relates a time-correlation function to a Fourier component of a complex kinetic coefficient, which is $\sigma_{\mu\nu}(\omega)$ in the present example. By further manipulations, Equation (45.22) can be expressed as a quantum-mechanical version of the fluctuation–dissipation theorem, Equations (42.17) and (42.18).[27]

Chapman–Kolmogorov Equation

We now proceed with the discussion of the master equation (45.3) and its microscopic basis for a large closed system. Here we stress that the approach to equilibrium in a macroscopic system, from a coarse-grained view, is a Markov process, that is, one without memory [see above Equations (42.12)]. We show first that the time development of Markoffian systems can be described by the Chapman–Kolmogorov, or Smoluchowski, equation.

We are interested in the time development of coarse-grained transition probabilities $P_I(t)$. To emphasize that these are coarse-grained quantities, as defined below Equation (45.3), we use capital-letter subscripts (I, J, ...) to represent, for example, Z_I microscopic states i. Thus, with $w_i(t)$ constant for all Z_i in the coarse-grained state I, we define

$$P_I(t) \equiv Z_I w_i(t) \tag{45.23}$$

The master equation (45.3) then takes the form

$$\frac{dP_I}{dt} = \sum_J (W_{IJ}P_J - W_{JI}P_I) \tag{45.24a}$$

where

$$W_{IJ} \equiv Z_J^{-1} \sum_{\substack{i\varepsilon I \\ j\varepsilon J}} W_{ij} \tag{45.24b}$$

An essential assumption of the present treatment is that the $w_i(t)$ in a given coarse-grained cell I are all constant, not only at an initial time $t = 0$, but for all times. It is this assumption that gives a system Markoffian behavior, and hence irreversibility, in the present development.

These last observations require further discussion. We recall that time development in a quantum-mechanical system is governed by the Schrödinger equation for wave functions, state vectors, or amplitudes. The associated transition matrix $\mathsf{U}(t_2, t_1)$ relates amplitudes $b_i(t_2)$, at time t_2, to those at an earlier time t_1,

$$b_i(t_2) = \sum_j \langle i|\mathsf{U}(t_2, t_1)|j\rangle b_j(t_1)$$

The probability $w_i(t)$ is then given by the expression

$$w_i(t_2) \equiv |b_i(t_2)|^2 = \sum_{j,j'} \langle i|\mathsf{U}(t_2, t_1)|j\rangle\langle i|\mathsf{U}(t_2, t_1)|j'\rangle^* b_j(t_1)b_{j'}^*(t_1)$$

For the coarse-grained probability $P_I(t_2)$ we then obtain

$$P_I(t_2) = \sum_{J,J'} \sum_{\substack{j\varepsilon J \\ j'\varepsilon J'}} \sum_{i\varepsilon I} \langle i|\mathsf{U}(t_2, t_1)|j\rangle\langle i|\mathsf{U}(t_2, t_1)|j'\rangle^* b_j(t_1)b_{j'}^*(t_1) \tag{45.25}$$

This equation governs the behavior of a single system. If we average it over many systems, a step that corresponds to our incomplete information about the detailed behavior of macroscopic systems, then we expect to get that

$$[b_j(t_1)b_{j'}^*(t_1)]_{\text{ave}} = \delta_{j,j'} Z_J^{-1} P_J(t_1) \tag{45.26}$$

Equation (45.25), without adjusting the notation to distinguish between $P_I(t_2)$ before and after the averaging process (for reasons of convenience only), then takes the form

$$P_I(t_2) = \sum_J T_{t_2-t_1}(I|J)P_J(t_1) \tag{45.27}$$

in which a coarse-grained transition probability $T_t(I|J)$ is defined by[28]

$$T_t(I|J) \equiv Z_J^{-1} \sum_{\substack{i\varepsilon I \\ j\varepsilon J}} |\langle i|\mathsf{U}(t, 0)|j\rangle|^2 \tag{45.28}$$

The crucial step, or assumption, that results in irreversible behavior is Equation (45.26).

Several observations must now be made concerning the results (45.27) and (45.28). To begin with, the time development achieved here is much more specific than that implied by the equation of motion (23.2) for the statistical matrix $\underline{\rho}(t)$, for even though the statistical matrix describes a quantum-mechanical mixed state, its equation of motion is practically equivalent to the Schrödinger equation itself. Though the basis for Equation (45.27) is also the Schrödinger equation, a coarse-graining in time has been achieved effectively by Equation (45.26). Thus it would not be meaningful to treat the coarse-grained states (I, J, ...) on a microscopic time scale. Definition (45.28), however, provides the transition probabilities $T_t(I|J)$ with a microscopic basis. Finally, the averaging process (45.26) has the consequence that Equation (45.27) gives the development in time of the probabilities $P_I(t)$ as a Markoff process. Time evolution of quantum-mechanical amplitudes at the (fine-grained) microscopic level has been replaced by an equation that governs the time evolution of coarse-grained probabilities.

Equation (45.27) can be used to define a Markoff process. The probabilities at a time t_2 are determined only by those at an earlier time t_1 and by a general single-step probability function $T_{t_2-t_1}(I|J)$.[29] Because the times t_1 and t_2 are arbitrary times, we may use Equation (45.27) to write

$$\begin{aligned} P_I(t_1 + s) &= \sum_J T_{t_1+s}(I|J)P_J(0) \\ &= \sum_{J,J'} T_{t_1}(I|J')T_s(J'|J)P_J(0) \end{aligned} \tag{45.29}$$

Inasmuch as the initial probabilities $P_J(0)$ are arbitrary, we then deduce a general form of the *Chapman–Kolmogorov*, or *Smoluchowski, equation*

$$T_{t_1+s}(I|J) = \sum_{J'} T_{t_1}(I|J')T_s(J'|J) \tag{45.30}$$

This equation is fundamental for the quantitative analysis of a Markoff process. It shows that for the Markoff process the development in time from an initial state J to a final state I, at time $t_1 + s$, proceeds independently of the distribution over intermediate states J' at time s.

Derivation of the Master Equation

The master equation in its form (45.24a) can be derived from the Chapman–Kolmogorov equation by setting $t_1 = \Delta t$, where Δt is a very small time interval on a macroscopic time scale. In this limit the transition probabilities $T_{\Delta t}(I|J)$ of Equation (45.28) must be related to those of Equation (45.24b) by the equation

$$T_{\Delta t}(I|J') = \delta_{I,J'}\left\{1 - \Delta t \sum_{J''} W_{J''J'}\right\} + \Delta t W_{IJ'} \qquad \textbf{(45.31)}$$

We must emphasize that this equation can only be meaningful if the small time interval Δt is large compared with any microscopic collision or relaxation times τ, because the basic transition probabilities W_{ij} are not defined for shorter time intervals. The coarse-graining in time, achieved effectively with Equations (45.27) and (45.28), is consistent with this observation.

On substituting Equation (45.31) into Equation (45.30), and relabeling s as the time t, we obtain in the limit $\Delta t \to 0$

$$\frac{d}{dt}T_t(I|J) = \sum_{J'} \{W_{IJ'}T_t(J'|J) - W_{J'I}T_t(I|J)\} \qquad (45.32)$$

That this equation is equivalent to the master equation (45.24a) can be seen by multiplying through by $P_J(0)$, summing over all macroscopic states J, and using finally the first equality of Equation (45.29). Thus the master equation follows from the coarse-grained description of the time evolution of systems as a Markoff process. We may use the master equation in the form (45.3), given originally, provided it is interpreted in the manner described there.

Results we have deduced in Sections 41 and 42 can also be derived starting from the master equation. In particular, it is possible to obtain the phenomenological relations (42.12) and the Onsager reciprocal relations (42.21) by using this approach.[30] The use of the master equation can be viewed as a method in the study of irreversible phenomena which, while yielding practical results, is closely related to the underlying microscopic theory. The master equation is a stepping-stone in nonequilibrium theory in the same sense as is the statistical matrix or partition function of equilibrium theory. Its use seems to be equivalent to the use of the fluctuation–dissipation theorem or of Kubo's method.

Finally, it is worth reiterating the viewpoint presented in Section 25 in connection with Figures V.3 and V.4 that irreversibility in macroscopic phenomena is partly a matter of defining an appropriate time scale, that is, an observer's time scale. Once this has been done, then what we observe to be irreversibility should follow from theory quite naturally. In fact, the key to all the derivations we have given in this chapter has been to assume that macroscopic (i.e., observer) time intervals are large compared with the time intervals during which microscopic processes take place.

We also must not take the arrow of time too seriously, as all of our analyses of irreversible phenomena have assumed a positive direction of time; when examined carefully, this direction could indeed have been chosen to be negative without altering the character of the irreversible phenomena. The basic symmetry relations (41.9) and (42.20) are the key to this conclusion, as one can verify by detailed examination of the derivations given.[31] Thus microscopic reversibility in no way contradicts the fact that in a macroscopic description all parameters tend to change in time so as to approach some equilibrium value, that is, to exhibit irreversible behavior.

Ehrenfests' Urn Model

The occurrence of large spontaneous fluctuations in a system, regardless of the direction of time, is extremely improbable. However, to observe what would happen if a system did have a large fluctuation, we need only bring the system to this condition by external intervention. On removal of the external constraint irreversibility occurs, as the system "practically always" returns to its equilibrium condition according to laws of the type we have discussed in this chapter. Point a in Figure V.2 is an example of a highly improbable fluctuation that has occurred during a system's approach to equilibrium.

The approach to equilibrium can be discussed in very precise terms by using a famous model attributable to the Ehrenfests.[32] See Figure IX.5. Suppose that $2N$ balls, numbered from 1 to $2N$, are distributed in two urns A and B, and suppose that a set of $2N$ cards is also numbered from 1 to $2N$ and mixed randomly. A card is drawn, and the ball whose number corresponds to that on the card is transferred from one urn to the other. After replacing the card in the pack, another is drawn, again randomly, and the whole procedure is repeated again and again. We denote by $n_A(s)$ and $n_B(s)$ the respective numbers of balls in urns A and B after s draws, so that

$$\begin{aligned} n_A(s) + n_B(s) &= 2N \\ n_A(s) - n_B(s) &= 2k(s) \end{aligned} \tag{45.33}$$

Then $n_A = N + k$ and $n_B = N - k$. We also define $\Delta(s) \equiv 2|k(s)|$.

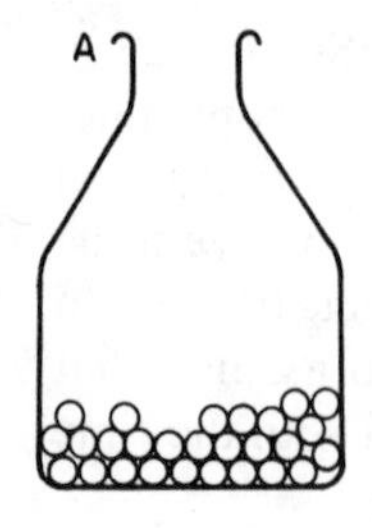

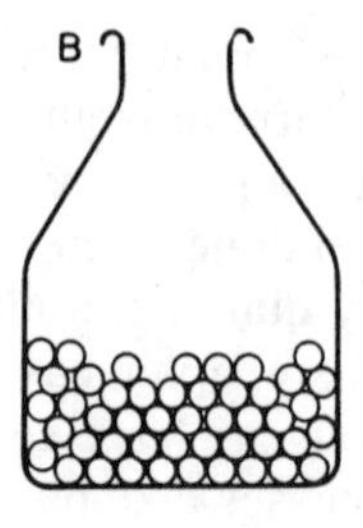

FIGURE IX.5 Ehrenfests' urn model. Numbered balls from 1 to $2N$ are placed in two urns with n_A balls in A and $n_B = 2N - n_A$ balls in B.

When the Ehrenfests' urn model is conducted experimentally it is instructive to plot $\Delta(s)$ as a function of s. One obtains a curve that "at equilibrium" fluctuates in the vicinity of $\Delta(s) = 0$. Moreover, if one starts with the "highly improbable" initial condition $n_A(0) = 2N$, then the curve of $\Delta(s)$ falls toward zero in a manner similar to that of the H function, Figure V.2. Thus the study, both experimentally and theoretically, of the Ehrenfests' urn model can serve as a model for much more complicated studies of the approach to equilibrium in macroscopic systems.

We outline here a first orientation in the analysis of the urn model and leave for the interested reader the pursuit of more sophisticated studies elsewhere.[32] We let $m \equiv n_A(s)$ denote the number of balls in urn A after s draws. Then on the next draw only the values $m \pm 1$ can occur for $n_A(s + 1)$. The corresponding "transition probabilities" $Q(m|m \pm 1)$ are calculated simply to be

$$Q(m|m + 1) = \frac{n_B}{2N} = \frac{2N - m}{2N}$$
$$Q(m|m - 1) = \frac{n_A}{2N} = \frac{m}{2N} \tag{45.34}$$

When $m > N$, these transition probabilities can be expressed in terms of $\Delta(s)$ as $Q(m|m \pm 1) = (2N \pm \Delta)/4N$.

From Equations (45.34) we can conclude (for the case $m > N$) that the probability $Q(m|m - 1)$ of a decrease in $\Delta(s)$ is larger when $\Delta(s)$ is larger, and conversely. Thus Equations (45.34) express the tendency of $\Delta(s)$ to decrease. For example, if initially $n_A(0) = 2N$, then it is clear that the first draw will involve a decrease of $\Delta = 2N$ by two units. With the second draw, there will still be an overwhelming probability $(1 - 1/2N)$ that Δ will continue to decrease and only a probability $1/2N$ that Δ will resume its original value. If $2N >>> 1$, it can be seen that the probability of Δ decreasing remains very large, until Δ has itself become quite small. An irreversible decrease of Δ can alnost certainly be expected to be observed when $2N >>> 1$.[33]

Let us now introduce the probability $P(n_0|m, s)$ that after s draws, starting from an initial value $n_A(0) \equiv n_0$, the value of $n_A(s)$ will be m. Since each step of this model process can be described as Markoff process, we will obtain an equation similar to (45.27) as it would be applied if each time interval were a unit time interval.

$$P(n_0|m, s) = \sum_{\ell} P(n_0|\ell, s - 1)Q(\ell|m) \tag{45.35}$$
$$\xrightarrow[s \to 0]{} \delta(n_0, m)$$

But according to Equations (45.34) we can write $Q(\ell|m)$ in the form

$$Q(\ell|m) = \left(\frac{\ell}{2N}\right)\delta(\ell, m+1) + \left(\frac{2N-\ell}{2N}\right)\delta(\ell, m-1) \tag{45.36}$$

Therefore, Equation (45.35) can be written as the recursion relation

$$\begin{aligned} P(n_0|m, s) &= \left(\frac{m+1}{2N}\right)P(n_0|m+1, s-1) \\ &\quad + \left(\frac{2N-m+1}{2N}\right)P(n_0|m-1, s-1) \end{aligned} \tag{45.37a}$$

If we recall that $m = n_A(s) = N + k$ and use the notation $P(n_0|m, s) \equiv P(k, s)$, then this last result can also be expressed as

$$\begin{aligned} P(k, s) &= \left(\frac{N+k+1}{2N}\right)P(k+1, s-1) \\ &\quad + \left(\frac{N-k+1}{2N}\right)P(k-1, s-1) \\ &\xrightarrow[s\to 0]{} \delta(k, n_0 - N) \end{aligned} \tag{45.37b}$$

To date equation (45.37b) has been solved only for the limiting case of large s values. As we discuss further below, this limit is indeed one of interest. However, one can extract from Equation (45.37b) the answers for various average values. Thus, in addition to verifying that $P(k, s)$ is correctly normalized,

$$\begin{aligned} \sum_{k=-N}^{+N} P(k, s) &= \sum_{\ell=-N}^{+N} P(\ell, s-1)\left[\left(\frac{N+\ell}{2N}\right) + \left(\frac{N-\ell}{2N}\right)\right] \\ &= \sum_{\ell=-N}^{+N} P(\ell, s-1) = 1 \end{aligned} \tag{45.38}$$

where one uses $P(N+1, s) = P(-N-1, s) = 0$ for all s values, one can also show that

$$\begin{aligned} \langle k(s)\rangle &\equiv \sum_{k=-N}^{N} kP(k, s) = (n_0 - N)\left(1 - \frac{1}{N}\right)^s \\ \langle k^2(s)\rangle &= (n_0 - N)^2\left(1 - \frac{2}{N}\right)^s + \frac{N}{2}\left[1 - \left(1 - \frac{2}{N}\right)^s\right] \end{aligned} \tag{45.39}$$

The proof of these two results is left for Problem IX.7. Equations (45.39) show that for very large s values, $\langle k\rangle \to 0$ and $\langle k^2\rangle = 1/4\langle\Delta^2\rangle \to N/2$, in agreement with our expectations for these two quantities.

An important problem from the viewpoint of physical analogies consists of studying the limiting case in which $s \to \infty$. It would be expected generally that the limit probability $\lim_{s\to\infty} P(n_0 \mid m, s)$ would be independent of the initial value n_0 and that in this case we could write $W(m)$ for this quantity. Unfortunately, this expectation is not fulfilled for the Ehrenfests' urn model, as has been shown by Kac.[32] One can, however, consider a "coarse-grained" model instead of the above "fine-grained" case by examining an ensemble of Ehrenfests' urn models, the various members of which have different initial values n_0. If we let $W(n_0)$ be the distribution function of the initial values, with $\Sigma_{n_0=0}^{2N} W(n_0) = 1$, then the condition for a stationary, or time-independent, distribution is that $W(m)$ satisfy Equation (45.35). Thus for a stationary ensemble we require that

$$W(m) = \sum_{n_0=0}^{2N} W(n_0)\, Q(n_0 \mid m) \tag{45.40}$$

If Equation (45.40) is satisfied for one step in the time evolution of the ensemble, then it will be satisfied for every step.

Exercise 45.3 Show that the binomial distribution (1.2)

$$W_{2N}(m) = 2^{-2N} \binom{2N}{m}$$

satisfies Equation (45.40).

Since the binomial distribution is normalized to unity, we may conclude from Exercise 45.3 that the solution to Equation (45.40) for the stationary-ensemble distribution function $W(m)$ is

$$W(m) = 2^{-2N} \binom{2N}{m} \xrightarrow[N>>1]{} \frac{1}{\sqrt{\pi N}} e^{-k^2/N} \tag{45.41}$$

where we have used Equations (2.11), (2.14), and (2.15) for the normal distribution and $k = m - N$. This result agrees with Equations (45.39) in the limit $s >>> 1$, as would be expected.

Finally, we inquire about the frequency with which various m values recur for a stationary ensemble of Ehrenfests' urn models. The answer is determined by the recurrence time T_r, for unit time intervals. In analogy with Equation (25.12) [see also Problem V.8(a)], we may argue that this recurrence time is given by

$$T_r = [W(m)]^{-1} \xrightarrow[N>>1]{} \sqrt{\pi N}\, e^{k^2/N} \tag{45.42}$$

The smallest recurrence time occurs for $k = 0$, and even in this case we obtain, when the unit time interval is 1 sec and $N = 40$, $T_r = \sqrt{40\pi} \cong 11.2$

sec. For the case $k = 10$ this shortest recurrence time is multiplied by $e^{2.5} = 12.2$, and so on.

The condition we call equilibrium is dominated at the microscopic level by those events that recur most frequently. When the recurrence time for an event is very long, then we will make the judgment once the event has occurred, or has been staged by external intervention, that what we observe subsequently is an irreversible progression away from that condition toward equilibrium.

Svedberg's Experiment

In Problem V.8(b) we asked for the probability of finding an evacuated region in a room. The corresponding recurrence time, as Equation (45.42) shows, is proportional to the reciprocal of this probability. This problem can be generalized to the question: What is the probability $P(m)$ for finding m and only m particles in a volume of interest $\Delta\Omega$? The answer is that $P(m)$ is given by the Poisson distribution (2.7) with $\lambda = n \cdot \Delta\Omega$, where n is the particle density in the room. The recurrence time for finding m and only m particles in $\Delta\Omega$ will be given by

$$T_r = \tau[P(m)]^{-1} = \tau(m!)(n\Delta\Omega)^{-m}\, e^{n\Delta\Omega} \tag{45.43}$$

where τ is the time interval between successive observations.

Consider now the application of Equation (45.43) to a colloidal solution of particles undergoing Brownian motion at diffusion equilibrium. In a classic experiment of this type by Svedberg, and analyzed in detail by M. von Smoluchowski,[34] it was it was found that the average $\overline{m}$ of particles that were observed in a field of view $\Delta\Omega$ was given by $\overline{m} = n\Delta\Omega = 1.55$. Observations were at time intervals τ, and average recurrence times for $m = 0, 1, 2, \ldots$ were recorded as shown in Table IX.1. The values calculated by using

Table IX.1 Recurrence Times T_r Versus the Number m of Brownian-Motion Particles Seen in a Volume $\Delta\Omega$. ($\overline{m} = n\Delta\Omega = 1.55$; $\tau \sim 1$ sec)

m	$\left(\frac{T_r}{\tau}\right)_{\text{obs}}$	$\left(\frac{T}{\tau}\right)_{\text{Smol.}}$	$\left(\frac{T_r}{\tau}\right)_{(45.43)}$
0	6.08	5.54	4.71
1	3.13	3.16	3.04
2	4.11	4.05	3.82
3	7.85	8.07	7.60
4	18.6	20.9	19.7

Equation (45.43) are also included in this table and the agreement is seen to be reasonably good. Agreement with the more accurate calculations of Smoluchowski, also tabulated, is quite good.

In Svedberg's experiment the value $m = 7$ was recorded only once in the sequence of 518 observations, in accord with the value $T_r/\tau \cong 1100$ predicted by using Equation (45.43). To observe the value $m = 18$, Svedberg would probably have had to continue his experiment for about 500,000 years!

Bibliography

Texts

A. Abragam, *The Principles of Nuclear Magnetism* (Oxford, Clarendon Press, 1961). The theory of spin relaxation in liquids and gases is treated in depth in Section II of Chapter VIII.

R. Balescu, *Equilibrium and Nonequilibrium Statistical Mechanics* (Wiley, 1975). An extensive range of topics is treated, mostly with more sophisticated discussion than in the present chapter. Recent developments in the field are emphasized.

H. B. Callen, *Thermodynamics* (Wiley, 1960), Chapters 16 and 17. An excellent development of the theory and application of irreversible thermodynamics.

S. R. de Groot, *Thermodynamics of Irreversible Processes* (North-Holland, 1951). Development and application of irreversible thermodynamics entirely from the macroscopic point of view. The thermomechanical effect (our Section 43) is treated in some detail in Chapter III.

C. Kittel, *Introduction to Solid State Physics,* 3rd edition (Wiley, 1968), Chapter 16. Basic concepts in the theory of magnetic resonance phenomena, background for our Section 44, are presented clearly from an experimental outlook.

F. Reif, *Fundamentals of Statistical and Thermal Physics* (McGraw-Hill, 1965), Chapter 15. The perspective from which we introduce the topic of irreversible processes, in Section 41, follows the development in this excellent advanced undergraduate text.

N. Wax, *Selected Papers on Noise and Stochastic Processes* (Dover, 1954). In general mathematical treatments random events (including those with short time correlations) are referred to as stochastic processes. In this book Professor Wax has collected six classic papers on the subject.

Original Article

R. Kubo, "Statistical Mechanical Theory of Irreversible Processes. I. General Theory and Simple Applications to Magnetic and Conduction Problems," *J. Phys. Soc. Japan,* **12** (1957), p. 570.

Review Articles in *Fundamental Problems in Statistical Mechanics,* edited by E. G. D. Cohen (North-Holland, 1962):

L. Van Hove, "Master Equation and Approach to Equilibrium for Quantum Systems."

N. G. Van Kampen, "Fundamental Problems in Statistical Mechanics of Irreversible Processes."

Notes

1. In the derivation of Equation (41.2) we have assumed that Equation (41.1) is valid for all times τ' in the interval $0 \leqslant \tau' \leqslant \tau$, an assumption that is certainly not justified when $\tau' \leqslant \tau_{\text{coll}}$. It is therefore important that we apply Equation (41.2) only when $\tau' \sim \tau >> \tau_{\text{coll}}$, or $t' \sim t + \tau$. Most of the integration region in Figure IX.2 satisfies this requirement.

2. See Equations (24.6).

3. The importance of the present result is that it can be used to relate the resistance R to the spectral density of the fluctuating voltage (Nyquist's theorem). See Problem IX.1(c).

4. See inequality (41.14).

5. Refer to the discussion associated with Equation (12.37).

6. The general theory of Markov processes is treated in W. Feller, *An Introduction to Probability Theory and Its Applications,* Vol. I, 3rd ed. (Wiley, 1968), Chapters XV–XVII; and A. T. Bharucha-Reid, *Elements of the Theory of Markov Processes and Their Applications* (McGraw-Hill, 1960).

7. The mean free path λ is related to the collision time by Equation (12.4), $\lambda = u\tau_{\text{coll}}$.

8. Because the pressure and temperature differences between the two compartments are considered small, adjustments of the total energy $(E + E')$ in order to hold T and T' constant will be negligible when the hole is small. Similarly, adjustments of the separate compartment volumes to maintain the two pressure values can be neglected. Alternatively, we can argue that even with no heat flow or volume adjustments in Figure IX.3, changes in T, T', $\mathcal{P}$, and $\mathcal{P}'$ can be neglected in first approximation.

9. Note that, although similar, process (a) is not the steady-state Joule–Thomson process of Exercise 8.1, because the enthalpy-per-particle w carried through the hole from one compartment is not precisely the same as that in the other compartment. Thus in process (a) we have $\Delta T = 0$ and $\Delta \mathcal{P} = 0$, whereas in the Joule–Thomson $\Delta w \equiv 0$ implies generally that both $\Delta T \neq 0$ and $\Delta \mathcal{P} \neq 0$.

10. F. London, *Superfluids,* Volume II, 2nd rev. ed. (Wiley, 1961), Section 12.

11. Treatment of more complicated two-dimensional examples, in which magnetic fields play a role, is given in the text by Callen, *op. cit.*

12. We omit average-value brackets in the notation for the macroscopic fluxes in this section because we do not consider the corresponding microscopic quantities.

13. Equation (43.14) is equivalent to Equation (42.7a). Recall that the chemical potential g is the average removal energy for a particle, computed for constant entropy and volume, which can be expressed as $g = me$.

14. L. I. Schiff, *Quantum Mechanics,* 3rd ed. (McGraw-Hill, 1968), p. 172.

15. See Equations (27.17) and (27.23).

16. The operators O_n are defined by the Taylor-series expression

$$\mathsf{O}_n = \left\{ \frac{\partial n}{\partial \lambda^n} (e^{\lambda \mathsf{A}} \mathsf{B}\, e^{-\lambda \mathsf{A}}) \right\}_{\lambda = 0}$$

which is equivalent to the cited recursion relation.

17. See A. Abragam, *op. cit.*, for calculations that take $\mathsf{H}'(t)$ to be the magnetic dipole–dipole interaction.

18. We make a crude estimate. The Heisenberg uncertainty principle $\Delta E \cdot \Delta t \gtrsim \hbar$, where ΔE is a nuclear spin energy, implies that the spin-correlation time τ_c must be $>\hbar/\Delta E$. Since liquid atomic-fields variations may occasionally be $\gtrsim 10^3$ gauss, $\Delta E_{\max} \sim 10^{-20}$ erg and therefore the values of τ_c that occur are roughly $\sim 10^{-7}$ sec.

19. More generally, $\tilde{\omega}$ and $\tilde{J}(\omega)$ are imaginary quantities that would be set equal to zero in the approximation $\beta\hbar\omega \ll 1$ of this section. See Problem IX.3.

20. C. Kittel, *op. cit.*, Chapter 16.

21. Pauli's derivation of the master equation is outlined in the article by Van Hove, *op. cit.*

22. Notice that the approximations made in the second and third equalities of Equation (44.20) can be viewed as the application of a coarse-graining procedure to the liquid states.

23. This requirement can easily be qualified, for nonconserved particles, by a discussion similar to that given below Equations (36.3). See also the derivation of Equation (40.9).

24. We have used here the detailed-balance theorem (and not the reciprocity theorem), which is valid whenever the system–reservoir interaction is weak. See, for example, J. M. Blatt and V. F. Weisskopf, *Theoretical Nuclear Physics* (Wiley, 1952), Section X.2.

25. Note that the first of Equations (27.5) gives $P_{o,z} = \tanh(\beta\mu\mathscr{H}_o) \cong \beta\mu\mathscr{H}_o$ for spin $S = 1/2$ particles. This result is also obtained as the answer to Problem V.1(c).

26. L. I. Schiff, *loc. cit.*, p. 29.

27. R. Kubo, *op. cit.*, Section 7.

28. Since at the microscopic level there can be no preferred origin of time, the transition matrix $\mathsf{U}(t_2, t_1)$ must depend on the times t_1 and t_2 only through the difference $t_2 - t_1$. Equation (45.27) is justified without reference to ensemble theory in Sections 24 and 25 of the article by N. G. Van Kampen, *op. cit.*

29. A general framework for classifying random, or stochastic, processes is outlined at the beginning of the classic article "On the Theory of the Brownian Motion. II" by M. C. Wang and G. E. Uhlenbeck in *Rev. Mod. Phys.* **17** (1945), p. 323; reprinted by N. Wax, *op. cit.* The Markoff process is the simplest one that involves correlations in time.

30. N. G. Van Kampen, *Physica* **23** (1957), pp. 707, 816.

31. For example, the damping coefficient α of Equation (41.12) actually reverses its sign with time, as can be verified by examination of the final equality in the equation preceding Equation (41.10).

32. This model is discussed in detail in Appendix III.A. of R. Jancel, *Foundations of Classical and Quantum Statistical Mechanics* (Pergamon, 1969), translated by W. E. Jones. See also the early analysis by F. Kohlrausch and E. Schrödinger, *Physik. Zeits.* **27** (1926), p. 306.

33. Note the similarity of this model to the free-expansion example discussed near the end of Section 25.

34. Smoluchowski's analysis and detailed references to the original papers are given in a famous review article, "Stochastic Problems in Physics and Astronomy," by S. Chandresekar in *Rev. Mod. Phys.* **15** (1943), p. 1; reprinted by N. Wax, *loc.*

cit. Smoluchowski's more exact expression for the recurrence time in Svedberg's experiment agrees with Equation (45.43) when $m >> \overline{m}$.

35. This ensemble is the "coarse-grained" microcanonical ensemble, discussed above Equations (26.3), in which the occupied closed-system energy states are all within a spread δE.

Problems

IX.1 (a) Define coefficients $C(\omega)$ of a real, random quantity $y(t)$ by

$$C(\omega) \equiv \frac{1}{2\pi}\int_{-\tau}^{\tau} dt\, y(t)\, e^{-i\omega t} = C^*(-\omega)$$

where $\tau >> \tau_c$ and τ_c is the correlation time of $y(t)$. Show that the quantity $y_\tau(t)$ defined by $y_\tau(t) \equiv \Theta(\tau - |t|)y(t)$, with $\Theta(x)$ a step function, is related to these $C(\omega)$ by Fourier transformation as

$$y_\tau(t) = \int_{-\infty}^{+\infty} d\omega\, C(\omega)\, e^{i\omega t}$$

(b) Argue that the correlation function $K(s)$ of Equation (41.5), defined for the variable y instead of F, can be written as

$$K(s) = \frac{1}{2\tau}\int_{-\infty}^{+\infty} dt\, y_\tau(t) y_\tau(t + s)$$

Then show that the spectral density $J(\omega)$ [second of Equations (44.23)] is given by

$$\langle y^2 \rangle J(\omega) = \frac{\pi}{\tau}|C(\omega)|^2$$

This result demonstrates that $J(\omega)$ is always a positive real quantity when $y(t)$ is real.

(c) Derive the result from Equation (41.12), for $\omega\tau_c \lesssim 1$,

$$\langle F^2 \rangle J(\omega) \cong \langle F^2 \rangle J(0) = \frac{\kappa T \alpha}{\pi} \qquad \text{(Nyquist's theorem)}$$

where the first approximate equality follows from the definition of $J(\omega)$. This result can be used together with Equation (44.26) to gain further insight into inequality (41.18).

(d) Argue that $T_2(\omega) < T_1(\omega)$, where T_1 and T_2 are the relaxation times of Section 44.

IX.2. (a) Starting from Equation (44.11), derive the expression

$$\frac{d\underline{\sigma}_I'(t)}{dt} = -\int_0^\infty ds \sum_\mu \langle\mu|\{[\mathsf{H}_I'(\mathrm{t}), [\mathsf{H}_I'(t-s), \underline{\tilde{\sigma}}_o + \underline{\tilde{\sigma}}_I']]\, \underline{\rho}_\ell + \mathsf{H}_I'(t)(\underline{\tilde{\sigma}}_o + \underline{\tilde{\sigma}}_I')[\mathsf{H}_I'(t-s), \underline{\rho}_\ell] - [\mathsf{H}_I'(t-s), \underline{\rho}_\ell](\underline{\tilde{\sigma}}_o + \underline{\tilde{\sigma}}_I')\mathsf{H}_I'(t)\}|\mu\rangle$$

where $\underline{\tilde{\sigma}}_I' \equiv \underline{\tilde{\sigma}}_I'(0)$.

(b) Take spin-state matrix elements of this result, and show that the effect of the last two terms is to remove $\underline{\tilde{\sigma}}_o$ from the first, thereby leading to Equation (44.12).
[*Suggestion:* (1) Off-diagonal matrix elements of the commutator bracket $[\mathsf{H}_I'(t-s), \underline{\rho}_\ell]$ give an exponential factor $e^{i\Delta E(t-s)/\hbar}$, where $(t-s) \lesssim \tau_c$, as argued below Equation (44.19). But liquid states have energy-level densities that are so large that this factor oscillates rapidly even for such small $(t-s)$ values. Argue that one can use

$$\int_0^\infty ds = \frac{1}{2}\int_{-\infty}^{+\infty} ds$$

without changing the result for

$$\frac{d\underline{\sigma}_I'(t)}{dt}$$

and, therefore, that the exponential factor leads to conservation of energy in the matrix elements considered. One can then re-express these matrix elements so that the commutator bracket $[\mathsf{H}_I'(t-s), \underline{\tilde{\sigma}}_o]$ is involved. (2) The approximation for nuclear-spin energies, $\beta\hbar\omega \sim 10^{-7}\mathcal{H}_o/T \ll 1$, where the external field $\mathcal{H}_o$ is measured in gauss, can then be used to discard uncanceled terms.]

IX.3. (a) Use the conservation-of-energy argument of Problem IX.2(b) to show that the spectral density is a real quantity that satisfies

$$J(-\omega) = e^{-\hbar\omega/\kappa T} \cdot J(\omega)$$

Verify that the time-correlation function $G(s)$ is not a real quantity when $J(-\omega) \neq J(\omega)$ and, in particular, that $\tilde{J}(\omega)$ of Equation (44.28) is an imaginary quantity.

(b) Show that the time-correlation function $G(s)$ can be written as

$$G(s) = 2\int_0^\infty d\omega[J_1(\omega)\cos\omega s - i\, J_2(\omega)\sin\omega s]$$

where the real quantities $J_1(\omega)$ and $J_2(\omega)$ are even and odd functions of ω respectively. Their sum is $J(\omega)$ and their ratio is given by

$$\frac{J_2(\omega)}{J_1(\omega)} = \tanh\left(\frac{\hbar\omega}{2\kappa T}\right) \underset{\hbar\omega << \kappa T}{\longrightarrow} 0$$

(c) Verify the integral relations

$$J_1(\omega) = \frac{1}{\pi}\int_0^{\infty} ds\, G(s) \cos \omega s$$

$$J_2(\omega) = \frac{i}{\pi}\int_0^{\infty} ds\, G(s) \sin \omega s$$

(d) Prove that

$$J_2(\omega) = -\frac{i}{\pi} P \int_{-\infty}^{+\infty} d\nu\, J_1(\nu) \left(\frac{1}{\nu - \omega}\right)$$

where P denotes principal value.

IX.4. (a) Start with definition (45.11) of the Gibbs H function and use the master equation (45.3) to derive the H theorem for a large closed system that includes a small system of interest in interaction with a temperature-fugacity reservoir. Show that at equilibrium there will be equal probability for occupation of all energy states in a microcanonical ensemble comprised of the large closed systems.[35] [*Suggestion:* Use Equation (45.6) and recall Exercise 25.2.]

(b) Show that if the reservoir probabilities $\tilde{\omega}_s$ are given by Equation (36.15), then at equilibrium the small-system probabilities w_i will also be given by this expression. This result implies, according to Equation (45.9) and as it must, that $dw_i/dt = 0$ when $\mathcal{H}(t)$ reaches its minimum value.

IX.5. *Fokker–Planck equation for Brownian motion*

We wish to describe the one-dimensional Brownian motion of a colloidal particle as a Markoff process, in which the probabilities for finding successive velocities are stochastically independent. Thus if $P_s(u \mid u_o)du$ is the probability distribution function, normalized to unity, that a particle starting with velocity u_o will have velocity u after a time s, then we expect that

$$P_{s+\tau}(u \mid u_o) = \int_{-\infty}^{+\infty} du_1\, P_\tau(u \mid u_1) P_s(u_1 \mid u_o)$$

which is the Chapman–Kolmogorov equation (45.30).

(a) Show for small time intervals τ that the above equation can be rewritten in the form

$$\frac{\partial P_s}{\partial s}\tau = -P_s(u \mid u_o) + \int_{-\infty}^{+\infty} d\xi\, P_s(u - \xi \mid u_o)P_\tau(u \mid u - \xi)$$

(b) Define velocity moments τM_n by the equation

$$M_n \equiv \frac{1}{\tau}\int_{-\infty}^{+\infty} d\xi\, \xi^n\, P_\tau(u + \xi \mid u)$$

and assume for macroscopically small τ values that the integrand of the result in part (a) is appreciable only for small ξ values. Show then that

$$\frac{\partial P_s}{\partial s} = \sum_{n=1}^{\infty} \frac{(-1)^n}{n!}\frac{\partial^n}{\partial u^n}[M_n P_s(u \mid u_o)]$$

Suggestion: Write

$$P_s(u - \xi \mid u_o)P_\tau(u \mid u - \xi) = \sum_{n=0}^{\infty} \frac{(-\xi)^n}{n!}\frac{\partial^n}{\partial u^n}[P_s(u \mid u_o)P_\tau(u + \xi \mid u)]$$

On neglecting all moments M_n for $n > 2$, we obtain the Fokker–Planck equation for $P = P_s(u \mid u_o)$,

$$\frac{\partial P}{\partial s} \xrightarrow[\tau \to 0]{} -\frac{\partial}{\partial u}(M_1 P) + \frac{1}{2}\frac{\partial^2}{\partial u^2}(M_2 P)$$

(c) Use Equations (41.11) and (41.17) for τM_1 and τM_2, respectively, with $F_{\text{ext}} = 0$. Then simplify the Fokker–Planck equation and show that the solution for $P_s(u \mid u_o)$ is a Maxwell–Boltzmann-like distribution

$$P_s(u \mid u_o) = \left[\frac{m'}{2\pi\kappa T}\right]^{1/2} \exp\left[-\frac{m'}{2\kappa T}(u - u_o')^2\right]$$

The quantity $u_0' = u_0 e^{-\gamma s}$ [see Equation (41.13)] and the mass $m' = m(1 - e^{-2\gamma s})^{-1}$. Interpret the $s \to 0$ and $s \to \infty$ limits of this result for $P_s(u \mid u_o)$.

IX.6. *Direct calculation of kinetic coefficients*

(a) Consider the thermomolecular pressure-difference effect of Figure IX.3 in the small-hole limit, with $w = 5/2\, \kappa T$ in Equations 43.6 for an ideal monatomic gas. Show first for this ideal gas that particle flow and energy flow from the left compartment to the right compartment are given, respectively, by (cf. Exercise 11.2):

$$\frac{dN_L}{dt} = A\mathcal{P}(2\pi m\kappa T)^{-1/2}; \qquad \frac{dE_L}{dt} = A\mathcal{P}(2\kappa T/\pi m)^{1/2}$$

where A is the cross-sectional area of either compartment. Then, for the case of small temperature and pressure differences between the two comparments, write down explicit forms of Equations (43.6) and identify the four kinetic coefficients L_{11}, L_{12}, L_{21}, and L_{22}. Verify the Onsager relation $L_{12} = L_{21}$ for this example and show that Equation (43.9) is readily obtained when $\dot{N} = 0$.

(b) For a second example, consider the flow of a monatomic gas in a horizontal uniform pipe in which there is a pressure gradient $d\mathcal{P}/dx$ and temperature gradient dT/dx along the pipe. In this case the phenomenological relations for the energy- and particle-current densities, J_E and J_N, are similar to Equations (43.6) with ΔT and ΔP replaced by dT/dx and $d\mathcal{P}/dx$ respectively. Argue that the particle-flux J_N, at the cross-sectional plane where $x = 0$ along the pipe, is given by

$$J_N = (A\lambda)^{-1} \int_{-\infty}^{-\infty} dx\, e^{-|x|/\lambda}(dN/dt)$$

where dN/dt is the answer given in part (a) evaluated at the location x, with cross-sectional area A, and λ is the mean free path for the (weakly interacting) gas particles. Evaluate this result, after integrating by parts, by neglecting the variation in $\mathcal{P}(x)$ and $T(x)$ over a mean free path. Repeat the calculation for the (kinetic) energy–current density J_E. Compare both results with the phenomenological relations for J_N and J_E in order to identify the kinetic coefficients, as in part (a).

IX.7. Derive Equations (45.39) using Equations (45.37b) and (45.38).

CHAPTER X

ONE- AND TWO-PARTICLE DISTRIBUTION FUNCTIONS

We consider details of a many-body system that can be surmised by studying it with a probe, such as a scattering particle. The knowledge gleaned from experiments with probes can be translated into information about the distribution of particles in the system and, in particular, about their correlations in position and time. This information can then be used in two different ways. First, one can attempt to make theoretical calculations of the few-particle distribution functions and to compare predictions with observation. Second, it is possible to calculate directly equilibrium thermodynamic properties of a system using one- and two-particle distribution functions, thereby giving an alternate procedure to the use of the partition function for this purpose.[1] In this chapter we first investigate generally properties of the one- and two-particle distribution functions in position and momentum space. Then focusing attention on the pair distribution function in position space, we examine its Fourier transform, called the structure factor, and its generalization to a space–time correlation function, whose Fourier transform is called the linear response function. These functions are then used in a study of long order in crystals (Section 49). Finally, in Section 50 we show explicitly how to calculate the one-particle thermodynamic Green's function.

46 One-Particle Distribution Functions

To demonstrate how a scattering experiment can provide information about a system, we consider first in this section a simplified example of the scattering of a low-energy particle by a fluid. We assume that the particle interacts only minimally with the fluid molecules so that multiple scattering within the fluid can be neglected. In other words, we assume that the scattered particle interacts with only a single fluid molecule as it traverses the system. On examination of the elastic scattering cross section for this process we are led to introduce the system *form factor,* which is a Fourier transform of the single-particle density function in position space. Following this treatment we investigate the momentum distribution in a system.

The differential cross section for scattering of a particle by a fluid is given by the well-known Golden Rule No. 2 formula,[2]

$$\frac{d\sigma}{d\Omega} = \frac{2\pi}{\hbar} \cdot \frac{mL^3}{\hbar k_o} \cdot \sum_{\mu,\nu} \langle \mu | \underline{\rho}_\ell | \mu \rangle |\langle \mathbf{k}, \nu | \mathsf{G} | \mathbf{k}_o, \mu \rangle|^2 \rho_f \tag{46.1}$$

where $d\Omega$ is an element of solid angle, L^3 is the volume of the scattering system (for box normalization), $\underline{\rho}_\ell$ is the statistical matrix for the N-particle fluid, G is the transition matrix for the scattering process, and ρ_f is the density of final states. We neglect here any effects due to the spin of the scattering particle, whose mass is m. The incident and final wave numbers of the particle are $\mathbf{k}_o$ and $\mathbf{k}$, respectively, and (μ, ν) represent fluid-state quantum numbers. Moreover, energy conservation in the scattering process dictates that

$$(E_\mu + \omega_o) = (E_\nu + \omega_f)$$

where

$$\omega_o = \hbar^2 k_o^2 / 2m$$

and

$$\omega_f = \hbar^2 k^2 / 2m$$

The density of final energy states for the scattered particle is calculated in a standard way to be[3]

$$\rho(\omega_f) = \frac{d^3 \mathbf{n}_f}{d\omega_f \, d\Omega} = \frac{mL^3}{(2\pi)^3} \frac{k}{\hbar^2} \tag{46.2}$$

We now neglect multiple scattering effects and assume that the transition matrix G can be approximated by a sum over N two-particle transition matrices G_α, that is, we assume

$$\mathsf{G} \cong \sum_{\alpha=1}^{N} \mathsf{G}_\alpha$$

for weakly interacting scattered particles.[4] Here G_α involves the Hamiltonian H_o, which is the kinetic energy of the incident particle plus the complete fluid-system Hamiltonian, and the interaction V_α between the incident particle and the αth fluid particle. We then consider the particle-fluid matrix elements of G_α needed in Equation (46.1), again neglecting any spin effects.

$$\langle \mathbf{k}, \nu|\mathsf{G}_\alpha|\mathbf{k}_o, \mu\rangle = L^{-3}\int d^3\mathbf{r}_p\, d^3\mathbf{r}'_p$$
$$\times \prod_{\beta=1}^{N} d^3\mathbf{r}_\beta\, d^3\mathbf{r}'_\beta \langle \nu \mid \mathbf{r}_1 \cdots \mathbf{r}_N\rangle \exp(-i\mathbf{k}\cdot\mathbf{r}_p)$$
$$\times \langle \mathbf{r}_p; \mathbf{r}_1 \cdots \mathbf{r}_N|\mathsf{G}_\alpha|\mathbf{r}'_p; \mathbf{r}'_1 \cdots \mathbf{r}'_N\rangle$$
$$\times \exp(i\mathbf{k}_o\cdot\mathbf{r}'_p)\langle \mathbf{r}'_1 \cdots \mathbf{r}'_N|\mu\rangle$$
$$= L^{-3}\int d^3\mathbf{r}_{p_\alpha}\, d^3\mathbf{r}'_{p_\alpha} \tag{46.3}$$
$$\times \prod_{\beta=1}^{N} d^3\mathbf{r}_\beta\, d^3\mathbf{r}'_\beta \langle \nu|\mathbf{r}_1 \cdots \mathbf{r}_N\rangle \exp[-i\mathbf{k}\cdot(\mathbf{r}_{p_\alpha} + \mathbf{r}_\alpha)]$$
$$\times \langle \mathbf{r}_{p_\alpha}; \mathbf{r}_1 \cdots \mathbf{r}_N|\mathsf{G}_\alpha|\mathbf{r}'_{p_\alpha}; \mathbf{r}'_1 \cdots \mathbf{r}'_N\rangle$$
$$\times \exp[i\mathbf{k}_o\cdot(\mathbf{r}'_{p_\alpha} + \mathbf{r}'_\alpha)] \times \langle \mathbf{r}'_1 \cdots \mathbf{r}'_N|\mu\rangle$$

where $\mathbf{r}_{p_\alpha} = \mathbf{r}_p - \mathbf{r}_\alpha$ and and similarly for $\mathbf{r}'_{p_\alpha}$. Thus in the first equality of this equation the particle coordinates all refer to an arbitrary origin, whereas in the second equality the origin for the incident particle's coordinates is at particle α. The relevance of this coordinates transformation is that in the second equality the fluid scattering process by particle α can readily be related to the corresponding two-particle scattering process. In this connection, we note that the kinetic-energy operator for the incident particle, when expressed in terms of the gradient operator $\partial/\partial\mathbf{r}_{p_\alpha}$, still involves the mass m and not a reduced mass.

At this point we make definite the condition that the system is not a gas by assuming that the αth particle does not recoil independently in the scattering process, because it is bound. Effectively then, the matrix elements of G_α in Equation (46.3) must be mostly diagonal, of the form[5]

$$\langle \mathbf{r}_{p_\alpha}; \mathbf{r}_1 \cdots \mathbf{r}_N|\mathsf{G}_\alpha|\mathbf{r}'_{p_\alpha}; \mathbf{r}'_1 \cdots \mathbf{r}'_N\rangle$$
$$= \left[\prod_{\beta=1}^{N} \delta^{(3)}(\mathbf{r}_\beta - \mathbf{r}'_\beta)\right]\langle \mathbf{r}_{p_\alpha}; \mathbf{r}_1 \cdots \mathbf{r}_N|\mathsf{G}_\alpha|\mathbf{r}'_{p_\alpha}; \mathbf{r}_1 \cdots \mathbf{r}_N\rangle$$

The consequence of this assumption is that we may simplify Equation (46.3) to

$$\langle \mathbf{k}, \nu|\mathsf{G}_\alpha|\mathbf{k}_o, \mu\rangle = \int \prod_{\beta=1}^{N} d^3\mathbf{r}_\beta$$
$$\times \langle \nu|\mathbf{r}_1 \cdots \mathbf{r}_N\rangle\langle \mathbf{k}; \mathbf{r}_1 \cdots \mathbf{r}_N|\mathsf{G}_\alpha|\mathbf{k}_o; \mathbf{r}_1 \cdots \mathbf{r}_N\rangle \tag{46.4}$$
$$\times \exp[i(\mathbf{k}_o - \mathbf{k})\cdot\mathbf{r}_\alpha]\langle \mathbf{r}_1 \cdots \mathbf{r}_N|\mu\rangle$$

The state labels **k** and $\mathbf{k}_o$ now refer specifically to an origin attached to the αth particle, which is why the exponential factor $\exp[i(\mathbf{k}_o - \mathbf{k}) \cdot \mathbf{r}_\alpha]$ has appeared.

The next step in the development is to utilize the scattering-length approximation for the matrix elements of G_α in Equation (46.4)

$$\langle \mathbf{k}; \mathbf{r}_1 \cdots \mathbf{r}_N | \mathsf{G}_\alpha | \mathbf{k}_o; \mathbf{r}_1 \cdots \mathbf{r}_N \rangle \cong \frac{2\pi\hbar^2 a_s}{mL^3} \tag{46.5}$$

where a_s is the scattering length for the (liquid particle)–(incident particle) interaction.[6] This approximation is valid for a weak interaction and low particle energies, such that $qr_o << 1$, where r_o is a measure of the range of the interaction and $\mathbf{q} = \mathbf{k}_o - \mathbf{k}$ is the momentum transfer to the liquid. Substitution of Equation (46.5) into (46.4) yields the important result

$$\langle \mathbf{k}, \nu | \mathsf{G}_\alpha | \mathbf{k}_o, \mu \rangle \cong \left(\frac{2\pi\hbar^2 a_s}{mL^3} \right) \langle \nu | e^{i\mathbf{q}\cdot\mathbf{r}_\alpha} | \mu \rangle \tag{46.6}$$

On summing Equation (46.6) over all N particles α and then substituting it and (46.2) into Equation (46.1), we obtain for the differential scattering cross section at low energies,

$$\frac{d\sigma}{d\Omega} \cong a_s^2 \sum_{\mu,\nu} \langle \mu | \rho_e | \mu \rangle \left| \left\langle \nu \left| \sum_{\alpha=1}^{N} e^{i\mathbf{q}\cdot\mathbf{r}_\alpha} \right| \mu \right\rangle \right|^2 \left(\frac{k}{k_o} \right) \tag{46.7}$$

We now examine the $\nu = \mu$ term, corresponding to elastic scattering, and set $k = k_o$ in this case since the recoil energy of the liquid will be negligible.

The influence of the liquid system in the low-energy scattering process is determined, according to Equation (46.7), by the many-body operator $\sum_{\alpha=1}^{N} e^{i\mathbf{q}\cdot\mathbf{r}_\alpha}$. For elastic scattering, the matrix elements of this operator can be rewritten as follows:

$$\begin{aligned} \left\langle \mu \left| \sum_{\alpha=1}^{N} e^{i\mathbf{q}\cdot\mathbf{r}_\alpha} \right| \mu \right\rangle &= \int d^3\mathbf{r}\, e^{i\mathbf{q}\cdot\mathbf{r}} \left\langle \mu \left| \sum_{\alpha=1}^{N} \delta^{(3)}(\mathbf{r} - \mathbf{r}_\alpha) \right| \mu \right\rangle \\ &= \int d^3\mathbf{r}\, e^{i\mathbf{q}\cdot\mathbf{r}} \langle \mu | \mathsf{n}(\mathbf{r}) | \mu \rangle \\ &= \int d^3\mathbf{r}\, e^{i\mathbf{q}\cdot\mathbf{r}} n_\mu(\mathbf{r}) \underset{\mathbf{q}=0}{\longrightarrow} N \end{aligned} \tag{46.8}$$

where the operator $\mathsf{n}(r)$ is given quantum-mechanically, as well as classically, by the first of Equations (24.2) and the density function $n_\mu(r)$ is the expectation value of $\mathsf{n}(r)$ for the fluid state μ. We next introduce a notation for the integral in the final equality of Equation (46.8); namely, we introduce the *form factor* $F_\mu(\mathbf{q})$ by the definition

$$F_{\mu}(\mathbf{q}) \equiv N^{-1} \int d^3\mathbf{r}\, e^{i\mathbf{q}\cdot\mathbf{r}} n_{\mu}(\mathbf{r}) \xrightarrow[\mathbf{q}=0]{} 1 \tag{46.9}$$

For a homogeneous fluid the density $n_{\mu}(\mathbf{r})$ will vary only near the surface and this fact explains the terminology form factor. If the system is large, then $F_{\mu}(\mathbf{q})$ will differ from zero only for small q values. In the thermodynamic limit we obtain for a homogeneous fluid

$$F_{\mu}(\mathbf{q}) \xrightarrow[\Omega,N\to\infty]{} (2\pi)^3\Omega^{-1}\, \delta^{(3)}(\mathbf{q}) \tag{46.9a}$$

In this case, we must interpret $\delta^{(3)}(0)$ as being equal to $\Omega/(2\pi)^3$.

After substituting Equations (46.8) and (46.9) into the $\nu = \mu$ term of Equation (46.7), we obtain for the elastic scattering cross section,

$$\left(\frac{d\sigma}{d\Omega}\right)_{\text{el}} = (Na_s)^2 \sum_{\mu} \langle\mu|\underline{\rho}_{e\ell}|\mu\rangle |F_{\mu}(\mathbf{q})|^2$$

$$\left(\frac{d\sigma}{d\Omega}\right)_{\text{el}} = (Na_s)^2 \langle F^2(\mathbf{q})\rangle \tag{46.10}$$

$$\xrightarrow[\Omega,N\to\infty]{} Na_s^2(2\pi)^3 n\, \delta^{(3)}(\mathbf{q})$$

We see that a measurement of the elastic scattering cross section can provide information about the density of matter in a finite system by inversion of Equation (46.9). However, for a large liquid system in thermodynamic equilibrium there is no elastic scattering except in the forward direction.[7] In Section 48 we return to study the more interesting case of inelastic scattering from a liquid.

Momentum Distribution

It will become clear in this chapter that the most measurable distribution functions for a system are those in position space, such as the density function $n_{\mu}(\mathbf{r})$ whose averaged Fourier transform can be determined from elastic scattering measurements. Thus although momentum-transfer measurements are made, their analysis reveals little information about the arrangement of a system's particles in momentum space. Of course, this fact does not diminish our inherent interest in momentum-space distribution functions, especially in the low-temperature region where, as for quantum fluids, momentum-space ordering prescribes particle motions. It is from this perspective that we discuss the momentum distribution and its calculation.

The momentum distribution, or one-particle reduced distribution function, has been defined by Equations (24.10) and (24.13). From the outset we must emphasize for a quantum fluid that the momentum distribution $\langle \mathsf{n}(k)\rangle$ is not the same quantity as the quasi-particle occupation number $\overline{n}'_{\mathbf{k}} \equiv \nu'(\mathbf{k})$, which is given by Equation (19.14) for the Fermi-statistics case. As explained above Equation (19.5), the latter quantity gives the distribution of quasi-particle energies in a Fermi fluid. Moreover, since these energies correspond to definite momentum states for large fluids, we may conclude that at $T = 0$ the distribution function $\nu'(\mathbf{k})$ is identical, in momentum space, with the free-particle distribution function $\nu(\mathbf{k})$ of Equation (32.1). But only for an ideal gas is $\langle \mathsf{n}(\mathbf{k}) \rangle$ equal to either $\nu(\mathbf{k})$ or $\nu'(\mathbf{k})$.

Consider an infinite Fermi fluid. In Problem IV.4 we have indicated a procedure for calculating $\langle \mathsf{n}(\mathbf{k}) \rangle$ using perturbation theory, and the result, which has the same qualitative behavior for all Fermi fluids, is shown schematically in Figure X.1. The existence of particle interactions causes the free-particle momentum distribution of Figure VII.1 to be rounded off, even at $T = 0$. The fact that the momentum distribution, as opposed to $\nu'(\mathbf{k})$, is not square at $T = 0$ simply reflects the fact that the many-body wave function of the system has high Fourier components—even though a single quasi-particle state continues to be given by the plane-wave function (17.27).[8]

The momentum distribution of a system is an indirectly measurable quantity at best. For heavy atomic nuclei, which may be considered to be finite Fermi fluids, a method for measuring $\langle \mathsf{n}(\mathbf{k}) \rangle$ indirectly was first alluded to by Serber.[9] He observed that in high-energy nucleon–nucleon collisions the most probable momentum transfers are $\lesssim \hbar/r_o$, where r_o is the range of the nuclear force. But r_o^{-1} is $\sim k_F$, where the Fermi momentum $\hbar k_F$ may be defined for a heavy nucleus by Equation (19.2) with $(2S + 1) \rightarrow 4$.[10] Assuming that $\langle \mathsf{n}(\mathbf{k}) \rangle$ does not deviate drastically from $\nu'(\mathbf{k})$, we may conclude that a sizable fraction of nucleon–nucleon collisions in heavy nuclei are forbidden by the Pauli principle. Therefore, high-energy nuclear reactions are sensitive to the momentum distribution. Unfortunately, collective effects in nu-

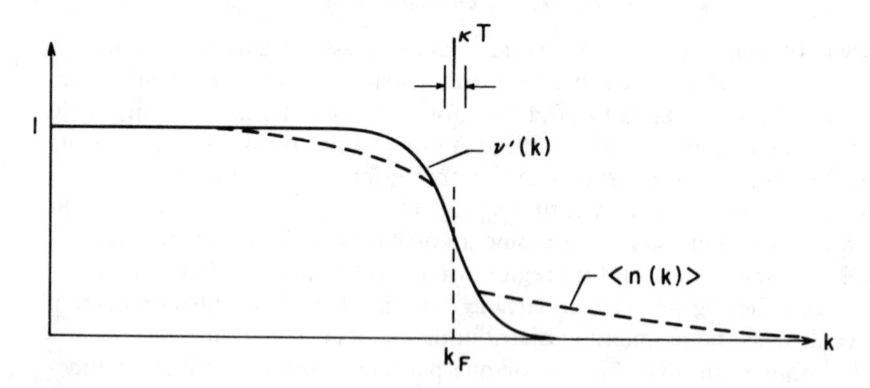

FIGURE X.1 The momentum distribution $\langle \mathsf{n}(k) \rangle$ and the average quasi-particle occupation number $\nu'(k)$ for a Fermi fluid near $T = 0$ (schematic).

clei make quantitative determinations of $\langle \mathsf{n}(\mathbf{k}) \rangle$, starting from a knowledge of free nucleon–nucleon collision cross sections, difficult if not impossible to extract from nuclear cross-section measurements.

Despite the fact that the momentum distribution tends to elude quantitative measurement, it is instructive to pursue its calculation as a prototype for the theoretical evaluation of distribution functions. The calculation of $\langle \mathsf{n}(\mathbf{k}) \rangle$ also provides an independent means for determining the density theoretically, using $n = \Omega^{-1} \Sigma_{\mathbf{k}} \langle \mathsf{n}(\mathbf{k}) \rangle$.

To calculate $\langle \mathsf{n}(k) \rangle$ for a single-component system, where $k = (\mathbf{k}, m)$, we substitute the first equality of Equation (24.10) into Equation (24.13) and use Equation (36.15a) for the statistical matrix. After referring to Equations (36.16) and (37.3), it is easy to verify that an explicit expression for $\langle \mathsf{n}(k) \rangle$ is given, for a large translationally invariant system, by

$$\langle \mathsf{n}(k) \rangle = e^{-\Omega f} \sum_{N=1}^{\infty} \frac{1}{N!} e^{\beta N g} \sum_{k_1 \cdots k_N} \times \exp\left(-\beta \sum_{\alpha=1}^{N} \omega_\alpha \right) \left(\sum_{\gamma=1}^{N} \delta_{k,k_\gamma} \right) W^{(S)} \begin{pmatrix} k_1 k_2 \cdots k_N \\ k_1 k_2 \cdots k_N \end{pmatrix} \tag{46.11}$$

where the $W_N^{(S)}$ function is defined by Equations (37.5) and (37.6). We next compare this expression with Equation (37.3) for the grand partition function $Z_G = e^{\Omega f}$ and conclude that $\langle \mathsf{n}(k) \rangle$ can be derived from Z_G by a simple partial differentiation, that is,

$$\langle \mathsf{n}(k) \rangle = -\beta^{-1} e^{-\Omega f} \left[\frac{\partial}{\partial \omega(k)} e^{\Omega f} \right]_{W^{(S)}} \tag{46.12}$$

Here it is important to emphasize that all $W_N^{(S)}$ functions must be held fixed when the derivative is taken with respect to the free-particle energy $\omega(k)$.

To transform the prescription (46.12) to a form that is amenable for calculation, we recall that the Ursell equations (37.7) were written independently of the free-particle factor $\exp(-\beta \Sigma_{\alpha=1}^{N} \omega_\alpha)$ in Equation (37.3). Furthermore, the grand potential f, as given by Equation (37.2), is written explicitly in terms of the quantum cluster functions $U_N^{(S)}$. Therefore, Equation (46.12) can be simplified to the form

$$\langle \mathsf{n}(k) \rangle = -\beta^{-1} \left[\frac{\partial}{\partial \omega(k)} \Omega f \right]_{U^{(S)}} \tag{46.13}$$

where now all $U_N^{(S)}$ functions are to be held fixed when taking the partial derivative. But according to Equation (37.16) the quantum Ursell functions can be written as linear combinations of products of T functions, which are in turn related to Boltzmann cluster functions by Equation (37.14). Thus we may also write

$$\langle \mathsf{n}(k) \rangle = -\beta^{-1}\left[\frac{\partial}{\partial\omega(k)}\Omega f\right]_T \tag{46.13a}$$

Exercise 46.1 Show, using Equations (46.13) and (38.10), that the momentum distribution in an ideal gas is given by

$$\langle\langle \mathsf{n}(k)\rangle\rangle = \nu(k) = (e^{\beta(\omega_k - g)} - \varepsilon)^{-1}$$

where $\mathfrak{z} = e^{\beta g}$, in agreement with Equation (32.1).

We conclude this section by deriving an expression for the momentum distribution of an imperfect gas. Upon substituting Equation (38.2) for $\Omega f = \mathcal{P}\Omega/\kappa T$ into Equation (46.13a), we obtain

$$\begin{aligned}
\langle\langle \mathsf{n}(k_1)\rangle\rangle &= -\beta^{-1}\left[\frac{(2S+1)\Omega}{\lambda_T^3}\right]\sum_{N=1}^{\infty} e^{\beta N g}\left[\frac{\partial c_N}{\partial\omega(k_1)}\right]_T \\
&= e^{\beta(g-\omega_1)} + \varepsilon e^{2\beta(g-\omega_1)} + e^{3\beta(g-\omega_1)} \\
&\quad + [e^{\beta(g-\omega_1)} + 2\varepsilon e^{2\beta(g-\omega_l)}] \\
&\quad \times \sum_{k_2}[e^{\beta(g-\omega_2)} + \varepsilon e^{2\beta(g-\omega_2)}]T\begin{pmatrix} k_1k_2 \\ k_1k_2\end{pmatrix} \\
&\quad + \frac{1}{2}\varepsilon e^{\beta(g-\omega_1)}\sum_{k_2k_3} e^{\beta(g-\omega_2)} \\
&\quad \times e^{\beta(g-\omega_3)}\,T\begin{pmatrix} k_1k_2k_3 \\ k_1k_2k_3\end{pmatrix} + \cdots
\end{aligned} \tag{46.14}$$

where the second equality is derived by using Equations (38.8) and the invariance of the T functions under column permutations. We note that the first line in this second equality gives the first three terms in the expansion of the free-particle momentum distribution $\nu(k)$. Further investigation of the momentum distribution for real systems requires a general analysis starting from the basic expression (46.13). At the end of Appendix F we indicate how such an analysis can proceed by writing $\langle\langle \mathsf{n}(k)\rangle\rangle$ as a diagrammatic expansion. Alternatively, a direct perturbation-series expansion is given by Equations (50.25) and (50.22) in Section 50.

47 Pair Distribution Functions

Two-particle, or pair, distribution functions are studied for systems at equilibrium. The position-space case, when normalized to unity for large particle separations, is called the radial distribution function $D(r)$, and this quantity is investigated first. Theoretical calculations and results of measurements of $D(r)$ are both discussed. An examination of the pair distribution function in momentum space is also included.

The pair distribution function in position space, $P(\tilde{r}_1, \tilde{r}_2)$ is defined by the third of Equations (24.6).[11] On using the first of Equations (24.2) for the operator $\mathsf{n}(\mathbf{r})$, we obtain

$$P(\tilde{\mathbf{r}}_1, \tilde{\mathbf{r}}_2) = \operatorname{Tr}\left(\underline{\rho} \sum_{\alpha \neq \beta = 1}^{N} \delta^{(3)}(\tilde{\mathbf{r}}_1 - \mathbf{r}_\alpha)\, \delta^{(3)}(\tilde{\mathbf{r}}_2 - \mathbf{r}_\beta)\right) \tag{47.1}$$

where if we wish to consider spin effects, then we must also include in this expression factors $\delta_{\xi_\alpha, \tilde{\xi}_1}\, \delta_{\xi_\beta, \tilde{\xi}_2}$ for the spin coordinates ξ_α. It is convenient to use the coordinates $\mathbf{R} \equiv \frac{1}{2}(\tilde{\mathbf{r}}_1 + \tilde{\mathbf{r}}_2)$ and $\mathbf{r} \equiv \tilde{\mathbf{r}}_1 - \tilde{\mathbf{r}}_2$, instead of the coordinates $\tilde{\mathbf{r}}_1$ and $\tilde{\mathbf{r}}_2$. Since

$$\delta^{(3)}(\tilde{\mathbf{r}}_1)\, \delta^{(3)}(\tilde{\mathbf{r}}_2) = \delta^{(3)}(\mathbf{R})\, \delta^{(3)}(\mathbf{r})$$

we may then write the expression for $P(\tilde{\mathbf{r}}_1, \tilde{\mathbf{r}}_2)$ in the form

$$P(\tilde{\mathbf{r}}_1, \tilde{\mathbf{r}}_2) = \operatorname{Tr}\left(\underline{\rho} \sum_{\alpha \neq \beta = 1}^{N} \delta^{(3)}(\mathbf{r} - \mathbf{r}_\alpha + \mathbf{r}_\beta)\, \delta^{(3)}[\mathbf{R} - \frac{1}{2}(\mathbf{r}_\alpha + \mathbf{r}_\beta)]\right)$$

For a large system at equilibrium this average value cannot depend on the coordinate system with which we choose to evaluate it, and hence there can be no dependence in the terms of the trace on $\frac{1}{2}(\mathbf{r}_\alpha + \mathbf{r}_\beta)$. Alternatively, Equation (24.21a) shows that $P(\tilde{\mathbf{r}}_1, \tilde{\mathbf{r}}_2)$ does not depend on the coordinate $\mathbf{R}$. Thus by either argument we conclude that the δ function that involves $\mathbf{R}$ and which gives unity when integrated over all space, can be replaced effectively by Ω^{-1}. We arrive thereby at the expression

$$P(\tilde{\mathbf{r}}_1, \tilde{\mathbf{r}}_2) = \frac{1}{\Omega} \operatorname{Tr}\left(\underline{\rho} \sum_{\alpha \neq \beta = 1}^{N} \delta^{(3)}(\mathbf{r} - \mathbf{r}_\alpha + \mathbf{r}_\beta)\right) \tag{47.2}$$

For a fluid we can expect that as $\mathbf{r} \to \infty$, $P(\tilde{\mathbf{r}}_1, \tilde{\mathbf{r}}_2)$ will approach the uncoupled value n^2 [see the first of Equations (24.8)]. We now remove this normalization factor from the pair distribution function; the resulting quantity is called the *radial distribution function* $D(\mathbf{r})$. Thus

$$\begin{aligned} D(\mathbf{r}) &\equiv n^{-2} P(\tilde{\mathbf{r}}_1, \tilde{\mathbf{r}}_1 - \mathbf{r}) \\ &= (n^2\Omega)^{-1} \operatorname{Tr}\left(\underline{\rho} \sum_{\alpha \neq \beta = 1}^{N} \delta^{(3)}(\mathbf{r} - \mathbf{r}_\alpha + \mathbf{r}_\beta)\right) \xrightarrow[r \to \infty]{} 1 \end{aligned} \tag{47.3}$$

We also introduce here a closely related function $G(\mathbf{r})$, which is essentially the first term in definition (24.6) of the pair distribution function. On referring to Equations (47.1) and (47.2), we find

$$G(\mathbf{r}) \equiv n^{-2}\,\mathrm{Tr}(\underline{\rho}\mathrm{n}(\tilde{\mathbf{r}}_1)\mathrm{n}(\tilde{\mathbf{r}}_1 - \mathbf{r}))$$

$$G(\mathbf{r}) = (n^2\Omega)^{-1}\,\mathrm{Tr}\left(\underline{\rho} \sum_{\alpha,\beta=1}^{N} \delta^{(3)}(\mathbf{r} - \mathbf{r}_\alpha + \mathbf{r}_\beta)\right) \tag{47.4}$$

$$= n^{-1}\,\delta^{(3)}(\mathbf{r}) + D(\mathbf{r})$$

When spin effects are of interest, then the normalization of $D(r)$ and $G(r)$ should be changed by the replacements $n^{-1} \to (2S + 1)/n$, corresponding to the introduction of the spin-coordinate factors $\delta_{\xi_\alpha,\tilde{\xi}_1}\,\delta_{\xi_\beta,\tilde{\xi}_2}$. We return to further consideration of the *space correlation function* $G(\mathbf{r})$ in the next section.

The fact that $D(\mathbf{r}) \neq 1$ for a fluid is associated with the occurrence of *local density fluctuations* in any real system.[12] Local density fluctuations in a fluid are due both to statistical degeneracy and to particle interactions, and the region over which $D(\mathbf{r})$ differs from unity, that is, the region in which two-particle correlations are appreciable, is a measure of the extent of these fluctuations. For a classical ideal gas $D(\mathbf{r}) \equiv 1$.

Calculation of Pair Distribution Function

To calculate the pair distribution function in position space we may apply Equation (47.1). In analogy with Equation (46.11) we write, including spin effects and using the notation $r_i = (\mathbf{r}_i, \xi_i)$,

$$\begin{aligned} P(\tilde{r}_1, \tilde{r}_2) = e^{-\Omega f} \sum_{N=2}^{\infty} \frac{1}{N!} e^{\beta N g} \sum_{\xi_1 \cdots \xi_N} \int d^3\mathbf{r}_1 \cdots d^3\mathbf{r}_N \\ \times \sum_{\alpha \neq \beta = 1}^{N} \delta^{(3)}(\tilde{\mathbf{r}}_1 - \mathbf{r}_\alpha)\,\delta_{\tilde{\xi}_1,\xi_\alpha} \\ \times\, \delta^{(3)}(\tilde{\mathbf{r}}_2 - \mathbf{r}_\beta)\,\delta_{\tilde{\xi}_2,\xi_\beta}\, W_H^{(S)}\!\left(\begin{matrix} r_1 \cdots r_N \\ r_1 \cdots r_n \end{matrix}\right)\Bigg] \end{aligned} \tag{47.5}$$

Here the N-particle (Heisenberg-picture) function $W_{H,N}^{(S)}$ is defined, similarly to Equation (37.6) for $W_N^{(S)}$, by

$$W_H^{(S)}\!\left(\begin{matrix} r_1 \cdots r_N \\ r_1' \cdots r_N' \end{matrix}\right) \tag{47.6}$$

$$\equiv \sum_{\mathcal{P}'} \varepsilon^{P'} U_{P'} \langle r_1 r_2 \cdots r_N | e^{-\beta \mathrm{H}} | r_2' \cdots r_N' \rangle$$

This quantity is merely an expression for symmetrized matrix elements of the unnormalized statistical matrix in the position representation. By using this notation, the expression (47.5) for $P(\tilde{r}_1, \tilde{r}_2)$ can also be written as a functional derivative,[13]

$$P(\bar{r}_1, \bar{r}_2) = e^{-\Omega f}\left[\frac{\delta}{\delta \bar{r}_1}\frac{\delta}{\delta \bar{r}_2} e^{\Omega f}\right]_{W_H^{(S)}} \tag{47.7}$$

where the grand partition function (36.16) evaluated in the position representation for a single-component system is

$$Z_G = e^{\Omega f} = \sum_{N=0}^{\infty} \frac{1}{N!} e^{\beta N g} \sum_{\xi_1 \cdots \xi_N} \times \int d^3\mathbf{r}_1 \cdots d^3\mathbf{r}_N \, W_H^{(S)}\begin{pmatrix} r_1 \cdots r_N \\ r_1 \cdots r_N \end{pmatrix} \tag{47.8}$$

The $W_H^{(S)}$ functions are to be held constant when the functional derivative is taken. The notation in Equation (47.7) is intended to exclude repeated functional differentiation of a single coordinate, which would give $\alpha = \beta$ terms in Equation (47.1).

As in Section 46 we next refer to the Ursell equations (37.7) whose position-space version results in

$$\Omega f = \sum_{N=1}^{\infty} \frac{1}{N!} e^{\beta N g} \sum_{\xi_1 \cdots \xi_N} \times \int d^3\mathbf{r}_1 \cdots d^3\mathbf{r}_N \, U_H^{(S)}\begin{pmatrix} r_1 \cdots r_N \\ r_1 \cdots r_N \end{pmatrix} \tag{47.9}$$

which is equivalent to Equation (37.2) for the grand potential f. Then it is straightforward to verify that Equation (47.7) for $P(r_1, r_2)$ can also be written as

$$P(r_1, r_2) = \langle \mathsf{n}(r_1)\rangle\langle \mathsf{n}(r_2)\rangle + \left[\frac{\delta^2}{\delta r_1 \, \delta r_2} \Omega f\right]_{U_H^{(S)}} \tag{47.10}$$

in which the Ursell functions $U_H^{(S)}$ are to be held constant during functional differentiation and where

$$\langle \mathsf{n}(r)\rangle = \langle \mathsf{n}(\mathbf{r}, \xi)\rangle = e^{-\Omega f}\left[\frac{\delta}{\delta r} e^{\Omega f}\right]_{W_H^{(S)}} = \left[\frac{\delta}{\delta r}\Omega f\right]_{U_H^{(S)}} \tag{47.11}$$

Since $\langle \mathsf{n}(r)\rangle = (2S + 1)^{-1} n$ for a homogeneous fluid we obtain for the radial distribution function (47.3), in this case, the expression

$$D(\mathbf{r}, \xi_1, \xi_2) = 1 + \left(\frac{2S + 1}{n}\right)^2 \left[\frac{\delta^2}{\delta r_1 \, \delta r_2} \Omega f\right]_{U_H^{(S)}} \tag{47.12}$$

Clearly, the second term on the right side of this expression provides, theoretically, a measure of the magnitude and extent of density fluctuations in a fluid. If we are not interested in spin effects, then instead of this equation we must use the prescription

$$D(\mathbf{r}) = 1 + n^{-2}\left[\frac{\delta^2}{\delta \mathbf{r}_1\, \delta \mathbf{r}_2}\Omega f\right]_{U_H^{(S)}} \tag{47.12a}$$

Imperfect Gas

We illustrate the use of Equation (47.12a) by calculating $D(\mathbf{r})$ for an imperfect gas. Inasmuch as $\Omega f = \Omega \mathcal{P}/\kappa T$, we obtain with the aid of Equations (38.2), (38.5), (38.6), and (38.18),

$$\begin{aligned} D(\mathbf{r}) &= 1 + \left(\frac{2S+1}{n^2 \lambda_T^3}\right)\Omega \sum_{N=2}^{\infty} \mathfrak{z}^N \left[\frac{\delta^2 c_N}{\delta \mathbf{r}_1\, \delta \mathbf{r}_2}\right]_{U_H^{(S)}} \\ &\cong 1 + D_2(\mathbf{r}) \end{aligned} \tag{47.13}$$

where the two-particle ($N = 2$) contribution to $D(\mathbf{r})$ is given by

$$D_2(\mathbf{r}) \equiv -\left(\frac{\lambda_T^3 \Omega}{2S+1}\right)\left[\frac{\delta^2}{\delta \mathbf{r}_1\, \delta \mathbf{r}_2}(a_2^{(0)} + a_2^{(V)}(\beta))\right]_{U_H^{(S)}} \tag{47.14}$$

For an imperfect gas we can expect generally that $D_2(\mathbf{r})$ will give an accurate measure of density fluctuations. Because of the functional differentiation that appears in this last definition we cannot use Equation (38.19) for the free-particle exchange term $a_2^{(0)}$. Instead, we must use the expression[14]

$$\begin{aligned} a_2^{(0)} &= -\frac{1}{2}\frac{\varepsilon}{\Omega}\left(\frac{\lambda_T^3}{2S+1}\right)\sum_{\xi_1,\xi_2} \\ &\quad \times \int d^3\mathbf{r}_1\, d^3\mathbf{r}_2\, \langle r_1|e^{-\beta \mathsf{K}}|r_2\rangle\langle r_2|e^{-\beta \mathsf{K}}|r_1\rangle \\ &= -\frac{1}{2}\left(\frac{\varepsilon}{\Omega \lambda_T^3}\right)\int d^3\mathbf{r}_1'\, d^3\mathbf{r}_2' \exp\left[-\left(\frac{2\pi}{\lambda_T^2}\right)(\mathbf{r}_1' - \mathbf{r}_2')^2\right] \end{aligned} \tag{47.15}$$

The integrand in this equation is the cluster-function product $\lambda_T^6 U_{H,1}\, U_{H,1}$, which must be held constant for the derivative in Equation (47.14).

In similar fashion we rewrite Equation (38.20) for $a_2^{(V)}$, with the aid of Equation (38.17) and Exercise 38.3, as

$$\begin{aligned} a_2^{(V)}(\beta) &= -\frac{1}{2}\left(\frac{\lambda_T^3}{2S+1}\right)\frac{1}{\Omega}\sum_{\xi_1,\xi_2}\int d^3\mathbf{r}_1\, d^3\mathbf{r}_2 \\ &\quad \times \{\langle r_1 r_2|e^{-\beta \mathsf{H}} - e^{-\beta \mathsf{H}_0}|r_1 r_2\rangle + \varepsilon\langle r_1 r_2|e^{-\beta \mathsf{H}} - e^{-\beta \mathsf{H}_0}|r_2 r_1\rangle\} \end{aligned} \tag{47.16a}$$

$$a_2^{(V)} = -\frac{\sqrt{2}}{\Omega}\int d^3\mathbf{r}_1' \, d^3\mathbf{r}_2' \times \{(2S+1)\langle\mathbf{r}_{12}'|(e^{-\beta\mathsf{H}} - e^{-\beta\mathsf{H}_o})_{\text{rel}}|\mathbf{r}_{12}'\rangle + \varepsilon\langle\mathbf{r}_{12}'|(e^{-\beta\mathsf{H}} - e^{-\beta\mathsf{H}_o})_{\text{rel}}|\mathbf{r}_{21}'\rangle\} \tag{47.16b}$$

where $\mathsf{H}_o = \mathsf{K}$. To derive the second line we have assumed a spin-independent two-particle interaction. We have also used the expression

$$\langle\mathbf{R}|(e^{-\beta\mathsf{K}})_{\text{CM}}|\mathbf{R}'\rangle = \frac{2\sqrt{2}}{\lambda_T^3}\exp\left[-\frac{2\pi}{\lambda_T^2}(\mathbf{R}-\mathbf{R}')^2\right] \xrightarrow[\mathbf{R}'=\mathbf{R}]{} \frac{2\sqrt{2}}{\lambda_T^3}$$

We note that the factor Ω^{-1} that appears in Equation (47.16), but not in Equation (38.20), is equivalent to a δ function involving the coordinate $\mathbf{R}$, as is discussed above Equation (47.2).

Exercise 47.1 Verify Equations (47.13) to (47.16).

We now substitute Equations (47.15) and (47.16) into Equation (47.14), use the fact that either of the primed variables $\mathbf{r}_1'$ and $\mathbf{r}_2'$ in the integrals can be labeled $\mathbf{r}_1$ or $\mathbf{r}_2$, and obtain the exact expression

$$\begin{aligned} D_2(\mathbf{r}) &= \left(\frac{\varepsilon}{2S+1}\right)\exp\left(-\frac{2\pi r^2}{\lambda_T^2}\right) \\ &\quad + 2\sqrt{2}\,\lambda_T^3\,\langle\mathbf{r}|(e^{-\beta\mathsf{H}} - e^{-\beta\mathsf{K}})_{\text{rel}}|\mathbf{r}\rangle \\ &\quad + \left(\frac{2\sqrt{2}\lambda_T^3\varepsilon}{2S+1}\right)\langle\mathbf{r}|(e^{-\beta\mathsf{H}} - e^{-\beta\mathsf{K}})_{\text{rel}}|-\mathbf{r}\rangle \\ &= 2\sqrt{2}\lambda_T^3\left\{\langle\mathbf{r}|(e^{-\beta\mathsf{H}})_{\text{rel}}|\mathbf{r}\rangle + \left(\frac{\varepsilon}{2S+1}\right)\langle\mathbf{r}|(e^{-\beta\mathsf{H}})_{\text{rel}}|-\mathbf{r}\rangle\right\} - 1 \end{aligned} \tag{47.17}$$

Here we have again used Equation (28.12), this time with $\lambda_T^2 \to 2\lambda_T^2$ for relative coordinates. If finally we consider the classical limit, as in Equation (38.21), then we obtain for $D(\mathbf{r}) \cong 1 + D_2(\mathbf{r})$ a well-known expression for imperfect gases,

$$D(\mathbf{r}) \cong e^{-\beta V(\mathbf{r})}\left[1 + \left(\frac{\varepsilon}{2S+1}\right)\exp\left(-\frac{2\pi r^2}{\lambda_T^2}\right)\right]$$

$$D(\mathbf{r}) \cong e^{-\beta V(\mathbf{r})} \qquad (\text{for } \lambda_T \ll r_o \ll r) \tag{47.18}$$

We previously used a multicomponent version of this result as Equation (40.17).

Exercise 47.2 Show, using Equations (38.18) to (38.20) and (47.17), that the second virial coefficient is given quite generally by an integral over $D_2(\mathbf{r})$, that is,

$$a_2(\beta) = -\left(\frac{2S+1}{2\lambda_T^3}\right)\int d^3\mathbf{r}\, D_2(\mathbf{r}) \tag{47.19}$$

Measurements of D(r) for Liquids

The scattering of X-rays and neutrons by liquids provides a means for measuring the radial distribution function, as we explain in the next section. In Figure X.2 we show curves of $D(r)$ at two different temperatures for liquid argon, a particularly simple substance to interpret because no interference effects due to ions or diatomic and polyatomic molecules occur.

The liquid argon curves show a behavior that one might expect. Thus they show that there is a high probability for two particles to be at a distance $r \cong \ell = 3.42$ Å, where ℓ is the interparticle spacing. Because of the tendency for this "lattice structure" to persist at larger distances, there is a second, less pronounced maximum in $D(r)$ corresponding to second nearest neighbors. However, since a classical liquid does not exhibit long-range order, we expect and observe that $D(r) \to 1$ as $r \to \infty$. We also see that there is zero probability that two argon atoms will occupy the same space; in fact, $D(r) \cong 0$ for $r \lesssim (0.8)(3.42) \cong 2.5$ Å. Finally, it is clear that these effects all tend to disappear as the temperature is increased.

The radial distribution function has also been measured for liquid helium

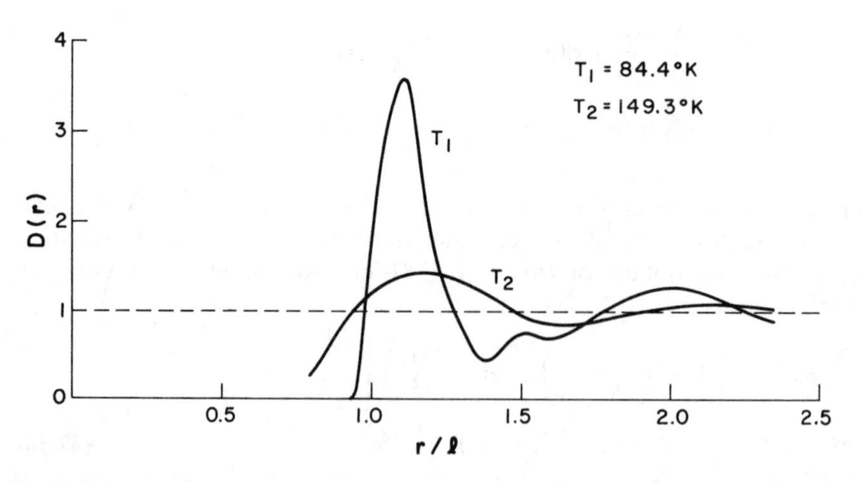

FIGURE X.2 Experimental curves for the radial distribution function of liquid argon at two different temperatures.[15] The interparticle spacing ℓ is 3.42 Å for liquid argon.

four over a range of temperatures above and below the λ point.[16] The results are quite similar to the lower-temperature liquid argon observations, and they show no essential difference between liquid He I and liquid He II. This fact is consistent with the interpretation, given at the end of Section 34, that the λ point is due to momentum-space ordering (that is, to Bose–Einstein condensation), and not to any peculiar behavior in position space.

Pair Distribution Function in Momentum Space

We now investigate the momentum correlations that occur between two particles in a system. In analogy with the derivation of Equation (46.12), we can derive from Equations (24.14) and (24.10) an expression for the pair distribution function in momentum space; namely,

$$
\begin{aligned}
P(k, k') &= e^{-\Omega f} \sum_{N=2}^{\infty} \frac{1}{N!} e^{\beta N g} \sum_{k_1 \cdots k_N} \\
&\quad \times \exp\left(-\beta \sum_{\alpha=1}^{N} \omega_\alpha\right)\left(\sum_{\gamma \neq \delta=1}^{N} \delta_{k,k_\gamma}\, \delta_{k',k_\delta}\right) W^{(S)}\begin{pmatrix} k_1 k_2 \cdots k_N \\ k_1 k_2 \cdots k_N \end{pmatrix} \qquad \textbf{(47.20)} \\
&= \beta^{-2} e^{-\Omega f} e^{-\beta\omega(k)} \left[\frac{\partial}{\partial \omega(k')} e^{\beta\omega(k)} \frac{\partial}{\partial \omega(k)} e^{\Omega f}\right]_{W^{(S)}} \\
&= \beta^{-2} e^{-\beta\omega(k)} \left[\frac{\partial}{\partial \omega(k')} e^{\beta\omega(k)} \frac{\partial}{\partial \omega(k)} \Omega f\right]_{U^{(S)}} \\
&\quad + \beta^{-2}\left[\frac{\partial}{\partial \omega(k')} \Omega f\right]_{U^{(S)}} \left[\frac{\partial}{\partial \omega(k)} \Omega f\right]_{U^{(S)}} \\
&= \langle \mathrm{n}(k)\rangle\langle \mathrm{n}(k')\rangle + \beta^{-2} e^{-\beta\omega(k)} \left[\frac{\partial}{\partial \omega(k')} e^{\beta\omega(k)} \frac{\partial}{\partial \omega(k)} \Omega f\right]_T
\end{aligned}
$$

where we have used Equation (46.13) in the final equality. Detailed examination of this expression can yield information about two-particle correlations in momentum space.

Let us consider generally the effect of the two partial derivatives in the final equality of Equation (47.20). On referring to Equations (37.2) and (37.15) or (37.16), we can see that two cases arise in general. In the first case the two partial derivatives involve $e^{-\beta\omega_i}$ factors associated with different T functions.[17] It can be shown, using the method described at the end of Appendix F to derive Equation (F.13), that this case results (for a translationally invariant system) in $\varepsilon\delta_{k,k'}$ times the first term in the final equality above. In the second case the two partial derivatives involve $e^{-\beta\omega_i}$ factors associated with the same T function.[18] This case is due primarily to momentum-space correlations associated with particle interactions, which vanish for an ideal gas and which, in the notation of Appendix F, are associated with primary 2 graphs. Here we denote this second case by the subscript $T^{(V)}$ and obtain thereby the general expression[19]

$$P(k, k') = \langle \mathsf{n}(k)\rangle\langle \mathsf{n}(k')\rangle(1 + \varepsilon\delta_{k,k'}) + \beta^{-2} e^{-\beta\omega(k)}\left[\frac{\partial}{\partial\omega(k')} e^{\beta\omega(k)}\frac{\partial}{\partial\omega(k)}\Omega f\right]_{T^{(V)}} \tag{47.21}$$

Exercise 47.3

(a) For an ideal gas show in analogy with the derivation of Equation (31.16) that

$$\begin{aligned}\langle[\Delta \mathsf{n}_i]^2\rangle &= \langle \mathsf{n}_i^2\rangle - \langle \mathsf{n}_i\rangle^2 = \beta^{-2}\frac{\partial^2}{\partial\omega_i^2}\ln Z \\ &= \langle \mathsf{n}_i\rangle(1 + \varepsilon\langle \mathsf{n}_i\rangle)\end{aligned} \tag{47.22}$$

(b) Use the answer to part (a) to show that Equation (47.21) and results of Problem V.3(a) and (b) are consistent for an ideal gas.

It is also straightforward to determine a general expression in momentum space for the reduced density matrix ρ_2 of Equations (24.15). We do this for the ideal-gas case only. If we recall Equation (24.17) for the diagonal case, then on examination of the result (47.21) it is quite easy to guess the correct expression

$$\langle k_1k_2|\rho_2|k_1'k_2'\rangle = \nu(k_1)\nu(k_2)[\delta_{k_1,k_1'}\,\delta_{k_2,k_2'} + \varepsilon\,\delta_{k_1,k_2'}\,\delta_{k_2,k_1'}] \tag{47.23}$$

where $\nu(k) = \langle \mathsf{n}_k\rangle$ is given by Equation (32.1). We now use this result to derive an explicit relation for the radial distribution function of a degenerate ideal gas. The procedure, which uses Equation (24.21), exhibits a second method for calculating $D(r)$.

We substitute Equation (47.23) into Equation (24.21a), with spin effects included so that $k_i = (\mathbf{k}_i, m_i)$, and obtain

$$\begin{aligned}D(\mathbf{r}, \xi_1, \xi_2) &= \left(\frac{2S + 1}{n\Omega}\right)^2 \sum_{k_1k_2k_1'k_2'} e^{i(\mathbf{k}_1 - \mathbf{k}_1')\cdot\mathbf{r}} \\ &\qquad \chi_{m_1}(\xi_1)\chi_{m_2}(\xi_2)\chi^\dagger_{m_1'}(\xi_1)\chi^\dagger_{m_2'}(\xi_2)\langle k_1k_2|\rho_2|k_1'k_2'\rangle \\ &\underset{V=0}{\rightarrow} \left(\frac{2S + 1}{n\Omega}\right)^2 \sum_{k_1k_2} \nu(k_1)\nu(k_2)\chi_{m_1}(\xi_1)\chi_{m_2}(\xi_2) \\ &\qquad \times [\chi^\dagger_{m_1}(\xi_1)\chi^\dagger_{m_2}(\xi_2) + \varepsilon\, e^{i(\mathbf{k}_1 - \mathbf{k}_2)\cdot\mathbf{r}}\,\chi^\dagger_{m_1}(\xi_2)\chi^\dagger_{m_2}(\xi_1)] \\ &= \left(\frac{2S + 1}{n\Omega}\right)^2 \sum_{\mathbf{k}_1,\mathbf{k}_2} \nu(k_1)\nu(k_2)[1 + \varepsilon\,\delta_{\xi_1,\xi_2}\, e^{i\mathbf{q}\cdot\mathbf{r}}]\end{aligned} \tag{47.24}$$

where the $\chi_m(\xi)$ are a complete set of spin wave functions and $\mathbf{q} \equiv \mathbf{k}_1 - \mathbf{k}_2$. The first term in the brackets of the third equality gives 1 after summing over all states. Therefore, in the thermodynamic limit this ideal-gas expression becomes

$$D(\mathbf{r}, \xi_1, \xi_2) = 1 + \varepsilon\, \delta_{\xi_1,\xi_2}\left(\frac{2S+1}{n}\right)^2 \left[(2\pi)^{-3}\int d^3\mathbf{k}\, \nu(\mathrm{k})\, e^{i\mathbf{k}\cdot\mathbf{r}}\right]^2$$
$$= 1 + \varepsilon\, \delta_{\xi_1,\xi_2}\left(\frac{2S+1}{n}\right)^2 \left[(2\pi^2 r)^{-1}\int_0^\infty k\, dk\, \nu(k) \sin kr\right]^2 \tag{47.25}$$

Exercise 47.4 Evaluate the expression (47.25) explicitly for an ideal Fermi gas at zero temperature and, using Equation (19.2), which gives the density n in terms of k_F, verify the result

$$D(\mathbf{r}, \xi_1, \xi_2) = 1 - 9\, \delta_{\xi_1,\xi_2}\left[\frac{\sin k_F r - k_F r \cos k_F r}{(k_F r)^3}\right]^2 \tag{47.26}$$
$$\xrightarrow[k_F r \ll 1]{} 1 - \delta_{\xi_1,\xi_2}$$

The small-r limit shows, in agreement with the Pauli exclusion principle, that free fermions at $T = 0$ experience an effective repulsion when they have the same spin coordinate. Free fermions with opposite spin coordinates are uncorrelated.

48 Structure Factor and Linear Response Function

In this section we introduce the Fourier transforms of the radial distribution function and of the space–time correlation function $G(\mathbf{r}, t)$. These transforms are called the structure factor $S(\mathbf{q})$ and the linear response function $R(\mathbf{q}, \omega)$, respectively, and we indicate how they are measured for a fluid by inelastic scattering experiments. The space–time correlation function is examined in some detail.

We begin by introducing the Fourier transform of the space correlation function $G(\mathbf{r})$ of Equation (47.4). We omit spin effects and include a factor n, so that the *structure factor* $S(\mathbf{q})$ is defined to be

$$S(\mathbf{q}) \equiv n\int d^3\mathbf{r}\, e^{i\mathbf{q}\cdot\mathbf{r}} G(\mathbf{r})$$
$$S(\mathbf{q}) = 1 + n\int d^3\mathbf{r}\, e^{i\mathbf{q}\cdot\mathbf{r}} D(\mathbf{r}) \tag{48.1}$$

We also introduce the so-called inelastic-part $S'(\mathbf{q})$ of the structure factor for a fluid by the definition

$$S'(\mathbf{q}) \equiv S(\mathbf{q}) - (2\pi)^3 n\, \delta^{(3)}(\mathbf{q}) \\ = 1 + n \int d^3\mathbf{r}\, e^{i\mathbf{q}\cdot\mathbf{r}}[D(\mathbf{r}) - 1] \tag{48.2}$$

Clearly, $[S'(\mathbf{q}) - 1]$ is a measure of the effect of two-particle correlations in a fluid, because $[D(\mathbf{r}) - 1]$ is such a measure.

For an isotropic fluid, inversion of Equation (48.2) gives

$$D(r) - 1 = \frac{1}{(2\pi)^3 n} \int d^3\mathbf{q}\, e^{-i\mathbf{q}\cdot\mathbf{r}}[S'(\mathbf{q}) - 1] \\ = (2\pi^2 nr)^{-1} \int_0^\infty q\, dq\, \sin qr[S'(q) - 1] \tag{48.3}$$

The radial distribution function for fluids is, in fact, determined from scattering experiments by using Eq. (48.3). After removal of the elastic scattering cross section, which corresponds (approximately) to the term $(2\pi)^3 n\, \delta^{(3)}(\mathbf{q})$ in Equation (48.2),[20] the remaining data, normalized to unity for large q values, give $S'(q)$. In these experiments $\hbar q$ is the momentum transfer to the fluid in the scattering process. In Figure X.3, we show results of neutron-scattering measurements of $S'(q)$ for liquid helium four under its normal vapor pressure. Fourier inversion of these curves, which show a small difference between He I and He II in the peak near $q = 2$ Å, gives the radial distribution function $D(r)$ discussed below Figure X.2 in Section 47.

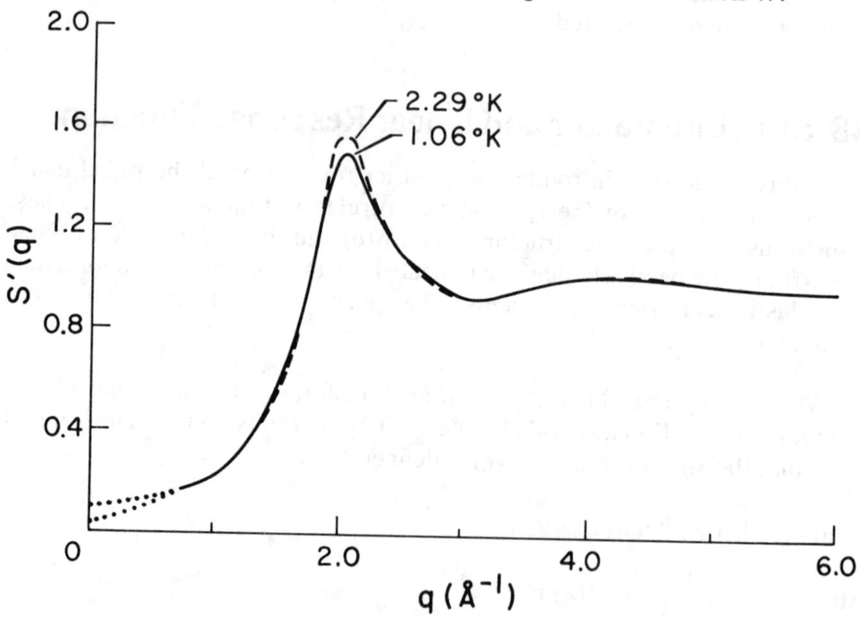

FIGURE X.3 The liquid structure factor $S'(q)$ for liquid helium four under its normal vapor pressure, above and below the λ point.[21]

When the second equality of Equation (47.4) is substituted into Equation (48.1), then we find for $S(q)$ the theoretical expression,

$$S(\mathbf{q}) = \frac{1}{\langle \mathsf{N} \rangle} \sum_{\mu} \langle \mu | \underline{\rho} | \mu \rangle \sum_{\alpha,\beta=1}^{N} \langle \mu | \exp[i\mathbf{q} \cdot (\mathbf{r}_\alpha - \mathbf{r}_\beta) | \mu \rangle]$$
$$= \frac{1}{\langle \mathsf{N} \rangle} \sum_{\mu,\nu} \langle \mu | \underline{\rho} | \mu \rangle \left| \langle \nu | \sum_{\alpha=1}^{N} \exp(i\mathbf{q} \cdot \mathbf{r}_\alpha | \mu \rangle \right|^2 \tag{48.4}$$

where μ and ν are many-body states for the system of interest. We call

$$\underline{\rho}_{\mathbf{q}} \equiv \sum_{\alpha=1}^{N} e^{i\mathbf{q} \cdot \mathbf{r}_\alpha} \tag{48.5}$$

the *density-fluctuation operator,* since it is the Fourier transform of the density operator $\mathsf{n}(r)$, Equation (24.2), and it is the operator whose matrix elements squared provide, via Equation (48.4), a measure of density fluctuations.

Exercise 48.1

(a) Use Equation (17.12) to show that the Fock-space representation of $\underline{\rho}_{\mathbf{q}}$ is given by

$$\underline{\rho}_{\mathbf{q}} = \sum_{k} \mathsf{a}^{\dagger}_{k+q} \mathsf{a}_k \tag{48.5a}$$

where $q = (\mathbf{q}, 0)$. The $\mathbf{q} = 0$ limit of both Equations (48.5) and (48.5a) gives $\underline{\rho}_0 = \mathsf{N}$, where N is the total-particle-number operator (17.6).

(b) For a fluid show that the $\mathbf{q} = 0$ limit of Equation (48.4) is

$$S(0) = \frac{\langle \mathsf{N}^2 \rangle}{\langle \mathsf{N} \rangle} = \langle \mathsf{N} \rangle + n\kappa T \cdot \kappa_T$$

where to derive the second equality we have used Equation (36.21). According to Equation (46.9a) we can also write

$$\langle N \rangle = (2\pi)^3 \, n \, \delta^{(3)} \, (0)$$

Equations (46.8) and (46.9) can be used to identify the $\nu = \mu$ term in Equation (48.4) for an infinite fluid. This leaves for $S'(q)$ the theoretical expression,

$$S'(\mathbf{q}) = \frac{1}{\langle \mathsf{N} \rangle} \sum_{\substack{\mu,\nu \\ (\mu \neq \nu)}} \langle \mu | \underline{\rho} | \mu \rangle | \langle \nu | \underline{\rho}_{\mathbf{q}} | \mu \rangle |^2 + (2\pi)^3 n \, \delta^{(3)}(\mathbf{q}) \left(\frac{\langle \mathsf{N}^2 \rangle}{\langle \mathsf{N} \rangle^2} - 1 \right)$$
$$\xrightarrow[q \to 0]{} n\kappa T \cdot \kappa_T = \frac{1}{\langle \mathsf{N} \rangle} (\Delta N)^2_{\text{rms}} \tag{48.6}$$

where the $\mathbf{q} = 0$ limit follows from Exercise 48.1(b) above. This limit is indicated for liquid helium four in Figure X.3 by the dashed lines, which are extrapolations of the neutron scattering data.

Linear Response Function

We return now to Equation (46.7) for the cross section of a scattering experiment in which a weakly interacting probe particle is used. It is easy to generalize this expression to the differential scattering cross section $d^2\sigma/d\Omega\, d\omega$ in which one measures both a momentum transfer $\hbar\mathbf{q}$ and an energy transfer ω to the fluid. Thus we find

$$\frac{d^2\sigma}{d\Omega\, d\omega} = \langle \mathrm{N}\rangle a_s^2\left(\frac{k}{k_o}\right)R(\mathbf{q}, \omega) \tag{48.7}$$

where the *linear response function* $R(\mathbf{q}, \omega)$ is defined by

$$R(\mathbf{q}, \omega) \equiv \frac{1}{\langle \mathrm{N}\rangle}\sum_{\mu,\nu} \langle \mu|\underline{\rho}|\mu\rangle\, \delta(E_\mu + \omega - E_\nu)|\langle \nu|\underline{\rho}_{\mathbf{q}}|\mu\rangle|^2 \tag{48.8}$$

E_μ and E_ν being fluid energies.[22] For low-energy neutron scattering experiments the energy ω is related to k and k_o by $\omega = (\hbar^2/2m)(k_o^2 - k^2)$.

According to the discussion of Equation (48.6), we can separate the elastic and inelastic parts of $R(\mathbf{q}, \omega)$ as

$$\begin{aligned} R(\mathbf{q}, \omega) &= (2\pi)^3 n\, \delta^{(3)}(\mathbf{q})\, \delta(\omega) \cdot \frac{\langle \mathrm{N}^2\rangle}{\langle \mathrm{N}\rangle^2} \\ &\quad + \frac{1}{\langle \mathrm{N}\rangle}\sum_{\mu\neq\nu} \langle \mu|\underline{\rho}|\mu\rangle\, \delta(E_\mu + \omega - E_\nu)|\langle \nu|\underline{\rho}_{\mathbf{q}}|\mu\rangle|^2 \end{aligned} \tag{48.8a}$$

It is also easy to see that $S(\mathbf{q})$ and $S'(\mathbf{q})$, of Equations (48.4) and (48.2), are related to $R(\mathbf{q}, \omega)$ by the simple integrals,

$$\begin{aligned} S(\mathbf{q}) &= \int_{-\infty}^{+\infty} d\omega\, R(\mathbf{q}, \omega) \\ S'(\mathbf{q}) &= \int_{-\infty}^{+\infty} d\omega\, [R(\mathbf{q}, \omega) - (2\pi)^3 n\, \delta^{(3)}(\mathbf{q})\, \delta(\omega)] \end{aligned} \tag{48.9}$$

Thus the linear response function is seen to be a convenient generalization of the structure factor. Moreover, Equation (48.7) shows that $S'(\mathbf{q})$ can indeed be determined from scattering cross-section measurements, as we have indicated previously in our discussions of $D(\mathbf{r})$ and $S(\mathbf{q})$.[23]

Space–Time Correlation Function

We next introduce the Fourier transform in both space and time of the linear response function by the equation

$$R(\mathbf{q}, \omega) = \frac{n}{2\pi\hbar}\int_{-\infty}^{+\infty} dt\, e^{-i\omega t/\hbar} \int d^3\mathbf{r}\, e^{i\mathbf{q}\cdot\mathbf{r}}\, G(\mathbf{r}, t) \qquad \mathbf{(48.10)}$$

The quantity $G(r, t)$ is the *space–time correlation function* for a many-body system, as we show after inverting this defining equation to obtain

$$\begin{aligned} n^2G(\mathbf{r}, t) &= \frac{n}{(2\pi)^3}\int_{-\infty}^{+\infty} d\omega\, e^{i\omega t/\hbar} \int d^3\mathbf{q}\, e^{-i\mathbf{q}\cdot\mathbf{r}}\, R(\mathbf{q}, \omega) \\ &= \frac{1}{(2\pi)^3\Omega}\sum_{\mu,\nu} \langle\mu|\underline{\rho}|\mu\rangle\, e^{i(E_\nu - E_\mu)t/\hbar} \int d^3\mathbf{q}\, e^{-i\mathbf{q}\cdot\mathbf{r}} |\langle\nu|\underline{\rho}_{\mathbf{q}}|\mu\rangle|^2 \\ &= \frac{1}{(2\pi)^3\Omega}\sum_{\mu} \langle\mu|\underline{\rho}|\mu\rangle \sum_{\alpha,\beta=1}^{N} \int d^3\mathbf{q}\, e^{-i\mathbf{q}\cdot\mathbf{r}}\langle\mu| e^{-i\mathbf{q}\cdot\mathbf{r}_\beta}\, e^{i\mathsf{H}t/\hbar}\, e^{i\mathbf{q}\cdot\mathbf{r}_\alpha}\, e^{-i\mathsf{H}t/\hbar}|\mu\rangle \end{aligned}$$

where H is the Hamiltonian of the system. Here the matrix element involving the density-fluctuation operators can be considered to be evaluated in the Heisenberg picture. In this picture we define time-dependent operators $\mathsf{O}(t)$ by the operator equation

$$\mathsf{O}(t) \equiv e^{i\mathsf{H}t/\hbar}\, \mathsf{O}\, e^{-i\mathsf{H}t/\hbar} \qquad (48.12)$$

It is then straightforward to show for an arbitrary time T that $n^2G(\mathbf{r}, t)$ can be written in the form

$$\begin{aligned} n^2G(\mathbf{r}, t) &= \frac{1}{(2\pi)^3\Omega}\sum_{\mu} \langle\mu|\underline{\rho}|\mu\rangle \sum_{\alpha,\beta=1}^{N} \\ &\quad \times \int d^3\mathbf{q}\, e^{-i\mathbf{q}\cdot\mathbf{r}}\langle\mu| e^{-i\mathbf{q}\cdot\mathbf{r}_\alpha(T)} e^{i\mathbf{q}\cdot\mathbf{r}_\beta(T+t)}|\mu\rangle \\ &= \frac{1}{(2\pi)^6\Omega}\int d^3\mathbf{R} \sum_{\mu} \langle\mu|\underline{\rho}|\mu\rangle \sum_{\alpha,\beta=1}^{N} \\ &\quad \times \int d^3\mathbf{q}_1\, d^3\mathbf{q}_2\, e^{-i\mathbf{q}_2\cdot\mathbf{r}}\, e^{i(\mathbf{q}_1-\mathbf{q}_2)\cdot\mathbf{R}}\langle\mu| e^{-i\mathbf{q}_1\cdot\mathbf{r}_\alpha(T)}\, e^{i\mathbf{q}_2\cdot\mathbf{r}_\beta(T+t)}|\mu\rangle \\ &= \frac{1}{\Omega}\int d^3\mathbf{R}\langle\mu|\underline{\rho}|\mu\rangle \sum_{\alpha,\beta=1}^{N} \\ &\quad \times \langle\mu|\delta^{(3)}(\mathbf{R1} - \mathbf{r}_\alpha(T))\, \delta^{(3)}(\mathbf{R1} + \mathbf{r1} - \mathbf{r}_\beta(T+t))|\mu\rangle \end{aligned}$$

where $\mathbf{r}_\beta(T + t)$ and $\mathbf{r}_\alpha(T)$ are noncommuting operators in general.[24]

Apart from the $\Omega^{-1}\int d^3\mathbf{R}$ integration, the final equality for $n^2G(\mathbf{r}, t)$ in Equation (48.13) gives the probability that if one particle is at position $\mathbf{R}$ at time T, then another (or the same) particle will be at position $(\mathbf{R} + \mathbf{r})$ at time $(T + t)$. Clearly, this probability, calculated quantum mechanically for a mixed state, cannot depend on either T or $\mathbf{R}$ for a large fluid system. Therefore, we can replace $\Omega^{-1}\int d^3\mathbf{R}$ by unity, and $G(\mathbf{r}, t)$ can be identified properly as a space–time correlation function, that is,

$$\mathbf{G}(\mathbf{r}, t) = \frac{1}{n^2}\left\langle \sum_{\alpha=1}^{N} \delta^{(3)}(\mathbf{R1} - \mathbf{r}_\alpha(T)) \cdot \sum_{\beta=1}^{N} \delta^{(3)}(\mathbf{R1} + \mathbf{r1} - \mathbf{r}_\beta(T + t)) \right\rangle \quad \textbf{(48.14)}$$

$$= \frac{1}{n^2}\left\langle \sum_{\alpha=1}^{N} \delta^{(3)}(\mathbf{R1} - \mathbf{r}_\alpha(0)) \cdot \sum_{\beta=1}^{N} \delta^{(3)}(\mathbf{R1} + \mathbf{r1} - \mathbf{r}_\beta(t)) \right\rangle$$

$$\xrightarrow[t\to 0]{} G(\mathbf{r}) = \frac{1}{n}\delta^{(3)}(\mathbf{r}) + D(\mathbf{r})$$

At $t = 0$ the operators $\mathbf{r}_\alpha(t)$ can be treated as classical position variables, and $G(\mathbf{r}, t)$ reduces to the space–correlation function $G(\mathbf{r})$ introduced by Equation (47.4).

Exercise 48.2 Show that

$$G(-\mathbf{r}, -t) = G^*(\mathbf{r}, t) \quad (48.15)$$

When $t \neq 0$ the space–time correlation function will not, in general, be a real quantity.

Some indication as to the difficulty of calculating the linear response function or the space–time correlation function in terms of the basic particle interactions can be given. We consider Equation (48.11) for $G(\mathbf{r}, t)$ of a fluid, use Equation (48.5a) for $\rho_\mathbf{q}$, and evaluate the sums over many-body states in the free-particle representation. As with the derivation of Equation (28.3) or (37.4), we obtain[25]

$$G(\mathbf{r}, t) = \langle \mathsf{N}\rangle^{-2} e^{-\Omega f} \sum_{N=0}^{\infty} \frac{1}{N!} e^{\beta N g} \sum_{k,k',\mathbf{q}} e^{-i\mathbf{q}\cdot\mathbf{r}} \sum_{k_1\cdots k_N} \sum_{\mathcal{P}} \varepsilon^P$$
$$\times \langle k_1 \cdots k_N | \mathsf{a}_k^\dagger \mathsf{a}_{k+\mathbf{q}}\, e^{i\mathsf{H}t/\hbar} \mathsf{a}_{k'+\mathbf{q}}^\dagger \mathsf{a}_{k'}\, e^{-(\beta + it/\hbar)\mathsf{H}} U_P | k_1 \cdots k_N\rangle$$
$$= \langle \mathsf{N}\rangle^{-2} e^{-\Omega f} \sum_{N=0}^{\infty} \frac{1}{(N!)^2} e^{\beta N g} \sum_{k,k',\mathbf{q}} e^{-i\mathbf{q}\cdot\mathbf{r}} \sum_{k_1\cdots k_N} \sum_{k'_1\cdots k'_N} \sum_{\mathcal{P},\mathcal{P}'} \varepsilon^{P+P'}$$
$$\times \langle k_1 \cdots k_N | \mathsf{a}_k^\dagger \mathsf{a}_{k+\mathbf{q}}\, e^{i\mathsf{H}t/\hbar} U_{P'} | k'_1 \cdots k'_N\rangle$$
$$\times \langle k'_1 \cdots k'_N | \mathsf{a}_{k'}^\dagger \mathsf{a}_{k'-\mathbf{q}}\, e^{-(\beta + it/\hbar)\mathsf{H}} U_P | k_1 \cdots k_N\rangle$$
$$= \langle \mathsf{N}\rangle^{-2} \sum_{k_1,k'_1,\mathbf{q}} e^{-i\mathbf{q}\cdot\mathbf{r}} \langle k_1 + \mathbf{q}, k'_1 - \mathbf{q} | \mathbf{R}_2(t) | k'_1 k_1\rangle$$

where we define a two-particle matrix $\mathsf{R}_2(t)$ by

$$\langle k'''_1 k''_1 | \mathsf{R}_2(t) | k'_1 k_1\rangle = e^{-\Omega f} \sum_{N=1}^{\infty} [(N-1)!]^{-2} e^{\beta N g} \sum_{k_2\cdots k_N} \sum_{k'_2\cdots k'_N} \sum_{\mathcal{P},\mathcal{P}'} \varepsilon^{P+P'}$$
$$\times \langle k'''_1 k_2 \cdots k_N | e^{i\mathsf{H}t/\hbar} \mathsf{U}_{P'} | \mathsf{k}'_1 \cdots \mathsf{k}'_N\rangle \quad (48.17)$$
$$\times \langle \mathsf{k}''_1 \mathsf{k}'_2 \cdots \mathsf{k}'_N | \mathsf{e}^{-(\beta + it/\hbar)\mathsf{H}} \mathsf{U}_P | \mathsf{k}_1 \cdots \mathsf{k}_N\rangle$$

We note that the selection of the variables k_1 and k_1' in the third equality of Equation (48.16) can each be made in N different ways. It can be seen that for $t \neq 0$ this expression (48.17) is considerably more complicated to evaluate than is the grand partition function (37.4).

Exercise 48.3 Show that

$$\sum_{k_1,k_1'}{}' \langle k_1 k_1' | \mathrm{R}_2(t) | k_1' k_1 \rangle = \langle \mathrm{N}^2 \rangle = \langle \mathrm{N} \rangle^2 [1 + \Omega^{-1} \kappa T \cdot \kappa_T] \tag{48.18}$$

where we have used a result from Exercise 48.1(b) to obtain the second equality.

On substituting Equation (48.18) into Equation (48.16), we find for $G(\mathbf{r}, t)$ the alternative expression

$$G(\mathbf{r}, t) = (1 + \Omega^{-1} \kappa T \cdot \kappa_T) + (n\Omega)^{-2} \sum_{\substack{k_1, k_1', \mathbf{q} \\ (\mathbf{q} \neq 0)}} \times e^{-i\mathbf{q}\cdot\mathbf{r}} \langle k_1 + \mathbf{q}, k_1' - \mathbf{q} | \mathrm{R}_2(t) | k_1' k_1 \rangle \tag{48.19}$$

Here, the first term gives the forward-scattering contribution to the linear response function (48.10).

Measurements of R(q, ω) for Liquid Helium

Neutron scattering experiments in liquid helium have revealed for He II that with a significant part of the inelastic scattering there is a unique energy transfer ω for each q value. Thus we may write for He II,

$$[R(\mathbf{q}, \omega) - (2\pi)^3 n\, \delta(\omega)\, \delta^{(3)}(\mathbf{q})] = Z(\mathbf{q})\, \delta[\omega - \varepsilon(\mathbf{q})] + R^{(1)}(\mathbf{q}, \omega) \tag{48.20}$$

where $R^{(1)}(\mathbf{q}, \omega)$ is a continuum contribution. Values (unnormalized) for $Z(q)$ have been measured by Henshaw and Woods[26] over a wide range of q values, and these show that $Z(q)$ is undoubtedly associated in Figure X.3 with the pronounced peak in $S'(q)$ around 2 Å^{-1}. This peak is therefore referred to as the single-excitation peak.

The measured single-particle-like excitations in liquid He II are identified as the quasi-particles of this quantum fluid.[27] It is believed that the quasi-particles in He II can be excited uniquely in the neutron scattering experiments because of the existence of a zero-momentum background in the liquid. To the extent that this interpretation is correct, it can be asserted that Bose–Einstein condensation occurs in liquid helium four. However, as we discussed at the end of Section 34, the zero-momentum background (for He II at rest), which is the superfluid component, is most certainly a very complicated entity. Thus, in addition to zero-momentum condensate particles in the superfluid, there is surely also an abundance of zero-momentum $(\mathbf{p}, -\mathbf{p})$

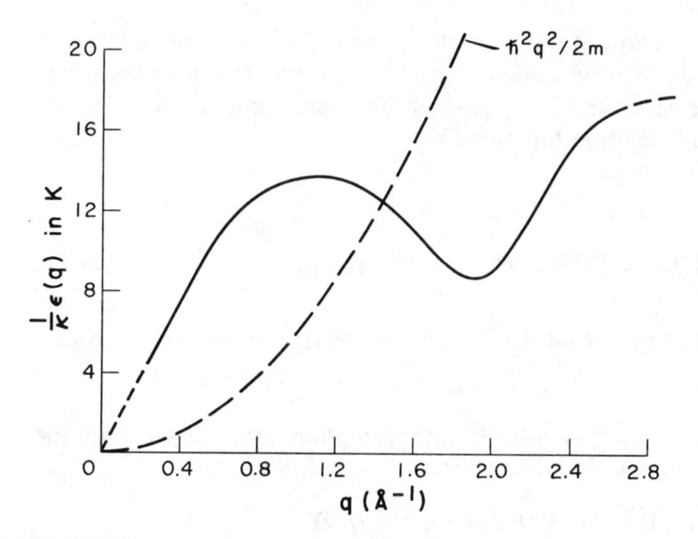

FIGURE X.4 Quasi-particle energy–momentum relation in He II, measured at $T = 1.12$K by the scattering of 4.04-Å neutrons. The phonon region below 0.3 Å^{-1} has been extrapolated from the data to the origin with a dashed line. Shown for comparison is the energy–momentum relation of a free helium atom.

boson pairs. The continuum contribution $R^{(1)}(\mathbf{q}, \omega)$ to the linear response function is probably due only to scattering from the quasi-particles in He II, as in an ordinary fluid.

In Figure X.4 we show the measured curve of $\varepsilon(q)$ for He II.[26] Recent measurements that reveal a second branch of $\varepsilon(q)$ at higher energies are not included. This curve was first predicted by Landau[28] in connection with his phenomenological theory of He II. The slope (in appropriate units) of the linear, or *phonon,* low-momentum region is equal to the velocity of sound (237 m/sec) in He II. The pronounced minimum around 2 Å^{-1} is referred to as the *roton* region of $\varepsilon(q)$.

49 Long-Range Order

Long-range spatial order means that two particles at a large distance from each other are correlated in their positions. The simplest example conceptually is a crystal, and in this section we demonstrate theoretically the long-range order of simple crystals. In particular, we show for crystals that $D(\mathbf{r})$ does not approach unity as $\mathbf{r} \to \infty$.[29] In a fluid there can be long-range correlations near the critical point where the density fluctuations increase without limit.[30] We investigate this case also following the original theory of Ornstein and Zernicke.

We initiate our study of long-range order in a crystal by considering the space–time correlation function $G(\mathbf{r}, t)$, Equation (48.11). On using the operator relation (48.12) this expression becomes

$$G(\mathbf{r}, t) = \frac{1}{(2\pi)^3 n^2 \Omega} \sum_{\alpha,\beta=1}^{N} \int d^3\mathbf{q}\, e^{-i\mathbf{q}\cdot\mathbf{r}} \langle e^{-i\mathbf{q}\cdot\mathbf{r}_\beta(0)}\, e^{i\mathbf{q}\cdot\mathbf{r}_\alpha(t)} \rangle$$

$$= \frac{1}{(2\pi)^3 nN} \sum_{\alpha,\beta=1}^{N} \int d^3\mathbf{q}\, e^{-i\mathbf{q}\cdot\mathbf{r}} \langle e^{-i\mathbf{q}\cdot(\mathbf{R}_\beta + \underline{\xi}_\beta(0))}\, e^{i\mathbf{q}\cdot(\mathbf{R}_\alpha + \underline{\xi}_\alpha(t))} \rangle$$

$$G(\mathbf{r}, t) = \frac{1}{(2\pi)^3 n} \sum_{\alpha=1}^{N} \int d^3\mathbf{q}\, e^{i\mathbf{q}\cdot(\mathbf{R}_\alpha - \mathbf{r})} \langle e^{-i\mathbf{q}\cdot\underline{\xi}_1(0)}\, e^{i\mathbf{q}\cdot\underline{\xi}_\alpha(t)} \rangle \tag{49.1}$$

where in the second equality we have introduced the notation $\mathbf{r}_\alpha = \mathbf{R}_\alpha + \underline{\xi}_\alpha$ of Equation (30.2). For a large crystal, the equilibrium position $\mathbf{R}_\beta$ of any one particle can be chosen as the origin of our coordinate system, and this particle can be labeled as particle 1. We shall assume, for reasons of simplicity, that there is only one atom per unit cell in the crystal.[31]

Mathematical Development

Our primary objective now is to evaluate the average value brackets in the third equality of Equation (49.1). One can surmise that this average value will entail a space–time correlation function between particles 1 and α in the crystal. Study of this microscopic correlation function will then enable us to elucidate the macroscopic property of long-range order in the crystal. The goal is achieved by utilizing a correct quantum-mechanical analysis of the coordinates $\underline{\xi}_\alpha(t)$ and corresponding momenta $\underline{\pi}_\alpha(t)$, with the normal-mode representation of these $6N$ variables also required.

It is sufficient for the purposes of this section to use the harmonic approximation (30.3) for the crystal Hamiltonian, except that here we must regard this Hamiltonian quantum mechanically. First we apply the operator expansion theorem of Exercise 44.2 to arrive at an iterated expression for the ith coordinate $\underline{\xi}_i(t)$ ($i = 1, 2, \cdots, 3N$). With the aid of the commutation relation $[\underline{\xi}_i, \underline{\pi}_j] = i\hbar\delta_{ij}\, 1$, we obtain[32]

$$\underline{\xi}_i(t) = e^{i\mathsf{H}t/\hbar}\underline{\xi}_i e^{-i\mathsf{H}t/\hbar} = \underline{\xi}_i + \frac{it}{\hbar}[\mathsf{H}, \underline{\xi}_i] + \frac{1}{2}\left(\frac{it}{\hbar}\right)^2 [\mathsf{H}, [\mathsf{H}, \underline{\xi}_i]] + \cdots$$

$$= \underline{\xi}_i + \left(\frac{t}{m}\right)\underline{\pi}_i - \frac{1}{2}\left(\frac{t}{m}\right)^2 \sum_{j=1}^{3N} a'_{ij}\underline{\xi}_j - \frac{1}{3!}\left(\frac{t}{m}\right)^3 \sum_{j=1}^{3N} a'_{ij}\underline{\pi}_j + \cdots \tag{49.2}$$

Clearly, in the harmonic approximation $\underline{\xi}_i(t)$ is merely a linear combination of all the $\underline{\xi}$'s and $\underline{\pi}$'s at $t = 0$. Therefore, $[\underline{\xi}_1(0), \underline{\xi}_\alpha(t)]$ is a c number. We can apply this conclusion, in conjunction with the operator identity[33]

$$e^{\mathsf{A}}e^{\mathsf{B}} = e^{\mathsf{A}+\mathsf{B}+\mathsf{C}}$$

$$\mathsf{C} = \frac{1}{2}[\mathsf{A},\mathsf{B}] + \frac{1}{12}\left(\Big[[\mathsf{A},\mathsf{B}],\mathsf{B}\Big] + \Big[[\mathsf{B},\mathsf{A}],\mathsf{A}\Big]\right) + \cdots \tag{49.3}$$

$$\longrightarrow \frac{1}{2}[\mathsf{A},\mathsf{B}] \qquad (\text{for } [\mathsf{A},\mathsf{B}] = c \text{ number})$$

to write the operator product in Equation (49.1) as

$$e^{-i\mathbf{q}\cdot\underline{\xi}_1(0)}\, e^{i\mathbf{q}\cdot\underline{\xi}_\alpha(t)} = e^{i\mathbf{q}\cdot[\underline{\xi}_\alpha(t)-\underline{\xi}_1(0)]}\, e^{1/2[\mathbf{q}\cdot\underline{\xi}_1(0),\mathbf{q}\cdot\underline{\xi}_\alpha(t)]} \tag{49.4}$$

This relation completes the first step of our mathematical analysis.

Inasmuch as the second factor on the right side of Equation (49.4) is a c number, we may take it outside of the averaging brackets when this result is substituted into Equation (49.1). This leaves the equilibrium average of the first factor, which we examine next, to be evaluated. On expanding the exponential of the first factor in Equation (49.4), we obtain for its average value the expression

$$\begin{aligned}\langle e^{i\mathbf{q}\cdot[\underline{\xi}_\alpha(t)-\underline{\xi}_1(0)]}\rangle &= \sum_{m=0}^{\infty}\frac{i^m}{m!}\left\langle\left[\sum_{j=1}^{3} q_j(\underline{\xi}_\alpha(t) - \underline{\xi}_1(0))_j\right]^m\right\rangle \\ &= \sum_{n=0}^{\infty}\frac{(-1)^n}{(2n)!}\left\langle\left[\sum_{j=1}^{3} q_j(\underline{\xi}_\alpha(t) - \underline{\xi}_1(0))_j\right]^{2n}\right\rangle\end{aligned} \tag{49.5}$$

In the second equality we have used the fact that the quantum-mechanical average over a product of an odd number of $\underline{\xi}$ and $\underline{\pi}$ factors, recalling Equation (49.2), must equal zero. To proceed, we need to use results from the following lengthy exercise.

Exercise 49.1 To evaluate the average value (49.5) explicitly, it is necessary to transform quantum mechanically to the normal-mode representation (30.4) of the Hamiltonian. A correct transformation is[34]

$$\begin{aligned}\underline{\xi}_\alpha &= \frac{1}{\sqrt{Nm}}\sum_k \mathsf{q}_k\, \hat{\boldsymbol{\varepsilon}}_k\, e^{i\mathbf{k}\cdot\mathbf{R}_\alpha} \\ \underline{\pi}_\alpha &= \sqrt{\frac{m}{N}}\sum_k \mathsf{p}_k\, \hat{\boldsymbol{\varepsilon}}_k\, e^{-i\mathbf{k}\cdot\mathbf{R}_\alpha}\end{aligned} \tag{49.6}$$

where $k = (\mathbf{k}, \lambda)$ with $\lambda = (1, 2, 3)$ being a polarization index, and where $\mathbf{k}$ is restricted to the first Brillouin zone.[35] Furthermore, since $\underline{\xi}_\alpha$ and $\underline{\pi}_\alpha$ are Hermitean operators we must have, with $\mathsf{q}_k = \mathsf{q}_\lambda(\mathbf{k})$ and so on,

$$\mathsf{q}_\lambda^\dagger(\mathbf{k}) = \mathsf{q}_\lambda(-\mathbf{k}), \quad \mathsf{p}_\lambda^\dagger(\mathbf{k}) = \mathsf{p}_\lambda(-\mathbf{k}), \quad \text{and} \quad \hat{\boldsymbol{\varepsilon}}_\lambda(-\mathbf{k}) = \hat{\boldsymbol{\varepsilon}}_\lambda(\mathbf{k})$$

The three unit polarization vectors $\hat{\varepsilon}_\lambda(\mathbf{k})$ are assumed to be orthogonal. In Equations (49.6) there are N wave numbers $\mathbf{k}$ in the first Brillouin zone. For a general wave number $\mathbf{q}$, we also have the condition

$$\sum_{\alpha=1}^{N} e^{i\mathbf{q}\cdot\mathbf{R}_\alpha} = N \sum_{(hk\ell)} \delta_{\mathbf{q},\mathbf{Q}(hk\ell)} \longrightarrow N\delta_{\mathbf{q},0} \quad \text{(for } \mathbf{q} \text{ in first Brillouin zone)} \tag{49.7}$$

The vector $\mathbf{Q}(hk\ell)$, with Miller indices $(hk\ell)$, is any reciprocal lattice vector for the crystal or the null vector.[36]

(a) Show that the quantum-mechanical form of Equation (30.4) is

$$\mathsf{H}(\mathsf{q}, \mathsf{p}) = \frac{1}{2}\sum_k (\mathsf{p}_{-k}\mathsf{p}_k + \omega_k^2 \mathsf{q}_{-k}\mathsf{q}_k) \tag{49.8}$$

where $-k$ denotes $(-\mathbf{k}, \lambda_k)$ and the frequencies ω_k, which cannot be dependent on the particle index α, are determined by the diagonalization condition

$$\omega_k^2\, \hat{\boldsymbol{\varepsilon}}_{k,i} = \frac{1}{m}\sum_{\beta=1}^{N}\sum_{j=1}^{3} a_{\alpha\beta,ij}\, \hat{\boldsymbol{\varepsilon}}_{k,j}\, e^{i\mathbf{k}\cdot(\mathbf{R}_\beta - \mathbf{R}_\alpha)} \tag{49.9}$$

Here i and j denote component directions of the polarization vectors, so that the sum over all $3N$ coordinates $\underline{\xi}_{\alpha,i}$, for example, is given by a sum over the N lattice locations α with three components at each site. Thus $a_{\alpha\beta,ij}$ is the coefficient in Equation (30.3) of $\underline{\xi}_{\alpha,i}\underline{\xi}_{\beta,j}$.

(b) Follow the procedure used in Equation (49.2) and show that

$$\mathsf{q}_k(t) = \mathsf{q}_k \cos\omega_k t + \left(\frac{\mathsf{p}_{-k}}{\omega_k}\right)\sin\omega_k t \tag{49.10}$$

where $[\mathsf{q}_k, \mathsf{p}_{k'}] = i\hbar\, \delta_{k,k'}\, \mathsf{1}$. [The commutation relation may be established after inverting Equations (49.6) with the aid of (49.7).]

We now return to an analysis of Equation (49.5). It is well-known that every product of two harmonic oscillator variables can be expressed, equivalently, in terms of products of two annihilation and creation operators.[37] But in the thermodynamic limit we must obtain, for every term in Equation (49.5), that all such products involve different sets of the variables k of Equations (49.6).[38] Moreover, there are $(2n - 1)!!$ ways of equating, or matching, n pairs of the k variables in the nth term of Equation (49.5), corresponding to the $2n$ factors of $[\underline{\boldsymbol{\xi}}_\alpha(t) - \underline{\boldsymbol{\xi}}_1(0)]$ whose average is to be taken. We conclude therefore, and quite remarkably, that the macrocanonical ensemble average (49.5) can be simplified, effectively, to

$$\langle e^{i\mathbf{q}\cdot[\underline{\xi}_\alpha(t)-\underline{\xi}_1(0)]}\rangle = \sum_{n=0}^{\infty}\frac{(-1)^n}{n!}\left\langle\frac{1}{2}[\mathbf{q}\cdot(\underline{\xi}_\alpha(t)-\underline{\xi}_1(0))]^2\right\rangle^n$$
$$= \exp\left(-\frac{1}{2}\langle[\mathbf{q}\cdot(\underline{\xi}_\alpha(t)-\underline{\xi}_1(0))]^2\rangle\right) \tag{49.11}$$

We next substitute Equations (49.4) and (49.11) into Equation (49.1), note that the commutator in Equation (49.4) can be replaced by its statistical average since Tr $\underline{\rho} = 1$, and obtain

$$G(\mathbf{r},t) = \frac{1}{(2\pi)^3 n}\sum_{\alpha=1}^{N}\int d^3\mathbf{q}\, e^{i\mathbf{q}\cdot(\mathbf{R}_\alpha-\mathbf{r})}\exp\left(-\frac{1}{2}\langle[\mathbf{q}\cdot(\underline{\xi}_\alpha(t)-\underline{\xi}_1(0))]^2\rangle\right)$$
$$\times\exp\left(\frac{1}{2}\langle[\mathbf{q}\cdot\underline{\xi}_1(0),\mathbf{q}\cdot\underline{\xi}_\alpha(t)]\rangle\right) \tag{49.12}$$

$$G(\mathbf{r},t) = \frac{1}{(2\pi)^3 n}\sum_{\alpha=1}^{N}\int d^3\mathbf{q}\, e^{i\mathbf{q}\cdot(\mathbf{R}_\alpha-r)}\exp\left(-\frac{1}{2}\sum_{i,j=1}^{3}q_iq_j[M_{ij}^{(1)}(0)-M_{ij}^{(\alpha)}(t)]\right)$$

Here the 3×3 matrix $M_{ij}^{(\alpha)}(t)$ is defined by

$$M_{ij}^{(\alpha)}(t) \equiv 2\,\langle\underline{\xi}_1(0)_i\,\underline{\xi}_\alpha(t)_j\rangle = M_{ji}^{(\alpha)}(t)$$
$$\xrightarrow[t\to 0]{} 2\langle\underline{\xi}_1(0)_i\,\underline{\xi}_\alpha(0)_j\rangle = 2\langle\underline{\xi}_1(t)_i\,\underline{\xi}_\alpha(t)_j\rangle \tag{49.13}$$

where the symmetry in i and j follows because these labels only specify component directions of the $\underline{\xi}$'s. Note that the equilibrium average $M_{ij}^{(\alpha)}(0)$ cannot depend on the origin of time [final equality of (49.13)]. The matrix $M_{ij}^{(\alpha)}(t)$ is the microscopic space–time correlation function mentioned above Equation (49.2).

Exercise 49.2 Show for a crystal structure with reflection invariance that

$$M_{ij}^{(\alpha)}(-t) = M_{ij}^{(\alpha)*}(t)\xrightarrow[t\to 0]{}\text{real}$$
$$\sum_{i,j=1}^{3}q_iq_j[M_{ij}^{(1)}(0)-\text{Re}\,M_{ij}^{(\alpha)}(t)]\rangle\, 0 \tag{49.14}$$

[*Suggestion*: Prove the inequality by using the fact that the argument of the exponential on the right side of Equation (49.11) is negative for arbitrary **q**.]

According to the inequality of Exercise 49.2, the real part of the factor quadratic in q_i in the exponential of Equation (49.12) is positive. Therefore, the **q** integral is convergent. Using standard procedures,[39] we can transform the $\mathbf{q}$ coordinates orthogonally by the complex transformation $q_i = \Sigma_{k=1}^{3} b_{ik}\zeta_k$ so that the matrix $[M_{ij}^{(1)}(0) - M_{ij}^{(\alpha)}(t)]$ is diagonalized, that is,

$$\sum_{i,j=1}^{3} b_{ki}^{-1}[M_{ij}^{(1)}(0) - M_{ij}^{(\alpha)}(t)]b_{j\ell} = \mathcal{M}_k^{(\alpha)}(t)\delta_{k,\ell} \tag{49.15}$$

On defining the inverse matrix of $[M_{ij}^{(1)}(0) - M_{ij}^{(\alpha)}(t)]$ by

$$N_{ij}^{(\alpha)}(t) \equiv \sum_{k=1}^{3} b_{ik} \frac{1}{\mathcal{M}_k^{(\alpha)}(t)} b_{kj}^{-1} \tag{49.16}$$

we find with straightforward integration in Equation (49.12) the result,

$$G(\mathbf{r}, t) = \sum_{\alpha=1}^{N} \frac{|N^{(\alpha)}(t)|^{1/2}}{(2\pi)^{3/2} n} \cdot \exp\left(-\frac{1}{2} \sum_{i,j=1}^{3} N_{ij}^{(\alpha)}(t)(r - R_\alpha)_i (r - R_\alpha)_j \right) \tag{49.17}$$

where $|N^{(\alpha)}(t)| = |\mathcal{M}^{(\alpha)}(t)|^{-1}$ is the determinant of the matrix $N_{ij}^{(\alpha)}(t)$.

Long-Range Order in Crystals

We can use Equation (49.17) to investigate long-range order in crystals. First we observe that in the long-range, or asymptotic, region $|\mathbf{R}_\alpha| \to \infty$, it must be so that $M_{ij}^{(\alpha)}(t) \to 0$ for all t. In this limit $N_{ij}^{(\alpha)}(t)$ is then independent of t and α, since it becomes the inverse of the real matrix $M_{ij}^{(1)}(0)$. We write N_{ij}^{∞} for this limiting case and obtain for $G(\mathbf{r}, t)$ the real function

$$\begin{aligned} G(\mathbf{r}, t) \xrightarrow[\mathbf{r}\to\infty]{} & \frac{|N^{(\infty)}|^{1/2}}{(2\pi)^{3/2} n} \sum_{\alpha_{\text{dist}}} \exp\left(-\frac{1}{2} \sum_{i,j=1}^{3} N_{ij}^{(\infty)}(r - R_\alpha)_i (r - R_\alpha)_j \right) \\ &= \lim_{\mathbf{r}\to\infty} D(\mathbf{r}) \end{aligned} \tag{49.18}$$

where α_{dist} are crystal atoms near the point $\mathbf{r}$ and located at a large distance from atom 1. Since this expression is independent of t, it must also be valid at $t = 0$ where, according to Equation (48.14), it reduces to the radial distribution function $D(\mathbf{r})$ as we have indicated. Surprisingly, only the equilibrium average $M_{ij}^{(1)}(0)$ occurs in the asymptotic region.

We see that the radial distribution function for a crystal is characterized asymptotically by a long-range order in which it takes maximum values at the atom sites $\mathbf{r} = \mathbf{R}_\alpha$. This is the expected result. Although $D(\mathbf{r}) \neq 1$ for $\mathbf{r} \to \infty$, it nevertheless is normalized so that $\int d^3\mathbf{r} D(\mathbf{r}) = N^{-1} N_{\text{dist}}\, \Omega \cong \Omega$, as for the normal fluid case and as one can readily check.[40]

We next compute the Fourier transform in space of $G(\mathbf{r}, t)$. Using Equation (49.12), we find

$$
\begin{aligned}
n\int d^3\mathbf{r}\, e^{i\mathbf{q}\cdot\mathbf{r}}G(\mathbf{r}, t) &= \sum_{\alpha=1}^{N} \exp(i\mathbf{q}\cdot\mathbf{R}_\alpha) \\
&\quad \cdot \exp\left(-\frac{1}{2}\sum_{i,j=1}^{3} q_i q_j [M_{ij}^{(1)}(0) - M_{ij}^{(\alpha)}(t)]\right) \\
&\xrightarrow[t\to 0]{} S(\mathbf{q})
\end{aligned}
\tag{49.19}
$$

where the $t = 0$ limiting form follows from Equations (48.1) and (48.14). Thus we have arrived at an expression for the crystal structure factor. Moreover, since the elastic part of the linear response function involves the factor $\delta(\omega)$, we can derive an expression for this function [see Equation (48.10)] by setting $t \to \infty$ in the first equality of Equation (49.19) above, where $M_{ij}^{(\alpha)}(\infty) = 0$.[41] Of course, this limit is independent of t, and therefore we obtain the elastic part of the linear response function simply by multiplying the right side of the limiting form of Equation (49.19) by $\delta(\omega)$. We obtain thereby,

$$
\begin{aligned}
R_{\text{el}}(\mathbf{q}, \omega) &= \delta(\omega)\sum_{\alpha=1}^{N} \exp(i\mathbf{q}\cdot\mathbf{R}_\alpha)\exp\left(-\frac{1}{2}\sum_{i,j=1}^{3} q_i q_j M_{ij}^{(1)}(0)\right) \\
R_{\text{el}}(\mathbf{q}, \omega) &= (2\pi)^3 n\,\delta(\omega)\exp\left(-\frac{1}{2}\sum_{i,j=1}^{3} q_i q_j M_{ij}^{(1)}(0)\right) \\
&\quad \cdot \sum_{(hk\ell)} \delta^{(3)}(\mathbf{q} - \mathbf{Q}(hk\ell))
\end{aligned}
\tag{49.20}
$$

where to derive the second equality we have used Equation (49.7).

We can see that long-range order in a crystal corresponds in the elastic part of the linear response function to scattering from all possible crystal planes, in addition to the scattering term $(2\pi)^3 n\ \delta^{(3)}(\mathbf{q})\ \delta(\omega)$, which occurs also for a normal fluid. The familiar $\mathbf{q} = 0$ term is due to scattering in the first Brillouin zone. Thus by this complicated procedure we obtain the well-known Bragg law for scattering from crystals.[42] We can also derive an expression for the inelastic part of $R(\mathbf{q}, \omega)$ using Equations (48.10) and (49.19). The exponential factor in Equation (49.20) is called the *Debye–Waller factor* and it governs the intensity, but not the angular resolution, of elastic scattering as a function of temperature.[43]

Fluctuations Near the Critical Point

The asymptotic approach to long-range order of the radial distribution function is itself an interesting question. Characteristically, this approach has a $1/r$ dependence. Rather than pursue this feature for a crystal, by further studying Equation (49.17) for $G(\mathbf{r}, t)$, we consider instead a simpler example, namely, the behavior of the radial distribution function in a classical dense gas as the critical point is approached. The analysis we give was due originally to Ornstein and Zernicke.[44]

Our starting point is the relation, which follows from Equations (48.2) and (48.6),

$$\frac{(\Delta N)^2_{\rm rms}}{\langle N \rangle} = n\kappa T \cdot \kappa_T = 1 + D \tag{49.21}$$

where

$$D \equiv n \int d^3\mathbf{r}\, [D(\mathbf{r}) - 1] \tag{49.22}$$

The quantity $\kappa_T = -\Omega^{-1}(\partial\Omega/\partial\mathcal{P})_T$ is the isothermal compressibility (9.2), which diverges for a dense gas as the temperature is lowered to the critical point ($T \rightarrow T_C^+$). Thus large density fluctuations occur in this region, and these may be associated with the occurrence of long-range order at the critical point. An observational consequence, quite distinct from the more general case of liquid condensation, is the phenomenon of critical opalescence.

Let us define: $d(\mathbf{r}) \equiv D(\mathbf{r}) - 1$. The quantity $d(\mathbf{r})$ differs from zero, as we remarked previously below Equation (47.4), because of local density fluctuations. Near the critical point the definition of "local" must be extended to comprise the large regions of macroscopic fluctuation, and we show that this fact enables one to deduce the long-range behavior of $d(\mathbf{r})$. We note that $d(\mathbf{r}_1 - \mathbf{r}_2)$ governs the correlated fluctuations of two particles at different positions $\mathbf{r}_1$ and $\mathbf{r}_2$, regardless of what happens in the rest of the gas. We can also define, for comparison, a function $f(\mathbf{r}_1 - \mathbf{r}_2)$, which governs the correlated fluctuations of two particles at different positions $\mathbf{r}_1$ and $\mathbf{r}_2$ when all the rest of the gas particles are held at fixed locations.

The essence of the Ornstein–Zernicke derivation is to relate $d(\mathbf{r}_{12})$ to $f(\mathbf{r}_{12})$ by an integral equation. In particular, we may argue (see Exercise 49.3 below), with $\mathbf{r}_1 \neq \mathbf{r}_2$, that

$$d(\mathbf{r}_{12}) = f(\mathbf{r}_{12}) + n \int_{\mathbf{r}_3 \neq (\mathbf{r}_1, \mathbf{r}_2)} d^3\mathbf{r}_3\, f(\mathbf{r}_{13})\, d(\mathbf{r}_{32}) \tag{49.23}$$

Substitution of this expression into Equation (49.22) yields, without difficulty,

$$D = F + FD = F(1 - F)^{-1} \tag{49.24}$$

where

$$F \equiv n \int d^3\mathbf{r}\, f(\mathbf{r}) = n \int d^3\mathbf{r}_2\, f(\mathbf{r}_{12}) \tag{49.25}$$

Equation (49.24) for D can be substituted into Equation (49.21) to give the result $n\kappa T\kappa_T = (1 - F)^{-1}$, which implies that $F \rightarrow 1$ as the critical point is approached.

Exercise 49.3 Show, using Equations (24.6) and (47.3), that for a homogeneous system

$$\frac{1}{n}\langle \Delta \mathsf{n}(\mathbf{r}_1)\Delta \mathsf{n}(\mathbf{r}_2)\rangle = n\, d(\mathbf{r}_1 - \mathbf{r}_2) + \delta^{(3)}(\mathbf{r}_1 - \mathbf{r}_2) \xrightarrow[\mathbf{r}_1 \neq \mathbf{r}_2]{} n\, d(\mathbf{r}_1 - \mathbf{r}_2) \tag{49.26}$$

where $\Delta\mathsf{n}(\mathbf{r}) \equiv [\mathsf{n}(\mathbf{r}) - \langle \mathsf{n}(\mathbf{r})\rangle]$, $\langle \mathsf{n}(\mathbf{r})\rangle = n$, and $\mathsf{n}(\mathbf{r})$ is defined by the first of Equations (24.2). This result can be used in the justification of the integral equation (49.23).[45] Thus we can argue that the correlation (or probability) of density fluctuations for particles at the points $\mathbf{r}_1$ and $\mathbf{r}_2$, that is, $nd(\mathbf{r}_{12})\, d^3\mathbf{r}_2$ has contributions (for a particle at $\mathbf{r}_1$) due to independent fluctuations at $\mathbf{r}_2$, that is, $nf(\mathbf{r}_{12})d^3\mathbf{r}_2$, plus all fluctuations at $\mathbf{r}_2$ due to independent fluctuations at points $\mathbf{r}_3 \neq \mathbf{r}_2$.

The "direct" correlation function $f(\mathbf{r})$ must be a short-range function, corresponding to the range r_o of orginary interparticle forces. We should not be surprised, therefore, that $F \sim 1$, even near the critical point. Moreover, since $d(\mathbf{r}_{12})$ must be of long range near the critical point, we can for large distances $\mathbf{r}_{12}$ neglect $f(\mathbf{r}_{12})$ in comparison with $d(\mathbf{r}_{12})$ in Equation (49.23). Also, we can legitimately expand $d(\mathbf{r}_{32})$, in this equation, about the point $\mathbf{r}_3 = \mathbf{r}_1$ and thereby obtain for an isotropic gas,

$$\begin{aligned} d(\mathbf{r}_{12}) &= n\int d^3\mathbf{r}_3 f(\mathbf{r}_{13})\, \{d(\mathbf{r}_{12}) + [(\mathbf{r}_3 - \mathbf{r}_1)\cdot \nabla_1]d(\mathbf{r}_{12}) \\ &\qquad + \frac{1}{2}\sum_{i,j=1}^{3} (r_3 - r_1)_i (r_3 - r_1)_j \frac{\partial^2}{\partial r_{1i}\partial f r_{1j}} d(\mathbf{r}_{12}) + \cdots\} \\ &\cong F[d(\mathbf{r}_{12}) + \varepsilon^2 \nabla_1^2\, d(\mathbf{r}_{12})] \qquad (r_{12} >> r_o) \end{aligned} \tag{49.27}$$

where the length ε is defined by

$$\varepsilon^2 \equiv \frac{n}{6F}\int d^3\mathbf{r}\, r^2 f(\mathbf{r}) \tag{49.28}$$

It is unnecessary to include further terms in Equation (49.27) when studying the long-range behavior of $d(\mathbf{r}_{12})$ near the critical point.

We now return to using the variable $\mathbf{r} = (\mathbf{r}_1 - \mathbf{r}_2)$ and solve the differential equation (49.27), which can be rewritten in the familiar form

$$(\nabla^2 - b^{-2})\, d(\mathbf{r}) = 0 \qquad (\mathrm{r} >> \mathrm{r}_o) \tag{49.29}$$

where

$$b^2 \equiv F\,\varepsilon^2(1 - F)^{-1} = D\,\varepsilon^2 \xrightarrow[T \to T_C^+]{} \infty \tag{49.30}$$

The spherically symmetric solution, well-defined for $r \to \infty$, of Equation (49.29) is

$$D(\mathbf{r}) - 1 = d(\mathbf{r}) = \left(\frac{D}{4\pi n b^2 r}\right) e^{-r/b} = \left(\frac{1}{4\pi n \varepsilon^2 r}\right) e^{-r/b} \tag{49.31}$$

where the normalization constant is determined by Equation (49.22). This is the desired result. Since the range $b \to \infty$ as the critical point is approached, it shows for this limit that $[D(\mathbf{r}) - 1] \to 0$ very slowly, that is,

$$\lim_{T \to T_C^+} D(\mathbf{r}) = 1 + \frac{1}{4\pi n \varepsilon^2 r} \tag{49.32}$$

where the short-range ε is defined by Equation (49.28). Equation (49.32) characterizes the long-range order in terms of $D(\mathbf{r})$ for a dense gas near the critical point.

Accurate experimental investigations of this theory have been made using the scattering of light by fluids. In the vicinity of the critical point, this scattering is referred to as critical scattering, or *critical opalescence*. A basis for analyzing critical scattering is provided by Equation (49.23), which is a key ingredient to the Ornstein–Zernicke theory. We first define the Fourier transforms of $d(\mathbf{r})$ and $f(\mathbf{r})$ by

$$s(\mathbf{q}) \equiv n \int d^3\mathbf{r}\; e^{i\mathbf{q}\cdot\mathbf{r}}\, d(\mathbf{r}) \tag{49.33}$$

$$c(\mathbf{r}) \equiv n \int d^3\mathbf{r}\; e^{i\mathbf{q}\cdot\mathbf{r}}\, f(\mathbf{r})$$

With these definitions, the Fourier transform of Equation (49.23) becomes

$$s(\mathbf{q}) = c(\mathbf{q})[1 + s(\mathbf{q})]$$

The structure factor $S'(\mathbf{q})$ of Equations (48.2) and (48.6), a quantity that is measured in critical-scattering experiments, can then be written in terms of $c(\mathbf{q})$ as

$$S'(\mathbf{q}) = [1 - c(\mathbf{q})]^{-1} \xrightarrow[\mathbf{q} \to 0]{} n\kappa T \kappa_T \tag{49.34}$$

It is particularly interesting to consider the small-q limit of $S'(\mathbf{q})$, because then the range parameters ε and b contribute to the expression. Thus one can use definitions (49.25) and (49.28) to write $[S'(\mathbf{q})]^{-1}$ in the approximate form

$$\frac{1}{S'(\mathbf{q})} \cong 1 - F + F\varepsilon^2 q^2 = F\varepsilon^2(q^2 + b^{-2}) \xrightarrow[q\to 0]{} \frac{F\varepsilon^2}{b^2} \tag{49.35}$$

where the second equality follows from definition (49.30). Now in the vicinity of the critical point $F \cong 1$ ε remains finite and b becomes very large. Therefore, as the critical point is approached, say at constant density and for $T \to T_C^{(+)}$, curves of $[S'(\mathbf{q})]^{-1}$ versus q^2 should yield straight lines with vanishingly small intercepts. The experimental data do indeed conform to this prediction, although the intercepts at $q^2 = 0$ seem not to agree with the limit in (49.34).[46]

The $q^2 \to 0$ limit in Equation (49.35) corresponds to a large-r limit of the radial distribution function $D(\mathbf{r})$. One can argue that it is precisely in this limit that the Ornstein–Zernicke theory must break down, a failure that can then be correlated with the above-mentioned discrepancy. Thus an equilibrium theory of fluctuation phenomena cannot be valid if the fluctuation in particle number, $nd(\mathbf{r}_{12})\delta\Omega_2$ of Exercise 49.3, evaluated for $r = |\mathbf{r}_1 - \mathbf{r}_2| \lesssim$ (long-range order parameter, b), is greater than unity when $\delta\Omega_2$ is a volume within which order occurs. Let us assume that a measure of this volume is $\delta\Omega_2 \sim \Omega^{-1}(\Delta\Omega)^2_{\text{rms}}$. We then set as a criterion for the validity of the Ornstein–Zernicke theory the condition

$$n\kappa T\kappa_T d(b) = \frac{\kappa T\kappa_T}{4\pi e\,\varepsilon^2 b} << 1 \tag{49.36}$$

where Equation (49.21) has been used for $(\Delta N)_{\text{rms}} = n(\Delta\Omega)_{\text{rms}}$. On equating the $\mathbf{q} = 0$ limits in Equations (49.34) and (49.35), and applying the resulting identity twice, this last condition can be used to establish a maximum range value b, or minimum b^{-1} value,

$$b^{-1} = (n\kappa T\kappa_T F\varepsilon^2)^{-1/2} >> (4\pi e n F\varepsilon^4)^{-1} \tag{49.36a}$$

The implication of these inequalities is that Equation (49.35), that is, Ornstein–Zernicke theory, may not be valid in the small-q limit when the inequalities fail, that is, for very small $(T - T_C)$ values.[47]

50 One-Particle Thermodynamic Green's Function

In Section 20 we studied two single-particle Green's functions: one that involves position and time variables and a second that involves position and temperature variables. The Fourier transforms of these two Green's functions were shown to be closely related by using their Lehmann representations. In this section we exhibit an explicit perturbation series expansion of the momentum-temperature, or thermodynamic, Green's function $G(k_2t_2; k_1t_1)$ for systems at equilibrium. The expansion procedure makes use of a generalization of a theorem due to Wick. The final result is expressible graphically, using diagrams similar to the Feynman graphs of field theory. Perturbation expansions of thermodynamic quantities are also derived.

We are interested in calculating the thermodynamic Green's function $G(k_2t_2; k_1t_1)$ of Equation (20.12) for systems at equilibrium,

$$G(k_2t_2; k_1t_1) = \langle P\{\mathsf{a}_H(k_2, t_2)\mathsf{a}_H^\dagger(k_1, t_1)\}\rangle \qquad \textbf{(50.1)}$$

where as indicated by Equation (20.14) the temperature dependence must be only on the difference $(t_2 - t_1)$. Although for translationally invariant (fluid) systems we must also have $\mathbf{k}_2 = \mathbf{k}_1$, where $k_i = (\mathbf{k}_i, m_i)$, in this section we permit the notation k_i to represent quantum numbers of the eigenstates of any appropriate unperturbed Hamiltonian $\mathsf{H}_o = \mathsf{K} + \mathsf{U}$ [see Equation (17.12)]. Moreover, the complete Hamiltonian $\mathsf{H} = \mathsf{H}_o + \mathsf{V}$ will include a perturbation Hamiltonian V assumed to be comprised of both V_2 [see Equation (17.19)] and any single-particle interaction potential $\tilde{\mathsf{U}}$, which is chosen to be excluded from H_o. Thus $\mathsf{V} = \mathsf{V}_2 + \tilde{\mathsf{U}}$. For simplicity we consider only single-component systems in this section.

We begin by reviewing formally the procedure established in Section 37 for calculating the grand partition function $Z_G = e^{\Omega f}$. Consider the free-particle statistical matrix $\underline{\rho}_o$, defined by the equation

$$\underline{\rho}_o \equiv e^{-\Omega f_o}\, e^{\beta(\mathsf{N}g - \mathsf{H}_o)} \qquad \textbf{(50.2)}$$

where g is the actual chemical potential for the system of interest and the free-particle grand potential f_o is given by Equation (38.10). Then Equation (37.4) for $e^{\Omega f}$ can be written concisely as

$$e^{\Omega(f-f_o)} = \langle \mathsf{W}(\beta)\rangle_o \equiv \mathrm{Tr}(\underline{\rho}_o \mathsf{W}(\beta)) \qquad \textbf{(50.3)}$$

in which $\mathsf{W}(\beta)$ is defined by Equation (37.5). The subscript o indicates use of the statistical matrix $\underline{\rho}_o$.

We next introduce a generalization of the operator $\mathsf{W}(\beta)$, namely,

$$\begin{aligned} \mathsf{W}(t_2, t_1) &\equiv e^{t_2\mathsf{H}'_o}\, e^{-(t_2-t_1)\mathsf{H}'}\, e^{-t_1\mathsf{H}'_o} \\ &= \mathsf{W}(t_2)\mathsf{W}^{-1}(t_1) \\ &\xrightarrow[t_1\to 0]{} \mathsf{W}(t_2) \end{aligned} \qquad (50.4)$$

where the effective Hamiltonian H' is defined by $\mathsf{H}' \equiv \mathsf{H} - g\mathsf{N}$ and similarly for H_o'.[48] Thus Equation (50.2) for $\underline{\rho}_o$ can also be written as $\underline{\rho}_o = e^{-\Omega f}e^{-\beta H_o'}$. By using the operator $W(t_2, t_1)$ we may rewrite Equations (50.1) for the equilibrium Green's function in the more complicated form

$$G(k_2t_2; k_1t_1) = e^{-\Omega(f-f_o)} \langle P\{W(\beta, t_2)a_{k_2}(t_2)W(t_2, t_1)a_{k_1}^\dagger(t_1)W(t_1)\}\rangle_o \tag{50.5}$$

where the statistical matrix $\underline{\rho}$ is given by Equation (36.15a) as $\underline{\rho} = \exp(-\Omega f) \times \exp(-\beta H')$. In this equation the annihilation and creation operators are now expressed in the interaction picture, as defined by the operator equations,[49]

$$\begin{aligned} a_k(t) &\equiv \exp(tH_o')\, a_k \exp(-tH_o') = a_k \exp(-t\varepsilon_k) \\ a_k^\dagger(t) &\equiv \exp(tH_o')\, a_k^\dagger \exp(-tH_o') = a_k^\dagger \exp(t\varepsilon k) \end{aligned} \tag{50.6}$$

Here the eigenvalues of H_o' have been denoted by ε_k, not to be confused with the quantity ε, which takes only the values $+1$ (for bosons) and -1 (for fermions). In particular, on using the notation of Equation (17.12) we have for eigenstates $|k_i\rangle$:

$$\langle k_i|H^{(1)}|\mathbf{k}_j\rangle - g\delta_{i,j} = \varepsilon_{k_i}\, \delta_{i,j}$$

A remarkable simplification of Equation (50.5) can now be achieved, for in view of the chronological ordering operator P we may use the second equality of (50.4), and Equation (50.3), to write:

$$G(k_2t_2; k_1t_1) = \frac{\langle P\{W(\beta)a_{k_2}(t_2)a_{k_1}^\dagger(t_1)\}\rangle_o}{\langle W(\beta)\rangle_o} \tag{50.7}$$

To justify, say, that

$$P\{a_{k_1}^\dagger(t_1)W(t_1)\} = P\{W(t_1)a_{k_1}^\dagger(t_1)\}$$

in Equation (50.5) we must use the iterated form of $W(t_1)$,

$$W(t) = \sum_{n=0}^{\infty} \frac{(-1)^n}{n!} \int_0^t ds_1 \cdots ds_n\, P[V(s_1)V(s_2) \cdots V(s_n)] \tag{50.8}$$

which relation follows from the integral equation[50]

$$W(\beta) = 1 - \int_0^\beta dt\, V(t)\, W(t)$$

where $V(t) \equiv \exp(tH_o')V \exp(-tH_o')$. Thus in the iterated form (50.8), every variable s_i satisfies $s_i < t$. Furthermore, for Fermi particles the operator V will always involve an even number of construction operators, because of particle-number conservation, so that the factor of ε in the definition of P

[see Equation (20.3)] never enters into Equation (50.8) when only operators $\mathsf{V}(s_i)$ are commuted.

Equation (50.7) provides a starting point for calculation of the one-particle thermodynamic Green's function. Because we must use the iterated form (50.8) for $\mathsf{W}(\beta)$, the calculation automatically results as a perturbation expansion in the interaction Hamiltonian V. It is the explicit formulation of this expansion to which we now turn.

Generalization of Wick's Theorem

When the expansion (50.8) is substituted into Equation (50.7), one obtains

$$G(k_2t_2; k_1t_1) = \langle \mathsf{W}(\beta)\rangle_o^{-1} \sum_{n=0}^{\infty} \frac{(-1)^n}{n!} \int_0^{\beta} ds_1 \cdots ds_n \times \mathrm{Tr}(\underline{\rho}_o P\{\mathsf{V}(s_1)\mathsf{V}(s_2) \cdots \mathsf{V}(s_n)\mathsf{a}_{k_2}(t_2)\mathsf{a}_{k_1}^{\dagger}(t_1)\}) \tag{50.9}$$

Our initial objective now is to evaluate the trace, or sum over all states, in this equation and for this purpose it is necessary only to focus attention on the construction operators that occur in the operators V. In particular, we are confronted with the evaluation of traces of the specific form

$$T \equiv \mathrm{Tr}(\underline{\rho}_o P\{\mathsf{b}(s_1)\mathsf{b}(s_2) \cdots \mathsf{b}(s_r)\}) \tag{50.10}$$

where the symbol $\mathsf{b}(s)$ will be used to denote either $\mathsf{a}_k(s)$ or $\mathsf{a}_k^{\dagger}(s)$ of Equations (50.6). Note that some of the temperature variables s_i, but not necessarily the associated state labels k_i, may be equal in Equation (50.10). It is in the evaluation of the trace (50.10) that a form of Wick's theorem[51] is required.

The chronological ordering operator in Equation (50.10) has the effect that it rearranges the operator product from the order $(s_1, s_2, \ldots, s_r)$ to some new order $(t_1, t_2, \ldots, t_r)$ with associated sign factor ε^P. Here P is the parity of the permutation of the t_i variables with respect to the s_i. Moreover, if we denote by z_i that $z_i = +1$ if b_{k_i} is a creation operation and $z_i = -1$ if b_{k_i} is an annihilation operator [see Equations (50.6)], then Equation (50.10) can be rewritten as

$$\begin{aligned} T &= \varepsilon^P \,\mathrm{Tr}[\underline{\rho}_o \mathsf{b}_{k_1}(t_1)\mathsf{b}_{k_2}(t_2) \cdots \mathsf{b}_{k_r}(t_r)] \\ &= \varepsilon^P \prod_{i=1}^{r} e^{z_i t_i \varepsilon_i} \,\mathrm{Tr}[\underline{\rho}_o \mathsf{b}_1\mathsf{b}_2 \cdots \mathsf{b}_r) \end{aligned} \tag{50.11}$$

With this result our task is simplified to the problem of evaluating $\mathrm{Tr}(\underline{\rho}_o\mathsf{b}_1 \cdots \mathsf{b}_r)$

To proceed we make use of the simple identity

$$\mathsf{b}_i\mathsf{b}_j = [\mathsf{b}_i, \mathsf{b}_j]_\varepsilon + \varepsilon \mathsf{b}_j\mathsf{b}_i$$

where $[\mathrm{b}_i, \mathrm{b}_j]_\varepsilon \equiv \mathrm{b}_i\mathrm{b}_j - \varepsilon \mathrm{b}_j\mathrm{b}_i$ and the commutation relations for the two cases $\varepsilon = +$ and $\varepsilon = -$ are given by Equations (17.7) and (17.8) respectively. At this point we observe further that r will always be an even number for Fermi-particle interactions. Therefore, upon using this last identity repeatedly in $\mathrm{Tr}(\underline{\rho}_o\mathrm{b}_1 \cdots \mathrm{b}_r)$ we obtain

$$\begin{aligned}\mathrm{Tr}(\underline{\rho}_o\mathrm{b}_1 \cdots \mathrm{b}_r) = {} & [\mathrm{b}_1, \mathrm{b}_2]_\varepsilon \,\mathrm{Tr}(\underline{\rho}_o\mathrm{b}_3\mathrm{b}_4 \cdots \mathrm{b}_r) \\ & + \varepsilon[\mathrm{b}_1,\mathrm{b}_3]_\varepsilon \,\mathrm{Tr}(\underline{\rho}_o\mathrm{b}_2\mathrm{b}_4 \cdots \mathrm{b}_r) \\ & + \cdots + [\mathrm{b}_1, \mathrm{b}_r]_\varepsilon \,\mathrm{Tr}(\underline{\rho}_o\mathrm{b}_2\mathrm{b}_3 \cdots \mathrm{b}_{r-1}) \\ & + \varepsilon \,\mathrm{Tr}(\mathrm{b}_1\underline{\rho}_o\mathrm{b}_2\mathrm{b}_3 \cdots \mathrm{b}_r)\end{aligned}$$

where in the last term we have made use of the cyclic property of the trace. In this last term we can apply Equations (50.2) and (50.6) to show that

$$\mathrm{b}_1\underline{\rho}_o = \underline{\rho}_o\mathrm{b}_1 \, e^{z_1\beta\varepsilon_1} \tag{50.12}$$

The final term then has the same form as the initial trace of interest, which can therefore be written as

$$\begin{aligned}\mathrm{Tr}(\underline{\rho}_o\mathrm{b}_1 \cdots \mathrm{b}_r) = {} & \mathrm{b}_1^{\cdot}\mathrm{b}_2^{\cdot} \,\mathrm{Tr}(\underline{\rho}_o\mathrm{b}_3\mathrm{b}_4 \cdots \mathrm{b}_r) \\ & + \varepsilon\mathrm{b}_1^{\cdot}\mathrm{b}_3^{\cdot} \,\mathrm{Tr}(\underline{\rho}_o\mathrm{b}_2\mathrm{b}_4 \cdots \mathrm{b}_r) \\ & + \cdots + \mathrm{b}_1^{\cdot}\mathrm{b}_r^{\cdot} \,\mathrm{Tr}(\underline{\rho}_o\mathrm{b}_2\mathrm{b}_3 \cdots \mathrm{b}_{r-1})\end{aligned} \tag{50.13}$$

where the *simple contraction* $\mathrm{b}_i^{\cdot}\mathrm{b}_j^{\cdot}$ is defined by

$$\mathrm{b}_i^{\cdot}\mathrm{b}_j^{\cdot} = [\mathrm{b}_i, \mathrm{b}_j]_\varepsilon(1 - \varepsilon \, e^{z_i\beta\varepsilon_i})^{-1} \tag{50.14}$$

This quantity is, of course, a c number and not an operator.

We now note that the effect of Equation (50.13) is to reduce the trace of a product of r operators to a sum over traces of products of $(r - 2)$ operators. It is therefore a simple matter, in principle, to iterate Equation (50.13) repeatedly in each of the traces on the right side. If we indicate a given simple contraction by the same number of multiple dots on each b operator, for example, $\mathrm{b}_k^{\cdots}\mathrm{b}_\ell^{\cdots}$, then Equation (50.13) can be expanded finally to the completely contracted form

$$\begin{aligned}\mathrm{Tr}(\underline{\rho}_o\mathrm{b}_1\mathrm{b}_2 \cdots \mathrm{b}_r) = {} & \mathrm{b}_1^{\cdot}\mathrm{b}_2^{\cdot}\mathrm{b}_3^{\cdot\cdot}\mathrm{b}_4^{\cdot\cdot} \cdots \mathrm{b}_r^{\cdots} \\ & + \mathrm{b}_1^{\cdot}\mathrm{b}_2^{\cdot\cdot}\mathrm{b}_3^{\cdot}\mathrm{b}_4^{\cdot\cdot} \cdots \mathrm{b}_r^{\cdots} \\ & + \mathrm{b}_1^{\cdot}\mathrm{b}_2^{\cdot\cdot}\mathrm{b}_3^{\cdot\cdot}\mathrm{b}_4^{\cdot} \cdots \mathrm{b}_r^{\cdots} \\ & + \cdots + \mathrm{b}_1^{\cdot}\mathrm{b}_2^{\cdot\cdot}\mathrm{b}_3^{\cdot\cdot}\mathrm{b}_4^{\cdots} \cdots \mathrm{b}_r^{\cdot} \\ & + \cdots \\ = {} & \Sigma \text{ (all possible fully contracted products)}\end{aligned} \tag{50.15}$$

where a fully contracted product includes ½ r simple contractions. Notice that no factors of ε appear in Equation (50.15), because the b_i's in every term have been reordered to the identity permutation, as defined by the left side of (50.15). This convention has been adopted for notational convenience, but in any application we must return to the ordering in which contracted pairs of operators are all side by side.

When Equation (50.15) is substituted into Equation (50.11), then in addition to the simple contractions there will be the exponential factors $e^{z_i t_i \varepsilon_i}$. These can be incorporated with the contraction symbol. Moreover, for the operators $\mathsf{b}_{k_i}(t_i)$ and $\mathsf{b}_{k_j}(t_j)$ it is important to distinguish the two cases $t_i > t_j$ and $t_i < t_j$. We are thus led to define a new contraction symbol

$$\begin{aligned} \overline{\mathsf{b}_{k_i}(t_i)\,\mathsf{b}_{k_j}}(t_j) &\equiv \mathrm{Tr}(\rho_o P\{\mathsf{b}_{k_i}(t_i)\mathsf{b}_{k_j}(t_j)\}) \\ &= e^{(z_i t_i \varepsilon_i + z_j t_j \varepsilon_j)}[\theta(t_i - t_j)\mathsf{b}_i\mathsf{b}_j + \varepsilon\theta(t_j - t_i)\mathsf{b}_j\mathsf{b}_i] \end{aligned} \tag{50.16}$$

But according to Equations (50.14), (17.7), and (17.8) the only nonvanishing contractions are those that involve both an annihilation and a creation operator. Therefore, definition (50.16) applies to only two cases, which can be exhibited explicitly as

$$\begin{aligned} \overline{\mathsf{a}_{k_i}(t_i)\,\mathsf{a}^\dagger_{k_j}}(t_j) &= \varepsilon\overline{\mathsf{a}^\dagger_{k_j}(t_j)\mathsf{a}_{k_i}}(t_i) \\ &= G_o(k_i t_i; k_j t_j) = S_o(k_i, t_i - t_j)\delta_{i,j} \\ &= e^{(t_j - t_i)\varepsilon_i}[\theta(t_i - t_j) + \varepsilon\nu(\varepsilon_i)]\delta_{i,j} \end{aligned} \tag{50.17}$$

where $\nu(\varepsilon_k) = (e^{\beta\varepsilon_k} - \varepsilon)^{-1}$.[52]

The case of equal temperature variables ($t_i = t_j$) requires special discussion. It is included in Equation (50.17) if one uses the convention [see below Equation (20.3)] that $\theta(0) = 0$.[53] Thus one observes first that such contractions only occur, essentially, for two construction operators within a given $\mathsf{V}_2(t)$ or $\tilde{\mathsf{U}}(t)$. Furthermore, within these interaction operators [see Equations (17.12) and (17.19)] the creation operators are always to the left of the annihilation operators. The only case ever encountered is, therefore,

$$\overline{\mathsf{a}^\dagger_{k_i}(t)\mathsf{a}_{k_j}}(t) = \nu(\varepsilon_k)\delta_{i,j}$$

We now substitute Equations (50.15) and (50.16) into Equation (50.11) for the trace T. If we note that the factor ε^P can be absorbed by rearranging the contracted b's according to the original identity permutation in Equation (50.10), then we finally obtain

$$\begin{aligned} T = {} & \overline{\mathsf{b}(s_1)\mathsf{b}}(s_2)\overline{\mathsf{b}(s_3)\mathsf{b}}(s_4) \cdots \quad \overline{\mathsf{b}}(s_r) \\ & + \mathsf{b}(s_1)\mathsf{b}(s_2)\mathsf{b}(s_3)\mathsf{b}(s_4) \cdots \quad \overline{\mathsf{b}}(s_r) \\ & + \mathsf{b}(s_1)\mathsf{b}(s_2)\mathsf{b}(s_3)\mathsf{b}(s_4) \cdots \quad \overline{\mathsf{b}}(s_r) \end{aligned} \tag{50.18}$$

$$+\cdots \qquad + \overbrace{\mathrm{b}(s_1)\overbrace{\mathrm{b}(s_2)\overbrace{\mathrm{b}(s_3)\mathrm{b}(s_4)}\cdots}\cdots \quad \mathrm{b}(s_r)} \tag{50.18}$$

$$+\cdots$$

$$= \Sigma \text{ (all possible fully contracted products)}$$

This result is the required generalization of Wick's theorem for use in Equation (50.9). Every term on the right side involves a product of ½(r) contractions, each of which, according to Equation (50.17), is a free-particle thermodynamic Green's function.

Cancellation of Unconnected Terms

The role of the factor $\langle \mathrm{W}(\beta)\rangle_o^{-1}$ in Equation (50.9) for the one-particle thermodynamic Green's function is to cancel many, many terms that arise when (50.18) is substituted. The canceled terms are those that are unconnected in the sense defined above Equation (37.16). Thus the Ursell equations (37.7) exhibit all possible ways in which unconnected products of terms can occur in the grand partition function $Z_G = e^{\Omega f}$ of Equation (37.4). The logarithm of this quantity, that is, Ωf of Equation (37.2), then involves only the connected products of Equation (37.16). A similar result occurs for the single-particle Green's function $G(k_2t_2; k_1t_1)$.

To appreciate better the notion of a connected term in the present development, we anticipate the diagrammatic notation to be introduced below by defining vertex symbols that correspond to the two interaction Hamiltonians $\mathrm{V}_2(t)$ and $\tilde{\mathrm{U}}(t)$. These are shown in Figure X.5, and the indicated bracket quantities, or vertex functions, are defined as follows:

$$
\begin{aligned}
{}^{t_1t_2}\!\begin{pmatrix} k_1\, k_2 \\ k_3\, k_4 \end{pmatrix}_t &\equiv -[\theta(t_1 - t) + \varepsilon\nu(\varepsilon_1)][\theta(t_2 - t) + \varepsilon\nu(\varepsilon_2)] \\
&\quad \times \langle k_1k_2|V^{(S)}|k_3k_4\rangle \exp[t(\varepsilon_1 + \varepsilon_2 - \varepsilon_3 - \varepsilon_4)]
\end{aligned} \tag{50.19a}
$$

$$
{}^{t_1}\!\begin{pmatrix} k_1 \\ k_0 \end{pmatrix}_{t_0} \equiv -\varepsilon[\theta(t_1 - t_0) + \varepsilon\nu(\varepsilon_1)]\langle k_1|\tilde{U}|k_0\rangle \exp[t_0(\varepsilon_1 - \varepsilon_0)] \tag{50.19b}
$$

where the matrix element in the first expression is given by Equation (17.20). The overall negative sign in each of these definitions arises because of the $(-1)^n$ factor in Equation (50.9), and the accompanying factor of ε in the second equation is explained below. The remaining factors are attributable to the contractions (50.17).

We now notice that the contractions (50.17) occur generally between different interactions, or vertices (except for the equal temperature case), and this suggests that we represent each contraction by a line between two vertices. In fact, the corresponding contraction expressions (50.17) have already been allocated (essentially) to the outgoing lines of the vertex functions (50.19). In this way diagrams, or graphs, that consist of interconnected vertices can be constructed.

Consider next the factor of 1/4, which, according to Equation (17.19b), occurs for each operator V_2 in Equation (50.9). Inasmuch as all of the construction operators are contracted by theorem (50.18), we will have two indistinguishable choices for connecting (or contracting) the outgoing lines from a vertex V_2, and similarly for the incoming lines. Therefore all of the factors of 1/4 will be canceled when theorem (50.18) is substituted into Equation (50.9). The one exception to this canceling will be when two lines connect the same two vertices. In each such special case a factor of 1/2 remains in the final expression; the corresponding double linkage is called a *double bond* between two vertices.

Although the line for every contraction in a graph is associated with a vertex at each end, it does not follow that the entire graphical structures, which occur when theorem (50.18) is substituted into Equation (50.9), are interconnected. Moreover, special attention must be given to the construction operators $a_{k_2}(t_2)$ and $a^{\dagger}_{k_1}(t_1)$, which are not associated with any interaction Hamiltonian. Brief reflection will convince one that the contractions associated with each of these two operators must be interconnected by tracing lines through vertices

Let us suppose that there are n_1 vertices associated with a connected graph that arises from contractions involving the two operators $a_{k_2}(t_2)$ and

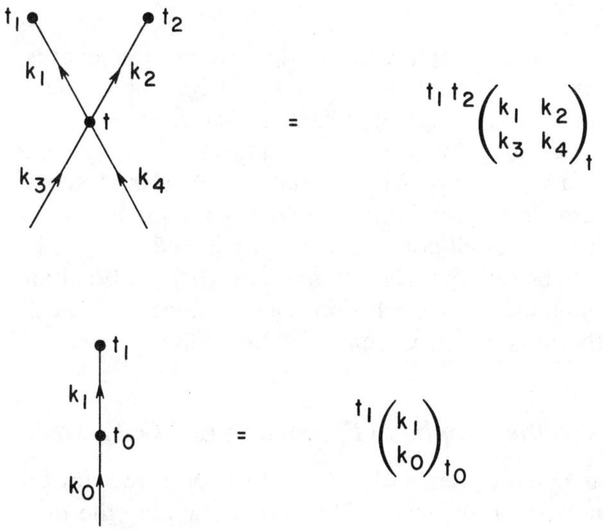

FIGURE X.5 Symbols for the vertices of connected ζ graphs. The bracket quantities are defined by Equations (50.19).

$a^\dagger_{k_1}(t_1)$. For the present we call these structural combinations type 1 graphs. With an n-vertex term arising from Equation (50.9) there will also be $n_2 = n - n_1$ other vertices connected (or not) in various ways to each other, but not (by assumption) connected to the first n_1 vertices. We call these n_2-vertex structural combinations type 2 graphs. Because all possible fully contracted products occur in theorem (50.18) and because the temperature variables s_i in Equation (50.9) are dummy variables that can always be relabeled, it must be so that

$$\binom{n}{n_1} = \frac{n!}{n_1!\, n_2!}$$

identical graphical structures will arise, corresponding to the assumed situation.[54] Every such case will therefore factor into two main graphical parts: (1) a type 1 graph that consists of n_1 interconnected vertices and is associated with a factor $(n_1!)^{-1}$, and (2) the remaining type 2 graph that is associated with a factor $(n_2!)^{-1}$. No factor of $n!$ remains for the corresponding product contribution to Equation (50.9).

Consider next the factor $\langle \mathsf{W}(\beta)\rangle_o^{-1}$ in Equation (50.9), where according to Equations (50.3) and (50.8) we may write

$$e^{\Omega(f-f_o)} = \langle \mathsf{W}(\beta)\rangle_o = \sum_{n=0}^{\infty} \frac{(-1)^n}{n!} \int_0^\beta ds_1 \cdots ds_2 \times \mathrm{Tr}(\rho_o\, P\{\mathsf{V}(s_1)\mathsf{V}(s_2) \cdots \mathsf{V}(s_n)\}) \tag{50.20}$$

When the above considerations are applied to the right side of this expression, then it becomes clear that $\langle \mathsf{W}(\beta)\rangle_o$ is also a sum over many, many graphical structures. In particular, every graphical structure in Equation (50.20) has the form of a type 2 graph as described. On turning the argument around, we see that every different type 1 graph that occurs in the trace of Equation (50.9) must be associated with (i.e., multiplied by) an infinite series of type 2 graphs whose sum, since all possible terms are included, must be $\langle \mathsf{W}(\beta)\rangle_o$. Therefore there will be two canceling factors of $\langle \mathsf{W}(\beta)\rangle_o$ in Equation (50.9), and we arrive at the important conclusion: Only connected (type 1) graphs contribute to the thermodynamic Green's function (50.9).

Connected ζ Graphs and the Graphical Expansion of $G(k_2t_2; k_1t_1)$

We now introduce connected ζ graphs ($\zeta = 0, 1, 2\cdots$) and give rules for their construction and analytic prescription. Then we show how the one-particle Green's function can be expressed as a sum over all different connected 1 graphs. We begin by establishing a sign convention.

We note that there is no overall factor of ε associated with the free-particle Green's function, or contraction (50.17). The definition of this quantity corresponds to the ordering $aa^\dagger$. But the initial configuration of construction operators in the interaction Hamiltonians of Equation (50.9) corresponds to the ordering $a^\dagger a^\dagger aa$ for V_2 and $a^\dagger a$ for $\tilde{U}$. Therefore, if we establish an identity permutation in which annihilation operators are to the left of creation operators in V_2 and $\tilde{U}$, then the sign convention of the contraction (50.17) will be satisfied automatically by the ordering of construction operators in theorem (50.18), which ordering derives from Equation (50.9). Of course, we do not actually interchange the construction operators as they occur, say, in Equation (17.19b) for V_2, but rather we are here only defining an identity permutation. There is, in the end, no consequence of this convention for the operators V_2, since $\varepsilon^2 = 1$, but for each of the operators $\tilde{U}$ there will occur an extra factor of ε, and this factor has already been included in Equation (50.19b).

An *nth-order connected ζ graph* ($\zeta = 0, 1, 2, \cdots$) consists of a collection of n vertices entirely interconnected by m directed lines, with ζ external incoming lines and ζ external outgoing lines, each of which is associated with a single-particle state variable and an (inverse) temperature variable as specified below. For a given set of external line labels, two connected ζ graphs are different if their topological structures, including line directions, are different.[55] Corresponding to each nth-order connected ζ graph is an analytic expression that is determined by the following procedures.

(a) Assign a temperature variable $t_\mu (\mu = 1, 2, \cdots, n + \zeta)$ to the head end of every line. Note that the heads of the ζ outgoing lines are not vertices.

(b) To each vertex ($\mu = 1, 2 \cdots n$), which must be one of the two types in Figure X.5, assign a vertex function [see Equations (50.19)].

(c) To each of the internal lines assign an integer $i (i = 1, 2, \cdots, m)$ and a state variable k_i. To each of the 2ζ external lines assign a pregiven state variable.

(d) Form a product of all the vertex functions and assign an overall factor of ε^P, where P is the sign of the permutation of the bottom row variables with respect to the top row variables. Here an identity permutation of the state variables of the ζ external incoming lines relative to those of the outgoing lines must be assigned.

(e) Assign a factor S^{-1} to the entire graph, where the *symmetry number* S is defined to be the total number of permutations of the m integers associated with the internal lines that leave the graph topologically unchanged (including the positions of these numbers relative to the lines).

(f) Sum over all m internal state variables.

(g) Integrate from 0 to β over each of the $n - \zeta$ temperature variables not attached to the head end of an external line.

Examples of connected 0 and 1 graphs are shown in Figure X.6.

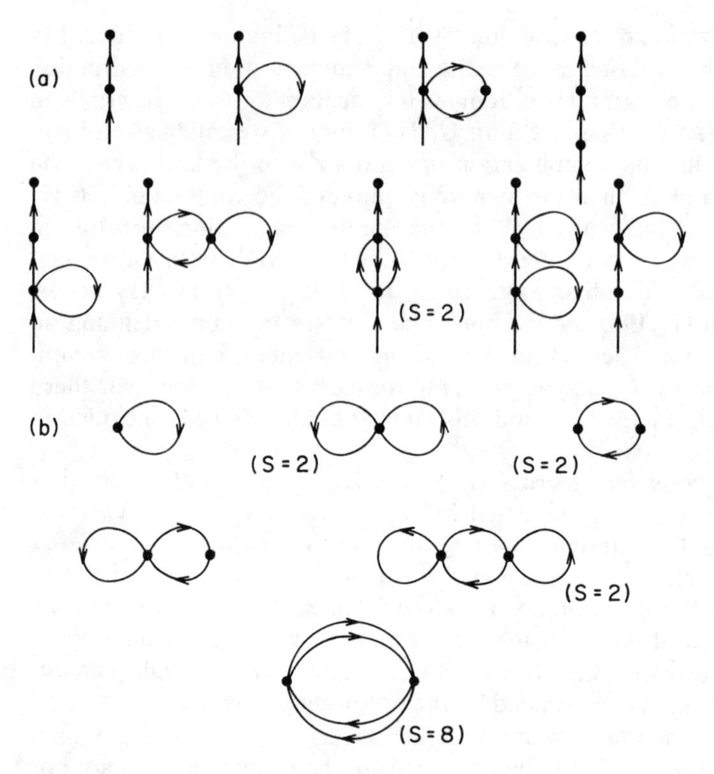

FIGURE X.6 (a) All of the 1- and 2-vertex connected 1 graphs; and (b) all of the 1- and 2-vertex connected 0 graphs. When it differs from unity the symmetry number S for a graph has been included.

We now define two quantities $\mathscr{G}(2, 1)$ and $\mathscr{L}(2; 1)$ by the equations:

$$\mathscr{G}(k_2t_2; k_1t_1) \equiv \delta(t_2 - t_1)\delta_{k_1,k_2} + \varepsilon\mathscr{L}(k_2t_2; k_1t_1) \tag{50.21}$$

$$\mathscr{L}(k_2t_2; k_1t_1) \equiv \sum_{n=1}^{\infty} [\text{all different } n\text{th-order connected 1 graphs}] \tag{50.22}$$

where (k_2, t_2) are the variables assigned to the outgoing external line of each 1 graph and (k_1, t_1) are the variables assigned to the incoming external lines. The inverse temperatures t_2 and t_1 are located at the head ends of these lines. With the aid of these definitions, the one-particle thermodynamic Green's function (50.9) can be written as

$$G(k_2t_2; k_1t_1) = e^{(t_1\varepsilon_1 - t_2\varepsilon_2)} \int_0^{\beta} ds\ \mathscr{G}(k_2t_2; k_1s)[\theta(s - t_1) + \varepsilon\nu(\varepsilon_1)] \tag{50.23}$$

In the absence of any interactions ($\mathscr{L}(2, 1) = 0$), this expression reduces to the free-particle Green's function, or contraction (50.17).

To verify Equation (50.23) we apply the method used in Appendix F to establish Equation (F.9). First we observe that after substitution into Equation (50.9) every interconnected term in Equation (50.18) corresponds to some connected 1 graph. Thus apart from overall numerical factors, the procedures given for writing down connected 1 graphs incorporate correctly the factors that result from the interaction Hamiltonians V_2 and $\tilde{U}$ and from the contractions (50.17). The overall sign in rule (d) follows directly from the convention for the identity permutation to be used in conjunction with Equation (50.18) [see paragraph preceding rules]. Here we recall that eventually the construction operators in every term of Equation (50.18) must be permuted until pairs of contracted operators are all side by side as in definition (50.16), with the annihilation operator to the left of the creation operator. The resulting sign will then be ε^P.[56] We may conclude that to prove Equation (50.23) we need only demonstrate that rule (e) for connected ζ graphs is correct when $\zeta = 1$.

Associated with each connected ζ graph, we now introduce *numbered ζ graphs* in which integers associated with the vertex labels are assigned to these vertices and may not be permuted. Two numbered ζ graphs are different if they have different topological structures, including the positions of these numbers relative to the vertices. We let D be the total number of different numbered ζ graphs that can be generated from a given connected ζ graph. The proof of Equation (50.23) is completed by the three steps of Exercise 50.1.[57]

Exercise 50.1

(a) Argue that the symmetry number S can be written as

$$S = 2^d r$$

where d is the total number of double bonds in a connected ζ graph [see below Equations (50.19)]. The number r is the number of vertex-label permutations that leave the graph topologically unchanged when, as above, integers are assigned to the n vertices instead of to the lines.

(b) For an n-vertex connected ζ graph show that the number D is given by

$$D = \frac{n!}{r}$$

(c) Prove that rule (e) for connected 1 graphs is correct by arguing that the number of terms in Equation (50.18) that yield the same connected 1 graph after substitution into Equation (50.9) is equal to

$$D \cdot 2^{2n_v - d}$$

where n_v is the number of V vertices and $(n - n_v)$ is the number of $\tilde{U}$ vertices.

We now use the notation (20.14) for translationally invariant systems at equilibrium, in which case both incoming and outgoing lines of the one-particle Green's function must be associated with the same momentum **k**.[58] Thus we can write, on referring to Equations (50.21) and (50.23),

$$G(kt_2; kt_1) = S(k, t_2 - t_1)$$
$$= e^{(t_1 - t_2)\varepsilon_k}\{[\theta(t_2 - t_1) + \varepsilon\nu(\varepsilon_k)] + \varepsilon \int_0^\beta ds[\theta(s - t_1) + \varepsilon\nu(\varepsilon_k)]\mathcal{L}(k, t_2, s)\} \quad \textbf{(50.23a)}$$

where we use the notation $\mathcal{G}(k, t_2, t_1)$ and $\mathcal{L}(k, t_2, t_1)$ for $\mathcal{G}$ and $\mathcal{L}$ when describing translationally invariant systems (see also Exercise 50.2).

Although the derivation of Equation (50.23) only provides rules for calculating $S(k, t)$ in the interval $(0 \leq t \leq \beta)$, we can extend the results of calculations to temperature regions outside of this interval by using Equation (20.35). This latter identity can also be applied to Equation (50.23a) for the special case $t = 0$; that is, we must have $S(k, 0) \equiv \varepsilon S(k, \beta)$. This identity gives

$$\int_0^\beta ds\ \mathcal{L}(k, 0, s) = \varepsilon e^{-\beta\varepsilon_k} \int_0^\beta ds\ \mathcal{L}(k, \beta, s) \quad (50.24)$$

where we have used the fact that $\theta(0) = 0$. One consequence of this new identity is that Equation (20.18) and the result of Problem IV.7(b) both yield for the momentum distribution the same result:

$$\langle \mathsf{n}(k) \rangle = \nu(k)[1 + \varepsilon \int_0^\beta ds\ \mathcal{L}(k, \beta, s)] \quad \textbf{(50.25)}$$

Thermodynamics

A by-product of the analysis in this section is a perturbation-series expansion for the grand potential f. By using Equation (50.20) in conjunction with the Ursell equations (37.7) one can show that

$$f(\beta, g, \Omega) = f_o(\beta, g, \Omega) + \Omega^{-1} \sum_{n=1}^{\infty} (\text{all different } n\text{th-order connected 0 graphs}) \quad \textbf{(50.26)}$$

where the free-particle grand potential f_o is given by Equation (38.10) as

$$f_o(\beta, g, \Omega) = -\frac{\varepsilon}{\Omega}\sum_k \ln(1 - \varepsilon e^{-\beta\varepsilon_k}) = \frac{\varepsilon}{\Omega}\sum_k \ln[1 + \varepsilon\nu(\varepsilon_k)] \tag{50.27}$$

To construct a proof of Equation (50.26) we must recall the definition of $\langle \mathsf{W}(\beta)\rangle_o$ by Equation (50.3) and then interpret each Ursell function $U_N^{(S)}$ in Equations (37.7) to be a sum over connected products of vertices as explained above Equation (50.20). Moreover, if instead of using Equation (37.10) for b_n we define all of the nth-order terms in Equation (50.26) to be the quantity b_n, then the development associated with Equations (37.11) and (37.12) can be used. The correct symmetry numbers of rule (e) are deduced for this case, in which $\zeta = 0$, by again using the concept of numbered ζ graphs and the results of Exercise 50.1.[59] With these observations the proof of Equation (50.26) becomes quite straightforward, and we do not consider it further here.

It is of interest to discuss the relationship between the two expressions (37.2) and (50.26) for the grand potential. We recall that in Section 37 the Ursell functions were studied in terms of effects due separately to particle statistics and to particle dynamics [see Equations (37.14) to (37.16) and Appendix F]. Thus single-particle clusters arose in Section 37 due only to the effects of particle statistics; a single-particle potential $\tilde{U}$ was not considered there. In the present treatment single-particle effects due to particle statistics are summarized by, or included in, the factors $\nu(\varepsilon_k)$ of Equations (50.17) and (50.19) and in the free-particle grand potential f_o. The effects of particle dynamics are exhibited explicitly in the vertex functions (50.19). Consequences of the dynamics and further effects due to particle statistics are completely intermingled in connected 0 graphs.

Exercise 50.2 The average energy of a system can be derived from any explicit calculation of the grand potential (50.26) by using the prescription (36.17). It can also be calculated for a large fluid system by using Equation (20.21) together with an explicit calculation of the thermodynamic Green's function (50.23a). Show, for this second method, that

$$\langle E\rangle = \sum_k [(\varepsilon_k + g + \frac{1}{2}\langle k|\tilde{U}|k\rangle)\,\langle \mathsf{n}_k\rangle - \frac{1}{2}\mathscr{L}(k, t_2, t_2^+)] \tag{50.28}$$

where we recall that ε_k is a single-particle eigenvalue of the effective Hamiltonian $\mathsf{H}_o' = \mathsf{H}_o - g\mathsf{N}$.[60] Since $\mathscr{L}(k, t_2, t_2^+)$ must be independent of the variable t_2, we may conclude as with $S(k, t_2 - t_1)$, that $\mathscr{L}(k, t_2, t_1)$ must depend only on the difference $(t_2 - t_1)$ of the temperature variables. For the quantity $\mathscr{L}$, however, this property is only true for a translationally invariant system.

Hartree–Fock Method

In the Hartree–Fock method of many-body theory, one attempts to find that single-particle potential U_{HF} that best represents the effect of particle interactions on a particle (or quasi-particle) in a given single-particle state.[61] The Hartree–Fock potential is an effective potential, and it must usually be found in some self-consistent manner. For a quantum fluid U_{HF} may be equivalent to the quasi-particle interaction energy U_i, which, for example, is defined in the Landau theory of a Fermi fluid by Equation (19.6). We discuss the method only briefly here, within the context of the present development.

The essence of the Hartree–Fock method is to add U_{HF} to the free-particle, or unperturbed, Hamiltonian H_o and correspondingly to subtract it within the perturbation V. We can incorporate this idea into the developments of this section by changing our point of view about the interaction potential $\tilde{\mathsf{U}}$, which was introduced below Equation (50.1). Thus in the Hartree–Fock method we set

$$\tilde{\mathsf{U}} = -\mathsf{U}_{HF} \tag{50.29}$$

and introduce thereby one further minus sign into Equation (50.19b). The Hamiltonian H_o becomes

$$\mathsf{H}_o = \mathsf{K} + \mathsf{U} + \mathsf{U}_{HF} \tag{50.30}$$

where the interaction U accounts for any external potentials acting on the system.

The choice of the Hartree–Fock potential is arbitrary, inasmuch as it has been both added and subtracted within the total Hamiltonian. It may be identified as a quasi-particle interaction energy, as indicated above, or it may be chosen so as to "minimize" the perturbation energy ΔE according to some criterion. A perturbation calculation of the one-particle Green's function in which the first choice is made is given as Problem X.5.

In closing this section we note that the internal structure of connected ζ graphs can be investigated quite generally. For example, one may study all internal structures that themselves have the form of connected 1 graphs. The sum over all such structures is then found to be the one-particle Green's function or, equivalently, the quantity $\mathcal{G}(k_2t_2; k_1t_1)$ of Equation (50.21). The details of such an analysis can then shed further light on the role of using a Hartree–Fock potential in the theory, or on what is also called the self-energy problem of many-body theory.[62]

The procedure that we have outlined could also be used to calculate the two-particle thermodynamic Green's function, which is closely related to the two-particle distribution functions discussed in Section 47 and 48. Other methods for calculating many-body Green's functions, and a wealth of formal studies of the properties of these functions, can be found in other texts and in the literature—to which the interested reader is referred.[63]

Bibliography

Texts

E. Feenberg, *Theory of Quantum Fluids* (Academic Press, 1969), Chapter I. Mathematical properties of the radial distribution function $D(\mathbf{r})$ and the structure factor $S(\mathbf{q})$ are deduced and summarized.

T. L. Hill, *Statistical Mechanics, Principles and Selected Applications* (McGraw-Hill, 1956), Chapter 6. A development of the theory and application of position-space distribution functions within the framework of classical statistical mechanics.

A. Isihara, *Statistical Physics* (Academic Press, 1971), Chapters 6 and 15. The theory of distribution functions and Green's functions is covered. Surveys of many related applications are also included.

Original Papers

T. Matsubara, "A New Approach to Quantum-Statistical Mechanics," *Prog. Theor. Phys.* **14,** 351 (1955); also published in *The Quantum Theory of Many-Particle Systems,* edited by H. L. Morrison (Gordon and Breach, 1962). The theory of thermodynamic Green's functions is developed and applications are given in quantum statistical mechanics.

L. Van Hove, "Correlations in Space and Time and Born Approximation Scattering in Systems of Interacting Particles," *Phys. Rev.* **95** (1954), p. 249. The space–time correlation function $G(\mathbf{r}, t)$ is introduced, and its importance in the analysis and interpretation of scattering experiments in matter is explained. Our treatment of long-range order in crystals (Section 49) is taken from this article.

Notes

1. See Problems X.1 and X.5 and Exercise 20.3.

2. L. I. Schiff, *Quantum Mechanics* 3rd ed. (McGraw-Hill, 1968), Equations (37.3) and (37.5).

3. Strictly speaking, ρ_f equals $\rho(\omega_f)$ times a small correction factor $[1 + dE_\nu/d\omega_f]^{-1}$ to account for the fact that ρ_f is the density of final states for both $\mathbf{k}$ and ν. We omit this factor, for convenience.

4. The general scattering formalism is examined in Section 2 of an article by A. Kerman, H. McManus, and R. M. Thaler, *Ann. Phys.* (*N.Y.*) **8** (1959), p. 551.

5. The assumption that particle α does not recoil independently of the rest of the fluid implies that these matrix elements do not involve the gradient operator $\partial/\partial\mathbf{r}_\alpha$. See, for example, Equations (17.14).

6. See Equations (37.21), (37.34), and (39.39) in L. I. Schiff, *loc. cit.*

7. For solid bodies with long-range order, that is, for crystals, elastic scattering occurs from all possible crystal planes. See Equation (49.20) and its discussion.

8. For a dilute gas of hard-sphere fermions, it has been shown that the (derivative of the) momentum distribution varies discontinuously from $\nu'(k)$ as shown in Figure X.1 [see F. Mohling, *Phys. Rev.* **122** (1961), p. 1062, Equations (110) and (111)]. In the zero-temperature limit this effect becomes a sharp discontinuity at the Fermi surface. The existence of such a sharp discontinuity has been analyzed rigorously for a one-dimensional model due to Luttinger by D. C. Mattis and E. H. Lieb, *J. Math. Phys.* **6** (1965), p. 304.

9. R. Serber, *Phys. Rev.* **72** (1947), p. 1114.

10. The density in a nucleus is due to both neutrons and protons, of essentially equal numbers; hence the value 4 instead of 2. Also, the range r_o and the internucleon spacing $\ell(\equiv n^{-1/3})$ in atomic nuclei are approximately the same.

11. Because we are interested only in equilibrium systems in this chapter, we have suppressed nonequilibrium time dependence in the general notation. The time dependence that occurs in the following sections is associated with time correlations, and such correlations occur even for systems at equilibrium.

12. This relationship can be readily established in analogy with the momentum–space case of Problem V.3(a). See also Exercise 49.3.

13. See footnote 15 in Chapter IV.

14. See Equations (38.8) for $c_2 = -a_2$, (28.10) and (28.12).

15. Curve T_1 is for a pressure of 0.8 atm, somewhat lower than that at the melting point, and curve T_2 is for a pressure of 46.8 atm, which is about at the critical point. See J. DeBoer, *Repts. Prog. Phys.* **12** (1949), p. 368.

16. D. G. Hurst and D. G. Henshaw, *Phys. Rev.* **100** (1955), p. 994.

17. For this first case it also must be possible, after the partial derivatives are taken, to factor the T-function product into two parts with no momentum variables other than k and k' in common.

18. This second case also includes those T-function products that do not factor, as explained in footnote 17.

19. We do not consider here complications due to Bose–Einstein condensation in a Bose fluid.

20. See Equation (46.10) or, for the precise $\mathbf{q} = 0$ limit, Exercise 48.1(b).

21. D. G. Henshaw, *Phys. Rev.* **119** (1960), p. 9. Considerable theoretical discussion of these measurements is given by A. Miller, D. Pines, and P. Nozières in *Phys. Rev.* **127** (1962), p. 1452.

22. Equations (46.7) and (48.7) do not include the possibility that the scattering length a_s may vary with the spin state or with the isotope or impurity of the scattering fluid particle. Such a variation gives an additional "incoherent" scattering cross section in the notation of Van Hove, *op. cit.*, Section III. Also, in the present development we are using a grand canonical ensemble, which implies, for example, that N^2 in Equation (46.10) becomes $\langle \mathsf{N}^2 \rangle$.

23. Note that $S'(\mathbf{q})$ is not precisely a factor in Equation (46.7), because $k = k(\nu)$. Additional discussion and calculations of $R(\mathbf{q}, \omega)$ are given in an article by A. J. Glick, *Ann. Phys. (N.Y.)* **17** (1962), p. 61.

24. R. Kubo [*J. Phys. Soc. Japan* **12** (1957), p. 570] has introduced a generalized linear response function

$$\phi_{BA}(t) \equiv (i\hbar)^{-1}\mathrm{Tr}(\underline{\rho}[\mathsf{A}, \mathsf{B}(t)])$$

Kubo's expression $\chi_{BA}(\omega)$ for a Fourier transform of this quantity for an isotropic system at equilibrium, and when $\mathsf{A} = \underline{\rho}_{\mathbf{q}}^{\dagger}$ and $\mathsf{B} = \underline{\rho}_{\mathbf{q}}$, is

$$\chi(\omega) \equiv \frac{1}{i\hbar}\int_0^{\infty} dt\, e^{-i\omega t/\hbar}\,\mathrm{Tr}(\underline{\rho}[\underline{\rho}_{\mathbf{q}}^{\dagger}, \underline{\rho}_{\mathbf{q}}(t)])$$

$$= \frac{2\langle \mathsf{N}\rangle}{\hbar}\int_{-\infty}^{\infty} d\omega' R(\mathbf{q}, \omega')\int_0^{\infty} dt\, e^{-i\omega t/\hbar}\sin\left(\frac{\omega' t}{\hbar}\right)$$

Therefore, although it is closely related to it, $R(\mathbf{q}, \omega)$ is not a special case of $\chi_{BA}(\omega)$. One example of $\chi_{BA}(\omega)$, with $\mathsf{A} = q\Sigma_{\alpha}\mathsf{x}_{\alpha,\nu}$ and $\mathsf{B} = \mathsf{j}_{\mu}$, is the electrical conductivity tensor $\sigma_{\mu\nu}(\omega)$ of Equation (45.20).

25. The notation in Equations (48.16) and (48.17) is somewhat mixed, inasmuch as we have not introduced state vectors in the number representation [see Equation (17.21) and Problem VIII.5(b)].

26. D. G. Henshaw and A. D. B. Woods, *Phys. Rev.* **121** (1961), p. 1266.

27. In Section 19 we introduced the concept of quasi-particles in a Fermi fluid, and this concept can also be appied to a Bose fluid.

28. L. D. Landau, *J. Phys. (USSR)* **11** (1947), p. 91.

29. Equation (47.12a) suggests that in the absence of long-range order $D(\mathbf{r}) \to 1$ as $\mathbf{r} \to \infty$. See also the results (47.18) and (47.26).

30. See the discussion at the end of Section 36. We focus attention on the critical point rather than on the more general liquid–vapor phase transition, because at this limit point we need to consider only a single phase.

31. An excellent summary of the fundamentals of crystal structure is given in Chapters 1 and 2 of C. Kittel, *Introduction to Solid State Structure,* 3rd ed. (Wiley, 1968).

32. All the atoms in the crystal are assumed to have the same mass m and, for convenience, we use the notation $a_{ij} = m^{-1}a'_{ij}$ in Equation (49.2).

33. See footnote 20 in Chapter VI.

34. D. Pines, *Elementary Excitations in Solids* (Benjamin, 1963), Section 2-2.

35. This restriction is equivalent to the restriction $\omega < \omega_m$ of Equation (30.15).

36. Equation (49.7) can be understood intuitively if one imagines summing the atoms by considering first all possible crystal planes that pass through atom 1 (at the origin). The exponential factor $\mathbf{q} \cdot \mathbf{R}_{\alpha}$ vanishes for any one such plane, with Miller indices $(hk\ell)$, if $\mathbf{q}$ is normal to it, as will be the case when $\mathbf{q} = \mathbf{Q}(hk\ell)$. Considering then this one plane, all other atoms can be assigned to parallel planes, for which $\mathbf{Q} \cdot \mathbf{R}_{\alpha} = 2\pi(\text{integer})$ (C. Kittel, *loc. cit.,* pp. 55 and 56). Thus Equation (49.7) is indeed true for all reciprocal lattice vectors and the null vector.

37. See Problem X.4(a).

38. According to Equations (49.6) each of the operators $\underline{\xi}_{\alpha}$ and $\underline{\pi}_{\alpha}$ is proportional to $\Omega/\sqrt{N}$, since the k sums are proportional to Ω (or N). A product of two $\underline{\xi}_{\alpha}$ and $\underline{\pi}_{\alpha}$ operators, linked together by one common k sum, will therefore be an intensive quantity in the thermodynamic limit. Note that the ensemble average of any product of two annihilation and creation operators that involves different k values is zero.

39. H. Goldstein, *Classical Mechanics* (Addison-Wesley, 1980), Sections 4.3 and 6.2.

40. The exponential factor in Equation (49.17) for $G(\mathbf{r}, t)$ is brought to diagonal form by the inverse orthogonal transformation $\Sigma_{i=1}^{3} b_{ki}^{-1}(R_\alpha - r)_i$; see also Equation (49.16).

41. Correspondingly, the $t \to \infty$ limit of the space–time correlation function itself is given by Equation (49.18) with all atoms α summed.

42. For clarification of this conclusion see C. Kittel, *loc. cit.*, p. 56.

43. See Problem X.4(c).

44. This version of the Ornstein–Zernicke theory is taken from L. Rosenfeld, *Theory of Electrons* (Dover, 1965), Chapter V, Section 6. See also M. E. Fisher, *J. Math. Phys.* **5** (1964), p. 944.

45. A detailed derivation of Equation (49.23) from Equation (49.26) is given on pages 82 and 83 of Rosenfeld's book, *loc. cit.*

46. A review of critical-scattering experiments of this type is given in Section 4 of the article by M. E. Fisher, *loc. cit.* The behavior of the isothermal compressibility κ_T and other thermodynamic properties near the critical points of fluids is discussed in Section 2 of this article. See also Problem X.7 and footnote 65.

47. For a detailed review of critical-point static phenomena, see L. P. Kadanoff *et al., Rev. Mod. Phys.* **39** (1967), p. 395. For a review of more recent theoretical developments on this topic, see M. E. Fisher, *Rev. Mod. Phys.* **46** (1974), p. 597.

48. See footnote 25 in Chapter IV.

49. See also the statement of Problem VIII.5(c). The second equalities of Equation (50.6) can be readily proved by using Equations (17.9) and the operator-expansion theorem of Exercise 44.2.

50. The derivation of this equation is indicated in Problem VIII.5(a).

51. G. C. Wick, *Phys. Rev.* **80** (1950), p. 268.

52. See Equations (20.14) and (20.33), and note also that the free-particle Green's function $S_o(k, t)$ is investigated in Problem IV.7.

53. More precisely, one can take the limit [cf. Equation (20.18)] $t_i = t_j - \eta$, with $\eta = 0+$, in Equations (50.16) and (50.17).

54. Often there will be additional ways in which identical terms can arise and these further possibilities will be dealt with below (see Exercise 50.1).

55. One can distort these graphs and stretch their lines without changing their topological structure.

56. Top- and bottom-row coordinates in the vertex factors correspond to creation and annihilation operators respectively. This fact does not alter the overall sign factor ε^P. There is, however, always one extra permutation associated with the (external lines) construction-operator product $\mathsf{a}_{k_2}(t_2)\mathsf{a}_{k_1}^\dagger(t_1)$ in (50.9). The extra factor of ε is included in definition (50.21).

57. The rules for connected ζ graphs can also be phrased so that every line carries a free-particle Green's function, or *propagator*, S_o. Our rules have been arranged so that the line factors are all unity.

58. We assume also that the one-particle Green's function is independent of any spin projections m_i, as will be the case when there is rotational and inversion invariance in the system.

59. In the present proof the numbers that occur in Equations (37.11) and (37.12) denote the number of vertices in a graph rather than the number of lines. Thus on using the notation of Exercise 50.1 a given connected 0 graph is associated with

$(n + n_v)$ internal lines. The corresponding number of terms in $U_N^{(S)}$, where $N = n + n_v$, is

$$2^{2n_v - d} \cdot D = 2^{2n_v - d}\left(\frac{n!}{r}\right) = \left(\frac{4^{n_v}}{S}\right)n!$$

The factors of 4 cancel those from the interaction Hamiltonians V_2 [see Equation (17.19b)].

60. A consequence of using the effective Hamiltonian H_o' is that $\langle E\rangle \rightarrow \langle E\rangle - \langle \mathsf{N}\rangle g$ on the left side of Equation (20.21) and $\omega_k \rightarrow \varepsilon_k$ on the right side (when $\tilde{\mathsf{U}} = 0$).

61. The Hartree–Fock method, in lowest approximation, was applied originally to the electrons of an atom. See L. I. Schiff, *loc. cit.*, pp. 431–433. Although historically the Hartree–Fock method is associated with use of the variational principle, this characterization of the method applies in the present context only to the leading-order perturbation-theory contribution to U_{HF}.

62. Such an analysis, applied to a fully ionized gas, is given by I. Rama Rao and C. R. Smith, *Phys. Rev.* **A2** (1970), p. 843.

63. See, for example, R. D. Mattuck, *A Guide to Feynman Diagrams in the Many-Body Problem* (McGraw-Hill, 1967); A. L. Fetter and J. D. Walecka, *Quantum Theory of Many-Particle Systems* (McGraw-Hill, 1971); and P. Nozières, *Theory of Interacting Fermi Systems* (Benjamin, 1964).

64. The quantity $\rho = mn$, and $\mathbf{U}$ is the local velocity vector (12.28).

65. The behavior of four simple fluids near their critical points is discussed both theoretically and experimentally in Section V of the review article by Kadanoff *et al.* (footnote 47).

Problems

X.1. (a) Show that the average energy of a many-particle system can be written in terms of reduced density matrices (24.15) as

$$\langle E\rangle = \mathrm{Tr}(\underline{\rho}_1 H^{(1)}) + \mathrm{Tr}(\underline{\rho}_2 V^{(s)})$$

where $H^{(1)}$ is the one-particle Hamiltonian defined by Equation (17.10a) and $V^{(s)}$ is given by Equation (17.20). Assume that the particle interaction energy is determined entirely by two-particle potential energies.

(b) Using classical statistical mechanics, show that the average energy for a system of N particles is given in terms of reduced distribution functions (23.20) by

$$\langle E\rangle = \int d\omega\left(\frac{p^2}{2m}\right)P_1(\mathbf{r}, \mathbf{p}, t) + \frac{1}{2}\int d\omega_1\, d\omega_2\, V_{12}P_2(\mathbf{r}_1, \mathbf{r}_2; \mathbf{p}_1, \mathbf{p}_2; t)$$

$$= N\left[\frac{3}{2}\kappa T + \frac{1}{2}n\int d^3\mathbf{r}\; V(\mathbf{r})\, D(\mathbf{r})\right]$$

where the second equality occurs at equilibrium. Here we assume that only two-particle interaction energies occur, and these are taken to be velocity-independent in the second equality.

(c) Show that the pressure $\mathscr{P}$ in a classical system at rest can be written as

$$\mathscr{P} = -\frac{1}{3\Omega}\left\{\sum_{\alpha=1}^{N}\langle \mathbf{r}\cdot\dot{\mathbf{p}}\rangle_\alpha - \frac{1}{2}\int d^3\mathbf{r}_\alpha\, d^3\mathbf{r}_\beta\, P(\mathbf{r}_\alpha, \mathbf{r}_\beta, t) r_{\alpha\beta}V'(r_{\alpha\beta})\right\}$$

where $V'(r) = \partial V/\partial r$.
[*Suggestion:* Apply Newton's second law to the αth particle, and distinguish between internal and external forces in the system.]
Consider the case in which there is local equilibrium in a fluid system and show that the preceding expression can be manipulated to give

$$\mathscr{P}(\mathbf{r}, t) = \frac{1}{3}[\rho\langle U^2\rangle + \mathrm{Tr}(\mathrm{B}_{ii})]$$

where the tensor $B_{ij}(\mathbf{r}, t)$ is defined by Equation (12.12) and the average value is given now by Equation (12.6). Here it must be assumed that the local coordinate system is inertial.[64] For fluids at thermodynamic equilibrium, verify that the first expression for $\mathscr{P}$ also yields

$$\mathscr{P} = n\kappa T - \frac{1}{6}n^2\int d^3\mathbf{r}\, D(\mathbf{r})\, rV'(r)$$

[*Suggestion:* Recall the virial theorem (28.15).]

X.2. Equation (48.9) is called a sum rule for the linear response function. In this problem a second sum rule is established.

(a) Show that

$$\int_{-\infty}^{+\infty}\omega\, d\omega\, R(\mathbf{q}, \omega) = \frac{1}{2\langle \mathsf{N}\rangle}\mathrm{Tr}(\rho\{\underline{\rho}_\mathbf{q}^\dagger[\mathsf{H}, \underline{\rho}_\mathbf{q}] + [\underline{\rho}_\mathbf{q}^\dagger, \mathsf{H}]\underline{\rho}_\mathbf{q}\})$$

where the two terms within the brackets of the trace are equal and $\underline{\rho}_\mathbf{q}$ is given by Equation (48.5a).

(b) Use the Fock-space notation of Section 17 to establish for local, spin-independent particle interactions, and in the absence of external potentials, that

$$[\mathsf{H}, \underline{\rho}_\mathbf{q}] = [\mathsf{K}, \underline{\rho}_\mathbf{q}] = \left(\frac{\hbar^2 q^2}{2m}\right)\underline{\rho}_\mathbf{q} + \mathsf{B}_\mathbf{q}$$

where

$$B_{\mathbf{q}} \equiv \frac{\hbar^2}{m} \sum_{k} (\mathbf{k} \cdot \mathbf{q}) a_{k+\mathbf{q}}^{\dagger} a_k$$

Show in this case, for an isotropic system, that

$$\int_{-\infty}^{+\infty} \omega \, d\omega \, R(\mathbf{q}, \omega) = \frac{1}{2\langle N \rangle} \mathrm{Tr}(\underline{\rho}[\underline{\rho}_{\mathbf{q}}^{\dagger}, B_{\mathbf{q}}])$$

(c) Use the result of part (b) to establish the second sum rule; namely,

$$\int_{-\infty}^{+\infty} \omega \, d\omega \, R(\mathbf{q}, \omega) = \frac{\hbar^2 q^2}{2m}$$

X.3. *Linear response function for ideal Fermi gas at $T = 0$.*

(a) Show for $q > 0$ that

$$R(q, \omega) = \frac{(2S+1)}{(2\pi)^3 n} \int d^3\mathbf{k}_1 \, d^3\mathbf{k}_2 \, \nu(k_1)[1 - \nu(k_2)]$$
$$\times \, \delta^{(3)}(\mathbf{k}_2 - \mathbf{k}_1 - \mathbf{q}) \, \delta(\omega_2 - \omega_1 - \omega) \qquad (\omega > 0)$$
$$= 0 \qquad (\omega < 0)$$

where $\nu(k)$ is the Fermi distribution (19.1).

(b) Integrate the result of part (a) and show (for $q > 0$) that

$$R(q, \omega) = \frac{3}{8yE_F} \begin{cases} x & \text{if } 0 < x < y(2-y) & (y < 2) \\ [1 - \frac{1}{4}\left(\frac{x}{y} - y\right)^2] & \text{if } y|y-2| < x < y(y+2) & (\text{all } y) \\ 0 & & \text{otherwise} \end{cases}$$

where $y = q/k_F$, $x = \omega/E_F$, and $E_F = \hbar^2 k_F^2/2m$.
(*Suggestion*: First write the three-dimensional δ function as a Fourier integral.)

(c) Show, using Equation (48.9) and the result of part (b), that the structure factor $S'(q)$ is given by

$$S'(q) = \begin{cases} \frac{1}{4} y \left(3 - \frac{1}{4} y^2\right) & (0 < y < 2) \\ 1 & (y > 2) \end{cases}$$

(d) Use Equation (48.3) together with the answer to part (c) to derive Equation (47.26). Note that $D(r)$ differs from the quantity $\Sigma_{\xi_1\xi_2} D(\mathbf{r}, \xi_1, \xi_2)$, when $S = ½$, by a normalization factor $(2S + 1)^2 = 4$.

(e) Verify the second sum rule of Problem X.2 using the result of part (b).

X.4. (a) Refer to Exercise 49.1 and show that the transformation

$$\mathsf{p}_k = i\left(\frac{\hbar\omega_k}{2}\right)^{1/2}(\mathsf{a}_k^\dagger - \mathsf{a}_{-k})$$

$$\mathsf{q}_k = \left(\frac{\hbar}{2\omega_k}\right)^{1/2}(\mathsf{a}_k + \mathsf{a}_{-k}^\dagger)$$

reduces the crystal Hamiltonian (49.8) to diagonal form in the number representation, that is, show that

$$\mathsf{H} = \sum_k\left(\mathsf{n}_k + \frac{1}{2}\right)\hbar\omega_k$$

where $\mathsf{n}_k = \mathsf{a}_k^\dagger\mathsf{a}_k$ and the annihilation and creation operators satisfy the Bose commutation relations (17.7).

(b) Derive for the quantity $M_{ij}^{(\alpha)}(t)$ of Equation (49.13) the expression

$$M_{ij}^{(\alpha)}(t) = \frac{\hbar}{(2\pi)^3 nm}\sum_{\lambda=1}^{3}\int\frac{d^3\mathbf{k}}{\omega_\lambda(\mathbf{k})}\varepsilon_{\lambda,i}(\mathbf{k})\varepsilon_{\lambda,j}(\mathbf{k})[1 - e^{-\beta\hbar\omega_\lambda(\mathbf{k})}]^{-1}$$
$$\times\{e^{-i[\mathbf{k}\cdot\mathbf{R}_\alpha - \omega_\lambda(\mathbf{k})t]} + e^{-\beta\hbar\omega_\lambda(\mathbf{k})}e^{i[\mathbf{k}\cdot\mathbf{R}_\alpha - \omega_\lambda(\mathbf{k})t]}\}$$

where the integration region is over the first Brillouin zone.

(c) Use the Debye model of Section 30 to show that the Debye–Waller factor in Equation (49.20) can be written in the form

$$\exp\left(-\frac{1}{2}\sum_{i,j=1}^{3} q_iq_jM_{ij}^{(1)}(0)\right) = \exp\left(-\frac{1}{3}q^2d^2(T)\right)$$

where

$$d^2(T) = \frac{9}{2}\left(\frac{\hbar^2}{m\kappa T_D}\right)\frac{1}{u^2}\int_0^u x\,dx\left(\frac{1+e^{-x}}{1-e^{-x}}\right) \qquad (u = T_D/T)$$

with the Debye temperature T_D defined by Equation (30.19). The quantity $d^2(T)$ is a measure of the mean-square displacement of an atom. Evaluate $d(T)$ in the high-temperature limit ($T >> T_D$) and in the low-temperature limit ($T << T_D$). Show for the high-temperature limit that the Debye–Waller factor decreases exponentially with increasing temperature.

X.5. (a) Write down explicit expressions for the nine connected 1 graphs of Figure X.6(a) using the Hartree–Fock method with the notation $U_{HF} = U$. Consider the general off-diagonal case for these graphs, and let their sum be denoted by $\mathscr{L}_2(kt_2; k_ot_1)$. Perform the temperature integration for the 1 graph whose symmetry number is S = 2 and simplify your answer using the identity $\nu_k = e^{-\beta\varepsilon_k}(1 + \varepsilon\nu_k)$.

(b) Choose the Hartree–Fock potential (to second order) to be

$$\langle k|U|k_o\rangle = \sum_{k_2} \nu_2 \langle kk_2|V^{(S)}|k_o k_2\rangle$$

$$+ \frac{1}{2} P \sum_{k_2 k_3 k_4} [\nu_2(1 + \varepsilon\nu_3)(1 + \varepsilon\nu_4) - \varepsilon(1 + \varepsilon\nu_2)\nu_3\nu_4]$$

$$\times \frac{\langle kk_2|V^{(S)}|k_3k_4\rangle\langle k_3k_4|V^{(S)}|k_o k_2\rangle}{(\varepsilon_k + \varepsilon_2 - \varepsilon_3 - \varepsilon_4)}$$

where P denotes that the principal value of the energy denominator is to be computed. Show that this choice yields for $\mathscr{L}_2$ the simple expression

$$\mathscr{L}_2(kt_2; k_o t_1) = \frac{1}{2} e^{t_1(\varepsilon_k - \varepsilon_o)} P \sum_{k_2 k_3 k_4} \frac{\langle kk_2|V^{(S)}|k_3k_4\rangle\langle k_3k_4|V^{(S)}|k_o k_2\rangle}{(\varepsilon_k + \varepsilon_2 - \varepsilon_3 - \varepsilon_4)}$$

$$\times e^{(t_2 - t_1)(\varepsilon_k + \varepsilon_2 - \varepsilon_3 - \varepsilon_4)}[\theta(t_2 - t_1)\varepsilon\nu_2(1 + \varepsilon\nu_3)(1 + \varepsilon\nu_4)$$

$$+ \theta(t_1 - t_2)(1 + \varepsilon\nu_2)\nu_3\nu_4]$$

(c) Consider the case of a large fluid at equilibrium and set $k_o = k$ in the results of part (b). Using $\varepsilon_k = \omega_k + \langle k|U|k\rangle - g$, show that Equation (50.28) gives for the average system energy,

$$\langle E\rangle = \sum_k \nu_k\omega_k + \frac{1}{2}\sum_{k_1k_2} \nu_1\nu_2\langle k_1k_2|V^{(S)}|k_1k_2\rangle$$

$$+ \frac{1}{4} P \sum_{k_1k_2k_3k_4} \nu_1\nu_2(1 + \varepsilon\nu_3)(1 + \varepsilon\nu_4) \frac{|\langle k_1k_2|V^{(S)}|k_3k_4\rangle|^2}{(\varepsilon_1 + \varepsilon_2 - \varepsilon_3 - \varepsilon_4)^2}$$

$$+ O(V^3)$$

To derive this result it is necessary to determine an expression for the momentum distribution using, say, Equation (50.25). [See also Problem IV.4(b).]

(d) Use the result of part (c) to show that the quasi-particle energy prescription of Equation (19.6), that is,

$$\omega_k + \langle k|U|k\rangle = \frac{\partial\langle E\rangle}{\partial\nu_k}$$

yields the Hartree–Fock potential of part (b) when $k_o = k$.

X.6. *Perturbation-theory calculation of Green's functions*

(a) Use the result of Problem X.5(b) to calculate the one-particle thermodynamic Green's function (50.23a) and its Fourier transform (20.27) to $O(V^2)$.
[*Suggestion*: Use the identity $\nu_k = e^{-\beta\varepsilon_k}(1 + \varepsilon\nu_k)$ and Equation (20.36).]

(b) Apply Equation (20.38) to identify the spectral density function as

$$\rho(k_1, \omega') = 2\pi\, \delta(\omega' - g - \varepsilon_1)$$
$$\times \{1 - \frac{1}{2} P \sum_{k_2k_3k_4} f(k_1k_2 \mid k_3k_4)(\varepsilon_1 + \varepsilon_2 - \varepsilon_3 - \varepsilon_4)^{-2}\}$$
$$+ \pi P \sum_{k_2k_3k_4} f(k_1k_2 \mid k_3k_4)\, \delta(\omega' - g + \varepsilon_2 - \varepsilon_3 - \varepsilon_4)$$
$$\times (\varepsilon_1 + \varepsilon_2 - \varepsilon_3 - \varepsilon_4)^{-2}$$
$$+ O(V^3)$$

where

$$f(k_1k_2 \mid k_3k_4) = |\langle k_1k_2|V^{(S)}|k_3k_4\rangle|^2$$
$$\times [\nu_2(1 + \varepsilon\nu_3)(1 + \varepsilon\nu_4) - \varepsilon(1 + \varepsilon\nu_2)\nu_3\nu_4]$$
$$\geqslant 0$$

(c) Substitute the answer to part (b) into Equation (20.32) for the Fourier transform of the space–time Green's function, and manipulate your result to the form

$$\hbar^{-1}\tilde{S}(k_1,\omega) = (1 + \varepsilon\nu_1)(\omega - \varepsilon_1 + i\eta)^{-1}\{1 - \frac{1}{2}(\omega - \varepsilon_1 + i\eta)^{-1}$$
$$\times \sum_{k_2k_3k_4} f(k_1k_2 \mid k_3k_4)[P(\varepsilon_1 + \varepsilon_2 - \varepsilon_3 - \varepsilon_4)^{-1}$$
$$- (\omega + \varepsilon_2 - \varepsilon_3 - \varepsilon_4 + i\eta)^{-1}]\}$$
$$- \varepsilon\nu_1(\omega - \varepsilon_1 - i\eta)^{-1}\{1 - \frac{1}{2}(\omega - \varepsilon_1 - i\eta)$$
$$\times \sum_{k_2k_3k_4} f(k_1k_2 \mid k_3k_4)[P(\varepsilon_1 + \varepsilon_2 - \varepsilon_3 - \varepsilon_4)^{-1}$$
$$- (\omega + \varepsilon_2 - \varepsilon_3 - \varepsilon_4 - i\eta)^{-1}]\}$$
$$- i\pi\varepsilon\, P \sum_{k_2k_3k_4} g(k_1k_2 \mid k_3k_4)\, \delta(\omega + \varepsilon_2 - \varepsilon_3 - \varepsilon_4)$$
$$+ O(V^3)$$

where

$$2\pi i\, \delta(x) = \lim_{\eta \to 0+} [(x - i\eta)^{-1} - (x + i\eta)^{-1}]$$

and

$$g(k_1k_2 \mid k_3k_4) = |\langle k_1k_2|V^{(S)}|k_3k_4\rangle|^2 \times [(1 + \varepsilon\nu_1)(1 + \varepsilon\nu_2)\nu_3\nu_4 - \nu_1\nu_2(1 + \varepsilon\nu_3)(1 + \varepsilon\nu_4)] \times (\varepsilon_1 + \varepsilon_2 - \varepsilon_3 - \varepsilon_4)^{-2}$$

Show that $\tilde{S}(k_1, \omega)$ can also be written approximately as

$$\hbar^{-1}\tilde{S}(k_1, \omega) \cong (1 + \varepsilon\nu_1)[\omega - \varepsilon_1 + i\eta - \Sigma(k_1, \omega + i\eta)]^{-1} - \varepsilon\nu_1[\omega - \varepsilon_1 - i\eta - \Sigma(k_1, \omega - i\eta)]^{-1} - i\pi\varepsilon P \sum_{k_2k_3k_4} g(k_1k_2 \mid k_3k_4)\, \delta(\omega + \varepsilon_2 - \varepsilon_3 - \varepsilon_4)$$

where

$$\Sigma(k_1, \omega \pm i\eta) = [\Sigma_R(k_1, \omega) \mp i\,\Sigma_I(k_1, \omega)]$$

and

$$\Sigma_R(k_1, \omega) \equiv \frac{1}{2} P \sum_{k_2k_3k_4} f(k_1k_2 \mid k_3k_4) \times [(\omega + \varepsilon_2 - \varepsilon_3 - \varepsilon_4)^{-1} - (\varepsilon_1 + \varepsilon_2 - \varepsilon_3 - \varepsilon_4)^{-1}] \xrightarrow[\omega \to \varepsilon_1]{} 0$$

$$\Sigma_I(k_1, \omega) \equiv \frac{\pi}{2} \sum_{k_2k_3k_4} f(k_1k_2 \mid k_3k_4)\, \delta(\omega + \varepsilon_2 - \varepsilon_3 - \varepsilon_4) \geqslant 0$$

(d) Calculate the momentum–time Green's function by using Equation (20.23) and the second answer to part (c). Consider only the case in which $\Sigma_I \ll \varepsilon_1$, so that the poles of $\tilde{S}(k_1, \omega)$ are dominated by their real parts at $\omega = \varepsilon_1$, and show that

$$i\tilde{S}(k_1, \tau) = e^{-i\varepsilon_1\tau/\hbar}[\theta(\tau) + \varepsilon\nu_1]\, e^{-|\tau|/\tau_e(k_1)} + \frac{1}{2}\varepsilon P \sum_{k_2k_3k_4} g(k_1k_2 \mid k_3k_4)\, e^{i(\varepsilon_2 - \varepsilon_3 - \varepsilon_4)\tau/\hbar} + O(V^3) \xrightarrow[\tau \to 0\pm]{} [\theta(0\pm) + \varepsilon\langle \mathsf{n}(k_1)\rangle]$$

where $\langle \mathsf{n}(k)\rangle$ is the momentum distribution and $\tau_e(k_1) \equiv \hbar/\Sigma_I(k_1, \varepsilon_1)$ can be interpreted as a (quasi-particle) lifetime.

X.7. *Behavior of a Van der Waals' gas near the critical point*
Express Van der Waals' equation of state in terms of the reduced variables T_r, $\mathcal{P}_r$, and $\rho_r = v_r^{-1}$ of Problem VIII.7(b). Then verify the following properties for a Van der Waals' gas near the critical point[65]:

(a) $(\mathcal{P}_r - 1) = \dfrac{3}{2}(\rho_r - 1)^3 \qquad \text{at } T_r = 1$

(b) $\kappa_T = \left(\rho \dfrac{\partial \mathcal{P}}{\partial \rho}\right)^{-1} \underset{\substack{\rho_r = 1 \\ T \to T_c^+}}{\longrightarrow} [6\mathcal{P}_C(T_r - 1)]^{-1}$

(c) At the end points of the liquid (ℓ)–vapor (g) transition region, that is, along the *coexistence curve,*

$$(\rho_{r,\ell} - 1) = (1 - \rho_{r,g}) = 2(1 - T_r)^{1/2}$$

[*Suggestion for part (c):* Assume $(4t - p) \propto t^2$, where $p \equiv 1 - \mathcal{P}_r$ and $t \equiv (1 - T_r)$.]

Note: If one is careful, then the Maxwell construction Equation (39.8) does not need to be used in order to derive result (c). However, Equation (39.8) is indeed required to justify the suggested assumption.

CHAPTER XI

REAL SYSTEMS IDEALIZED

In this chapter important applications of statistical mechanics not treated earlier in the text are included. The emphasis is on real systems, suitably idealized so that the predominant physical description can be calculated theoretically by straightforward analysis. Two sections are devoted to magnetism—paramagnetism and ferromagnetism—and one to chain molecules. Sections 53 and 54 introduce the subject of relativistic statistics and its application to the electrons in a white dwarf star.

51 Paramagnetism

Magnetic effects in ordinary matter are due both to paramagnetism, in which the angular momentum of particles tends to line up parallel to the direction of applied external magnetic fields, and to diamagnetism, in which the external fields induce microscopic currents with associated magnetism. Because the angular momentum of particles is both orbital and spin in nature, we can give a satisfactory account of these magnetic effects only by using quantum statistical mechanics. In fact, we first prove a remarkable theorem, due to Van Leeuwin and also to Bohr, which states that the magnetic susceptibility of a classical system of charges is identically zero. The main topic of this section is a general quantum-mechanical description of paramagnetism, which in special circumstances becomes ferromagnetism—the subject of the following section. Also treated is a unique effect called Van Vleck paramagnetism.

We begin by proving the Bohr–Van Leeuwin theorem, and to this end we consider a collection of N charges, the ith having orbital angular momentum $\mathbf{L}_i$ measured relative to some origin. The system is assumed to be in thermodynamic equilibrium without macroscopic motion. The magnetic moment $\boldsymbol{\mu}_i$ of the ith charge is then $e_i\mathbf{L}_i/2m_ic$, in Gaussian units, so that the total magnetization of the system is

$$\langle \mathbf{M} \rangle = \frac{\int d\Gamma \left[\sum_{i=1}^{N} \frac{e_i}{2m_ic} \mathbf{L}_i \right] e^{-\beta E}}{\int d\Gamma \, e^{-\beta E}} \tag{51.1}$$

in which both paramagnetic and diamagnetic effects are included. The integration element is the volume element (23.12) of Γ space, and the total energy E is given by

$$E = \sum_{i=1}^{N} \left\{ \frac{1}{2m_i} \left[\mathbf{p}_i - \frac{e_i}{c} \mathbf{A}_i(\mathbf{r}_i) \right]^2 + e_i \, \phi_i(\mathbf{r}_i) \right\} + V \tag{51.2}$$

Here $\mathbf{A}_i(\mathbf{r}_i)$ and $\phi_i(\mathbf{r}_i)$ are vector and scalar potentials acting on the ith particle at $\mathbf{r}_i$, and these may be due to any external fields as well as to the rest of the system.[1] The quantity V is the nonelectromagnetic interaction energy, which is assumed to be velocity independent.

Now the orbital angular momentum is associated with mechanical motion, that is, $\mathbf{L}_i = \mathbf{r}_i \times \mathbf{p}_i'$, where $\mathbf{p}_i' = m\mathbf{v}_i$ is related to the canonical momentum $\mathbf{p}_i$ by

$$\mathbf{p}_i' = \mathbf{p}_i - \frac{e_i}{c} \mathbf{A}_i(\mathbf{r}_i) \tag{51.3}$$

We can therefore make a transformation of momentum coordinates from $\mathbf{p}_i$ to $\mathbf{p}_i'$, with Jacobian equal to unity, and obtain from Equation (51.1) the magnetization expression

$$\langle \mathbf{M} \rangle = \frac{\int d\Gamma' \left[\sum_{i=1}^{N} \frac{e_i}{2m_ic} \mathbf{L}_i \right] e^{-\beta E'}}{\int d\Gamma' \, e^{-\beta E'}} \tag{51.1a}$$

where now

$$E' = \sum_{i=1}^{N} \left(\frac{p_i'^2}{2m_i} + e_i\phi_i \right) + V$$

But this is precisely the expression that one would write down for $\mathbf{A}_i = 0$, that is, in the absence of any magnetic fields, and therefore the magnetization calculated classically is identically zero.

Physically, the Bohr–Van Leeuwin theorem can be visualized as the vector cancellation of all possible magnetic orbits in a system, where in classical statistics all possible magnetic orbits can occur. Thus at any given point of a system there is, on the average, no net electric current. The theorem fails, primarily, because it does not take into account the intrinsic spins of particles.

Quantum Statistical Theory

We assume that there is only one kind of atom (or ion) in the system of interest and that, in so far as paramagentic effects are concerned, the atoms may be considered to be noninteracting.[2] Furthermore, we assume ordinary temperatures such that the atoms are essentially all in their ground states, that is, we assume

$\kappa T \ll$ [excitation energy of first excited state]

where $\Delta E_{\text{exc}} = \Delta(\hbar\omega) = (\hbar c)\,\Delta k \cong (2 \times 10^{-5} \cdot \Delta k)$ eV with Δk in cm^{-1}. Typically, $\Delta k \gtrsim 10^4\ \text{cm}^{-1}$ for first excited states; at $T = 300\text{K}$, $\kappa T \sim 0.025$ eV.

Now the splitting in magnetic fields of an atomic energy level into its $(2J + 1)$ magnetic substates involves energy differences $\sim\mu_B\mathscr{H} \cong (6 \times 10^{-5})\mathscr{H}$ eV, with $\mathscr{H}$ measured in units of 10^4 gauss and where μ_B is the Bohr magneton. We can therefore also expect

$\mu_B\mathscr{H} \ll$ [excitation energy of first excited state]

Finally, since the nuclear magneton is $\sim(10^{-3}\ \mu_B)$ we can for all but the very lowest temperatures neglect any hyperfine splitting of the $(2J + 1)$ magnetic substates. Thus we are led to consider only the magnetic partition function $(Z_{\mathscr{H}})^N$, where[3]

$$Z_{\mathscr{H}} \equiv \sum_{m=-J}^{+J} e^{-\beta m g \mu_B \mathscr{H}} = \sum_{m=-J}^{+J} e^{-mx/J} \tag{51.4}$$

with the magnetic field direction $\hat{\mathscr{H}}$ chosen as the axis of quantization for angular momentum. The quantity g is the *Lande g factor,* or spectroscopic splitting factor, for an atom in its ground state, where for electrons $g = 2.0023 \cong 2.00$. The useful parameter x is defined by

$$x \equiv Jg\mu_B\mathscr{H}/\kappa T \tag{51.5}$$

For fluids we neglect here any effects due to statistical degeneracy.

To calculate the magnetization of the system, we now use the equation

$$\langle M\rangle = N\langle\mu\rangle \tag{51.6}$$

where the average value $\langle\mu\rangle$ of the atomic magnetic moment is given by

$$\begin{aligned}\langle\mu\rangle &= -\frac{1}{Z_{\mathcal{H}}}\sum_{m=-J}^{+J}(mg\mu_B)\,e^{-\beta mg\mu_B\mathcal{H}} \\ &= (Jg\mu_B)\left(\frac{d}{dx}\ln Z_{\mathcal{H}}\right)\end{aligned} \tag{51.7}$$

Also the magnetic susceptibility per unit volume, χ, is defined by the equation

$$\chi \equiv \frac{1}{\Omega}\left(\frac{\langle M\rangle}{\mathcal{H}}\right) = n\frac{\langle\mu\rangle}{\mathcal{H}} \tag{51.8}$$

We next calculate $\langle\mu\rangle$ and χ explicitly for several cases of interest.

The partition function (51.4) can be readily expressed in closed form as

$$Z_{\mathcal{H}} = \frac{\sinh\left(\dfrac{2J+1}{2J}\right)x}{\sinh\left(\dfrac{x}{2J}\right)} \tag{51.4a}$$

Substitution of this result into Equation (51.7) yields for $\langle\mu\rangle$ the general expression

$$\langle\mu\rangle = \left(\frac{J}{J+1}\right)^{1/2}\mu_o\,B_J\,(x) \tag{51.9}$$

where $\mu_o \equiv \sqrt{J(J+1)}\,g\mu_B$ is the magnetic moment of a free atom, calculated quantum mechanically as $\langle\boldsymbol{\mu}^2\rangle^{1/2}$. The quantity $B_J(x)$, called the *Brillouin function,* is defined by

$$\begin{aligned}B_J(x) &= \frac{d}{dx}\ln Z_{\mathcal{H}} = \left(\frac{2J+1}{2J}\right)\coth\left(\frac{2J+1}{2J}\right)x - \frac{1}{2J}\coth\left(\frac{x}{2J}\right) \\ &\xrightarrow[x\to 0]{} \frac{1}{3}\left(\frac{J+1}{J}\right)x\left[1 - \frac{1}{30}\left(\frac{2J^2+2J+1}{J^2}\right)x^2 + O(x^4)\right] \\ &\xrightarrow[J=1/2]{} \tanh x \\ &\xrightarrow[J\to\infty]{} L(x)\end{aligned} \tag{51.10}$$

Here the *Langevin function* $L(x)$, for the limiting case $J \to \infty$, is defined by

$$L(x) \equiv \coth x - \frac{1}{x} \tag{51.11}$$

The case $J = 1/2$ gives for the average atomic moment the value $\langle \mu \rangle = ((1/2) g\mu_B)\tanh x$.[4] The Brillouin function is plotted for various J values in Figure XI.1.

The Langevin function applies to the classical limit in which all possible angular-momentum projections onto the magnetic field direction are allowed for the ground state of an atom. This limit occurs when we set $(2J + 1) \rightarrow \infty$, provided that the intrinsic magnetic moment μ_o is kept finite.[5] In this case

$$\langle \mu \rangle = \mu_o L(\mu_o \mathcal{H}/\kappa T) \qquad \text{(classical limit)} \tag{51.9a}$$

which is a result of Problem VI.2

In the high-temperature or weak-field limit $x << 1$, Equations (51.9) and (51.10) give for the average atomic moment,

$$\langle \mu \rangle = \frac{1}{3} \frac{\mu_o^2 \mathcal{H}}{\kappa T} \qquad (x << 1) \tag{51.12}$$

a result that can also be derived, using classical statistics, from Equations (51.9a) and (51.11). In this limit, the magnetic susceptibility is

$$\chi = \frac{1}{3} \frac{n\mu_o^2}{\kappa T} \tag{51.13}$$

an expression known from experiment as *Curie's law*.

Finally we consider the low-temperature strong-field limit $x >> 1$, where the Brillouin function approaches the number unity. In this case we find for

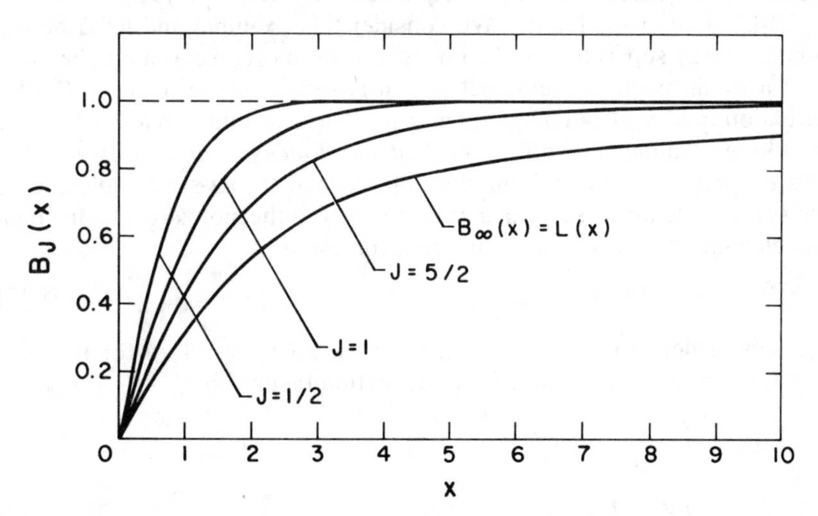

FIGURE XI.1 Dependence of the Brillouin function $B_J(x)$ on its argument x for various values of J.

$\langle\mu\rangle$ the limiting value,

$$\langle\mu\rangle = Jg\mu_B \qquad (x >> 1) \tag{51.14}$$

Thus at very low temperatures it becomes possible to line up all the atoms in the direction of the magnetic field. In strong fields ($\mathcal{H} \sim 50{,}000$ gauss) magnetic saturation has been achieved with various paramagnetic salts for $T \sim 1\text{K}$.[6]

The g factor introduced in Equation (51.4) is defined for atomic states by the equation,

$$\langle Jm|\underline{\mu}_z|Jm\rangle \equiv -mg\mu_B \tag{51.15}$$

This matrix element can be evaluated for various coupling models. For example, in Russell–Saunders (or LS) coupling the magnetic-moment operator is $\underline{\mu}_z = -(\mathsf{L}_z + 2\mathsf{S}_z)\mu_B = -(\mathsf{J}_z + \mathsf{S}_z)\mu_B$. One then obtains the well-known result,[7]

$$\begin{aligned} g &= 1 + \frac{1}{m}\langle LSJm|\mathsf{S}_z|LSJm\rangle \\ &= 1 + \frac{J(J+1) + S(S+1) - L(L+1)}{2J(J+1)} \end{aligned} \tag{51.16}$$

Van Vleck Paramagnetism

Some organic molecules have a triplet ($J = 1$) excited state not far above a singlet ($J = 0$) ground state. We consider this example and let Δ be the zero-field energy separation of the levels. Furthermore, we assume the presence of a weak magnetic field such that it is sufficient to use second-order perturbation theory to calculate the magnetic susceptibility. When $\kappa T << \Delta$, the resulting paramagnetic effect is called *Van Vleck paramagnetism*.

We use definition (51.15) for the unperturbed triplet-state molecular g factor (g) and define a second g factor (g'), for the nonzero off-diagonal matrix elements in this example, by the expression

$$\langle 10|\underline{\mu}_z|00\rangle = \langle 00|\underline{\mu}_z|10\rangle = -g'\mu_B \tag{51.17}$$

It is assumed here that $g' \neq 0$. Then the state vectors $|\Psi_{Jm}\rangle$ for the four states (J,m) are given in first-order perturbation theory by

$$\begin{aligned} |\Psi_{00}\rangle &= |00\rangle - \left(\frac{g'\mu_B\mathcal{H}}{\Delta}\right)|10\rangle \\ |\Psi_{10}\rangle &= |10\rangle + \left(\frac{g'\mu_B\mathcal{H}}{\Delta}\right)|00\rangle \\ |\Psi_{11}\rangle &= |11\rangle,\ |\Psi_{1,-1}\rangle = |1,-1\rangle \end{aligned} \tag{51.18}$$

The energies of these four states are, to second order in perturbation theory,

$$\begin{aligned} E_{00} &= -\frac{(g'\mu_B\mathcal{H})^2}{\Delta} \\ E_{10} &= \Delta + \frac{(g'\mu_B\mathcal{H})^2}{\Delta} \\ E_{11} &= \Delta + g\mu_B\mathcal{H} \\ E_{1,-1} &= \Delta - g\mu_B\mathcal{H} \end{aligned} \qquad (51.19)$$

where the zero of energy is chosen to be the unperturbed ground-state energy. The corresponding magnetic moments are

$$\begin{aligned} \langle\Psi_{00}|\underline{\mu}_z|\Psi_{00}\rangle &= -\langle\Psi_{10}|\underline{\mu}_z|\Psi_{10}\rangle = \frac{2(g'\mu_B)^2\mathcal{H}}{\Delta} \\ \langle\Psi_{11}|\underline{\mu}_z|\Psi_{11}\rangle &= -\langle\Psi_{1,-1}|\underline{\mu}_z|\Psi_{1,-1}\rangle = -g\mu_B \end{aligned} \qquad (51.20)$$

To calculate the average molecular magnetic moment we use the equations:

$$\begin{aligned} \langle\mu\rangle &= \frac{1}{Z_{\mathcal{H}}}\sum_{J,m}\langle\Psi_{Jm}|\underline{\mu}_z|\Psi_{Jm}\rangle\, e^{-\beta E_{Jm}} \\ Z_{\mathcal{H}} &= \sum_{J,m} e^{-\beta E_{Jm}} \end{aligned} \qquad (51.21)$$

In the low-temperature limit $\kappa T << \Delta$, we then perceive that only the $J = 0$ term contributes to $\langle\mu\rangle$ and $Z_{\mathcal{H}}$. We obtain for $\langle\mu\rangle$, in this case, the Van Vleck paramagnetism result $\langle\mu\rangle = \langle\Psi_{00}|\underline{\mu}_z|\Psi_{00}\rangle$. According to Equation (51.8), the magnetic susceptibility is

$$\chi = \frac{2n(g'\mu_B)^2}{\Delta} \qquad (\kappa T << \Delta) \qquad (51.22)$$

Exercise 52.1 Show, using Equations (51.8) and (51.19) to (51.21), that when $\kappa T >> \Delta$ one obtains a Curie-law form for the susceptibility, namely,

$$\chi = \frac{n}{2\kappa T}(g^2 + g'^2)\,\mu_B^2 \qquad (\kappa T >> \Delta) \qquad (51.22a)$$

Note that the energy splitting Δ does not appear in this result.

52 Ferromagnetism

The theory of ferromagnetism in its simplest form can be developed starting directly from the considerations given at the end of Section 1. Alternatively, one can choose the analysis of Section 51 as a point of departure. We adopt this latter approach here, developing thereby the quantum-mechanical version of a molecular-field theory due originally to Pierre Weiss (1907). The underlying cause of ferromagnetism is shown to be the Heisenberg exchange interaction. Using this mechanism the internal magnetic field can be identified by a self-consistent procedure, due initially to Peter R. Weiss, which we apply in lowest approximation. The section is concluded with a study of magnetization near the critical temperature.

The salient feature of the molecular-field theory of ferromagnetism is the introduction of an internal magnetic field. Thus we note first that the magnetic field $\mathcal{H}$ (or B), appearing in Equation (51.4) and thereafter, should properly be interpreted as being equal to the sum of an external field $\mathcal{H}_o$ and an internal field $\mathcal{H}_{\text{int}}$. If we assume that the internal field is proportional to the magnetization M of the material, that is, $\mathcal{H}_{\text{int}} = \lambda M/\Omega = \lambda n\langle\mu\rangle$, then the quantity x of Equation (51.5) becomes

$$x = Sg\mu_B(\mathcal{H}_o + \lambda n\langle\mu\rangle)/\kappa T \tag{52.1}$$

where we henceforth use the letter S instead of J to denote angular momentum. Thus it is the spins of outer atomic electrons that are responsible primarily for ferromagnetism. Equation (51.9) then takes the form

$$\langle\mu\rangle = Sg\mu_B B_S(x) \tag{52.2}$$

If we solve Equation (52.2) for $B_S(x)$ and substitute for $\langle\mu\rangle$ from Equation (52.1), then we obtain the relation

$$B_S(x) = \frac{\kappa T}{(Sg\mu_B)^2\lambda n}\left(x - \frac{Sg\mu_B\mathcal{H}_o}{\kappa T}\right) \tag{52.3}$$

where the Brillouin function $B_S(x)$ is given by Equation (51.10). It is the solution $x = x_o$ to Equation (52.3) that provides us with initial theoretical insight into the nature of ferromagnetism. We show below that the proportionality constant λ is a large number for ferromagnetic materials. Moreover, the critical temperature T_C below which magnetization occurs spontaneously, when $\mathcal{H}_o = 0$, is determined by the equation

$$\kappa T_C = \frac{1}{3}S(S+1)(g\mu_B)^2\lambda n \tag{52.4}$$

as we also demonstrate. We next introduce the notation

$$\eta \equiv (\kappa T)^{-1} S g \mu_B \mathcal{H}_o$$
$$\xrightarrow[T \to T_C]{} \left[\frac{1}{3}(S+1) g \mu_B \lambda n\right]^{-1} \mathcal{H}_o \equiv \eta_C \tag{52.5}$$

so that Equation (52.3) can be written more concisely as

$$B_S(x) = \frac{1}{3}\left(\frac{S+1}{S}\right)\frac{T}{T_C}(x - \eta) \tag{52.3a}$$

A graphical solution $x_o = x_o(\mathcal{H}_o)$ to this equation is displayed in Figure XI.2.

The first thing to note about Figure XI.2. is that when $\lambda = 0$, then the straight line, corresponding to the right side of Equation (52.3a), goes through the x-axis parallel to the y-axis at the point $x = \eta$. It therefore intersects the Brillouin function at the point $x_o = \eta$ and the magnetization is then determined, as in Section 51, directly by Equation (52.2). When $\lambda \neq 0$ one cannot, in general, solve Equation (52.3a) analytically, and the graphical solution must be used. Spontaneous magnetization occurs whenever the

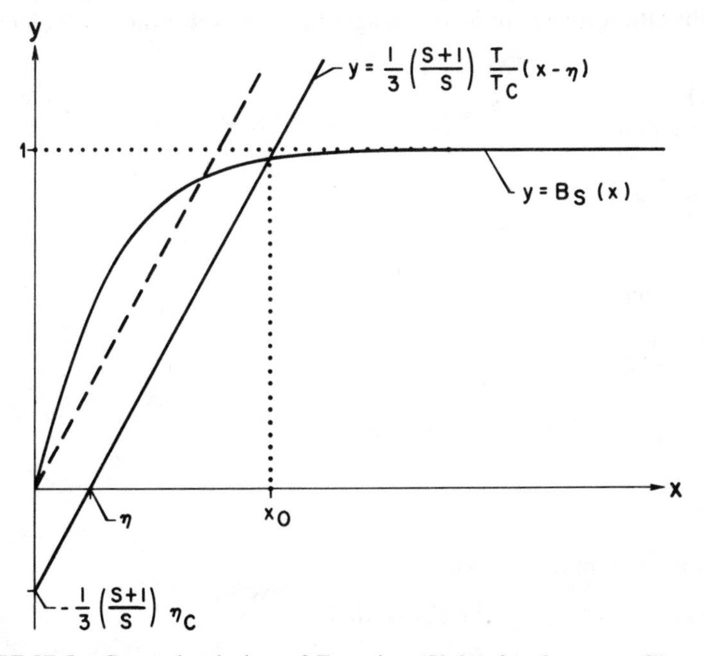

FIGURE XI.2 General solution of Equation (52.3a) for the cases $\mathcal{H}_o = 0$ (dashed line) and $\mathcal{H}_o \neq 0$, as determined by the intersection at x_o of the straight line with the $y = B_S(x)$ curve. Spontaneous magnetization (for $\mathcal{H}_o = 0$) occurs only when the dashed line intersects the Brillouin-function curve.

slope of the dashed line, for the case $\mathcal{H}_o = 0$, is less than the slope of the $y = B_S(x)$ curve at the origin. Now, from Equation (51.10) we can derive an expression for this latter slope as follows:

$$\frac{dB_S(x)}{dx} = \left(\frac{1}{2S}\right)^2 \operatorname{csch}^2\left(\frac{x}{2S}\right) - \left(\frac{2S+1}{2S}\right)^2 \operatorname{csch}^2\left(\frac{2S+1}{2S}\right)x$$

$$\xrightarrow[x\to 0]{} \frac{1}{3}\left(\frac{S+1}{S}\right)\left[1 - \frac{1}{10}\left(\frac{2S^2+2S+1}{S^2}\right)x^2 + O(x^4)\right] \tag{52.6}$$

$$\xrightarrow[S=1/2]{} \operatorname{sech}^2 x$$

$$\xrightarrow[S\to\infty]{} \frac{dL(x)}{dx} = \frac{1}{x^2} - \operatorname{csch}^2 x$$

Therefore, magnetization can occur spontaneously whenever

$$\left.\frac{dB_S(x)}{dx}\right|_{x=0} = \frac{1}{3}\left(\frac{S+1}{S}\right) > \frac{1}{3}\left(\frac{S+1}{S}\right)\frac{T}{T_C}$$

that is, whenever $T < T_C$. We may conclude that the *critical temperature* T_C has been defined correctly by (52.4). The following exercise demonstrates that this condition for spontaneous magnetization is not only sufficient, but also necessary.

Exercise 51.2

(a) Show that

$$\frac{dB_S(x)}{dx} > 0$$

for all x.

(b) Show that

$$\frac{d^2B_S(x)}{dx^2} < 0$$

for all $x \geqslant 0$ and, therefore, that

$$\frac{dB_S(x)}{dx}$$

has its maximum value at $x = 0$.

[*Suggestion for part (b):* Prove that

$$f(y) \equiv \left(\frac{\sinh y}{y}\right)^3 \Big/ \cosh y$$

is a monotonically increasing function of y for $y > 0$.]

We now discuss the constant λ. If the internal field $\mathcal{H}_{\text{int}}$ of iron were due to magnetic dipoles, then we would have $\lambda = 4\pi$, as is well-known.[8] In this case, and with $S = \frac{1}{2}$, $g \cong 2$, and $n \sim (2\text{Å})^{-3}$, the critical temperature would be determined by Equation (52.4) to be $T_C \sim 1\text{K}$. The corresponding internal magnetic field would be $\mathcal{H}_{\text{int}} \sim 4\pi n\mu_B \sim 10^4$ gauss. In fact, certain paramagnetic salts do reveal critical temperatures for spontaneous magnetization, at very low temperatures, which can be attributed to magnetic dipole interactions. But for iron the critical temperature for spontaneous magnetization is $T_C = 1043\text{K}$, corresponding to a value of $\lambda \sim 10^4$. The magnetization in a single domain for $T < T_C$ is known to be $M/\Omega \sim 1700$ gauss, so that $\mathcal{H}_{\text{int}} \sim 10^7$ gauss. Obviously, the magnetic dipole interaction cannot be invoked to explain ferromagnetism.

Heisenberg Exchange Interaction

The mechanism that seems to explain the large internal magnetic fields of ferromagnets is the electron Coulomb exchange interaction, first proposed by Heisenberg. The atoms in an ionic crystal are, of course, bound by Coulomb forces between the ions and electrons. The direct Coulomb energy term $K_{\mu\nu} = \langle \mu\nu | V_{\text{coul}} | \mu\nu \rangle$, which arises in the first-order perturbation theory calculation for two electrons in atomic states μ and ν, gives a repulsive contribution to this binding. There is, moreover, an exchange energy term

$$\pm J_{\mu\nu} = \pm \langle \nu\mu | V_{\text{coul}} | \mu\nu \rangle$$

Now the magnetic dipole energy is roughly $\mu_B \mathcal{H}_{\text{int}} \sim 4\pi n \mu_B^2 \sim 10^{-4}$ eV, whereas single- (outer) electron electrostatic exchange energies in crystals are ~ 0.1 eV. Therefore, the exchange interaction, for electrons in different atoms, can have the order of magnitude necessary to explain ferromagnetism. For electrons, with $V_{\text{coul}} > 0$, this exchange interaction is repulsive $(+J_{\mu\nu})$ for singlet states and attractive $(-J_{\mu\nu})$ for triplet states.[9] Thus the exchange interaction favors spontaneous magnetization.

Consider two electrons in states μ and ν with corresponding spin operators $\mathbf{s}_\mu = \frac{1}{2}\underline{\boldsymbol{\sigma}}_\mu$ and $\mathbf{s}_\nu = \frac{1}{2}\underline{\boldsymbol{\sigma}}_\nu$, where $\underline{\sigma}_{\mu,i}$ is a Pauli spin matrix. The sign $(\pm)$ of the Coulomb exchange interaction for these two electrons can be represented by the operator $-(\frac{1}{2} + 2\mathbf{s}_\mu \cdot \mathbf{s}_\nu) = -\frac{1}{2}(1 + \underline{\boldsymbol{\sigma}}_\mu \cdot \underline{\boldsymbol{\sigma}}_\nu) = \pm 1$, in which the $+$ sign corresponds to the singlet state and the $-$ sign to the triplet states. Then, apart from a spin-independent factor, the exchange interaction for these two electrons can be represented as $-2\mathbf{s}_\mu \cdot \mathbf{s}_\nu \langle \nu\mu | V_{\text{coul}} | \mu\nu \rangle$. Therefore, if we represent the spin operator for the outer electrons of the αth atom by $\mathbf{S}_\alpha = \Sigma_{\mu\in\alpha}\, \mathbf{s}_\mu$, then the effective exchange interaction Hamiltonian for two atoms in a ferromagnetic domain will be

$$\mathsf{H}'_{\alpha\beta} = -2J_{\alpha\beta}\, \mathbf{S}_\alpha \cdot \mathbf{S}_\beta \qquad (52.7)$$

We have assumed here, for reasons of simplicity, that all relevant outer electrons (i.e., those not in closed shells) in an atom have the same exchange interaction $J_{\alpha\beta}$.[10] This expression for $\mathsf{H}'_{\alpha\beta}$ is the *Heisenberg exchange interaction* used in the theory of ferromagnetism. If we sum over all pairs of atoms in the domain, then we obtain the complete "spin-interaction Hamiltonian" due to Coulomb exchange,

$$\mathsf{H}' = -\sum_{\alpha\neq\beta} J_{\alpha\beta}\,\mathbf{S}_\alpha\cdot\mathbf{S}_\beta \tag{52.7a}$$

The theory of ferromagnetism, as we have outlined it so far, was due originally to Pierre Weiss. We consider next attempts to derive a self-consistent internal magnetic field $\mathcal{H}_{\text{int}}$ based on the Heisenberg exchange interaction, following a method first developed by Peter R. Weiss.[11] This latter theoretical development can be understood in first approximation by the following simplified analysis. We assume, for atom α, that $J_{\alpha\beta}$ is appreciable only for the z nearest-neighbor atoms β. We then define the internal field $\mathcal{H}_{\text{int}}$ by equating $-\underline{\boldsymbol{\mu}}_\alpha\cdot\mathcal{H}_{\text{int}}$ to the average interaction energy experienced by the αth atom. Upon using Equation (52.7) and replacing $J_{\alpha\beta}$ by an average interaction energy J, we obtain

$$g\mu_B\mathbf{S}_\alpha\cdot\mathcal{H}_{\text{int}} = -2J\,\mathbf{S}_\alpha\cdot\sum_{\beta=1}^{z}\langle\mathbf{S}_\beta\rangle$$

which implies that

$$\mathcal{H}_{\text{int}} = -\frac{2J}{g\mu_B}\sum_{\beta=1}^{z}\langle\mathbf{S}_\beta\rangle \tag{52.8}$$

For a simple crystal, $\langle\mathbf{S}_\beta\rangle$ must be independent of β and, moreover, it is related to the magnetic moment $\langle\underline{\boldsymbol{\mu}}\rangle$ by $\langle\underline{\boldsymbol{\mu}}\rangle = -g\mu_B\langle\mathbf{S}\rangle$. We can therefore use the relation $M = N\langle\underline{\mu}\rangle$ to derive the result

$$\begin{aligned}\mathcal{H}_{\text{int}} &= \frac{2zJ}{(g\mu_B)^2}\cdot\frac{M}{N} = \frac{2zJ}{n(g\mu_B)^2}\cdot\frac{M}{\Omega}\\ &\equiv \lambda\,M/\Omega\end{aligned} \tag{52.9}$$

Thus in the simplest quantum-mechanical version of the Weiss molecular-field approximation we find that

$$\lambda = \frac{2zJ}{n(g\mu_B)^2} \tag{52.10}$$

which, for $zJ \sim 0.2$ eV, $g \cong 2$ and $n \sim (2\ \text{Å})^{-3}$, gives $\lambda \sim 10^4$.

The subject of ferromagnetism also includes the study of ferromagnetic domains.[12] At low temperatures actual ferromagnetic specimens often appear to be composed of a number of randomly oriented small regions called domains within each of which the local magnetization is saturated according to the above theory, or its generalization. Magnetic saturation at $\mathcal{H}_o = 0$ corresponds in Equation (52.3a) to the values $B_S(x_o) = 1$ and $\eta = 0$, in which case one obtains

$$x_o = \left(\frac{Sg\mu_B}{\kappa T}\right)\lambda n\langle\mu\rangle = \left(\frac{3S}{S+1}\right)\cdot\frac{T_C}{T}$$

or equivalently, and as one would expect, $\langle\mu\rangle = Sg\mu_B$. This condition corresponds to a ground state for the magnetic energy of a simple ferromagnet, or single domain. The elementary excitations above this ground state are quantized *spin waves,* called *magnons,* whose investigation is yet another interesting subject.[13] We conclude our treatment here by investigating the magnetization and its fluctuation in a simple ferromagnet near its critical temperature.

Behavior Near the Critical Temperature

We first inquire how it can be that a simple ferromagnet will ever exhibit spontaneous magnetization, because, in the absence of an externally provided axis in space, the system would have to be symmetric with respect to arbitrary rotations. The thermodynamic potential would then be a minimum at thermodynamic equilibrium for zero magnetization [inequality (9.15)], and indeed it is above the critical temperature. How then can the rotational symmetry of the system be "broken" below the critical temperature? The answer is that although spontaneous magnetization apparently destroys rotational symmetry for $T < T_C$, it must do so in an arbitrary direction. The minimum of G that occurs at nonzero magnetization when $T < T_C$ occurs with equal probability in all directions. We now examine the onset of this magnetization, within the framework of the molecular-field approximation, as the temperature is lowered to T_C.

When the temperature is close to or greater than the critical temperature and the external magnetic field $\mathcal{H}_o$ is weak ($\eta \ll 1$), then we can expect that the solution x_o, as determined graphically in Figure XI.2, will be a small number. In this case we can use the small-parameter expansion of $B_S(x)$, Equation (51.10), to obtain from Equation (52.3a) the result

$$x_o\left(\frac{T}{T_C} - 1\right) + \frac{\alpha_S}{3}x_o^3 + O(x_o^5) = \eta_C \tag{52.11}$$

where $\eta_C \sim \mathcal{H}_o/\mathcal{H}_{\text{int}} \ll 1$ is defined in the second line of (52.5), $\mathcal{H}_{\text{int}} = \lambda n\langle\mu\rangle$

and

$$\alpha_S \equiv \frac{1}{10}\left(\frac{2S^2 + 2S + 1}{S^2}\right) \xrightarrow[S=1/2]{} 1 \xrightarrow[S\to\infty]{} \frac{1}{5} \tag{52.12}$$

We can solve Equation (52.11) to find for x_o the limiting values,

$$\begin{aligned} x_o &\cong \frac{\eta_C}{\left(\frac{T}{T_C} - 1\right)} \ll 1 \qquad \text{for } \left(\frac{T}{T_C} - 1\right) \gg \eta_C^{2/3} \\ &\cong \left(\frac{3}{\alpha_S}\cdot \eta_C\right)^{1/3} \ll 1 \qquad \text{for } \left|\frac{T}{T_C} - 1\right| \ll \eta_C^{2/3} \\ &\cong \left(\frac{3}{\alpha_S}\right)^{1/2}\left(1 - \frac{T}{T_C}\right)^{1/2}\left[1 + \frac{1}{2}\left(\frac{\alpha_S}{3}\right)^{1/2}\frac{\eta_C}{(1 - T/T_C)^{3/2}} + O\left(1 - \frac{T}{T_C}\right)\right] \\ &\qquad \text{for } 1 > \left(1 - \frac{T}{T_C}\right) \gg \eta_C^{2/3}. \end{aligned} \tag{52.13}$$

The corresponding expressions for $B_S(x_o)$ are

$$\begin{aligned} B_S(x_o) &\cong \frac{1}{3}\left(\frac{S+1}{S}\right)\frac{\eta_C}{\left(\frac{T}{T_C} - 1\right)} \qquad \text{for } \left(\frac{T}{T_C} - 1\right) \gg \eta_C^{2/3} \\ &\cong \frac{1}{3}\left(\frac{S+1}{S}\right)\left(\frac{3}{\alpha_S}\eta_C\right)^{1/3} \qquad \text{for } \left|\frac{T}{T_C} - 1\right| \ll \eta_C^{2/3} \\ &\cong \frac{1}{3}\left(\frac{S+1}{S}\right)\left(\frac{3}{\alpha_S}\right)^{1/2}\left(1 - \frac{T}{T_C}\right)^{1/2} \qquad \text{for } \left(1 - \frac{T}{T_C}\right) \gg \eta_C^{2/3} \end{aligned} \tag{52.14}$$

We now use this last result, together with Equations (51.8) and (52.2), to find for the magnetic susceptibility, when $T > T_C$,

$$\begin{aligned} \chi \equiv \frac{1}{\Omega}\left(\frac{M}{\mathcal{H}_o}\right) &= \frac{C}{T - T_C} \qquad \text{for } \left(\frac{T}{T_C} - 1\right) \gg \eta_C^{2/3} \\ &= \left(\frac{3}{\alpha_S \eta_C^2}\right)^{1/3}\frac{C}{T_C} \propto \mathcal{H}_o^{-2/3} \qquad \text{for } \left(\frac{T}{T_C} - 1\right) \ll \eta_C^{2/3} \end{aligned} \tag{52.15}$$

where

$$C \equiv \frac{T_C}{\lambda} = \frac{1}{3}S(S+1)\frac{n}{\kappa}(g\mu_B)^2 \tag{52.16}$$

The first line of Equation (52.15) gives the well-known *Curie–Weiss law* for the magnetic susceptibility above the critical point T_C of a ferromagnet.[14] According to the second line of Equation (52.15) a plot of χ^{-1} deviates from linear dependence on $(T - T_c)$ toward an intercept that gives $\chi^{-1} = 0$ at an actual critical temperature less than T_C, except in the limit $\mathscr{H}_o \rightarrow 0$ when this second case does not apply. This behavior is qualitatively similar to that observed experimentally (see Figure XI.3). However, the present theory is in two respects especially inadequate near the critical temperature.

On the one hand, the molecular-field theory breaks down near the critical temperature, because spin fluctuations, to be discussed further below, become extremely large when $|T - T_C| \ll T_C$. The present description of the long-range order, or spontaneous magnetization, then becomes incorrect. In this respect, the mean-field theory discussed in this section is quite similar to the Ornstein–Zernicke theory of density fluctuations near the critical point of a fluid, discussed at the end of Section 49. There it was demonstrated, in connection with inequalities (49.36), that the description of long-range order by the Ornstein–Zernicke theory is not valid when $|T - T_C|$ becomes extremely small. But the molecular-field theory also fails to describe short-range order correctly near the critical temperature, because nearest-neighbor interactions are not treated with sufficient care in this temperature region. Of course, long- and short-range orders in a substance are not totally independent effects.[15]

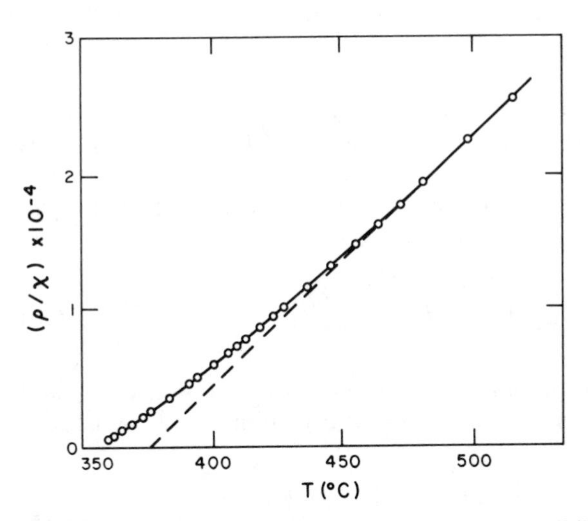

FIGURE XI.3 Reciprocal of the magnetic susceptibility per gram (ρ = mass density) of nickel in the neighborhood of the Curie–Weiss temperature (354°C = 627K). The dashed line is a linear extrapolation from high temperatures. Observed values deviate from linearity over a temperature scale that is ~10 times that suggested by Equation (52.15).[16]

We conclude this section by applying our present results to a calculation of magnetization fluctuations in a simple ferromagnet. We focus attention on one domain that is part of a larger magnetic medium, all in a fixed weak external field $\mathcal{H}_o$. Furthermore, we assume that the total magnetization ($M_{\text{med}} + M$) of this entire system can be held constant. Now when the magnetization of a single domain changes in a fixed external field $\mathcal{H}_o$, then the corresponding energy change is $dE = \mathcal{H}_o\, dM$.[17] The probability that a spontaneous fluctuation $(\Delta M)^2$ will occur is

$$P(\Delta M) \propto \exp\left\{-\frac{1}{2\kappa T}\left(\frac{\partial^2 E}{\partial M^2}\right)(\Delta M)^2\right\} \tag{52.17}$$

Exercise 52.2 Use the method of Problem III.7 to prove Equation (52.17).

From Equation (52.17) we can deduce that the rms fluctuation in the magnetization of a system is given by

$$(\Delta M)_{\text{rms}} = \left[\frac{1}{\kappa T}\left(\frac{\partial^2 E}{\partial M^2}\right)\right]^{-1/2} = \left[\frac{1}{\kappa T}\left(\frac{\partial \mathcal{H}_o}{\partial M}\right)\right]^{-1/2} = \left(\kappa T\frac{\partial M}{\partial \mathcal{H}_o}\right)^{1/2} \tag{52.18}$$

This expression can be rewritten, using Equation (52.2), as

$$\frac{1}{\sqrt{N}}(\Delta M)_{\text{rms}} = \left[\kappa T S g \mu_B \frac{dB_S(x_o)}{dx_o}\frac{\partial x_o}{\partial \mathcal{H}_o}\right]^{1/2} \tag{52.19}$$

Upon referring to Figure XI.2 and Equation (52.6), one can readily check that the fluctuations are well-behaved in the limit $\mathcal{H}_o \to 0$ at all temperatures except possibly near the critical temperature T_C. Near the critical temperature, where $x_o << 1$, we may use the second line of Equation (52.6) to simplify Equation (52.19) and obtain the limiting expression.

$$\frac{1}{\sqrt{N}}(\Delta M)_{\text{rms}}\Bigg|_{\mathcal{H}_o=0} \xrightarrow[T\to T_C]{} \left[\frac{1}{3}(S+1)g\mu_B\kappa T_C\right]^{1/2}\left(\frac{\partial x_o}{\partial \mathcal{H}_o}\right)^{1/2}_{\mathcal{H}_o=0} \tag{52.19a}$$

We next determine the partial derivative $(\partial x_o/\partial \mathcal{H}_o)$ for T near T_C and $\mathcal{H}_o \to 0$. According to the first and third lines of Equation (52.13) we have

$$\begin{aligned}\lim_{\mathcal{H}_o\to 0}\left(\frac{\partial x_o}{\partial \mathcal{H}_o}\right) &= \left[\frac{1}{3}(S+1)g\mu_B\lambda n\right]^{-1}\left(\frac{T}{T_C}-1\right)^{-1} \qquad (T > T_C) \\ &= \frac{1}{2}\left[\frac{1}{3}(S+1)g\mu_B\lambda n\right]^{-1}\left(1-\frac{T}{T_C}\right)^{-1} \qquad (T \leqslant T_C)\end{aligned} \tag{52.20}$$

Substitution of this result into Equation (52.19a) yields a final expression for the magnetization, or spin, fluctuation in a ferromagnetic domain near the critical temperature, and according to the molecular-field theory.

$$\frac{1}{\sqrt{N}}(\Delta M)_{\text{rms}}\bigg|_{\mathcal{H}_o=0} \xrightarrow[|1-(T/T_C)|<<1]{} \left[\frac{1}{3}S(S+1)\right]^{1/2}(g\mu_B)\left|1-\frac{T}{T_C}\right|^{-1/2} \times \begin{cases} 1 & \text{if } T > T_C \\ \dfrac{1}{\sqrt{2}} & \text{if } T < T_C \end{cases} \quad \textbf{(52.21)}$$

Thus the spin fluctuations in a ferromagnet become infinite at the critical temperature, just as do the density fluctuations in an ordinary fluid near its liquid–vapor phase transition.[18] The spin fluctuations near the critical temperature, of course, are associated with the onset of a long-range order, which is their spontaneous alignment when $T < T_C$.

Finally, we give an indication of the range of validity of the molecular-field theory in the vicinity of the critical temperature. Our validity criterion will be that the magnetization fluctuation $\langle\Delta\mu(\mathbf{r}_1)\Delta\mu(\mathbf{r}_2)\rangle$ for two spins separated by a distance over which long-range order is established (the "coherence length," ξ) must be small compared with $\langle\mu\rangle^2$ itself. Next we appeal to the corresponding analysis for the Ornstein–Zernicke theory given at the end of Section 49 [see also Problem X.7(b)]. There we saw that whereas the (approximate) theory gives $(\Delta N)^2_{\text{rms}} \propto (T - T_C)^{-1}$ for $T \to T_C^{(+)}$, the corresponding fluctuation $d(\mathbf{r})$ for $r \sim b$ has $d(b) \propto (T - T_C)^{1/2}$. By analogy, and after referring to Equation (52.21), we may expect that $\langle\Delta\mu(\mathbf{r}_1)\Delta\mu(\mathbf{r}_2)\rangle \propto (T_C - T)^{1/2}$ for $|\mathbf{r}_1 - \mathbf{r}_2| \sim \xi$ and $T \lesssim T_C$. On the other hand, Equations (52.2) and (52.14) show that $\langle\mu\rangle^2 \propto (T_C - T)$ for $T \lesssim T_C$. Therefore our validity criterion must give a minimum condition for the parameter $\varepsilon \equiv (1 - T/T_C) << 1$; that is, the molecular-field theory is not valid for very small $|\varepsilon|$ values.[19]

53 Relativistic Statistics

In Chapter VII we discussed degenerate gases and included applications to several real and ideal systems. Further examples in nature occur in astrophysics, and concepts that emerge from the study of degenerate gases serve usefully in starting models for certain stars. Thus the white dwarf star and the neutron star are examples of highly degenerate Fermi gases, at least in part. We begin this section by presenting numerology associated with a typical white dwarf star to demonstrate that its electrons must be treated using relativistic statistics. We also establish that the primary active constituent, thermodynamically, in a white dwarf star is its electron gas. Then we outline the mathematical treatment of an ideal gas of relativistic fermions in both the degenerate and nondegenerate limits.

It is believed that the densities of matter in white dwarf stars range around $\rho \sim 10^6$ gm/cm^3 and that very little of this matter is hydrogen. The electron number density, which is approximately half the nucleon number density, is than $n_e \sim 3 \times 10^{29}$ electrons/cm^3. We use Equation (19.2) to define the Fermi momentum, even for relativistic statistics, and therefore we obtain for white dwarf stars, $k_F \sim 2 \times 10^{10}$ cm^{-1}. The corresponding relativistic Fermi energy is

$$
\begin{aligned}
E_F &\equiv m_e c^2 \left[\left(1 + \left(\frac{\hbar k_F}{m_e c} \right)^2 \right)^{1/2} - 1 \right] \\
&\sim \frac{1}{2}(m_e c^2) \cong \frac{1}{4}\,\text{MeV} \sim \kappa(3 \times 10^9\,\text{K})
\end{aligned}
\tag{53.1}
$$

These numbers show that the electrons in a white dwarf star involve relativistic, but not extreme-relativistic, energies.

Now the interior temperature of a typical white dwarf star is $T \sim 10^7$K. We can then conclude from the above numbers that its electrons form a highly degenerate, relativistic Fermi gas and, in fact, calculations based on this model are in agreement with the known observational facts about these stars. Finally, we notice that the average kinetic energy of the ions in the star is $\kappa T \sim 1$ keV $\ll (E_F)_{\text{el}}$. In the case of a white dwarf star composed entirely of helium, for which the ion-number density would be $\frac{1}{2}n_e$, we can then estimate that the electron pressure is ~200 times the ion pressure.[20] Therefore, in first approximation we can consider that the thermodynamic properties of these stars are due to the electron gas, which is essentially an ideal gas because of the complete ionization that must occur.

Degenerate Fermi Gas

We investigate the thermodynamic functions of a highly degenerate ideal Fermi gas, using relativistic statistics, in some detail. Equation (32.1) for the mean occupation number $\nu(k)$ and Figures VII.1 and VII.2 continue to be valid for this case, except that now for the single-particle energies we use the expression

$$
\begin{aligned}
\omega(k) &= mc^2 \left\{ \left[1 + \left(\frac{\hbar k}{mc} \right)^2 \right]^{1/2} - 1 \right\} \xrightarrow[k \to k_F]{} E_F \\
&\xrightarrow[\hbar k \ll mc]{} (\hbar k)^2/2m \qquad \text{(nonrelativistic limit)} \\
&\xrightarrow[\hbar k \gg mc]{} \hbar c k \qquad \text{(extreme-relativistic limit)}
\end{aligned}
\tag{53.2}
$$

An important consequence of this expression is that the relation (31.18), that is, $\mathscr{P}\Omega = ⅔\langle E\rangle$, holds only in the nonrelativistic limit; in the extreme relativistic limit we obtain Equation (35.10), $\mathscr{P}\Omega = ⅓\langle E\rangle$.

The zero of energy in Equation (53.2) does not include the rest-mass energy mc^2, because mass-energy conversion, such as electron–positron pair production, is not significant in a white dwarf star. Thus, as indicated above, $\kappa T \ll mc^2$ for the electrons. Nevertheless, two parameters that include the rest-mass energy, that is, $(g + mc^2)$ and $(E_F + mc^2)$, where g is the chemical potential of the Fermi gas, enter into the calculations below. Corresponding dimensionless parameters are η_o and Γ_o, defined to be

$$\begin{aligned} \eta_o &\equiv 1 + \frac{g}{mc^2} \\ \Gamma_o &\equiv 1 + \frac{E_F}{mc^2} = \left[1 + \left(\frac{\hbar k_F}{mc}\right)^2\right]^{1/2} \end{aligned} \tag{53.3}$$

Equation (19.2) can be used to relate the number density $n = N/\Omega$ to Γ_o as

$$n \equiv \frac{(2S + 1)}{6\pi^2} k_F^3 = \frac{(2S + 1)}{6\pi^2}\left(\frac{mc}{\hbar}\right)^3 (\Gamma_o^2 - 1)^{3/2} \tag{53.4}$$

We now wish to write down relativistic generalizations of Equations (32.7) and (32.8). Starting from Equations (31.13) and (31.15) to (31.17), it is straightforward to derive general expressions for the density, energy, and pressure in the thermodynamic limit. We use the variable $x = \beta[\omega(k) - g]$ and define $\eta(x)$ by the relation

$$\eta(x) \equiv 1 + \frac{g}{mc^2}\left(1 + \frac{x}{\beta g}\right) \xrightarrow[x\to 0]{} \eta_o \tag{53.5}$$

Then $\nu(x) = (e^x + 1)^{-1}$, $\hbar k/mc = (\eta^2 - 1)^{1/2}$, and we obtain:

$$\begin{aligned} n &= \frac{(2S + 1)}{2\pi^2}\int_0^\infty k^2\,dk\,\nu(k) \\ &= \frac{(2S + 1)}{6\pi^2}\left(\frac{mc}{\hbar}\right)^3 \int_{-\beta g}^{\infty} dx(\eta^2 - 1)^{3/2}\left(-\frac{d\nu}{dx}\right) \end{aligned} \tag{53.6}$$

$$\begin{aligned} \frac{\langle E\rangle}{\Omega} &= \frac{(2S + 1)}{2\pi^2}\int_0^\infty k^2\,dk\,\omega(k)\,\nu(k) \\ &= \frac{(2S + 1)}{16\pi^2}\left(\frac{mc}{\hbar}\right)^3 mc^2 \int_{-\beta g}^{\infty} dx\left\{\eta(2\eta^2 - 1)\sqrt{\eta^2 - 1} - \frac{8}{3}(\eta^2 - 1)^{3/2}\right. \\ &\qquad \left. - \ln(\eta + \sqrt{\eta^2 - 1})\right\}\left(-\frac{d\nu}{dx}\right) \end{aligned} \tag{53.7}$$

$$\mathcal{P} = \frac{(2S+1)}{6\pi^2}\int_0^\infty k^3\frac{d\omega(k)}{dk}\,dk\,\nu(k)$$

$$= -\frac{\langle E\rangle}{\Omega} + \frac{(2S+1)}{6\pi^2}\left(\frac{mc}{\hbar}\right)^3 mc^2\int_{-\beta g}^{\infty} dx(\eta-1)(\eta^2-1)^{3/2}\left(-\frac{d\nu}{dx}\right)$$

$$= \frac{(2S+1)}{16\pi^2}\left(\frac{mc}{\hbar}\right)^3 mc^2\int_{-\beta g}^{\infty} dx\left\{-\eta(2\eta^2-1)\sqrt{\eta^2-1} + \frac{8}{3}\eta(\eta^2-1)^{3/2} + \ln(\eta+\sqrt{\eta^2-1})\right\}\left(-\frac{d\nu}{dx}\right) \tag{53.8}$$

$$\xrightarrow[g\ll mc^2]{} \frac{2}{3}\frac{\langle E\rangle}{\Omega} \qquad \text{(nonrelativistic limit)}$$

$$\xrightarrow[g\gg mc^2]{} \frac{1}{3}\frac{\langle E\rangle}{\Omega} \qquad \text{(extreme-relativistic limit)}$$

We now use the analysis following Equation (32.8) to evaluate the above integrals in the low-temperature limit where $\beta g \gg 1$.

According to Equations (32.11) to (32.14), the general low-temperature expansion for the integral $I(F)$, defined by (32.10), is

$$I(F) = F(0) + \frac{\pi^2}{6}F^{(2)}(0)\left(\frac{\kappa T}{g}\right)^2 + \frac{7\pi^4}{360}F^{(4)}(0)\left(\frac{\kappa T}{g}\right)^4 + O\left(\frac{\kappa T}{g}\right)^6 \tag{53.9}$$

If we recall that F is defined to be a function of the variable $(\kappa T/g)x$, then from Equations (53.5) to (53.8) we can derive the expressions:

$$n = \frac{(2S+1)}{6\pi^2}\left(\frac{mc}{\hbar}\right)^3 \times\left\{(\eta_o^2-1)^{3/2} + \frac{\pi^2}{2}\frac{(2\eta_o^2-1)}{\sqrt{\eta_o^2-1}}\left(\frac{\kappa T}{mc^2}\right)^2 + O\left(\frac{\kappa T}{mc^2}\right)^4\right\} \tag{53.10a}$$

$$\frac{\langle E\rangle}{\Omega} = \frac{(2S+1)}{16\pi^2}\left(\frac{mc}{\hbar}\right)^3 \times mc^2\left\{\eta_o(2\eta_o^2-1)\sqrt{\eta_o^2-1} - \frac{8}{3}(\eta_o^2-1)^{3/2} - \ln(\eta_o+\sqrt{\eta_o^2-1}) + \frac{4\pi^2}{3}\left(\frac{\eta_o-1}{\eta_o+1}\right)^{1/2} \times(3\eta_o^2+\eta_o-1)\left(\frac{\kappa T}{mc^2}\right)^2 + O\left(\frac{\kappa T}{mc^2}\right)^4\right\} \tag{53.10b}$$

$$\mathscr{P} = \frac{(2S+1)}{16\pi^2}\left(\frac{mc}{\hbar}\right)^3 \times mc^2\left\{-\eta_o(2\eta_o^2 - 1)\sqrt{\eta_o^2 - 1} + \frac{8}{3}\eta_o(\eta_o^2 - 1)^{3/2} + \ln(\eta_o + \sqrt{\eta_o^2 - 1}) + \frac{4\pi^2}{3}\eta_o\sqrt{\eta_o^2 - 1}\left(\frac{\kappa T}{mc^2}\right)^2 + O\left(\frac{\kappa T}{mc^2}\right)^4\right\} \tag{53.10c}$$

where $\eta_o = 1 + (g/mc^2)$. We see from these expressions that in the low-temperature limit for a relativistic Fermi gas, the thermodynamic functions become expansions in the parameter $(\kappa T/mc^2)^2 << 1$.

To determine the unknown parameter η_o, or the chemical potential g, we invert the first of Equations (53.10) with the aid of Equations (53.3) and (53.4). After some algebraic manipulations we find

$$\eta_o = \left(1 + \frac{g}{mc^2}\right) = \Gamma_o\left\{1 - \frac{\pi^2}{6}\frac{(2\Gamma_o^2 - 1)}{\Gamma_o^2(\Gamma_o^2 - 1)}\left(\frac{\kappa T}{mc^2}\right)^2 + O\left(\frac{\kappa T}{mc^2}\right)^4\right\} \tag{53.11}$$

In the low-temperature limit the parameters η_o and Γ_o are equal, which implies, as we would expect, that $g = E_F$ in this limit. Also, in the nonrelativistic limit Equation (53.11) is equivalent to Equation (32.17).If we substitute Equation (53.11) into the second and third of Equations (53.10), we obtain the results:

$$\frac{\langle E\rangle}{\Omega} = \frac{(2S+1)}{16\pi^2}\left(\frac{mc}{\hbar}\right)^3 \times mc^2\left\{\Gamma_o(2\Gamma_o^2 - 1)\sqrt{\Gamma_o^2 - 1} + \frac{8}{3}(\Gamma_o^2 - 1)^{3/2} - \ln(\Gamma_o + \sqrt{\Gamma_o^2 - 1}) + \frac{4\pi^2}{3}\Gamma_o\sqrt{\Gamma_o^2 - 1}\left(\frac{\kappa T}{mc^2}\right)^2 + O\left(\frac{\kappa T}{mc^2}\right)^4\right\} \tag{53.12}$$

$$\mathscr{P} = \frac{(2S+1)}{16\pi^2}\left(\frac{mc}{\hbar}\right)^3 \times mc^2\left\{-\Gamma_o(2\Gamma_o^2 - 1)\sqrt{\Gamma_o^2 - 1} + \frac{8}{3}\Gamma_o(\Gamma_o^2 - 1)^{3/2} + \ln(\Gamma_o + \sqrt{\Gamma_o^2 - 1}) + \frac{4\pi^2}{9}\sqrt{\Gamma_o^2 - 1}\left(\frac{\Gamma_o^2 + 1}{\Gamma_o}\right)\left(\frac{\kappa T}{mc^2}\right)^2 + O\left(\frac{\kappa T}{mc^2}\right)^4\right\}$$

where the density is related to Γ_o by Equation (53.4).

Exercise 53.1

(a) Use $\Gamma_o = 1 + (E_F/mc^2)$ and Equation (53.4) to show that Equations (53.12) reduce to Equations (32.18) and (31.18) in the nonrelativistic limit.

(b) With the aid of the thermodynamic identity

$$T\frac{\langle S\rangle}{\Omega} = \mathcal{P} + \frac{\langle E\rangle}{\Omega} - ng = \mathcal{P} + \frac{\langle E\rangle}{\Omega} - nmc^2(\eta_o - 1)$$

show that the entropy of a relativistic Fermi gas at very low temperatures is given by the expression

$$\frac{\langle S\rangle}{\kappa\Omega} = \frac{1}{6}(2S+1)\left(\frac{mc}{\hbar}\right)^3\left(\frac{\kappa T}{mc^2}\right)\Gamma_o\sqrt{\Gamma_o^2 - 1}\left\{1 + O\left(\frac{\kappa T}{mc^2}\right)^2\right\} \qquad (53.13)$$

which reduces to Equation (32.19) in the nonrelativistic limit.

Nondegenerate Fermi Gas

We now examine the thermodynamic expressions for a relativistic Fermi gas in the nondegenerate region, where $\kappa T >> mc^2$. For simplicity, however, we do not treat quantum effects or the associated radiation field, which must be included in any complete analysis of this high-energy region. For this high-temperature case, we include the rest-mass energy mc^2 in the definition of ω_k, which implies that we must also include it in the chemical potential g'. Thus we set

$$\begin{aligned} g' &= g + mc^2 \\ \omega_k &= [(mc^2)^2 + (\hbar ck)^2]^{1/2} \end{aligned} \qquad (53.14)$$

where g is the chemical potential used in the above treatment of the highly degenerate region. We note that g' is related to the parameter η_o of Equation (53.3) by $g' = \eta_o mc^2$.

We substitute Equation (32.1) for $\nu(k)$ into the first lines of Equations (53.6) to (53.8), and use the integration variable $x = (\omega_k/mc^2)$, to obtain the expanded forms for n, $\langle E\rangle/\Omega$, and $\mathcal{P}$:

$$\begin{aligned} n &= \frac{(2S+1)}{2\pi^2}\left(\frac{mc}{\hbar}\right)^3 \sum_{s=1}^{\infty} \varepsilon^{s-1} e^{s\beta g'} \int_1^{\infty} x\sqrt{x^2-1}\, e^{-s\alpha x}\, dx \\ \frac{\langle E\rangle}{\Omega} &= \frac{(2S+1)}{2\pi^2}\left(\frac{mc}{\hbar}\right)^3 mc^2 \sum_{s=1}^{\infty} \varepsilon^{s-1} e^{s\beta g'} \int_1^{\infty} x^2\sqrt{x^2-1}\, e^{-s\alpha x}\, dx \\ \mathcal{P} &= \frac{(2S+1)}{6\pi^2}\left(\frac{mc}{\hbar}\right)^3 mc^2 \sum_{s=1}^{\infty} \varepsilon^{s-1} e^{s\beta g'} \int_1^{\infty} (x^2-1)^{3/2}\, e^{-s\alpha x}\, dx \\ &= \frac{(2S+1)}{2\pi^2}\left(\frac{mc}{\hbar}\right)^3 \kappa T \sum_{s=1}^{\infty} \frac{\varepsilon^{s-1}}{s} e^{s\beta g'} \int_1^{\infty} x\sqrt{x^2-1}\, e^{-s\alpha x}\, dx \end{aligned} \qquad (53.15)$$

where $\alpha = mc^2/\kappa T$ and $\varepsilon = -1$ for Fermi statistics. These integrals can be evaluated exactly in terms of Bessel functions after making the change-of-variable substitution $x = \cosh\theta$. We do not follow this general procedure, but rather we consider only the nondegenerate limit with relativistic statistics in which $\alpha \ll 1$. We then obtain the following asymptotic expansions of the integrals in Equations (53.15):

$$n = \frac{(2S+1)}{2\pi^2}\left(\frac{mc}{\hbar}\right)^3 \sum_{s=1}^{\infty} \varepsilon^{s-1} e^{s\beta g'} \int_1^{\infty} x^2\, dx\, e^{-s\alpha x}\left(1 - \frac{1}{2x^2} + \cdots\right)$$

$$= \frac{(2S+1)}{\pi^2}\left(\frac{\kappa T}{\hbar c}\right)^3 \sum_{s=1}^{\infty} \frac{\varepsilon^{s-1}}{s^3} e^{s\beta g}\left[1 + s\alpha + \frac{1}{4}s^2\alpha^2 + \cdots\right]$$

$$\frac{\langle E\rangle}{\Omega} = \frac{3(2S+1)}{\pi^2}\left(\frac{\kappa T}{\hbar c}\right)^3 \kappa T \sum_{s=1}^{\infty} \frac{\varepsilon^{s-1}}{s^4} e^{s\beta g} \tag{53.16}$$

$$\times\left[1 + s\alpha + \frac{5}{12}s^2\alpha^2 + \frac{1}{12}s^3\alpha^3 + \cdots\right]$$

$$\mathcal{P} = \frac{(2S+1)}{\pi^2}\left(\frac{\kappa T}{\hbar c}\right)^3 \kappa T \sum_{s=1}^{\infty} \frac{\varepsilon^{s-1}}{s^4} e^{s\beta g}\left[1 + s\alpha + \frac{1}{4}s^2\alpha^2 + \cdots\right]$$

We note that in the final expressions (53.16) the chemical potential $g = g' - mc^2$ has again appeared. In the temperature–density region

$$n \ll \left(\frac{\kappa T}{\hbar c}\right)^3 \tag{53.17}$$

we need to keep only the $s = 1$ terms in Equations (53.16), inasmuch as this condition implies that $e^{\beta g} \ll 1$. On solving the first of these equations for $e^{\beta g}$, we obtain the approximate expression

$$e^{\beta g} = \frac{\pi^2}{(2S+1)} n\left(\frac{\hbar c}{\kappa T}\right)^3\left[1 - \alpha + \frac{3}{4}\alpha^2 + \cdots\right] + O\left(\frac{\hbar c}{\kappa T}\right)^6 n^2$$
$$\cong \frac{\pi^2}{(2S+1)} n\left(\frac{\hbar c}{\kappa T}\right)^3 < 1 \tag{53.18}$$

This equation then shows that the chemical potential g is large and negative, that is, $-g > \kappa T \gg mc^2$, in this region. Condition (53.17) is equivalent to the condition on the interparticle spacing $\ell = n^{-1/3}$, $\ell \gg (\hbar c/\kappa T) = (\hbar/mc)\alpha$, where $\hbar/mc$ is the Compton wavelength. When this condition holds we get the familiar ideal-gas law $\mathcal{P} = n\kappa T$ from the first and third of Equations (53.15). The first and second of Equations (53.16), when combined, yield the following result

$$\frac{\langle E\rangle}{\Omega} = 3n\kappa T\left[1 + \frac{1}{6}\alpha^2 + \cdots\right] \cong 3n\kappa T \qquad (\kappa T \gg mc^2) \tag{53.19}$$

which is equivalent to the expected expression for this relativistic region, namely, $\langle E\rangle \cong 3\mathcal{P}\Omega$. The results (53.16) provide a relativistic generalization of the weak-degeneracy case treated at the end of Section 31.

54 White Dwarf Star Model

We derive thermodynamic results for a white dwarf star model at zero temperature. We have already discussed in Section 53 that in first approximation we need consider only its degenerate electron gas for the thermodynamic description of a white dwarf star, and we begin by citing appropriate limiting expressions for this model. Of course, the ion background determines the mass density of the star as well as its dynamics via gravitational attraction. We include these considerations in a derivation of the general equation of state. This equation determines a density–radius relation for the star, which is then investigated qualitatively. The maximum, or limiting, mass M_3 for white dwarf stars is studied.

The main thermodynamic results we need from Section 53 are Equations (53.12) for the energy density and pressure of a degenerate free-electron gas. We first repeat these results here for the limit $T = 0$, using both the parameters $y = \hbar k_F/m_e c$ and

$$\Gamma_o = 1 + \frac{E_F}{m_e c^2} = (1 + y^2)^{1/2} \tag{54.1}$$

where m_e is the electron mass. On setting $S = ½$, we obtain

$$\begin{aligned} \mathcal{P}_e &\xrightarrow[T\to 0]{} A\, f(\Gamma_o) \\ u_e \equiv \frac{\langle E_e\rangle}{\Omega} &\xrightarrow[T\to 0]{} A\, g(\Gamma_o) \end{aligned} \tag{54.2}$$

where

$$A \equiv (\pi/3)\left(\frac{m_e c}{h}\right)^3 m_e c^2$$

and

$$\begin{aligned} f(\Gamma_o) &= \Gamma_o(2\Gamma_o^2 - 5)\sqrt{\Gamma_o^2 - 1} + 3\ln\left(\Gamma_o + \sqrt{\Gamma_o^2 - 1}\right) \\ &= y(2y^2 - 3)\sqrt{1 + y^2} + 3\ln\left(y + \sqrt{1 + y^2}\right) \end{aligned} \tag{54.3}$$

$$f(\Gamma_o) \xrightarrow[y \ll 1]{} \frac{8}{5} y^5 - \frac{4}{7} y^7 + O(y^9) \qquad \text{(nonrelativistic limit)}$$

$$\xrightarrow[y \gg 1]{} 2y^4 - 2y^2 + 3 \ln 2y - \frac{7}{4} + O(y^{-2})$$

(extreme-relativistic limit)

$$\begin{aligned} g(\Gamma_o) &= 8(\Gamma_o - 1)(\Gamma_o^2 - 1)^{3/2} - f(\Gamma_o) \\ &= 8y^3\left(\sqrt{1 + y^2} - 1\right) - f(\Gamma_o) \end{aligned} \tag{54.4}$$

$$\xrightarrow[y \ll 1]{} \frac{12}{5} y^5 - \frac{3}{7} y^7 + O(y^9) \qquad \text{(nonrelativistic limit)}$$

$$\xrightarrow[y \gg 1]{} 6y^4 - 8y^3 + 6y^2 - 3 \ln 2y + \frac{3}{4} + O(y^{-2})$$

(extreme-relativistic limit)

The electron-density expression (53.4) can be expressed in terms of the mass density ρ of the star, so as to include the ion background. To accomplish this objective, we note that the density ρ_α of the αth nuclear component in the star, with molecular weight A_α in gm/mole and atomic number Z_α, corresponds to an electron-density contribution

$$n_\alpha = Z_\alpha N_o A_\alpha^{-1} x_\alpha \rho \tag{54.5}$$

where N_o is Avogadro's number. The quantity x_α is the relative mass abundance of the αth nuclear component, that is, $x_\alpha \rho$ is its mass density in gm/cm^3. Therefore the total electron density is given by the sum of Equation (54.5) over all components α, as well as by Equation (53.4).

$$\begin{aligned} n_e &= \frac{1}{3\pi^2}\left(\frac{m_e c}{\hbar}\right)^3 y^3 \\ &= N_o \rho \sum_\alpha \frac{Z_\alpha x_\alpha}{A_\alpha} = \frac{N_o \rho}{\mu_e} \end{aligned} \tag{54.6}$$

where

$$\mu_e \equiv \left(\sum_\alpha \frac{Z_\alpha x_\alpha}{A_\alpha}\right)^{-1} \tag{54.7}$$

is called the *mean molecular weight per free electron* in the star. For a white dwarf star with only a small amount of hydrogen, $\mu_e \cong 2$ gm/mole.

Equation (54.6) can now be solved for the mass density ρ in terms of the parameter y, giving

$$\rho = By^3 \tag{54.8}$$

where $B \propto \mu_e$. The quantities A and B in Equations (54.2) and (54.8), respectively, have the magnitudes

$$A = \frac{\pi}{3}\left(\frac{m_e c}{h}\right)^3 m_e c^2 = 6.01 \times 10^{22} \text{ erg/cm}^3$$
$$B = \frac{8\pi}{3}\left(\frac{m_e c}{h}\right)^3 \frac{\mu_e}{N_o} = (9.74 \times 10^5)\mu_e \text{ gm/cm}^3 \quad (54.9)$$

with μ_e measured in gm/mole. Inasmuch as $\rho \sim 10^6$ gm/cm^3 in white dwarf stars, we see quite explicitly from these last two equations that $y = \hbar k_F/m_e c \sim 1$ for these stars and, therefore, that relativistic statistics must be used for the electrons.

Equation of State

The equation of state in the model white dwarf star is the pressure–density relation $\mathcal{P}_e(\Gamma_o)$ of Equations (54.2), where $\Gamma_o = (1 + y^2)^{1/2}$ is related to ρ by Equation (54.8). This equation must be supplemented by a relation that gives the parameter Γ_o as a function of radius r. To deduce this latter relation we first derive a differential equation that involves both Γ_o and r by following a standard procedure. The basic idea is to equate the change with r in free-electron pressure $\mathcal{P}_e$ to the change with r in force per unit area of gravitational attraction.

Consider a spherical shell of radius r and thickness dr in the star, assumed spherical and of uniform composition. Note that the electrons, which behave locally as an ideal gas in first approximation, have a density variation on a larger scale that because of electrical attraction follows that of the ions. Let $d\mathcal{P}_e(<0)$ be the change in the electron pressure as one goes outward from r to $r + dr$ in the star. See Figure XI.4. The corresponding inward force on the spherical shell, needed to balance this pressure change, is $-4\pi r^2 d\mathcal{P}_e$. This force is provided by gravitational attraction on the shell due to the matter of mass $M(r)$ that is interior to r, that is,

$$-4\pi r^2 d\mathcal{P}_e = \frac{GM(r)\rho(r)}{r^2}(4\pi r^2 dr) \quad (54.10)$$

where

$$M(r) = \int_0^r \rho(r')(4\pi r'^2 dr') \quad (54.11)$$

When we solve Equation (54.10) for $M(r)$, substitute the result into the left side of Equation (54.11), and differentiate both sides of the resulting equa-

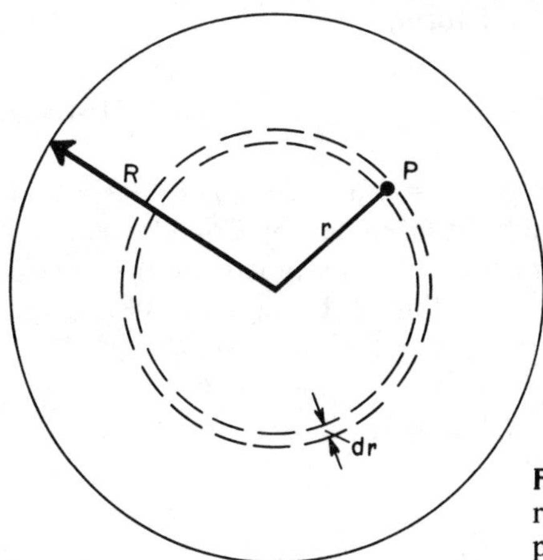

FIGURE XI.4 White dwarf star model of radius R. At point P there is an electron pressure $\mathscr{P}_e(r)$ and a mass density $\rho(r)$.

tion with respect to r, we obtain the fundamental equation of equilibrium,

$$\frac{1}{r^2}\frac{d}{dr}\left(\frac{r^2}{\rho}\frac{d\mathscr{P}_e}{dr}\right) = -4\pi G\rho \tag{54.12}$$

The solution to this equation gives the desired $\Gamma_o(r)$ relation.

To proceed, we substitute Equations (54.2) and (54.8), for $\mathscr{P}_e$ and ρ, into Equation (54.12) to obtain a differential equation entirely in terms of Γ_o and r,

$$\frac{1}{r^2}\frac{d}{dr}\left[r^2(\Gamma_o^2 - 1)^{-3/2}\frac{df(\Gamma_o)}{dr}\right] = -\left(\frac{4\pi GB^2}{A}\right)(\Gamma_o^2 - 1)^{3/2}$$

On substituting the first equality of Equation (54.3) into this equation, we find the result

$$\frac{1}{r}\frac{d^2}{dr^2}(r\Gamma_o) = \frac{1}{r^2}\frac{d}{dr}\left(r^2\frac{d\Gamma_o}{dr}\right) = -\left(\frac{\pi}{2}\frac{GB^2}{A}\right)(\Gamma_o^2 - 1)^{3/2} \tag{54.13}$$

We next define the value of Γ_o at $r = 0$ to be Γ_c (>1) and introduce the function $\Phi(r)$ by

$$\Phi(r) \equiv \frac{\Gamma_o(r)}{\Gamma_c} \xrightarrow[r\to 0]{} 1 \tag{54.14}$$

We also scale the radius r by the length parameter α as $r = \alpha\zeta$, where

$$\begin{aligned}\alpha \equiv \frac{1}{\Gamma_c B}\left(\frac{2A}{\pi G}\right)^{1/2} &= \frac{N_o}{\Gamma_c}\left(\frac{\hbar}{\mu_e c}\right)\left(\frac{3\pi}{4}\frac{\hbar c}{Gm_e^2}\right)^{1/2} \\ &= 7.77 \times 10^8\,(\mu_e\Gamma_c)^{-1}\ \text{cm}\end{aligned} \tag{54.15}$$

Equation (54.13) then takes the alternative form,

$$\frac{1}{\zeta}\frac{d^2}{d\zeta^2}(\zeta\Phi) = -(\Phi^2 - \Gamma_c^{-2})^{3/2} \xrightarrow[\zeta\to 0]{} -(1 - \Gamma_c^{-2})^{3/2} \quad \textbf{(54.13a)}$$

with boundary conditions at the origin: $\Phi(0) = 1$ and $d\Phi(\zeta)/d\zeta \xrightarrow[\zeta\to 0]{} 0$. Numerical solutions to this equation have been tabulated by Chandrasekhar.[21] These solutions depend, of course, on the value assumed for the central density $\rho_c = \rho(0)$ of the star, which, according to Equation (54.8), determines the parameters $y_c = y(0)$ and $\Gamma_c = (1 + y_c^2)^{1/2}$.

Exercise 54.1 Show from Equation (54.13a) that

$$\Phi(\zeta) \xrightarrow[\zeta\to 0]{} 1 + O(\zeta^2)$$

which implies that

$$\frac{d\Phi(\zeta)}{d\zeta} \xrightarrow[\zeta\to 0]{} 0$$

Radius and Mass of Star

The radius $R = \alpha\zeta_1$ of the white dwarf star model is determined by the value ζ_1 at which the density, or $y(\zeta)$, in the solution to Equation (54.13a) first goes to zero. At this point $\Gamma_o(R) = 1$, and from Equation (54.14),

$$\Phi(\zeta_1) = \Gamma_c^{-1} \quad \textbf{(54.16)}$$

Finite values of ζ_1, which are of order 4 to 5, occur in all solutions to Equation (54.13a) for which $\Gamma_c > 1$. Therefore, we can conclude from Equation (54.15) that the radius of a typical white dwarf star is only slightly greater than the radius of the earth ($= 6.37 \times 10^8$ cm).

The mass interior to the radial distance $r = \alpha\zeta$ is given by Equation (54.11) as

$$\begin{aligned} M(r) &= 4\pi\alpha^3 \int_0^{\zeta} \rho\zeta'^2\, d\zeta' \\ &= 4\pi B(\alpha\Gamma_c)^3 \int_0^{\zeta} (\Phi^2 - \Gamma_c^{-2})^{3/2}\, \zeta'^2\, d\zeta' \\ &= 4\pi B(\alpha\Gamma_c)^3 \int_0^{\zeta} \left[-\frac{d}{d\zeta'}\left(\zeta'^2 \frac{d\Phi}{d\zeta'}\right)\right] d\zeta' \end{aligned}$$

We have used Equations (54.8) and (54.14), and $y = \sqrt{\Gamma_o^2 - 1}$, to derive the second equality of this equation. Substitution of the differential equation

(54.13a) then leads to the third equality. The total mass $M(R)$ of the star is determined by the relation

$$M(R) = 4\pi B(\alpha\Gamma_c)^3\left(-\zeta^2 \frac{d\Phi}{d\zeta}\right)_{\zeta=\zeta_1} \quad \textbf{(54.17)}$$

where B is given by the second of Equations (54.9) and

$$(\alpha\Gamma_c) = (7.77 \times 10^8)\mu_e^{-1} \text{ cm}$$

We note that

$$\left(-\zeta^2 \frac{d\Phi}{d\zeta}\right)_{\zeta=\zeta_1}$$

in Equation (54.17) depends on Γ_c, as does the star radius $R = \alpha\zeta_1$. Therefore, by using either graphical or numerical methods one can eliminate the dependence on Γ_c between R and M to find a direct relation $R = R(M, \mu_e)$ for the white dwarf star model. The result is displayed as Figure XI.5.

We can understand Figure XI.5 by the following simple considerations. Instead of a star whose density varies with radius, as above, imagine the star to be of uniform density. Then the work that would be required to compress the free-electron gas from infinite dilution to the star of radius R, in the absence of any gravitational forces, would be

$$-\int_{\infty}^{R} \mathcal{P}_e(r) 4\pi r^2 \, dr$$

This work can be equated to minus the gravitational potential energy of such a star, which is equal to $\frac{3}{5}\, GM^2/R$. On differentiation of this equality at

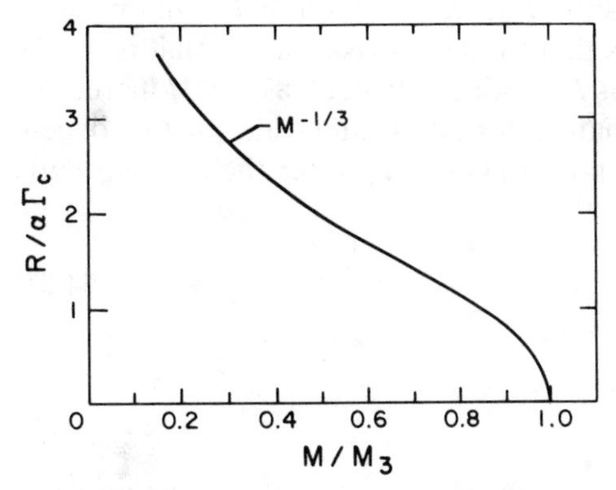

FIGURE XI.5 The radius R of the white dwarf star model as a function of its mass M, with both quantities expressed in dimensionless units. The quantity M_3 is the limiting mass of Equation (54.21), which occurs for $\Gamma_c \rightarrow \infty$.

constant M, we obtain

$$\mathcal{P}_e(R) = \frac{3}{20\pi} \cdot \frac{GM^2}{R^4} \tag{54.18}$$

Next we substitute Equations (54.3) and (54.8) into Equation (54.2) to obtain for the free-electron pressure the limiting equations

$$\begin{aligned} \mathcal{P}_e = Af(\Gamma_o) &\xrightarrow[y\ll 1]{} \frac{8}{5} A \left(\frac{\rho}{B}\right)^{5/3} && \text{(nonrelativistic limit)} \\ &\xrightarrow[y\gg 1]{} 2A\left[\left(\frac{\rho}{B}\right)^{4/3} - \left(\frac{\rho}{B}\right)^{2/3}\right] && \text{(extreme-relativistic limit)} \end{aligned} \tag{54.19}$$

where

$$\rho = M\left(\frac{4\pi}{3}R^3\right)^{-1}$$

On equating the preceding two equations, and using Equation (54.15), we can derive an expression for R as a function of M in each of the two limits:

$$\begin{aligned} \frac{R}{\alpha\Gamma_c} &= 3\left[\frac{4\pi}{3}\frac{B}{M}\right]^{1/3} (\alpha\Gamma_c) && \text{(nonrelativistic limit)} \\ \left(\frac{4\pi}{3}\frac{B}{M}\right)^{2/3} R^2 &= 1 - \frac{4}{15}\left(\frac{M}{\frac{4\pi}{3}B}\right)^{2/3} (\alpha\Gamma_c)^{-2} && \text{(extreme-relativistic limit)} \end{aligned} \tag{54.20}$$

The first of Equations (54.20) shows that for small mass values, that is, in the nonrelativistic limit, the radius R decreases with increasing mass as $M^{-1/3}$. The second equation shows that for large mass values, that is, in the extreme-relativistic limit, the radius R continues to decrease as M increases. A surprising result, however, is that in this latter limit, and for this degenerate electron-gas model, the star radius goes to zero for the *limiting mass value* M_3, where

$$\begin{aligned} M_3 &\equiv 4\pi KB(\alpha\Gamma_c)^3 \\ &= (5.74 \times 10^{33})K\,\mu_e^{-2}\ \text{gm} \end{aligned} \tag{54.21}$$

and

$$K = \frac{1}{3}(15/4)^{3/2} = 2.42$$

The more accurate constant K, as predicted by Equation (54.17) for the extreme-relativistic limit where $\Gamma_c \to \infty$, is

$$K = \lim_{\Gamma_c \to \infty} \left(-\zeta^2 \frac{d\Phi}{d\zeta} \right)_{\zeta = \zeta_1} = 2.02$$

In terms of solar masses, and with $K = 2.02$, we then find $M_3 = (5.89 \mu_e^{-2}) M_\odot$. Thus white dwarf stars with $\mu_e \cong 2$ are limited to a maximum mass value only slightly in excess of the solar mass $M_\odot \cong 2 \times 10^{33}$ gm, a fact that is in agreement with observations on these stars.[22]

55 Chain Molecules

In Section 1 we mentioned the subject of polymer chains in connection with our discussion of probability distributions. In this section we treat in some detail three problems involving molecular chains.[23] In the first problem we consider a one-dimensional chain of N elements, each of fixed length a, which can fold upon each other in all possible ways. We assume that this chain is in a temperature bath T and calculate the tension F necessary to maintain the effective length L of the chain. This molecular model is the simplest one that embodies the basic characteristic of rubber elasticity. In the second problem we allow the chain to deform in three dimensions. Analogies with the paramagnetic alignment of spins are discussed. Finally, in the third problem a model of wool elasticity is analyzed.

The first chain molecule that we discuss is the one-dimensional folding model of Section 1. We assume that there is no significant configurational energy for the molecular chain. Moreover, the independent variables for the system are L and T. Then to calculate the tension for a given temperature we must consider the Helmholtz free energy Ψ and use the formula

$$F = \left(\frac{\partial \Psi}{\partial L} \right)_T \quad \textbf{(55.1)}$$

where $\Psi = E - TS$ and F is the force of tension acting on the chain. Since the energy is independent of L in this problem, we see immediately that the tension is due entirely to the entropy S.

To specify a given configuration for the molecule we assume a positive direction to the right and start at the left end of the molecule (see Figure I.3). Then each successive element is directed either to the right $(+)$ or to the left $(-)$. The total configuration of the molecule chain is determined, in part, by the numbers n_+ and n_- of $(+)$ and $(-)$ orientations. But the length L of the chain does not depend upon the ordering in which the (+) and (−) orientations occur. Therefore, the only configurational quantity of interest in this problem is the unnormalized probability $W(n_+, n_-)$ for finding the numbers n_+ and n_-, where

$$W(n_+, n_-) = \frac{N!}{n_+!n_-!} \tag{55.2}$$

Then the entropy of the chain molecule is given, for large $N = n_+ + n_-$, by Equation (4.24) as

$$\begin{aligned} S &= \kappa \ln W(n_+, n_-) \\ &\xrightarrow[N>>1]{} \kappa[N \ln N - n_+ \ln n_+ - n_- \ln n_-] \end{aligned} \tag{55.3}$$

where Stirling's formula (2.6) has been used.

Now the length L is calculated as $L = (n_+ - n_-)a$, which together with the equation $N = n_+ + n_-$ gives for n_+ and n_- the expressions,

$$n_{\pm} = \frac{1}{2}(N \pm L/a) \tag{55.4}$$

The entropy expression (55.3) then becomes

$$\frac{1}{N\kappa} S = -\frac{1}{2}\left(1 + \frac{L}{Na}\right)\ln\left(\frac{1}{2} + \frac{L}{2Na}\right) - \frac{1}{2}\left(1 - \frac{L}{Na}\right)\ln\left(\frac{1}{2} - \frac{L}{2Na}\right) \tag{55.5}$$

which is a special case of Equation (4.25). Therefore, from Equation (55.1) we can derive for the tension F, when $N >> 1$, the result

$$\begin{aligned} F &= \frac{\kappa T}{2a} \ln\left(\frac{1 + L/Na}{1 - L/Na}\right) \\ &\xrightarrow[L<<Na]{} \left(\frac{\kappa T}{Na^2}\right)L \end{aligned} \tag{55.6}$$

in which the second line corresponds to Hooke's law.

We can solve the first equality of Equation (55.6) for L/N to obtain

$$\frac{L}{N} = a \tanh\left(\frac{Fa}{\kappa T}\right) \tag{55.7}$$

Let us now compare this result for a one-dimensional molecular chain with that obtained in Section 51 for the magnetism of a system of N spin-½ atoms. On referring to Equations (51.5), (51.9), and (51.10), we may deduce the following set of analogies:

$$\langle \mu \rangle \leftrightarrow L/N$$

$$\frac{1}{2} g\mu_B \leftrightarrow a \qquad \left(\text{for } J = \frac{1}{2}\right)$$

$$\mathscr{H} \leftrightarrow F$$

This set of analogies could have been used as an alternative, although heuristic, method for deriving Equation (55.6). Inasmuch as the final result (55.6) is determined uniquely by the three quantities N, L, and T, it is unnecessary to make any distinction between a microscopic result and an average value in this calculation.

Three-Dimensional Folding

In the three-dimensional problem the *a priori* probabilities ½ for orientations (+) and (−) are replaced by unnormalized continuous probabilities, or solid angle differential elements

$$d\omega_\alpha = \sin \Theta_\alpha d\Theta_\alpha d\phi_\alpha \tag{55.8}$$

where $\alpha(=1, 2, \ldots N)$ labels the αth element in the molecular chain. It is convenient to choose the z-axis of the coordinate system to lie along the direction between the end points of the chain. Then the effective length L of the chain is given by

$$L = a\sum_{\alpha=1}^{N} \cos \Theta_\alpha \tag{55.9}$$

and the probability for finding the chain with length L is proportional to the total "N-particle volume":

$$\Gamma(L)dL \equiv a\int \cdots \int \prod_{\alpha=1}^{N} d\omega_\alpha \qquad \left(\text{with } L < a \sum_{\alpha=1}^{N} \cos \Theta_\alpha < L + dL\right) \tag{55.10}$$

We can solve the three-dimensional problem by introducing a macrocanonical ensemble in which the tension F, instead of the length L, is held constant. In this ensemble the energy E is replaced by $E - LF$, where L is given by Equation (55.9)

Exercise 55.1 In the macrocanonical ensemble in which all possible L values are allowed we introduce, for the constraint of "known" $\langle L\rangle$, an undetermined Lagrange multiplier βF and use a classical limit in N-particle position space of the method of the most probable distribution (Section 27).

(a) Verify that the most probable value $\langle L\rangle$ in this $T - F$ ensemble is given by

$$\langle L\rangle = \beta^{-1}\left(\frac{\partial \ln Z}{\partial F}\right)_T \tag{55.11}$$

where Z is the macrocanonical partition function,

$$Z = \int \cdots \int \left(\prod_{\alpha=1}^{N} d\omega_\alpha\right) e^{-\beta(E-LF)} \tag{55.12}$$

in which L is given by Equation (55.9).
[*Suggestion:* Argue that in Equation (27.1), $g_i \propto (\Pi_{\alpha=1}^{N} \delta\omega_\alpha)_i$ for the ith set of N directions.]

(b) Use the method of Sections 27 and 36 to show that $\beta = (\kappa T)^{-1}$ and that the thermodynamic energy $\Phi \equiv -\beta^{-1} \ln Z$ is related to the Helmholtz free energy by

$$\Phi = \Psi - F\langle L\rangle = \langle E\rangle - T\langle S\rangle - F\langle L\rangle \tag{55.13}$$

[*Suggestion:* Note that $-\partial \ln Z/\partial\beta = \langle E\rangle - F\langle L\rangle$ and that for a constraint of known $\langle L\rangle$, $đQ = d\langle E\rangle_L + đW$, where $d\langle E\rangle_L = d\langle E\rangle - F' d\langle L\rangle$ and F' is the external tension. It is necessary to show that $F = F'$. This equality can be verified by using the identity $(\partial\Phi/\partial\langle L\rangle)_{T,F} \equiv 0$, for the $T - F$ ensemble at equilibrium, together with Equation (55.1) for F'.[24]]

In our present problem we may set $E = 0$ in the partition function Z of Exercise 55.1. On using Equation (55.9) for L, we readily obtain for Z the expression

$$\begin{aligned} Z &= \prod_{\alpha=1}^{N} \int d\omega_\alpha\, e^{\beta Fa\cos\Theta_\alpha} \\ &= \left[\int_0^{2\pi} d\phi \int_0^{\pi} \sin\Theta\, e^{\beta Fa\cos\Theta}\, d\Theta\right]^N \\ &= \left[\frac{4\pi}{\beta Fa} \sinh \beta Fa\right]^N \end{aligned} \tag{55.14}$$

Equation (55.11) then gives for $\langle L\rangle$ the result, valid when $N >> 1$,

$$\langle L\rangle = N\kappa T \frac{\partial}{\partial F} \ln\left(\frac{\sinh \beta Fa}{\beta Fa}\right)$$

$$\langle L\rangle = Na \cdot L\left(\frac{Fa}{\kappa T}\right) \tag{55.15}$$

where the Langevin function $L(x)$ is defined by Equation (51.11). In this three-dimensional case Hooke's law, for $\langle L\rangle << Na$, takes the form

$$F = 3\left(\frac{\kappa T}{Na^2}\right)\langle L\rangle$$

The analogy established below Equation (55.7), between the magnetization of atomic spins and a one-dimensional molecular chain, also applies in the three-dimensional case. We set the correspondences for this continuous case by considering the limit $J \to \infty$. On referring to Equation (51.9a) we find: $\langle\mu\rangle \leftrightarrow \langle L\rangle/N$, $\mu_o \leftrightarrow a$, and $\mathcal{H} \leftrightarrow F$.

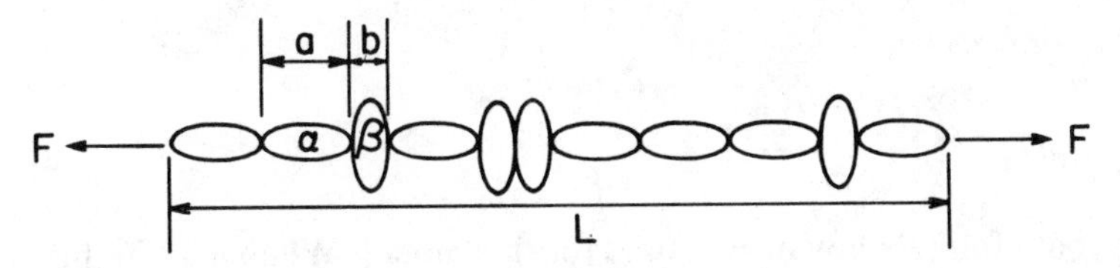

FIGURE XI.6 Simple chain-molecule model for keratin molecules in wool.

Wool Elasticity

The freely folding chain molecule in three dimensions serves as a simple realistic model for describing some polymers. In our third problem we treat a simplified model for keratin molecules in wool. In this model for wool elasticity, N monomeric units are arranged along a straight line to form a chain molecule. Each monomeric unit is assumed to be capable of being either in a state α, with length a and energy E_α, or in a state β, with length b and energy E_β, as shown in Figure XI.6. We use the $T - F$ ensemble of Exercise 55.1 for this one-dimensional model to derive the relation between the most probable length $\langle L \rangle$ and the tension F.

The partition function (55.12) for the wool model becomes, for this case with discrete energy states,

$$Z = \sum_{N_\alpha=0}^{N} \frac{N!}{N_\alpha! N_\beta!} e^{-\beta[N_\alpha E_\alpha + N_\beta E_\beta - F(N_\alpha a + N_\beta b)]} \qquad (55.16)$$
$$= [e^{-\beta(E_\alpha - Fa)} + e^{-\beta(E_\beta - Fb)}]^N$$

Here we have used the fact that the number of states with energy $E = N_\alpha E_\alpha + N_\beta E_\beta$ and length $L = N_\alpha a + N_\beta b$ is the binomial coefficient (55.2). Equation (55.11), for the dependence of the most probable length $\langle L \rangle$ on the tension F, then gives for $N >> 1$

$$\frac{\langle L \rangle}{N} = [e^{\beta(Fa - E_\alpha)} + e^{\beta(Fb - E_\beta)}]^{-1}[ae^{\beta(Fa - E_\alpha)} + be^{\beta(Fb - E_\beta)}]$$
$$= (e^x + e^\Delta)^{-1}(ae^x + be^\Delta) \qquad (55.17)$$
$$\xrightarrow[x<<1]{} \left(\frac{a + be^\Delta}{1 + e^\Delta}\right) + \frac{Fe^\Delta(a - b)^2}{\kappa T(1 + e^\Delta)^2} + \cdots$$

where $x = \beta F(a - b)$ and

$$\Delta \equiv (E_\alpha - E_\beta)/\kappa T \qquad (55.18)$$

If we define $\langle L \rangle_o \equiv N(1 + e^\Delta)^{-1}(a + be^\Delta)$ to be the value of $\langle L \rangle$ when $F = 0$, then Equation (55.17) can also be written, for small distortions from equilibrium and when $a \neq b$, as

$$F \cong \left(\frac{1 + e^{\Delta}}{a - b}\right)^2 \kappa T e^{-\Delta}\left(\frac{\langle L\rangle - \langle L\rangle_o}{N}\right) \tag{55.17a}$$

which shows that Hooke's law also results for this model. When $a = b$, then $\langle L\rangle = \langle L\rangle_o$ is independent of F.

The second equality of Equation (55.16) shows that we may effectively consider the energy of a monomeric unit to be $(E_\alpha - Fa)$ or $(E_\beta - Fb)$, according to whether the unit is in the state α or β respectively. By this reasoning one can use the concept of the internal partition function, Equation (28.25), to arrive directly at the result (55.16).

Bibliography

Texts

S. Chandrasekhar, *An Introduction to the Study of Stellar Structure* (Dover, 1957), Chapters X and XI. A classic and extremely readable treatment of the subject. Our reference is to the analysis of the relativistic electron gas and the application of this model to white-dwarf-star calculations.

J. P. Cox and R. T. Giuli, *Principles of Stellar Structure* (Gordon and Breach, 1968), Chapters 24 and 25. The quantum statistical theory is explored exhaustively in Chapter 24, and expansions of thermodynamic functions are tabulated for many different temperature–density regimes. Chapter 25 elaborates in comprehensive detail on the theory of white dwarf stars.

P. J. Flory, *Statistical Mechanics of Chain Molecules* (Interscience, 1969), Chapter VIII. In which the subject is developed and applied in great detail.

C. Kittel, *Introduction to Solid State Physics,* 3rd ed. (Wiley, 1968), Chapters 14 and 15. Paramagnetism, ferromagnetism, and other magnetic phenomena are treated theoretically, and the relation of the theories to a wealth of experimental knowledge is elaborated abundantly. An analysis of Van Vleck paramagnetism is given in Appendix G.

R. Kubo, *Statistical Mechanics* (Interscience, 1965). Among the many instructive problems listed and solved in this unique text are three that relate to chain molecules and that have been used as a basis for our Section 55. These three problems are Problem 13 in Chapter 1 and Problems 13 and 17 in Chapter 2.

J. E. Mayer and M. G. Mayer, *Statistical Mechanics* (Wiley, 1940). A brief and clear development of the theory of paramagnetism is given in Sections h and i of Chapter 15.

F. Reif, *Fundamentals of Statistical and Thermal Physics* (McGraw-Hill, 1965). A simplified treatment of ferromagnetic theory is given in Sections 6 and 7 of Chapter 10.

J. H. Van Vleck, *The Theory of Electric and Magnetic Susceptibilities* (Oxford, Clarendon, 1932), Chapter IV. The classical theory of magnetic susceptibilities is treated. Special attention is given to explaining why classical statistical mechanics alone cannot be applied to atomic systems.

Review Article
J. H. Van Vleck, "A Survey of the Theory of Ferromagnetism", *Rev. Mod. Phys.* **17**, 27 (1945). Although obviously limited in its coverage to developments before 1945, this excellent article is still relevant today because of its careful and thorough analysis of the basic theories of ferromagnetism.

Notes

1. Because interparticle potential energies in Equation (51.2) occur twice for each pair of particles, internal potentials $\phi_i(\mathbf{r})$ must be multiplied by ½. The external fields may also vary slowly in time.

2. In first approximation magnetic interactions between the atoms may be represented by an internal magnetic field $\mathcal{H}_{\text{int}}$. When $\mathcal{H}_{\text{int}}$ becomes significant, as it does, say, for ferromagnetic materials (Section 52), then we must include the effects of magnetic interactions explicitly.

3. The paramagnetism of atoms is due to motion of their electrons, with charges $-e$.

4. This result can also be written in terms of the spin polarization as $\langle\mu\rangle = \frac{1}{2} g\mu_B \times P_{o,z}$; see footnote 25 in Chapter IX.

5. In other words, when we set $J \to \infty$ to obtain the classical limit, then the parameter x becomes $\mu_o\mathcal{H}/\kappa T$ and it is assumed that μ_o is a given atomic magnetic moment.

6. C. Kittel, *op. cit.*, p. 433.

7. M. E. Rose, *Elementary Theory of Angular Momentum* (Wiley, 1957), p. 98.

8. See, for example, J. D. Jackson, *Classical Electrodynamics* 2nd ed. (Wiley, 1975), Sections 5.6–5.8.

9. Many-electron wave functions must be totally antisymmetric, which implies that a pair of electrons in a singlet spin state must be in a spatially symmetric state, corresponding to $+J_{\mu\nu}$. Similarly, a pair of electrons in a triplet spin state must be associated with $-J_{\mu\nu}$. By contrast, the effective exchange interaction for free fermions is zero for singlet states and repulsive for triplet states. (See Exercise 47.4.)

10. We have also ignored orbital angular-momentum contributions to the atomic spin S_α. Having completed our discussion of the underlying justification for Equation (52.7) in terms of electron–electron interactions, we can now return to our treatment of ferromagnetism in terms of interacting atoms for which, say, $n = N/\Omega$ represents the density of atoms in the material.

11. P. R. Weiss, *Phys. Rev.* **74** (1948), p. 1493. See also the review article by J. H. Van Vleck, *op. cit.*

12. C. Kittel, loc. cit., pp. 488–496.

13. *Ibid.*, pp. 464–471.

14. Notice that the constant C is also the coefficient of T^{-1} in Curie's law, Equation (51.13).

15. One consequence of a correct analysis of ferromagnetic substances is that the heat capacity for zero external field exhibits a sharply discontinuous behavior at the critical temperature and a high-temperature tail, whereas the present theory predicts

a less abrupt discontinuity with no high-temperature tail. The solution of the so-called "Ising model" of ferromagnetism in two dimensions demonstrates the importance of short-range correlations for explaining this behavior; see K. Huang, *Statistical Mechanics* (Wiley, 1967), Chapters 16 and 17.

16. The data are from P. Weiss and R. Forrer, *Ann. Phys.* (*Paris*) **5** (1926), p. 153, and have been discussed and analyzed extensively by J. S. Kouvel and M. E. Fisher, *Phys. Rev.* **136A** (1964), p. 1626. See also Section IV of the review article by L. P. Kadanoff *et al., Rev. Mod. Phys.* **39** (1967), p. 395.

17. J. D. Jackson, *loc. cit.*, p. 215.

18. See the discussion at the end of Section 36.

19. Details relating to this analysis are given in Section II of the review article by Kadanoff *et al., op. cit.*

20. The relations given below Equation (53.2) can be used to estimate these pressures.

21. S. Chandrasekhar, *op. cit.*, App. IV.

22. For further, and extensive, discussion of white dwarf stars, the interested reader is referred to the text by Cox and Giuli.

23. In addition to the texts by Flory and Kubo cited in the bibliography, an additional reference on polymers is D. A. McQuarrie, *Statistical Thermodynamics* (Harper & Row, 1973), Chapter 14.

24. Maximizing the entropy subject to the constraints of known $\langle E \rangle$ and $\langle L \rangle$ is equivalent to finding the unrestricted minimum of $\beta\Phi$.

APPENDIX

A
DARWIN–FOWLER METHOD OF MEAN VALUES

In Exercise 4.1 we showed how to calculate the most probable number $\bar{n}_i$ of particles in the ith μ-space cell of a classical ideal gas. In this appendix we show how the so-called Darwin–Fowler method of mean values provides an expression for the average value $\langle n_i \rangle$. Explicit results when the total particle number N is very large are obtained by appeal to the method of steepest descents, the subject of Appendix B.

Application to the determination of the average occupation numbers $\langle N_i \rangle$ in a macrocanonical ensemble is also given, and the results obtained are compared with those of Section 27. In both applications of the method the fluctuations in occupation numbers are examined and shown to be not large ordinarily.

There are two related questions that can be answered by the Darwin–Fowler method of mean values. One is the determination of the average occupation number $\langle n_i \rangle$ for particles in the ith μ-space cell Z_i of a classical ideal gas,[1] and the other concerns the normalization of the probabilities $W_N(n_1, n_2, \dots n_i, \dots)$ of Equation (4.18). In this appendix we set $C = 1$, so that the sum of these probabilities is definitely not normalized to unity. Individually, with $C = 1$ each W_N becomes the number of ways of choosing a given set $\{n_i\}$ of occupation numbers in the gas, as required for calculation of the statistical entropy (4.24).

$$W_N(n_i) = N! \prod_i \frac{Z_i^{n_i}}{n_i!} \tag{A.1}$$

To calculate the average, or mean, values $\langle n_i \rangle$ we must consider all possible $W_N(n_i)$ values, corresponding to all possible sets $\{n_i\}$. Thus we must consider the ensemble of all possible ideal-gas systems, that is, an ensemble of ensembles, [see above Equation (4.21)] in which each ideal-gas system is subject to the constraints (4.19) and (4.20)

$$\sum_i n_i = N$$
$$\sum_i n_i \, \omega_i = E \tag{A.2}$$

The definition of $\langle n_i \rangle$ is then given by

$$\langle n_i \rangle = \frac{\sum_{\{n_i\}} n_i W_N(n_i)}{\sum_{\{n_i\}} W_N(n_i)} \tag{A.3}$$

As mentioned above, we shall be interested in the normalization of the $W_N(n_i)$, for example, in the calculation of $\langle n_i \rangle$ by Equation (A.3). Therefore we define the quantity P_N by

$$P_N \equiv \sum_{\{n_i\}} W_N(n_i) \tag{A.4}$$

which is to be evaluated subject to constraints (A.2). Now in the absence of the energy constraint, we would obtain for P_N simply

$$P_N = (Z_1 + Z_2 + \cdots + Z_i + \cdots)^N \qquad \text{(incorrect)}$$

since the expansion of the right side of this expression is, for this case, the multinomial expansion (A.4). The *Darwin–Fowler method* achieves the explicit evaluation of P_N, in the limit $N \to \infty$, subject to both constraints (A.2); at the same time it avoids the use of Stirling's formula (2.6) for the $n_i!$ that are not large.

The mean values $\langle n_i \rangle$ can be calculated as simple partial derivatives of P_N with respect to the Z_i. Thus Equations (A.1), (A.3), and (A.4) can be combined to give the result

$$\langle n_i \rangle = Z_i \frac{\partial \ln P_N}{\partial Z_i} \tag{A.5}$$

Exercise A.1 Show that the fluctuation $\langle (\Delta n_i)^2 \rangle$ in the value of n_i is given by

$$\langle (\Delta n_i)^2 \rangle \equiv \langle n_i^2 \rangle - \langle n_i \rangle^2 = Z_i \frac{\partial}{\partial Z_i}\left(Z_i \frac{\partial \ln P_N}{\partial Z_i} \right)$$
$$\langle (\Delta n_i)^2 \rangle = Z_i \frac{\partial \langle n_i \rangle}{\partial Z_i} \tag{A.6}$$

The method of mean values permits calculation of the fluctuations in n_i using this equation; these fluctuations cannot be investigated using the method of the most probable distribution.

Generating Function for P_N

At the end of Section 5 we introduced the concept of generating functions to evaluate difficult sums in probability theory. We now introduce a generating function $[u(z)]^N$ to evaluate the sum (A.4). We assume first that the energy levels ω_i of the system are all expressed in terms of some energy unit ε. With this convention, a function $u(z)$ is introduced with the definition

$$u(z) \equiv \sum_i Z_i z^{\omega_i} \tag{A.7}$$

where the summation is over all single-particle states (cells). This quantity is essentially the partition function for a single particle, or molecule.[2]

Further study of the observations made below Equation (A.4) reveals that the energy constraint (A.2) can be included in a determination of P_N for $u(z)$ by using the following prescription:

$$P_N = \text{coefficient of } z^E \text{ in } [u(z)]^N$$

In this prescription, both constraints (A.2) are taken into account. The Darwin–Fowler method achieves an implementation of this prescription.

We may assume that the energy levels of the identical particles in the ideal gas, or ensemble, are ordered in some way, such as

$$\omega_1 \leqslant \omega_2 \leqslant \omega_3 \leqslant \cdots \leqslant \omega_i \leqslant \cdots$$

We can assume further that the ideal gas itself has finite extension so that, apart from degeneracies, the energy levels ω_i are all discretely separated. We now place a simple restriction on these energy levels by requiring that the unit of energy ε be chosen small enough so that, to desired accuracy, every energy level ω_i can be written as an integral multiple of ε. Thus for free particles in a one-dimensional box we can write $\omega_j\varepsilon = n_j^2\varepsilon$. We also choose the energy unit ε to be large enough so that not all of the levels ω_i have a common divisor other than unity.[3] If, finally, we let the quantity z take on complex values, then the above prescription is equivalent to the complex integral

$$P_N = \frac{1}{2\pi i} \oint \frac{[u(z)]^N}{z^{E+1}}\, dz \tag{A.8}$$

where the contour is any closed contour, within the circle of convergence of $u(z)$, about the origin. Note that for the purposes of evaluating this contour integral we can set $\omega_1 = 0$, since this choice only affects the zero of energy that can be canceled in the numerator and denominator of the integrand.

It is necessary to study the integrand,

$$I(\mathfrak{z}) \equiv \frac{[u(\mathfrak{z})]^N}{\mathfrak{z}^{E+1}} \equiv e^{v(\mathfrak{z})} \tag{A.9}$$

of the contour integral (A.8). We show for $\mathfrak{z}$ real and positive that $I(\mathfrak{z})$ is always positive and exhibits only one very sharp minimum. We can then use the *method of steepest descents* to evaluate P_N in terms of derivatives of the function $v(\mathfrak{z})$, defined by the second equality in (A.9).[4]

We let $\mathfrak{z}_o = x_o$ be the point along the real axis where $I(\mathfrak{z})$ takes its minimum value, and $\mathfrak{z}_s > \mathfrak{z}_o$ be any singular point of $u(\mathfrak{z})$ and $I(\mathfrak{z})$.[5] Now $x^{-(E+1)}$ decreases monotonically from ∞ to 0 as $\mathfrak{z} = x$ increases along the positive real axis. Similarly, $[u(x)]^N$ increases monotonically from Z_1^N to ∞ as x increases from 0 to x_s. Therefore, $I(x)$ is always positive and exhibits at least one minimum, as shown in Figure A.1.

Now, the derivative of $I(\mathfrak{z})$ is given by

$$\frac{dI(\mathfrak{z})}{d\mathfrak{z}} = I(\mathfrak{z})v'(\mathfrak{z}) \tag{A.10a}$$

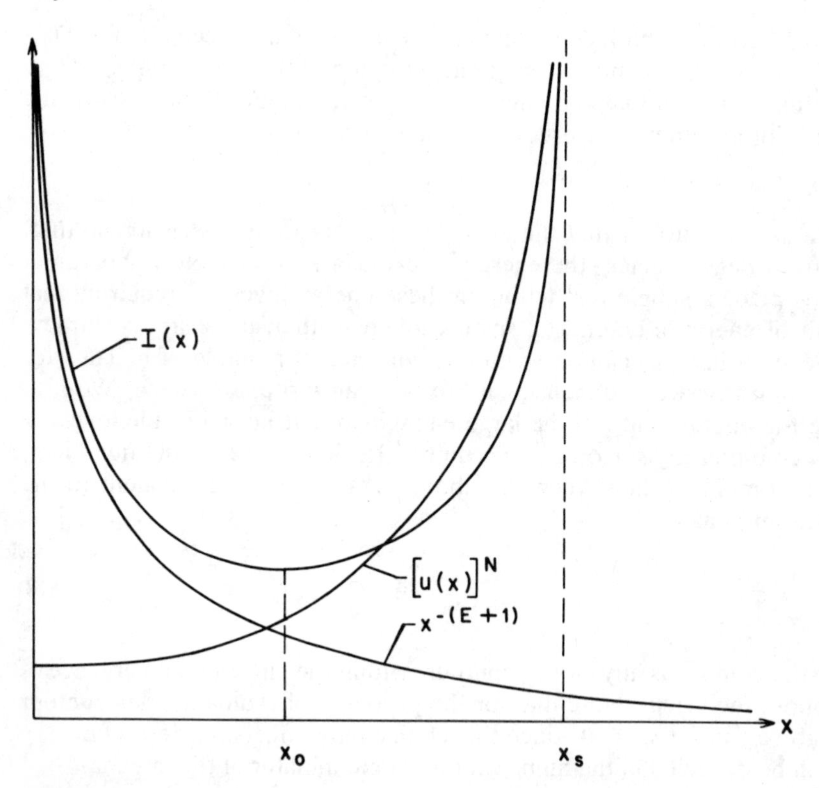

FIGURE A.1 The integrand of the contour integral (A.8), plotted as a function of the real variable x (schematic).

where $v(z)$ is defined by Equation (A.9) and $v'(z)$ is given by

$$v'(z) = N\frac{u'(z)}{u(z)} - \frac{E+1}{z} \tag{A.10b}$$

The minimum point $z_o = x_o$ is, of course, determined by the condition $v'(z_o) = 0$. Furthermore, the second derivative of $I(z)$ is derived as

$$\begin{aligned}\frac{d^2I(z)}{dz^2} &= I(z)\{[v'(z)]^2 + v''(z)\} \\ &= I(z)\{[v''(z) + z^{-1}v'(z)] + v'(z)[v'(z) - z^{-1}]\}\end{aligned} \tag{A.11a}$$

where

$$\begin{aligned}v''(z) + z^{-1}v'(z) &= \frac{N}{[u(z)]^2}\{u''(z)u(z) - [u'(z)]^2 + z^{-1}u(z)u'(z)\} \\ &= \frac{N}{[zu(z)]^2}\sum_i Z_i z^{\omega_i}\left\{\omega_i\left(\sum_k Z_k z^{\omega_k}\right)^{1/2} - \frac{\sum_i Z_j \omega_j z^{\omega_j}}{\left(\sum_k Z_k z^{\omega_k}\right)^{1/2}}\right\}^2 \\ &= O(N)\end{aligned} \tag{A.11b}$$

which is always positive for $z = x > 0$. Thus the second derivative of $I(x)$ is always positive along the positive real axis, and therefore the first derivative of $I(x)$ is a monotonically increasing function.[6] We conclude that $I(z)$ can have only one minimum along the positive real axis. That the minimum is very sharp follows from the fact that E and N are both very large numbers, which implies in turn that $v''(z)$ is also a very large quantity.

In Appendix B the above considerations about $I(z)$ are used and extended to the evaluation of the contour integral (A.8) by the method of steepest descents. The final result (B.10),[7] which is exact in the limit $N \to \infty$, can be written with the aid of Equation (A.9) as

$$\begin{aligned}\ln P_N &= v(z_o) - \frac{1}{2}\ln[2\pi\, v''(z_o)] \\ &= -(E+1)\ln z_o + N\ln[u(z_o)] - \frac{1}{2}\ln[2\pi\, v''(z_o)] \\ \ln P_N &\cong -E\ln z_o + N\ln u(z_o)\end{aligned} \tag{A.12}$$

The logarithm of $v''(z_o)$ has been neglected in the third line of this last equation, because, according to Equation (A.11b), the function $v''(z_o)$ is proportional to N when $N \ggg 1$ and $N^{-1}\ln N \to 0$ in this limit.

Interpretation of Results

Substitution of result (A.12) into Equation (A.5) yields the expression

$$\frac{\langle n_i\rangle}{N} = \frac{Z_i \mathfrak{z}_o^{\omega_i}}{\sum_i Z_i \mathfrak{z}_o^{\omega_i}} \tag{A.13a}$$

in which the zero of energy [see below Equation (A.8)] could now be re-adjusted. If we replace the unknown quantity $\mathfrak{z}_o$ by a different unknown, β, by means of the definition

$$\mathfrak{z}_o \equiv e^{-\beta\varepsilon} \tag{A.14}$$

and incorporate the energy unit ε into ω_i, thereby redefining $\varepsilon\omega_i$ as ω_i, then Equation (A.13a) becomes

$$\frac{\langle n_i\rangle}{N} = \frac{Z_i e^{-\beta\omega_i}}{\sum_i Z_i e^{-\beta\omega_i}} \tag{A.13b}$$

We next make the identification [see above Equation (4.18)]

$$Z_i = \delta^3\mathbf{n}_i = \frac{\Omega\delta^3\mathbf{k}_i}{(2\pi\hbar)^3} \tag{A.15}$$

so that we can compare result (A.13b) with Equation (4.16) for the most probable occupation numbers $\bar{n}_i/N$. Inasmuch as Equation (4.16) or (4.17) for $\bar{n}_i$ can be derived using the method of Exercise 4.1, the formal proof that $\langle n_i\rangle = \bar{n}_i$ in the limit $N \to \infty$ is now complete.

With the aid of Equation (A.10b) and $(E + 1)\varepsilon \cong E\varepsilon \to E$, it is easy to show that the condition $v'(\mathfrak{z}_o) = 0$ is equivalent to the second of the constraint equations (A.2) when $n_i \to \langle n_i\rangle$. The integrals [see Equations (2.4)]

$$\sum_i Z_i e^{-\beta\omega_i} = \frac{\Omega}{(2\pi\hbar)^3}\int d^3\mathbf{k}\, e^{-\beta k^2/2m} = \frac{\Omega}{\lambda_T^3}$$

$$\sum_i Z_i\omega_i e^{-\beta\omega_i} = \frac{\Omega}{(2\pi\hbar)^3}\int d^3\,\mathbf{k}\left(\frac{k^2}{2m}\right)e^{-\beta k^2/2m} = \frac{3}{2\beta}\frac{\Omega}{\lambda_T^3} \tag{A.16}$$

where $\lambda_T = (2\pi\hbar^2\beta/m)^{1/2}$, can then be used to verify that $E = \frac{3}{2}N\beta^{-1}$, from which we may conclude that $\beta = (\kappa T)^{-1}$. Finally, substitution of the first of Equations (A.16) into Equation (A.13b) gives Equation (4.17) explicitly.

Equation (A.13a) can now be substituted into Equation (A.6) for the fluctuation in particle number $\langle(\Delta n_i)^2\rangle$. Since $\langle n_i\rangle/N$ is a small quantity for a classical gas, we find the result

$$\frac{\langle(\Delta n_i)^2\rangle}{N} = \frac{\langle n_i\rangle}{N} - \left(\frac{\langle n_i\rangle}{N}\right)^2 \cong \frac{\langle n_i\rangle}{N} \ll 1 \qquad \textbf{(A.17)}$$

We may conclude that in a classical ideal gas the fluctuation in occupation number, $\langle(\Delta n_i)^2\rangle \cong \langle n_i\rangle$, is "normal," that is, no larger than the occupation numbers themselves.

The maximum entropy of a classical ideal gas is given formally by Equation (4.26),

$$\overline{S} = \kappa \ln W_N(\overline{n}_i) \qquad \textbf{(A.18a)}$$

With $C = 1$, this expression can be evaluated explicitly (see Exercise 4.3) to give

$$S = \kappa \ln\left[N!\left(\frac{e^{5/2}}{n\lambda_T^3}\right)^N\right] \qquad \text{(A.18b)}$$

Alternatively, we may calculate the entropy for an ideal gas by referring to the general considerations at the end of Section 4 about the entropy of a system. We can argue that the entropy of the ensemble of all possible ideal-gas systems must be given by the expression $\kappa \ln P_N$, where P_N is given by Equation (A.4). In other words, the entropy S for this ensemble is given by

$$S = \kappa \ln\left[\sum_{\{n_i\}} W_N(n_i)\right] \qquad \textbf{(A.19)}$$

where the summation is subject to constraints (A.2). A remarkable fact, demonstrated in Exercise A.2, is that in the limit $N \rightarrow \infty$, this last expression also gives Equation (A.18b) when evaluated explicitly. Equations (A.18a) and (A.19) are equivalent because the number $\mathcal{N}$ of large W_N in the sum over all W_N is such that $\mathcal{N} \ll W_N(n_i)$, even though $\mathcal{N}$ itself is a large number. The entropy of a system may indeed be defined in many ways.[8]

Exercise A.2 Use Equation (A.12) to calculate the normalization of the $W_N(n_i)$ subject to constraints (A.2); that is, show

$$P_N = \sum_{\{n_i\}} W_N(n_i) \xrightarrow[N\rightarrow\infty]{} N!\left(\frac{e^{5/2}}{n\lambda_T^3}\right)^N$$

Application to a Macrocanonical Ensemble

It is straightforward to apply the Darwin–Fowler method, using Equations (A.1) to (A.12), to a macrocanonical ensemble. Upon referring to the notation of Section 27 we can make the following analogies with quantities used in the ideal-gas example:

$$\begin{aligned}
&N \to M \text{ (number of systems in ensemble)}\\
&E \to E_M \text{ (total ensemble energy)}\\
&\omega_i \to E_i \text{ (system energy level)}\\
&n_i \to N_i \text{ (occupation number in ensemble)}\\
&Z_i \to p_i \text{ (\textit{a priori} probability for \textit{i}th state)}
\end{aligned} \tag{A.20}$$

With regard to this last identification, we here treat every state of the system separately, so that the degeneracy g_i of the ith state is unity. Instead, the p_i's are introduced as variables, to permit the writing of equations such as (A.5) and (A.6), with the understanding that eventually we shall set $p_i = 1$ for all i.

Our task now is to interpret thermodynamically the statistical results that proceed from Equation (A.12). The condition that the right side of Equation (A.10b) vanishes at $\mathfrak{z} = \mathfrak{z}_o$ determines the average system energy $\langle E \rangle$ to be

$$\langle E \rangle \equiv \frac{\varepsilon E_M}{M} = \frac{\varepsilon \mathfrak{z}_o\, u'(\mathfrak{z}_o)}{u(\mathfrak{z}_o)} \tag{A.21}$$

where ε is the energy unit and the function $u(\mathfrak{z}_o)$, with $p_i = 1$, is found from Equations (A.7) and (A.14) to be

$$\begin{aligned}
u(\mathfrak{z}_o) &= \sum_i \mathfrak{z}_0^{E_i}\\
&\xrightarrow[\varepsilon E_i \to E_i]{} \sum_i e^{-\beta E_i}
\end{aligned} \tag{A.22}$$

Thus the function $u(\mathfrak{z}_o)$ is the partition function Z, of Equation (27.6), for a macrocanonical ensemble. Equations (A.13b) and (A.21), with substitutions (A.20), become identical to Equations (27.5).

Next we write $F' = \ln u(\mathfrak{z}_o)$, with $u(\mathfrak{z}_o)$ as given by Equation (A.22). Then the proof that $\beta = (\kappa T)^{-1}$ follows precisely as it does with Equations (27.9) to (27.16). Equation (27.17) gives $F' = -\beta\Psi$, where $\Psi = \langle E \rangle - T\langle S \rangle$ is the Helmholtz free energy. Here the system entropy $\langle S \rangle$ is the quantity introduced thermodynamically by Equation (27.13). Internal consistency, in the present development, can be demonstrated finally by considering the average system entropy that is obtained statistically from Equations (A.19), (A.4), and (A.12). With the further aid of Equations (A.14) and (A.21) we obtain

$$\begin{aligned}
\langle S \rangle &= \frac{\kappa}{M} \ln P_M\\
&= \kappa[\beta\langle E \rangle + \ln u(\mathfrak{z}_o)]
\end{aligned} \tag{A.23}$$

which with $\beta = (\kappa T)^{-1}$ gives $\ln(\mathfrak{z}_o) = -\beta\Psi$, as above.

We may conclude that ensemble average values are indeed equal to corresponding most probable values, and this identification is verified conclusively by computing the fluctuations $\langle(\Delta N_i)^2\rangle$. We use the analogies (A.20) and substitute Equation (A.13a) for $\langle N_i\rangle$ into Equation (A.6). With the aid of Equations (A.14) and (A.21), we can then derive the expression

$$\langle(\Delta N_i)^2\rangle = \langle N_i\rangle\left\{1 - \frac{\langle N_i\rangle}{M} - (E_i - \langle E\rangle)\left(\frac{\partial\beta}{\partial p_i}\right)_{p_j=1}\right\} \tag{A.24}$$

in which the E_i now include the energy unit ε. Here we have considered, as we did not do in the derivation of Equation (A.17), that β may depend on the variable p_i before setting $p_j = 1$ for all j. (Clearly, the distribution of system energies in a temperature bath will vary with the *a priori* probabilities p_j, which implies that $\partial\beta/\partial p_i \neq 0$ when all other parameters are held constant.) The following exercise shows how the partial derivative in (A.24) may be evaluated to give a more explicit expression for the fluctuations $\langle(\Delta N_i)^2\rangle$.

Exercise A.3

(a) Equation (A.21) for $\langle E\rangle$ may be written in the form

$$\langle E\rangle = \frac{s_1}{s_o}$$

where

$$s_\ell \equiv \sum_i p_i E_i^\ell e^{-\beta E_i}$$

Use this expression for $\langle E\rangle$ to derive the identity

$$\frac{ds_1}{s_1} - \frac{ds_o}{s_o} = 0$$

for arbitrary variations in the p_i.
[*Suggestion:* Recall that Equation (A.21) was derived by evaluating Equation (A.10b) at the point $\mathfrak{z} = \mathfrak{z}_o$.]

(b) Use the identity of part (a) to evaluate the partial derivative

$$(\partial\beta/\partial p_i)_{p_j=1}$$

and show that Equation (A.24) for $\langle(\Delta N_i)^2\rangle$ can be rewritten as

$$\langle(\Delta N_i)^2\rangle = \langle N_i\rangle\left\{1 - \frac{\langle N_i\rangle}{M}\left[1 + \frac{(E_i - \langle E\rangle)^2}{\langle(\Delta E)^2\rangle}\right]\right\} \tag{A.25}$$

where

$$\langle(\Delta E)^2\rangle = \langle E^2\rangle - \langle E\rangle^2 = \frac{s_2}{s_o} - \left(\frac{s_1}{s_o}\right)^2$$

This result is in agreement with the general statement (26.10) about the fluctuations of extensive variables, for an ensemble in this case. Therefore [see Equation (26.9)], the rms deviations $\langle(\Delta N_i)^2\rangle^{1/2}$ are indeed small in the ensemble, and the most probable values $\overline{N}_i$ $(= \langle N_i \rangle)$ are extremely probable. The "unexpected" term in (A.25), which involves energy fluctuations, contributes to making the fluctuations $\langle(\Delta N_i)^2\rangle$ smaller than in the "normal" case (A.17).

Bibliography for Appendix A

R. H. Fowler, *Statistical Mechanics* (Cambridge, 1936), Chapter II. In which the method of mean values, as developed originally by Darwin and Fowler, is applied to a number of physical systems, both classical and quantum mechanical.

E. Schrödinger, *Statistical Thermodynamics* (Cambridge, 1967), Chapter VI. The source for the development in this appendix; Schrödinger's treatment of the Darwin–Fowler method is both elegant and simple.

Notes

1. As in Section 4, $Z_i = \delta^3\mathbf{n}_i$ is the size, or degeneracy, of a cell in the phase space of a single molecule (μ space).

2. See Equation (28.22). In this appendix we omit the factor $(N!)^{-1} \cong (e/N)^N$ from $W_N(n_i)$, which factor, as explained at the end of Section 4, accounts for particle indistinguishability.

3. This restriction on ε is used in Equation (A.8) and in the evaluation of P_N at the end of Appendix B. The restriction becomes somewhat artifical when it is applied, at the end of this appendix, to the energy levels of systems in a macrocanonical ensemble.

4. Because the minimum of the integrand $I(\mathfrak{z})$ is extremely sharp, we expand $v(\mathfrak{z}) \equiv \ln I(\mathfrak{z})$, rather than $I(\mathfrak{z})$ itself, about the minimum value $\mathfrak{z}_o$. See the discussion of Equation (2.5) for further clarification of this point.

5. Along the real axis $\mathfrak{z}_s = x_s$ is the point at infinity in the present application. For an application of this development to degenerate ideal gases see Section 31. In particular, see footnote 3 of Chapter VII in which the distinction between the region of analyticity of $I(\mathfrak{z})$ and the radius of convergence of a series expansion is clarified.

6. In the second equality of Equation (A.11a) the second term is negative when $0 < v'(x) < x^{-1}$. However, in the small region of x values where this occurs, near x_o, this second term is O(1) and therefore negligible compared with the first O(N) term.

7. In Appendix B, the quantity P_N is called $J(\mathfrak{z}_o)$.

8. See also Equation (4.25) and Problem II.4.

APPENDIX

B METHOD OF STEEPEST DESCENTS

The method of steepest descents, due originally to P. Debye, is constructed for a contour integral whose integrand has one sharp minimum along the positive real axis. Such an integrand, when analytic, must have a saddle point at the minimum point. The integral is then evaluated by choosing the contour to pass through this point along the paths of steepest descent, normal to the real axis.

We wish to evaluate a contour integral $J(z_o)$ of the type (A.8) whose integrand can be written in the exponential form (A.9).

$$J(z_o) \equiv \frac{1}{2\pi i} \int_C e^{v(z)}\, dz \tag{B.1}$$

It is assumed that $v(z)$ has one sharp minimum along the positive real axis at $z = z_o$ (see Figure A.1). The contour C, which passes around the origin, is specified further by requiring that it cross the real axis normally at z_o, as shown in Figure B.1. This choice of the contour makes it possible to evaluate $J(z_o)$ by the method of steepest descents.[1]

We begin by summarizing properties of the function $v(z)$ as we shall require them for the development of this appendix. Thus since $v(z)$ has a sharp minimum along the real axis at the point $z = z_o$, we have in addition to the condition $v'(z_o) = 0$, the expansion

$$v(z) = v(z_o) + \frac{1}{2} v''(z_o)(z - z_o)^2 + O(z - z_o)^3 \tag{B.2}$$

where $v''(z_o)$ is large, real, and positive.[2] We proceed now by developing, in the vicinity of the point z_o, the analytic properties of a function $v(z)$ of the type (B.2).

Let the function $v(z)$ be written in terms of its real and imaginary parts, f and g, respectively, as

$$v(z) = f(x, y) + i\, g(x, y) \tag{B.3}$$

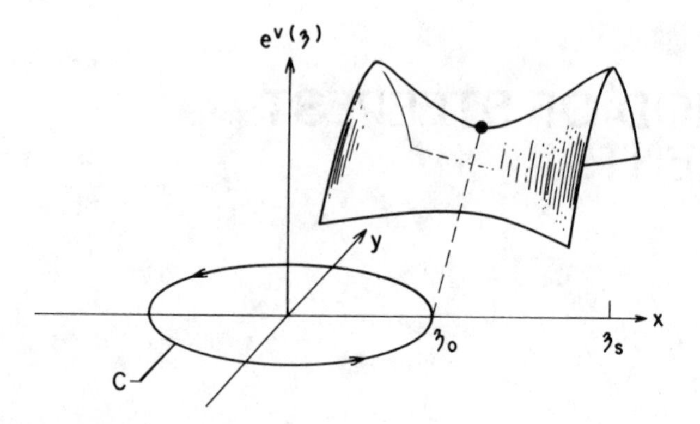

FIGURE B.1 The choice of contour C in the method of steepest descents. Shown also is the saddle point at z_o.

where $z = x + iy$. We assume that $v(z)$ is analytic everywhere along and in the vicinity of the contour C. Then the functions f and g satisfy the Cauchy–Riemann equations,

$$\begin{aligned}\frac{\partial f}{\partial x} &= \frac{\partial g}{\partial y} \\ \frac{\partial f}{\partial y} &= -\frac{\partial g}{\partial x}\end{aligned} \tag{B.4}$$

in this region. Also, they each satisfy the two-dimensional Laplace's equation and are therefore *harmonic functions,* where a function that is harmonic in a region can have neither an absolute maximum nor an absolute minimum in that region.[3]

Along the positive real axis the imaginary part of $v(z)$ is zero, and we have $v(z) = f(x, 0)$. Moreover, $f(x, 0)$ has a sharp minimum at z_o, where, since $v(z)$ is analytic at z_o, this minimum cannot be an absolute minimum. We may conclude, therefore, that $f(x, y)$ must be a sharp maximum in some direction away from the real axis.[2] Thus the point z_o is a *saddle point,* as is shown in Figure B.1.

We are now in a position to argue that we should choose the contour C so that it "drops" along the paths of steepest descent from z_o, for along a path of steepest descent the imaginary part of $v(z)$ is zero (as it also is along the real axis), and therefore rapid oscillations in the value of the integrand in Equation (B.1) cannot occur. We prove this fact first and then establish that the two paths of steepest descent are perpendicular to the real axis at z_o.

The variation of $f(x, y)$ along any path is given by

$$\frac{\partial f}{\partial s} = \cos\Theta \frac{\partial f}{\partial x} + \sin\Theta \frac{\partial f}{\partial y} \tag{B.5}$$

where $ds = (dx^2 + dy^2)^{1/2}$ is a differential path length and the angle Θ is the direction of the path at the point (x, y) measured relative to the x-axis. Now anywhere along a path of steepest descent the function $f(x, y)$ must be at a minimum point relative to a direction perpendicular to the path. In other words, since this path lies along the lowest points of a valley, we must have

$$\left.\frac{\partial f}{\partial s}\right|_{\Theta \to \Theta + \pi/2} = 0 \tag{B.6}$$

If we combine Equations (B.5) and (B.6), and use the Cauchy–Reimann equations (B.4), then we can establish easily that along a path of steepest descent

$$\frac{\partial g}{\partial s} = 0 \tag{B.7}$$

Therefore, the imaginary part of $v(\mathfrak{z})$ is constant, in fact zero, along the two paths of steepest descent from $\mathfrak{z}_o$. Thus at the point $\mathfrak{z} = \mathfrak{z}_o$ there are two intersecting lines along which $g(x, y) = 0$.

Consider next the function

$$\tilde{v}(\mathfrak{z}) \equiv v(\mathfrak{z}) - v(\mathfrak{z}_o) = \tilde{\phi}(x, y) + i\,g(x, y)$$

which vanishes at $\mathfrak{z}_o$. Inasmuch as the point $\mathfrak{z}_o$ is a saddle point, where the real function $\tilde{\phi}$ vanishes, we can readily argue that there must be two intersecting lines at the point $\mathfrak{z}_o$ along which $\tilde{\phi}(x, y) = 0$. We have now proved that both the real and imaginary parts of $\tilde{v}(\mathfrak{z})$ have this latter property, and one can next prove that at $\mathfrak{z}_o$ the four curves $\tilde{\phi}\,(x, y) = 0$ and $g(x, y) = 0$ all intersect at angles of $\pi/4$ radians.[4] Therefore, the paths of steepest descent are normal to the x-axis at the saddle point. We choose the contour C to follow these two paths for a ways before closing back around the origin, as shown in Figure B.1.

We can now write the contour integral (B.1) in the following manner:

$$J(\mathfrak{z}_o) = J_o(\mathfrak{z}_o) + \frac{1}{2\pi i}\int_{C'} e^{v(\mathfrak{z})}\, d\mathfrak{z} \tag{B.8a}$$

$$J_o(\mathfrak{z}_o) \equiv \frac{1}{2\pi i}\int_{\mathfrak{z}_o - ia}^{\mathfrak{z}_o + ia} e^{v(\mathfrak{z})}\, d\mathfrak{z} \tag{B.8b}$$

where a is a small real number and C' is the open contour that consists of C with the small path length from ($\mathfrak{z}_o$ − ia) to ($\mathfrak{z}_o$ + ia) removed. We next evaluate the integral $J_o(\mathfrak{z}_o)$ approximately with the aid of expansion (B.2).

First we rewrite this Taylor series as an expansion about the maximum point $\mathfrak{z}_o$ in the direction of the imaginary axis:

$$v(\mathfrak{z}) = v(\mathfrak{z}_o) - \frac{1}{2} v''(\mathfrak{z}_o)\, \zeta^2 + O(\zeta^3) \tag{B.2a}$$

where $\mathfrak{z} - \mathfrak{z}_o \equiv i\zeta$ and $v''(\mathfrak{z}_o)$ is large, real, and positive.

Since the important contribution to $J_o(\mathfrak{z}_o)$ occurs in the region of ζ values,

$$\zeta \lesssim [v''(\mathfrak{z}_o)]^{-1/2} << 1$$

and we may choose the small number a to lie outside this region, the integral $J_o(\mathfrak{z}_o)$ may now be approximated as follows:

$$\begin{aligned} J_o(\mathfrak{z}_o) &= \frac{1}{2\pi} e^{v(\mathfrak{z}_o)} \int_{-a}^{+a} d\zeta\, e^{-(1/2)v''(\mathfrak{z}_o)\zeta^2 + O(\zeta^3)} \\ &\cong \frac{1}{2\pi} e^{v(\mathfrak{z}_o)} \int_{-\infty}^{+\infty} d\zeta\, e^{-(1/2)v''(\mathfrak{z}_o)\zeta^2} \\ &= [2\pi\, v''(\mathfrak{z}_o)]^{-1/2}\, e^{v(\mathfrak{z}_o)} \end{aligned} \tag{B.9}$$

where we have used the second of Equations (2.4) to derive the final equality. The correction terms to this result form an asymptotic series.[5] That these correction terms are small relies on the fact that $v''(\mathfrak{z}_o)$ is large, that is, on the fact that $\exp[v(\mathfrak{z}_o)]$ is sharply peaked in the imaginary direction.

We now return to Equation (B.8a) and consider the contribution to $J(\mathfrak{z}_o)$ from the second integral. One normally argues, in any particular application, that this second integral can be neglected. For example, in the Darwin–Fowler method as it was applied in Appendix A, we must argue that oscillations due to $\mathscr{I}m[v(\mathfrak{z})]$ cause the integrand (A.9) to be small everywhere on C', as compared with the value it takes at $\mathfrak{z} = \mathfrak{z}_o$. The proof that this indeed occurs uses the restriction placed on the energy levels ω_i, above Equation (A.8), that not all of the ω_i values have a common divisor other than unity. With this restriction it is possible to show[6] that the overwhelming contribution to $J(\mathfrak{z})$ occurs along that part of the contour C where the function $v(\mathfrak{z})$ has its maximum value, namely, at $\mathfrak{z} = \mathfrak{z}_o$. We obtain then from Equations (B.8a) and (B.9), with $J(\mathfrak{z}_o) = P_N$ in the case of the Darwin–Fowler method,

$$J(\mathfrak{z}_o) \cong [2\pi\, v''(\mathfrak{z}_o)]^{-1/2}\, e^{v(\mathfrak{z}_o)} \tag{B.10}$$

Bibliography for Appendix B

E. T. Copson, *Asymptotic Expansions* (Cambridge, 1967), Chapters 7 and 8. In which the method of steepest descents is developed and applied rigorously and with mathematical perspective.

H. Jeffries and B. S. Jeffries, *Methods of Mathematical Physics* (Cambridge, 1972), pp. 503–506.
P. M. Morse and H. Feshbach, *Methods of Theoretical Physics, Part I* (McGraw-Hill, 1953), Section 4.6.

Notes

1. It is always permissible to translate the origin, which implies that the choice of contour C is really quite general. The coordinate axes may also be rotated arbitrarily. For example, when $v(x)$ has a sharp maximum, instead of minimum, along the positive (real) x-axis, then the integration technique (B.9) of the steepest-descents method can be applied directly using the variable x. Such is the case for the $\Gamma(z)$ function (2.1) when $z >> 1$ [see the derivation (2.6) of Stirling's formula for $N!$].

2. See Equation (A.11b), for application to the Darwin–Fowler method, to verify that the minimum is very pronounced, or sharp.

3. E. C. Titchmarsh, *The Theory of Functions* (Oxford, 1952), Section 5.14.

4. E. C. Titchmarsh, *loc. cit.*, Section 3.52.

5. A systematic method for deriving this series, whose terms are due to the $O(\zeta^3)$ neglected terms in the approximate expression (B.2a) for $v(\mathfrak{z})$, is presented in the text by Morse and Feshbach, *op. cit.*

6. See E. Schrödinger, *Statistical Thermodynamics* (Cambridge, 1967), pp. 31–33.

APPENDIX C

CARNOT CYCLE AND ABSOLUTE TEMPERATURE SCALE

In this appendix we introduce the Carnot cycle as that process which gives the maximum work output from a thermally isolated pair of temperature reservoirs. Following a discussion of Carnot's theorem, which relates to the maximum possible efficiency of heat engines, we then introduce the thermodynamic (or Kelvin) temperature scale and explain its equivalence to the absolute temperature scale, defined by Equation (6.3). Finally, we give a brief discussion of the second law of thermodynamics.

Consider a thermally isolated system that consists initially of several bodies not in thermal equilibrium with each other. This system can be allowed to approach equilibrium in such a way that it does work on external objects. The amount of work extracted, and hence the final equilibrium state of the system, depends on the process by which thermal equilibrium is approached.

The question arises as to which process can produce the maximum possible work. Here we are only interested in the work produced by virtue of the nonequilibrium condition. This means that we must ignore any work that might be done by a general expansion of the system, or by other mechanical processes, since this work could also be done by the system after it is in thermal equilibrium. Thus we assume that the total volume Ω and other external parameters λ of the system remain constant during the process; that is, we restrict our discussion to heat engines.

Let the initial energy of the system be E_0 and the energy of a final equilibrium state as a function of its entropy be $E(S)$. Inasmuch as the system is thermally isolated, the work W done by it is given simply by $W = E_0 - E(S)$. On differentiating this expression with respect to S we obtain, with the aid of Equation (6.3),

$$\frac{\partial W}{\partial S} = -\left(\frac{\partial E}{\partial S}\right)_{\Omega,\lambda} = -T \tag{C.1}$$

where T is the temperature of the final state. Since this derivative is negative, we may conclude that W increases as S decreases. But the entropy of a thermally isolated system can only increase or remain constant. Therefore the maximum possible work done will be obtained if the total entropy S remains constant during the process; that is, the process must be reversible.

Carnot Cycle

We now consider the special case of two bodies at different temperatures T_1 and T_2, such that $T_2 > T_1$. We can calculate the maximum work δW_{max} that can be extracted from this system during the exchange of a small quantity of energy between the two bodies by arguing as follows: First, we must emphasize that if the transfer of energy were to be accomplished by direct contact of the bodies, then no work whatsoever would be done. In this case the process would be an irreversible one, and the entropy of the system would increase by $|\delta E_2|(1/T_1 - 1/T_2)$, where $-\delta E_2$ is the energy transfer or heat taken from body two. [Equation (7.12) applies to each body, because their temperatures do not change during the small energy transfer.]

We can see that to bring about a reversible transfer of energy between the two bodies, and hence to obtain the maximum work output, it is necessary to introduce some auxiliary body (the working body) that performs a definite reversible cyclic process. This process must take place in such a way that a direct exchange of energy between two bodies occurs only when they are both at the same temperature. Thus the working body at temperature T_2 must be brought into contact with the system body (reservoir) at temperature T_2, to gain isothermally an energy $-\delta E_2$. Then it must be adiabatically cooled to the temperature T_1, at which temperature it gives up energy δE_1 to the reservoir at temperature T_1. Finally, the working body is returned adiabatically to its initial state at temperature T_2. The entire reversible process is called the *Carnot cycle* (see Figure C.1). During the expansions associated with this cycle, positive work is done by the auxiliary body on external objects. Clearly, the Carnot cycle is the only possible reversible cycle with which a heat engine can operate between two temperature reservoirs.

We now calculate the (maximum) external work done during a Carnot cycle, noting that the auxiliary body need not be considered, since it returns to its initial state. On referring again to Equation (7.12), we see that this work is given by

$$\delta W_{max} = (-\delta E_2) - \delta E_1 = -T_2\delta S_2 - T_1\delta S_1$$

But the total entropy change $(\delta S_1 + \delta S_2)$ for a reversible process is zero, which implies that

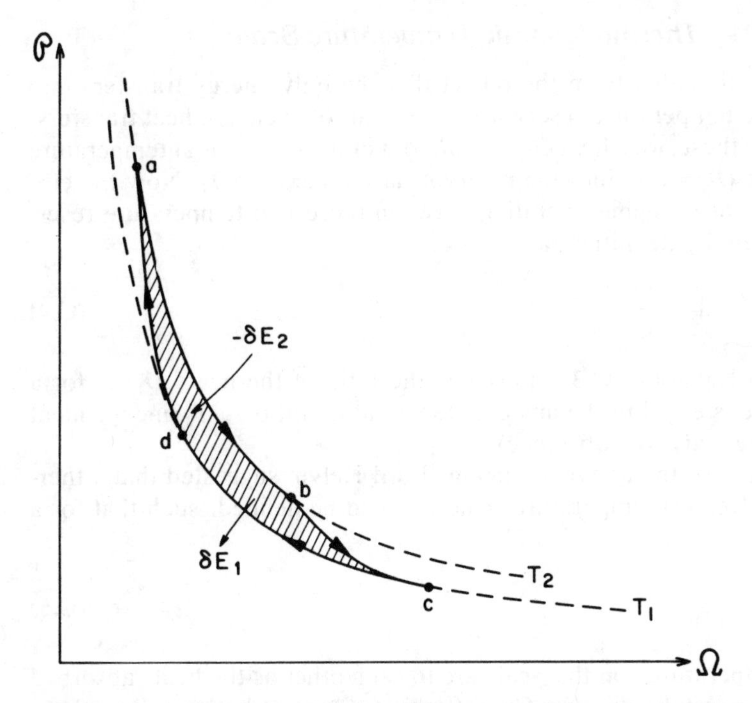

FIGURE C.1 The Carnot cycle on a $\mathcal{P}$–Ω diagram for a working body (an ideal gas in this case) operating between two temperature reservoirs. Curves *b–c* and *d–a* represent adiabatic processes. The net external work per cycle, $\oint \mathcal{P}\, d\Omega$, done by this ideal heat engine is given by the shaded area.

$$\begin{aligned}\delta W_{\max} &= (T_2 - T_1)(-\delta S_2) \\ &= \left(\frac{T_2 - T_1}{T_2}\right)(-\delta E_2)\end{aligned} \tag{C.2}$$

Therefore, the efficiency $e \equiv \delta W/|\delta E_2|$ of a heat engine operating on a Carnot cycle is given by

$$e_{\max} = \frac{\delta W_{\max}}{|\delta E_2|} = 1 - \frac{T_1}{T_2} \tag{C.3}$$

Since a Carnot cycle gives the maximum possible work output for a given energy transfer $-\delta E_2$ from the reservoir at temperature T_2, we may conclude that: "No heat engine operating between two temperature reservoirs can have an efficiency greater than that of a Carnot cycle." This theorem was due originally to Sadi Carnot.

Thermodynamic Temperature Scale

We have stipulated from the outset that the only energy transfers into and out of the temperature reservoirs of the Carnot cycle are heat transfers. For one cycle these are a heat $Q_2 = -\delta E_2$ out of the reservoir at temperature T_2 and a heat $Q_1 = \delta E_1$ into the reservoir at temperature T_1. Now the efficiency of any heat engine operating between these two temperature reservoirs is written, by definition, as

$$e = \frac{Q_2 - Q_1}{Q_2} = 1 - \frac{Q_1}{Q_2} \tag{C.4}$$

According to Equation (C.3), however, the ratio of the heats Q_1/Q_2 for a Carnot engine is equal to the absolute temperature ratio T_1/T_2, independent of the working body (or substance).

On the basis of the above argument, Lord Kelvin suggested that a thermodynamic (Kelvin) temperature scale Θ could be defined, such that for a Carnot cycle

$$\frac{\Theta_2}{\Theta_1} \equiv \frac{Q_2}{Q_1} \tag{C.5}$$

Thus two temperatures on this scale are to each other as the heats absorbed and rejected, respectively, by a Carnot engine operating between these temperatures. To complete the definition of the thermodynamic scale (in degrees Kelvin), the arbitrary value $\Theta_{tr} = 273.16\text{K}$ is assigned to the triple point of water.

Now we make two observations relative to the thermodynamic temperature scale, as defined above.

(1) The absolute temperature scale T can be identified with the thermodynamic temperature scale Θ. Therefore, since Equation (C.3) follows from Equation (7.12), which is equivalent to definition (6.3) in Section 6, we may conclude that Equation (6.3) in is, indeed, a consistent way to define the absolute temperature scale. Thus both Equations (6.3) and (C.5) can be used to define the same absolute temperature scale to within an arbitrary scaling constant.

(2) Since definition (C.5) [as well as (6.3)] is independent of the working substance, one can imagine the use of an ideal gas as the working body in a Carnot cycle. Moreover, the most convenient absolute temperature scale to define operationally is the ideal gas scale T', which is based on constant-volume thermometer measurements.[1] On using the ideal-gas law, one can show independently that Equation (C.5) holds for the temperature scale T', that is, $Q_2/Q_1 = T_2'/T_1'$. Therefore, on setting $T_{tr}' = 273.16°$, we obtain that $\Theta = T = T'$, and the thermodynamic temperature scale agrees with the ideal-gas scale.

Exercise C.1 It is well known[2] that application of the ideal-gas law $\mathcal{P}\Omega = N\kappa T'$ to an adiabatic process yields $\mathcal{P}\Omega^{\gamma} =$ constant, where γ is the heat-capacity ratio, C_P/C_V. Use this result to show that $Q_2/Q_1 = T_2'/T_1'$ for an ideal gas operating in a Carnot cycle.

Second Law of Thermodynamics

In Section 6 we used the maximum entropy principle to describe the macroscopic state of thermodynamic equilibrium. We also used this principle to characterize the approach to equilibrium as a condition of increasing entropy. The law of increasing entropy is stated formally as the *second law of thermodynamics,* in the form: "The entropy of a closed system tends to increase."[3]

There are other, nonstatistical, statements of the second law of thermodynamics. These relate to cyclic heat engines and refrigerators, such as the Carnot engine and its reverse. In particular, there is the Clausius statement concerning refrigerators and the Kelvin–Planck statement concerning heat engines which can be phrased as follows. *Clausius:* "It is impossible to operate a cyclical machine whose sole effect is to convey heat continuously from one body to another at a higher temperature." *Kelvin–Planck:* "It is impossible to extract heat from a system at one uniform temperature and to convert it into work with no other effect occurring in the system or its environment."

Exercise C.2 Prove that both the Clausius and Kelvin–Planck statements of the second law of thermodynamics follow from the entropy statement.

Bibliography for Appendix C

L. D. Landau and E. M. Lifshitz, *Statistical Physics* (Addison-Wesley, 1958), translated by E. Peierls and R. F. Peierls, Section 19. This lucid section on maximum work has been used as a basis for the first part of the present appendix.

R. Resnick and D. Halliday, *Physics, Part I* (Wiley, 1966), Chapter 25. A thorough exposition at the elementary level of the Carnot cycle and its relation to entropy and the second law of thermodynamics.

Notes

1. R. Resnick and D. Halliday, *op. cit.*, Section 21-5.
2. *Ibid.*, p. 586, Example 5.

3. E. T. Jaynes [*Am. J. Phys.* **33** (1965), p. 391] has given a precise operational statement of the second law of thermodynamics using the definition of an adiabatic process as one (reversible or irreversible) for which there is no external heat exchange: "The experimental entropy cannot decrease in a reproducible adiabatic process that starts from a state of complete thermodynamic equilibrium."

APPENDIX

D REMARKS RELATED TO THE COLLISION INTEGRAL

The collision integral plays a significant role in many aspects of classical kinetic theory. In its fundamental form as derived in Section 23 it changes sign under time reversal, a property that is consistent with the microscopic laws of physics from which it is derived. The approximate expressions (11.10) and (11.14) we used in Chapter III also change sign under time reversal, as we have argued in Section 25 below Equation (25.11). Implicitly required for this argument is the time-reversal invariance of the classical transition probability (11.11).

In this appendix we first discuss underlying symmetries of the classical transition probability. Then we cite some consequences for the hydrodynamic equations (Section 12) of the exact form (23.26) of the collision integral.

In Section 11 we derived expression (11.10) for the collision integral, namely,

$$\left[\frac{\partial}{\partial t} P(\mathbf{r}_1, \mathbf{k}_1, t)\right]_{\text{coll}} = \int d^3\mathbf{k}_2\, d\omega_q [P_2(\mathbf{r}_1, \mathbf{r}_2'; \mathbf{k}_1', \mathbf{k}_2'; t) - P_2(\mathbf{r}_1, \mathbf{r}_2; \mathbf{k}_1, \mathbf{k}_2; t)] \times R(\mathbf{k}_1, \mathbf{k}_2; \hat{\mathbf{q}}) \tag{D.1}$$

where $R(\mathbf{k}_1, \mathbf{k}_2, \hat{\mathbf{q}})$ is the transition probability for the scattering process $(\mathbf{k}_1, \mathbf{k}_2) \rightarrow (\mathbf{k}_1', \mathbf{k}_2')$. The function R is related to the differential scattering cross section σ by the first equality of Equation (11.11). Moreover, symmetry properties of the collision integral (D.1), or of its related form (11.14), follow from symmetry properties of the microscopic physical quantity R. In fact, one property, the second equality of Equation (11.11), was used to obtain the expression (D.1) itself. We now discuss some of the symmetries of R.

Symmetries of the Transition Probability

According to Equation (11.4) the momentum-transfer vector $\mathbf{q}$ is given by

$$\mathbf{q} = \mathbf{k}_2' - \mathbf{k}_2 = \mathbf{k}_1 - \mathbf{k}_1' \tag{D.2}$$

where the second equality is due to conservation of momentum for the two-body collision process. Clearly, the three vectors $\mathbf{k}_1$, $\mathbf{k}_2$, and $\mathbf{q}$ determine completely both $\mathbf{k}_1'$ and $\mathbf{k}_2'$. Similarly, for the inverse process $\mathbf{k}_1'$, $\mathbf{k}_2'$, and $\mathbf{q}$ determine $\mathbf{k}_1$ and $\mathbf{k}_2$. For elastic scattering, the only case of interest in our present considerations, the magnitude of the relative momenta (11.3) is conserved and we may replace $\mathbf{q}$ by the unit vector $\hat{\mathbf{q}}$ in the preceding two statements. Thus there is no change in functional meaning if we write, here, $R(\mathbf{k}_1, \mathbf{k}_2; \mathbf{k}_1', \mathbf{k}_2')$ instead of $R(\mathbf{k}_1, \mathbf{k}_2, \hat{\mathbf{q}})$. With this notation we may conveniently cite various symmetry properties of R for classical point particles without internal degrees of freedom.

(a) Invariance under time reversal: Equations (25.9) to (25.11) define the meaning of time reversal for microscopic physical quantities. Thus the scattering process $(\mathbf{k}_1, \mathbf{k}_2) \rightarrow (\mathbf{k}_1', \mathbf{k}_2')$ becomes, under time reversal, $(-\mathbf{k}_1', -\mathbf{k}_2') \rightarrow (-\mathbf{k}_1, -\mathbf{k}_2)$. Corresponding invariance of the transition probability means that

$$R(\mathbf{k}_1, \mathbf{k}_2; \mathbf{k}_1', \mathbf{k}_2') = R(-\mathbf{k}_1', -\mathbf{k}_2'; -\mathbf{k}_1, -\mathbf{k}_2) \tag{D.3}$$

(b) Invariance under rotation: This symmetry expresses the fact that the scattering process must be invariant under any rotation $\mathcal{R}^{-1}$ of the coordinate system. Equivalently, we may rotate the scattering process itself by $\mathcal{R}$, and invariance of the transition probability under rotations can be stated as

$$R(\mathbf{k}_1, \mathbf{k}_2; \mathbf{k}_1', \mathbf{k}_2') = R(\mathcal{R}\mathbf{k}_1, \mathcal{R}\mathbf{k}_2; \mathcal{R}\mathbf{k}_1', \mathcal{R}\mathbf{k}_2') \tag{D.4}$$

(c) Invariance under reflections or inversion: Reflection invariance means that if the scattering process (or coordinate system) is reflected through any plane, then there is no effect on the transition probability R.[1] We may represent a reflection operation by the letter $\mathcal{I}$, and write

$$R(\mathbf{k}_1, \mathbf{k}_2; \mathbf{k}_1', \mathbf{k}_2') = R(\mathcal{I}\mathbf{k}_1, \mathcal{I}\mathbf{k}_2; \mathcal{I}\mathbf{k}_1', \mathcal{I}\mathbf{k}_2') \tag{D.5a}$$

Inversion is a generalization of the reflection operation in which the entire scattering process is reflected about a point, so that we get for invariance of R under inversion,

$$R(\mathbf{k}_1, \mathbf{k}_2; \mathbf{k}_1', \mathbf{k}_2') = R(-\mathbf{k}_1, -\mathbf{k}_2; -\mathbf{k}_1', -\mathbf{k}_2') \tag{D.5b}$$

Inversion I is equivalent to a rotation through 180 degrees followed by a reflection [see Figure D.1(d)]. It is equivalent to the change from physical

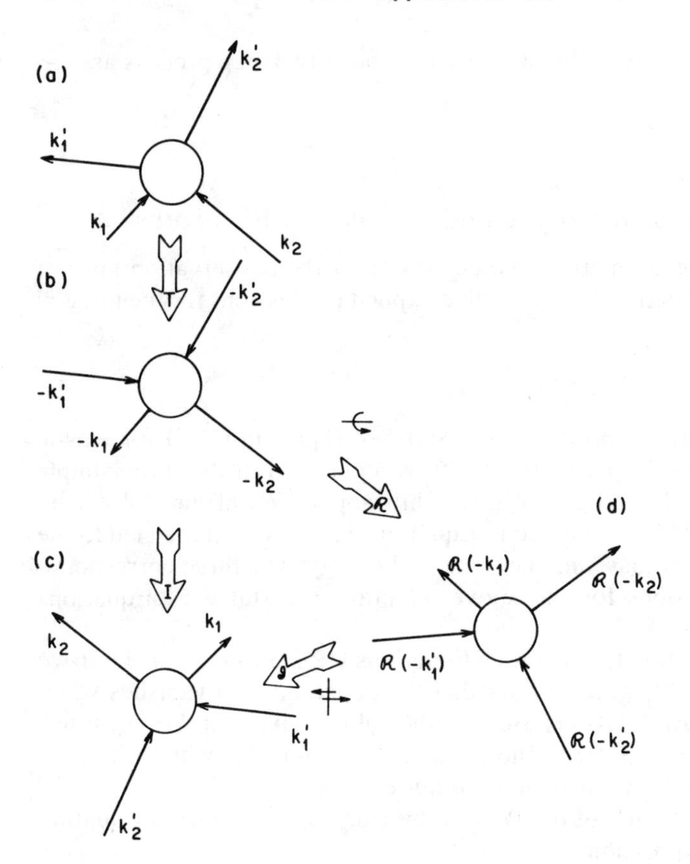

FIGURE D.1 Symmetries of the classical scattering of two point particles: (a) The basic process $(\mathbf{k}_1, \mathbf{k}_2) \to (\mathbf{k}_1', \mathbf{k}_2')$. (b) The time-reversed (T) process, followed by (c) an inversion (I). (d) An intermediate step between (b) and (c), which consists of a rotation through 180 degrees ($\mathscr{R}$) followed by a reflection ($\mathscr{I}$). Note that, in this case, $\mathscr{R}(-\mathbf{k}_i) = \mathscr{I}(\mathbf{k}_i)$.

description within a right-handed coordinate system to one within a left-handed system.

If the two-body scattering process is subjected (theoretically) to a time reversal followed by an inversion, then invariance of the transition probability under both of these operations [Equations (D.3) and D.5b)] yields the result

$$R(\mathbf{k}_1, \mathbf{k}_2; \mathbf{k}_1', \mathbf{k}_2') = R(\mathbf{k}_1', \mathbf{k}_2'; \mathbf{k}_1, \mathbf{k}_2) \tag{D.6}$$

or, equivalently, $R(\mathbf{k}_1, \mathbf{k}_2, \hat{\mathbf{q}}) = R(\mathbf{k}_1', \mathbf{k}_2', -\hat{\mathbf{q}})$. In the center-of-momentum system, this last result is equivalent to the second equality of Equation

(11.11). Figure D.1 shows schematically the above two-step process associated with Equation (D.6).

Exact Collision Integral and the Conservation Laws

To give a complete microscopic derivation of the conservation laws of hydrodynamics in Section 12, we needed explicit expressions for the integral

$$-\int d^3\mathbf{k}_1\, \chi(\mathbf{r}_1, \mathbf{k}_1, t)\left(\frac{\partial P}{\partial t}\right)_{\text{coll}}$$

where χ is a conserved property that satisfies Equation (12.5) for elastic molecular collisions. In particular, we were interested in the three simple cases: $\chi = m$, $\chi = \mathbf{k}$, and $\chi = k^2/2m$. Thus application of the Boltzmann transport equation (11.27) resulted in Equation (12.21), which then led to the conservation laws for mass, momentum, and energy. The three corresponding explicit expressions for the above integral were stated as Equations (12.23) to (12.25).

We now outline the derivation of Equations (12.23) to (12.25). First we recall theorem (12.22), which shows that the above integral vanishes when the approximate form (11.14) is used for the collision integral. We must use the exact expression (23.26) in the present derivation, for which the conserved property (12.5) of χ is then not required.

On using the definition of the Poisson bracket, we obtain from Equation (23.26) the exact expression

$$-\int d^3\mathbf{k}_1\, \chi(\mathbf{r}_1, \mathbf{k}_1, t)\left(\frac{\partial P}{\partial t}\right)_{\text{coll}} = -\int d^3\mathbf{k}_1\, d^3\mathbf{r}_2\, d^3\mathbf{k}_2\, \chi(\mathbf{r}_1, \mathbf{k}_1, t) \times \frac{\partial}{\partial \mathbf{r}_1} V(r_{21}) \cdot \frac{\partial}{\partial \mathbf{k}_1} P_2(\mathbf{r}_1, \mathbf{r}_2; \mathbf{k}_1, \mathbf{k}_2; t) \tag{D.7}$$

in which we have considered only the case of a velocity-independent central molecular interaction $V(r_{12})$. It is assumed that the distribution function P_2 vanishes for large momentum values, so that an integration by parts with respect to $\mathbf{k}_1$ yields, quite generally,

$$-\int d^3\mathbf{k}_1\, \chi(\mathbf{r}_1, \mathbf{k}_1, t)\left(\frac{\partial P}{\partial t}\right)_{\text{coll}} = \int d^3\mathbf{k}_1\, d^3\mathbf{r}_2\, d^3\mathbf{k}_2\, P_2(\mathbf{r}_1, \mathbf{r}_2; \mathbf{k}_1, \mathbf{k}_2; t) \times \frac{\partial}{\partial \mathbf{k}_1} \chi(\mathbf{r}_1, \mathbf{k}_1, t) \cdot \frac{\partial}{\partial \mathbf{r}_1} V(r_{21}) \tag{D.8}$$

This result shows that Equation (12.23) for the mass case $\chi = m$ is correct.

Equation (D.8) can be rewritten in an alternate form by using $d\omega = d^3\mathbf{r}\, d^3\mathbf{k}$, $\mathbf{r}_{21} = \mathbf{r}_2 - \mathbf{r}_1$ and the fact that we are only interested here in position- and time-independent χ. Thus we obtain

$$-\int d^3\mathbf{k}\, \chi(\mathbf{k}) \left(\frac{\partial P}{\partial t}\right)_{\text{coll}} = -\int d\omega_1\, d\omega_2\, P_2(\mathbf{r}_1, \mathbf{r}_2; \mathbf{k}_1, \mathbf{k}_2; t)\, \delta^{(3)}(\mathbf{r}_1 - \mathbf{r}) \times V'(r_{21})\hat{\mathbf{r}}_{21} \cdot \frac{\partial}{\partial \mathbf{k}_1} \chi(\mathbf{k}_1) \tag{D.9}$$

where $V'(r)$ denotes the derivative of the potential V and $\hat{\mathbf{r}}_{21}$ is a unit vector. The left side is now a function of $\mathbf{r}$ and t, instead of $\mathbf{r}_1$ and t.

We consider next the special case $\chi(\mathbf{k}) = \mathbf{k}$, for which Equation (D.9) becomes

$$\begin{aligned} -\int d^3\mathbf{k}\, \mathbf{k} \left(\frac{\partial P}{\partial t}\right)_{\text{coll}} &= -\int d\omega_1\, d\omega_2\, P_2(\mathbf{r}_1, \mathbf{r}_2; \mathbf{k}_1, \mathbf{k}_2; t)\, \delta^{(3)}(\mathbf{r}_1 - \mathbf{r}) \\ &\quad \times V'(r_{21})\hat{\mathbf{r}}_{21} \qquad \text{(D.10)} \\ &= \frac{1}{2}\int d^3r_1\, d^3\mathbf{r}_2\, P(\mathbf{r}_1, \mathbf{r}_2, t)\, V'(r_{21})\hat{\mathbf{r}}_{21} \\ &\quad \times [\delta^{(3)}(\mathbf{r}_2 - \mathbf{r}) - \delta^{(3)}(\mathbf{r}_1 - \mathbf{r})] \end{aligned}$$

To derive the second equality of this equation we have used definition (12.13) of the pair distribution function $P(\mathbf{r}_1, \mathbf{r}_2, t)$ and the fact that this function must be symmetric in the position coordinates. The difference in δ functions can be written formally as a Taylor-series expansion, in powers of the "small" distance $\mathbf{r}_{21}$, as follows:

$$\begin{aligned} [\delta^{(3)}(\mathbf{r}_1 - \mathbf{r} + \mathbf{r}_{21}) - \delta^{(3)}(\mathbf{r}_1 - \mathbf{r})] &= \left[\sum_{n=1}^{\infty} \frac{1}{n!}(-\mathbf{r}_{21} \cdot \nabla_r)^n\right] \delta^{(3)}(\mathbf{r}_1 - \mathbf{r}) \qquad \text{(D.11)} \\ &= -(\nabla_r \cdot \mathbf{r}_{21}) \left[\sum_{n=1}^{\infty} \frac{1}{n!}(-\mathbf{r}_{21} \cdot \nabla_r)^{n-1}\right] \delta^{(3)}(\mathbf{r}_1 - \mathbf{r}) \end{aligned}$$

Substitution of the formal expansion (D.11) into Equation (D.10) gives a result that can be written, on switching to the independent integration variables $\mathbf{r}_1$ and $\mathbf{r}_{21}$, as follows:

$$-\int d^3\mathbf{k}\mathbf{k}\left(\frac{\partial P}{\partial t}\right)_{\text{coll}} = \nabla\cdot\left\{-\frac{1}{2}\int d^3\mathbf{r}_1\, d^3\mathbf{r}_{21}\mathbf{r}_{21}\hat{\mathbf{r}}_{21} \times V'(r_{21})P(\mathbf{r}_1, \mathbf{r}_1+\mathbf{r}_{21}, t) \times\left[\sum_{n=1}^{\infty}\frac{1}{n!}(\mathbf{r}_{21}\cdot\nabla_{r_1})^{n-1}\right] \times \delta^{(3)}(\mathbf{r}_1-\mathbf{r})\right\} \tag{D.12}$$

$$= \nabla\cdot\left\{-\frac{1}{2}\int d^3\mathbf{r}_1\, d^3\mathbf{r}_{21}\,\delta^{(3)}(\mathbf{r}_1-\mathbf{r})\mathbf{r}_{21}\hat{\mathbf{r}}_{21} \times V'(r_{21}) \times\left[\sum_{n=1}^{\infty}\frac{1}{n!}(-\mathbf{r}_{21}\cdot\nabla_{r_1})^{n-1}\right] \times P(\mathbf{r}_1, \mathbf{r}_1+\mathbf{r}_{21}, t)\right\}$$

But the variation of $P(\mathbf{r}_1, \mathbf{r}_1 + \mathbf{r}_{21}, t)$ with $\mathbf{r}_1$ is expected to be extremely small over the region of integration of $\mathbf{r}_{21}$, that is, over the range r_o of the interaction potential V. In other words, the effective expansion parameter in the series summation is $\sim r_o/L \ll 1$, where L is a macroscopic length that is characteristic of density fluctuations in the fluid. We may therefore neglect all terms in n beyond the first one in Equation (D.12) and obtain, to excellent approximation,

$$-\int d^3\mathbf{k}\,\mathbf{k}\left(\frac{\partial P}{\partial t}\right)_{\text{coll}} \cong \nabla\cdot\left\{-\frac{1}{2}\int d^3\mathbf{r}_1\, d^3\mathbf{r}_2\,\delta^{(3)}(\mathbf{r}_1-\mathbf{r})\mathbf{r}_{21}\hat{\mathbf{r}}_{21} \times V'(r_{21})\, P(\mathbf{r}_1, \mathbf{r}_2, t)\right\} \tag{D.13}$$

$$= \nabla\cdot\overleftrightarrow{B}(\mathbf{r}, t)$$

where the dyad $\overleftrightarrow{B}$ is defined by Equation (12.12). This completes the proof of Equation (12.24).

To derive Equation (12.25) we need the result of Problem V.7 for $\partial\phi/\partial t$, where $\phi(\mathbf{r}, t)$ is defined by Equation (12.15). If we subtract $\nabla\cdot(\phi\mathbf{u})$ and the expression for $\partial\phi/\partial t$ from Equation (D.9), with $\chi(\mathbf{k}) = k^2/2m$, then we obtain

$$-\left\{\int d^3\mathbf{k}\,(k^2/2m)\left(\frac{\partial P}{\partial t}\right)_{\text{coll}} + \frac{\partial \phi}{\partial t} + \nabla\cdot(\phi\mathbf{u})\right\} \tag{D.14}$$

$$= \frac{1}{2}\nabla\cdot\int d\omega_1\, d\omega_2\, P_2(\mathbf{r}_1, \mathbf{r}_2; \mathbf{k}_1, \mathbf{k}_2; t)\,\delta^{(3)}(\mathbf{r}_1 - \mathbf{r})V(r_{21})\left(\frac{\mathbf{k}_1}{m} - \mathbf{u}\right)$$

$$- \frac{1}{2}\int d\omega_1\, d\omega_2\, P_2(\mathbf{r}_1, \mathbf{r}_2; \mathbf{k}_1, \mathbf{k}_2, t)\,\delta^{(3)}(\mathbf{r}_1 - \mathbf{r})V'(r_{21})\hat{\mathbf{r}}_{21}\cdot\left(\frac{\mathbf{k}_1}{m} + \frac{\mathbf{k}_2}{m}\right)$$

$$= \frac{1}{2}\nabla\cdot\int d\omega_1\, d\omega_2\, P_2(\mathbf{r}_1, \mathbf{r}_2; \mathbf{k}_1, \mathbf{k}_2; t)\,\delta^{(3)}(\mathbf{r}_1 - \mathbf{r})V(r_{21})\left(\frac{\mathbf{k}_1}{m} - \mathbf{u}\right)$$

$$+ \frac{1}{2}\int d\omega_1\, d\omega_2\, P_2(\mathbf{r}_1, \mathbf{r}_2; \mathbf{k}_1, \mathbf{k}_2; t)V'(r_{21})\left(\hat{\mathbf{r}}_{21}\cdot\frac{\mathbf{k}_1}{m}\right)$$

$$\times [\delta^{(3)}(\mathbf{r}_2 - \mathbf{r}) - \delta^{(3)}(\mathbf{r}_1 - \mathbf{r})]$$

$$= \frac{1}{2}\nabla\cdot\int d\omega_1\, d\omega_2\, P_2(\mathbf{r}_1, \mathbf{r}_2; \mathbf{k}_1, \mathbf{k}_2; t)\,\delta^{(3)}(\mathbf{r}_1 - \mathbf{r})V(r_{21})\left(\frac{\mathbf{k}_1}{m} - \mathbf{u}\right)$$

$$- \frac{1}{2}\nabla\cdot\int d\omega_1\, d^3\mathbf{r}_{21}\, d^3\mathbf{k}_2\,\delta^{(3)}(\mathbf{r}_1 - \mathbf{r})V'(r_{21})\mathbf{r}_{21}\left(\hat{\mathbf{r}}_{21}\cdot\frac{\mathbf{k}_1}{m}\right)$$

$$\times\left[\sum_{n=1}^{\infty}\frac{1}{n!}(-\mathbf{r}_{21}\cdot\nabla_{r_1})^{n-1}\right]P(\mathbf{r}_1, \mathbf{r}_1 + \mathbf{r}_{21}; \mathbf{k}_1, \mathbf{k}_2; t)$$

We have used the symmetry of the distribution function P_2 under interchange of all the coordinates for particles 1 and 2 to deduce the second equality above. To arrive at the final equality we have used Equation (D.11) and the arguments associated with Equation (D.12). As with Equation (D.13), we now drop all of the terms in n beyond the first one ($n = 1$) in this last expression and obtain, to excellent approximation,

$$-\left\{\int d^3\mathbf{k}(k^2/2m)\left(\frac{\partial P}{\partial t}\right)_{\text{coll}} + \frac{\partial \phi}{\partial t} + \nabla\cdot(\phi\mathbf{u})\right\} \tag{D.15}$$

$$\cong \frac{1}{2}\nabla\cdot\int d\omega_1\, d\omega_2\, P_2(\mathbf{r}_1, \mathbf{r}_2; \mathbf{k}_1, \mathbf{k}_2; t)\,\delta^{(3)}(\mathbf{r}_1 - \mathbf{r})$$

$$\times\left[V(r_{21})\left(\frac{\mathbf{k}_1}{m} - \mathbf{u}\right) - V'(r_{21})\mathbf{r}_{21}\left(\hat{\mathbf{r}}_{21}\cdot\frac{\mathbf{k}_1}{m}\right)\right]$$

$$= \nabla\cdot\left[\overleftrightarrow{B}(\mathbf{r}, t)\cdot\mathbf{u}(\mathbf{r}, t) + \mathbf{q}_V(\mathbf{r}, t)\right]$$

where the quantities $\overleftrightarrow{B}$ and $\mathbf{q}_V$ are defined by Equations (12.12) and (12.18), respectively. Q.E.D.[2]

Bibliography for Appendix D

Text

K. Huang, *Statistical Mechanics* (Wiley, 1965). Our presentation of the symmetries of the transition probability is based on the detailed exposition given in Section 3.2 of this text.

Original Paper

J. H. Irving and J. G. Kirkwood, "The Statistical Mechanical Theory of Transport Processes. IV. The Equations of Hydrodynamics," *J. Chem. Phys.* **18** (1950), p. 817.

Notes

1. For the purposes of Chapter III and this appendix, only the electromagnetic interactions of atomic and molecular physics need be considered. Of course, the transition probability is not invariant under reflections when the weak interaction of β decay phenomena is included.

2. Note that for the equilibrium distributions of large fluid systems $P_2(\mathbf{r}_1, \mathbf{r}_1 + \mathbf{r}_{21}; \mathbf{k}_1, \mathbf{k}_2; t)$ must be independent of $\mathbf{r}_1$ (as well as t) because of Galilean invariance. In this limit, the expressions (12.12) and (12.18), for $\overleftrightarrow{B}$ and $\mathbf{q}_V$, respectively, become exact.

APPENDIX E

WIGNER DISTRIBUTION FUNCTION

The Wigner distribution function $f_w(q_i, p_i, t)$ is defined in terms of off-diagonal position-space matrix elements of the statistical matrix $\underline{\rho}$. Conditions are shown under which average values calculated from f_w agree with the prescription (21.9) for their calculation using $\underline{\rho}$. The rather complicated equation of motion that f_w satisfies is derived and shown to reduce to Liouville's theorem in the limit $h \to 0$. One concludes, therefore, that the Wigner distribution function is a suitable quantum-mechanical analog to the classical distribution function. For this reason one also obtains a rigorous derivation of Liouville's theorem.

For convenience, but without loss of generality, we consider a system of N identical particles with no internal degrees of freedom. Then the Wigner distribution function $f_w(\mathbf{q}_1 \cdots \mathbf{q}_N; \mathbf{p}_1 \cdots \mathbf{p}_N; t)$ is constructed from the statistical matrix $\underline{\rho}$ by the prescription

$$f_w(\mathbf{q}_1 \cdots \mathbf{q}_N; \mathbf{p}_1 \cdots \mathbf{p}_N; t) \equiv (2\pi)^{-3N} \int \prod_{\alpha=1}^{N} (d^3\boldsymbol{\tau}_\alpha \, e^{i\boldsymbol{\tau}_\alpha \cdot \mathbf{p}_\alpha}) \times \langle \mathbf{q}_j - \tfrac{1}{2}\hbar\boldsymbol{\tau}_j | \underline{\rho} | \mathbf{q}_j + \tfrac{1}{2}\hbar\boldsymbol{\tau}_j \rangle \tag{E.1}$$

where $|\mathbf{q}_j + \frac{1}{2}\hbar\boldsymbol{\tau}_j\rangle$ is a shorthand notation for $|\mathbf{q}_1 + \frac{1}{2}\hbar\boldsymbol{\tau}_1, \ldots, \mathbf{q}_N + \frac{1}{2}\hbar\boldsymbol{\tau}_N\rangle$. We note that for the off-diagonal set of coordinates $\{\mathbf{q}_\alpha - \frac{1}{2}\hbar\boldsymbol{\tau}_\alpha, \mathbf{q}_\alpha + \frac{1}{2}\hbar\boldsymbol{\tau}_\alpha\}$, $\mathbf{q}_\alpha$ represents a center-of-mass and $\hbar\boldsymbol{\tau}_\alpha$ a relative coordinate. It is the relative coordinate $\hbar\boldsymbol{\tau}_\alpha$ that is used to generate momentum dependence in the Wigner distribution function.

It is easy to verify from definition (E.1) that the Wigner distribution function is real, although not necessarily positive. The fact that it is not definitely a positive quantity stems from its quantum-mechanical origins; despite this apparent shortcoming, f_w gives the average values of physical quantities correctly, as we now demonstrate, and this is really all that can be required of a distribution function.

We first prove a general theorem for the position-space matrix elements of $\mathsf{A}\underline{\rho}$, where A is any (Hermitian) physical quantity that may be a function of both coordinates $\mathbf{q}_\alpha$ and momenta $\mathbf{p}_\alpha$. We write this theorem, using the notation of definition (E.1), as follows:

$$\langle \mathbf{q}_j - \tfrac{1}{2}\hbar\boldsymbol{\tau}_j|\mathsf{A}\underline{\rho}|q_j + \tfrac{1}{2}\hbar\boldsymbol{\tau}_j\rangle = (2\pi)^{-3N} \times \int \prod_{\alpha=1}^{N}\left[d^3\mathbf{y}_\alpha\, d^3\mathbf{p}_\alpha\, d^3\ell_\alpha\, e^{-i\tau_\alpha\cdot\mathbf{p}_\alpha}\, e^{i\ell_\alpha\cdot(\mathbf{q}_\alpha-\mathbf{y}_\alpha)}\right] \times \tilde{\mathbf{A}}(\mathbf{q}_j - \tfrac{1}{2}\hbar\boldsymbol{\tau}_j,\, \mathbf{p}_j + \tfrac{1}{2}\hbar\ell_j)\cdot f_w(\mathbf{y}_j, \mathbf{p}_j) \tag{E.2}$$

where the real quantity $\tilde{A}(\mathbf{q}_j, \mathbf{p}_j)$ is defined by the relation for position-space matrix elements,

$$\begin{aligned}\langle \mathbf{q}_j|\mathsf{A}|\mathbf{q}_j'\rangle\, e^{i\Sigma_\alpha \mathbf{k}_\alpha\cdot\mathbf{q}_\alpha'} &= \left[\prod_{\alpha=1}^{N}\delta^{(3)}(\mathbf{q}_\alpha - \mathbf{q}_\alpha')\right]\tilde{A}\left(\mathbf{q}_j', \frac{\hbar}{i}\frac{\partial}{\partial \mathbf{q}_j'}\right) e^{i\Sigma_\alpha \mathbf{k}_\alpha\cdot\mathbf{q}_\alpha'} \\ &= \left[\prod_{\alpha=1}^{N}\delta^{(3)}(\mathbf{q}_\alpha - \mathbf{q}_\alpha')\right]\tilde{A}(\mathbf{q}_j, \hbar\mathbf{k}_j)\, e^{i\Sigma_\alpha \mathbf{k}_\alpha\cdot\mathbf{q}_\alpha}\end{aligned} \tag{E.3}$$

Care must be exercised when this equation is applied to functions involving products of position and momentum variables. We assume here that such functions are written as symmetrized products, and this is the significance of the notation $\tilde{A}$.[1] For example, the electromagnetic-interaction term $\mathbf{p}\cdot\mathbf{A}$ occurs from the expansion of $(\mathbf{p} - (e/c)\mathbf{A})^2/2m$ as $\tfrac{1}{2}(\mathbf{p}\cdot\mathbf{A} + \mathbf{A}\cdot\mathbf{p})$, where in quantum mechanics $\mathbf{p} = -i\hbar\nabla$.

Proof of Theorem (E.2). We consider the matrix element on the left side of Equation (E.2) and proceed as follows:

$$\begin{aligned}\langle \mathbf{q}_j - \tfrac{1}{2}\hbar\boldsymbol{\tau}_j|\mathsf{A}\underline{\rho}|\mathbf{q}_j + \tfrac{1}{2}\hbar\boldsymbol{\tau}_j\rangle &= \int\prod_{\alpha=1}^{N}(d^3\mathbf{q}_\alpha')\langle \mathbf{q}_j - \tfrac{1}{2}\hbar\boldsymbol{\tau}_j|\mathsf{A}|\mathbf{q}_j'\rangle \\ &\quad\times \langle \mathbf{q}_j'|\underline{\rho}|\mathbf{q}_j + \tfrac{1}{2}\hbar\boldsymbol{\tau}_j\rangle \\ &= \frac{1}{(2\pi)^{6N}}\int\prod_{\alpha=1}^{N}[d^3\mathbf{q}_\alpha'\, d^3\mathbf{x}_\alpha\, d^3\mathbf{x}_\alpha'\, d^3\mathbf{k}_\alpha\, d^3k_\alpha' \\ &\quad\times e^{i\mathbf{k}_\alpha'\cdot(\mathbf{q}_\alpha'-\mathbf{x}_\alpha')}\, e^{-i\mathbf{k}_\alpha\cdot(\mathbf{q}_\alpha+(1/2)\hbar\tau_\alpha-\mathbf{x}_\alpha)}] \\ &\quad\times \langle \mathbf{q}_j - \tfrac{1}{2}\hbar\boldsymbol{\tau}_j|\mathsf{A}|\mathbf{q}_j'\rangle\langle \mathbf{x}_j'|\underline{\rho}|\mathbf{x}_j\rangle\end{aligned}$$

$$= \frac{\hbar^{3N}}{(2\pi)^{6N}}$$

$$\times \int \prod_{\alpha=1}^{N} [d^3\mathbf{y}_\alpha \, d^3\mathbf{z}_\alpha \, d^3\mathbf{k}_\alpha \, d^3\mathbf{k}'_\alpha \, e^{i\mathbf{k}'_\alpha \cdot (\mathbf{q}_\alpha - (1/2)\hbar\tau_\alpha - \mathbf{y}_\alpha + (1/2)\hbar\mathbf{z}_\alpha)}$$

$$\times e^{-i\mathbf{k}_\alpha \cdot (\mathbf{q}_\alpha + (1/2)\hbar\tau_\alpha - \mathbf{y}_\alpha - (1/2)\hbar\mathbf{z}_\alpha)}]$$

$$\times \tilde{\mathbf{A}}(\mathbf{q}_j - \tfrac{1}{2}\hbar\boldsymbol{\tau}_j, \hbar\mathbf{k}'_j)\langle \mathbf{y}_\alpha - \tfrac{1}{2}\hbar\mathbf{z}_\alpha | \rho | \mathbf{y}_\alpha + \tfrac{1}{2}\hbar\mathbf{z}_\alpha \rangle$$

where to derive the third equality we have used Equation (E.3) and the change of variables ($\mathbf{x}_\alpha = \mathbf{y}_\alpha + \tfrac{1}{2}\hbar\mathbf{z}_\alpha$, $\mathbf{x}'_\alpha = \mathbf{y}_\alpha - \tfrac{1}{2}\hbar\mathbf{z}_\alpha$). Next we use definition (E.1) of the Wigner distribution function to obtain

$$\langle \mathbf{q}_j - \tfrac{1}{2}\hbar\boldsymbol{\tau}_j | \mathsf{A}\rho | \mathbf{q}_j + \tfrac{1}{2}\hbar\boldsymbol{\tau}_j \rangle = \left(\frac{\hbar}{2\pi}\right)^{3N}$$

$$\times \int \prod_{\alpha=1}^{N} [d^3\mathbf{y}_\alpha \, d^3\mathbf{k}_\alpha \, d^3\mathbf{k}'_\alpha \, e^{i(\mathbf{k}'_\alpha - \mathbf{k}_\alpha)\cdot(q_\alpha - \mathbf{y}_\alpha)}$$

$$\times e^{-i(\hbar/2)(\mathbf{k}'_\alpha + \mathbf{k}_\alpha)\cdot\tau_\alpha}]$$

$$\times \tilde{A}(\mathbf{q}_j - \tfrac{1}{2}\hbar\boldsymbol{\tau}_j, \hbar\mathbf{k}'_j) f_w\left(\mathbf{y}_j, \frac{\hbar}{2}(\mathbf{k}_j + \mathbf{k}'_j)\right)$$

Theorem (E.2) now follows, after making the change of variables [$\mathbf{p}_\alpha = (\hbar/2)(\mathbf{k}'_\alpha + \mathbf{k}_\alpha)$, $\ell_\alpha = \mathbf{k}'_\alpha - \mathbf{k}_\alpha$]. The Hermitian conjugate of this theorem can be written, after changing the signs of both $\boldsymbol{\tau}_\alpha$ and ℓ_α, as

$$\langle \mathbf{q}_j - \tfrac{1}{2}\hbar\boldsymbol{\tau}_j | \rho\mathsf{A} | \mathbf{q}_j + \tfrac{1}{2}\hbar\boldsymbol{\tau}_j \rangle = (2\pi)^{-3N} \int \prod_{\alpha=1}^{N} [d^3\mathbf{y}_\alpha \, d^3\mathbf{p}_\alpha \, d^3\ell_\alpha \, e^{-i\tau_\alpha\cdot\mathbf{p}_\alpha} e^{i\ell_\alpha\cdot(\mathbf{q}_\alpha - \mathbf{y}_\alpha)}]$$

$$\times \tilde{A}(\mathbf{q}_j + \tfrac{1}{2}\hbar\boldsymbol{\tau}_j, \mathbf{p}_j - \tfrac{1}{2}\hbar\ell_j) f_w(\mathbf{y}_j, \mathbf{p}_j) \quad \textbf{(E.4)}$$

Calculation of Average Values

The trace of either Equation (E.2) or (E.4) gives the (real) average value $\langle \mathsf{A} \rangle$ as

$$\langle \mathsf{A} \rangle \equiv \int \prod_{\alpha=1}^{N} (d^3\mathbf{q}_\alpha)\langle \mathbf{q}_j | \mathsf{A}\rho | \mathbf{q}_j \rangle$$

$$\langle \mathsf{A} \rangle = (2\pi)^{-3N} \int \prod_{\alpha=1}^{N} [d^3\mathbf{q}_\alpha \, d^3\mathbf{y}_\alpha \, d^3\mathbf{p}_\alpha \, d^3\ell_\alpha \, e^{i\ell_\alpha \cdot (\mathbf{q}_\alpha - \mathbf{y}_\alpha)}]$$

$$\times \tilde{A}(\mathbf{q}_j, \mathbf{p}_j + \tfrac{1}{2}\hbar\ell_j) \, f_w(\mathbf{y}_j, \mathbf{p}_j) \tag{E.5}$$

$$\underset{\hbar \to 0}{\longrightarrow} \int \prod_{\alpha=1}^{N} (d^3\mathbf{q}_\alpha \, d^3\mathbf{p}_\alpha) A(\mathbf{q}_j, \mathbf{p}_j) \, f_w(\mathbf{q}_j, \mathbf{p}_j)$$

Thus the average value $\langle \mathsf{A} \rangle$, as defined in quantum mechanics by Equation (21.9), reduces in the classical limit to an average calculated by using the Wigner distribution function. (Note that the distinction between A and $\tilde{A}$ disappears when $\hbar = 0$.) This property of the Wigner distribution function can be stated more strongly, for when one considers physical quantities A that are functions of position only or momentum only, then the third line of Equation (E.5) holds quite generally and not just in the limit $\hbar \to 0$.

Exercise E.1

(a) Show for physical quantities A, which are momentum independent, that

$$\langle \mathsf{A} \rangle = \int \prod_{\alpha=1}^{N} (d^3\mathbf{q}_\alpha \, d^3\mathbf{p}_\alpha) A(\mathbf{q}_j) f_w(\mathbf{q}_j, \mathbf{p}_j) \tag{E.6a}$$

(b) Show for physical quantities A, which are position independent, that

$$\langle \mathsf{A} \rangle = \int \prod_{\alpha=1}^{N} (d^3\mathbf{q}_\alpha \, d^3\mathbf{p}_\alpha) \, A\,(\mathbf{p}_j) \, f_w(\mathbf{q}_j, \mathbf{p}_j) \tag{E.6b}$$

One consequence of these two results, readily verified by setting A = 1, is that the Wigner distribution function is normalized correctly to unity.

Equation of Motion for f_w

We now use Equations (E.2) and (E.4) in conjunction with the quantum-mechanical equation of motion (23.2) for $\underline{\rho}$ to derive an equation of motion for the Wigner distribution function. We begin by considering general off-diagonal position-space matrix elements, as in Equations (E.1) and (E.2), of Equation (23.2).

$$i\hbar\frac{\partial}{\partial t}\langle \mathbf{q}_j - \tfrac{1}{2}\hbar\boldsymbol{\tau}_j | \underline{\rho} | \mathbf{q}_j + \tfrac{1}{2}\hbar\boldsymbol{\tau}_j \rangle = \langle \mathbf{q}_j - \tfrac{1}{2}\hbar\boldsymbol{\tau}_j | \mathsf{H}\underline{\rho} - \underline{\rho}\mathsf{H} | \mathbf{q}_j + \tfrac{1}{2}\hbar\boldsymbol{\tau}_j \rangle \tag{E.7}$$

Substitution of this expression into the time derivative of Equation (E.1), followed by the use of Equations (E.2) and (E.4), then yields the desired equation of motion for f_w.

$$\frac{\partial}{\partial t} f_w(\mathbf{q}_j, \mathbf{p}_j) = -\frac{i}{\hbar}(2\pi)^{-6N} \int \prod_{\alpha=1}^{N} [d^3\mathbf{y}_\alpha\, d^3\boldsymbol{\tau}_\alpha\, d^3\mathbf{s}_\alpha\, d^3\ell_\alpha\, e^{i\tau_\alpha\cdot(\mathbf{p}_\alpha - \mathbf{s}_\alpha)}\, e^{i\ell_\alpha\cdot(\mathbf{q}_\alpha - \mathbf{y}_\alpha)}]$$
$$\times [\tilde{H}(\mathbf{q}_j - \tfrac{1}{2}\hbar\boldsymbol{\tau}_j, \mathbf{s}_j + \tfrac{1}{2}\hbar\ell_j) - \tilde{H}(\mathbf{q}_j + \tfrac{1}{2}\hbar\boldsymbol{\tau}_j, \mathbf{s}_j - \tfrac{1}{2}\hbar\ell_j)]$$
$$\times f_w(\mathbf{y}_j, \mathbf{s}_j) \tag{E.8}$$

We now show that this equation of motion reduces to Liouville's equation (23.19) for f_w in the limit $\hbar \to 0$. To this end we perform Taylor-series expansions in powers of $\hbar$ of the Hamiltonian functions in Equation (E.8). We define the notation

$$\frac{\partial^\lambda}{\partial \mathbf{q}_j^\lambda} \equiv \frac{\partial^\lambda}{\partial \mathbf{q}_1^{\lambda_1}\, \partial \mathbf{q}_2^{\lambda_2} \cdots \partial \mathbf{q}_N^{\lambda_N}} \tag{E.9}$$

and similarly for $\partial^\lambda/\partial \mathbf{p}_j^\lambda$, where $\lambda \equiv \Sigma_{\ell=1}^{N} \lambda_i$, and write Equation (E.8) as

$$\frac{\partial}{\partial t} f_w(\mathbf{q}_j, \mathbf{p}_j) = -\frac{i}{\hbar}(2\pi)^{-6N} \int \prod_{\alpha=1}^{N} [d^3\mathbf{y}_\alpha\, d^3\boldsymbol{\tau}_\alpha\, d^3\mathbf{s}_\alpha\, d^3\ell_\alpha\, e^{i\tau_\alpha\cdot(\mathbf{p}_\alpha - \mathbf{s}_\alpha)}\, e^{i\ell_\alpha\cdot(\mathbf{q}_\alpha - \mathbf{y}_\alpha)}]$$
$$\times f_w(\mathbf{y}_j, \mathbf{s}_j) \sum_{\{\lambda_i\}} \prod_{k=1}^{N} \left[\frac{(\tfrac{1}{2}\hbar\boldsymbol{\tau}_k)^{\lambda_k}}{\lambda_k!}\right]$$
$$\times \frac{\partial^\lambda}{\partial \mathbf{q}_j^\lambda}[(-1)^\lambda \tilde{H}(\mathbf{q}_j, \mathbf{s}_j + \tfrac{1}{2}\hbar\ell_j) - \tilde{H}(\mathbf{q}_j, \mathbf{s}_j - \tfrac{1}{2}\hbar\ell_j)]$$
$$= -\frac{i}{\hbar}(2\pi)^{-6N} \int \prod_{\alpha=1}^{N} [d^3\mathbf{y}_\alpha\, d^3\boldsymbol{\tau}_\alpha\, d^3\mathbf{s} d^3\ell_\alpha\, e^{i\ell_\alpha\cdot(\mathbf{q}_\alpha - \mathbf{y}_\alpha)}] f_w(\mathbf{y}_j, \mathbf{s}_j)$$
$$\times \sum_{\{\lambda_i\}} \prod_{k=1}^{N} (\lambda_k!)^{-1} \left(-\frac{\hbar}{2i}\right)^\lambda \frac{\partial^\lambda}{\partial \mathbf{s}_j^\lambda} e^{i\Sigma_\beta \tau_\beta\cdot(\mathbf{p}_\beta - \mathbf{s}_\beta)} \tag{E.10}$$
$$\times \frac{\partial^\lambda}{\partial \mathbf{q}_j^\lambda}[(-1)^\lambda \tilde{H}(\mathbf{q}_j, \mathbf{s}_j + \tfrac{1}{2}\hbar\ell_j) - \tilde{H}(\mathbf{q}_j, \mathbf{s}_j - \tfrac{1}{2}\hbar\ell_j)]$$
$$= -\frac{i}{\hbar}(2\pi)^{-3N} \int \prod_{\alpha=1}^{N} [d^3\mathbf{y}_\alpha\, d^3\ell_\alpha\, e^{i\ell_\alpha\cdot(\mathbf{q}_\alpha - \mathbf{y}_\alpha)}] \sum_{\{\lambda_i\}} \prod_{k=1}^{N} (\lambda_k!)^{-1} \left(\frac{\hbar}{2i}\right)^\lambda$$
$$\times \frac{\partial^\lambda}{\partial \mathbf{p}_j^\lambda} f_w(\mathbf{y}_j, \mathbf{p}_j) \cdot \frac{\partial^\lambda}{\partial \mathbf{q}_j^\lambda}[(-1)^\lambda \tilde{H}(\mathbf{q}_j, \mathbf{p}_j + \tfrac{1}{2}\hbar\ell_j) - \tilde{H}(\mathbf{q}_j, \mathbf{p}_j - \tfrac{1}{2}\hbar\ell_j)]$$

In the first equality of Equation (E.10) $\Sigma_{\{\lambda_i\}}$ refers to the infinite sums over all λ_i, $i = 1, 2, \ldots, N$, and in the third equality we have integrated by parts before performing the N-dimensional $\boldsymbol{\tau}_\alpha$ and $\mathbf{s}_\alpha$ integrals. A further set of expansions yields, in analogy with the development of Equation (E.10),

$$\begin{aligned}
\frac{\partial}{\partial t} f_w(\mathbf{q}_j, \mathbf{p}_j) &= \frac{1}{2} \sum_{\{\lambda_i\}} \sum_{\{\mu_i\}} \prod_{k=1}^{N} (\lambda_k!)^{-1} \prod_{\ell=1}^{N} (\mu_\ell!)^{-1} \left(\frac{\hbar}{2i}\right)^{\lambda+\mu-1} [(-1)^\mu - (-1)^\lambda] \\
&\quad \times \left[\frac{\partial^\mu}{\partial \mathbf{p}_j^\mu} \frac{\partial^\lambda}{\partial \mathbf{q}_j^\lambda} \tilde{H}(\mathbf{q}_j, \mathbf{p}_j)\right] \cdot \left[\frac{\partial^\mu}{\partial \mathbf{q}_j^\mu} \frac{\partial^\lambda}{\partial \mathbf{p}_j^\lambda} f_w(\mathbf{q}_j, \mathbf{p}_j)\right] \\
&= \sum_{\substack{\{\lambda_i\} \\ \lambda=\text{odd}}} \sum_{\substack{\{\mu_i\} \\ \mu=\text{even}}} \prod_{k=1}^{N} (\lambda_k!)^{-1} \prod_{\ell=1}^{N} (\mu_\ell!)^{-1} \left(\frac{\hbar}{2i}\right)^{\lambda+\mu-1} \\
&\quad \times \left\{\left[\frac{\partial^\mu}{\partial \mathbf{p}_j^\mu} \frac{\partial^\lambda}{\partial \mathbf{q}_j^\lambda} \tilde{H}(\mathbf{q}_j, \mathbf{p}_j)\right] \cdot \left[\frac{\partial^\mu}{\partial \mathbf{q}_j^\mu} \frac{\partial^\lambda}{\partial \mathbf{p}_j^\lambda} f_w(\mathbf{q}_j, \mathbf{p}_j)\right] \right. \\
&\quad \left. - \left[\frac{\partial^\mu}{\partial \mathbf{p}_j^\mu} \frac{\partial^\lambda}{\partial \mathbf{q}_j^\lambda} f_w(\mathbf{q}_j, \mathbf{p}_j)\right] \cdot \left[\frac{\partial^\mu}{\partial \mathbf{q}_j^\mu} \frac{\partial^\lambda}{\partial \mathbf{p}_j^\lambda} \tilde{H}(\mathbf{q}_j, \mathbf{p}_j)\right]\right\} \\
&= \{\tilde{H}, f_w\} + O(\hbar^2)
\end{aligned} \tag{E.11}$$

where we have interchanged μ_i and λ_i labels to arrive at the second term in the second equality. In the limit $\hbar \to 0$, the Wigner distribution theorem satisfies Liouville's theorem (23.19), inasmuch as the distinction between H and $\tilde{H}$ disappears in this limit. In connection with the third equality of this last result (E.11) we must emphasize that there is still dependence on $\hbar$ in the first, Poisson-bracket term, since f_w is the quantum-mechanically defined function (E.1).

Exercise E.2

(a) Derive the first equality of Equation (E.11).

(b) Write the Hamiltonian $\tilde{H}$ as the sum of a kinetic-energy part $K = \Sigma_{\alpha=1}^{N}(p_\alpha^2/2m)$ and a potential-energy part V and show for velocity-independent interactions that

$$\begin{aligned}
\frac{\partial}{\partial t} f_w(\mathbf{r}_j, \mathbf{p}_j) &= -\sum_{\alpha=1}^{N} \left(\frac{\mathbf{p}_\alpha}{m}\right) \cdot \frac{\partial}{\partial \mathbf{r}_2} f_w(\mathbf{r}_j, \mathbf{p}_j) \\
&\quad + \sum_{\substack{\{\lambda_i\} \\ \lambda=\text{odd}}} \prod_{k=1}^{N} (\lambda_k!)^{-1} \left(\frac{\hbar}{2i}\right)^{\lambda-1} \\
&\quad \left[\frac{\partial^\lambda}{\partial \mathbf{r}_j^\lambda} V(\mathbf{r}_j)\right] \cdot \left[\frac{\partial^\lambda}{\partial \mathbf{p}_j^\lambda} f_w(\mathbf{r}_j, \mathbf{p}_j)\right]
\end{aligned} \tag{E.12}$$

We have shown that the quantum-mechanically defined Wigner distribution function (E.1) has properties that can be expected of it. It produces average values correctly, by Equations (E.5) and (E.6), and it satisfies Liouville's theorem in the classical limit. It is therefore a suitable quantum-mechanical analog of the classical distribution function (23.13). But if this is the case, then we may turn the argument around and say that the final equality of Equation (E.11), evaluated at $\hbar = 0$, constitutes a rigorous proof of Liouville's theorem.

Bibliography for Appendix E

Original Papers

E. Wigner, "On the Quantum Correction for Thermodynamic Equilibrium," *Phys. Rev.* **40** (1932), p. 749. In which the Wigner distribution function was originally introduced, with some of its properties and consequences examined.

W. E. Brittin and W. R. Chappell, "The Wigner Distribution Function and Second Quantization in Phase Space," *Rev. Mod. Phys.* **34** (1962), p. 620.

Note

1. Rules that associate classical quantities with quantum-mechanical operators are called correspondence rules; writing a function as a symmetrized product is only one such rule. A general discussion of rules that have been proposed has been given by L. Cohen, *J. Math. Phys.* **7** (1966), p. 781. A useful generalization of the so-called Weyl correspondence rule and investigation of its consequences for the Wigner distribution function have been given by G. C. Summerfield and P. F. Zweifel, *J. Math. Phys.* **10** (1969), p. 233.

APPENDIX F

GRAPHICAL SOLUTION TO THE PERMUTATION PROBLEM OF QUANTUM STATISTICS

The permutation problem of quantum statistics is to separate the clustering effect due to particle identity, or statistics, from that due to particle dynamics. Its analytic solution is accomplished at the end of Section 37 with Equation (37.16). In this appendix we show, after substitution into Equation (37.2) for the grand potential, that this solution can be greatly simplified by expressing it graphically in terms of primary 0 graphs. The appendix is introduced by defining, more generally, primary ζ graphs for any positive integer ζ. It is ended by showing that the momentum distribution can be expressed in terms of primary 1 graphs.

The permutation problem arises when one attempts to find an explicit expression for the general cluster integral (37.10) of quantum statistics. Although the permutation problem may be bypassed in the derivation of such an explicit expression, the problem itself is of interest, as we discussed at the end of Section 37. Moreover, the graphical solution we give in this appendix demonstrates a general technique that may be applied in many other contexts. We begin, then, by defining primary ζ graphs for $\zeta = 0, 1, 2, \ldots$. These graphs were first introduced by Lee and Yang, but the rules for them presented here are a modified set.[1]

A *primary ζ graph* ($\zeta = 0, 1, 2, \ldots$) consists of a collection of vertices connected by directed lines, with ζ external incoming lines and ζ external outgoing lines. Each vertex is characterized by a number $\alpha(\alpha = 1, 2, \ldots)$, which gives the number of incoming and of outgoing lines at that vertex. Two primary graphs are different if their topological structures are different.[2] Primary graphs with $\zeta \neq 0$ can be used in the calculation of distribution functions, as we show for the case $\zeta = 1$.

To each primary ζ graph a term is assigned that is determined by the following procedures.

(1) Associate with each *internal line* a different integer $i(i = 1, 2, \ldots, m)$ and a corresponding momentum (and spin) k_i.[3]

(2) If $\zeta \neq 0$, then associate the external lines with certain pregiven momenta.

(3) To each α vertex, assign a factor.

$$T\begin{pmatrix} k_{A_1} & \cdots & k_{A_\alpha} \\ k_{B_1} & \cdots & k_{B_\alpha} \end{pmatrix}$$

where $k_{A_1} \cdots k_{A_\alpha}$ are the momenta associated with its outgoing lines, $k_{B_1} \cdots k_{B_\alpha}$ are the momenta associated with its incoming lines, and the T_α function is defined by Equation (37.14).

(4) Assign a factor $\varepsilon e^{\beta(g-\omega_k)}$ for each internal line.

(5) Assign a factor S^{-1} to the entire graph, where S = symmetry number. The symmetry number is defined to be the total number of permutations of the m integers associated with the internal lines that leave the graph topologically unchanged, including the positions of these numbers relative to the lines.

(6) Assign an overall sign factor $\varepsilon^{P'}$ to the entire graph, where P' is the permutation of the $(m + \zeta)$ bottom-row coordinates, in the product of T functions, with respect to the $(m + \zeta)$ top-row coordinates. An identity permutation for the momenta of the ζ external incoming lines relative to the momenta of the ζ external outgoing lines must be assigned.

(7) Finally, sum over all internal momenta $k_1, k_2, \cdots, k_m$.

Figure F.1 shows some examples of primary 0 graphs along with their associated symmetry numbers.

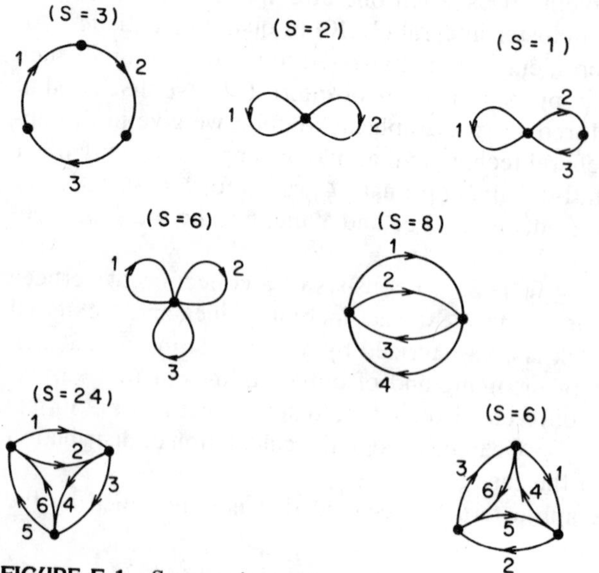

FIGURE F.1 Some primary 0 graphs with their associated symmetry numbers, S.

Exercise F.1 Verify each of the symmetry numbers given in Figure F.1.

Grand Potential

For simplicity we consider only a large, translationally invariant single-component system in this appendix. For this case the grand potential (37.2) becomes[3]

$$\Omega f(\beta, \Omega, g) = \sum_{N=1}^{\infty} \frac{1}{N!} \sum_{k_1 \cdots k_N} \left[\prod_{i=1}^{N} \varepsilon e^{\beta(g-\omega_i)} \right] \varepsilon^N U^{(S)}\begin{pmatrix} k_1 \cdots k_N \\ k_1 \cdots k_N \end{pmatrix} \tag{F.1}$$

To demonstrate the diagrammatic method, we begin by indicating the correspondence between leading terms in this expression and lowest-order primary 0 graphs. We use the fact [see Equation (38.7)] that

$$T\begin{pmatrix} k \\ k' \end{pmatrix} = \varepsilon\, \delta_{k,k'} \tag{F.2}$$

Then the result of substituting Equations (37.15) into Equation (F.1) can readily be shown to agree exactly with the sum of the expressions given for the following primary 0 graphs:

$$= \varepsilon \sum_{k} \left[\varepsilon\, e^{\beta(g-\omega_k)} \right] \tag{F.3}$$

$$\begin{aligned} &= \frac{\varepsilon}{2} \sum_{k_1,k_2} \left[\varepsilon\, e^{\beta(g-\omega_1)} \right] \left[\varepsilon\, e^{\beta(g-\omega_2)} \right] T\begin{pmatrix} k_1 \\ k_2 \end{pmatrix} T\begin{pmatrix} k_2 \\ k_1 \end{pmatrix} \\ &= \frac{\varepsilon}{2} \sum_{k} \left[\varepsilon\, e^{\beta(g-\omega_k)} \right]^2 \end{aligned} \tag{F.4}$$

$$= \frac{\varepsilon}{3} \sum_{k} \left[\varepsilon\, e^{\beta(g-\omega_k)} \right]^3 \tag{F.5}$$

$$= \frac{1}{2} \sum_{k_1 k_2} \left[\varepsilon\, e^{\beta(g-\omega_1)} \right] \left[\varepsilon\, e^{\beta(g-\omega_2)} \right] T\begin{pmatrix} k_1\, k_2 \\ k_1\, k_2 \end{pmatrix} \tag{F.6}$$

$$\begin{aligned} &= \varepsilon \sum_{k_1 k_2 k_3} \left[\varepsilon\, e^{\beta(g-\omega_1)} \right] \left[\varepsilon\, e^{\beta(g-\omega_2)} \right] \left[\varepsilon\, e^{\beta(g-\omega_3)} \right] T\begin{pmatrix} k_1\, k_2 \\ k_1\, k_3 \end{pmatrix} T\begin{pmatrix} k_3 \\ k_2 \end{pmatrix} \\ &= \sum_{k_1 k_2} \left[\varepsilon\, e^{\beta(g-\omega_1)} \right] \left[\varepsilon\, e^{\beta(g-\omega_2)} \right]^2 T\begin{pmatrix} k_1\, k_2 \\ k_1\, k_2 \end{pmatrix} \end{aligned} \tag{F.7}$$

$$= \frac{1}{3!} \sum_{k_1 k_2 k_3} \left[\varepsilon\, e^{\beta(g-\omega_1)} \right] \left[\varepsilon\, e^{\beta(g-\omega_2)} \right] \left[\varepsilon\, e^{\beta(g-\omega_3)} \right] T\begin{pmatrix} k_1\, k_2\, k_3 \\ k_1\, k_2\, k_3 \end{pmatrix} \tag{F.8}$$

Notice that there are two terms of the type (F.5) and six terms of the type (F.7) in the third of Equations (37.15).

The expressions (F.3) to (F.8) suggest that Equation (F.1) for the grand potential can be rewritten simply as

$$f(\beta, \Omega, g) = \Omega^{-1} \sum (\text{all different primary 0 graphs}) \tag{F.9}$$

If this equation is a correct generalization, and it is, then clearly the primary 0 graph notation brings considerable order and simplicity to the complicated array of terms that occur when equations of the type (37.15) are substituted into Equation (F.1).

We now use Equation (37.16) to give a general proof[4] that Equation (F.9) is correct. We first assign, instead of associate, a different integer $i(i = 1, 2, \ldots, m)$ to each of the m internal lines of a 0 graph. The resulting 0 graph, generated from the original primary 0 graph, is called a *numbered 0 graph*. Two numbered 0 graphs are different if they have different topological structures, including the positions of the numbers relative to the lines. Thus permutations of these numbers relative to the lines, as in rule (5) above, are not allowed for a numbered 0 graph and the numerical factor associated with each such graph is defined to be 1.

Let D be the total number of different numbered 0 graphs that can be generated from the same primary 0 graph. From the definition of the symmetry number S it is easy to verify that

$$D = (N!)\, S^{-1} \tag{F.10}$$

We next observe, after substituting Equation (37.16), that different terms in Equation (F.1) can give identical results after summing over all the internal momentum variables. Graphically, such terms correspond to the same primary 0 graph, but to different numbered 0 graphs. On reflection, one can conclude that the total number of such terms is identical with the total number D of corresponding different numbered 0 graphs. Since according to Equation (F.1) for f we must divide every term by $N!$, the resulting factor for each primary 0 graph is S^{-1}, and rule (5) is correct.

We observe, finally, that there is a one-to-one correspondence between every different term in Equation (F.1) and some primary 0 graph. Moreover, the sign factor given in rule (6) agrees with that assigned in Equation (37.16), because the part ε^N in (37.16) has been distributed to the N line factors of rule (4). Therefore, Equation (F.9) is indeed correct.

The grand potential f_o for an ideal gas can be deduced immediately from Equation (F.9) and the special cases (F.3) to (F.5). It is

$$\begin{aligned} f_o(\beta, \Omega, g) &= \varepsilon\Omega^{-1} \sum_{N=1}^{\infty} \frac{1}{N} \sum_k \left[\varepsilon\, e^{\beta(g-\omega_k)} \right]^N \\ &= -\frac{(2S+1)\varepsilon}{\Omega} \sum_k \ln\left[1 - \varepsilon\, e^{\beta(g-\omega_k)} \right] \end{aligned} \tag{F.11}$$

The factor N^{-1} in the first equality is the symmetry number of the N-vertex free-particle primary 0 graph. According to Equation (36.11), this result agrees with Equations (31.13) and (31.15), which were derived earlier for the pressure of an ideal degenerate gas.

Momentum Distribution

We can also deduce an expression, similar to Equation (F.9), for the momentum distribution $\langle \mathsf{n}(k) \rangle$. It is convenient to use prescription (46.13a), which, with the aid of Equation (F.9), becomes

$$\langle \mathsf{n}(k) \rangle = -\beta^{-1} \left[\frac{\partial}{\partial \omega(k)} \sum \text{(all different primary 0 graphs)} \right]_T \tag{F.12}$$

Here the T functions are represented by vertices in the primary 0 graphs, and these must be held constant during the indicated partial differentiation; by contrast, the internal lines are associated with the free-particle factors $\varepsilon e^{\beta(g-\omega_k)}$ whose partial derivative may be taken. The result of performing this derivative is to give for $\langle \mathsf{n}(k) \rangle$ the expression

$$\langle \mathsf{n}(k) \rangle = \varepsilon\, e^{\beta(g-\omega_k)} \sum \text{(all different primary 1 graphs)} \tag{F.13}$$

It is instructive to show that the terms written down in Equation (46.14) [see also Equation (F.2)] are in agreement with this general formula.

Exercise F.2

(a) Write down primary 1 graphs corresponding to each of the terms in the second equality of Equation (46.14) and show graphically that these can be derived from the set of primary 0 graphs (F.3) to (F.8). [*Suggestion:* Consider first that the primary 1 graph that corresponds to Equation (F.3) is $\uparrow = \varepsilon$.

(b) Use Equations (F.11) and (F.12) to show that the momentum distribution for an ideal gas is

$$\langle \mathsf{n}(k) \rangle_o = [e^{\beta(\omega_k - g)} - \varepsilon]^{-1}$$

in agreement with Equation (32.1).

To construct a general proof of Equation (F.13), we observe first that the partial derivative in Equation (F.12) corresponds to breaking open, one at a time, each of the lines in every primary 0 graph and writing the corresponding internal line factor explicitly, as in Equation (F.13). If there are n_0 lines in a given 0 graph, with symmetry number S_0, then one generates by this procedure n_0 terms, each of which is equal to $S_0^{-1} S_1$ times a corresponding primary 1 graph, with symmetry number S_1. We define the number n_1 to be the number of identical primary 1 graphs of any certain kind derived from the given 0 graph by cutting each of its n_0 lines. On using the definition of the symmetry number given in rule (5), it is straightforward to verify for each different 1 graph that

$$S_0 = n_1 S_1 \tag{F.14}$$

We conclude that the symmetry numbers of the primary 1 graphs in Equation (F.13) are always correctly reproduced by taking the indicated partial derivative in Equation (F.12). Since every primary 1 graph corresponds uniquely to one and only one primary 0 graph, the proof of Equation (F.13) from Equation (F.12) is complete.

Bibliography for Appendix F

Original Paper

T. D. Lee and C. N. Yang, "Many-Body Problem in Quantum Statistical Mechanics IV. Formulation in Terms of Average Occupation Number in Momentum Space," *Phys. Rev.* **117** (1960), p. 22.

Notes

1. F. Mohling, *Phys. Rev.* **122** (1961), p. 1043.
2. One can distort these graphs and stretch their lines without changing their topological structure.
3. The state labels k represent both a momentum value $\mathbf{k}$ and a magnetic spin quantum number m. The state variables k can be replaced, more generally, by the quantum numbers μ corresponding to the general single-particle eigenstate of H_o.
4. Adapted from T. D. Lee and C. N. Yang, *op. cit.*, App. A.

INDEX